Managing Plant Genetic Diversity

Managing Plant Genetic Diversity

edited by

Johannes M.M. Engels

International Plant Genetic Resources Institute (IPGRI)
Rome, Italy

V. Ramanatha Rao

IPGRI Regional Office for Asia, the Pacific and Oceania
Serdang, Malaysia

Anthony H.D. Brown

CSIRO
Canberra, Australia

and

Michael T. Jackson

International Rice Research Institute
Makati City, The Philippines

CABI *Publishing*

CABI *Publishing* is a division of CAB *International*

CABI Publishing
CAB International
Wallingford
Oxon OX10 8DE
UK

Tel: +44 (0)1491 832111
Fax: +44 (0)1491 833508
Email: cabi@cabi.org
Web site: www.cabi-publishing.org

CABI Publishing
10 E 40th Street
Suite 3203
New York, NY 10016
USA

Tel: +1 212 481 7018
Fax: +1 212 686 7993
Email: cabi-nao@cabi.org

The International Plant Genetic Resources Institute (IPGRI) is an autonomous international scientific organization, supported by the Consultative Group on International Agricultural Research (CGIAR). IPGRI's mandate is to advance the conservation and use of genetic diversity for the well-being of present and future generations. IPGRI has its headquarters in Rome, Italy, and offices in another 22 countries worldwide. It operates through three programmes: (i) the Plant Genetic Resources Programme; (ii) the CGIAR Genetic Resources Support Programme; and (iii) the International Network for the Improvement of Banana and Plantain (INIBAP).

IPGRI, Via dei Tre Denari, 472/a, 00057 Maccarese (Fiumicino), Rome, Italy.
Tel: +39 06 61181; fax: +39 06 6197 9661
E-mail: ipgri@cgiar.org; Web site: www.ipgri.cgiar.org

A catalogue record for this book is available from the British Library, London, UK.

Library of Congress Cataloging-in-Publication Data
Managing plant genetic resources / edited by J.M.M. Engels ...[et al.].
p. cm.
Includes bibliographical references (p.).
ISBN 0-85199-522-5 (alk. paper)
1. Germplasm resources, Plant–Management. 2. Crops–Germplasm resources–Management. I. Engels, J. M. M.
SB123.3 .M35 2001
333.95′34--dc21 00-054709

ISBN 0 85199 522 5

Typeset in Garamond by Columns Design Ltd, Reading
Printed and bound in the UK by Biddles Ltd, Guildford and King's Lynn

Contents

Contributors

G.A. Abrigo, Philippine Rice Research Institute (PhilRice), Maligaya, Muñoz, Nueva Ecija, 3119 Philippines

A.H. Alexandrov, Forest Research Institute, Sofia, Bulgaria

T.A. Al-Turki, KACST Herbarium, Natural Resources and Environmental Research Institute, King Abdulaziz City for Science and Technology, PO Box 6086, Riyadh-11442, Kingdom of Saudi Arabia

G.H. Badaloo, Mauritius Sugar Industry Research Institute, Reduit, Republic of Mauritius

P. Balakhrishna, Regional Biodiversity Programme, Asia, IUCN-World Conservation Union, 48 Vajira Road, Colombo 5, Sri Lanka

D. Barrantes, Escuela de Biologia, Universidad de Costa Rica, Ciudad Universitaria 'Rodrigo Facio', San Pedro de Montes de Oca, San José, Costa Rica

N. Barve, Ashoka Trust for Research in Ecology and the Environment (ATREE), Hebbal, Bangalore 560 024, India

K.S. Bawa, Department of Crop Physiology, University of Agricultural Sciences, GKVK, Bangalore 560 065, India

F. Begemann, German Centre for Documentation and Information in Agriculture (ZADI), Information Centre for Genetic Resources (IGR), Villichgasse 17, 53177 Bonn, Germany

S. Biber-Klemm, Faculty of Law, University of Basel, Maiengasse 51, CH-4056 Basel, Switzerland

I. Blada, Forest Research and Management Institute, Bucharest, Romania

A.H.D. Brown, Centre for Plant Biodiversity Research, CSIRO Plant Industry, GPO Box 1600, Canberra ACT, Australia

C.L. Brubaker, Centre for Plant Biodiversity Research, CSIRO Plant Industry, GPO Box 1600, Canberra ACT, Australia

M.A. Calibo, Genetic Resources Center, International Rice Research Institute (IRRI), MCPO Box 3127, 1271 Makati City, Philippines

E. Castro, Escuela de Biologia, Universidad de Costa Rica, Ciudad Universitaria 'Rodrigo Facio', San Pedro de Montes de Oca, San José, Costa Rica

A. Chaudhury, CSIRO Plant Industry, GPO Box 1600, Canberra ACT 2601, Australia

P.C. Chew, Plant Biotechnology Laboratory, Faculty of Science and Technology, Universiti Kebangsaan Malaysia, 43600 Bangi, Selangor, Malaysia

E.A. Chiwona, Malawi Plant Genetic Resources Centre, Chitedze Agricultural Research Station, Lilongwe, Malawi

M.M. Clyde, Plant Biotechnology Laboratory, Faculty of Science and Technology, Universiti Kebangsaan Malaysia, 43600 Bangi, Selangor, Malaysia

J. Degreef, National Botanic Garden of Belgium, Domaine de Bouchout, 1860 Meise, Belgium

M. Denich, Centre for Development Research (ZEF), University of Bonn, Bonn, Germany

N. Donita, Forest Research and Management Institute, Bucharest, Romania.

F. Engelmann, International Plant Genetic Resources Institute (IPGRI), Via dei Tre Denari 472a, 00057 Maccarese (Fiumicino), Rome, Italy

J.M.M. Engels, International Plant Genetic Resources Institute (IPGRI), Via delle Sette Chiese 142, Rome, Italy

L. Frese, Federal Centre for Breeding Research on Cultivated Plants (BAZ), Gene Bank, Bundesallee 50, 38116 Braunschweig, Germany

S. Fukuoka, National Institute of Agrobiological Resources (NIAR), Tsukuba 305–8602, Japan

F. Gandara, Instituto de Pesquisas e Estudos Florestais, Av. Padua Dias 11, Piracicaba, SP Brazil

K.N. Ganeshaiah, Department of Genetics and Plant Breeding, University of Agricultural Sciences, GKVK, Bangalore 560 065, India

P.L. Gautam, National Bureau of Plant Genetic Resources, New Delhi 110 012, India

D. Godden, Department of Agricultural Economics, A04, University of Sydney, NSW 2006, Australia

Tadesse Woldermariam Gole, Centre for Development Research (ZEF), University of Bonn, Bonn, Germany

S.L. Greene, USDA ARS, Washington State University, 24106 N. Bunn Road, Prosser, WA 99350, USA

M. Gritsenko, USDA ARS, Washington State University, 24106 N. Bunn Road, Prosser, WA 99350, USA

L. Guarino, International Plant Genetic Resources Institute (IPGRI), Regional Office for the Americas, c/o CIAT, AA 6713, Cali, Colombia

N. Haq, International Centre for Underutilized Crops (ICUC), Southampton, UK

S. Harrer, German Centre for Documentation and Information in Agriculture (ZADI), Information Centre for Genetic Resources (IGR), Villichgasse 17, 53177 Bonn, Germany

J.G. Hawkes, School of Continuing Studies, University of Birmingham, Edgbaston, Birmingham B15 2TT, UK

Th. Hazekamp, International Plant Genetic Resources Institute (IPGRI), Via dei Tre Denari 472a, 00057 Maccarese (Fiumicino), Rome, Italy

R.J. Hijmans, International Potato Center (CIP), Lima, Peru

L. Hipolito, Philippine Rice Research Institute (PhilRice), Maligaya, Muñoz, Nueva Ecija, 3119 Philippines

T. Hodgkin, International Plant Genetic Resources Institute (IPGRI), Via dei Tre Denari 472a, 00057 Maccarese (Fiumicino), Rome, Italy

M.T. Jackson, Genetic Resources Center, International Rice Research Institute (IRRI), MCPO Box 3127, 1271 Makati City, Philippines

T.V. Jacobs, Department of Botany, University of Transkei, Private Bag XI, UNITRA, Umtata 5117, South Africa

A. Jarvis, International Plant Genetic Resources Institute (IPGRI), Regional Office for the Americas, c/o CIAT, AA 6713, Cali, Colombia

J.D. Jiménez Krause, German Centre for Documentation and Information in Agriculture (ZADI), Information Centre for Genetic Resources (IGR), Villichgasse 17, 53177 Bonn, Germany

R.C. Johnson, USDA ARS WRPIS, 59 Johnson Hall, Washington State University, Pullman, WA 99164, USA

J. Jose, Central Plantation Crops Research Institute, Kasaragod-671124, Kerala, India

V. Joshi, National Bureau of Plant Genetic Resources, New Delhi 110 012, India

P. Kageyama, Instituto de Pesquisas e Estudos Florestais, Av. Padua Dias 11, Piracicaba, SP Brazil

R. Kambuou, National Agricultural Research Institute, Port Moresby NCD, Papua New Guinea

A. Karp, IACR-Long Ashton Research Station, University of Bristol, Long Ashton, Bristol BS41 9AF, UK

J. Kennedy, School of Business, La Trobe University, Bundoora, VIC 3083, Australia

M.C. Kiran, Ashoka Trust for Research in Ecology and the Environment (ATREE), Hebbal, Bangalore 560 024, India

S. Kochhar, National Bureau of Plant Genetic Resources, New Delhi 110 012, India

R.J. Kohel, Crop Germplasm Research Unit, 2765 F&B Road, College Station, TX 77845, USA

M.P. Koshy, Department of Forest Sciences, University of British Columbia, Vancouver, BC V6T 1Z4, Canada

S. Kresovich, Institute for Genomic Diversity and Department of Plant Breeding, Cornell University, Ithaca, NY 14853–2703, USA

C. Lanaud, CIRAD, TA 40/03, Av. Agropolis, 34398 Montpellier Cédex 5, France

A.J. Luongo, Institute for Genomic Diversity and Department of Plant Breeding, Cornell University, Ithaca, NY 14853–2703, USA

G. Macaya, Centro de Investigación en Biologia Celular y Molecular, Universidad de Costa Rica, Ciudad Universitaria 'Rodrigo Facio', San Pedro de Montes de Oca, San José, Costa Rica

Bhag Mal, International Plant Genetic Resources Institute (IPGRI), New Delhi 110 012, India

R. Manimekalai, Central Plantation Crops Research Institute, Kasaragod-671124, Kerala, India

N. Maxted, School of Biological Sciences, University of Birmingham, Birmingham, UK

M. Milne, Department of Agricultural Economics, A04, University of Sydney, NSW 2006, Australia

S.R. Morin, Genetic Resources Center, International Rice Research Institute (IRRI), MCPO Box 3127, 1271 Makati City, Philippines

J.C. Motamayor, FUNDACITE, Av. Las Delicias, Maracay, Edo. Aragua, Venezuela

M. Nageswara Rao, Department of Crop Physiology, University of Agricultural Sciences, GKVK, Bangalore 560 065, India

G. Namkoong, Department of Forest Sciences, University of British Columbia, Vancouver, BC V6T 1Z4, Canada

W.A. Neves do Amaral, International Plant Genetic Resources Institute (IPGRI), Via dei Tre Denari 472a, 00057 Maccarese (Fiumicino), Rome, Italy

M.N. Normah, Plant Biotechnology Laboratory, Faculty of Science and Technology, Universiti Kebangsaan Malaysia, 43600 Bangi, Selangor, Malaysia

K. Okuno, Hokkaido National Agricultural Experiment Station, Sapporo, 062-8555, Japan

R. Ortiz, Crop Improvement Division, International Institute of Tropical Agriculture (IITA), c/o L.W. Lambourn & Co., Carolyn House, 26 Dingwall Road, Croydon CR9 3EE, UK

S. Padulosi, International Plant Genetic Resources Institute (IPGRI), Regional Office for Central and West Asia and North Africa, c/o ICARDA, PO Box 5466, Aleppo, Syria

V.A. Parthasarathy, Central Plantation Crops Research Institute, Kasaragod-671124, Kerala, India

J. Peacock, CSIRO Plant Industry, GPO Box 1600, Canberra ACT 2601, Australia

P. Perrino, National Research Council, Germplasm Institute, Bari, Italy

J.-L. Pham, Centre IRD, BP 5045, 34032 Montpellier Cédex, France

G. Postolache, Institute of Botany, Chisinău, Moldova

S.M. Quilloy, Genetic Resources Center, International Rice Research Institute (IRRI), MCPO Box 3127, 1271 Makati City, Philippines

N. Rajanaidu, Malaysian Palm Oil Board, No. 6 Persiaran Institusi, Bandar Baru Bangi, 43400 Kajang, Selangor, Malaysia

V. Ramanatha Rao, International Plant Genetic Resources Institute (IPGRI), Regional Office for Asia, the Pacific and Oceania, Serdang, 43400 Selangor, Darul Ehsan, Malaysia

K. Ramdoyal, Mauritius Sugar Industry Research Institute, Reduit, Republic of Mauritius

G. Ravikanth, Department of Crop Physiology, University of Agricultural Sciences, GKVK, Bangalore 560 065, India

O.J. Rocha, Escuela de Biologia, Universidad de Costa Rica, Ciudad Universitaria 'Rodrigo Facio', San Pedro de Montes de Oca, San José, Costa Rica

S. Roscher, German Centre for Documentation and Information in Agriculture (ZADI), Information Centre for Genetic Resources (IGR), Villichgasse 17, 53177 Bonn, Germany

G. Saavedra, Instituto de Investigaciones Agropecuarias, CRI La Platina, PO Box 439, Santiago, Chile

N.R. Sackville Hamilton, Institute of Grassland and Environmental Research (IGER), Aberystwyth, UK

I. Salma, Strategic, Environment and Natural Resources Research Center, MARDI, PO Box 12301, GPO 50774, Kuala Lumpur, Malaysia

I. Santini de Siegel, German Centre for Documentation and Information in Agriculture (ZADI), Information Centre for Genetic Resources (IGR), Villichgasse 17, 53177 Bonn, Germany

S.D. Sastrapradja, Indonesian Institute for Sciences, Center for Research in Biotechnology, Raya Bogor, Km 46, Cibinong, Bogor, Indonesia

G.T. Scarascia-Mugnozza, National Academy of Sciences, Rome, Italy

S.J. Schloss, Institute for Genomic Diversity and Department of Plant Breeding, Cornell University, Ithaca, NY 14853–2703, USA

L.S. Sebastian, Philippine Rice Research Institute (PhilRice), Maligaya, Muñoz, Nueva Ecija, 3119 Philippines

R. Uma Shaanker, Department of Crop Physiology, University of Agricultural Sciences, GKVK, Bangalore 560 065, India

G.D. Sharma, National Bureau of Plant Genetic Resources, New Delhi 110 012, India

B.W.S. Sobral, Virginia Bioinformatics Institute, Virginia Tech (0477), 1750 Kraft Drive, Suite 1400, Blacksburg, VA 24061, USA

W. Spoor, Scottish Agricultural College, The King's Buildings, Agriculture Building, West Mains Road, Edinburgh EH9 3JG, UK

A. Stella, Instituto de Pesquisas e Estudos Florestais, Av. Padua Dias 11, Piracicaba, SP Brazil

M.S. Swaminathan, M.S. Swaminathan Research Foundation, 3 Cross Street, 600113 Chennai (Madras), India

A. Tan, Aegean Agricultural Research Institute, PO Box 9, Menemen, Izmir, Turkey

A.S. Tan, Aegean Agricultural Research Institute, PO Box 9, Menemen, Izmir, Turkey

M. Taylor, TaroGen, Secretariat of the Pacific Community (SPC), Suva, Fiji

D. Teketay, Ethiopian Agricultural Research Organization, Addis Ababa, Ethiopia

J. Turok, International Plant Genetic Resources Institute (IPGRI) Via dei Tre Denari 472a, 00057 Maccarese (Fiumicino), Rome, Italy

A. Upadhyay, Central Plantation Crops Research Institute, Kasaragod-671124, Kerala, India

Th.J.L. van Hintum, Centre for Genetic Resources, The Netherlands (CGN), Plant Research International, Wageningen, The Netherlands

R. van Treuren, Centre for Genetic Resources, The Netherlands (CGN), Plant Research International, Wageningen, The Netherlands

G. Vandemark, USDA ARS, Washington State University, 24106 N. Bunn Road, Prosser, WA 99350, USA

G. Venturella, Università degli Studi di Palermo, Dipartimento di Scienze Botaniche, Via Archirafi 38, I-90123 Palermo, Italy

P.L.G. Vlek, Centre for Development Research (ZEF), University of Bonn, Bonn, Germany

J.T. Williams, International Centre for Underutilized Crops (ICUC), Southampton, UK

J. Yu, Crop Germplasm Research Unit, 2765 F&B Road, College Station, TX 77845, USA

G. Zervakis, National Agricultural Research Foundation, Institute of Kalamata, Lakonikis 85, 24100 Kalamata, Greece

Foreword

It has been many years since leading members of the plant genetic resources community got together to critically review the latest scientific and technological developments and collectively map out action for the future. Thus, after wide consultation with many of our partners, in June 2000 in Kuala Lumpur, Malaysia, IPGRI, together with the Malaysian Palm Oil Board (MPOB), convened the International Conference on Science and Technology for Managing Plant Genetic Diversity in the 21st century (SAT21).

The conference brought together more than 230 participants from over 60 countries. Most of the papers presented at the conference are included in these proceedings. The final chapter provides a synthesis of the major outcomes and recommendations of the conference and their implications for future research. I believe that the value of SAT21 was enormous, extending well beyond the conference's assessment of the state of knowledge and exploration of priorities for the future. As always at such events, a major benefit was undoubtedly the informal sharing of information among participants, and the forging of friendships and networks with the potential to lead to new, productive research partnerships. An international steering committee was established in Kuala Lumpur to explore various options for maintaining the momentum achieved at the conference. One such option is the creation of an international genetic resources society which would organize future conferences at regular intervals. For IPGRI, SAT21 has provided new understandings and helped us forge new partnerships that will be invaluable in our efforts to achieve our vision, as described in our strategy, *Diversity for Development*. This vision foresees collective, concerted action by farmers, forest dwellers, pastoralists, scientists, development workers and political leaders in harnessing the full potential of the earth's plant genetic diversity to eradicate poverty, to achieve food security and to protect the environment. The conference addressed the key scientific and technical aspects of this vision: the wise management of plant genetic resources, which is the material to which farmers and scientists apply their skills in the service of development.

I would like to thank here all the individuals and institutions involved in organizing the conference, in particular our co-organizers, the Malaysian Palm Oil Board (MPOB), and co-sponsors: the Ministry of Science, Technology and Environment, Government of Malaysia (MOSTE), the Food and Agriculture Organization of the United Nations (FAO), the Canadian International Development Agency (CIDA), the German Federal Ministry for Economic Cooperation and Development (BMZ), the Forest Research Institute of Malaysia (FRIM), the Malaysian Agricultural Research and Development Institute (MARDI), the Universiti Putra Malaysia (UPM), the Australian Agency for International Development (AusAID), the US Agency for International Development (USAID), the Industrial Credit and Investment Corporation of India (ICICI) and the Technical Centre for Agricultural and Rural Cooperation (CTA).

I would also like to thank all the members of the International Programme Committee for their valuable input to the programme, the IPGRI Task Force and the Local Organizing Committee for all their hard work and dedication in handling the logistics, and a special word of thanks must go to Ms Patti Sands who worked so tirelessly to help make the conference a success. Finally, I would like to express my thanks to the editors of this volume who have put together what I believe will be, for many years to come, an invaluable reference text on the science and technology of managing plant genetic resources.

Geoffrey Hawtin
Director General, IPGRI

Jack Hawkes: Plant Collector, Researcher, Educator and Visionary

M.T. Jackson
Genetic Resources Center, International Rice Research Institute (IRRI), Makati City, Philippines

In 1938, Jack Hawkes persuaded his superiors at the Empire Potato Breeding Station in Cambridge, UK, to sanction a visit to Leningrad to meet with Russian scientists working on potatoes and their taxonomy. Jack was scheduled to join an expedition late that same year to collect potatoes in the Americas. While in Leningrad, he met N.I. Vavilov and I suppose that was the start of a life-long interest in genetic resources and their use.

Jack graduated with first class honours from the University of Cambridge in 1937, and received his PhD in 1942 from the same university. His thesis was one of the first studies on the diversity and taxonomy of potatoes, based on the materials he had collected during 1938 and 1939. He continued to work in Cambridge until 1948 when he moved with his wife Barbara and two daughters to Colombia, where his twin sons were born. In Colombia, he helped establish a national potato programme and, with his younger colleague Nelson Estrada, he developed a breeding strategy involving crosses between tetraploid and diploid potatoes that remains until today an important method to broaden diversity in potato breeding. In 1952, he accepted a Lectureship in the Department of Botany at the University of Birmingham. In the 1960s he was first given a Personal Chair, and then appointed Mason Professor of Botany and Head of Department. Jack remained in the department (although it changed its name) until his retirement in 1982.

Jack's contribution to the taxonomy and biosystematics of potatoes has been enormous. Over several decades he returned to Central and South America several times to collect wild potatoes. With his Danish colleague, J.P. Hjerting, he published important monographs on the potatoes of Argentina, Brazil, Paraguay and Uruguay (in 1969) and Bolivia (in 1989). He established crop plant evolution and taxonomy studies as important disciplines at Birmingham. He led a major project that culminated in the publication of a computer-mapped flora of Warwickshire, the very first venture of this type. With colleagues in the university's Medical School, he applied serology to understand species relationships among potato species, another pioneering approach.

I first met Jack in early 1970 when, as an undergraduate student, I applied for a place on the Master's course on genetic resources conservation and use that he had just initiated. Like many of his students, I was infected by Jack's enthusiasm for the study of crop plants and their evolution, in which he brought together so many of his interests, particularly archaeology. The early 1970s were heady days; we were pioneers in the emerging discipline of genetic resources conservation. Jack invited many of the leading lights of the genetic resources conservation movement, among them Erna Bennett and Jack Harlan, to Birmingham, to interact with his students. Through Jack I later met Sir Otto Frankel.

I interacted with Jack professionally on our work and interests in potatoes over 20 years before moving on to rice. I made only one small collecting trip with him, in the Andes of central Peru in 1975. It was an eye-opening experience to collect with someone so knowledgeable about potatoes, their taxonomy and ecology. It seemed as though Jack could almost smell the potatoes without seeing them. We always found small populations in just the places he expected.

Jack's contribution to genetic resources conservation and use has been outstanding. His leading contribution as an educator and visionary has been the training he provided to students, most coming from developing countries. Today, many of these former students occupy important positions in national and international programmes, as both administrators and researchers. There now exists a cadre of trained scientists who understand the technical issues and challenges of genetic conservation.

It has been my pleasure and privilege to work with him; I am proud to have him as my mentor and friend.

The Evolution of Plant Genetic Resources and the Work of O.H. Frankel

J.G. Hawkes
School of Continuing Studies, University of Birmingham, Edgbaston, Birmingham, UK

It is a curious aspect of human thought processes that until we find the right words to encapsulate them, we do not seem to make progress with them. Let me try to explain this.

If we consider the collection of plant diversity we automatically cast our minds back to the genius of the Russian scientist, N.I. Vavilov. Before his time plant breeders made selections and hybrids from the old landraces of their own countries. Now, of course, we realize that a wider range of genetic diversity might well be much more useful in providing genes for disease and pest resistance. Until Vavilov this concept was hardly considered. We have not only come to recognize the need for conserving the genetic diversity of our ancient crop plants, we have also recognized the value of the genetic diversity of related wild species. Some crops, such as wheat and barley, do not have many wild ancestral species related to them. Others, like potatoes, have almost too many related wild species. The cultivated potato, *Solanum tuberosum*, is 'backed up' by some 200 wild species, many of them carrying genes for adaptation and disease resistance that are not known in the cultivated species. This is just one example of the value of wild species related to crops and bearing extremely valuable resistance genes unknown in the crop itself.

By the 1950s breeders were providing farmers with high-yielding and more pest- and disease-resistant varieties. This was of course excellent, but the breeders did not at first realize that these new varieties were replacing the very genetic diversity that they would need as a basis for further advances in the future. Indeed, it was not until the late 1960s and the early 1970s that this paradox became evident and the importance of conserving what were then called 'plant genetic resources' or 'crop plant genetic resources' was recognized.

The leader of this movement, to which I had the honour to belong at its early stages, was undoubtedly O.H. Frankel. I had the greatest good fortune to meet Otto Frankel and to work with him at various Food and Agriculture Organization (FAO) conferences, the first being in 1967 at FAO's headquarters in Rome, Italy. At FAO, Otto had met Erna Bennett, and they worked together to organize the 1967 conference. I took part in that conference and was asked to join their newly formed Panel of Experts to organize a world network of genetic resources institutions. Another panel member was Jack Harlan, whom I also got to know very well.

What were my reactions to these three people? Otto was fiery, complex, practical and a whole mixture of other attributes. He did not suffer fools gladly and said exactly what he thought to whoever he talked to. He was, at the same time, loyal, honest and friendly. I admired him tremendously. He and Erna Bennett used to fight like cat and dog, but she, again, was honest, fierce and outspoken. Jack Harlan, on the other hand, was quiet and much less outspoken but, like the others, was a really excellent scientist. I liked and respected all three of them and I believe the feelings were mutual. I learned a great deal from their experiences and ways of thinking. In his excellent 1999 obituary of Otto, Lloyd Evans quoted Otto, who, when asked how he and his wife got on together in Canberra, Australia, after his retirement, replied, 'We don't mind. We make our own environment.' And this is what Otto had done all his life.

Aside from his interactions with Jack Harlan, Erna Bennett and me, Otto really invented the concepts of genetic conservation of plants useful to man, something that had not previously been clearly thought out. It was also a concept of real value to humanity at a time when the old varieties and landraces of crops were being quickly replaced by new high-yielding varieties throughout the world, and particularly in developing countries.

One aspect of this concept of plant genetic resources conservation struck me very forcefully at our FAO meetings. We knew what ought to be done, but who was going to carry it out? Perhaps some four or five of us knew about genetic resources but this was a worldwide problem, needing many trained scientists from many countries, particularly from the developing world. We needed geneticists, plant breeders, botanists and seed physiologists at least, and in large numbers, but at that time there were not enough of them to carry out these tasks. What could be done?

This was where I believed I could help, by establishing in my Department of Botany at Birmingham a Master's course in the 'Conservation and Utilization of Plant Genetic Resources'. I talked to the Faculty of Science about this and it was agreed to, as long as I did not ask for more money! I assured them that I would use my own departmental resources, which I did to begin with. When the course flourished, I was then able to obtain funds from FAO and other bodies. The first intake of postgraduate students took place in 1969. Otto thought I would not be able to continue for more than a year or two for lack of students. How wrong he was! The course still continues, long after my retirement, and is now in its 31st year. Counting the students who come for only short parts of the course, nearly 1000 students have graduated from the course. So much for Otto's predictions!

I feel very proud that I was able to set in motion such a training course from which these students, mostly from the developing world, have benefited. More importantly, the course has played an important role in conserving the genetic diversity of crop plants and their wild relatives. I know that Otto and Erna would have valued that. And Otto, who was born in 1900 and at 98 survived throughout almost the whole of the 20th century, was, I know, happy that by his efforts and the efforts of those of us who have been associated with him, our task has largely been successful.

Genetic resources exploration, conservation and use are now well known throughout the world, and have been of the utmost value to humanity. The value and use of wild and cultivated genetic resources to breeders have come of age.

1 The History of *ex situ* Conservation and Use of Plant Genetic Resources

G.T. Scarascia-Mugnozza[1] and P. Perrino[2]
[1]*National Academy of Sciences, the Forty, Rome, Italy;* [2]*National Research Council, Germplasm Institute, Bari, Italy*

Introduction

Information on the history of plant genetic resources conservation and use can be found in several reports, catalogues and newsletters published by different institutions (Food and Agriculture Organization (FAO), International Board for Plant Genetic Resources (IBPGR), International Plant Genetic Resources Institute (IPGRI), Consultative Group on International Agricultural Research (CGIAR) and others). Recent developments in the maintenance, management and sustainable use of biodiversity, in general, and of plant genetic resources in particular led to the publication of special issues on the history of conservation and agricultural use of plant germplasm (Scarascia-Mugnozza, 1995, 1998; Pistorius, 1997).

Plants have travelled, during human migrations and along the ancient caravan routes, from continent to continent. Moving from the Old to the New World and vice versa, they have made many important contributions to agricultural and eating habits around the planet (FAO, 1959).

Movement of plants from place to place and from people to people implies the use of germplasm as a food but also for improving agricultural production and for increasing diversification. Starting from the beginning of agriculture, man has stored plants and seeds from one cycle of cultivation to the next in different ways, some of which are known to us and are still used today. Storage of germplasm also took place during migration. As agriculture progressed and human population increased, the need to store plants and seeds *ex situ* grew and involved even longer distances and lapses of time than the short break between seasons of cultivation or time needed to migrate. In a broad sense, *ex situ* conservation of germplasm is a practice that humans have used since the beginning of agriculture, to expand cultivation and/or to colonize new lands and to ensure the spread of agriculture around the world.

This chapter aims to trace a historic profile of *ex situ* plant germplasm conservation with a main emphasis on the aspects most relevant to protection of genetic diversity and its use for agricultural development.

History of *ex situ* Plant Germplasm Conservation

From the beginning of agriculture, farmers have domesticated hundreds of plant species and within them genetic variability has increased owing to migration, natural mutations and crosses, and unconscious or conscious selection. This gradual and continuous expansion of genetic diversity within crops went on for several millennia, until scientific principles and techniques influenced the development of agriculture. This happened at the beginning of the 20th century, when Mendel's laws

(eds J.M.M. Engels, V. Ramanatha Rao, A.H.D. Brown and M.T. Jackson)

were used. The spread of new and more productive crop varieties, which were genetically less heterogeneous than primitive populations, paradoxically started the well-known process of 'genetic erosion'.

In fact, in the 1920s and 1930s, N.I. Vavilov and Jack Harlan began to notice that traditional crop varieties, or landraces, were being lost from cultivated fields around the world. Since then, scientific efforts to conserve plant genetic diversity also focused on collecting material and placing it in *ex situ* storage. In this chapter, we have divided all of the efforts made by humans on *ex situ* germplasm conservation into the following sections:

- conservation and use of crop germplasm before the 1967 FAO/IBP (International Biological Programme) Technical Conference;
- plant genetic resources developments since 1967;
- ex-situ collections and genebanks;
- genetic resources: maintenance, use and regulation.

Conservation and Use of Crop Germplasm before the 1967 FAO/IBP Technical Conference

Before 1967, the international exchange of genetic resources functioned mainly among the network of plant introduction stations in western Europe, the USA, Australia, New Zealand and eastern Europe (mainly the Soviet Union). There were only a few genebanks (introduction stations) that exchanged genetic material. The oldest and most famous of them, with worldwide scope and adequate evaluation facilities, were:

1. The All-Union Institute for Plant Industry, in Leningrad (now St Petersburg), Russia (1920).
2. The Commonwealth Potato Collection at Cambridge, UK (before the Second World War).
3. The collections for research programmes of the Rockefeller Foundation in the USA (1943).
4. The National Seed Storage Laboratory (NSSL) at Fort Collins, Colorado, USA (1958).

Most of the other seed banks were inadequate with regard to the later requirements of international agricultural research and most collections were considered erratic or unreliable. Moreover, most collections required frequent regeneration. In spite of the circumstances under which most of the collections, of truly international scope, were held in the 1950s, the overall picture on a regional level showed promising initiatives.

For example, in West Africa, plant quarantine regulations were initiated in the late 1950s. The Inter-African Phytosanitary Convention of 1954 was the umbrella of the Organisation of African Unity. Ghana established a Plant Exploration and Introduction Service. In the 1960s, universities in Nigeria and Côte d'Ivoire started collecting activities. In Latin America, plant exploration started in the 1950s and 1960s. Argentina started a National Service, Venezuela began maintenance of papaya and oil crops, Colombia focused on potato and grasses, and Costa Rica and Mexico on cacao. Most of these activities had a strong interaction with the Rockefeller Foundation. In Asia, India started collecting pulses, cruciferous and forage crops in the late 1960s. Other collections were started by the International Rice Research Institute (IRRI) in the Philippines and other countries. In Japan, the National Seed Storage Laboratory was opened in 1966 (Ito, 1972). Australia introduced much germplasm in the late 1960s, although a Plant Introduction Service, under the Australian Commonwealth Scientific and Industrial Research Organization (CSIRO), was established in 1930.

However, the collection of plant genetic resources was more significant in North America and Europe as well as the activities undertaken by FAO.

North America

Official government recognition of the importance of agricultural development first came in 1827, with President John Adams. Only in 1898 was a great impetus provided through the creation of the Office of Foreign Seed and Plant Introduction. In the early 1960s, this office was renamed the New Crop Research Branch, with its headquarters located at the USDA Plant Industry Station in Beltsville, Maryland.

Owing to increasing demands for food and fibre and the industrialization of agriculture, crop improvement between 1900 and 1930 was mostly concerned with adaptation and yield factors of new varieties. However, even before the Second World War, breeders had problems related to disease resistance, quality, planting and harvesting methods, and reactions to plant protection or weed control practices. Therefore, germplasm material in US genebanks was mainly collected for short-term use in specific breeding programmes or for tests regarding agricultural diversification.

Once the demand of researchers was satisfied, the germplasm was stored at the NSSL at Fort Collins, Colorado. This laboratory was created in 1958 and was the first genebank with long-term seed storage equipment.

Advanced and well-organized collection of germplasm was coordinated by the Rockefeller Foundation in the USA in the 1940s and 1950s. A programme for the improvement of basic crops, primarily maize, wheat and potato was started. This was called the Mexican Agricultural Program, and was guided mainly by N.H. Borlaugh (1970 Nobel Peace Prize winner). Later on, this programme gave rise to the so-called 'Green Revolution'. Similar projects were carried out in Guatemala, El Salvador, Venezuela, Brazil, Uruguay, Argentina, Costa Rica, Cuba, Colombia, Peru and Chile, under the auspices of US and American Land-Grant Universities. These collections, according to some scientists, formed the basis for a global network recommended by the FAO Panel of Experts in the early 1970s.

An important initiative of the USA was, then, the establishment of four Regional Plant Introduction Stations, at Ames (Iowa, 1947), Geneva (New York, 1948), Experiment (Georgia, 1949) and Pullman (Washington, 1952) and one inter-regional programme on potato. The stations were coordinated by the New Crops Research Branch at the Agricultural Research Stations (ARS) in Beltsville. Thus, the NSSL at Fort Collins, Colorado, was used for long-term storage and preservation of valuable plant germplasm propagated by seed, while the four smaller Regional Stations were used to maintain stocks for ongoing breeding work.

Europe

In the 1920s the USSR went through a rapid industrialization of its agricultural sector. Vavilov made successful collecting expeditions during the 1920s and 1930s; not only in the USSR, but also in over 50 countries in Asia, the Americas, Northern Africa, Europe and the Mediterranean basin. In all, 50,000 seed samples, mainly of wheat, rye, oat, pea, lentil, chickpea and maize, were collected and provided the basis for the establishment of modern genebanks in the USSR (Plucknett *et al.*, 1987).

Vavilov's ideas led him to establish the well-known concept of the 12 Vavilovian centres of crop diversity. Plant breeding approaches after Vavilov's death urged the USSR to collect new material. In the early 1960s, Zhukovsky (from the N.I. Vavilov All-Union Scientific Research Institute of Plant Industry, VIR, Leningrad) complained about the lack of a continuous introduction of new material of cotton, maize, potato, bean, pumpkin, tomato, pepper, tobacco and groundnut. In fact, despite Vavilov's efforts the institute's collections had considerable gaps (especially from Latin America, Australia, the Balkans and the Iberian Peninsula). In the 1960s, Zhukovsky was the only Russian contacting FAO in order to organize expeditions to Latin America to introduce potato species resistant to virus degeneration, races of *Phytophthora*, nematodes, Colorado beetle, *Epilachna* and other pests (Whyte, 1958).

Germplasm activities were also initiated at the Institut für Kulturpflanzenforschung, founded in 1942 at the Tuttenhof domain, near Vienna. The first director was Hans Stubbe. During the Second World War, in 1945, the Institute was moved to Quedlinburg and in 1946 to Gatersleben (German Democratic Republic). In 1948, it was integrated into the German Academy of Science in Berlin, renamed in 1968 Zentralinstitut für Genetik und Kulturpflanzenforschung, under the direction of Rudorf Mansfeld (1949–1960) and S. Danert (1961–1970).

In Western Europe, there was a general need for an international central organization. An attempt to solve the problem of maintaining genetic stocks of potato species and varieties in their original integrity and free from diseases was made by the British Commonwealth, which had established the Commonwealth Potato Collection (and by the USA with its IR-1, Inter-Regional Potato Introduction Project at Wisconsin). However, for several reasons, such as lack of updated reports, and because breeders had discarded more promising lines, this initiative did not work (Hawkes, 1961).

The aforementioned situation was typical for other crops. For example, in Italy, since the beginning of the century, the improvement of *Triticum* species was based, by N. Strampelli and later on by others, on the collection and use of the wide genetic diversity available in landrace populations and also on distant germplasm, for example from Japan (Scarascia-Mugnozza and Porceddu, 1972; Porceddu, 1972). This European experience led to the conclusion that an international station could be the solution to the problem of conservation.

The European Society for Research and Plant Breeding (EUCARPIA) was established in 1956 in the Netherlands. In the early 1960s it was the first organization to promote a collecting network. It started off with a more generalist ecogeographical orientation (ecoregional genebanks). In 1962, the third EUCARPIA General Congress, in Paris, emphasized the danger of loss of genetic resources. In 1966, EUCARPIA delegates advised the European plant breeding institutes to start a collaboration through regional genebanks on the continent.

The proposal resulted in the establishment of the following four sub-regional genebanks:

- North Western Europe. The bank was established at the former Institute of Crop Science and Seed Research of the Federal Agricultural Research Centre (FAL) at Braunschweig-Völkenrode, West Germany. Today, the genebank is part of the Federal Centre for Breeding Research of Cultivated Plants (BAZ). Dieter Bommer was the main initiator.
- Central and Eastern Europe. In the original plan, more than one genebank would be needed, for example, in Leningrad (already existing) and Gatersleben (already existing). Hans Stubbe (1942), O. Schwarz (1946), R. Mansfeld (1949), S. Danert (1961) and C.O. Lehmann (1970) were the main actors.
- Southern Europe, including the Mediterranean region. The genebank was established at Bari, Italy. The main instigator was G.T. Scarascia-Mugnozza.
- Scandinavia. The bank was established at Lund, Sweden, as a cooperative effort among the Nordic countries (Denmark, Finland, Iceland, Norway and Sweden). The main instigators were Ebbe Kjellqvist and Stig Blixt.

Although none of the genebanks acted as a sub-regional centre, they were all very active and successful. In the 1970s, a special section of EUCARPIA, the Genebank Committee, formally linked to the section Wild Species and Primitive Forms, was established (Hawkes was the first Chairman). There were annual meetings of members and genebank directors from Eastern Europe.

FAO

In the 1950s and early 1960s, the other major actor in the conservation of genetic resources was FAO. Several World Catalogues of Genetic Stocks (wheat, rice, maize, barley) were set up in the late 1950s, while the FAO Plant Introduction Newsletter was seen as a key initiative and intermediary between breeders of the world. In the 1960s, the FAO Plant and Protection Service dealt with a continuous stream of enquiries for samples of seed or vegetative material for use by breeders. FAO genetic resources policies show a significant distinction between planning and actual programming. A strong call for immediate action in conservation, particularly for landraces and wild relatives, had already been made during the 10th Session of the FAO Conference in Rome, November 1959. The need for a truly intergovernmental initiative to streamline germplasm conservation and distribution was recognized during the 1961 Technical Meeting on Plant Exploration and Introduction, but was not worked out during later years. The dominance of breeders in the 1960s had a double impact: (i) conservation and use were closely linked; and (ii) storage in the first instance took place in industrialized countries and was tied to plant breeding institutes. Naturally, the history of FAO in the collection and exchange of germplasm is much richer in events (Pistorius, 1997).

The 1961 Technical Meeting on Plant Exploration and Introduction was the first initiative on a multilateral basis with the aim of extending initiatives in the field of plant introduction. For this purpose it was suggested that: (i) National and Regional Introduction Stations should be set up under the aegis of FAO; and (ii) Exploration Centres should be built in regions of greatest genetic diversity and serve as centres for research on environmental interaction and add knowledge on landraces and wild relatives (Rudorf, 1961).

A pilot Exploration Centre was established in 1964, at Izmir in Turkey, which started under a joint project between the Turkish government, the UN Development Programme/Special Fund and FAO. It also acted as Regional Centre for Afghanistan, Iran, Iraq, Pakistan, Syria and Turkey, and organized germplasm collection, conservation and evaluation. For some reason the Izmir Centre began to work successfully only later, in the middle of the 1970s, and played its main role at national level (FAO/UNDP, 1970; Frankel, 1985). A FAO-Unit of Crop Ecology and Genetic Resources was established in Rome, in 1967, under the guidance of R. Pichel and E. Bennett.

Plant Genetic Resources Developments since 1967

1967 FAO/IBP Conference

The 1967 FAO/IBP Technical Conference on the Exploration, Utilization and Conservation of Plant Genetic Resources, held in Rome and organized by the International Biological Programme (IBP) and FAO, was an early occasion for a group of experts, later referred to as the Panel of Experts on Plant Genetic Resources, to define a global strategy for the conservation of plant genetic resources.

In particular, the rising concern about possible genetic erosion of landraces and wild relatives due to modern agriculture, and the more general, increasing need of the agro-industry for a steady flow of new germplasm, convinced the members of the Conference to give more consideration to the generalist approach of conservation. New cold-storage techniques,[1] developed in the 1960s, made long-term *ex situ* storage possible.

The alternative of *in situ* conservation was brought forward as well, although this met considerable opposition from the scientists who emphasized the direct use of genetic resources in mainstream breeding programmes.

The scientific arguments for *in situ* conservation, as complementary and as an alternative to *ex situ*, were mostly based on genecological premises. In any case, soon after the 1967 Conference, *in situ* conservation did not materialize and remained merely a theoretical dispute among breeders and geneticists. In general, the debates dealing with arguments on genetics, single-gene resistance versus polygenic resistance, socio-economic implications of single-gene resistance, genecological ideas and breeding strategies, led to conclusions in favour of *ex situ* conservation (FAO, 1969).

The report of the 1967 Conference ('Genetic Resources in Plants – Their Exploration and Conservation') confirmed a general consensus on the need for more facilities and efforts for both *in situ* and *ex situ* conservation. During the following decade, as already mentioned, *ex situ* conservation became the dominant conservation strategy.

The 1967 FAO/IBP Technical Conference generated some important guidelines for the establishment of a global network for *ex situ* long-term conservation. The conversion of these guidelines into practical action took place quickly. A Panel of Experts on Plant Exploration and Introduction (established in 1965) generated a Plan of Action presented during the 1973 FAO/IBP Technical Conference on Plant Genetic Resources (Rome, 12–16 March), published under the title *Crop Genetic Resources for Today and Tomorrow* (Frankel and Hawkes, 1975).

The most important achievement of the Panel of Experts (which met six times from 1966 to 1975) was the formulation of basic criteria for conservation and use of genetic material. These were: (i) that plant material was to be made available immediately and without restriction to all breeders requesting it; and (ii) that genetic variability had to be maintained for future generations in long-term storage under conditions of maximum physical and genetic security (FAO, 1969).

During the Third Session of the Panel of Experts, in Rome, 1969 (FAO, 1969), the Panel pointed out a few regions in the world and the corresponding native crops that needed immediate attention: the Near East (wheat, etc.), the Sudanian zone of Africa (*Oryza glaberrima*, etc.), southern and eastern Africa (forage grasses), Ethiopia (various local varieties), South and Central America (cotton, etc.), South-East Asia (tropical fruits, etc.) and Oceania (yam).

This priority list was very broad and not practical. In its last formal meeting (March 1975), the Panel made a modified ranking. It recommended that the FAO first had to look for cooperation with the Germplasm Laboratory at Bari, to coordinate further exploration in the west and central Mediterranean regions. Ethiopia was given second priority, and the third was for some tropical crops. These modified criteria illustrate a shift from a crop-oriented to a region-oriented approach.

A third important result of the Panel was a categorization of *ex situ* collections: base collections (for long-term conservation), active collections (for research and distribution) and working collections (usually maintained at plant breeding institutions).

These developments led to the planning in 1973 of another joint FAO/IBP Technical Conference on Crop Genetic Resources. This differed from the one in 1967 in that it did not formulate scientific parameters for the conservation and use of plant genetic resources. However, the book that came out of it (Frankel and Hawkes, 1975) formulated practical

[1] There is a rich literature on research and results of methods for sampling, conservation techniques in low temperature chambers, for engineering aspects of units for long-term conservation of germplasm, for methods of rejuvenation, and so on.

action plans on: optimal sampling strategies in genetic conservation, sampling techniques for *ex situ* collections, methods of exploration in seed crops, vegetatively propagated crops and tree species, and long-term storage of seed and pollen.

Genetic erosion issue reaches the global scale

Although the Panel of Experts had been very concerned about the negative consequences of genetic erosion, the issue received little public attention in the 1960s. The early 1970s, however, saw an unexpected recognition of the issue as the world witnessed two serious consequences of genetic erosion. In 1970, a serious outbreak of southern corn-leaf blight in the USA reminded the scientific world that genetic variability was not always enough if the cytoplasm is entirely of one kind. In the same year, a catastrophic outbreak of coffee rust caused great losses in Brazil with higher coffee world market prices as a consequence. These cases provoked publicity on a global scale, generating a flow of information on other, ongoing cases of genetic erosion. The Panel presented all these facts.

In 1972, within the agricultural community of the USA, the Agricultural Board and the National Research Council published a report, *Genetic Vulnerability of Major Crops*, which attracted much attention, within both scientific circles and circles of agricultural non-governmental organizations (NAS, 1972). The key lesson of the 1970s was that genetic uniformity is the basis of vulnerability to epidemics and, more generally, to biotic and abiotic stresses. The situation poses substantial challenges to scientists and to the nation involved (NAS, 1972).

Clearly the market wants uniformity. The irony is that the uniformity of crops facilitates epidemics and could be a synonym of vulnerability and crop damage; unfortunately the scientist, not the market, tends to receive the blame.

The United Nations Conference on Human Environment (UNCHE, 1972)

The UNCHE was held in Stockholm in June 1972. The influence of the ideas of the FAO Panel of Experts became significant in the clear division of labour between *in situ* and *ex situ* conservation. It was stated that both are needed, but it became clear that genetic resources with an agricultural value had to be conserved in 'national or regional genetic conservation centres' *ex situ*, such as the NSSL, USA or the VIR, USSR.

Wild relatives of crop species, on the other hand, would be maintained in their 'natural environment' (*in situ*) for which the UNESCO Man and Biosphere Programme was recommended to fulfil the important role. To establish a network of *ex situ* genetic resources, the UNCHE recommended 'that the appropriate UN agency establish an international liaison unit for plant genetic resources' and 'to provide the secretariat for periodic meetings of international panels and seminars on the subject; a conference on germplasm might be convened to follow up the successful FAO/IBP conference of 1967'.

Plans for a global network of genebanks

After Stockholm 1972 and the 1973 FAO/IBP Technical Conference there was no direct impact in terms of more collections and storage activities. But, there was a rapidly growing network of several International Agricultural Research Centres (IARCs), belonging to the CGIAR network. The 'establishment of a mechanism to encourage, coordinate and support action to conserve genetic resources and make them available for use' suggested during a meeting (1972) of the Panel in Beltsville, Maryland, was submitted to the CGIAR Technical Adviser Committee (TAC). This move had tremendous consequences for the position of scientists, governments and international agencies (FAO, first of all) in the conservation of genetic resources.

The plan embraced four elements:

- To create a coordinating centre (which later became the International Board for Plant Genetic Resources – IBPGR).
- To stimulate the establishment of genebanks in international centres already existing in developing countries (those already established were: IRRI, in 1960, International Maize and Wheat Improvement Centre (CIMMYT), in 1966, International Centre for Tropical Agriculture (CIAT), in 1967, and International Institute for Tropical Agriculture (IITA), in 1968).
- To establish genebanks in new international centres: the West African Rice Development

Association (WARDA, 1971), the International Potato Centre (CIP, 1971), and the International Crops Research Institute for the Semi-Arid Tropics (ICRISAT, 1972). After 1972, to these organizations already existing, some others, like the International Livestock Centre for Africa (ILCA, 1974) and the International Centre for Agricultural Research in the Dry Areas (ICARDA, 1976) were also included in the network.

- To establish new 'regional' genebanks in the Vavilovian centres of crop diversity; following an 'ecogeographical approach', the centres would serve as a global division of labour in collection efforts.

While the scientific orientation of the FAO Panel embraced general incentives to conserve genetic resources, the institutional orientation of CGIAR was to extend its mandate in considering genetic erosion as a threat to the further development of the Green Revolution.

The establishment of IBPGR

The IBPGR, established in 1974, was administratively integrated into FAO, but technically it was autonomous. The establishment of IBPGR accelerated the formation of a world network of genebanks. In this context, IBPGR had a precise and important task: 'to promote and assist in the worldwide effort to collect and conserve the plant germplasm needed for future research and production' (TAC, 1972). IBPGR replaced the activities of the Panel of Experts.

However, IBPGR might never have existed, or at least not in its present form, if FAO had not seen the need for conserving the genetic resources of crops and had not set up a Panel of Experts to provide the necessary scientific basis for this to be undertaken (Frankel, 1985). IBPGR, on the other hand, offered FAO the financial support and manpower to build up and coordinate a new world network of genebanks.

IBPGR also managed to create and maintain standard criteria for the upkeep of collections. The institutions in the network that preserve genetic material on a long-term basis, form the core of IBPGR's international network. By 1984 it consisted of 40 collections in about 30 countries. Under the impulse for the preservation of their genetic resources and the need to establish ad hoc structures, several developing countries were pushed to reorganize their national agriculture systems, including institutes or research groups for the attainment and maintenance of genetic resources collections.

The direct link with CGIAR implied that conservation strategies were defined as direct requirements for international agricultural research, in its need for landraces (and wild relatives) threatened by replacement by major crops, instead of those for local populations in developing countries. This 'supply function' also implied that conservation tended to focus on major crops (most of all, rice, maize, potato, wheat and sorghum) and constituted a clear break with the FAO Panel of Experts' initial view.

Political and scientific issues and the establishment of the FAO Commission on Plant Genetic Resources

The term 'genetic resources' refers to the genetic information contained in the genes, the terms 'biological diversity/biodiversity' encompass all species of plants, animals and microorganisms, including ecosystems and ecological processes. In the late 1960s and 1970s a group of 77 developing countries proclaimed a New International Economic Order (NIEO) in several fora, including FAO. Other driving forces were non-governmental organizations (NGOs) and the Rural Advancement Foundation International (RAFI). The NGOs' constant lobbying within FAO against the North's dominance in the exchange and use of genetic resources and against the neglect of the rights and needs of small farmers, helped to shape the genetic resources issue as it stood throughout the 1980s. The International Coalition for Development Action (ICDA), renamed Genetic Resources Action International (GRAIN), was also an important source of information for NGOs. In the late 1980s, GRAIN published *New Hope or False Promise? Biotechnology and Third World Agriculture* (Hobbelink, 1987).

Scientists within the CGIAR network tried to convince NGO critics that the results from the research mostly benefited developing countries. RAFI and GRAIN reacted by saying that the CGIAR centres were the main driving forces behind the Green Revolution, the creators of monocrop cultures and the main contributors to genetic erosion.

During the 20th Conference, in 1979, a number of developing countries started to ask for information about the following issues (Esquinas-Alcázar, 1989):

- Who owns the genetic resources collected with international money and stored in countries other than those in which they were collected? Who will guarantee their long-term security?
- What guarantee is there for continued free exchange of material in *ex situ* collections?
- How can countries benefit from the plant genetic resources that their farmers have produced, improved and conserved over millennia, as they currently lack the technical and financial capacity to use these resources for their own benefit?

Throughout the 1980s, FAO remained the principal forum in which developing countries tried to pursue their interests, which comprised attempts to support the establishment within FAO of an international legal framework to set global standards and rules for the conservation and exchange of genetic resources, with the aim that genetic resources would remain available in the public domain.

During the 21st Conference, on 25 November 1981, Resolution 6/81 was approved, which in its statement (e) emphasizes the lack of an 'international agreement for ensuring the conservation, maintenance, and free exchange of genetic resources of agricultural interest contained in existing germplasm banks'. This resolution became one of the most hotly debated in the history of FAO. The genetic resources issue, having been a matter of discussion by experts during most of the 1970s, suddenly in 1981 received worldwide press coverage. NGOs were actively in favour, while industrialized countries, particularly the USA, UK and Australia heavily opposed the resolutions.

Nevertheless, the 22nd FAO Conference (1983) approved the proposal for an International Undertaking, which includes 11 articles and the establishment of the FAO Commission on Plant Genetic Resources (CPGR). In spite of the limited support from industrialized countries for the Undertaking, the 22nd FAO Conference was considered a major victory for developing countries. The first session of the CPGR took place in March 1985 and since then it has met every 2 years. The discussion on the alternative FAO genebank network (point 2 of Resolution 6/81) proved that the idea of establishing a physical genebank at FAO headquarters was not feasible. The proposal for an international network of storage facilities, in the frame of CGIAR appeared more workable.

Global System on Plant Genetic Resources (1983)

The adoption of the FAO International Undertaking on Plant Genetic Resources (1983) confirmed and strengthened the establishment of an international network. Article 7 of the Undertaking asked for coordinated action by all institutions involved in the conservation of genetic resources to develop a Global System on Plant Genetic Resources for Food and Agriculture (Fig. 1.1), whose objective is to ensure the safe conservation, and promote the availability and sustainable utilization of plant genetic resources for present and future generations, by providing a flexible framework for sharing the benefits and burdens (FAO, 1995b).

The Global System is a means to effectuate the decisions of CPGR, but would also consist of *in situ* conservation areas and an *ex situ* network of active and base collections, and form a core to strengthen FAO's political position in the collection, conservation, evaluation and documentation, as well as the use of genetic resources. The CPGR and the Undertaking are the main support for the Global System. This falls into three main elements:

1. International agreements: (i) Code of Conduct for Plant Germplasm Collection and Transfer; (ii) Code of Conduct on Biotechnology; and (iii) Basic agreements on genebanks.
2. Three network systems (global mechanisms): (i) the world Information and Early Warning System on Plant Genetic Resources; (ii) a network of *ex situ* genebanks (since 1994, also including the CGIAR collections); and (iii) a network of *in situ* and on-farm conservation areas.
3. Global Instruments: (i) State of the World's Plant Genetic Resources; (ii) Global Plan of Action for the Conservation and Sustainable Utilization of Plant Genetic Resources for Food and Agriculture (GPA); and (iii) International Fund. (FAO, 1996b).

The establishment of IPGRI

In 1989–1990, the CGIAR, following a proposal by its Technical Advisory Committee, supported 'the establishment of IBPGR as an international organization independently managed and preferably located near FAO headquarters in Italy' (TAC, 1986). The International Plant Genetic Resources Institute (IPGRI), as successor of IBPGR, was for-

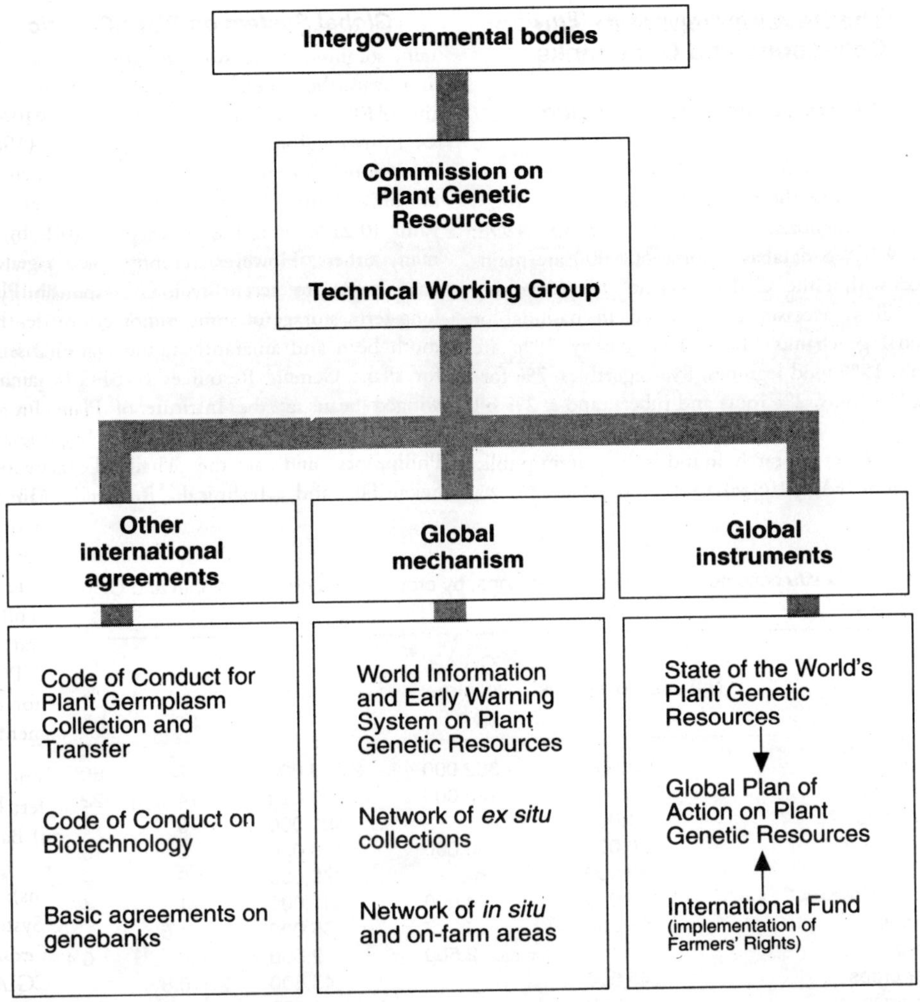

Fig. 1.1. The global system for the conservation and utilization of plant genetic resources for food and agriculture (source: FAO, 1995b).

mally established on 9 October 1991, and ratified by the Italian Parliament in March 1994.

IPGRI does not have research facilities but operates primarily as a catalyst and facilitator, contracting most of its research to partner organizations. In this way, IPGRI works to enhance strategic and adaptive research aimed at solving key genetic resources problems. IPGRI is also a specialized development agency that provides direct technical support to national plant genetic resources programmes. Its way of working is based on strong linkages with many partners, a proactive, bottom-up approach and needs-driven objectives. The flexibility of this approach allows the Institute to respond to changing needs and circumstances and to take a broad view of biodiversity issues in general and the conservation and use of plant genetic resources in particular. The Institute's partners are found at all levels of genetic resources work: national plant genetic resources programmes, research institutes (international and regional), organizations, universities, herbaria and botanical gardens, the private sector, non-governmental organizations, community-based organizations including farmers and women's groups (various IPGRI newsletters, Scarascia-Mugnozza, 1999).

The Development of *ex situ* Collections and Genebanks

Size of collections worldwide

At present, over 6 million accessions are stored *ex situ* throughout the world (Plucknett *et al.*, 1987; Scarascia-Mugnozza, 1995, 1998; FAO, 1998; FAO WIEWS database): some 600,000 are maintained within the CGIAR system, the remaining 5.4 million accessions are stored in national or regional genebanks (Table 1.1). Nearly 39% are cereals, 15% food legumes, 8% vegetables, 7% forages, 5% fruits, 2% roots and tubers, and *c.* 2% oil crops. Spices and medicinal, aromatic and ornamental species are rarely found in long-term public collections (FAO, 1996a).

For wheat, rice, potato, cassava, banana/plantain, sorghum, yam, sweet potato, chickpea, lentil and bean, the largest *ex situ* collections are held by the IARCs; for other crops the largest collections are in national institutions. Minor crops are poorly represented in *ex situ* collections; for example, there are only about 12,000 accessions of all species of yams (0.21%) and less of coconut (0.16%) and many others. However, recently some genebanks have begun to accept regional responsibility for long-term storage of some minor crops: rice bean, moth bean and amaranth at the National Bureau for Plant Genetic Resources (NBPGR) in India; winged beans at the Institute of Plant Breeding College of Agriculture (IPB-UPLB), Laguna, in the Philippines and at the Thailand Institute of Scientific and Technical Research (TISTR),

Table 1.1. *Ex situ* collected and stored accessions, by crop, maintained in national and CGIAR genebanks.[a]

Crop	National collections (a)	CGIAR centres (b)	Total[b] (c)	% a/g	% b/g	% c/g
Cereals	1,971,000	362,000	2,333,000	37	60	39
Food legumes	758,000	132,000	890,000	14	22	15
Vegetables	481,000	—	481,000	9	—	8
Forages	350,000	58,000	408,000	6	10	7
Fruit	279,000	—	279,000	5	—	5
Roots and tubers	77,000	24,000	101,000	1	4	2
Oil crops	95,000		95,000	1.8		1.7
Banana		2,500	2,500		0.4	< 0.1
Sugar crops	45,500		45,500	0.8		0.7
Beverages	43,000		43,000	0.8		0.5
Condiments	17,700		17,700	0.3		0.3
Cacao crops	9,400		9,400	0.2		0.2
Rubber	31,000		31,000	0.6		0.5
Fibre crops	76,300		76,300	1.4		1.3
Narcotics and drugs	28,000		28,000	0.5		0.5
Shelter crops	10,000		10,000	0.2		0.2
Ornamentals	23,200		23,200	0.4		0.4
Medicinal plants	2,300		2,300	< 0.1		< 0.1
Dyes	1,000		1,000	< 0.1		< 0.1
Perfume crops	600		600	< 0.1		< 0.1
Building materials	400		400	< 0.1		< 0.1
Others	1,100,600	21,500	1,122,100	20	4	19
Total (g)	5,400,000	600,000	6,000,000	100	100	100

[a] *Ex situ* collections consist of seed genebanks, field genebanks and *in vitro* genebanks.
[b] Seed genebanks (5,435,000 accessions); field genebanks (527,000 accessions); *in vitro* genebanks (38,000 accessions). Considering duplicates within and between collections, the total number of accessions is estimated to be from 1 to 2 million.
Sources: Plucknett *et al.*, 1987; Scarascia-Mugnozza, 1995; FAO WIEWS database, 1996; FAO, 1998.

Bangkok; faba beans at the International Centre for Agricultural Research in Dry Areas (ICARDA), Aleppo, in Syria; and adzuki beans at the National Institute of Agrobiological Research (NIAR), Tsukuba, in Japan. It is interesting to note that, for many export crops and commodities, large percentages of the global collections are concentrated in a few countries: oil palm accessions in Zaire (83%), most rubber accessions (76%) in Malaysia and coconut (22%) in Sierra Leone (FAO, 1998).

According to the type of genetic variation within *ex situ* collections, about 32% are advanced cultivars, 24% are landraces or old cultivars, 5% are advanced landraces and 8% are wild or crop relatives (FAO, 1998).

Some countries are consolidating national collections (*in situ* and *ex situ*) of indigenous genetic resources, which are of potential importance largely to the country itself. Quantitative estimates for 24 countries (Table 1.2) show that only a few countries (for example, Ethiopia, Cyprus and China) store high percentages of indigenous plants, of sizeable collections, although it is impossible to say how representative current *ex situ* collections are of total diversity in the respective countries. However, landraces of cereals are probably better represented than those of pulses, root crops, fruit and vegetables, excluding potato and tomato (FAO, 1996a).

Coverage of wild relatives is limited, and that of forest ornamentals, aromatic and medicinal species is minimal. Worldwide, there are gaps in the collections of minor crops and under-utilized species, especially for landraces and wild relatives from their centres of diversity and cultivation. Targeted collection of selected species and assessments of the genetic diversity of landraces are, therefore, priorities (Padulosi, 1996).

Table 1.2. Indigenous accessions in national genebanks. (From 24 Country Report, FAO, 1998.)

Country	Percentage
Europe	
Bulgaria	12
Czech Republic	16
Republic of Moldova	40
Romania	71
Slovakia	8
Belgium	75
Africa	
Cameroon (roots and tubers)	75
Cameroon (fruits)	25
Ethiopia	100
Mauritius	100
Angola	100
Malawi	100
Namibia	100
Senegal	10
Near East	
Islamic Republic of Iran	95
Cyprus	100
Iraq	22
America	
Brazil	24
Colombia	55
Ecuador	52
United States	19
Asia	
China	85
Democratic People's Republic of Korea	20
Republic of Korea	18
Sri Lanka	67

Storage facilities

Methods for germplasm conservation are determined by a number of factors: purpose (programmes to conserve genetic diversity for posterity or for present and short-term use), storage behaviour (type of seeds: orthodox, recalcitrant or intermediate) and resources (financial, human and institutional capacities and technologies) (Table 1.3) (FAO, 1996a).

By the end of the 1970s, there were about 54 seed stores. Today, besides the genebanks of the IARCs, there are over 1300 national and regional genebanks and seed storages on field collections, 397 of which are maintained under long- or medium-term storage conditions. In fact, many countries have more than one *ex situ* facility or collection (70 in India), although in many cases these are active or research collections (FAO, 1998). Many of the collections are breeders' or working collections and are likely to be short term. Of the 1308 genebanks registered in the World Information and Early Warning System (WIEWS) database, 496 (38%) are located in Europe, 328 (25%) in the Americas, and 293 (22%) in Asia (Table 1.4). The 15 largest national base collections together hold about 1.7 million accessions or 34% of the national seed collections (Table 1.5).

A total of 75 countries have facilities for medium- and long-term storage, but in many cases the major

Table 1.3. Technologies for *ex situ* conservation according to the type of plant genetic resources. (From FAO WIEWS database, 1996; FAO, 1998.)

Storage technology	Type of genetic material	Adequate storage
Desiccated seeds at low temperature (−18°C) and 3–7% moisture content	Orthodox seeds	Long-term conservation (base collection); provision of accessions for use (active collection)
Desiccated seeds at cool temperature	Orthodox seeds	Provision of accessions for use (active and working collections); medium-term conservation (base collections)
Ultra-dry seeds at room temperature	Orthodox seeds	Medium- to long-term conservation
Storage of dried seeds at room temperature	Some long-lived orthodox seeded species	Provision of accessions for use (active and working collection)
Cultivation of entire plants in field genebank	Vegetative species and some non-orthodox seeded species	Short or medium-term conservation (base collections); provision of accessions for use (active collections)
Slow growth under serial *in vitro* propagation	Vegetative species and some non-orthodox seeded species	Medium-term conservation; provision of accessions for use (active collections)
Cryopreservation at −196°C in liquid nitrogen or −154°C to −196°C in N_2 vapour	Seeds, pollen, tissue, cells, embryos of species capable of *in vitro* regeneration after drying and freezing	Long-term conservation
Freeze-dried seeds or tissue	Seed or plant tissue	Medium- to long-term conservation, depending on the species

constraints to reliable storage are: equipment, suitable drying facilities and electricity supplies. Based on the Country Reports, secure, long-term seed storage facilities are found in 35 countries: 15 in Europe, seven in the Americas, five in Asia, four in Africa and four in Near East regions. A further 56 countries have facilities for short- to medium-term storage only (FAO, 1998).

According to FAO WIEWS database, only about 56% of the accessions are stored in medium- or long-term conditions. Nearly 8% are stored in short-term, 1% in short/medium and 10% in field collections, *in vitro* or under cryopreservation. No information is available for 25% of the accessions.

Field genebanks and in vitro *facilities*

Plant species that are vegetatively propagated, have long life cycles or produce short-lived (recalcitrant) seed, are usually maintained in field genebanks, existing in at least 103 countries. These include crops like fruit trees of temperate zones, potato, cassava, banana/plantain, yam and tropical fruits, rubber, coffee, cocoa and coconut. Nevertheless, improvement and development of appropriate conservation technologies for vegetatively propagated plants, as well as for species with non-orthodox seeds, are needed (FAO, 1998).

Approximately 527,000 accessions are stored worldwide in field genebanks: 284,000 in Europe, 10,000 in the Near East, 84,000 in Asia and the Pacific, 16,000 in Africa and 117,000 in the Americas (FAO, 1998). More than 60 countries have *in vitro* conservation facilities. *In vitro* storage is an alternative or complementary method for conserving vegetatively propagated plants, or those with long life cycles and with recalcitrant seeds, which has been developed up to now for a small number of species. The technology requires expen-

Table 1.4. Number of genebanks and accessions in *ex situ* collections, by region. (From FAO WIEWS database, 1996; FAO, 1998.)

	Accessions		Genebanks	
Region	Number	%	Number	%
Africa	353,523	6	124	10
Latin America, the Caribbean	642,405	12	227	17
North America	762,061	14	101	8
Asia	1,533,797	28	293	22
Europe	1,934,574	35	496	38
Near East	327,963	6	67	5
Total	5,554,505	100	1,308	100
CGIAR Total	593,191		12	

sive equipment and skilled staff. The total number of *in vitro* accessions is about 38,000 (cassava, potato, sweet potato, Andean root and tuber crops, yam, banana/plantain, cocoyam, grasses, etc.); they are stored in about 63 *in vitro* genebanks, mostly at IARCs. There are, however, good examples at National Agricultural Research Systems; for instance the papaya collection at the Malaysian Agricultural Research and Development Institute.

Botanical gardens

Worldwide, 1500 botanical gardens (11% private) maintain living collections of plants (Fig. 1.2). About 10% of them also have seed banks, and 2% *in vitro* collections. Usually, vegetatively propagated species, forest trees, medicinal and ornamental species, as well as plant genetic resources for food and agriculture of local significance are well represented. In this way they can fill an important gap in *ex situ* conservation programmes. Because their mandate extends to all plant species, botanical gardens are characterized by few accessions per taxon and, consequently, by a high interspecific but low intraspecific genetic diversity.

Most of the botanical gardens (915) are situated in Europe, the Former Soviet Union, and the United States, where about 75% of the total germplasm of botanical gardens is conserved. Worldwide, 698 (47%) have germplasm collections, 80% of which are preserved as living collections, outdoors or in greenhouses. In particular the situation is as follows: 410 botanical gardens conserve ornamental species or wild native endangered species, many of them related to major crops; about 60% of these collections are in Europe, the USA, Japan, South Africa and Mexico; 169 also conserve medicinal and forest species, particularly in China, Japan, India and Brazil; 119 conserve germplasm of cultivated species, including landraces, semi-cultivated species and other wild species locally utilized. These collections can be found in Asia (India and China), Mesoamerica (Mexico) and Canada.

The *Index Seminum* Commission has been the main mechanism for germplasm exchange for 300 years, and about 2 million accessions are offered every year, but the feedback on how the germplasm is used is very poor.

Characterization and evaluation of the collections

In general, genebanks were installed in the 1970s and 1980s in response to genetic erosion. The urgency of the moment led to the rescue and amassing of a huge amount of plant genetic resources, which was not followed by extended evaluation and utilization. Only 56 and 55 countries, respectively, provided information on characterization and evaluation activities, and only small percentages of the collections have been subjected to some form of characterization or evaluation (FAO, 1998).

The situation for IARC's managed collections is much better for characterization and, in general, also for evaluation, in particular for resistance to pests and diseases. Most of the centres perform multi-site evaluation of accessions for desirable characteristics in different agroecological regions. So far, there are a few examples of germplasm being

Table 1.5. *Ex situ* storage facilities and regeneration situation in the world's largest national base collections. (From FAO, 1998.)

Country and institute	Accessions stored	Storage facilities	Status of regeneration
China Institute of Crop Germplasm	300,000	Long-term storage space available	Not yet needed since genebank is only 8 years old
United States National Seed Storage Laboratory	268,000	Long-term storage capacity of 1 million accessions	9% requires regeneration; main constraints are lack of human resources and facilities for regeneration of cross-pollinated crops
Russian Federation VIR	177,680	No long-term facilities	Regeneration required frequently
Japan NIAR	146,091	Long-term facilities	4% requires regeneration; no specific problems reported
India NBPGR	144,109	New genebank capacity for 600,000 accessions	63% requires regeneration; no specific problems reported
Republic of Korea RDA	115,639	Long-term facilities total capacity 200,000 accessions	50% requires regeneration; main problems are with cross-pollinated species
Canada PGRC	100,000	Long-term facilities	No specific problems reported
Germany IPK, Gatersleben	103,000	Long-term facilities	Main constraint is lack of staff resources
Italy Germplasm Institute, Bari	80,000	Long-term facilities	No specific problems reported
Brazil CENARGEN	60,000	Long-term facilities capacity for 100,000 accessions	64% requires regeneration; main constraints are funds, infrastructure and human resources
Germany FAL, Braunschweig	57,000	Long-term facilities	Main constraint is lack of staff resources
Ethiopia Biodiversity Institute	54,000	Long-term facilities	8% requires regeneration; main constraints are lack of funds, land and human resources
Hungary Institute for Agrobotany	45,833	Long-term facilities	40% requires regeneration; no specific problems reported
Poland Plant Breeding and Acclimatization Institute	44,883	Long-term facilities	3% requires regeneration; no specific problems reported
Philippines NPGRL	32,446	Long-term facilities	No specific problems reported
Total[a]	1,728,681		

[a] About 34% of the world collection: 6,000,000.

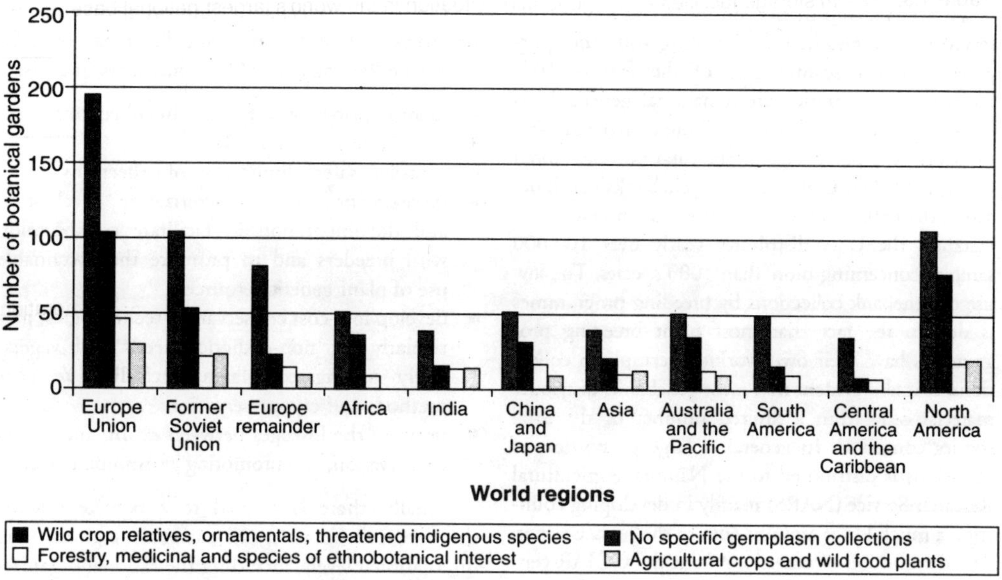

Fig. 1.2. Conservation of plant biodiversity and plant genetic resources in botanical gardens, by groups of species and/or information (1500 botanical gardens worldwide) (source: FAO, 1998).

systematically evaluated using a network approach, at national and international level. For instance the International Network for Genetic Evaluation of Rice (INGER), coordinated by IRRI and operating for 20 years, has greatly contributed to the increase in rice germplasm. Similarly, the International Network for the Improvement of Bananas and Plantains (INIBAP), working mainly through NARS, has developed the International Musa Testing Programme for the evaluation of banana and plantain collections. In the USA, several public sector and private institutions are collaborating with the United States Germplasm Enhancement Maize project (GEM), which is broadening the genetic base of maize hybrids. It is evident that international cooperation, in-depth documentation and information, and a more systematic evaluation and characterization in relevant environments are needed (FAO, 1998).

Distribution of germplasm from genebanks

In general, germplasm from genebanks and botanical gardens is freely available to bona fide users on request; however, increasingly, access restrictions are being introduced, partly caused by quarantine regulations. More recently, restrictions are being triggered by IPRs, the not yet concluded International Undertaking (IU), and restrictive access legislation (J. Engels, personal communication).

A fundamental achievement is represented by the legal status of the CGIAR genebanks. The legal status of CGIAR and other *ex situ* collections was left unresolved until after the implementation of the UNCED Convention on Biological Diversity (CBD), in 1993. The CBD offers the signing states international rights over their national resources (art. 3). The CBD does not cover *ex situ* collections, which were acquired prior to the implementation, i.e. the majority of the accessions in the CGIAR collections. The importance of the issues of *ex situ* collections and farmers' rights was further emphasized during the Nairobi Conference for the Adoption of the Agreed Text of the CBD in 1993. Soon after, the FAO adopted Resolution 7/93, which called for intergovernmental negotiations on: 'the issue of access on mutually agreed terms to plant genetic resources, including *ex situ* collections, not addressed by the Convention, and the issue of the realization of Farmers' Rights' (FAO, 1995a). On 26 October 1994, the 12 CGIAR centres holding plant genetic resources placed them under the auspices of the FAO International Network of *ex situ* collections, part of the FAO Global System (FAO, 1995b; Pistorius, 1997).

At present, information on the utilization of conserved genetic resources for breeding and other purposes is rather scanty. One of the few available indicators of utilization from national genebanks is the number of accessions distributed each year and expressed as a percentage of the total number stored in the genebank. Only very few genebanks distribute more than 10% of their accessions annually. For instance, the USA distributes yearly over 100,000 samples concerning more than 3000 species. The low use of genebank collections by breeding programmes is due to the fact that most plant breeding programmes have their own working germplasm collections. It is also evident that most genebanks distribute material only from a limited number of the total species conserved. In general, a large proportion of accessions is distributed to the National Agricultural Research Service (NARS) mainly in developing countries; a much lower amount goes to the private sector. The distribution of accessions from the CGIAR centres varies widely (Table 1.6). Most centres distribute at least 10% of their total accessions every year; a rate higher than that for most national genebanks (WRI, IUCN, UNEP, 1992; FAO, 1998).

Remarks and improvement of the system

It has become increasingly clear that the *ex situ* collections and the system of genebanks need to be strengthened. Concerns are based on:

- gaps in the collections of minor crops and under-utilized species, especially landraces and wild relatives from their centres of diversity and cultivation;
- a large and increasing number of accessions that need regeneration;
- several important species with non-orthodox seeds cannot be stored in seed genebanks;
- many countries lack resources to maintain germplasm that they or the international community have already paid to collect;
- few genebanks operate in conformity with the highest international standards, while the genetic integrity of most of the germplasm accessions in the remaining genebanks is threatened;
- duplication of accessions is far from complete, while there is a significant duplication of samples;
- documentation is variable and incomplete;
- access to information is limited;
- coordination between genebanks, breeders and other users is insufficient (FAO, 1998).

On such premises, improvement of the genebank system for *ex situ* collections could be attained following a number of measures, such as:

- identify priorities to fill gaps in collections;
- increase regeneration efforts;
- complete safety duplication of collections;
- increase primary characterization, evaluation and documentation to facilitate collaboration with breeders and to promote the sustainable use of plant genetic resources;
- develop low-cost conservation technologies, particularly for non-orthodox seeds and vegetatively propagated plants, including *in vitro* methods and cryopreservation;
- improve the linkages between *ex situ* and *in situ* conservation, also promoting a common database.

Finally, there is a need to maximize synergy through collaboration at national, regional and international levels, including the rational organization of base, active and working collections. This might include mechanisms to allow countries to place their material in secure storage facilities outside their borders, without compromising their sovereign rights. Financing mechanisms may be required to facilitate such rationalization of activities.

Genetic Resources: Maintenance, Use and Regulation

Ex situ *versus* in situ *and on-farm conservation of agrobiodiversity*

What are the real problems and the prospective, matured in the last decade, what are the implications, the possible approaches, in relation to collection, maintenance and evaluation of agrobiodiversity?

Of course, collecting and conserving *ex situ* samples in genebanks is of undoubted technical and economic advantage to both holders and users of germplasm, and it has been instrumental in the success of many national and international plant improvement programmes, leading to significant increases in productivity, as did the Green Revolution. However, while genebanks will continue to play their specific role, biological evolution, that is the continuous creation of biodiversity, cannot take place in stored material. It can only occur in nature, through the dynamics of continuous contact and interaction among life forms in ecosystems or, for crop plants and domestic animals, in agroecosystems.

Table 1.6. Percentage of germplasm samples distributed annually by CGIAR centres, by sector (1992–1994). (From FAO, 1998.)

Centres	Other international agricultural research centres (%)	Developing country national agricultural research system (%)	Developed country national agricultural research system (%)	Private sector (%)	Total number of samples distributed outside the centres
CIAT					
Phaseolus	0	54	46	0	1,979
Manihot	0	59	40	1	422
Forage legumes	16	51	27	6	1,655
Total	7	53	37	3	4,056
CIMMYT					
Maize	0	20	72	8	2,234
Wheat	0	69	28	3	2,372
Total	0	45	49	6	4,606
WARDA					
Total	25	75	0	0	1,872
ICARDA					
Total	5	63	32	0	13,013
CIP					
Potato		93	7		3,929
Sweet potato		95	5		1,023
Total		93	7		4,952
IITA					
Total	13	66	21	0	3,895
ICRISAT					
Total	0	91	2	7	19,570
IRRI					
Total	7	52	39	2	7,207
ILRI					
Total	9	64	7	20	1,071
INIBAP					
Total	3	64	33	0	371
Total	4	72	21	3	60,613

During recent years, national and international programmes, following considerable public calls towards nature conservation and 'to protect nature from man', have increased *in situ* biodiversity conservation by increasing the number of protected areas and reserves, which is also supported through the UNESCO-MAB programme, the Global Environment Facility of UNDP and UNEP projects and according to the multifaceted strategy developed at the UNESCO Seville Conference in 1995 (Scarascia-Mugnozza, 1995).

The need to allow such processes to continue has prompted an increasing engagement in *in situ* programmes of biodiversity conservation. The Convention on Biological Diversity (CBD) itself, in Article 8, promotes *in situ* conservation when it explicitly calls on signatory parties to 'establish a system of protected areas, or areas where special measures need to be taken, to conserve the biological diversity', with the aim of ensuring the conservation of ecosystems and agrobiodiversity, and of guaranteeing the sustainable utilization of the latter. Moreover, outside gene parks and protected areas, *in situ* conservation is often carried out at the farm level, 'on-farm conservation', where landraces and locally improved material are grown, utilized and conserved as components of traditional farming systems, and where they also evolve in response to their dynamics. Agricultural populations have, over millennia, conserved plant and animal genetic resources and practised selection for yield, quality, resistance to biotic and abiotic stresses and for medicinal and other economic prospects. Therefore, it is right and fair that *in situ* conservation, again according to the CBD, should aim to

> respect, preserve and maintain knowledge, innovations and practices of indigenous and local communities embodying traditional lifestyles relevant for the conservation and sustainable use of biological diversity ... and encourage the equitable sharing of the benefits arising from the utilization of such knowledge, innovations and practices.

However, as most *in situ* management practices have been directed towards habitat preservation and have focused on ecological rather than genetic considerations, technical expertise on conservation strategies and techniques focused on *in situ* conservation of useful species should be intensified.

It may be concluded that the *in situ* or, under specific conditions, on-farm conservation and cultivation of crop, domestic animal and agroforestry species may play a significant role not only in the effective maintenance of agrobiodiversity, but also as a component of sustainable development programmes, as recognized by Agenda 21. Measures such as the establishment of a multilaterally agreed funding mechanism should therefore be taken to promote, encourage and implement *in situ* and on-farm conservation.

Use of genetic resources

Use and exploitation of genetic resources postulate knowledge and evaluation of the characters expressed by the genome of the samples and the identification of desirable traits. Research in many scientific disciplines of relevance to plant genetic resources has allowed the development of more and more refined methodologies which allow geneticists and biotechnologists to pursue their programmes more effectively. In particular, the new scientific and technical tools should be progressively applied to scan the genome of specific valuable accessions.

An important project fundamental for the preservation, knowledge, evaluation and utilization of all biological diversity, and consequently basic for the use and valorization of agrobiodiversity, was designed, during 1999, in Paris meetings of the OECD Megascience Forum. It was decided to establish – in the frame of the UNEP Global Environment Facility – a Global Biodiversity Information Facility (GBIF), corresponding to the aims of the CBD.

Like biodiversity itself, information on conservation, preservation and evaluation of biodiversity and genetic resources is already distributed worldwide, and contained in electronic databases, including information on the distribution of organisms around the globe, information on physiological functions of organisms in relation to ecosystems, detailed genomic maps, etc., often compiled during unrelated and independent projects.

GBIF should strengthen preservation and utilization of knowledge and expertise on biodiversity by providing access and links to existing or new databases, synchronizing and planning for interoperability between them, developing novel interface designs and protocols for indexing and validation of documentation, and so on. Of primary importance will be the role of GBIF in the exchange of data, information and resources. Therefore, a clear agreement should be reached in which intellectual property rights, as well as breeders' and farmers' rights will be equitably protected and guaranteed.

In recent years the epochal event represented by

the development of biological sciences, especially molecular biology, and animal and plant genomics, and the results of biotechnological research and applications, made it increasingly evident that biotechnology can add value to genetic resources and biodiversity at large. The successful use of genetic engineering methods has raised the economic value and increased the potential of much biological diversity as a resource in breeding and research. Consequently, it has widened enormously the scope and boundaries of initiatives to protect and conserve biodiversity.

There may be unique opportunities for real international cooperation among developing and industrialized countries in building local capacities in developing countries, for the application of agrobiotechnologies to agrobiodiversity. Equitable and effective cooperation through technology transfer constitutes one of the major mechanisms by which resources of biodiversity can be conserved, managed and used sustainably for human welfare. Equity demands that developing countries receive benefits and compensation for the use of these resources, by improving their capacity to maintain their agrobiodiversity *in situ* or *ex situ*, to identify and evaluate useful genetic traits for plant and animal improvement and to apply the relevant biotechnologies for the optimal use of genetic resources, for the benefit of their populations.

In any case, it is important to underline 'that the realisation of benefits from such technologies depends on the continued conservation and availability of the biological resources. Investments in technology development should be made hand-in-hand with investment to protect the resource base' (FAO, 1996c).

Legal issues and rights

Since the feedstock for the biotechnology industry is biological diversity, it is necessary to formulate and adopt some basic ground rules which could regulate questions concerning, above all, conservation, use, ownership, access, benefit-sharing, indigenous rights, farmers' rights and intellectual property rights. These issues are technically and legally complex, but they can be efficiently faced by the international community if our actions proceed along the path traced by the principles of the CBD: conservation of biodiversity, sustainable use of its components, access to and transfer of genetic resources accumulated in *ex situ* genebanks or maintained *in situ*, and equitable sharing of the benefits of the use of genetic resources. This was confirmed by the FAO during the last decade[2] in the Leipzig Declaration (June 1996) and in the Global Plan for the Conservation and Sustainable Utilization of Plant Genetic Resources for Food and Agriculture. Also the Madras Declaration of 1996 confirmed this (Scarascia-Mugnozza and Swaminathan, 1996). The FAO World Food Summit (Rome, November 1996) declared, among the objectives of its third commitment, solemn support to the Leipzig Plan of Action through 'inter alia appropriate *in situ* and *ex situ* approaches, systematic survey and inventorying to broaden the genetic basis of crops, and fair and equitable sharing of benefits arising from the use of such resources'. The same questions were debated at the World Conference on Science (UNESCO, Budapest, 1999), which had to recognize that the expansion of IPR-controlled science is inevitable. Among the easily imaginable consequences of this are the increase in secrecy in scientific work and decreased attention towards the scientific problems related to the health and livelihoods of the poor.

In the light of the above, it is extremely urgent that a timely agreement be reached among holders of germplasm throughout the world in order to regulate the status and access to genetic resources and that an efficient mechanism be identified to assure equitable sharing of the benefits derived from their use.

Very recently, the FAO Commission on Plant Genetic Resources for Food and Agriculture, during its last session (April 2000), made further progress in searching for the development and establishment of an international plant genetic resources system, which provides recognition, and protection of farmers' and traditional communities' rights for an equitable share of the benefits.

Apart from the fundamental need for strengthening international cooperation to achieve suitable participatory systems for sharing benefits in a wide range of approaches and actions, we think that key issues within the ongoing international debate on plant genetic resources are: (i) the problems related to the realization of farmers' rights confronted to the patent system, being them complementary and not opposed to breeders' rights; and (ii) the status of the *ex situ*

[2] In 1995, the introductory McDougall Memorial Lecture to the general FAO Conference was dedicated to the potentialities and perspectives of biological diversity for food and agriculture (Scarascia-Mugnozza, 1995).

collections prior to the approval of CBD in 1993, remained out of the reach of the provisions of the Convention itself. Thus, the continuing uncertainty on the legal status and related aspects of access and maintenance of germplasm collections represents a threat to conservation of plant genetic resources.

The FAO-CGIAR Agreements of 1994 should be regarded as a major contribution to an open system of germplasm conservation and exchange. Further development of the Agreements should be open for consideration and consultation among all concerned parties of the plant genetic resources community with a view, for example, to create the most open and effective system possible for the conservation, exchange and enhancement of genetic material, presumably under the aegis of a supranational authority. This model might be open to all public or private, national or international institutes prepared to abide by the terms of the agreement and to provide a list of germplasm accessions to be exchanged under the conditions stipulated by the agreement. The excellent relationships of IARCs with National Agricultural Research Systems and their governments, and also with NGOs, could be a favourable pathway to reach agreements on the model of the FAO-CGIAR agreement. Countries could agree, for example, to place their genetic resources into such a framework on the basis of prior informed consent. Access to samples and information on these resources could be unrestricted (subject to a legal mechanism) for all countries that are parties to the agreement.

To achieve sustainable conservation efforts, it is of utmost importance to link conservation to use. Without strong linkages the value cannot be realized and eventually even conservation will be threatened. The issue of use, however, carries with it the problems connected with sharing rights (benefits) derived from plant genetic resources. According to the principles embodied in the International Undertaking and reiterated by the CBD and FAO World Food Summit, there is a need to ensure full respect of the rights of the countries in which agricultural biodiversity is found, and of their farmers and farming community, so as to avert the grave continuing danger of erosion of plant genetic resources and the irreparable loss of these resources.

As matter of fact, no concrete proposals have emerged up to now on giving practical meaning and content to the concept of farmers' rights as described by the International Undertaking as 'an obligation to compensate farmers for their past, present and future contribution conserving and making available plant genetic resources' (FAO, 1995a).

That is why the role of farmers' rights, a concept introduced by M.S. Swaminathan in 1979 in the FAO debates, and endorsed by FAO since 1989, is crucial (Swaminathan, 1995). They aim at reconciling the view of 'technology rich' and 'gene rich' countries in order to ensure the availability of plant genetic resources for biotechnological industry within an equitable system. They aim at providing some balance to 'formal' intellectual property rights (IPR), such as breeders' rights and patents, invented to reward innovation acquired from the advanced research and resources invested in the industrial countries. Farmers should be rewarded no less than institutions and private companies.

Farmers' rights and IPR need to be harnessed, in the interests of maintenance and use of biodiversity, of scientific research for advantages for industries and farmers, and above all for human welfare, by developing agreements that promote the equitable sharing of benefits coming from a proper use of genetic resources.

In our opinion, UN agencies like FAO, UNESCO, UNEP and UNDP, with the technical support of IPGRI and of CGIAR, might jointly develop ad hoc guidelines in order to harmonize the provision of Article 27(3)b of the World Trade Agreement (regarding IPR) with the ethics and equity provisions of the CBD (Articles 8(j) and 15 of CBD). Such a revision of IPR will help to foster symbiotic bio-partnerships, to protect biodiversity and eliminate fears of biopiracy.

In this context, it is appropriate to remember that, in the frame of the patent system of inventions, which is mandatory for the World Trade Organization (WTO), plant variety protection in industrialized countries is regulated by the International Convention for the Protection of New Varieties of Plants and implemented by the International Union for the Protection of New Varieties of Crop Plants (UPOV), which rewards only scientific plant breeders (UPOV, 1991).

Therefore, it is timely to put forward consistent and sound elements for enhancing mutual supportiveness among the international fora, especially the CBD and the FAO Commission for Genetic Resources for Food and Agriculture on one side, and the trade and IPR policies of WTO, the Trade Related Intellectual Property System (TRIPS) Agreement, the UPOV and the World Intellectual Property Rights Organization (WIPO), on the other.

Specular to such inter-foral cooperation, the best efforts should be made in order to pursue effective national systems and coordination mechanisms, by which negotiators in trade and IP conventions, while responding to their respective ministries and agencies, are aware and supportive of domestic policies and international positions embodied by officials negotiating in agricultural and environmental fora.

Conclusions

International cooperation in the field of genetic resources is imperative without delay. One of the most urgent issues to be solved, for political, environmental, scientific, economic and ethical reasons is the realization of farmers' rights within the context of the revision of the International Undertaking and the development of a mechanism for its implementation. The revised International Undertaking could then become a protocol of the CBD.

Equity means equilibrium between public good and private profit, and in particular demands that developing countries receive benefit and compensation for the use of genetic resources that they host or hosted and have already partly transferred to *ex situ* collections in developed countries, which are used by companies and breeders with beneficial results. Equitable and effective cooperation constitutes one of the major mechanisms by which resources of biodiversity can be conserved, managed and used sustainably all over the world, under legal agreements.

References

Esquinas-Alcàzar, J. (1989) The role of national, regional and international institutions in germplasm conservation. FAO, Rome, Italy (unpublished).

FAO (1959) *World Catalogue of Genetic Stocks: Barley.* FAO, Rome, Italy.

FAO (1969) *Report of the Third Session of the FAO Panel of Experts on Plant Exploration and Introduction. 25–28 March, Rome.*

FAO (1995a) Progress report on Resolution 3 of the Nairobi Final Act: ex situ collections and farmers' rights. FAO report presented at the *First Session of the Conference of Parties to the Convention on Biological Diversity, Nassau, The Bahamas, 28 November–9 December 1994.* FAO, Rome, Italy.

FAO (1995b) Progress report on the global system for the conservation and utilization of plant genetic resources for food and agriculture. *Report for the Sixth Session of the Commission on Plant Genetic Resources,* CPGR-6/95/4, April 1995. FAO, Rome, Italy.

FAO (1996a) *Report on the State of the World's Plant Genetic Resources for Food and Agriculture.* FAO, Rome, Italy.

FAO (1996b) *Global Plan of Action for the Conservation and Sustainable Utilization of Plant Genetic Resources for Food and Agriculture.* FAO, Rome, Italy.

FAO (1996c) The fourth international technical conference in the context of the FAO global system for the conservation and utilization of plant genetic resources for food and agriculture. Report prepared for the *International Technical Conference on Plant Genetic Resources (June 1996),* ICTPGR/96/INF/2, May 1996. FAO, Rome, Italy.

FAO (1998) *The State of the World's Plant Genetic Resources for Food and Agriculture.* FAO, Rome, Italy.

FAO/UNDP (1970) Turkey: Technical Report. Report prepared by the FAO Agricultural Research and Introduction Centre, Izmir for the Government of Turkey. FAO, Rome, Italy.

Frankel, O.H. (1985) Genetic resources: the founding years. *Diversity* 7.

Frankel, O.H. and Hawkes, J.G. (eds) (1975) *Crop Genetic Resources for Today and Tomorrow.* IBS series Vol. 2. Cambridge University Press, Cambridge, UK.

Hawkes, J.G. (1961) EUCARPIA: proposal for the establishment of a European potato introduction station. Paper submitted to the *FAO Technical Meeting on Plant Exploration and Introduction, Rome, 10–20 July 1961.* FAO, Rome, Italy.

Hobbelink, H. (1987) *New Hope or False Promise? Biotechnology and Third World Agriculture.* International Coalition for Development Action, Brussels, Belgium.

Ito, H. (1972) *Organization of the National Seed Storage Laboratory for Genetic Resources in Japan.* Chapman and Hall, Norfolk, UK.

National Academy of Sciences (NAS) (1972) *Genetic Vulnerability of Major Crops.* Washington, DC, USA.

Padulosi, S. (1996) Le risorse genetiche delle specie minori nel Mediterraneo e l'iniziativa IPGRI a sostegno della loro conservazione [Genetic resources of minor species in the Mediterranean and the initiative of IPGRI to encourage their maintenance. In: *The Minor Cereals*]. In: Porfiri, O., Castagna, R., Padulosi, S. and Codoni, D. (eds) *I Cereali Minori.* pp. 11–21.

Pistorius, R. (1997) *Scientists, Plants and Politics – A History of the Plant Genetic Resources Movement.* International Plant Genetic Resources Institute, Rome, Italy.

Plucknett, D.L., Smith, N.J.H., Williams, J.T and Anishetti, N.M. (1987) Crop germplasm conservation and developing countries. *Science* 220, 163–169.

Porceddu, E. (1972) Germplasm laboratory – Bari: aims and methods of genetic conservation in Italy. *Plant Genetic Resources Newsletter* No. 28. FAO, Rome, Italy, pp. 12–16.

Rudorf, W. (1961) Exploration centers within the areas of gene centers, and introduction centers in remote countries with large area of cultivated plants. Contribution to the *1961 FAO Technical Meeting on Plant Exploration and Introduction.* FAO, Rome, Italy (unpublished).

Scarascia-Mugnozza, G.T. (1995) The protection of biodiversity and the conservation and use of genetic resources for food and agriculture: potential and perspectives. *FAO 19th McDougall Memorial Lecture, Rome, Italy, 20 October 1995.*

Scarascia-Mugnozza, G.T. (1998) Diversità biologica: riflessioni sul multiforme tema. *Atti dei convegni Lincei 145, XV Giornata dell'Ambiente. Convegno: Conservazione e valorizzazione della biodiversità.* [Biological diversity: consideration on the multiform theme. *Proceedings of the Lincei Conferences 145, XV Day of the environment. Conference: Maintenance and valorisation of biodiversity*]. Rome, 5 June 1997.

Scarascia-Mugnozza, G.T. (1999) Italian efforts and Italy–IPGRI cooperation in the safeguard and development of plant genetics. Lecture held during the XXVth IPGRI-IBPGR Anniversary Celebration at FAO (Rome, 9 March 1999).

Scarascia-Mugnozza, G.T. and Porceddu, E. (1972) Activities and programs of the Italian Germplasm Laboratori. Paper presented at the *EUCARPIA Conference on European and Regional Genebanks, Izmir.* Tip. del Sud, Bari, pp. 3–16.

Scarascia-Mugnozza, G.T. and Swaminathan, M.S. (1996) Appello agli Scienziati di tutto il mondo per la conservazione e l'utilizzazione della biodiversità e delle risorse genetiche essenziali per l'agricoltura e la produzione agroalimentare. [Appeal to the scientists of the world for the maintenance and use of biodiversity and genetic resources important for food and agriculture.] *Memoria di Scienze Fisiche e Naturali*, vol. 114. Rendiconti della Accademia delle Scienze, detta dei XL, Serie V, Vol. XX, parte II.

Swaminathan, M.S. (1995) *Reaching the Unreached: Farmers' Rights and Plant Genetic Resources: Recognition & Reward: A Dialogue.* MacMillan India Limited, India.

Technical Advisory Committee (TAC) (1972) The Collection, Evaluation and Conservation of Plant Genetic Resources. Report of TAC ad Hoc Working Group held in Beltsville, USA, 20–25 March 1972.

TAC (1986) *Report of the Second External Programme and Management Review of the International Board for Plant Genetic Resources (IBPGR).* TAC, Rome, Italy.

UPOV (1991) *International Convention for the Protection of New Varieties of Plants.* International Union for the Protection of New Varieties of Crop Plants, Geneva, Switzerland.

Whyte, R.O. (1958) *Plant Exploration, Collection and Introduction.* FAO Agricultural Studies No. 41. FAO, Rome, Italy.

WRI, IUCN, UNEP (1992) *Global Biodiversity Strategy: Guidelines for Action to Save, Study, and Use Earth's Biotic Wealth Sustainably and Equitably.* WRI, IUCN, UNEP, Washington, DC.

2 The Past, Present and Future Contributions of Farmers to the Conservation and Development of Genetic Diversity

M.S. Swaminathan
M.S. Swaminathan Research Foundation, Chennai (Madras), India

Introduction

The significant contributions of rural people, particularly women, to the conservation and enhancement of agrobiodiversity remained unrecognized until only a few decades ago. Studies by Darwin, de Candolle, Vavilov, Harlan, Wellhausen and many others have shown a strong positive correlation between cultural diversity and genetic diversity, demonstrating the impact of human communities on the conservation and use of biodiversity. Agrobiodiversity is largely the result of human–nature interaction.

Ethnobiologists and social anthropologists have long recognized the role of rural people in the domestication and improvement of economic plants. However, it is only in the last 25 years that serious efforts have been made to develop ways to recognize and reward the invaluable contributions of farm families to the conservation and development of genetic diversity. A serious debate on this issue started in the forum of the Food and Agriculture Organization (FAO) in 1981 when I was Independent Chairman of the FAO Council. In November 1983, the FAO Council meeting under my Chairmanship set up an International Commission on Plant Genetic Resources (now renamed as the FAO Commission on Genetic Resources for Food and Agriculture, so as to include within its scope animal genetic resources also). Soon debate started in meetings of this Commission on the concept of farmers' rights. Although there has been a near consensus for nearly two decades on the need for recognizing and rewarding the contributions of farm families to genetic resources conservation, no internationally agreed methodology exists even now for this purpose.

The Convention on Biological Diversity (CBD), which came into force in December 1993, has for the first time in a legally binding document given explicit recognition to the role of indigenous communities to both accumulation and conservation of knowledge and information (Article 8j). The prior informed consent and benefit-sharing provisions of CBD have not become operational so far.

Conservation Methods and Economic Stakes

Sharing of benefits and the concomitant increase in the recognized value of the resources are the most effective ways to foster conservation and to ensure the continued availability of plant genetic resources (PGR). Today, there is a growing economic stake in the exploitation of habitats rich in plant genetic resources. That this will be disastrous is evident from the loss of nearly 3.5% of the world's forests since UNCED in Rio de Janeiro in 1992. *There is therefore an urgent need to create an economic stake in conservation.* In general, the issue of benefit-sharing

(eds J.M.M. Engels, V. Ramanatha Rao, A.H.D. Brown and M.T. Jackson)

has not received the detailed attention it deserves among those formulating legal measures for operationalizing the CBD. If FAO's revised International Understanding on plant genetic resources (PGR) incorporating provisions for farmers' rights becomes a protocol of the legally binding CBD, follow-up action will become mandatory. The adoption of the Cartagena Protocol on Biosafety is a sign of hope.

The different systems of conservation in vogue today are:

- *in situ* conservation;
- *ex situ* preservation;
- *in situ* on-farm conservation.

Unlike cryogenic preservation in genebanks, conservation involves both preservation and evolution. Therefore, *ex situ* preservation alone cannot provide the lasting benefits that accrue from the conservation of habitats and ecosystems rich in biodiversity. The different systems of conservation are shown in Fig. 2.1, which also indicates the agencies actively involved in the different methods of conservation/preservation.

There is need to develop ways to end the prevailing dichotomy in which the primary conservers remain poor while those using their material and information become rich. In the case of *in situ* and *ex situ* conservation, public funds are provided to establish biosphere reserves, protected areas, botanical gardens, genebanks, etc. However, in the case of *in situ* on-farm conservation of agrobiodiversity, rural people conserve landraces or folk varieties for public good at personal cost. Equity demands that their work be recognized and compensated. Thus, we should develop a framework for promoting benefit-sharing under the different conditions of conservation depicted in Fig. 2.1.

At a dialogue organized in January 1990 at Madras by the M.S. Swaminathan Research Foundation (MSSRF) in collaboration with the Keystone Center in the United States, the participants stated,

> we agree on the concept of Farmers' Rights and we agree that the contributions to a fund in recognition of these rights should not be voluntary. Practically speaking, a voluntary fund is a fund without resources. Thus, there should be a compulsory funding mechanism. This would ensure that Farmers' Rights are recognized in a real way.

The concept of farmers' rights includes recognition of the fact that farmers have developed and continue to help develop and maintain genetic diversity. In many cases, farmers engage in conscious and creative practices as they select and breed their crops. Ethnobotanists around the world have chronicled the invaluable contributions of tribal and rural women and men in the conservation and enhancement of genetic diversity in plants (Swaminathan and Kochhar, 1989).

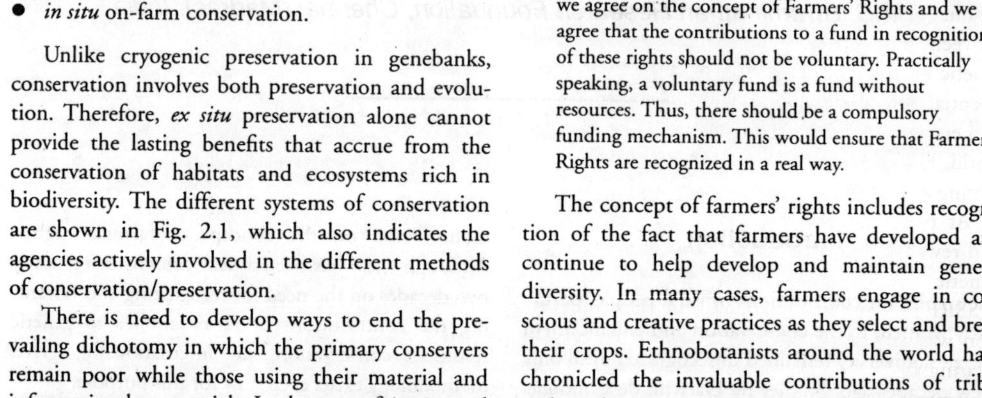

Fig. 2.1. Role of community conservation in integrated gene management.

The Plant Variety Protection Acts in vogue in developed countries generally conform to the provisions of the Conventions agreed to from time to time under the forum of the International Union for the Protection of New Varieties of Plants (UPOV). They are designed to promote invention and investment in plant breeding. However, they do not afford recognition to the conservation and enhancement by farm families of genetic diversity, which is the basic feedstock for plant breeding and biotechnology enterprises. By introducing provisions for farmers' rights, Plant Variety Protection Acts will concurrently foster innovation and conservation. They will thus help to strike a balance between homogeneity and heterogeneity in the genetic make-up of new cultivars. Such a balance is essential for sustainable agriculture. Breeders and farmers are allies in our struggle for a hunger-free world. Hence, their rights should be mutually reinforcing and not projected as antagonistic.

At two dialogues on methods of recognizing and rewarding informal innovations in the area of genetic resources conservation organized by MSSRF in 1994 and 1996, a detailed draft legislation was developed for converting the know-how relating to farmers' rights into field level do-how (Swaminathan, 1995, 1996). On the basis of these drafts, the Government of India has introduced in Parliament a Plant Variety Protection and Farmers' Rights Act.

Equity in Benefit-sharing

The need for internationally agreed methodologies for giving effect to the equity provisions of the CBD is now widely recognized. The issue of benefit-sharing has received considerable attention during the last 15 years in the forum of the Commission on Plant Genetic Resources of the FAO in relation to PGR. Although agreed methodologies for implementing farmers' rights at the field level have yet to emerge, several institutions have developed their own voluntary code of conduct in the matter of benefit-sharing in the commercial exploitation of agrobiodiversity.

Article 15 of CBD recognizes that 'States have sovereign rights over their own biological resources'. It also recognizes:

> the close and traditional dependence of many indigenous and local communities embodying traditional lifestyles on biological resources, and the desirability of sharing equitable benefits arising from the use of traditional knowledge, innovations and practices relevant to the conservation of biological diversity and the sustainable use of its components.

Implementation of the above provisions of CBD will require both material and information transfer agreements within and among nations, which incorporate provisions for prior informed consent and equity in benefit-sharing between the primary conservers and those utilizing their knowledge and material in plant breeding, genetic engineering and pharmaceutical and other biological enterprises. The Consultative Group on International Agricultural Research (CGIAR) has developed well-defined procedures for material and knowledge transfer.

The Trade Related Intellectual Property Rights (TRIPS) provisions of the World Trade Agreement call for either a patenting or a *sui generis* system of plant variety protection. This is for the purpose of providing incentives for innovation and investment in the use of plant genetic resources. Such Plant Variety Protection Acts have been in existence in several industrialized countries for over 50 years. Since 1961, UPOV has been prescribing guidelines for harmonizing the provision of such legislation in different countries. The UPOV convention, however, provides for recognizing and rewarding only the contributions of breeders and not of those who have conserved the basic raw material used by breeders. The time has come to end the sad irony of the poverty of the conservers in contrast to the prosperity of those utilizing the fruits of their knowledge and conservation ethics. Several developing countries in South America, Africa and Asia are currently in the process of enacting legislation incorporating provisions for benefit-sharing. Individual institutions and commercial companies have also been practising their own procedures for sharing benefits with the communities that provided the genetic material. These different steps illuminate the path towards an internationally agreed protocol on benefit-sharing under the auspices of FAO or CBD.

Equity in benefit-sharing is fundamental to the retention and revitalization of the *in situ* on-farm conservation traditions of rural and tribal families. Material and Information Transfer agreements should safeguard the interests of those providing the concerned material/information. CGIAR institutions are already adopting a Material Transfer Agreement procedure that will prevent the monopolistic exploitation of public-funded research for commercial profit. Benefit-sharing procedures will have to be developed at the individual and community levels. At the level of an individual farmer–conserver–innovator the same procedures for seeking recognition and reward as those available to professional breeders can be used. Only they may have to be helped in obtaining patents/plant variety protection in accordance with the prescribed national legislation. The problem is more complex in the case of benefit-sharing with entire communities. Procedures are available for identifying the area from which critical genes responsible for the commercial success of a new variety came. Thanks to molecular techniques, this possibility also extends to genes controlling quantitative traits like yield and quality. Therefore, appropriate reward can be given from the Community Biodiversity and Gene Funds proposed to be established under Biodiversity and Plant Variety Protection Acts in several developing countries. Breeders will have to be requested to disclose the full pedigrees of their new varieties and indicate to the extent possible the area from where the critical genes, including QTLs (quantitative trait loci), came. The communities concerned can decide how to use the funds provided. Obviously they should be used for community benefits, including the funds needed for strengthening on-farm conservation of landraces and seed technology.

The benefit-sharing methods appropriate to different systems of genetic resources conservation are summarized below.

In situ *conservation*

This largely takes the form of biosphere reserves, national parks and protected areas. Here, many opportunities are available for benefit-sharing, such as:

- Participatory forest management, involving benefit-sharing with reference to wood and non-wood forest products.
- Bioprospecting based on equity in sharing commercial profit.
- Ecotourism, which strengthens the livelihood security of local communities.
- Symbiotic social contracts between tribal and rural families and the corporate sector with reference to sourcing of material based on assured buy-back arrangements.

Ex situ *preservation*

Here, methods of sharing royalties with the providers of the critical information and material can be developed. Community Gene Funds can be established to provide recognition and reward to entire communities based on the analysis of the pedigrees of successful varieties.

In situ *on-farm conservation*

This is an area of vital importance to global food and health security. Community Biodiversity and Gene Funds could be used to compensate local communities for the yield and income loss they may incur by continuing to conserve landraces and folk varieties of economic plants, instead of abandoning them in favour of high-yielding varieties. In addition, both individual and community benefits can be conferred on the basis of transparent procedures. Technical Resource Centres for the implementation of the equity provisions of CBD, like the one existing at MSSRF, will be of help in this process. Thus, benefits can be in cash and kind. Social recognition and material benefits are both important. There are several informal models of such recognition and reward developed and operated by non-governmental organizations, academia, business and industry. Through a combination of political will, professional skill and people's participation, fair and equitable methods of benefit-sharing can now be introduced in all systems of conservation and use of plant genetic resources.

The FAO Global Plan of Action developed at Leipzig provides an excellent blueprint for plant genetic resources conservation. In the ultimate analysis, the success of conservation efforts will depend on people's participation. Equity in benefit-sharing is the trigger for fostering a people-centred conservation movement. Implementation of benefit-sharing procedures – rooted in the principles of ecology, social and gender equity and economics – will be an important step in achieving a better common future for humankind. This is why industrialized countries and biotechnology industries, for which biodiversity constitutes the basic feedstock, should contribute to a Global Community Gene Fund designed to reward the tribal and rural families who are today protecting public good at personal cost.

The three fundamental concepts of Article 15 of CBD, namely sovereign rights over genetic resources (article 15.1), access subject to prior informed consent (article 15.5) and access subject to mutually agreed terms (article 15.4), provide a framework for a new partnership between holders and providers of knowledge and genetic material and users, based on equity and ethics in sharing benefits.

Unilateral exploitation of biodiversity for commercial purposes is now widely regarded as biopiracy. This method of exploitation should give way to biopartnership based on procedures like co-patenting and equitable benefit-sharing. Such biopartnerships will help to foster symbiotic relationships between biodiversity providers and users and will foster the causes of conservation and sustainable use.

The recent initiative of the World Intellectual Property Rights Organization (WIPO) to consider questions relating to according recognition to traditional knowledge systems and informal innovations is a welcome first step in enlarging the concept of intellectual property rights (IPR). The World Trade Organization (WTO) should also initiate steps to enlarge the concept of TRIPS, so as to evolve a new 'TRIPS Plus' paradigm, where the 'plus' refers to equity and ethics in the sharing of benefits between traditional knowledge holders and conservers and commercial companies. Finally, UPOV should also restructure itself so as to become an 'International Union for the Protection of Breeders and Farmers' Rights'. Such action at the international level and appropriate legislative and non-legislative measures at the national level will help to give meaning and content to the concept of equitable benefit-sharing.

Global plan of action

At an International Technical Conference on Plant Genetic Resources held at Leipzig, Germany, from 17 to 23 June 1996, under the auspices of FAO, a global plan of action was developed for both *in situ* and *ex situ* conservation of PGR. Although at this Conference, benefit-sharing methodologies were not specifically discussed, the following areas that have a bearing on benefit-sharing were given consideration:

- development and commercialization of underutilized crops and species;
- supporting seed production and distribution;
- developing new markets for local varieties and 'diversity-rich' products;
- expanding and improving education and training.

The Leipzig declaration called for 'a new and more productive partnership between scientists and farmers to build upon the ongoing efforts of farmers to manage and improve their plant genetic resources, especially in marginal areas'.

Such partnership will help to address the twin needs for:

- revitalization of the *in situ* on-farm conservation of agrobiodiversity by rural families;
- creating an economic incentive system that provides for both compensation for loss of yield and reward for contributions to genetic resources conservation and enhancement.

The FAO Global Plan of Action represents the most comprehensive strategy for the conservation and sustainable use of PGR.

In a recent publication entitled *Business and Biodiversity* published by the World Business Council for Sustainable Development and IUCN (Stone *et al.*, 1997) the need for sharing benefits was reiterated in the following statement.

> Company management should ensure that local communities are closely involved with decisions affecting the use of natural resources. Of particular importance in some areas is the indigenous peoples' knowledge and the specific issues related to their use of, and relationship to, the land and biological resources. This relationship needs to be harnessed in a positive way, both for the company and the peoples, by developing agreements which promote the equitable sharing of resources.

Thus, the pathway suggested is bilateral negotiations between companies and local communities.

We also have information on both megabiodiversity areas and 'hot spot' locations with reference to threats to biodiversity (see Heywood, 1995). Many of these areas are also characterized by high population density (Cincotta *et al.*, 2000). Hence, extending support to the communities conserving and improving agrobiodiversity is not difficult, if there is a will to do so. UNEP,

FAO, UNESCO and the International Plant Genetic Resources Institute (IPGRI) of CGIAR could jointly help to develop internationally acceptable guidelines for according support from a Global Biodiversity Fund.

Community food and water security system

It is gratifying that several developing countries in South America, Africa and Asia are in the process of institutionalizing benefit-sharing procedures in legal measures relating to CBD. Experience in operating such legislation will lead to the refinement of benefit-sharing procedures by the beginning of the new millennium. This will then provide the basis for the adoption of an agreed protocol on benefit-sharing by the Conference of Parties to CBD. Inaction is not the answer to imperfections in currently available benefit-sharing methodologies. Learning by doing is the pathway to perfection.

In future, the contributions of farmers to *in situ* on-farm and *ex situ* on-farm conservation will depend much on the social recognition and economic incentives provided to the primary conservers. While the older generations conserved as a part of tradition and ethics, the younger generation is likely to continue the earlier traditions only if a recognition and reward system is put in place.

In addition to implementing the ethics and equity principles enshrined in CBD, it will be important to integrate the revitalization of the *in situ* on-farm conservation traditions with a Community Food and Water Security System of the type described below.

There is need for a community-centred food security system based on attention to all the links in the conservation–cultivation–consumption chain. One male and one female member of a panchayat (village or town council) should be trained to serve as a member of a Food and Water Security Corps. Such a system will include the following four major components.

1. *Field Genebank.* This involves the *in situ* on-farm conservation of landraces and local varieties of crops, through the revitalization of the conservation traditions of rural and tribal families, particularly women.
2. *Village Seed Bank.* The rural families often lose their seed stocks because of drought, flood and other natural calamities. Therefore, in each village a Village Seed Bank will be established through a Seed Security Self-help Group, supported by microcredit.
3. *Village Water Bank.* Conservation of rainwater, sustainable management of ground water and the conjunctive use of surface, ground and recycled water are important components of the Village Water Bank.
4. *Area Grain Bank.* It is important to maintain grain reserves of local staples to meet emergencies like drought and natural calamities. For this purpose, an Area Grain Bank will be established at a suitable location, each to serve about 25,000 families. The Grain Bank will be operated by a self-help group supported by a revolving fund.

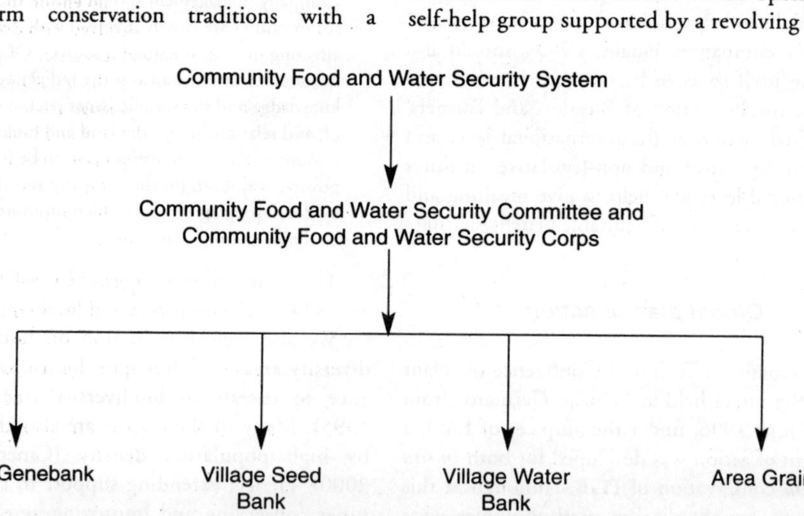

Fig. 2.2. Components of a community food and water security system.

Thus the Community Food and Water Security System will foster a sustainable people-centred and people-controlled method of ending food and drinking water insecurity at the level of each individual. It will help to ensure both food security and genetic resources conservation.

Contributions of Farming Families: A Case Study in Kolli Hills, Tamil Nadu

Past: subsistence mode of production

The local tribal community is known as Malayali. Agriculture is the mainstay of the Malayali economy. For a long time subsistence farming has been the predominant pattern and continued to be so until two decades ago. It was not just for survival; it was a way of life for them. Their traditional management practices enabled them to have a sustainable food system. The traditional cultivated food crops include minor millets such as Ragi (*Eleusine coracana*), Samai (*Panicum miliare*), Thinai (*Setaria italica*), Thirivaragu (*Paspalum scrobiculatum*), Panivaragu (*Panicum miliaceum*), rice (*Oryza sativa*), both wet land and dry land varieties, and pulses.

Almost all the food crops, with one or two exceptions, had a minimum of three to four varieties with different agronomic characters. Given their forefather's knowledge and their own experience, the Malayalis were motivated to adopt different crop varieties to suit local environmental, socio-economic and cultural conditions.

They adopted varieties based on a wide range of criteria. The Malayali farmers, both men and women, pointed out that different food crops and different varieties formed their agricultural system, because of their high yield, cooking qualities, desirable maturity period or ease of harvest. Further, some of the varieties provided good fodder and thatching material for their house roofs. Another significant aspect is the compatibility of different varieties with climatic and edaphic factors, resources, and cropping patterns (e.g. mixed cropping) of the Malayali farming system.

Many varieties were adopted for different reasons, which may be explained by giving a few examples.

- Sadan samai, a medium-duration variety, gives good yield. The straw serves as good fodder material for their cattle, but it is difficult to process.
- Malliya samai matures in 90 days and gives good yield, but the straw does not have fodder value. The straw is good thatching material.
- While the above two varieties are cultivated in terraced uplands, the long-duration variety of Perun samai is cultivated in the sloped terraces because it fits seasonal characteristics better.

There are cultural reasons for cultivation of particular food crops. A rice variety known as Karu Nellu even today is grown by the households worshipping the goddess 'Kongayi Amman', a local deity, for ritual offering. Similarly, Thinai flour is also offered as ritual offerings to local female deities. The delicacies prepared out of this grain are considered very tasty, and the gruel prepared out of it is given to women after childbirth. Thinai is also considered soodu (hot) by the locals. So the food prepared out of it is taken once a week, particularly during the winter season. This grain is quite easy to pound, compared with other millets. The concept of 'hot' and 'cold' foods is also inherent in tribal families, the 'hot' food being taken during winter months and the 'cold' food being consumed during summer.

Ragi was cultivated on a large scale until recently as it is considered very healthy (nutritious) and can be consumed year round. Ragi is followed by samai, Puzhudikar Nellu (a dry land rice variety) and other crops.

The Malayalis practise mixed cropping (mainly Ragi based) where at least six varieties of crops are sown simultaneously and cultivated in a single plot. It is a risk-minimizing device as well as a means of extending the cropping season.

Present: transformation from food production to commodity production

In recent times, i.e. in the late 1970s and early 1980s, a trend towards a market-oriented economy set in which altered their food and production system. This transformation took place as a result of various developmental activities such as improving accessibility through construction of roads, and setting up cooperative societies and schools. Entry of outsiders and exposure to urban life brought changes in socio-economic and cultural values. Monetary considerations favoured a shift from a subsistence economy to a market-oriented one.

A subsistence-based food production system has

largely been replaced by the cultivation of cassava in the terraced uplands and pineapple in the sloped and rocky lands. Between 1970 and 1984, the area under cassava and pineapple increased more than 50 times. Changes in agricultural practices and access to education forced people to reduce their livestock.

Another crucial factor is the increase in the human population, from approximately 19,000 in 1951 to 32,130 in 1991.

Their food security is gradually declining, making them dependent on the market. Most importantly, the agrobiodiversity of the Kolli Hills is at great risk. Another crucial factor which has a direct bearing on Kolli Hills agrobiodiversity is land alienation, which has been taking place at an alarming rate.

Disturbances in the exchange rates may affect the food supply of the Malayalis. To avoid any such disturbances, an alternative system of the kind described earlier, which will ensure their food security, is a must.

Future: revitalization of traditional agricultural and food systems

If the present trend continues there is more scope for greater commodification in future. Those who continue with farming will concentrate on the commercial crops that have economic value such as tapioca, pineapple and plantation crops. Those who are away for education or because of new opportunities have drifted towards non-farm activities. Such people and households would be completely diverted from their traditional agricultural activity. Unless they derive economic benefit from their conservation traditions, many of the local crops and varieties will soon become 'lost crops'. If this scenario is to be changed it needs to have a positive and effective intervention with people's involvement and participation.

A two-pronged strategy can help to mitigate the problem: one deals with cultural aspects and the other with the economic aspect. There should be a cultural motivation for the people whereby they will feel proud about their traditional crops. People should be encouraged to rediscover the cultural value of raising their traditional millets and other crops, by according social prestige to such traditions.

Value addition in terms of monetary gains is another crucial factor. Value addition in terms of both cultural and monetary returns would help to reinforce conservation of traditional millet crops and agrobiodiversity of Kolli Hills. The implementation of the concept of farmers' rights will also help. The fast-changing scenario in Kolli Hills is depicted in Fig. 2.2.

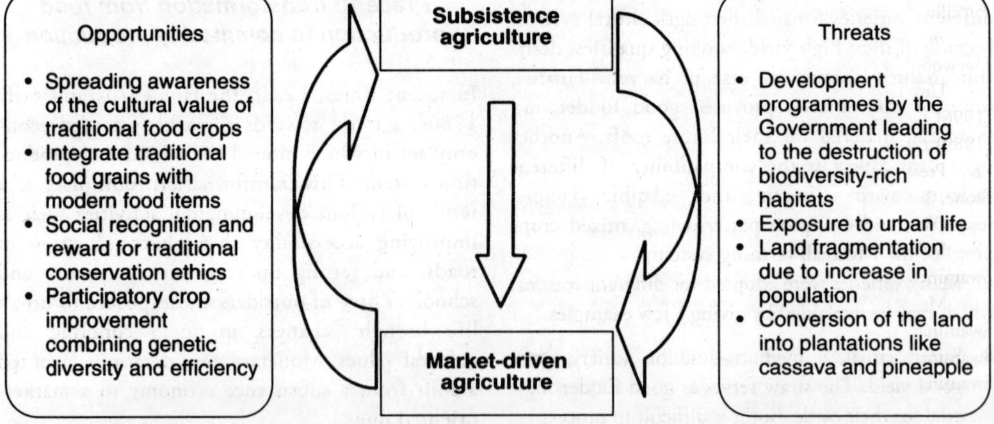

Fig. 2.3. Analysis of strengths, weaknesses, threats and opportunities (SWOT) in Kolli Hills.

Gender Dimensions of Biodiversity Management

Most of the studies on contributions of farming families are not gender disaggregated. MSSRF, in collaboration with the FAO Regional Office in Bangkok, has initiated a series of studies on the gender dimensions of biodiversity management (Swaminathan, 1998). More recently, detailed studies have been undertaken by Dr Virendra Kumar and Dr Hemal Kanvinde of MSSRF on gender roles in biodiversity conservation and management in Sri Lanka and the Maldives. The results so far stress the need to recognize the specific role of women in genetic resources conservation and enhancement. This is particularly important in the context of the equitable benefit-sharing provisions of CBD.

Conclusion

The books *Lost Crops of the Incas* and *Lost Crops of Africa* relate very clearly how ancient civilizations depended on a wide range of food and medicinal plants, and *Biodiversity in Trust* (Fuccillo *et al*., 1997) reports the enormous range of intraspecific diversity maintained in the genebanks of International Agricultural Research Centres (IARCs) belonging to the CGIAR system. We owe such rich variability to the conservation ethos of farm and tribal families. Often the conservation habits were not due to any commitment to conservation *per se*, but rather to the understanding that genetic variability diminishes vulnerability to biotic and abiotic stresses. Farming families in the past were motivated by a desire to minimize risks rather than to maximize profits. Risk aversion agronomy involved growing mixtures of crops and crop varieties.

Modern agriculture involves monoculture and the cultivation of the same genetic strain over large areas. There is a fear that when genetically modified varieties covered by proprietary science and released by commercial companies become popular, genetic homogeneity will increase in farmers' fields. The challenge lies in integrating genetic efficiency with genetic diversity. The procedure adopted by MSSRF scientists for this purpose is the integration of pre-breeding with participatory breeding. By adopting such procedures, we can conserve agrobiodiversity and at the same time enhance it through participatory breeding with farm families. This will help to select location-specific varieties adapted to the local agroecological and sociocultural conditions. If we adopt such scientific procedures and at the same time introduce legal measures for social recognition and economic reward, farm families will continue to conserve, nurture and enhance agrobiodiversity.

References

Cincotta, R.P., Wisnewski, J. and Engleman, R. (2000) Human population in the biodiversity hotspots. *Nature* 404 (27 April), 990–992.

Fuccillo, D., Sears, L. and Stapleton, P. (eds) (1997) *Biodiversity in Trust: Conservation and Use of Plant Genetic Resources in CGIAR Centres*. Cambridge University Press, Cambridge, UK.

Heywood, V.H. (ed.) (1995) *Global Biodiversity Assessment*, United Nations Environment Programme, Cambridge University Press, Cambridge, UK.

(1996) *Lost Crops of Africa*. A report of an ad hoc advisory panel, BOSTID. National Academy Press, Washington, DC.

(1989) *Lost Crops of the Incas. Little Known Plants of the Andes with Promise for Worldwide Cultivation*. National Academy Press, Washington, DC.

Stone, D., Ringwood, K. and Vorhies, F. (1997) *Business and Biodiversity – a Guide for the Private Sector*. WBCSE (World Business Council for Sustainable Development), IUCN (The World Conservation Union), Gland, Switzerland.

Swaminathan, M.S. (ed.) (1995) *Farmers' Rights and Plant Genetic Resources: Recognition and Reward: A Dialogue*. MacMillan India Ltd, Madras, India.

Swaminathan, M.S. (ed.) (1996) *Agrobiodiversity and Farmers' Rights*. Konark Publishers, Delhi, India.

Swaminathan, M.S. (ed.) (1998) *Gender Dimensions in Biodiversity Management*. Konark Publishers, Delhi, India.

Swaminathan, M.S. and Kochhar, S.L. (eds) (1989) *Plants and Society*. Macmillan Publishers, London.

3 The Impact of Gene Technologies on the Use of Genetic Resources

J. Peacock and A. Chaudhury
CSIRO Plant Industry, Canberra, Australia

We can rightly expect this to be the Century of Biology. We have enormously increased our power to ask questions about how living organisms grow, develop and function. The principal driver of this increase in research capacity, which is leading to vast increases in our knowledge base, is gene technology, with its associated tools and related technologies.

Gene technology has provided a quantum increase in our ability to probe the molecular and cellular bases of the biological processes of plants and animals, and of microorganisms for that matter. We now have new levels of understanding of the development, growth and responses of plants to changing environmental conditions.

Underlying development, growth and environmental responses is the control or regulation of gene expression and each component of the plant life cycle can be understood in terms of patterns of gene expression.

Along with the new knowledge have come major, new concepts in our understanding. For example, repression of gene activity is a major strategy of regulation of gene action, especially in the development of an organism. For many genes, the usual condition is that of inactivity; particular controls turn genes on at certain times and in certain cells. Another surprise is that dominant mutations, which were once supposed to be extremely rare, are in fact relatively common.

We have discovered some unexpected mechanisms of great importance in plants. For example, in the last 2 years, research has shown that plants have an immune system protecting them against virus infections. This system operates by interfering with RNA molecules that are necessary in the multiplication cycle of the virus. It is a powerful protection mechanism, which can now be harnessed in agriculture to protect crop plants from many viruses. It can also be harnessed to modulate the level of expression of any particular gene. Another recent finding is that epigenetic mechanisms operate along with genetic (DNA sequence-based) mechanisms of gene regulation, so chemical modification of DNA sequences plays a role in regulation of gene activity along with the coded linear DNA sequence itself. All this new knowledge enriches our germplasm resource collections and increases our capacity to use the genetic variation they contain.

The increase in our understanding of how plants develop and operate is already contributing to modern agriculture, enhancing our abilities to improve the performance of crops. Transgenic crop plants are already grown on more than 40 Mha of agricultural land. The new biology is also improving conventional plant breeding. We need high-performing cultivars of both types – transgenic and conventional – to meet the world's food needs.

Dr Swaminathan has convincingly put the case for the urgent need to increase food production and also for the need to achieve this increase with minimal harm to the environment. We need sustainable agricultural systems. He also emphasized the need to provide enhanced genotypes for subsistence farming, not just for the high-input, large-scale agricultural systems of developed countries.

The two keys to plant improvement are genetic variation and selection. The new biological sciences have increased the supply of genetic variation but the need to conserve existing variation and make it available for use is unquestioned. The new gene technologies are also helping us to meet this particular challenge by increasing our capacity to use genetic resources.

Most, but not all, genetic variation is a consequence of changes in the linear sequence of DNA nucleotides, either in the coding or in the controlling regions of a gene. The coding region of a gene determines what protein (gene product) will be produced. The controlling regions determine where, when and how much of that product will be produced. Changes in the linear sequence of nucleotides can alter the coding message and can change the interactions of regulatory proteins with the protein-binding motifs in the promoter and other control regions of a gene (Fig. 3.1).

We have powerful methods to detect changes in DNA sequence and thus characterize the extent of genetic diversity and populations of any given species. DNA sequence markers have proved to be very powerful in following particular examples of sequence in genetic resource and crop breeding programmes. For example, we are using this technique to monitor variation in the breeding material of macadamia, an Australian tree species, the only species in the Australian flora to enter the commercial food systems. DNA technology allows us to compare the genetic variation in wild populations with that currently in breeding populations and in genetic resource collections. The analysis shows where we must collect further material for the germplasm collection and will give clues as to what genotypes should be introduced into breeding programmes.

Our capacity to work with particular genes, for example self-incompatibility genes, is putting more rationality into decisions on population sizes for conservation of endangered species. The correlation of population size, number of incompatibility alleles and seed set in the plants of a population is a key factor in developing a strategy for *in situ* conservation of the species (Fig. 3.2).

Gene Technology Enhances Detection of Genetic Variation and Precision of Selection

The new gene technologies are having great impact on the use of genetic variation at all levels of relatedness of germplasm to crops (Fig. 3.3). We can illustrate this by considering the genetic resources available for bread wheats. The primary genepool of bread wheats provided the dwarfing genes that have been massively important in Australian wheat production and everywhere else in the world. In Australia, we have used two dwarfing genes in all of our present varieties. They have been instrumental in increasing agricultural performance but they do have some disadvantages under Australian conditions. The dwarf coleoptile of the semi-dwarf varieties can lead to poor stands when the seed is sown deeply, a practice that Australian farmers employ in some water-limited environments and in minimum tillage agriculture. Australian scientists scanned germplasm accessions for dwarfing genes that do not reduce coleoptile length even though the internodes of the growing plant are still shortened. A new dwarfing gene with these characteristics was identified and is being introduced into breeding populations. Crop establishment is improved considerably. The breeding process is being speeded up

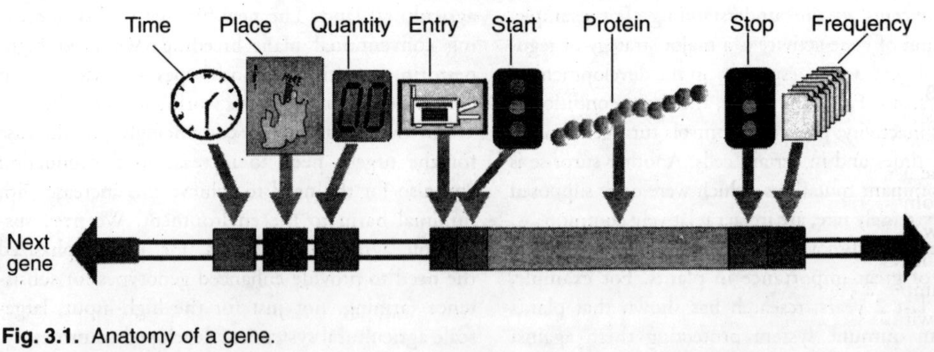

Fig. 3.1. Anatomy of a gene.

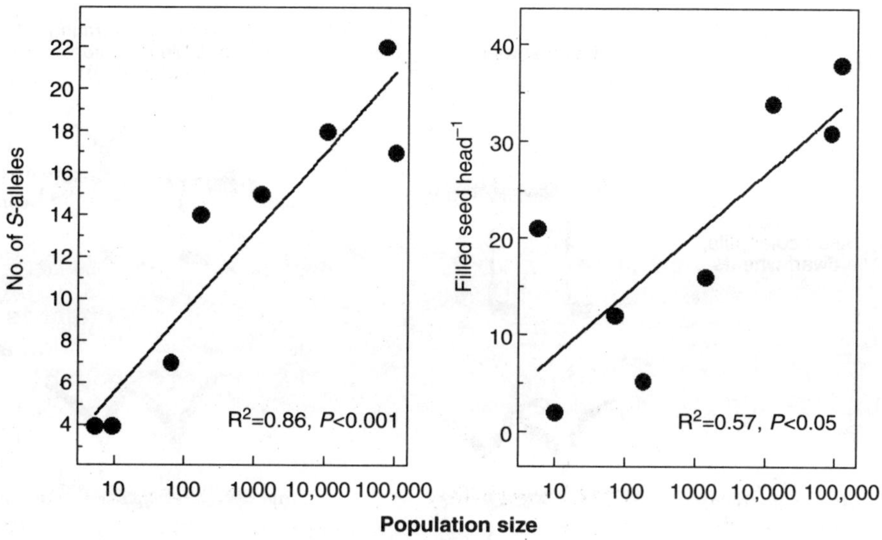

Fig. 3.2. Correlation of population size, *S* alleles and seed set.

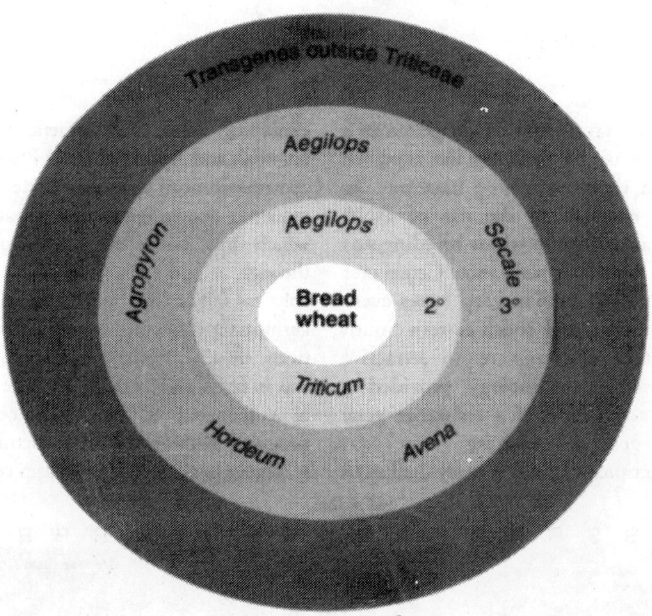

Fig. 3.3. Bread wheat genepool.

with the use of DNA sequence markers and it looks very promising indeed (Fig. 3.4).

New biological knowledge is also giving breeders the means to break barriers to increasing yield. An example in wheat agriculture is our ability to select, within the primary genepool, for increased efficiency in the use of the two naturally occurring carbon isotopes in the atmosphere. This is done using a mass spectrometer and has resulted in spectacular increases in yield in putative cultivars. In this case the new technologies provided a new selection method for the breeder.

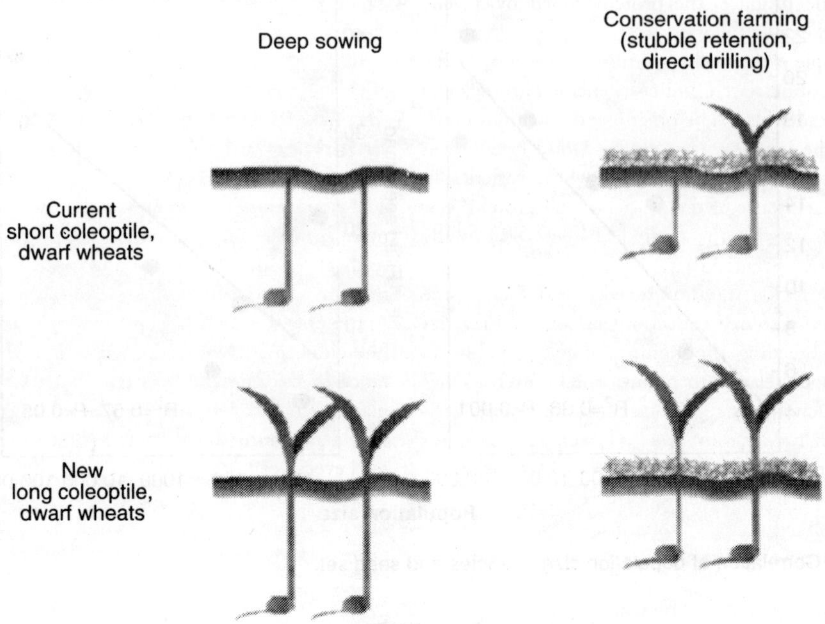

Fig. 3.4. Wheat dwarfing genes.

DNA markers can save breeding programmes a lot of time and expense by reducing the need for costly, difficult and time-consuming bioassays. In Australia the first example of the use of DNA markers for a resistance gene in wheat breeding was the gene for cereal nematode resistance. Cereal cyst nematodes cause AU$80 million crop losses every year in western Victoria and south-eastern South Australia, so resistant cultivars are an attractive prospect. DNA marker technology provided a powerful tool for acquisition of a resistance gene from the secondary genepool of wheat.

Screening to meet precise needs is really the key to extracting value from genetic resource collections. Lagudah and Appels (CSIRO Plant Industry, personal communication) were able to identify a DNA marker sequence closely linked to a nematode resistance gene which had been transferred to hexaploid wheat through a *Triticum tauschii* cross. Apart from being able to follow the specific gene from the *tauschii* germplasm collection in subsequent breeding populations, they were able to show that the DNA marker was in fact a perfect marker for the resistance gene as it is a sequence in the resistance gene itself. This provided a screen for other members of this class of resistance gene in the germplasm collection (Fig. 3.5).

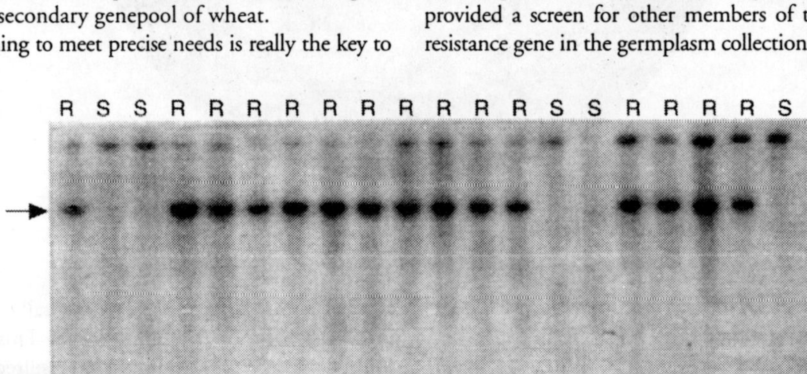

Fig. 3.5. DNA from a subset of *T. tauschii* individuals in an F_2 population in which *Cre3* is segregating (three independent populations).

A gene product, the protein coded by a gene sequence, can be followed in screening programmes. An example is the null condition (a deletion) of the granule-bound starch synthase gene on chromosome 4A of bread wheat. The organoleptic quality important to the Japanese taste in the Udon noodle was found to be present in certain wheat varieties in Australia. Absence of the 4A gene (and protein) was shown to provide the desired characteristic in the noodle dough (Fig. 3.6).

In the 1950s mutation breeding stood alongside germplasm resource collections as sources of variation for breeding programmes, but it has always been a poor relative to nature which, after all, has had a much longer time to accumulate genetic variations. In nature mutation has been coupled with the natural processes of selection so that the variation present in germplasm resource collections has a high probability of being useful. Now there are powerful new techniques for screening thousands of plants for sequence variance in any particular gene which is known to be of importance in a breeding programme. Detection of single nucleotide polymorphisms can reduce a collection of many thousands of accessions to some tens of plants with sequence changes in the gene of interest. These can then be screened for their phenotypic characteristics and used where appropriate.

Mutagenesis Programmes can Interface with Genetic Resource Use

The new technologies can also be used in tackling more complex goals. Apomixis is a process in which seeds are produced without any gametic union and in many cases without any role of the male gamete. Apomictic seeds are strictly matroclinal: a female line clone. Apomixis could be of immense importance in freezing the hybrid vigour of F_1 progeny plants so that farmers in developing countries would not need to buy new hybrid generations of seed every year.

Apomixis is present in many plants and occurs by different mechanisms. However, it has not yet been possible to integrate it into any of the major crops of the world. The species of *Tripsacum*, related to maize, contain characteristics of apomixis but so far it has not been possible to introgress the traits into maize. Isolation of genes that control apomixis and their introduction into the crops of the world would make a dramatic change in agriculture.

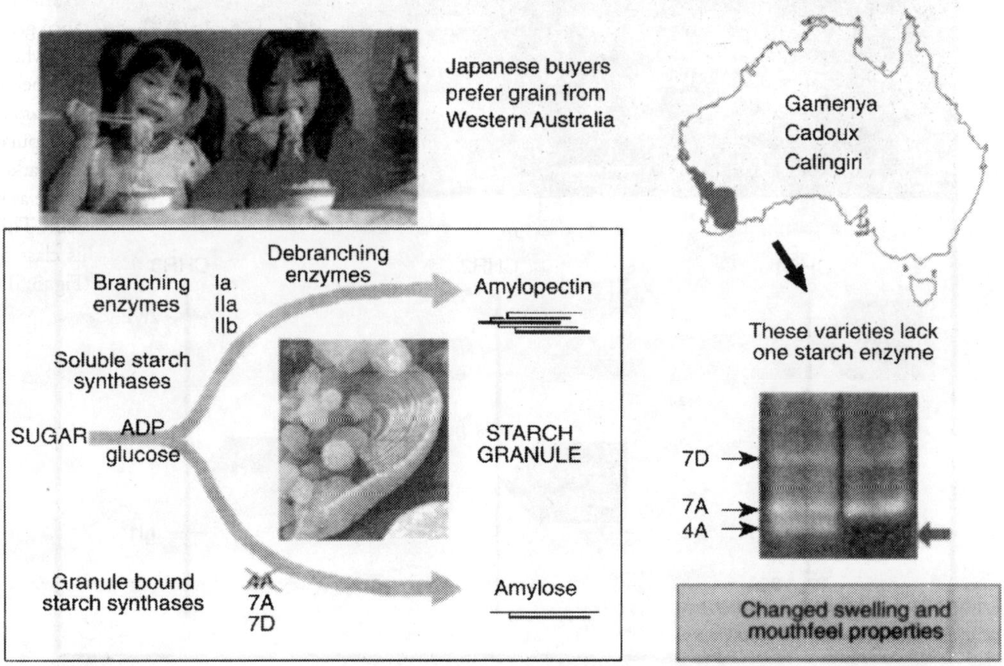

Fig. 3.6. A gene for taste.

We mounted a mutagenesis programme in *Arabidopsis* to identify genes that are important in the development of seed and in which a mutation permitted seed development without pollination taking place (Fig. 3.7). We have identified three genes that code for proteins that act as repressors of seed development in the absence of pollination. The absence of any one of the key repressors presumably allows a whole cascade of patterns of gene expression to occur so that seed development is initiated, but the seed development fails to proceed to viable seed formation. Even if these mutant plants are pollinated, seed development is arrested. The embryo develops only to the heart stage and then development fails.

There was a startling exception to this rule and that was when the pollen used was from a plant in which DNA methylation levels had been much reduced. In this case pollination provided for normal double fertilization and viable seed production (Fig. 3.8). This observation provides the basis for a new screen for genes in genetic resource collections that allow embryo development to occur without pollination but through the same paths as occurred when pollination in our experiments was from low-methylation plants. If such alleles can be identified then we would be very close to having a functional apomictic system.

In some ways the approach that I have described suggests that mutagenesis of genetic material could bypass screening of existing genetic resources but in reality it emphasizes that with increased knowledge we would have the ability to screen genetic resources for precisely the allelic variance required. In this case available germplasm resources could be screened for the class of mutation that generates the embryo development in the manner that we discovered with low methylation level pollen.

Transfer of Genes from Genetic Resource Accessions

A major difficulty in the use of variation from germplasm collections applies to the situation of sexual isolation of tertiary genepools. For example cotton, *Gossypium hirsutum*, is susceptible to *Fusarium* wilt, a disease presenting a major challenge

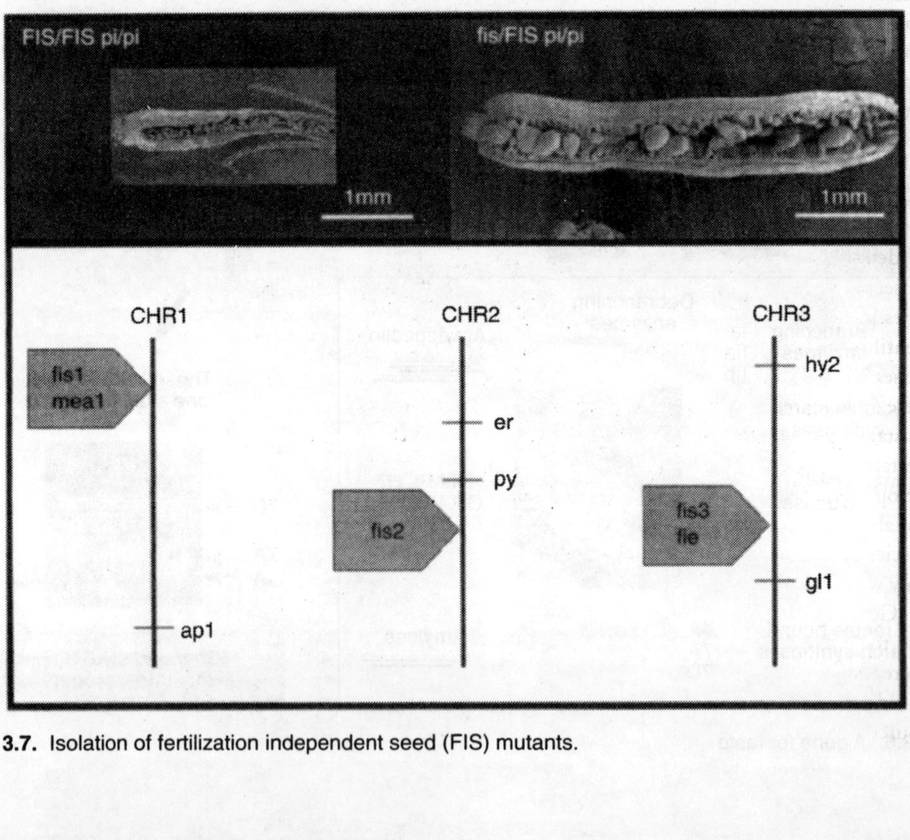

Fig. 3.7. Isolation of fertilization independent seed (FIS) mutants.

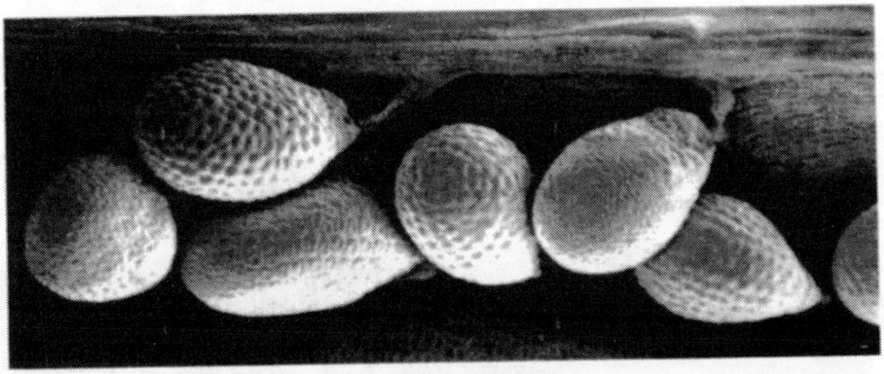

Fig. 3.8. Rescue of FIS mutants with low methylation pollen.

to the Australian cotton industry at present. Breeders have made some progress in selecting for tolerance but research has shown that a native Australian *Gossypium* species, *G. sturtianum*, appears to be immune to the *Fusarium* pathogen. It is, however, extremely difficult to transfer genetic material from *G. sturtianum* to *G. hirsutum*. Brubaker and Brown (CSIRO Plant Industry, personal communication) have made some progress using sophisticated technologies of chromosome doubling and embryo rescue, and by searching for appropriate homologous recombination in the hybrid plants. However, this would be a perfect opportunity to apply gene cloning of the resistance gene with subsequent transfer of the gene into *G. hirsutum* by transformation technology.

The latter approach is becoming a possibility now because of the advances in mapping the cotton genome and the fact that many resistance genes can be identified through their possession of particular conserved sequence motifs. Cloning of resistance genes in rice, in *Arabidopsis* and in other plants has identified common sequence features of many classes of resistance genes, and hybridization for these motifs will greatly facilitate acquisition of resistance genes from germplasm collections. Certainly in the case of the quaternary genepool of a crop plant – i.e. plants or other organisms that are not able in any circumstances to cross with the crop species – gene technology and gene transfer are the only ways to introduce new genetic material.

Gene transfer depends on two things: our ability to put genes into any target species, and increasingly we are able to do this, and our ability to identify and clone genes from a germplasm source (Fig. 3.9).

The Era of Genomics

Already the complete genome sequence of *Arabidopsis* is available and it is likely that the complete genome sequence of rice will be available before the end of 2001.

Once a specific gene has been cloned from *Arabidopsis* there is a high probability of being able to use the *Arabidopsis* gene sequence to isolate homologous gene sequences from other plants, particularly dicotyledonous plants. Similarly, homologous gene sequences from the rice genome will provide the launching platform for acquisition of genes from a range of monocotyledonous plants. So we will soon have available a high probability for the isolation and use of genes across the plant kingdom. In the case of cereal crops, the major food crops of the world, another feature of considerable value is that the gene sequence is highly conserved throughout the different species. This synteny of the cereal species will be a useful property in the exchange of information among the cereal plants (Fig. 3.10).

Along with sequences of genomes, many other techniques are developing which will be of great value in accessing genetic resources. Microarray technology allows the simultaneous monitoring of thousands of expressed genes and in this way specific genes, which are likely to be of value in a given environmental or biotic challenge situation, can be identified and then a process of screening can be mounted in a germplasm collection. The seed banks of the world contain lines that have many unique characteristics needed for successful crop production, resistance to stresses, resistance to diseases, to toxic minerals and so on, and already breeding efforts have used these resources in crop

Fig. 3.9. Genes can be inserted into most crops.

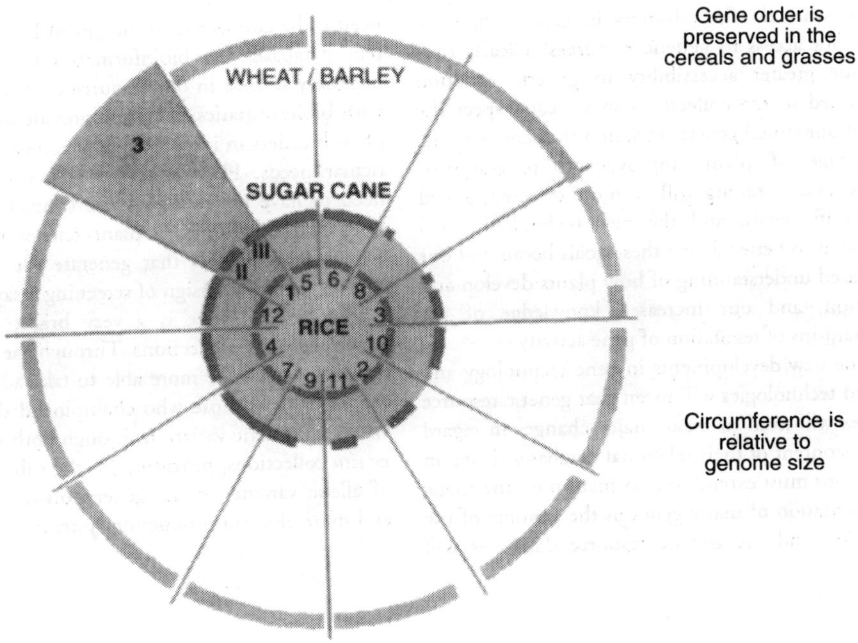

Fig. 3.10. Rice is the key to the genes of cereals.

improvement. But what has happened so far is only the tip of the iceberg compared with what promises to come in the next few years. Genes will be transferred not only across species but also from one genus to another and between quite unrelated families. One of the most successful and wide-spread transgenic situations of the present day is the transfer of the bacterial insecticide gene from *Bacillus thuringiensis* to crop plants where it provides protection against Lepidopteran pests.

Gene transfers across species have raised concerns in the community but as our knowledge of genomes increases we see that cross-species gene transfers will mostly be the introduction of different allelic forms of genes that exist in a wide range of species. There are even many genes shared across the different kingdoms of organisms. For example, genes that code for haemoglobin, previously thought to exist primarily in the animal kingdom, are now known to exist in all species of plants and in many microorganism groups. Transferring a haemoglobin gene from an animal to a plant could therefore be considered comparable to the transfer of a haemoglobin gene from one plant to another in an intraspecific cross.

In plant improvement programmes the ultimate objective is to develop plants that best fit our needs, whether they be our need for compatibility with the agricultural environment or our need to provide food products optimized for our health needs. Ideally we will assemble an array of coded genetic information in the genome of our production species which accomplishes all of these objectives.

Conclusions

The world's human population is now increasing by about 160 people per minute so more than ever before there is a requirement to provide for increases in food production. This has to be achieved in a world where no more land is available for cultivation and water and other essential resources are becoming limited in crop production. More than ever before there is a need to harness the genetic potential of crop plants and in particular the three major cereals – wheat, rice and maize – to the fullest possible extent. It is fortunate that this compelling need has arisen at the time of unprecedented progress in the plant sciences. The developments in our understanding of genetics and the molecular biology of plants have allowed the identification and characterization of many genes that will help to increase yield and protect plants from biotic and abiotic stresses. These developments are opening possibilities for the characterization and use of genetic resources that were not foreseen in earlier times.

So what do the advances in gene technology mean for access to genetic resources? Clearly they provide greater accessibility to genetic variation contained in the collections so we can expect the use of our stored genetic variation to increase to the advantage of plant improvement programmes. Access and screening will be more directed, aimed at specific goals, and the new technologies will enable us to better define these goals because of our increased understanding of how plants develop and function, and our increased knowledge of the mechanisms of regulation of gene activity.

The new developments in gene technology and related technologies will mean that genetic resource centres will need to make major changes in regard to the content of their relational databases. Data on accessions must extend to information on the allelic representation of many genes in the genome of the accession and the genetic resource databases will need to be connected to the global DNA and protein databases. So bioinformatics will become a necessary science in our resource centres. Scientists with bioinformatics skills will provide assistance to plant breeders in identifying key accessions for particular needs. Plant genetic resource centres will need to have increasingly close contact with basic research programmes in plant science since these are the programmes that generate the knowledge providing for the design of screening assays.

We believe there is a very bright future for genetic resource collections. Through the new technologies we will be more able to take advantage of the wisdom of those who championed the conservation of genetic variation through both *in situ* and *ex situ* collections, providing for the subsequent use of allelic variants in the generation of sustainable and more efficient production systems.

4 The New Genetic Era: Will it Help us in Managing Genetic Diversity?

A. Karp
IACR-Long Ashton Research Station, University of Bristol, Long Ashton, Bristol, UK

Introduction

The advances made in the science of genetics, from the rediscovery of Mendel's laws in the early 1900s to the development of DNA chips at the end of the last century have been simply astonishing. As the 21st century unfolds, the advancing field of plant genomics heralds promise of a new potential for analysing plant diversity at the level of single base-pair differences and for identifying the genetic basis of traits underlying the agronomic performance and environmental adaptation of plants.

This chapter examines the contribution already possible from the application of existing molecular genetic techniques to the field of plant genetic resource management and the gaps and challenges that remain to be addressed. It then seeks to assess the potential of genomics to enhance our capabilities for managing plant genetic diversity in the 21st century.

Addressing Key Issues in Genetic Resource Management

From the standpoint of assessing the role of molecular technology, the main issues are as follows. For *ex situ* genetic resources, the key issues are in essence the acquisition, maintenance, characterization and utilization of genetic diversity. For *in situ* genetic resources, they are the identification of populations for conservation, their management, access to the genetic resources and utilization (Karp *et al.*, 1997a).

Within all these key issues it is possible to identify numerous specific questions that continuously have to be faced in making decisions on genetic diversity. These questions vary from those that are relatively easy to define, for example concerning whether two individuals are identical or not, to those where definition is more difficult, such as problems relating to 'genetic erosion', or identifying 'useful variation'. Nevertheless, a common essential feature that is shared by all is the requirement for information on diversity, and it is here that the most fundamental contribution of molecular techniques is made.

The first major attraction of molecular techniques was their ability to detect genetic diversity at levels of resolution that exceeded those achievable by other methods, including the use of biochemical markers such as isozymes. The nature of the DNA-assay (originally restriction fragment length polymorphism (RFLP)) also meant that markers were more robust and independent of environmental conditions. Additional attractions, which emerged with the development of the PCR, were the possibility of obtaining information from tiny amounts of plant material at any stage of development, in addition to the increasing speed with which the assays could be performed and the ability to increase throughput substantially by automation.

There is little doubt that for all these reasons DNA-based assays have revolutionized our ability to characterize genetic diversity. The large number of articles describing their use for the characterization of crop germplasm and plant populations of many species are testimony to this success. As indicated above, however, managing genetic diversity requires more than just characterization. It requires information that can be used to help address the key issues of both *ex situ* and *in situ* genetic diversity management and to assist in the process of decision-making. For example, for assisting evaluation, for improving utilization and for the development, or validation, of genebanking strategies, such as in sampling strategies, multiplication, management and the development of core collections.

In this respect, it can be argued that the potentially infinite number of 'character differences' that can be revealed using DNA-based assays is only a minor part of the contribution of molecular techniques of relevance in the context of genetic diversity management. Of greater importance is the kind of information on genetic diversity that can be derived, together with the quality, speed and efficiency with which it can be obtained. Ultimately, it is only if this information can be used confidently by genetic resource managers to help improve practices, or assist in decision-making, that the benefits of molecular techniques could be said to be realized.

Molecular Techniques and Information on Genetic Diversity

The information on genetic diversity that can be derived using molecular techniques can be summarized as follows:

1. The amount of diversity (a quantification of the diversity present at intra- and interspecific levels).
2. The structure of the diversity (how the diversity is partitioned among and within populations in space and over time).
3. Evolutionary history (population history, genetic lineages, gene genealogies and distinguishing between differentiation and divergence, i.e. between current and past events).
4. Genetic distance (a measure of the extent of difference between genes, sequences, populations or species).
5. Relatedness (kinship, the degree of relatedness among relatives).
6. Identity (whether individuals are genetically identical or distinct).
7. Gene flow (the extent to which genes are exchanged).
8. Linkage disequilibria and the inter-locus correlation of allelic variation.

A large number of different molecular techniques are able to provide information of this kind, but despite recent advances and our increasing knowledge of plant genomes, no single technique is currently available that can provide all of the information listed above. Molecular techniques differ in their informational content. Some techniques produce dominant markers only, while others are co-dominant and permit detection of the different alleles. Some techniques yield multilocus profiles, while others detect variation at a single locus. Some markers provide genealogical information (i.e. on evolutionary history), while others provide information of phenetic relationships (i.e. based on overall similarities). Understanding exactly what type of information will be needed on genetic diversity to address specific management issues is, thus, at the crux of being able to determine whether any given technique will be a suitable means of obtaining the information. To add another complexity, the quality of the information obtained for any given marker system will also depend on the sampling strategy used and the way(s) in which the data have been scored and analysed. Finally, techniques also differ in resource requirements, in their reproducibility, in their amenability for automation and with respect to whether or not their data can be easily input into databases.

For all these reasons, choosing the most appropriate technique can be difficult. More often than not, a combination of techniques is needed to cover all the requirements of information that may be needed, which increases the cost of applying a molecular approach. Alternatively, a compromise has to be made which reduces either the type of information that can be obtained or the number of samples that can be analysed, and therefore places a limit on the potential gains from the application of molecular techniques.

Despite these difficulties, over the past 5 years the field of molecular genetic diversity screening has advanced significantly and general consensus is being reached as to the most appropriate marker techniques (Karp *et al.*, 1997b). Of the many different techniques for detecting polymorphisms, only the main types will be discussed here, with reference to their general suitability for genetic diversity management. The newer technology of DNA microarrays will be discussed later as their impact on the field is not yet known.

1. Probe hybridization (no PCR)

RFLPs were the original successful molecular marker technology because they are co-dominant and, thus, can detect heterozygotes (Evola *et al.*, 1986). They are also highly reproducible and can be exchanged between laboratories. Data obtained from RFLPs can be used in the characterization of genetic diversity in cultivars, accessions and natural populations and can assist genetic diversity management in many of the key areas discussed earlier. Although polymorphic probes have to be identified from cDNA or genomic DNA libraries, informative probes can be exchanged between laboratories and syntenic relationships among plant genomes, for example among the *Gramineae*, or *Solanaceae*, have further extended the wide usability of RFLP markers. These advantages have been increasingly outweighed, however, by the disadvantage of the lengthy technical procedure (which involves Southern blotting and probe hybridization), the difficulty of converting the method to an automated system and the low level of discrimination achievable in some crops.

Higher levels of polymorphism, capable of distinguishing even the closest relatives can be achieved using probes for hypervariable regions of the genome comprising tandemly repeating sequences. There are two main classes of these: minisatellites, which are typically 16–100 bp in length and microsatellites, or simple sequence repeats (SSRs), which are 2–8 bp in length. Minisatellites and microsatellites are found in abundance and the use of one motif as a probe will typically detect several hybridization sites resulting in a multi-locus profile. The high mutation rates of these tandem repeats compared to other nuclear sequences results in highly discriminate profiles that can distinguish individual members of a family and fingerprint cultivars (Vosman *et al.*, 1992). Although restriction enzyme digestion and probe hybridization are used in the procedure, the polymorphism observed results from variation in the number of tandem repeats present at each locus and is, thus, commonly described as variation in the number of tandem repeats, or VNTRs. VNTRs provided the first method of DNA fingerprinting for individual assignment (Jeffreys *et al.*, 1985; Bruford *et al.*, 1992) and have been extensively applied in animals and humans (including forensic science). In those cases where they have been identified in plants, they have been useful for germplasm characterization, particularly through their capacity to discriminate closely related cultivars in crops with low polymorphism levels (Vosman *et al.*, 1992). Applications in genetic diversity management have been limited, however, by the multi-locus nature of the profile obtained and the length of the procedure required to perform the assay. For both VNTRs and RFLPs the technical limitations reduced their usage once PCR had been developed and the Southern blotting, probe hybridization steps could be avoided.

2. PCR with generic or arbitrary primers

PCR brought with it a whole wave of new techniques, the most immediately applicable of which require no prior knowledge of sequences within the genome but rely on the use of either random arbitrary primers, or generic primers. In randomly amplified polymorphic DNAs (RAPDs) (Williams *et al.*, 1990), a single random primer is added to the template DNA and subjected to PCR. This simple but effective method of revealing polymorphisms is also cheap and universally applicable and was extensively applied for characterization of plant genetic diversity after its initial development. RAPDs are dominant markers. More problematic, however, is the fact that the profiles amplified are very sensitive to any change in laboratory conditions, consumables and equipment used to perform them. RAPDs are not robust enough to exchange among laboratories (Jones *et al.*, 1997), nor to withstand the passage of time and the continual change in equipment and enzymes that occur as techniques advance. Similar limitations are true for many other techniques which work on a comparable basis to RAPDs and which include AP-PCR (arbitrary-primed PCR), DAF (DNA amplification fingerprinting), directed amplification of minisatellite region DNA, ISSR (inter-SSR amplification) and RAMP (randomly amplified microsatellite polymorphism) (see Karp and Edwards, 1998). Even though the limitations of RAPDs and other similar techniques are well recognized, they are still useful for the characterization of genetic diversity, for estimating genetic similarities and genetic distances, for identification and classification and for population studies, although they are increasingly becoming restricted to situations where resources limit the choice of markers available or only quick, basic results are required.

The procedure of amplified fragment length polymorphism (AFLP) (Vos *et al.*, 1995) depends on the use of generic primers and is more robust than techniques such as RAPDs for two main reasons. Firstly, it involves restriction digestion and secondly, two primers are used in more stringent PCR conditions. In this technique the DNA is first digested with a rare cutter and a frequent cutter enzyme. Adaptors are added to the cut ends and then PCR is performed in two successive rounds. In the first round, the primers are based on the adaptor and restriction site sequences plus one additional nucleotide. In the second round between two and three additional nucleotides are added to the primers. The additional nucleotides act as selective bases because only fragments containing the complementary bases internal to the restriction site and adaptor sequences will be amplified. The resultant multi-locus profile is highly reproducible (Jones *et al.*, 1997) and extremely informative.

AFLPs provide data on many different sites within the genome in a single assay and are currently among the most informative of marker systems available with respect to polymorphic index (see Powell *et al.*, 1996). They have made a considerable impact on the characterization of crop germplasm by providing an extremely efficient means of identifying duplicates and contaminants, of determining genetic similarities and genetic distances and of discriminating between closely related accessions. The AFLP technique has also been adapted for use with primers that target specific sequences within the genome, or sequence-specific amplification polymorphism (SSAP) (Ellis *et al.*, 1998). Specific modifications, which utilize knowledge of gene sequences, will be described later. Here, two main classes are mentioned: SAMPL (selectively amplified microsatellite polymorphic loci), where one of the AFLP primers is substituted by a primer based on the sequence of a compound microsatellite (Witsenboer *et al.*, 1997); and assays that detect polymorphisms associated with the insertion of retrotransposons, such as inter-retrotransposon amplification polymorphism (IRAP) (Kalendar *et al.*, 1999), retrotransposon-microsatellite amplification polymorphism (REMAP) (Kalendar *et al.*, 1999) and retrotransposon-based insertional polymorphism (RBIP) (Flavell *et al.*, 1998). These techniques provide the means of profiling diversity within specific regions of the genome, which can be helpful in crops where polymorphisms are limited, by extending the range of markers that can be found. In the case of RBIP, PCR products can also be detected on gels or by dot blotting. The latter process can be automated, permitting potentially very high throughput screening and is thus of interest to breeding companies as well as curators of germplasm as a means of screening large numbers of samples. The use of this approach for screening genetic resources is currently being investigated (A. Flavell, University of Dundee, Scotland, UK, personal communciation).

3. PCR with sequence tagged primers

Multi-locus profiling techniques are an effective means of obtaining data at many loci in a single assay, but the information available at a single locus is limited (Karp *et al.*, 1997b). This is a particularly important shortcoming to recognize if gene frequency or genealogical information is required for genetic diversity management. Information on heterozygosity is also usually limited, which is a particular disadvantage for outcrossing species.

The alternative approach to multi-locus profiling is PCR with primers that target a single known site, such as a gene. In plants there are three sources of potential gene sequences: the chloroplast (cpDNA), mitochondrial (mtDNA), and nuclear (nDNA) genomes which differ in their mode of inheritance, evolutionary rates and recombination. Gene sequences are the most informative of markers available and are therefore very important for diversity studies from species to population level. They are especially important when information on evolutionary history (e.g. allele or gene genealogies, phylogenies, population history, etc.) is required. Sequence data are theoretically capable of providing all the types of information on diversity listed earlier. In practice, however, polymorphism rates can be too low to attain this goal at the intraspecific level and particularly at the level of individuals.

Microsatellite and minisatellite loci mutate at higher rates than gene sequences. If they are cloned and sequenced, primers to the flanking regions can be designed to produce a locus-specific, sequence-tagged marker. Minisatellites are generally difficult to clone by virtue of their size and have not been used extensively in plants. In contrast, microsatellites tagged in this way are extremely useful for genetic diversity studies because they provide single-locus, co-dominant, highly polymorphic markers (Morgante and Olivieri, 1993).

Throughput and information content can be increased by multiplexing and using automation, which means that information can be obtained from several different microsatellite loci simultaneously on large numbers of samples. Microsatellites are also robust markers and therefore suitable for exchange among laboratories and they provide data that are more amenable for input and storage in database formats. For these reasons microsatellites are being increasingly used to characterize genetic diversity in crop genepools and to establish databases of crop varieties. Together with sequences they are the predominant marker system for population genetic analysis and have also replaced classic DNA fingerprinting for identity and forensic applications. One of their main limitations is the cost and time required for their development, which led to their slower uptake as molecular markers in plants, compared with animals. Procedures for microsatellite isolation from plants have now been developed by several researchers and an increasing number of primer sets are being published on web sites and in journals. A second problem with microsatellites is that their mode of mutation does not conform well to the classical mutation models derived for gene sequences, although specific analytical techniques have been designed to overcome this problem.

Examples of the Use of Molecular Marker Techniques

The subject of genetic marker techniques and their appropriateness for genetic diversity analysis has only been given cursory treatment here compared with the depth to which this subject can be discussed. There are numerous reviews which cover this subject (e.g. Bachmann, 1994; Karp *et al.*, 1997b, 1998; Parker *et al.*, 1998; Sunnucks, 2000), and it is not my intention to reiterate these. Instead, there are three main points that I would like to emphasize:

1. The more clearly the question(s) being addressed are defined, the easier it will be to select the best marker system(s) and the best strategies for applying them.
2. A basic understanding of the different techniques with respect to what they can offer in terms of genetic information is essential for their successful application.
3. It is also important to consider how the information will be gathered (and by whom) and how (and by whom) it will be used.

Two collaborative studies on genetic diversity should serve to exemplify these points.

The use of molecular markers to assess genetic diversity in coconut genepools

Coconut (*Cocos nucifera* L.) is a key plantation crop of the tropics. Almost every part of the tree is used and its food and industrial products, such as coconut oil, copra and desiccated coconut, play an important role for the rural communities and economies of many developing countries. The current geographic distribution of coconut spans the coastal areas between 20 degrees north and 20 degrees south of the equator, with an altitude range from sea level to 1000 m. The wide geographic distribution of coconut and its diverse growth habit have led to the identification of some 900 accessions by the International Coconut Genetic Resources Network (COGENT). However, the crop is now in relative decline in many countries, due largely to the impact of diseases on yields and strong competition from other oil crops such as oil palm. COGENT is in the process of collecting threatened and useful germplasm to augment existing collections with a view to establishing regional genebanks. Specific questions that need to be addressed in this endeavour include: How can genetic diversity in populations of coconut growing *in situ* be assessed rapidly and efficiently in order that palms can be selected for inclusion in regional genebanks? How much diversity is present in coconuts of the different regions and can differences be detected between different growing regions? Can individual palms be adequately identified?

To assist in these efforts, in a succession of two transnational collaborative projects, appropriate molecular marker techniques were developed and their potential in helping to resolve these questions was demonstrated. At the outset, it was recognised that of the DNA-based methods that could be used to assess coconut germplasm, only a few would be suitable for application in networked efforts spanning several continents, consistent with the COGENT strategy. Results must be reproduced by different laboratories. Assays must be capable of being scored and analysed using standardized methods and the data obtained must be suitable for entry into databases. It was also important to consider that there are two main types of coconut, tall varieties, which are fast growing and outcrossing and dwarf varieties, which are mostly selfing. While the latter can be considered to be essentially fixed lines, tall varieties exist either as populations maintained under natural pollination or in the form of inter-population hybrids. Techniques that detect heterozygotes directly (i.e. without progeny testing) and give data on allelic differences were therefore preferred.

Microsatellites were considered to fit all the requirements for the coconut genetic diversity study, although these were not available at the time of starting the work. It was also felt necessary to complement the microsatellite analysis with AFLPs in order to investigate whether a multi-locus fingerprinting method would be needed to discriminate between closely related palms.

The objectives of the first collaborative project were to develop highly informative microsatellite markers in coconut and characterize their usefulness using a limited set of cultivars. In total, 41 primer pairs were identified and tested of which 38 were polymorphic, with an average of five alleles per locus. Analysis of similarity matrices based either on shared alleles at each locus (simple matching coefficient), or on allele bands across all loci (Jaccard coefficient), showed that Dwarfs grouped separately from Talls and that less genetic diversity was present in Dwarfs compared with Talls, as expected. In a wider test on 40 samples, eight microsatellites detected 64 alleles, giving an average of eight alleles per microsatellite. This pilot study indicated the potential of these markers for detecting diversity in coconut populations (Rivera *et al.*, 1999). In order to be useful in the context of the COGENT network it was necessary to ascertain that the microsatellites would be equally informative in coconut material from across the geographic range. In the second collaborative project, the microsatellites were tested on a wide range of ecotypes and the data gained from this study compared with those obtained using AFLPs. Genetic diversity was assessed in 31 individuals from 14 coconut populations across the geographic range using 37 of the microsatellites (one of the 38 original microsatellites failed to give a clear pattern when tested further) and 12 AFLP primer combinations. The microsatellites reveal high levels of genetic diversity, with between two and 16 alleles per locus and a total of 339 alleles in the 14 populations. Two of the four Dwarf populations were homozygous at all the loci, as expected, but Niu Leka Dwarf, which is known to be cross-pollinating, showed high levels of heterozygosity. Three microsatellites gave distinct genotypes for 12 out of the 14 populations suggesting that it would be possible to identify all 14 populations using only a small set of microsatellites (Teulat *et al.*, 2000).

The microsatellites revealed important information concerning the genetic diversity of the coconut populations. Genetic diversity was generally higher in populations from the South Pacific and South-East Asia. The East African populations had higher heterozygosity levels than those from West Africa and the populations from Tonga and Fiji generally had distinct alleles from those of the South Pacific. Data of this kind were not as apparent from the AFLPs. The AFLP analysis with 12 primer combinations gave a total of 1106 bands, of which 303 were polymorphic (27%). Similarity matrices were constructed from the two datasets using the proportion of shared alleles, in the case of the microsatellites, and Jaccard coefficient, in the case of the AFLPs. Cluster and principal coordinates analyses gave dendrograms and plots that showed similar relationships with both marker types, although AFLPs resolved tighter clusters. There was generally good agreement between groupings on the dendrogram and the geographic distribution of the coconuts (Teulat *et al.*, 2000).

The results indicated that both microsatellites and AFLPs provided valuable information on the genetic diversity of coconut which could address the questions posed but that the microsatellites, in particular, also meet the requirements relating to how the information will be gathered and how it will be used. The microsatellites now provide a valuable toolkit for genetic resource managers as they can be used in the laboratories of the different countries to identify populations with distinct alleles or high heterozygosity and for other practical problems such as detecting pollinations from non-parental types in breeding programmes. Efforts are currently underway to identify a small sub-set of microsatellites which could be used as a standard tester set to identify all the different populations. It may still be necessary, however, to use AFLPs in certain circumstances, for example, to check whether two individuals are identical or not, since a large number of microsatellites would be needed to achieve the same genome coverage possible with a single AFLP profile.

The use of molecular markers to improve genebank management

The coconut example is just one of many that illustrate the information gained when molecular techniques are chosen carefully to address specific questions on genetic diversity. Like other examples of its kind, the study is still, however, at the relatively small scale in terms of plant numbers. The application of molecular techniques at realistic scales equating to those of *ex situ* plant genetic resource collections, for example, might therefore still be questioned.

One example of where this issue has been tackled can be found in an EU-funded demonstration project on the lettuce collection held at the Centre for Plant Genetic Resources (CGN). This project specifically targeted the requirements of genebank managers and has sought to address the usefulness of molecular markers from a genebank perspective. Genebanks conserve *ex situ* genetic diversity of crops and their wild relatives by first sampling the diversity and then storing the diversity as efficiently as possible. The project centred around the numerous questions that arise throughout these practices relating to:

1. Which samples to include (or exclude, or bulk).
2. The identification of gaps in the collection.
3. The identification of duplicates.
4. How to choose material from the germplasm collection for utilization.
5. How to multiply samples without loss of diversity.

To demonstrate the possibility of applying molecular markers for improving genebank management relating to these issues, the project aimed to achieve two main goals: (i) a description of the genetic structure of the genebank and establishing possible correlations with useful traits by characterizing it with molecular markers; and (ii) the use of molecular markers to monitor and validate the multiplication techniques currently used for lettuce at the CGN. It was hoped that information gained from achieving the first goal would help address issues 1–4 while that gained from achieving the second goal would help address issue 5.

The lettuce collection at CGN comprises some 2118 accessions. It was chosen as a suitable test model because the collection is well documented (which facilitates the investigation of relationships between molecular markers and other characteristics), it is of medium-size compared with other genebanks, both cultivated and wild species are present, including selfing and outcrossing species, and the collection is extensively used.

In order to perform the demonstration, the molecular work was organized into two main activities corresponding to the two goals listed above. To obtain a description of the genetic structure, DNA was extracted from 6471 lettuce plants (two samples for selfing accession and five samples for outcrossing accessions). All individuals were genotyped using ten microsatellites (that had been previously mapped in lettuce and were known to occur on different linkage groups) and three AFLP primer combinations. In order to validate multiplication techniques, DNA was extracted from 900 plants and genotyping was performed as described above but for samples taken within accessions before and after standard multiplication procedures.

The project was a major undertaking involving five participating organizations. The biotech company Keygene (Netherlands), which invented the AFLP technology, performed all the DNA extractions and also the AFLPs for the whole collection. This took almost two years, of which two months were needed for DNA extractions. The microsatellites were developed by Plant Research International (The Netherlands), which performed all the automated assays required for achieving the large screen. Together with the scoring and cross-checking, this took over 2 years. Two other groups, at IACR-Long Ashton Research Station and the Botanisk Institut in Copenhagen, were responsible for achieving the second goal, which took 18 months. Contact with the genetic resources community was maintained through the involvement of the International Plant Genetic Resources Institute (IPGRI).

The results obtained have indicated that the molecular data will be useful in the management of the lettuce genebank, particularly with regard to: (i) assisting in the identification of duplicates; (ii) assisting in classification and identifying any misclassification; (iii) validating different core collection strategies that could be used; and (iv) validating the multiplication strategies used at the CGN.

A number of experiences were also gained from attempting to apply the techniques to such a large set of samples.

In the case of the AFLPs, the large number of accessions in a single species, particularly in cultivated lettuce, *Lactuca sativa* and its close wild relative *Lactuca serriola*, meant that low-density AFLP profiles had to be used in order to avoid saturation of the band profiles when all samples were compared. As a consequence, although it was possible to compare all accessions for one species in one dataset, it was not possible to resolve all the accessions using this approach. When similarity matrices were constructed from the data and input into clustering algorithms, the resultant dendrograms grouped numerous accessions together, including some known to be of distinct origin. AFLPs with high-density profiles would discriminate these and resolve relationships more clearly. A recommendation emerging for screening large sample sizes, such as whole genebanks is, thus, to use a two-tiered approach: first low density profiles to compare all samples and resolve the main clusters and then high density profiles to resolve samples within clusters. It was not possible to compare the different *Lactuca* species directly as the AFLP profiles were too dissimilar.

In the case of the microsatellites, one problem was that they only amplified well in *L. sativa* (in which they were developed) and the wild relative *L. serriola*. In the other species, amplification only occurred in occasional samples and the datasets were too incomplete for use in the analysis. If the entire genebank is to be genotyped with microsatellites, it would thus be necessary to isolate microsatellites from the wild species.

A number of issues arose for both marker types during the course of the lettuce project. After assays had been completed a considerable amount of time was required to cross-check all data points from the gels to the datasets. For the microsatellite data obtained on the ABI 377, more time was needed for this than foreseen. There is also a clear requirement for new software that is able to manage the massive sizes of the datasets. Similarly, new visualization techniques are needed. A new project is currently addressing some of these issues (in particular, the bioinformatics aspects).

The Major Gap still to be Bridged and Ways of Meeting this Challenge

The two examples described above serve to illustrate the gains achievable from application of molecular marker techniques to address genetic diversity management questions, but there are still challenges to be faced. In essence, these relate to the gap that exists between the diversity revealed by DNA assays and phenotypic or functional diversity. Microsatellites, AFLPs and other such DNA markers are successful because they are highly polymorphic but they are also generally employed in an anonymous fashion. This has invited criticism of the functional relevance of the diversity revealed or, more explicitly, whether or not DNA markers are really providing the type of information that is needed to make valued judgements on diversity.

The rapidly increasing information available from genome mapping and gene sequencing projects is providing an expanding resource which may help to bridge this gap. High-resolution genetic maps now mean that markers known to be very closely linked to traits of interest can be used and the availability of sequence data for an increasing number of important genes enables the development of screens for detecting diversity specifically at these loci.

Adapting the current PCR-based techniques to target functional diversity

As mentioned earlier, it is possible to adapt multilocus profiling procedures such as AFLPs to target specific sequences within the genome, for example relating to microsatellites, or the insertion of transposons. This approach has the potential to be adapted for profiling diversity within genes and gene families in plants.

One example where this is being explored is in an EU-funded project in which different approaches based on SSAP (Ellis *et al.*, 1998) and the use of primers for conserved regions are being used to investigate diversity in genes encoding key events in development, metabolism and intra-cellular signalling. Different specific target genes are being investigated by different participating groups including protein kinases and signal transduction genes, disease resistance genes, genes involved in synthesis of, or response to, phytohormones and genes coding for cytoskeletal proteins. All partners are testing the techniques initially on a panel of 16 genotypes each of potato, tomato and barley (ten cultivars, two exotics and four accessions). Any promising functional biodiversity markers are then being tested further on an extended range of cultivars and ecotypes. In order to confirm that at least some of the polymorphisms observed reflect functional variation, the expressions of a selected number of the sequences are also being analysed. The project is only mid-way but the results obtained so far for one of these gene targets illustrate the potential usefulness of the approach.

The SnRK1 gene family (Halford *et al.*, 1992) codes for protein kinases involved in the regulation of carbon metabolism and are implicated in stress responses. These enzymes are related to the sucrose non-fermenting kinase-1 of yeast (hence SNF-1 related kinase = SnRK-1), which is found to have a crucial role in the biochemical signalling leading to the switch from utilization of glucose to other sugars when glucose is unavailable. It therefore plays a central role in carbon catabolite derepression, and the related genes found in plants are similarly very important in nutrient utilization. SnRK1s have been found to phosphorylate and inactivate sucrose phosphate synthase (in cauliflower and barley), nitrate reductase and HMG-CoA reductase (in spinach), and to activate the gene expression of sucrose synthase in potato (using antisense technology; Halford, 1999). SnRK1s from rye, barley, potato, oat, *Arabidopsis* and sugarbeet have been characterized at the molecular and biochemical level. Amino acid sequence similarities between the plant, yeast and mammalian sequences range from 44 to 46%. The barley gene subfamily comprises approximately 10–20 members per haploid genome, while that of potato comprises approximately 5–10 and that of *Arabidopsis* 2. The genes from all three species show a high degree of conservation of gene structure in that they share around 68% sequence identity and the positions of introns within the coding regions are identical.

Primers for conserved regions within SnRK1s have been used to produce amplification products which have been screened for differences in order to develop more specific sequence-specific marker assays for these polymorphic regions. Initial experiments focused on using the Clontech Genome Walker kit (Anon, 1996) for amplification of promoter regions. This kit amplifies products from a specific primer in the known coding region of the gene to an adaptor primer which binds to an adaptor ligated to a restriction site in the promoter region. A choice had to be made as to where to design the conserved primer. Primers from exon 1 that extend into the promoter region would allow detection of possible changes in transcriptional control and experiments first concentrated on this area of potential importance.

Using this approach two amplified products were obtained (A and B) of which one (B) was found to be polymorphic among barley cultivars. Sequencing of the products indicated that they are highly related to the Bkin12 gene members of the SnRK1 family in barley and have identical coding regions, differing only in the promoter sequence. A wider screen using this approach, confirmed that the B product is polymorphic in barley cultivars. When the same approach was used in another barley species (*Hordeum jubatum*) a third product (C) was identified which was not observed in cultivated barleys. The sequence also shows very high homology with Bkin12.

These results indicate that the approach has potential for revealing variation in important genes. Normal PCR has confirmed that absence of variant C in *Hordeum vulgare* is not a false negative, and RT-PCR has shown that variant C is expressed at the RNA level in *H. jubatum*. Evidence for differential expression has been found, confirming that variant C is likely to be of functional significance, an aspect which is now being studied further (J.S.C. Clark and N.G. Halford, IACR-Long Ashton Research Station, 2000, personal communication).

Using genome maps and closely linked markers

An example of how molecular map information can be used to help identify variation of use to breeders can be found in Tanksley and McCouch (1997) where a strategy termed the advanced backcross QTL (quantitative trait loci) method is described for identifying superior genes that have been transferred from wild species into cultivated lines. In this strategy, crosses are made between elite parents and landraces or wild species, followed by two or three backcrosses to the elite parent. The molecular linkage map is used to identify the location of genes that have been transmitted to the progeny after backcrossing to the elite cultivated parent. In this way it is possible to establish which chromosomal segments of the wild species are associated with superior performance of the lines and to purify lines so that they contain only a specific QTL in an elite genetic background. This approach, which has been successfully demonstrated in tomato, should improve the utilization of diversity residing in secondary genepools but it is resource intensive and its limitations include the unknown frequency of alleles of potential value in the wild germplasm.

Identification of the genes controlling a trait and knowledge of their DNA sequence would facilitate the classification of variation in the germplasm pool. Classification of the sequence variants at a targeted locus would substantially reduce the amount of work needed to assess their potential for breeding and lead to the identification of superior alleles (Sorrells and Wilson, 1997).

An example where both map and sequence knowledge was used to analyse genetic diversity in an agronomically important trait has been described by Sicard *et al.* (2000) for disease resistance variation in lettuce. Resistance to downy mildew in lettuce, caused by the fungus *Bremia lactucae*, is determined by a large number of different genes. The genes that have been characterized so far are clustered in four linkage groups. The major cluster has been saturated with molecular markers. Resistance gene candidates (RGCs) encoding nucleotide binding sites (NBS) and leucine rich repeat (LRR) regions have been identified and one family of over 24 members is localized in the major disease resistance cluster. A microsatellite marker and two primer pairs that amplify sequences encoding a region in the middle of the LRR region were used as markers to screen accessions of the cultivated lettuce, *L. sativa*, the wild lettuces, *L. serriola*, *Lactuca saligna* and *Lactuca virosa* and natural populations of *L. serriola* from Israel (near the centre of diversity) and California (a recently colonized area). All the material studied had been characterized against all known pathotypes of *Bremia*. Polymorphism at the microsatellite locus was highly correlated with resistance to the pathogen *B. lactucae*, with different microsatellite haplotypes diagnostic for different resistance alleles, suggesting a strong potential for use of this marker in screening wider germplasm and in selection programmes. Very little length variation was observed in the amplified products of the LRR regions using either of the two primer sets but restriction site polymorphisms were detected. The LRR markers also correlated with resistance phenotypes. All three markers were able to discriminate between accessions that had been previously shown to be resistant to all known isolates of *B. lactucae.* A very high level of genetic diversity in resistance genes was revealed in the wild germplasm in this study, indicating that there is little redundancy and that wild accessions will be a rich source of new resistance genes.

DNA chips and microarrays: what do they offer for managing genetic diversity?

The enormous and ever-increasing number of sequences that continually appear in databases provides what is presently a largely untapped resource for identifying allelic diversity of individual characterized genes as a means of searching for those specific alleles that will confer trait improvement. One problem with using this resource, however, is that the majority of the sequences are cDNAs (complementary DNAs to sequences from RNA libraries) or ESTs (expressed sequence tagged sites, or cDNAs which have been at least partially sequenced for PCR primers) and for the majority of these sequences, the function of them is not yet known. This has led to an increasing demand for functional assignment which the development of the DNA microarray (or DNA chip) now promises to help meet.

It is easy for a casual observer to be left with the impression that DNA chips are a completely novel and somewhat remote technology but, in fact, they represent the latest development in the progression of hybridization arrays that started with the Southern blot over a quarter of a century ago (Southern, 1975). In the Southern blot, labelled probes are hybridized to DNA immobilized on a nitrocellulose or nylon membrane. Each track is a DNA sample which has been cut with restriction endonucleases and the fragments have been separated by gel electrophoresis and then transferred to the membrane by diffusion, where they bind strongly. The probes hybridize to DNA fragments only where complementary sequences are present and this basic principle still holds for all further developments. The next progression was filter-based screening of whole libraries, such as cDNA libraries spotted on to nitrocellulose filters. Usually several different filters were needed to accommodate the number of different colonies present in the library. Through the use of robotic workstations it then became possible to stamp large numbers of colonies on to single filters, for example entire cDNA or genomic DNA libraries. Such macroarrays have extensive applications in molecular biology and are often used, for example, in the screening of microsatellite-enriched libraries.

Increasing miniaturization accompanied this progression from Southern blot to macroarray, which culminated in the DNA chip. Two significant innovations led to the development of the DNA microarray. The first was the use of non-porous solid supports, such as glass, which improves the efficiency of hybridization and post-washing steps by allowing for unimpeded contact between probe and target. The second was the development of fluorescent-based detection systems. The result has been an explosion in development of different microarrays, pioneered by Stanford University (Stanford, California, USA) and by commercial companies, such as Affymetrix. I will only deal briefly with this developing field, whose major applications up till now have been mostly in humans and animals, concentrating the discussion around its potential use for detecting polymorphisms in plants. A number of recent reviews cover the subject of microarrays more comprehensively (e.g. Castellino, 1997; Chakravati, 1999; Gerhold *et al.*, 1999; Lander, 1999; Lipshutz *et al.*, 1999; Southern *et al.*, 1999).

In DNA microarrays, different cloned sequences from libraries are spotted on the glass slide and information is gained by determining at which of these sequences hybridization occurs. In contrast with earlier hybridization systems, the DNA on the solid support is, thus, normally referred to as the probe, and the fluorescent RNA or DNA, which is available in solution for hybridization, is referred to as the target. There are basically two types of microarray: expression arrays and arrays for identifying and genotyping mutations and polymorphisms.

Expression arrays are being used to monitor gene expression levels in different tissues and under different situations as a first step towards functional characterization of genes. The potential offered by microarrays is enormous here. DNA expression chips generated by Affymetrix, for example, can display from 65,000 to 400,000 oligonucleotides, representing up to 9000 genes on a 1.6 cm^2 glass surface. A set of four chips allows the entire 6000 genes of yeast to be assayed and in mammals 5–10,000 genes are already in common use for expression screening. Expression arrays are making an impact in the plant field and large functional genomics projects are being pursued by consortia of academics and industry. It is difficult to imagine, however, how expression arrays might be used directly in genetic diversity studies, although benefits will clearly be gained as sequences conferring important functions are identified. In contrast, genotyping arrays have the potential to make a direct contribution to genetic diversity studies in plants since they offer the possibility of locating, identifying and cataloguing genotypic differences in large sets of known DNAs simultaneously (Lipshutz *et al.*, 1999).

There are a number of different types of genotyping arrays of which only two will be described briefly here. In the first, a tiling array is designed to interrogate all base pair positions in a gene sequence. Overlapping 25-mers are used to cover the sequence. For each 25-mer stretch, four probes are spotted on to the solid support. Each probe has an identical sequence except at one base position, which is the same position in each probe. At this position the probes differ in one of the four possible nucleotides (A, G, C, T). When the target DNA is added, the brightest hybridization signal will occur at the correct probe only. In the tiling array, the next set of four probes differs in the next base-pair position, and so on, until the entire length of the sequence has been interrogated at every base-pair position. Commercially available Affymetrix chips on which are arrayed 350,000 oligonucleotides allow representation of 40,000 bp of sequence per chip, including both strands and each possible nucleotide mismatch.

A second approach to genotyping is to build an array containing the previously identified allelic variants of a gene. In this case the probe alternatives consist of a perfect match, or a single base mismatch, for the different alleles. This approach has the potential to permit the genotyping of vast numbers of loci simultaneously and could provide a promising way of screening for genetic diversity in important genes, such as disease resistance genes. Single nucleotide polymorphisms (SNPs) are the most common mutation in sequences. Once SNPs have been identified, large numbers can be arrayed on chips ('SNP chips') for detecting the presence of these mutations. As more and more genes are uncovered by sequencing, and increasing numbers of alleles are identified it may be possible to build arrays for all kinds of important traits, but the sequencing and identification has to be achieved first.

DNA microarray approaches are certainly impressive but there are some negative considerations to be taken into account. Some of these are general to all array types. For example, normally only SNPs can be detected and not insertions or deletions (unless the probe is provided in the array). Cross-hybridization between related sequences can cause problems and it can be difficult to perform the multiple PCR amplifications when several gene sequences are present on the array. Handling the information obtained is also no small task and requires sophisticated software. Finally, the costs are extremely high. Commercially available chips are in the US$1000 range and if a complete set of instruments (hybridization system, scanner and software) is purchased, the costs escalate to nearer US$250,000 (Granjeaud *et al.*, 1999).

This might seem to place the technology out of the reach of academic researchers and the plant genetic resource community. It is worth remembering, however, that the costs may well come down in the future and that, in the meantime, the principle of arrays can be profited from at a lower technological level. In *Arabidopsis*, for example, high-speed robotic printing of cDNAs on glass allowed expression studies to be carried out on 45 genes simultaneously (Schena *et al.*, 1995). It is within reach to construct similar arrays of oligonucleotides representing the different alleles of important genes.

The real limitation here is that the sequences to be placed on the array for genotyping polymorphisms in plant genes, in the main, still have to be identified. It will take some time to achieve this for many of the traits of interest in plants. This requirement is likely to be met first in those few crops of highest economic importance and one might question, when, or indeed if, it will be achieved in the majority of plant species of interest as genetic resources.

Somewhat provocatively, one might also question what advantages are really offered by these powerful but expensive tools for genetic resource managers. Using arrays of the tiling kind, for example, it would be possible to detect every single polymorphism that is present in a given sequence between different samples. It would still not be possible, however, to know what the functional significance of all the polymorphisms detected are, even if they occur in genes. Apart from the level of resolution and the scale and capacity for screening several different sequences simultaneously, it is therefore not so easy to see what additional benefits this might offer for genetic diversity studies. In the case of designing arrays for genotyping allelic difference at several loci, this does look attractive, as it is allelic variation that is most of interest for genetic diversity management. However, how many samples could be reasonably screened in this way, and who would be able to afford such an approach in the field of plant genetic resources?

Concluding Remarks

Since the application of RFLPs in the early 1980s, molecular genetic techniques have made a substantial impact on the characterization of plant genetic diversity in crop germplasm, genepools and natural populations. Information derived from their use is being successfully applied in many aspects of genetic diversity management, for example, identification of duplicates, developing sampling strategies and validating multiplication techniques.

In contrast, understanding the 'significance' or assessing the 'value' of the diversity revealed remains a tantalizing challenge. One reason for this is the gap existing between our understanding of diversity at the genetic (or molecular) and phenotypic (or functional levels). Challenges that remain concern the identification of traits of importance, and particularly being able to identify alleles that may confer an improvement. Advances here would enhance utilization of the resources. The technology is moving in this direction. High resolution genetic maps enable closely linked markers to be used and the increasing numbers of ESTs and SNPs provide routes for more targeted sequence-based approaches. DNA chips allow the detection of every mutation in a sequence and genotyping of several genes to be achieved simultaneously. The main problem, however, will still be to find which of the many sequences, and which of the many polymorphisms uncovered, are of interest, and worth the cost of their detection.

Acknowledgements

The author thanks Jacqueline H. A. Barker, Chloe Aldam, Keith J. Edwards and Rachel Trehin (IACR-Long Ashton Research Station) who were involved in the coconut work, together with Ramon Rivera (Philippine Coconut Authority), Beatrice Teulat and Francois Rognon (Burotrop, France), Patricia Lebrun, Luc Baudouin (CIRAD, France) and Toby Hodgkin (IPGRI). Chloe Aldam and Toby Hodgkin are also involved in the lettuce genebank project. Thanks are also given to the following collaborators in the EU projects (BIO4–96–2062 and BIO4–98–0332): Theo van Hintum, Ben Vosman, Clemens van der Wiel, Gerard van der Linden (Plant Research International, The Netherlands), Johan Peleman (Keygene, The Netherlands), Ole Seberg (Botanisk Institut, Copenhagen, Denmark), Marcello Buiatti (University of Florence, Italy), Mauro Cresti (University of Siena, Italy), Maria Dani (Tecnogen, Italy), Robbie Waugh (Scottish Crop Research Institute, UK). Special thanks to Jeremy Clark at IACR-Long Ashton Research Station for the results on Bkin12.

For the EU project on Functional Biodiversity, seeds of tomato cultivars were provided by Plant Research International (The Netherlands). The seeds from ten cultivars of barley were provided by the Scottish Crop Research Institute which also provided tubers from ten cultivars of potato. Seeds from *H. jubatum* were also obtained from Herbiseed Ltd, UK. Additionally, seeds from 29 species of *Hordeum* have been kindly provided by Dr Fredrik Ottosson (Swedish University of Agricultural Sciences), and seeds from 100 cultivars and landraces from Dr Helmut Knüpffer (IPK Gatersleben, Germany) for testing the markers further.

IACR receives grant-aided support from the Biotechnology and Biological Sciences Research Council, UK.

References

Anon. (1996) Clontech Universal GenomeWalker Kit User Manual (PT3042–1). Clontech Laboratories Inc., USA.

Bachmann, K. (1994) Tansley Review No. 63. Molecular markers in plant ecology. *New Phytologist* 126, 403–418.

Bruford, M.W., Hanotte, O., Brookfield, J.F.Y. and Burke, T. (1992) Single-locus and multilocus DNA fingerprinting. In: Hoezel, A.R. (ed.) *Molecular Genetic Analysis of Populations: A Practical Approach*. IRL Press at Oxford University Press, Oxford, UK, pp. 225–269.

Castellino, A.M. (1997) When the chips are down. *Genome Research* 7, 943–946.

Chakravati, A. (1999) Population genetics – making sense out of sequence. *Nature Genetics* Suppl. 21, 56–60.

Ellis, T.H.N., Poyser, S.J., Knox, M.R., Vershinin, A.V. and Ambrose, M.J. (1998) *Ty1*-copia class retrotransposon insertion site polymorphism for linkage and diversity analysis of pea. *Molecular and General Genetics* 260, 9–9.

Evola, S.V., Burr, F.A. and Burr, B. (1986) The suitability of restriction fragment length polymorphisms as genetic markers in maize. *Theoretical and Applied Genetics* 71, 765–771.

Flavell, A.J., Knox, M.R., Pearce, S.R. and Ellis, T.H.N. (1998) Retrotransposon-based insertion polymorphisms (RBIP) for high-throughput marker analysis. *Plant Journal* 16, 643–650.

Gerhold, D., Rushmore, T. and Caskey, T.C. (1999) DNA chips: promising toys have become powerful tools. *Trends in Biochemical Science* 24,168–173.

Granjeaud, S., Bertucci, F. and Jordan, B.R. (1999) Expression profiling: DNA arrays in many guises. *BioEssay* 21, 781–790.

Halford, N.G. (1999) Metabolic signalling and the partitioning of resources in plant storage organs. *Journal of Agricultural Science* 133, 243–249.

Halford, N.G., Vicente-Carbajosa, J., Sabelli, P.A., Shewry, P.R., Hannappel, U. and Kreis, M. (1992) Molecular analyses of a barley multigene family homologous to the yeast protein kinase gene SNF1. *Plant Journal* 2, 791–797.

Jeffreys, A.J., Wilson, V. and Thein, S.L. (1985) Hypervariable 'minisatellite' regions in human DNA. *Nature* 314, 67.

Jones, C.J., Edwards, K.J., Castaglione, S., Winfield, M.O., Sala, F., Van der Wiel, C., Biedemeyer, G., Vosman, B., Matthes, M., Daly, A., Brettschneider, R., Bettini, P., Buiatti, M., Maestri, E., Malcevschi, A., Marmiroli, N., Aert, R., Volckaert, G., Rueda, J., Linacero, R., Vazquez, A. and Karp, A. (1997) Reproducibility testing of RAPD, AFLP & SSR markers in plants by a network of European Laboratories. *Molecular Breeding* 3, 381–390.

Kalendar, R., Grob, T., Suoniemi, A. and Schulman, A.H. (1999) IRAP and REMAP: Two new retrotransposon-based DNA fingerprinting techniques. *Theoretical and Applied Genetics* 98, 704–711.

Karp, A. and Edwards, K.J. (1998) *DNA markers: a global overview*. In: Caetano-Anollés, G. and Gresshoff, P.M. (eds) *DNA Markers: Protocols, Applications and Overviews*. Wiley-VCH, Weinheim, Germany, pp. 1–13.

Karp, A., Kresovich, S., Bhat, K.V., Ayad, W.G. and Hodgkin, T. (1997a) Molecular tools in plant genetic resources conservation: a guide to the technologies. *IPGRI Technical Bulletin* No. 2. IPGRI, Rome, Italy.

Karp, A., Edwards, K.J., Bruford, M., Funk, S., Vosman, B., Seberg, O., Kremer, A., Boursot, P., Arctander, P., Tautz, D. and Hewitt, G.M. (1997b) Newer technologies for biodiversity evaluation: Opportunities and challenges. *Nature Biotechnology* 15, 625–628.

Karp, A., Isaac, P.G. and Ingram, D.S. (1998) *Molecular Tools for Screening Biodiversity: Plants and Animals*. Chapman & Hall, London, UK.

Lander, E.S. (1999) Array of hope. *Nature Genetics* Suppl. 21, 3–4.

Lipshutz, R.J., Fodor, P.A., Gingeras, T.T. and Lockhart, D.J. (1999) High density synthetic oligonucleotide arrays. *Nature Genetics* Suppl. 21, 20–24.

Morgante, M. and Olivieri, A.M. (1993) PCR-amplified microsatellites as markers in plant genetics. *Plant Journal* 3, 175–182.

Parker, P.G., Snow, A.A., Schug, M.D., Booton, G.C. and Fuerst, P.A. (1998) What molecules can tell us about populations: choosing and using a molecular marker. *Ecology* 79, 361–382.

Powell, W., Morgante, M., Andre, C., Hanafey, M., Vogel, J., Tingey, S.S. and Rafalski, J.A. (1996) The comparison of RFLP, RAPD, AFLP and SSR (microsatellite) markers for germplasm analysis. *Molecular Breeding* 2, 225–238.

Rivera, R., Edwards, K.J., Barker, J.H.A., Arnold, G.M., Ayad, G., Hodgkin, T. and Karp, A. (1999) Isolation and characterisation of polymorphic microsatellites in *Cocos nucifera* L. *Genome* 42, 668–675.

Schena, M., Shalon, D., Davis, R.W. and Brown, O.W. (1995) Quantitative monitoring of gene expression patterns with a complementary DNA microarray. *Science* 270, 467–470.

Sicard, D., Woo, S-S., Arroya-Garcia, R., Ochoa, O., Nguyen, D., Korol, A., Nevos, E. and Michelmore, R. (2000). Molecular diversity at the major cluster of disease resistance genes in cultivated and wild *Lactuca* spp. *Theoretical and Applied Genetics* 99, 405–418.

Sorrells, M. and Wilson, W.A. (1997) Direct classification and selection of superior alleles for crop improvement. *Crop Science* 37, 691–697.

Southern, E.M. (1975) Detection of specific sequences among DNA fragments separated by gel electrophoresis. *Journal of Molecular Biology* 98, 503–517.

Southern, E.M., Mir, K. and Shchepinov, M. (1999) Molecular interactions on microarrays. *Nature Genetics* Suppl. 21, 5–9.

Sunnucks, P. (2000) Efficient genetic markers for population biology. *Trends in Ecology and Evolution* 15, 199–203.

Tanksley, S.D. and McCouch, S.R. (1997) Seed banks and molecular maps: unlocking genetic potential from the wild. *Science* 277, 1063–1066.

Teulat, B., Aldam, C., Trehin, R., Lebrun, P., Barker, J.H.A., Arnold, G.M., Karp, A., Baudouin, L. and Rognon, F. (2000) An analysis of genetic diversity in coconut (*Cocos nucifera*) populations from across the geographic range using sequence-tagged microsatellites (SSRs) and AFLPs. *Theoretical and Applied Genetics* 100, 764–771.

Vos, P., Hogers, R., Bleeker, M., Rijans, M., Van De Lee, T., Hornes, M., Frijters, A., Pot, J., Peleman, J., Kuiper, M. and Zabeau, M. (1995) AFLP: a new technique for DNA fingerprinting. *Nucleic Acids Research* 23, 4407–4414.

Vosman, B., Arens, P., Rus-Kortekaas, W. and Smulders, M.J.M. (1992) Identification of highly polymorphic DNA regions in tomato. *Theoretical and Applied Genetics* 85, 239–244.

Williams, J.G.K., Kubelik, A.R., Livak, K.J., Rafalski, J.A. and Tingey, S.V. (1990) DNA polymorphisms amplified by arbitrary primers are useful as genetic markers. *Nucleic Acids Research* 18, 6531–6535.

Witsenboer, H., Vogel, J. and Michelmore, R.W. (1997) Identification, genetic localization and allelic diversity of selectively amplified microsatellite polymorphic loci in lettuce and wild relatives (*Lactuca* spp.). *Genome* 40, 923–936.

5 DNA Polymorphisms in Accessions of *Nephelium lappaceum* L.

P.C. Chew[1], M.M. Clyde[1], M.N. Normah[1] and I. Salma[2]

[1]Plant Biotechnology Laboratory, Faculty of Science and Technology, Universiti Kebangsaan Malaysia, Selangor, Malaysia; [2]Strategic, Environment and Natural Resources Research Centre, MARDI, Kuala Lumpur, Malaysia

Introduction

Nephelium is one of 130 genera in the family *Sapindaceae*, with the centre of origin for this genus thought to be the Indo-Malaysian region. According to Soepadmo (1979) there are 30 species of *Nephelium* in South-east Asia, 14 of which occur in Peninsular Malaysia. *Nephelium* spp. are evergreen trees and produce edible fruit, either sweet or sour. Examples of edible species are *N. lappaceum* (rambutan), *N. rambutan-ake* (pulasan), *N. cuspidatum*, *N. hypoleucum*, *N. maingayi* and *N. uncinatum* (Siebert, 1991). Apart from *N. lappaceum* and *N. rambutan-ake*, which are cultivated for their fruits, the other *Nephelium* species are only rarely and locally traded (Verheij and Coronel, 1991).

In Malaysia previous studies on *Nephelium* have been on botanical aspects (Lim, 1984), morphological characterization and agronomic traits (Whitehead, 1959; Salma, 1986). Both outcrossing and selfing can occur in rambutan. Some trees produce only male flowers, others produce hermaphrodite flowers with non-functional anthers, whereas other hermaphrodite trees function as both male and female. This last type of tree produces most fruit, and budwood is taken from them for commercial (clonal) propagation. Pollination is carried out by *Trigona* bees. Although the species is an outbreeder, the genetic diversity of cultivated rambutan is narrowing because clonal propagation of a few cultivars in nurseries has replaced home propagation (using seedlings).

Information on genetic diversity is still lacking, even though it is important and could be used to enhance selection and breeding programmes. The significance of genetic variability as a natural resource and the increased threats to crop genepools due to urbanization and deforestation had led MARDI to carry out systematic collection and conservation of the major fruit tree species which included *Nephelium* species. In 1984 MARDI, in collaboration with the International Board for Plant Genetic Resources (IBPGR), organized an expedition to collect and conserve the accessions of *N. lappaceum* (rambutan) and *N. rambutan-ake* (pulasan) from 85 districts throughout the country including the main islands such as Tioman, Langkawi and Pangkor. Budwoods from a total of 323 accessions were collected and vegetatively propagated in the nursery. Only 235 accessions survived and the plants are planted at Kemaman. At present there are 211 accessions still living.

Fruit characters show the main differences among the accessions. However their differences based on fruit characters as well as other morphological characters such as leaf and flower are not able to differentiate all the accessions. As such DNA markers would be useful.

Randomly amplified polymorphic DNA (RAPD) has been found to be a suitable technique for molecular characterization and determination of phenetic relationships, as shown in a similar study on *Lansium domesticum* (Song, 1999). In this study RAPD was chosen based on its reasonable cost and

(eds J.M.M. Engels, V. Ramanatha Rao, A.H.D. Brown and M.T. Jackson)

relative simplicity. The main objectives of this study are to estimate genetic diversity and to characterize available accessions using molecular markers, and to study genetic relatedness among accessions via comparison of genetic polymorphisms. In addition to the RAPD work, genetic diversity within the collection is also currently studied by MARDI using isozyme polymorphism.

Materials and Methods

Materials

The study material is centred on the MARDI-IPGRI (International Plant Genetic Resources Institute) collection of *N. lappaceum* at MARDI Station, Kemaman Terengganu. These accessions were collected from all states throughout Malaysia, including islands, in 1984 and planted out in 1986. The trees, now approximately 15 years old, have been fruiting since 1990. A total number of 211 accessions are available. Voucher specimens are deposited in the Herbarium of Universiti Kebangsaan Malaysia.

DNA extraction

Total DNA is isolated from fresh mature leaves using the modified CTAB method of Doyle and Doyle (1990). Genomic DNA extracted using 3–5 g of fresh leaves from each accession gave yields ranging from 5 ng μl^{-1} to 500 ng μl^{-1}. The difficulty in DNA extraction due to high content of polysaccharides and proteins was partly overcome by incubating the extraction mixture with potassium acetate in an ice bath, after which most of the proteins and polysaccharides are removed as a complex with the insoluble potassium dedecyl sulphate precipitate. The DNA samples are further subjected to a purification step using a High Pure PCR Template Preparation Kit (Boehringer Mannheim) according to the manufacturer's instructions. The DNA is diluted with sterile distilled water to give a concentration of 1 ng μl^{-1}.

Amplification conditions

PCR reactions were performed in a 25 µl reaction mixture containing 3 ng template DNA, 2.5 µl PCR buffer 10× (100 mM Tris-HCl, 15 mM $MgCl_2$ and 500 mM KCl, pH 8.3), 0.2 mM each of dATP, dCTP, dGTP and dTTP, 0.5 unit *Taq* polymerase (Boehringer Mannheim) and 10 pmol of single decamer primer (Operon Technologies Inc.). Amplification was performed using the GeneAmp PCR System 2400 (Perkin Elmer) thermal cycler with an initial 2 min of denaturation at 93°C; followed by 45 cycles of 45 s at 92°C, 2 min 45 s at 35°C, 1 min 45 s at 72°C, and ending with 10 min at 72°C to ensure that the primer extensions proceed to completion. After amplification is completed, 5 µl of PCR product mixture is loaded and electrophoresed on 1,5% agarose gel in 1× TAE buffer. The gel is stained with ethidium bromide (1µl ml^{-1}) for 20 min, viewed under ultraviolet light and photographed using Polaroid 665 film. Molecular weights of the fragments are estimated using a 100 bp DNA step Ladder marker (GibcoBRL).

Selection of primers

Primer screening was carried out using eight Operon Primer Kits, i.e. OPA, OPB, OPC, OPD, OPK, OPS, OPT and OPU, each containing 20 decamer primers. Of the 12 primers that yielded clear and reproducible bands, six (OPK-07, OPK-15, OPK-16, OPK-19, OPS-03 and OPS-18) have been used for the 175 accessions.

Homology test

A homology test was performed to confirm whether the fragments that co-migrated are homologous within and among populations and different fragment positions represented different loci. One of the amplified fragments from each of the primers is eluted, labelled and hybridized to Southern blots of RAPD gels.

Data analysis

The DNA amplification was repeated at least twice to ascertain the reproducibility of bands. Bands on RAPD gels were scored as present (1) or absent (0) for each DNA sample and the data analysed using NTSYS-pc (Rohlf, 1993). The data matrix was analysed to generate Jaccard similarity coefficient values which can be used to construct a dendrogram via UPGMA (Unweighted Pair Group Method with Arithmethic Mean). Shannon diversity indices (H_0) were calculated to measure the level of polymorphism between all accessions using the formula $H_0 = -\sum \pi_i \ln \pi_i$ where π_i is the frequency of the phenotype (fragment) *i* present.

Results and Discussion

Of the 160 primers tested, 12 were found to be suitable. Results of the present study are based on the use of six selected primers (OPK-07, OPK-15, OPK-16, OPK-19, OPS-03 and OPS-18). The sequence of each primer, and the number of bands that are polymorphic are given in Table 5.1. Overall, a total of 81 scorable bands (fragments) were generated, of which 73 showed polymorphism (90.12%).

Only clearly visible and reproducible bands representing fragments with molecular size in the range 400–2072 bp were used in the data analysis. Homology tests performed via Southern blotting and hybridization for some of the RAPD products confirmed that the fragments which co-migrated are homologous among the accessions and that different fragment positions represent different loci, i.e. non-allelic. The fragments used as probes for each of the six primers were $K07.2_{1356}$, $K15.5_{1487}$, $K16.5_{652}$, $K19.5_{1023}$, $S03.5_{1173}$ and $S18.7_{861}$ (Table 5.1).

The degree of genetic variation among the accessions (H_0) calculated for each primer (Table 5.2) showed OPS-03 with the highest value (4.02) and OPK-07 with the lowest (1.24). The mean value for Shannon index was 2.25. This is comparable to the value of 2.32 obtained in a similar study on 85 accessions of *L. domesticum*, another tropical fruit species (Song, 1999). However the value is likely to change when the results from the other accessions and several more primers are incorporated.

The results of the clustering analysis by the UPGMA method using datasets of each primer and primer combinations are shown in Table 5.2. Primer OPS-03 was able to distinguish the highest number of subsets (86) and OPK-07 the lowest (21). The combination of datasets from four primers (OPS-03, OPS-18, OPK-19 and OPK-15) gave the maximium number of 121 subsets (Table 5.3).

Table 5.2. The degree of genetic variation among the accessions (H_0) for each primer.

Primer	Number of distinguishable subsets	H_0
OPS-03	86	4.02
OPS-18	48	2.02
OPK-19	45	2.12
OPK-15	43	2.09
OPK-16	38	2.03
OPK-07	21	1.24
		Mean: 2.25

Table 5.1. Primer sequence, size (bp) and frequency of amplified bands.

Primer	OPK-07		OPK-15		OPK-16		OPK-19		OPS-03		OPS-18	
5'-3'	AGCGAGCAAG		CTCCTGCCAA		GAGCGTCGAA		CACAGGCGGA		CAGAGGTCCC		CTGGCGAACT	
P[a]	8/10 (80%)		15/16 (93.75%)		11/11 (100%)		11/12 (91.67%)		15/17 (88.24%)		13/15 (86.67%)	
1	1460	0.16	2050	0.04	1230	0.67	1720	1.00	1580	0.52	1400	1.00
2	1360	0.99	1930	0.95	880	0.94	1550	0.97	1450	1.00	1160	0.99
3	1280	0.90	1790	0.95	780	0.08	1480	0.97	1380	1.00	1040	0.86
4	1140	0.97	1600	0.06	730	0.05	1130	0.31	1240	0.39	990	0.98
5	1070	0.16	1490	0.99	650	0.97	1020	0.64	1170	0.85	940	0.10
6	1030	0.65	1440	0.61	610	0.16	1000	0.31	1130	0.20	890	0.76
7	930	0.96	1280	1.00	590	0.06	710	0.14	1030	0.80	860	0.60
8	880	0.08	1090	0.85	560	0.90	670	0.88	970	0.81	830	0.02
9	800	0.99	1070	0.71	520	0.37	650	0.05	910	0.44	760	0.05
10	680	1.00	870	0.72	450	0.96	510	0.96	850	0.91	700	1.00
11			810	0.86	430	0.45	480	0.20	820	0.44	660	0.98
12			780	0.30			420	0.05	800	0.60	540	0.99
13			710	0.97					710	0.34	490	0.18
14			600	0.06					670	0.18	450	0.73
15			560	0.95					600	0.65	410	0.58
16			470	0.97					550	0.93		
17									520	0.38		

[a] Number (%) of polymorphic bands.

Table 5.3. Primer combinations and the number of subsets they distinguish.

Primer combination	Number of RAPD fragments	Number of subsets distinguished
OPS-03	17	86
OPS-03, OPS-18	32	115
OPS-03, OPS-18, OPK-19	44	119
OPS-03, OPS-18, OPK-19, OPK-15	60	121
OPS-03, OPS-18, OPK-19, OPK-15, OPK-16	71	117
OPS-03, OPS-18, OPK-19, OPK-15, OPK-16, OPK-07	81	111

Jaccard's coefficient of genetic similarity between pairs of accessions ranged from 0.492 between accession 094 and accession 200, to 0.963 between accession 073 and accession 071. Such values are consistent with degrees of genetic relatedness within the same species, with the highest values possibly indicative of a common source for the planting material (e.g. previous releases by the Department of Agriculture which were planted in orchards and home gardens), and the lower values representing genetic variation between trees planted from seeds. Comparison of these results with morphological characters currently undertaken should yield a clearer understanding of the accessions in the collection.

Our preliminary results highlight the need for local germplasm collections to be studied before accessions may be lost. Since there is a high level of genetic diversity in the collection, and other methods of germplasm conservation are not well developed for rambutan it is therefore very important to maintain as many of these trees as possible.

Acknowledgement

The authors thank Che Rashid Endut, Head of MARDI Station, Kemaman, for his assistance. This study is funded by IRPA Grant No. 01-02-02-0026 from the Malaysian government.

References

Doyle, J.J. and Doyle, J.L. (1990) Isolation of plant DNA from fresh tissue. *Focus* 12, 13–15.

Lim, A.L. (1984) The reproductive biology of rambutan, *Nephelium lappaceum* L. (Sapindaceae). *Gard. Bull. Sing.* 37, 181–192.

Rohlf, F.J. (1993) NTSYS-pc, Numerical Taxonomy and Multivariate Analysis System. Version 1.80. Applied Biostatistics Inc., New York.

Salma, I. (1986) Rambutan (*Nephelium lappaceum* L.) clones and their classification. MARDI Report No. 107.

Siebert, B. (1991) *Nephelium* L. In: Verheij, E.W.M. and Coronel, R.E. (eds) *Plant Resources of South-East Asia* No. 2: *Edible Fruits and Nuts*. Pudoc, Wageningen, The Netherlands, pp. 233–235.

Soepadmo, E. (1979) Genetic resources of Malaysia – fruit trees. *Malaysian Applied Biology* 8, 33–42.

Song, B.K. (1999) Kepelbagaian genetik dalam *Lansium domesticum* Corr. MSc thesis, Universiti Kebangsaan Malaysia.

Verheij, E.W.M. and Coronel, R.E. (eds) (1991) *Plant Resources of South-East Asia* No. 2: *Edible Fruits and Nuts*. Pudoc, Wageningen, The Netherlands.

Whitehead, C. (1959) The rambutan. A description of the characteristics and potential of the more important varieties. *Malaysian Agricultural Journal* 42, 53.

6 Molecular Analysis of Phylogenetic Relationships among Coconut Accessions

A. Upadhyay, J. Jose, R. Manimekalai and V.A. Parthasarathy
Central Plantation Crops Research Institute, Kerala, India

Introduction

Coconut (*Cocos nucifera* L.), a member of family *Arecaceae* is an important oil crop, grown in coastal humid tropics. It provides subsistence to millions of families in coconut growing countries. The origin of evolution of this crop is considered to be the Asia–Pacific region (Harries, 1978). Coconut accessions can be broadly classified into two types: Talls and Dwarfs. Besides other differences, the Talls are preferentially cross-pollinated whereas the Dwarfs are mainly self-pollinated. At the Central Plantation Crops Research Institute (CPCRI), India, a large collection of coconut germplasm is being maintained. India is the site for the International Coconut Gene Bank for South Asia (ICGB-SA) and extensive germplasm collecting is underway to enrich the coconut germplasm centre. Characterization and cataloguing of coconut germplasm are important for any breeding programmes to improve its productivity. At present, morphological characters are used for evaluation and classification of collected germplasm (Kumaran *et al.*, 1998). However, germplasm characterization based on morphological traits is time-consuming, expensive and sometimes unsatisfactory because of environmental effects.

Molecular markers, which detect variation at the DNA level, provide a way to characterize germplasm accurately at a faster rate. Recently, an array of molecular marker techniques has been developed. Molecular markers like restriction fragment length polymorphism (RFLP) (Lebrun *et al.*, 1998a), randomly amplified polymorphic DNA (RAPD) (Ashburner *et al.*, 1997; Rodriguez *et al.*, 1997), amplified fragment length polymorphism (AFLP) (Perera *et al.*, 1998) and microsatellites or simple sequence repeats (SSRs) (Perera *et al.*, 2000) have been successfully employed for assessing genetic diversity in coconut. Although newer techniques like AFLP and SSRs are gaining importance due to their ability to detect more polymorphism (SSRs) and high multiplex ratio (AFLP) (Powell *et al.*, 1996), RAPD markers remain popular because of their simplicity and low development cost. The RAPD technique is an efficient tool for identifying variation and estimating diversity in different biological systems (Tingey and Tufo, 1993). RAPD markers are generated by PCR amplification of random genomic segments with a single primer of arbitrary sequence (Williams *et al.*, 1990). Since no a priori knowledge of genome structure is needed, they are specially useful for analysis of less studied genomes like coconut.

In the present study, RAPD markers were used to establish the genetic similarity among some indigenous and exotic coconut accessions maintained in the coconut germplasm centre at CPCRI, Kasaragod, India and widely used in the institute's ongoing breeding programme.

Materials and Methods

Plant material

Fourteen coconut accessions (nine Tall, four Dwarf and one intermediate type) maintained in the CPCRI germplasm collection at Kasaragod were used for the study. The details of these accessions are given in Table 6.1.

DNA extraction

DNA was extracted from newly emerged leaf using the protocol standardized earlier in our laboratory (Upadhyay *et al.*, 1999), as follows. Leaf tissue (5 g) was ground to fine powder in liquid N_2. The powdered tissue was transferred to a 50 ml polypropylene tube containing 25 ml DNA extraction buffer (100 mM Tris, 20 mM EDTA, 1% SDS, 0.2% β-mercaptoethanol, 5% PVP) and incubated at 65°C for 1h with intermittent mixing. After incubation, 15 ml phenol:chloroform:isoamyl alcohol (25:24:1) was added and mixed for 10 min by swirling. The solution was centrifuged at 20,000 g for 20 min. The supernatant was re-extracted with chloroform:isoamyl alcohol (24:1), followed by centrifugation. The supernatant was transferred to a fresh tube and DNA was precipitated by adding 2/3 volume of isopropanol. The precipitated DNA was spooled out with a microtip, transferred to a microtube and washed twice with 76% ethanol (containing 10 mM ammonium acetate). The pellet was dried and dissolved in 1 ml TE buffer. The DNA concentration was estimated spectrophotometrically as well as by comparing the band intensity with a known quantity of DNA on ethidium bromide stained 0.8% TAE agarose gel. The average yield of high molecular weight DNA was approximately 300 μg g^{-1} FW tissue.

RAPD analysis

Polymorphic primers were identified by screening 100 random decamer primers (Operon Technologies, USA) with DNA of WCT and COD. Amplification was carried out using PCR parameters as follows. The PCR reaction contained 25 ng DNA, 10 mM Tris-HCl (pH 9), 4.0 mM $MgCl_2$, 50 mM KCl and 0.01% gelatin, 100 μM each of dNTPs, 25 pmole of primer and 1.5 U *Taq* DNA polymerase in 25 μl reaction volume. All biochemicals were procured from M/s Bangalore Genei Pvt. Ltd, Bangalore, India. DNA was amplified in DNA Engine-PTC 200 (MJ Research), programmed for denaturation at 94°C for 5 min followed by 40 cycles of 1 min at 94°C, 1 min at 55°C and 2 min at 72°C. Cycling was concluded with a final extension at 72°C for a further 8 min. Amplification products were resolved by electrophoresis in a 1.2% agarose, 1X TAE gel at 60 V for 4 h. Controls lacking template DNA were included for each primer reaction mix. Amplification products were stained with ethidium bromide and visualized under UV light. Each band was considered as a RAPD marker.

Table 6.1. Details of coconut accessions.

Sl. no.	Accession	Abbreviation	Type	Place of collection
1.	West Coast Tall	WCT	Tall	Kerala, India
2.	Benaulim	BEN	Tall	Goa, India
3.	Laccadive Micro	LCM	Tall	Lakshadweep Islands, Arabian sea
4.	Laccadive Ordinary	LCO	Tall	Lakshadweep Islands, Arabian sea
5.	Kappadam	KAP	Tall	Kerala, India
6.	Andaman Ordinary	ADO	Tall	Andaman Islands
7.	Philippine Ordinary	PHO	Tall	Philippines
8.	San Ramon	SNR	Tall	Philippines
9.	Java	JVT	Tall	Indonesia
10.	Chowghat Orange Dwarf	COD	Dwarf	Kerala, India
11.	Chowghat Green Dwarf	CGD	Dwarf	Kerala, India
12.	Malayan Yellow Dwarf	MYD	Dwarf	Malaysia
13.	Malayan Orange Dwarf	MOD	Dwarf	Malaysia
14.	Gangabondam	GBD	Intermediate	Andhra Pradesh, India

Six polymorphic primers (OPA-10, OPA-11, OPB-1, OPB-5, OPC-5 and OPD-7) were selected to amplify DNA from 14 coconut accessions. The presence or absence of each PCR product was recorded.

Statistical analysis

Genetic diversity or heterogeneity was calculated according to Nei's (1975) formula. Heterogeneity was calculated for each marker and then averaged out for the total measure. Heterogeneity for Tall and Dwarf accessions was calculated by considering the markers present only in those respective groups of accessions.

The presence–absence data were entered into a binary data matrix as discrete variables (1 for the presence and 0 for the absence of a homologous band). Pairwise genetic distance was calculated based on the Nei and Li coefficient (Nei and Li, 1979) using computer package RAPDistance (Armstrong *et al.*, 1994). Genetic distance data were subjected to cluster analysis by the UPGMA method using PHYLIP software (Phylogeny Inference Package, version 3.5c, J. Felsenstein, Department of Genetics, University of Washington, Seattle).

Results and Discussion

Level of polymorphism

Of the 100 primers tested, only 54 primers amplified coconut DNA. Thirty-four primers detected at least one polymorphic band between one Tall (WCT) and one Dwarf (COD) accession. The number of polymorphic bands per primer ranged from 1 to 16. A total of 245 bands were generated by 34 polymorphic primers, of which 116 (47%) were polymorphic. The average number of polymorphic bands per primer was 2.2 (3.4 when only polymorphic primers were considered).

The number of polymorphic bands detected by each primer depends on the primer sequence, hence a variable number of polymorphic bands per primer was obtained. These results are consistent with earlier reports on RAPD analysis (Connolly *et al.*, 1994; Powell *et al.*, 1996; Ashburner *et al.*, 1997). The percentage (47%) of polymorphic bands between one Tall and one Dwarf coconut accession indicated a moderate level of polymorphism and was comparable with earlier reports on RAPD analysis in coconut (Ashburner *et al.*, 1997; Rodriguez *et al.*, 1997). The level of polymorphism in terms of the number of polymorphic bands per primer was also moderately high and found to be consistent with earlier reports on *Arabidopsis thaliana* (0.3), wheat (0.38) (Tingey and Tufo, 1993), soybean (1.56) (Powell *et al.*, 1996) and sweet potato (3.7) (Connolly *et al.*, 1994). These results indicate that RAPD markers can be a useful technique for germplasm characterization in coconut.

Polymorphism among accessions

Six primers generated 51 bands in 14 accessions, of which 35 (69%) bands were polymorphic. Among Tall accessions 50 bands were present, of which 33 (66%) were polymorphic. In contrast, Dwarf accessions had 30 bands and only 14 (47%) were polymorphic. RAPD markers were detected that were unique for LCM, LCO, BEN and WCT. The total heterogeneity among 14 accessions was 0.49 whereas that for Tall and Dwarf accessions was 0.46 and 0.40, respectively.

The pairwise genetic distance varied from 0.189 (CGD and MOD) to 0.62 (between WCT and MOD). The average genetic distance among Dwarf (0.31) was much less than that among Tall (0.45) accessions. Table 6.2 lists the genetic distance matrix for the 14 accessions.

When subjected to cluster analysis, all Dwarf accessions were grouped together (Fig. 6.1) whereas Tall accessions formed three groups. The two accessions from the Philippines, namely PHO and SNR, grouped together and along with ADO were closer to Dwarf than to Tall. KAP, a large nut sized accession from the west coastal region of India, was grouped with LCM, a small nut sized accession from the Lakshadweep Islands in the Arabian sea. Gangabondam, an intermediate type showed more genetic similarity to Dwarf accessions.

The data on genetic distance and heterogeneity indicated that more variance exists among Tall accessions than among Dwarf accessions. These results are comparable with earlier studies, which demonstrated higher variation in Tall than Dwarf (Ashburner and Rohde, 1994; Ashburner *et al.*, 1997; Perera *et al.*, 1998). This has been attributed to cross-pollination among Talls. Dwarf accessions from geographically distant regions (MOD and CGD) were genetically related (genetic distance, 0.186). Lebrun *et al.* (1998b) hypothesized that all the Dwarfs might have appeared at the same time

Table 6.2. Genetic distance matrix for 14 coconut accessions.

	WCT	PHO	BEN	ADO	LCO	JVT	KAP	SNR	LCM	COD	MOD	MYD	CGD
PHO	0.50	0.0											
BEN	0.40	0.50	0.0										
ADO	0.55	0.45	0.45	0.0									
LCO	0.49	0.53	0.43	0.44	0.0								
JVT	0.44	0.41	0.38	0.39	0.40	0.0							
KAP	0.48	0.48	0.38	0.43	0.47	0.42	0.0						
SNR	0.55	0.35	0.46	0.37	0.51	0.40	0.40	0.0					
LCM	0.45	0.55	0.39	0.53	0.54	0.50	0.32	0.51	0.0				
COD	0.60	0.45	0.48	0.30	0.47	0.38	0.49	0.40	0.55	0.0			
MOD	0.62	0.46	0.52	0.44	0.48	0.50	0.47	0.45	0.57	0.31	0.0		
MYD	0.61	0.49	0.52	0.40	0.48	0.46	0.49	0.37	0.59	0.30	0.25	0.0	
CGD	0.60	0.47	0.53	0.41	0.46	0.48	0.48	0.42	0.58	0.36	0.19	0.26	0.0
GBD	0.49	0.46	0.43	0.41	0.41	0.40	0.40	0.41	0.51	0.40	0.37	0.37	0.33

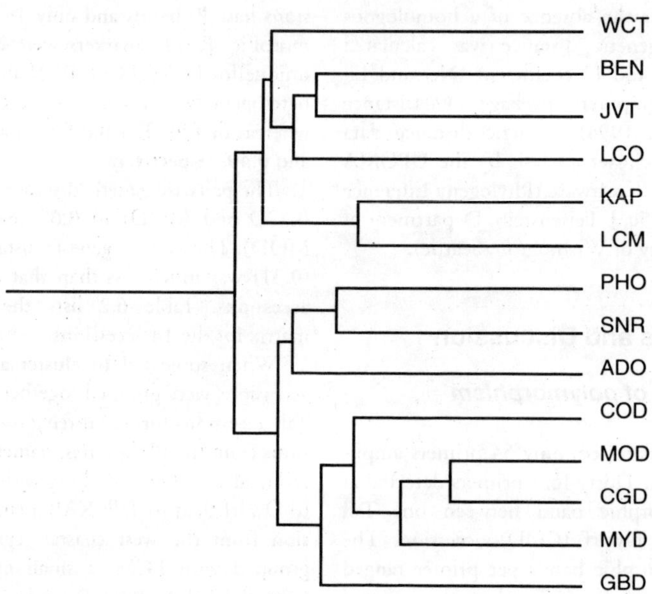

Fig. 6.1. Dendrogram showing cluster analysis of RAPD similarities among 14 coconut accessions.

and due to autogamy a major part of the genetic structure was conserved subsequently. Also Dwarf cultivars imported into new regions tend to remain genetically isolated from the local population. These reasons may explain why similar genotypes are found in distant regions.

The genetic diversity among these accessions was quite high (0.49). Ashburner *et al.* (1997) also observed high inter-population diversity among South Pacific accessions. They suggested that although the differentiation of these accessions might have arisen due to establishment of a population by a few individuals, the founder effect had not reduced diversity. The inclusion of accessions from distant regions might have resulted in relatively higher gene diversity in this study.

The cluster analysis placed two accessions with diverse fruit characters, namely KAP and LCM, together. Similar results have been reported earlier based on RFLP (Lebrun *et al.*, 1998) and phenol (Jay *et al.* 1989) analysis. The three exotic Tall collections (PHO, SNR, JVT) did not occupy a distinct position, rather, they were grouped with the indigenous accessions. Therefore, to widen genetic diversity of germplasm collections, estimation of genetic distance along with morphological data is also important.

Three Tall accessions, namely ADO, PHO and SNR, were grouped closer to Dwarf accessions. Similar results were obtained by Everard (1999) after analysing one Tall, one Dwarf and SNR. Rohde *et al.* (1995), using ISTR (Inverse Sequence Tagged Repeats) markers with 21 accessions, also found that some Tall accessions were grouped with Dwarfs. It could be assumed that grouping of the Tall accessions in this study may be due to the fact that these three accessions have some level of self-pollination due to interspadix overlapping of the mature male and female phase (Ratnambal *et al.*, 1995). The interspadix overlapping for ADO, PHO and SNR is 7.0, 4.2 and 4 days, respectively, which is comparable with intraspadix overlapping in Dwarfs. Thus the observed relationship of these accessions with Dwarfs may be due to their breeding behaviour.

In conclusion, this study has established the ability of RAPD markers to distinguish coconut accession with high efficiency. This information will form the base for analysis of intra-population variation. Extensive use of this technique and other molecular markers for characterization of coconut germplasm is envisaged. Such a study will help in planning future germplasm collecting and the selection of parents for breeding programmes.

Acknowledgements

The authors thank the Department of Biotechnology, Government of India, New Delhi, for financial support and Dr M.J. Ratnambal, Dr V. Niral and Dr V. Arunachalam for their help.

References

Armstrong, J.S., Gibbs, A.J., Peakall, R. and Weiller, G. (1994) The RAPDistance package: http:/ life.anu.edu.au/ molecular/ software/ rapd.html

Ashburner, G.R. and Rohde, W. (1994) Coconut germplasm characterization using DNA marker technology. *ACIAR Proceedings* 43, 44–46.

Ashburner, G.R., Thompson, W.K. and Halloran, G.M. (1997) RAPD analysis of South Pacific coconut palm populations. *Crop Science* 37, 992–997.

Connolly, A.G., Godwin, I.D., Cooper, M. and DeLacy, I.H. (1994) Interpretation of randomly amplified polymorphic DNA marker data for fingerprinting sweet potato (*Ipomoea batatas* L.) genotypes. *Theoretical and Applied Genetics* 88, 332–336.

Everard, J.M.D.T. (1999) An investigation towards developing a molecular approach to improve the efficiency of coconut breeding by RAPD-marker assisted selection. *CORD* XV(2), 115–130.

Harries, H.C. (1978) The evolution, dissemination and classification of *Cocos nucifera* L. *Botanical Review* 44, 205–317.

Jay, P., Bourdex, R., Potier, F. and Sanlaville, C. (1989) Note on polymorphism of coconut leaf polyphenols. *Oleagineux* 44, 151–161.

Kumaran, P.M., Koshi, P.K., Arunachalam, V., Niral, V. and Parthasarathy, V.A. (1998) Biometric clustering of coconut populations of three Indian Ocean Islands. (Abs. No. 18). PLACROSYM XIII, held at Coimbator, 16–18 December 1998, UPASI, Valparai, Tamil Nadu, India.

Lebrun, P., N'cho, Y.P., Seguin, M., Grivet, L. and Baudouin, L. (1998a) Genetic diversity in coconut (*Cocos nucifera* L.) revealed by restriction fragment length polymorphism (RFLP) markers. *Euphytica* 101, 103–108.

Lebrun, P., Grivet, L. and Baudouin, L. (1998b) The spread and domestication of the coconut palm in the light of RFLP markers. *Plantations, Recherché, Developpement* 5, 241–245.

Nei, M. (1975) *Molecular Population Genetics and Evolution. Frontiers of Biology*, vol. 40. North-Holland Publishing Company Ltd, Oxford, UK.

Nei, M. and Li, W.H. (1979) Mathematical model for studying genetic variation in terms of restriction endonucleases. *Proceedings of the National Academy of Sciences USA* 76, 5269–5273.

Perera, L., Russell, J.R., Proven, J., NcNicol, J.W. and Powell, W. (1998) Evaluating genetic relationship between indigenous coconut (*Cocos nucifera* L.) accessions from Sri Lanka by means of AFLP profiling. *Theoretical and Applied Genetics*, 96, 545–550.

Perera, L., Russell, J.R., Proven, J. and Powell, W. (2000) Use of microsatellite DNA markers to investigate the level of genetic diversity and population genetic structure of coconut (*Cocos nucifera* L.). *Genome* 43, 15–21.

Powell, W., Morgante, M., Andre, C., Hanafey, M., Voger, J., Tingey, S. and Rafalski, A. (1996) The comparison of RFLP, RAPD, AFLP and SSR (microsatellite) markers for germplasm analysis. *Molecular Breeding* 2, 225–238.

Ratnambal, M.J., Nair, M.K., Muralidharan, K., Kumaran, P.M., Bhaskar Rao, E.V.V. and Pillai, R.V. (1995) *Coconut Descriptors*. Part 1. Central Plantation Crops Research Institute, Kasaragod, Kerala, India.

Rodriguez, M.J.B., Estioko, L.P., Namia, T.I. and Soniega, J.A. (1997) Analysis of genetic diversity in coconut by RAPD. *The Philippine Journal of Coconut Studies* XXII, 1–7.

Rohde, W., Kullaya, A., Rodriguez, J. and Ritter, E. (1995) Genome analysis of Cocos nucifera L. by PCR amplification of spacer sequences separating a subset of copia-like *Eco*RI repetitive elements. *Journal of Genetics and Breeding* 49, 179–186.

Tingey, S.V. and del Tufo, J.P. (1993) Genetic analysis with random amplified polymorphic DNA markers. *Plant Physiology* 101, 349–352.

Upadhyay, A., Parthasarathy, V.A., Seema, G. and Karun, A. (1999) An efficient method of DNA extraction from coconut. *Agrotropica* 11(1), 35–38.

Williams, J.G.K., Kubelik, A.R., Livak, K.J., Rafalski, J.A. and Tingey, S.V. (1990) DNA polymorphism amplified by arbitrary primers are useful as genetic markers. *Nucleic Acids Research* 18, 6531–6535.

7 Molecular Characterization of *Gossypium* Germplasm for Cotton Improvement

R.J. Kohel and J. Yu
USDA, ARS, Crop Germplasm Research Unit, College Station, Texas, USA

Introduction

Cotton (*Gossypium* spp.) is the leading natural fibre crop and it is also an important oilseed crop. As one of the first crop plants to which Mendelian principles were applied, cotton is an important genetic system in collection, evaluation and utilization of germplasm. Improvement of cotton production has been successful over the past half century due to the utilization of exotic cotton germplasm in addition to refined cultivation practices. The traditional approach to the introduction of new genes into the elite cultivars is to screen entries from a cotton germplasm collection for a well-defined character, usually controlled by a single, dominant gene. Many of the easy improvements in the germplasm have already been made this way. However, most traits important to cotton production, such as lint yield and fibre strength are not monogenic, but polygenic. Many beneficial genes may have been left behind because of the genetic bottleneck imposed on cotton plants during early domestication processes and through modern breeding practices (Xiao *et al.*, 1996; Tanksley and McCouch, 1997). Although wild and exotic cotton germplasm may be less valuable for improvement of most traits based on phenotypic examination, it is possible that some additional useful genes remain undetected among thousands of *Gossypium* accessions. The difficult challenges for future improvement will require the integration of tools developed via biotechnology with those of conventional breeding. DNA-based genetic descriptors will revolutionize germplasm evaluation in the new century, and shift our attention more directly toward novel genes or new genotypes. Recent applications of genome mapping confirm that superior genes may be derived from the phenotypically inferior germplasm. In this chapter, we present our strategies for the molecular characterization of *Gossypium* germplasm. First, we will provide a little background and describe the *Gossypium* species, review the germplasm that is represented in the cotton germplasm collection, and then how we plan to characterize germplasm and how these efforts can be used to exploit the variability of this germplasm.

Distribution and Diversity

The eventual recognition of the differences in species and ploidy level led to the discovery of the rich and diverse evolutionary differentiation that has occurred in *Gossypium* with over 50 species currently recognized (Percival and Kohel, 1990; Stewart and Percival, unpublished). Cotton of the contemporary world comprises four cultivated species: diploid species (*Gossypium arboreum* and *Gossypium herbaceum*) of the A genome and known as Asiatic cottons and tetraploid species (*Gossypium barbadense* and *Gossypium hirsutum*) of the AD genome and known as the New World cottons

 Managing Plant Genetic Diversity
(eds J.M.M. Engels, V. Ramanatha Rao, A.H.D. Brown and M.T. Jackson)

(Lee, 1984). The diploid species are subdivided into eight genomic groups, A–K, skipping letters H, I and J. The two Asiatic cottons, *G. herbaceum* and *G. arboreum*, are the only members of the A genomic group with no true wild progenitor species identified. There are five tetraploid species, two of which are cultivated. The tetraploids are putatively of common origin from A genome and D genome parents, but the common progenitor tetraploid species does not exist. Of the cultivated cottons, the tetraploid *G. hirsutum* represents about 90% of world commerce, *G. barbadense* represents almost 10% and the Asiatic cottons represent a small portion (Lee, 1984). Much of the Asiatic cottons are consumed in home industry, and they are not included in the textile industry. An up-to-date review of the species is presented by Brubaker *et al.* (1999).

Germplasm Collection

The USDA cotton germplasm collection currently contains 7157 accessions, and it represents 2211 obsolete *G. hirsutum* cultivars, 2102 exotic *G. hirsutum* races, 1333 *G. barbadense* that are not clearly distinguished into obsolete and exotic types, 969 diploid *G. herbaceum* and *G. arboreum* from Asia, and 542 wild species. Extensive plant explorations have been made to acquire this germplasm; however, current efforts are focused on seed exchanges with other collections (Percival and Kohel, 1990). *Gossypium* is a long-lived perennial found in tropical and subtropical arid climates characterized by monsoon rainfall patterns. Individuals found in their native habitats occur as small isolated groups or as individuals, and not in dense or extended populations. The consequence of this and its relevance to the material in the germplasm collection, is that the accessions collected were uniform and not phenotypically heterogeneous or heterozygous and each accession is unique. This means that the task of characterization of the accessions is focused on distinguishing among them and not one of identifying and eliminating redundancy. It is when accessions from other collections such as a recent Russian collection are added that questions of duplication or redundancy become important considerations.

The Asiatic (diploid) cottons were the original cottons of world commerce that were subsequently replaced by *G. hirsutum* (Lee, 1984). It is, therefore, of no surprise that breeders of cotton are primarily interested in the *G. hirsutum* accessions and that they represent the greater number of accessions. To characterize the germplasm collection, we will have to develop strategies that are capable of differentiating among these *G. hirsutum* accessions. Detecting differences among the species is technically less complex, due to greater polymorphism, and a system that adequately detects intraspecific variability within *G. hirsutum* will have the power to make the more gross separations among the species.

Early Attempts or Methods of Characterization

Characterization of *Gossypium* is no new concept or endeavour. With the scientific discovery of different ploidy levels of *Gossypium*, cytological techniques were used to separate the diploid and the tetraploid genomes (Beasley, 1940, 1942; Percival *et al.*, 1999). Details of genomic groups and species are discussed in Percival *et al.* (1999).

The early attempts to differentiate among the tetraploid *G. hirsutum* accessions considered plant morphological differences combined with geographic origins to separate *G. hirsutum* into seven geographic races (Hutchinson, 1951). These distinctions remain today as the best criteria to describe the variability within *G. hirsutum* but many of the accessions have not been so described.

With advancing technology, the electrophoresis of seed proteins was used to separate species. We applied this procedure to the Asiatic and Upland cottons and were successful in separating these materials into various groups, or clusters. Fifteen such groups were defined within *G. hirsutum* that were not always consistent with known geographic origins and observed morphological differences. The results supported our concerns that geographic and morphological race classifications were inadequate, but protein banding alone was not adequate to interpret the variability. Again, technological advances resulted in the use of isozyme analyses. These were first applied to *Gossypium* as a taxonomic tool to distinguish differences at the species level (Cherry *et al.*, 1972; Wendel and Percy, 1990). We applied these isozyme protocols to the variability in *G. hirsutum*. However, the various isozyme assays resolved little or no variability within the species of real use.

Genomic Tools

The development of DNA markers provides new and more powerful tools that can be effectively used in discriminating among genebank accessions. Genetic linkage maps based on DNA markers offer new opportunities in evaluation and characterization of crop germplasm resources for their better utilization (Tanksley and McCouch, 1997). The new approach is to look directly for the genes of interest, removing the blocking factors from other parts of the genome (Xiao *et al.*, 1995). In cotton, we began a genomics programme that would provide the markers for germplasm characterization. Our and other previous research with isozymes and preliminary use of DNA markers established the low level of polymorphism in intraspecific *G. hirsutum*. Our in-house genomics programme has developed a cotton genetic map comprising 868 markers, 256 restriction fragment length polymorphisms (RFLPs), 152 randomly amplified polymorphic DNA (RAPDs), 198 simple sequence repeats (SSRs), 267 amplified fragment length polymorphisms (AFLPs), and five morphological markers assembled into 50 linkage groups that cover about 5000 cM of the cotton genome (Reddy *et al.*, 1997; Yu *et al.*, 1997; Yu and Kohel, 2000) (Fig. 7.1). The cotton genetic map is based on the interspecific cross of *G. hirsutum* (TM-1) and *G. barbadense* (3-79) (Yu and Kohel, 2000), which are the genetic standards for their respective species (Kohel *et al.*, 1970; Kohel, 1973). Two thirds of the linkage groups have been assigned to their respective genomic origin (A vs. D) or chromosomal identity (1–26) by use of diploid and aneuploid cottons. Thirteen quantitative trait loci (QTLs) for fibre quality properties (strength, length and fineness) have been identified in 3-79, an extra long staple cotton (Kohel *et al.*, 2000). These QTLs explained 35% to 50% of the total genetic variance for fibre characteristics in the F_2 population (Yu *et al.*, 1998). Characterization of these QTLs in intraspecific *G. hirsutum* crosses involving Pee Dee (PD6992) and Acala (HS427-10) cottons is in progress. Molecular tags for major genes have been identified for glandless cotton, photoperiod sensitivity and other important traits. From this base, diagnostic DNA markers can be identified that are capable of detecting polymorphism in intraspecific populations. We have observed that the level of polymorphism within intraspecific material is between 10 and 20% of that between our interspecific mapping parents.

In general, cotton genome research lags behind that of other major crops. To accelerate progress and leverage resources in cotton, we have initiated efforts to bridge cotton with other plant models such as *Arabidopsis*. Like most higher plants, cotton has about 50,000 functional genes per genome. It would be a very tedious task, and not practical to study individual genes or their combination for cotton improvement. The tools, including markers, maps, and gene transfer procedures, are still rather limited in cotton genomics. Bridging cotton with the advanced model plant systems would enhance cotton genomics and provide additional publicly available resources. *Arabidopsis* is an established model system for all higher plants. With increased investment, complete nucleotide sequences of the entire *Arabidopsis* genome will be available within the year. To make substantial progress into the area of cotton functional genomics, the availability of *Arabidopsis* genomic tools and their use in the cotton genome is currently under exploration. High levels of homology between *Gossypium* and model plant genomes and polymorphism among *Gossypium* germplasm were detected by using the conserved gene sequences. About 10% of the 40,000 *Arabidopsis* expressed sequence tagged sites (ESTs) may be readily identified in the cotton genome (Yu *et al.*, unpublished). These *Arabidopsis* ESTs would be complemented with unique cotton ESTs and genomic sequence information from cotton bacterial artificial chromosomes (BACs) to study cotton functional genomics. Such new genomic tools will facilitate cotton improvement programmes to eventually discover, manipulate and utilize thousands of valuable genes otherwise buried in *Gossypium* germplasm.

Preliminary Characterization

To begin the process of molecular characterization, we selected a subset of germplasm lines that represent what we consider the range of diversity of *G. hirsutum* and outliers of *G. barbadense* and Asiatic cottons. In a pilot experiment, we screened 280 accessions in subsets with 43 marker loci. The 280 accessions were selected based on their representation from different genepools or sources. The sampling process emphasized genetic composition more than the physical appearance. Among them, 151 were *G. hirsutum* accessions comprising 30 exotic, 41 obsolete and 80 Russian cottons. The 126 *G.*

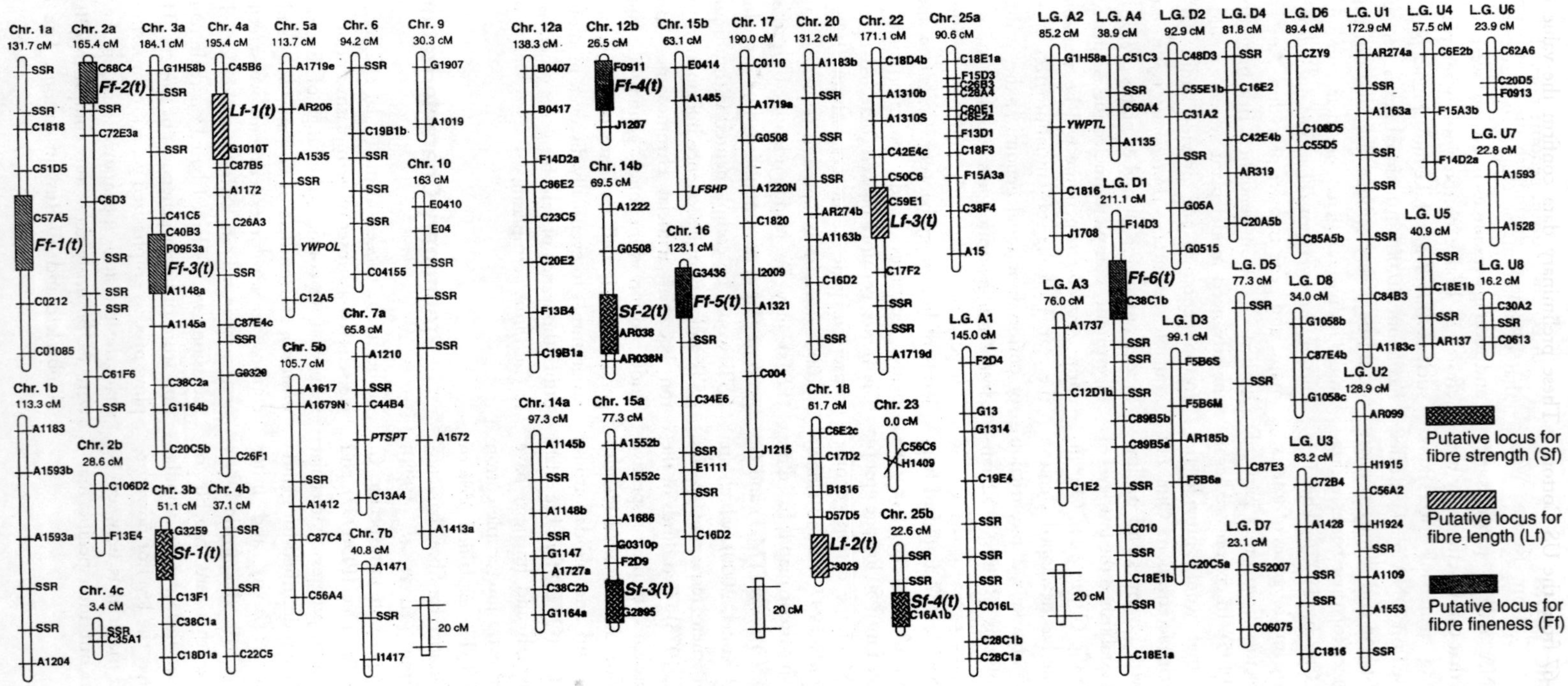

Fig. 7.1. A genetic map based on an F_2 population derived from the interspecific cross TM1 × 3-79. Thirteen QTLs for fibre quality traits are indicated on the linkage groups.

barbadense accessions were 67 from the US Cotton Germplasm Collection and 59 from Russia. We also included three Asiatic cottons together with TM-1 and 3-79, two genetic standards of AD genomes. The 43 DNA markers, mostly SSRs, were selected from the cotton molecular genetic map including gene-linked tags and representatives from each chromosome or linkage group. Based on our previous mapping surveys, some of these selected DNA markers were capable of detecting polymorphism among *G. hirsutum* accessions. DNA profiles of 280 *Gossypium* accessions were generated in the lab, and each major DNA fragment was recorded as either presence (score 1), absence (score 0), or unknown (score ?) for each of the 280 accessions. These preliminary data confirm the value of our DNA markers to distinguish the known variation (Fig. 7.2). The results of this cluster analysis indicated not only the separation of three major groups (*G. hirsutum*, *G. barbadense* and Asiatic cottons), but also the separation of each of two major groups (*G. hirsutum* and *G. barbadense*) into more than a dozen subgroups.

These preliminary data show us the strengths of DNA markers and the deficiencies. The strengths are rather obvious in the ability to separate among the accessions with greater detail than previously possible. But to be effective, we have to identify and overcome the deficiencies. The first deficiency is still the limited total number of markers. The second is the unknown

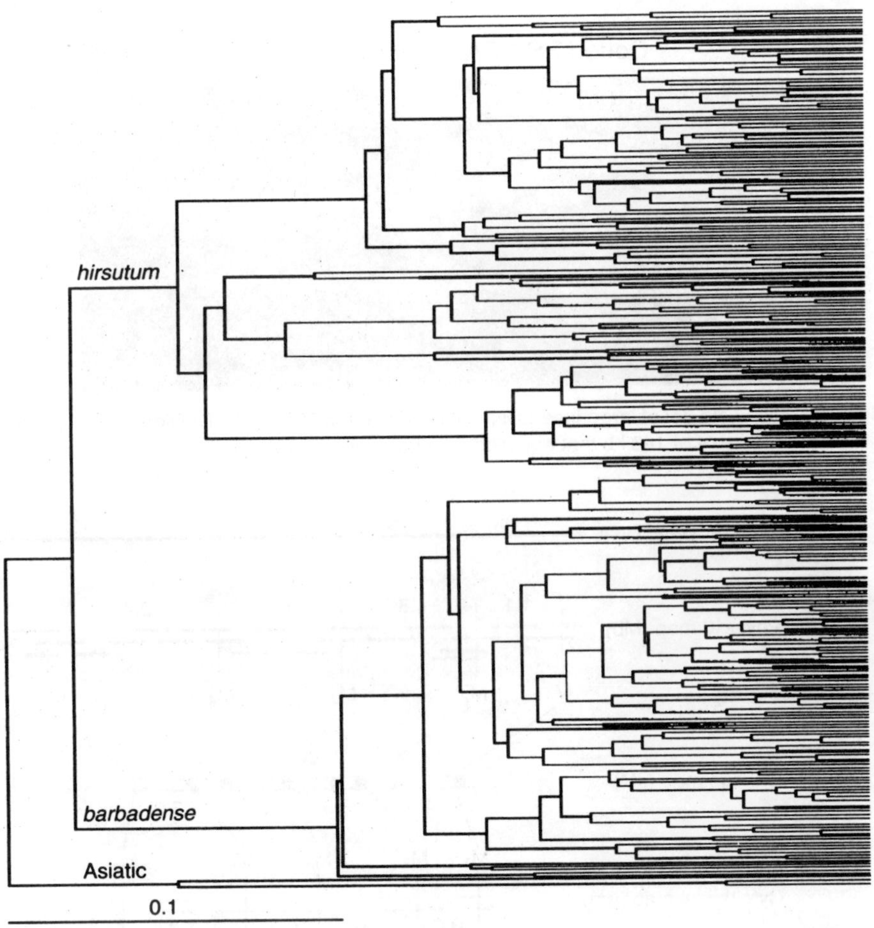

Fig. 7.2. Cluster analysis of 280 accessions. Major groups of *G. hirsutum*, *G. barbadense* and Asiatic cottons are clearly distinguished. Scale is the similarity index.

chromosomal location and representation of these markers within the A and D subgenomes. The third is not so much a deficiency as a requirement that we should place on the markers of choice. The utility of DNA markers for evaluation of the germplasm will depend on their portability among accessions within a specific *Gossypium* species. To this end, we have initiated a project that will resolve the deficiencies directly. We constructed a 7.3× genome equivalent BAC library with 115,584 clones with an average 143 kb insert size (Fig. 7.3) (Yu *et al.*, 2000). The success of this effort led to our decision to initiate a major effort to develop an integrated genetic and physical map (Fig. 7.4). A cotton chromosome would be reconstructed by overlapping such BAC insert DNA, and anchored with DNA markers selected from a genetic map that corresponds to a cytological map. Any BAC contigs or the extended contigs may serve as a resource for marker saturation, in addition to positional cloning of genes of agronomic importance.

At the second phase of this project, the genetic uniqueness of each accession will be determined by the two criteria: (i) a high number of unique DNA polymorphism relative to TM-1 and 3-79, which are the modern genetic standards; and (ii) new accessions that are dissimilar genetically (DNA profiles) to all other accessions. The resulting unique *Gossypium* accessions will be subject to further characterization with an additional 100 DNA markers once we overcome the limitation of sufficient DNA markers with the best ability to describe the germplasm. Accessions with DNA profiles mostly distinct from those of modern cultivars are likely to possess a large number of new genes that

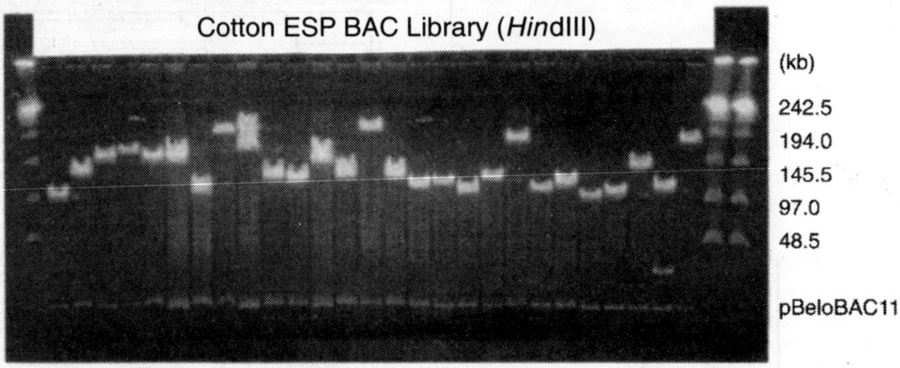

Fig. 7.3. Pulse-field gel of large-insert DNA clones from a BAC library. The library had 115,584 clones with an average size of 143 kb that equalled 7.3x genome equivalents.

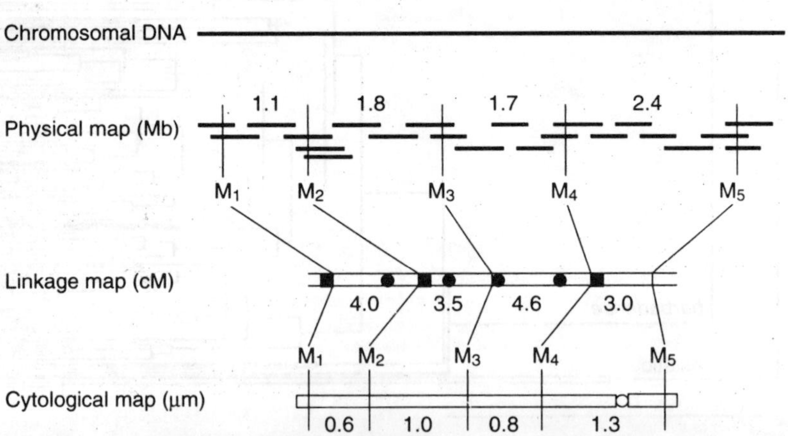

Fig. 7.4. Schematic illustration of integrative physical and genetic mapping: horizontal bars, BAC clone; M, DNA markers; filled squares, major genes; filled circles, QTLs.

lie buried but are potentially useful in crop improvement (Tanksley and McCouch, 1997). Recommendations for the further utilization of *Gossypium* germplasm resources will be made according to the uniqueness of their DNA profiles.

Resources for Characterization

We began the construction of a TM-1 BAC library to give an integrated genetic and physical map that will effectively resolve the 50 linkage groups to the 26 chromosome linkage groups. TM-1 is the widely used genetic standard for *G. hirsutum*, and it is a parent of our genetic map (Kohel *et al.*, 1970). We are constructing a 10× genome equivalent library, and with improved protocols, we expect 150,000 clones with an average 150 kb insert size. We are using a binary vector with one of the restriction enzymes that will add utility for future research projects (Hamilton *et al.*, 1999). From the positive BACs, we are producing at least 1000 SSR markers that will be placed on the integrated map. With the level of polymorphism within intraspecific material between 10 and 20%, this will mean that of the approximately 2000 markers on the interspecific genetic map, we will have 200 to 400 polymorphic markers among intraspecific accessions. Theoretically, this should give coverage of a marker every 12+ cM. A subset of about 100 markers selected from these should cover the entire genome with markers no more than 50 cM apart (Fig. 7.5). This core reference subset of markers will be publicly available so that all evaluations of germplasm can be shared and pooled into a common database for analysis and interpretation. Therefore, the core markers will serve as the standard to characterize workable sets of *Gossypium* accessions across different genepools or germplasm sources. Other marker sets can also be combined with the core set of markers and characterized in subsets of accessions. Information from our genomics programme is entered into CottonDB http://ars-genome@cornell.edu, in the genomics workbook http://algodon.tamu.edu and BAC information in http://hbz.tamu.edu (Chen *et al.*, 2000).

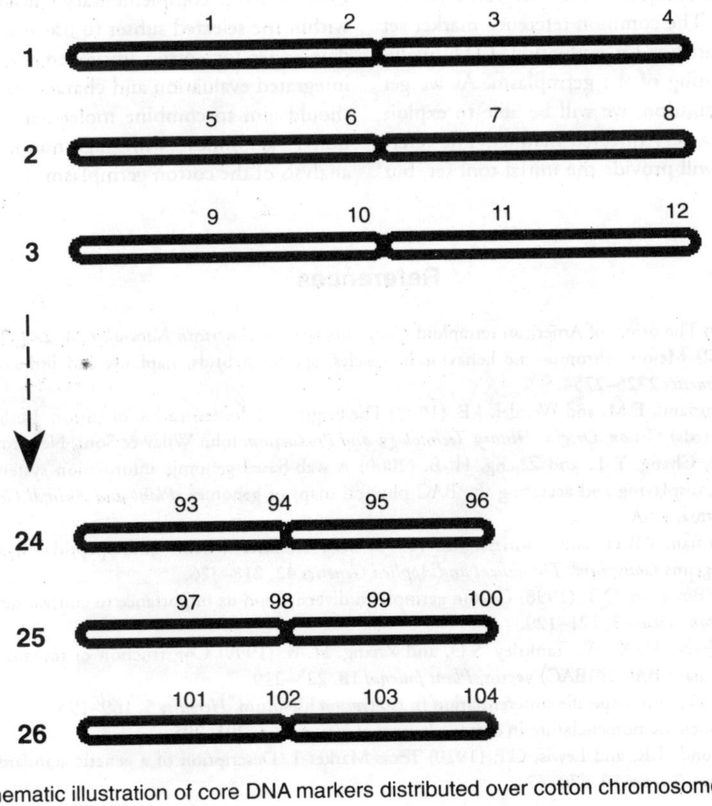

Fig. 7.5. Schematic illustration of core DNA markers distributed over cotton chromosomes.

Discussion

The need for and importance of characterizing *Gossypium* genetic resources is readily recognized. Currently genetic vulnerability is an increased concern as less than 1% of cotton germplasm accessions have been explored (Esbroeck and Bowman, 1998). The evaluation and utilization of methodologies for analysis of the germplasm collection have been limited mostly to morphological descriptors with masking negative effects such as yield reduction. Neutral molecular descriptors have recently been initiated for characterizing the cotton germplasm. However we are concerned that the emergent efforts will produce data and information that are fragmentary and cannot be readily combined into an integrated whole. Our proposed strategy and efforts are analogous to the establishment of a standard set of descriptors. A portable set of markers, such as SSRs, that can used in both rather simple labs equipped for gel-based PCR analysis or more sophisticated labs with automated genotyping instruments will allow the analysis of germplasm subsets from which the data and interpretations can be combined and integrated for the entire collection. The common reference marker set will provide an anchor for evaluation, QTL analysis, and gene mining of the germplasm. As we get beyond characterization, we will be able to exploit this variation in a very directed manner. The reference marker set will provide the initial tool set, but the integrated map and BAC clones will be powerful tools for the eventual application of functional genomics and gene mining to the germplasm.

What we propose is a major undertaking of time and resources. We have initiated the preliminary phases that can be completed with our in-house resources. With adequate funding the tools for characterization will be available in 3 years. Characterization will require additional funds and participants, so that the timely completion of the characterization will be resource driven. We also recognize that plans do not always evolve as envisioned; however, with a plan and specific goals, we will remain flexible in adapting the methodologies and technologies that will allow realization of the goal. Using the standard set of molecular descriptors (selected DNA markers), genetic relationships among cotton accessions can be determined as shown in our pilot study. It is always necessary to start with a subset of genetic outliers to maximize the potential success in the characterization process. A subset of genotypes will then be selected for further characterization of desirable genes. Available ESTs of the model plants such as *Arabidopsis* can be used to detect complementary conserved sequences within the selected subset to use as potential candidate genes for cotton improvement. By and large, integrated evaluation and characterization strategies should aim to combine molecular tools and functional genomics with agronomic performance analysis of the cotton germplasm.

References

Beasley, J.O. (1940) The origin of American tetraploid *Gossypium* species. *American Naturalist* 74, 285–286.

Beasley, J.O. (1942) Meiotic chromosome behavior in species, species hybrids, haploids and induced polyploids of *Gossypium. Genetics* 2725–2754.

Brubaker, C.L., Bourland, E.M. and Wendel, J.E. (1999) The origin and domestication of cotton. In: Smith, C.W and Cothren, J.T. (eds) *Cotton: Origin, History, Technology, and Production.* John Wiley & Sons, New York, pp. 3–31.

Chen, H., Tao, Q., Chang, Y.-L. and Zhang, H.-B. (2000) A web-based genomic information system for efficiently manipulating, displaying and accessing the BAC physical maps of genomes. *Plant and Animal Genome* VIII, San Diego, California, USA.

Cherry, J.P., Katterman, F.R.H. and Endrizzi, J.E. (1972) Seed esterases, leucine aminopeptidases, and catalases of species of the genus *Gossypium. Theoretical and Applied Genetics* 42, 218–226.

Esbroeck, G.V. and Bowman, D.T. (1998) Cotton germplasm diversity and its importance to cultivar development. *The Journal of Cotton Science* 2, 121–129.

Hamilton, C.M., Frary, A., Xu, Y., Tanksley, S.D. and Zhang, H.-B. (1999) Construction of tomato genomic DNA libraries in a binary-BAC (BIBAC) vector. *Plant Journal* 18, 223–229.

Hutchinson, J.B. (1951) Intra-specific differentiation in *Gossypium* hirsutum. *Heredity* 5, 169–193.

Kohel, R.J. (1973) Genetic nomenclature in cotton. *Journal of Heredity* 64, 291–295.

Kohel, R.J., Richmond, T.R. and Lewis, C.F. (1970) Texas Marker-1. Description of a genetic standard for *Gossypium hirsutum* L. *Crop Science* 10, 670–671.

Kohel, R.J., Yu, J., Park, Y.-H. and Lazo, G.R. (2000) Molecular mapping and characterization of genes controlling fiber quality in cotton. *Euphytica* (in press).

Lee, J.A. (1984) Cotton as a world crop. In: Kohel, R.J. and Lewis, C.F. (eds) *Cotton*. American Society of Agronomy, Madison, Wisconsin, pp. 1–25.

Percival, A.E. and Kohel, R.J. (1990) Distribution, collection, and evaluation of *Gossypium*. *Advances in Agronomy* 44, 225–256.

Percival, A.E., Wendel, J.E. and Stewart, J.M. (1999) Taxonomy and germplasm resources. In: Smith, C.W. and Cothren, J.T. (eds) *Cotton: Origin, History, Technology, and Production*. John Wiley & Sons, New York, pp. 33–63.

Reddy, A.S., Haisler, R.M., Yu, Z.H. and Kohel, R.J. (1997) AFLP mapping in cotton. *Plant Genome* V, San Diego, California.

Tanksley S.D. and McCouch, S.R. (1997) Seed banks and molecular maps: unlocking genetic potential from the wild. *Science* 277, 1063–1066.

Wendel, J.F. and Percy, R.G. (1990) Allozyme diversity and introgression in the Galapagos Island endemic *Gossypium darwinii* and its relationship to continental *G. barbadense*. *Biochemistry and Systematic Ecology* 18, 517–528.

Xiao J., Grandillo, S., Ahn, S., McCouch, S., Tanksley, S. and Yuan, L. (1996) Genes from wild rice improve yield. *Nature* 384, 223–224.

Yu, J., Park, Y.-H., Lazo, G.R. and Kohel, R.J. (1997) Molecular mapping of the cotton genome. *Proceedings of the 5th International Congress of Plant Molecular Biology*. Singapore.

Yu J., Park, Y.-H., Lazo, G.R. and Kohel, R.J. (1998) Molecular mapping of the cotton genome: QTL analysis of fiber quality characteristics. *Proceedings of Plant and Animal Genome VI, the International Conference on the Status of Plant and Animal Genome Research*. San Diego, California.

Yu, J., Kohel, R.J., Zhang, H.-B., Dong, J.-M. and Decanini, L.I. (2000) Construction of a cotton BAC library and its applications to gene isolation. *Proceedings of the 8th International Conference on the Status of Plant and Animal Genome Research*. San Diego, California.

Yu, Z.H. and Kohel, R.J. (2001) Cotton genome research in the United States. In: Jenkins, J.N. and Saha, S. (eds) *Emerging Technologies in Cotton Breeding*. Oxford University Press, New York.

8 Molecular Analysis of the Origin and Domestication of *Theobroma cacao* L.

J.C. Motamayor[1] and C. Lanaud[2]
[1]*FUNDACITE, Av. Las Delicias, Maracay, Edo. Aragua, Venezuela;* [2]*CIRAD, TA 40/03, Av. Agropolis, Montpellier, France*

Introduction

Molecular markers offer new tools for investigating the evolution and domestication of crop plant species, and the origin of today's commercial cultivars. Different markers have been employed in such studies, and several have compared the results from different techniques (in cacao for example, see N'Goran *et al.*, 1994; Lerceteau *et al.*, 1997). This chapter shows that additional factors like adequacy of sampling can lead to new evolutionary hypotheses emerging.

Cacao trees belong to the genus *Theobroma* and to the family *Malvaceae* (Alverson *et al.*, 1999). *Theobroma cacao* L. ($2n = 2x = 20$) is a tree native to wet tropical regions of the northern part of South America and, in some reports, of Central America. Indeed, there is a controversy about the origin and domestication of cacao.

For some authors cacao was introduced in Central America. Although the first home of domestication and culture is placed in Central America, Van Hall (1914) stated that the most probable home country of cacao is the region of the Orinoco and the Amazon basins, with the valleys of their tributary streams. Cheesman (1944) considered that the centre of origin of the cacao tree is the Upper Amazon near the Colombian–Ecuadorian border on the eastern flanks of the Andes. Cheesman's hypothesis is based on Pound's report (1938) that the greatest morphological diversity occurs in this area. Pound (1945), Desrosiers (1954) and Purseglove (1968) also placed the centre of origin of this species in South America, either in the lower eastern equatorial slopes of the Andes or in the upper waters of the Amazon River (Upper Amazon). Cheesman (1944) and these latter authors argued that although cacao has been cultivated in Mexico and Central America for over 2000 years, no truly wild populations were present in this region; they proposed that cacao was introduced into Central America and Mexico. Schultes (1984) hypothesized that once the cacao spread throughout the Amazon Valley, it could have dispersed along two routes: one leading north and the other west. In this way, the domestication of cacao occurred in South America and was then spread to Central America and Southern Mexico by migrating Indians (Schultes, 1984).

In contrast, however, other authors suggest that cacao was domesticated from wild plants from Central America. Wickizer (1951) argued that after the wild species evolved in either the Orinoco or Amazon basin, it spread in a wild state along the Gulf coast as far north as southern Mexico. Cuatrecasas (1964) suggested that before the domestication of cacao, natural populations occurred in the central part of Amazon-Guyana into southern Mexico. Stone (1984) stated that although the homeland of cacao is usually placed in the Amazon, this plant also occurs in the wild in Central America. Allen (1982) mentioned that cacao occurs as an apparently wild tree from southern Mexico to Bolivia and Brazil. Indeed, there

 Managing Plant Genetic Diversity (eds J.M.M. Engels, V. Ramanatha Rao, A.H.D. Brown and M.T. Jackson)

exist reports about wild cacao populations in southern Mexico (Miranda, 1962; Cuatrecasas, 1964; Gomez-Pompa *et al.*, 1990; De la Cruz *et al.*, 1995). Furthermore, Mora-Urpi (1958) found greater cacao variability in Mexico and Central America than in South America.

Cuatrecasas' (1964) hypothesis has been sustained by most authors (Cope, 1976; Wood and Lass, 1985; Gomez-Pompa *et al.*, 1990; Laurent *et al.*, 1994; N'Goran *et al.*, 1994; De la Cruz *et al.*, 1995; Whitkus *et al.*, 1998). Based on the great morphological diversity observed in Central as well as in South America, he proposed that north and south cacao populations developed into two forms separated geographically by the Panama Isthmus. Both populations evolved independently and developed a number of consistent distinguishing characteristics that led to their recognition as subspecies (*T. cacao* ssp. *cacao* and *T. cacao* spp. *sphaerocarpum*).

Krickeberg (1946) proposed that the Mayan people domesticated cacao. Miranda (1962) was the first to report wild cacao in southern Mexico as a completely indigenous cacao tree in the Lacandon rain-forest. Cuatrecasas (1964) claimed that wild plants from the Lacandon rainforest were possibly the ancestors of the domesticated cacao. It was even proposed that cultivated cacao was introduced into South America in pre-Columbian times (Mora-Urpi, 1958).

Cuatrecasas' subspecies correspond to the two morphogeographic groups proposed by Cheesman (1944). Subspecies *cacao* and *sphaerocarpum* belong to the Criollo and Forastero groups respectively. A third hybrid group originated from crosses between Criollo and Forastero was called Trinitario. The Forastero group is subdivided according to the geographical origin of the trees (Upper Amazon, Lower Amazon and Guyana Forastero).

Isozyme diversity

Several biochemical and molecular studies have analysed genetic diversity in the Criollo and Forastero groups. For instance, Lanaud (1987) analysed the diversity of 296 genotypes for nine polymorphic isozyme loci. A total of 30 alleles were identified in the study, and all the alleles except PAC1 (a common allele in Criollo and Trinitario populations) were present in the native populations of the Upper Amazon. The populations that had the most alleles per locus were those from the Upper Amazon (from 1.8 to 2.2). The populations from Venezuela and Guyana were the least variable with 1 to 1.5 alleles per locus. Ronning and Schnell (1994) studied the isozyme diversity at eight loci in 86 cacao clones stemming from the groups Forastero, Criollo, Trinitario and a hybrid undefined group. Allelic frequencies and genetic distances confirmed genetic differentiation between Forastero and Criollo.

Molecular diversity revealed by RFLP markers

Nuclear diversity

Laurent *et al.* (1994) analysed 201 genotypes belonging to various morphogeographic groups. Diversity at the total nuclear DNA level was analysed with 31 cDNA probes that revealed 87 polymorphic fragments. In spite of continuous variation between the groups, due in particular to a great number of hybrids (Trinitario), a clear structure appeared on axis 1 of an FAC among Forastero on the one hand and Criollo and Trinitario on the other. The populations of the Upper Amazon Forastero and Criollo were highly diverse. The diversity of Criollo overlapped the pool of Trinitario. Based on the high genetic diversity found in Forastero as well as in the Criollo group, Laurent *et al.* (1994) supported Cuatrecasas's (1964) hypothesis for a Central American origin of the Criollo group.

Lerceteau *et al.* (1997) analysed the genetic diversity of Ecuadorian Nacional clones (59), and also analysed Forastero (29), Trinitario (29), and Criollo (9) clones. Forty-three genomic probes, coded in terms of alleles, were used for this study. Within-group genetic diversity was almost identical between Forastero, Trinitario and Criollo (respectively 0.33; 0.31; 0.31); the Nacional group shown a value of 0.19. The highest mean number of heterozygous loci based on 31 restriction fragment length polymorphism (RFLP) probes was found in Trinitario and Criollo, and the lowest between the Nacional clones. Among them, some genotypes, sampled from very old Ecuadorian plantations, were almost totally homozygous.

Cytoplasmic DNA diversity

Laurent *et al.* (1993) studied DNA diversity of 177 genotypes of cacao trees for organellar DNA using heterologous mitochondrial probes (ATP synthetase of sunflower, cytochrome oxidase of wheat) and chloroplastic probes (rubisco from spinach). The mitochondrial probes revealed 44 mitotypes. Thirty-five of them were found in fewer than six clones. These minor types were essentially made up of Criollo and Trinitario. Among the nine types remaining, type 1 grouped two-thirds of the Forastero clones, comprising genotypes from Guyana, Venezuela, Brazil, Peru, Colombia and Ecuador. The other major type (mitotype 2) grouped the majority of Criollo and Trinitario (26 clones), and some Lower Amazon genotypes. In this study the mitochondrial DNA diversity was represented by an FAC. In the graphic analysis, it is interesting to note that the mitochondrial DNA diversity was much greater in Criollo than in Forastero, in contrast to the nuclear DNA diversity.

Molecular diversity revealed by RAPD markers

Figueira *et al.* (1994) analysed the diversity of genotypes belonging to the three groups, Criollo, Trinitario and Forastero, by means of randomly amplified polymorphic DNA (RAPD) markers (128 amplified fragments resulting from 23 primers). They observed continuous variation between these groups and no differences between them. On the other hand, they noticed a clear difference between wild and cultivated clones. Ribosomal DNA analysis confirmed this structure and the authors thus proposed a new classification for cacao trees not based on the three traditional groups, Criollo, Trinitario and Forastero, but on the groups 'wild cacao trees' and 'cultivated cacao trees'.

N'Goran *et al.* (1994) analysed the genetic diversity of 106 genotypes belonging to the various morphogeographic groups within Criollo, for 49 repeatable polymorphic RAPD products. The Hierarchic Ascendant Classification (HAC), established from the factorial coordinates of an FAC, showed a clear structure among Forastero and Criollo, as well as a clear differentiation between Upper and Lower Amazon Forastero (N'Goran *et al.*, 1994). After Southern blot analyses of 36 fragments, 12 corresponded to highly repeated and scattered sequences, 12 to unique sequences and 12 to low copy number sequences. If one considers only fragments corresponding to the unique sequences, no clear structuring appears. On the other hand, a genetic structure appears between Criollo and Forastero when only fragments corresponding to the highly repeated sequences are considered (while these do not differentiate Lower from Upper Amazon Forastero). From these results, N'Goran *et al.* (1994) supported Cuatrecasas's (1964) hypothesis about a natural dispersion of the species in Central America.

De la Cruz *et al.* (1995) analysed by RAPD 42 genotypes corresponding to what they called wild plants either from Mexico (sinkholes of Yucatan and rainforest of Chiapas) or from the Upper Amazon, to Criollo cultivars, Forastero and Trinitario, and to a representative of a nearby genus *Herrania*. The calculations of a dissimilarity index allowed the construction of a neighbour-joining tree. In this tree there was a greater similarity between cultivated Criollo and wild cacao trees of South America than between cultivated Criollo and 'wild' plants from Mexico. The authors suggested that the wild plants found in the ancient sacred groves of Maya probably do not exist in the germplasm collections and could be the genotypes closest among existing genotypes to the Maya cultivars.

In a more recent study, Whitkus *et al.* (1998) analysed by RAPD a wider sample (86 individuals), which included 26 wild genotypes collected in the Lacandona forest of Chiapas, Mexico, five individuals from sinkholes in the Yucatan and wild and cultivated clones of South America. Contrary to the populations and cultivars of South America where the intra-population diversity has been seen to be higher than the inter-population diversity, the genetic diversity found between the two Mexican populations was greater than that found in each region. This reflects the low level of polymorphism found in these two populations. A structure of the genetic diversity similar to that of De la Cruz *et al.* (1995) was observed. The two populations from Mexico appeared well differentiated and showed their originality. From these results, Whitkus *et al.* (1998) also supported Cuatrecasas' (1964) hypothesis. The authors underlined, however, the lack of affinity between cultivated Criollo and the wild plants observed in Mexico. According to these authors, cacao trees collected in the Lacandon forest could represent true wild cacao trees, and plants sampled in the sinkholes could have been introduced by the Maya from the wild populations. In that case, they would represent a subsample of the present populations in the Lacandon rainforest that could have diverged through isolation by distance since the time of the introductions (thus explaining the genetic differentiation between the two Mexican populations).

To analyse the relationship between Central and South American cacao further, we assayed variation in a range of molecular markers on Forastero and Criollo individuals already analysed in previous researches (Lanaud, 1987; Ronning and Schnell, 1994; Laurent *et al.*, 1994; N'Goran *et al.*, 1994; Figueira *et al.*, 1994; Lerceteau *et al.*, 1997). Most of them come from international germplasm collections or modern plantations. We also sampled other individuals from sites where it could be assumed that Criollo had not crossed with individuals from the other groups.

Material and Methods

Plant material

The samples were classified as Ancient Criollo, Modern Criollo, Trinitario, Amelonado individuals (Lower Amazon Forastero), Guyana Forastero, Upper Amazon Forastero and hybrids with at least one Upper Amazon Forastero parent. A complete list of individuals used in the study and their geographic origin is available on request.

The ancient Criollo individuals (n = 92) consisted of trees sampled from places where gene flow between Criollo and Trinitario or Forastero trees was absent or limited because introductions of Trinitario or Forastero material were unlikely. Most samples came from trees on old or abandoned farms and from private gardens in towns difficult to access. Sampling was conducted in Venezuela, Colombia, Nicaragua and Mexico. In Mexico, samples of Criollo were also collected in the rainforest where wild Criollo trees have been reported (Miranda, 1962; Cuatrecasas, 1964) and in places where Mayan people cultivated cacao (Yucatan sinkholes: Gomez-Pompà *et al.*, 1990 and in the Pacific coast of Mexico: Lopez-Mendoza, 1987). Samples from the rainforest of Belize that were associated with Mayan ruins (Mooledhar *et al.*, 1995) were also included for analysis.

Modern Criollo individuals (n = 70) were defined as those coming from modern farms or from farms where an important introduction of Trinitario or Forastero seedlings or clones was suspected. This class included material from germplasm collections of Costa Rica, Ivory Coast, Mexico, Venezuela and France. Modern Criollo represents the genotypes studied as Criollo in previous biochemical and molecular studies (Lanaud, 1987; Ronning and Schnell, 1994; Laurent *et al.*, 1994; N'Goran *et al.*, 1994; Figueira *et al.*, 1994; De la Cruz *et al.* 1995; Lerceteau *et al.*, 1997; Whitkus *et al.*, 1998).

Trinitario (n = 66) and Guyana (n = 5), Lower and Upper Amazon Forastero individuals (LAF, n = 9 and UAF, n = 27) and hybrids with at least one Upper Amazon Forastero parent (n = 14), were studied to compare the structure of their genetic diversity with that of Criollo and Modern Criollo.

DNA isolation, RFLP and microsatellite analyses

DNA was isolated as described in Risterucci *et al.* (2000). Two hundred and eighty-nine individuals were analysed for RFLPs, using procedures previously described (Lanaud *et al.*, 1995). Seventeen cDNA and eight genomic DNA probes, chosen for their coverage of the genetic map of *T. cacao* L. (Lanaud *et al.*, 1995), were used to screen individuals. DNA fragments were extracted from low melting agarose gel and hybridized, without cloning, on to the restricted DNA.

Sixteen microsatellites (Lanaud *et al.*, 1999) were used to screen the genetic diversity of 98 individuals already analysed by RFLPs. Microsatellites were chosen based on their position in the cacao genetic map (Risterucci *et al.*, 2000). Microsatellite allele sizes were scored by comparison of PCR product lengths to the sequence of the genomic clone from where primers were designed.

For RFLP data a factorial analysis of correspondences (FAC, Benzeckri, 1973), was carried out using the software GENETIX 4.0 (Laboratoire Génome et Populations, Université de Montpellier II, Montpellier). The gene diversity statistics (Nei, 1978), the mean allele number per locus, the percentage of polymorphic loci at 95% levels of significance, and the observed heterozygosity were calculated using the software GENETIX 4.0.

For microsatellite data the shared allele distance (D_{AS}) (Chakraborty and Jin, 1993) was calculated. This distance is equal to 1 minus the proportion of shared alleles:

$$D_{AS} = 1 - (a/2n)$$

where a is the number of alleles common to individuals i and j, and n the number of loci studied. This distance was computed by averaging the values over all available loci between two individuals. To determine the relatedness between Criollo and Forastero individuals a neighbour-joining tree (Saitou and Nei, 1987) was constructed from the shared allele distance between individuals obtained from microsatellite data using the program TREEMAKER (Cornuet J.-M. and Piry, S., 2000, personal communication).

Results

RFLP analyses

After hybridization of the 25 probes, 66 alleles were detected. The unbiased gene diversity (Nei, 1978) was higher for Forastero than Criollo (Table 8.1). The average number of alleles was the highest for the Forastero group, as well as the percentage of polymorphic loci and the observed heterozygosity. This last statistic was very low for Ancient Criollo, indicating homozygosity as a characteristic genetic factor for this group.

For RFLP markers, all Ancient Criollo individuals (92) mapped in the left half of the FAC (Fig. 8.1), with a cluster of homogeneous clones with several unresolved, near identical individuals in the fourth quadrant. Only eight Ancient Criollo geno-

Table 8.1. RFLP diversity within Criollo (ssp. *cacao*), Forastero (ssp. *sphaerocarpum*), Modern Criollo and Trinitario. Hnb, unbiased gene diversity (Nei, 1978); N, sample size; Hobs, observed heterozygosity; P(0.95), proportion of polymorphic loci when most frequent allele does not exceed 95%; A, mean number of alleles per locus.

	N	Hnb	Hobs	P(0.95)	A
ssp. *cacao*	92	0.002 (0.008)	0.002 (0.008)	0.00	1.08
ssp. *sphaerocarpum*	37	0.38 (0.17)	0.17 (0.11)	0.92	2.32
Modern Criollo	70	0.39 (0.18)	0.47 (0.23)	0.84	2.04
Trinitario	66	0.42 (0.18)	0.43 (0.19)	0.84	2.12

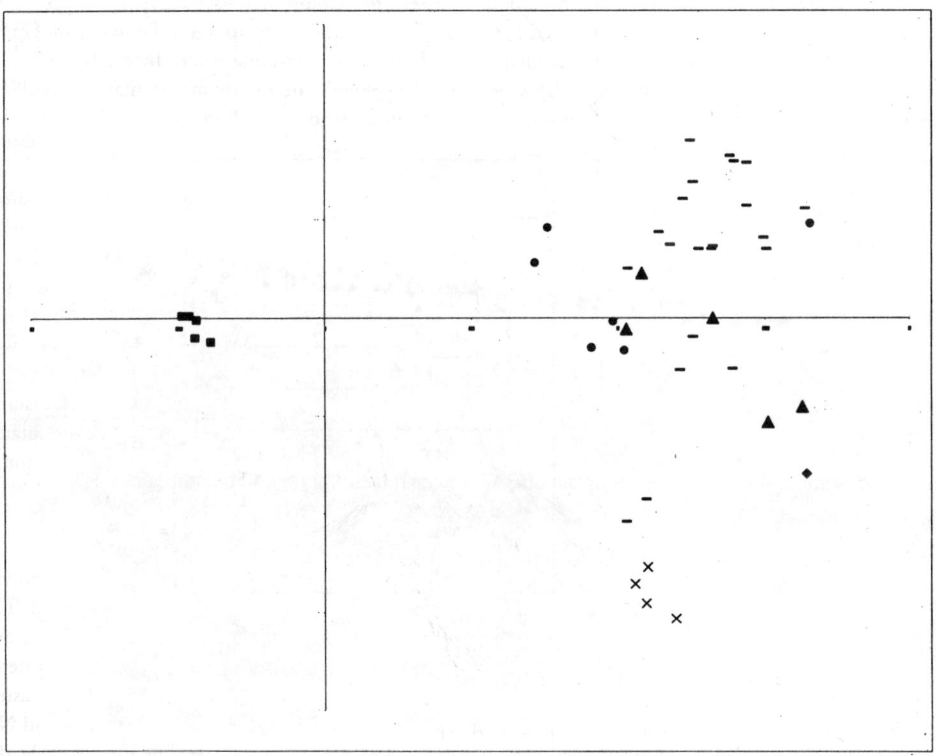

Fig. 8.1. Factorial analysis of correspondences (FAC). Squares, Ancient Criollo individuals; circles, Colombian–Ecuadorian individuals; crosses, Guyane individuals; triangles, Venezuelan–Brazilian individuals; diamonds, Amelonado type; and bars, Peruvian individuals.

types were observed in 92 individuals whereas each Forastero individual had a unique RFLP genotype. Some Criollo individuals shared their RFLP genotype among different morphotypes or across diverse geographical areas (Venezuela, Colombia, Nicaragua, Belize and Mexico). Upper Amazon Forastero was represented in the right half of the figure.

Consequently, Table 8.1 shows that the Ancient Criollo group (comprising individuals from the Lacandon rainforest) had very low diversity. Cacao trees from the Lacandon rainforest were identical at the RFLP level to those putatively cultivated by the Maya (those found in the sinkholes of Yucatan, in the Pacific Coast of Mexico and in Belize) as well as to individuals cultivated today in South America.

In Fig. 8.2, which excludes the samples from Upper Amazon Forastero, Modern Criollo individuals (black filled circles) are superimposed on Trinitario (black triangles). Furthermore they form a continuous line between Ancient Criollo and Amelonado Forastero, and are much less diverse than hybrids having an Upper Amazon Forastero parent.

The genetic diversity statistics for Modern Criollo were similar to those of Trinitario. The observed heterozygosity of Modern Criollo was significantly higher than that of Ancient Criollo. Equivalence between Modern Criollo and Trinitario is expected considering that the distinction based on morphological traits between Modern Criollo and Trinitario is subjective.

Microsatellite analyses

The 16 microsatellite loci detected 150 alleles. Despite the increased number of alleles, the genetic diversity of the Ancient Criollo group observed from microsatellite data was still very low compared with that observed for the Forastero group (Table 8.2). The gene diversity for the Ancient Criollo group was 0.04 in contrast to that for the Forastero group (0.78). Within geographic regions, the gene diversity values for 13 and five individuals from the Peru and Colombia–Ecuador regions were similarly high (0.70). The observed heterozygosity was 0.00 for the Ancient Criollo and 0.34 for Forastero, and other genetic statistics similarly contrasted. Figure 8.3 is the neighbour-joining tree calculated from the shared allele distances (D_{AS}) (Chakraborty and Jin, 1993). In this tree, Criollo individuals are closer to Colombian–Ecuadorian clones (EBC, LCT) than the latter to some Peruvian or Guyana or Lower Amazon Forastero individuals.

In general, the microsatellite markers evidenced the same trends regarding the genetic background

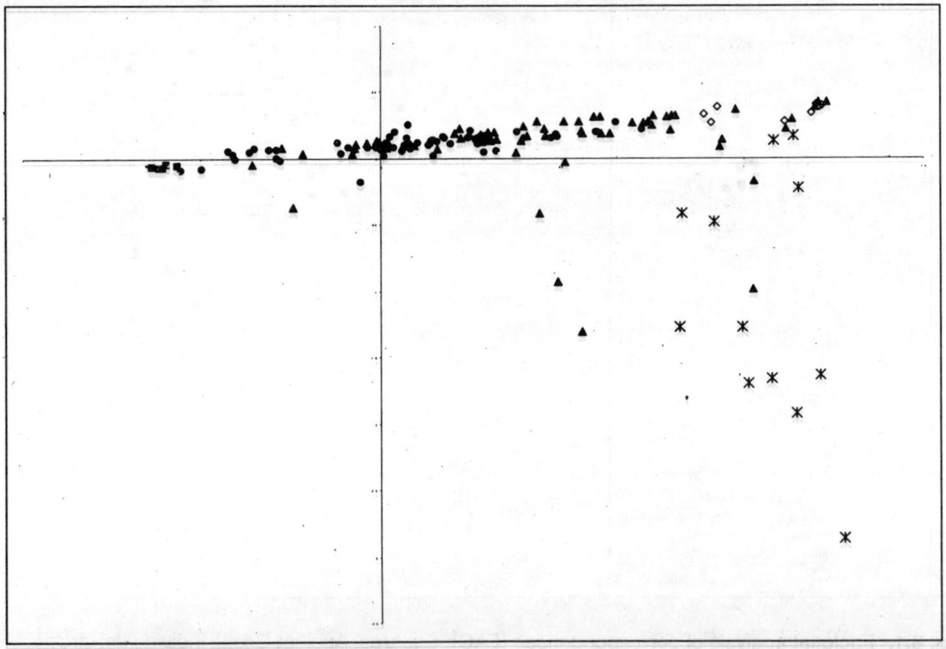

Fig. 8.2. Factorial analysis of correspondences (FAC). Squares, Ancient Criollo; circles, Modern Criollo; triangles, Trinitario; diamonds, Amelonados; cross + bar, hybrids having an Upper Amazon parent.

of Modern Criollo and Trinitario as those found using RFLPs. Modern Criollo had equivalent values for population genetic statistics to those obtained for Trinitario group (Table 8.2). The gene diversity for the Modern Criollo group was 0.52 while for the Trinitario group it was 0.50. The observed heterozygosity was 0.61 for Modern Criollo and 0.69 for Forastero.

Table 8.2. Genetic diversity for microsatellite markers within Criollo (ssp. *cacao*), Forastero (ssp. *sphaerocarpum*), Modern Criollo and Trinitario. Genetic diversity within Forastero from Peru and Ecuador is shown. N, sample size; Hnb, unbiased gene diversity (Nei,1978); Hobs, observed heterozygosity; P(0.95), proportion of polymorphic loci when most frequent allele does not exceed 95%; A, mean number of alleles per locus.

	N	Hnb	Hobs	P(0.95)	A
ssp.*cacao*	40	0.04 (0.13)	0.003 (0.01)	0.06	1.19
ssp. *sphaerocarpum*	28	0.78 (0.13)	0.34 (0.13)	1.00	8.69
Peru	13	0.70 (0.18)	0.48 (0.25)	1.00	5.50
Colombia–Ecuador	5	0.70 (0.13)	0.39 (0.19)	1.00	3.94
Modern Criollo	14	0.52 (0.03)	0.61 (0.14)	1.00	2.44
Trinitario	13	0.50 (0.05)	0.69 (0.16)	1.00	2.13

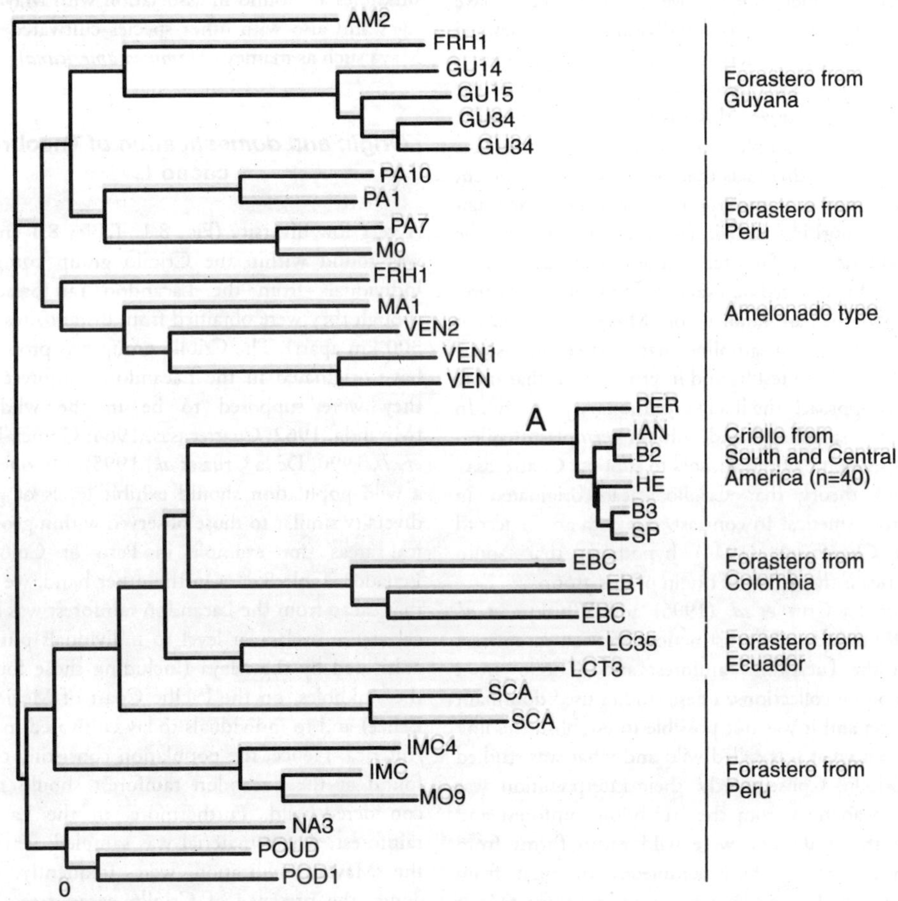

Fig. 8.3. Neighbour-joining tree of Forastero and Criollo individuals based on the shared allele distance calculated from microsatellite data. All Criollo individuals (N = 40) cluster under node A.

Discussion

The analyses of RFLPs and of microsatellite markers described here have shed new light on patterns of genetic diversity and genetic relationships within *T. cacao* L. Both techniques have yielded equivalent results, despite the fact that the number of alleles detected per locus was significantly higher for microsatellite loci than for RFLP loci. It is of interest, then, to ask why our results differ from those of others obtained using RFLPs (Laurent *et al.*, 1994; Lerceteau *et al.*, 1997), or isozymes (Lanaud, 1987; Ronning and Schnell, 1994) or RAPD markers (Figueira *et al.*, 1994; N'Goran *et al.*, 1994; Lerceteau *et al.*, 1997). Both our RFLP and microsatellite analyses clearly distinguished Ancient Criollo individuals from those introgressed with Forastero genes particularly from the Lower Amazon. As mentioned above, previous studies have taken introgressed Criollo individuals (those defined in this study as Modern Criollo) as valid representatives of the Criollo group. The present study is the first that emphasizes this distinction, and it arises from a proper sampling of individuals of each class. In this way, individuals that are classified as Ancient Criollo constitute the true Criollo group; that originally comprising individuals cultivated before the introduction of Forastero clones to cacao plantations, which caused natural hybridization, and subsequently the appearance of Modern Criollo or Trinitario. Thus a sampling that is based on historic data can lead to results and interpretations that more closely approach the likely evolutionary pathways. In this case, sampling based only on germplasm collections led some of the authors to support Cuatrecasas' (1964) theory that Criollo cacao originated in Central America. In contrast, our data are in accord with Cheesman's (1944) hypothesis that South America is the region of origin of this group.

De La Cruz *et al.* (1995) and Whitkus *et al.* (1998) found a clear distinction between cacao trees from the Lacandon rainforest and 'Criollo' from germplasm collections. These studies used dominant markers and it was not possible to establish the link between what was called wild and what was studied as Criollo. Consequently, their interpretation was that cacao trees from the Lacandon rainforest and from the 'sinkholes' were wild cacao forms from Central America, and completely different from domesticated cacao. Furthermore, contrary to our results obtained they found differentiation between samples from the Lacandon rainforest and those from the sinkholes. Regarding this last result, they proposed that cacao from the 'sinkholes', although 'wild', represented ancient introductions of cacao trees probably from the Lacandon rainforest, which, through isolation, diverged from the source material. Their results also allowed them to support Cuatrecasas' (1964) hypothesis.

For the present study we analysed three different field samplings, of eight individuals from three sinkholes from Yucatan and did not find any genetic difference from nine of the 13 individuals from the Lacandon rainforest in Mexico studied using all markers. All eight individuals showed exactly the same genotype for 25 RFLP and 16 microsatellite loci. Genetic evidence regarding the misclassification of Criollo trees from Central America as wild will be further discussed below. However it should be noted that those from the sinkholes are found in association with Mayan vestiges and also with other species cultivated by the Maya such as mamey (*Mammea americana*).

Origin and domestication of Theobroma cacao *L.*

A very low diversity (Fig. 8.1, Tables 8.1 and 8.2) was found within the Criollo group comprising individuals from the Lacandon rainforest even though they were obtained from distant sites (up to 300 km apart). The Criollo group was proposed to have originated in the Lacandon rainforest where they were supposed to be in the wild state (Miranda, 1962; Cuatrecasas, 1964; Gomez-Pompa *et al.*, 1990; De la Cruz *et al.*, 1995). On one hand, a wild population should exhibit levels of genetic diversity similar to those observed within geographical areas (for example in Peru or Colombia–Ecuador, Table 8.2). On the other hand, we found that cacao from the Lacandon rainforest was identical at the molecular level to individuals putatively cultivated by the Maya (including those found in the sinkholes, on the Pacific Coast of Mexico and Belize) and to individuals today cultivated in South America. Hence, the population consisting of trees found at the Lacandon rainforest should not be considered wild. Furthermore, in the Lacandon rainforest, where material was sampled, vestiges of the Mayan civilization were frequently found. Thus, the presence of Criollo cacao trees in the Lacandon rainforest might be the remnant of cacao cultivation by the Mayan civilization.

Our results support Cheesman's (1944) hypothesis that cacao was introduced into Central America and contradict Cuatrecasas's (1964) hypothesis that Criollo is a separate subspecies that had evolved independently from the South American individuals. If this latter were the case, all wild Forastero individuals should form a cluster independent of those individuals from Central America. In contrast Fig. 8.3 shows Criollo individuals closer to Forastero from Colombia and Ecuador than to other Forastero from Guyana or some from Peru. Therefore, the Criollo group is not forming a separate subspecies (*cacao*) to that comprising individuals from South America (*sphaerocarpum*). We propose that the Criollo group had a South American origin and a small number of individuals were transported by humans. This led to a genetic bottleneck as an important factor in Criollo differentiation from Forastero individuals.

Thus the Criollo group probably originates from a South American population where domestication commenced. But, from where? The South American origin of the Criollo group is controversial because of the lack of evidence of its presence in this region of the continent. Pre-Columbian South American civilizations did not develop such advanced cultures as Maya or Aztec, and did not leave as precise traces of their activities as the latter. However, there exist some historical data that report cacao cultivation at the time of the Conquest and its utilization for beverages and ceremonies by local Indians (Simon, 1882; Febres Cordero, 1892; Altolaguirre y Duvale, 1908; Jahn, 1927; Vazquez de Espinosa, 1948; Arellano Moreno, 1950). These reports concern the south-western region of Venezuela where only Criollo trees are found today. Although archaeological evidence of cacao domestication in South America is lacking, historic and genetic evidence seems to argue for this view.

The high morphological diversity found in the Criollo group could be due to mutations in relatively few genes participating in pod morphology, which have been deliberately selected by humans. Indeed, such different pod types as Porcelana and Pentagona (one very rough and the other very smooth) are contrasting. A special human interest could lie in the collection, maintenance and use of these cacao types. For example, Pentagona type has the finest pod cortex, increasing the ratio of beans weight to pod weight and facilitating the extraction of the beans from the fruit. Characteristic traits of Criollo trees such as the sugared pulp of its beans and the fact that it needs a shorter fermentation time, may be prized as a target of human selection through more than 2000 years of cultivation.

Acknowledgements

This study was partially financed by IPGRI through the Vavilov-Frankel Fellowship 1998, by the National Scientific Research Council (CONICIT) of Venezuela and by the International Centre for Agricultural Research and Development (CIRAD, France).

References

Allen, J.B. (1982) Collecting wild cocoa at its centre of diversity. *Proceedings of the Eighth International Cocoa Research Conference, Cartagena, Colombia.* Cocoa Producers' Alliance, Lagos, Nigeria, pp. 655–662.

Altolaguirre y Duvale, A. (1908) *Relaciones geograficas de la gobernacion de Venezuela (1767–1768).* Real Sociedad Geografica. Imp. del Patronato de Huérfanos de Admon. Militar, Madrid, Spain.

Alverson, W.S., Whitlock, B.A., Nyffeler, R., Bayer, C. and Baum, D.A. (1999) Phylogeny of the core Malvales: Evidence from ndhF sequence data. *American Journal of Botany* 86, 1474–1486.

Arellano Moreno, A. (1950) Fuentes para la historia economica de Venezuela (Siglo XVI). Comité Ejecutivo. *Tercera Conferencia Interamericana de Agricultura.* Serie Nal. 83 (Cuadernos Verdes). Caracas, Venezuela. Tip. 'El Compas'.

Benzeckri, J.P. (1973) *L'analyse des données.* 2. *L'analyse des correspondances.* Dunod, Paris, France.

Chakraborty, R. and Jin, L (1993) A unified approach to study hypervariable polymorphisms: statistical considerations of determining relatedness and populations distances. In: Pena, S.D.J., Chakraborty, R. Epplen, J.T. and Jeffreys, A.J. (eds) *DNA Fingerprinting: State of the Science.* Birjhauser Verlag, Basel, Switzerland, pp. 153–175.

Cheesman, E.E. (1944) Notes on the nomenclature, classification possible and relationships of cocoa populations. *Tropical Agriculture* 21, 144–159.

Cope, F.W. (1976) Cacao. *Theobroma cacao* L. (*Sterculiaceae*). In: Simmonds, N.W. (ed.) *Evolution of Crop Plants.* Longman, London, pp. 207–213.

Cuatrecasas, J. (1964) Cacao and its allies: a taxonomic revision of the genus *Theobroma*. *Contributions from the United States Herbarium* 35, 379- 614.

De la Cruz, M., Whitkus, R., Gomez-Pompa, A. and Mota-Bravo, L. (1995) Origins of cacao cultivation. *Nature* 375, 542–543.

Desrosiers, R. (1954) Diversidad genética del cacao como base en la seleccion de la resistancia a la enfermedad de la escoba de bruja. *Turrialba* 4, 131–134.

Febres Cordero, L. (1892) El chocolate y el chorote. Estudio historico. In: *Estudios sobre etnografia americana. Memorias escritas para ser presentadas al Congreso Internacional de Americanistas y al Congreso Geografico Hispano-portugués-americano, en sus sesiones de 1892*. Imp. Centenario, Mérida, Venezuela, pp. 55–71.

Figueira, A., Janick, J., Levy, M. and Goldsbrough, P. (1994) Re-examining the classification of *Theobroma cacao* L. using molecular markers. *Journal of the American Society for Horticultural Science* 119, 1073–1082.

Gomez-Pompa, A., Flores, J.S. and Fernandez, M.A. (1990) The sacred cacao groves of the Maya. *Latin American Antiquity* 1, 247–257.

Jahn, A. (1927) *Los aborigenes del occidente de Venezuela. Su historia, etnografia, y afinidades linguisticas*. Con un mapa etnologico y 33 planchas. Lit.y Tip. del Comercio, Caracas, Venezuela.

Krickeberg, W. (1946) *Etnologia de America*. Fondo de Cultura Economica, Mexico.

Lanaud, C. (1987) Nouvelles données sur la biologie du cacaoyer (*Theobroma cacao* L.): diversités de populations, systèmes d'incompatibilité, haploïdes spontanés; leurs conséquences pour l'amélioration génétique de cette espèce. PhD thesis, Université Paris XI, Orsay, France.

Lanaud, C., Risterucci, A.M., N'Goran, A.K.J., Clément, D., Flament, M.H., Laurent, V. and Falque, M. (1995) A genetic linkage map of *Theobroma cacao* L. *Theoretical and Applied Genetics* 91, 987–993.

Lanaud, C., Risterucci, A.M., Pieretti, I., Falque, M., Bouet, A. and Lagoda, P.J.L (1999) Isolation and characterization of microsatellites in *Theobroma cacao* L. *Molecular Ecology* 8, 2141–2152.

Laurent, V., Risterucci, A.M. and Lanaud, C. (1993) Chloroplast and mitochondrial DNA diversity in *Theobroma cacao*. *Theoretical and Applied Genetics* 87, 81–88.

Laurent, V., Risterucci, A.M. and Lanaud, C. (1994) Genetic diversity in cocoa revealed by cDNA probes. *Theoretical and Applied Genetics* 88, 193–198.

Lerceteau, E., Robert, T., Pétiard, V. and Crouzillat, D. (1997) Evaluation of the extent of genetic variability among *Theobroma cacao* accessions using RAPD and RFLP markers. *Theoretical and Applied Genetics* 95, 10–19.

Lopez Mendoza, R. (1987) *El cacao en Tabasco*. Coleccion Cuadernos Universitarios. Serie Agronomia No. 13. Universidad Autonoma de Chapingo, Mexico.

Miranda, F. (1962) Wild cacao in the Lacandona Forest, Chiapas, Mexico. *Cacao* (*Turrialba*) 7, 7.

Mooledhar, V., Maharaj, W. and O'Brien, H. (1995) The collection of Criollo cocoa germplasm in Belize. *Cocoa Grower's Bulletin* 49, 26–40.

Mora-Urpi, J. (1958) Notas sobre el posible origen y la variabilidad del cacao cultivado en América tropical. *Turrialba* (Costa Rica) 8, 34–43.

Nei, M. (1978) Estimation of average heterozygosity and genetic distance from a small number of individuals. *Genetics* 89, 583–590.

N'Goran, J.A.K., Laurent, V. Risterucci, A.M. and Lanaud, C. (1994) Comparative genetic diversity of *Theobroma cacao* L. using RFLP and RAPD markers. *Heredity* 73, 589–597.

Pound, F.J. (1938) Cocoa and Witches' Broom disease (*Marasmius perniciosus*) of South America with notes on other species of *Theobroma*. Yuille's Printery, Port of Spain, Trinidad and Tobago.

Pound, F.J. (1945) A note on the cacao population of South America. *Proceedings of the Cocoa Research Conference*. London, UK.

Purseglove, J.W. (1968) *Theobroma* L. In: Purseglove, J.W. (ed.) *Tropical Crops. Dycotyledons* 2. John Wiley & Sons, New York, pp. 571–599.

Risterucci, A.M., Grivet, L., N'Goran, J.A.K., Pieretti, I., Flament, M.H. and Lanaud, C. (2000) A high-density linkage map of *Theobroma cacao* L. *Theoretical and Applied Genetics* 101, 948–955.

Ronning, C.M. and Schnell, R.J. (1994) Allozyme diversity in a germplasm collection of *Theobroma cacao* L. *Journal of Heredity* 85, 291–295.

Saitou, N. and Nei, M. (1987) The neighbor-joining method: a new method for reconstructing phylogenetic trees. *Molecular Biology and Evolution* 4, 685–691.

Schultes, R.E. (1984) Amazonian cultigens and their northward and westward migrations in pre-Columbian times. In: Stone, D. (ed.) *Pre-Columbian Plant Migration*. Papers of the Peabody Museum of Archaeology and Ethnology, vol. 76. Harvard University Press, Cambridge, Massachusetts, pp. 69–83.

Simon, P. (1882) *Noticias historiales de las conquistas de Tierra Firme en las Indias Occidentales*. Primera parte. Edicion hecha sobre la de Cuenca de 1626. Imprenta de Medardo Rivas, Bogota, Colombia.

Stone, D. (1984) Pre-Columbian migration of *Theobroma cacao* L. and *Manihot esculenta* C. from Northern South America into Mesoamerica: a partially hypothetical view. *In:* Stone, D. (ed.) *Pre-Columbian Plant Migration.* Papers of the Peabody Museum of Archaeology and Ethnology, vol. 76. Harvard University Press, Cambridge, Massachusetts, pp. 69–83.

Van Hall, C.J.J. (1914) *Cocoa.* Macmillan, London.

Vazquez de Espinosa, A. (1948) *Compendium and Description of the West Indies,1628*, transl. by Charles Upson Clark, Smithsonian Miscellaneous Collections, vol. 108. Washington, DC.

Whitkus, R., De la Cruz, M. and Mota-Bravo, L. (1998) Genetic diversity and relationships of cocoa (*Theobroma cacao* L.) in southern Mexico. *Theoretical and Applied Genetics* 96, 621–627.

Wickizer, V.D. (1951) *Coffee, Tea and Cocoa, an Economic and Political Analysis.* Stanford University, California.

Wood, G.A.R. and Lass, R.A. (1985) *Cocoa.* Longman, London.

9 Technologies and Strategies for *ex situ* Conservation

F. Engelmann and J.M.M. Engels
International Plant Genetic Resources Institute (IPGRI), Rome, Italy

Introduction

Two basic conservation strategies, each composed of various techniques, are employed to conserve genetic diversity: *in situ* and *ex situ* conservation. Article 2 of the Convention on Biological Diversity provides the following definitions for these categories (UNCED, 1992). *Ex situ* conservation means the conservation of components of biological diversity outside their natural habitat. *In situ* conservation means the conservation of ecosystems and natural habitats and the maintenance and recovery of viable populations of species and, in the case of domesticated or cultivated species, in the surroundings where they have developed their distinctive properties.

There is an obvious fundamental difference between these two strategies: *ex situ* conservation involves the sampling, transfer and storage of target taxa from the collecting area, whereas *in situ* conservation involves the designation, management and monitoring of target taxa where they are encountered (Maxted *et al.*, 1997). Another difference lies with the more dynamic nature of *in situ* conservation and the more static nature of *ex situ* conservation. These two basic conservation strategies are further subdivided into specific techniques including seed storage, *in vitro* storage, DNA storage, pollen storage, field genebank and botanical garden conservation for *ex situ*, and protected area, on-farm and home garden conservation for *in situ*, each technique presenting its own advantages and limitations (Engels and Wood, 1999). *Ex situ* conservation techniques are particularly appropriate for the conservation of crops and their wild relatives, while *in situ* conservation is especially appropriate for wild species and for landrace material on farm.

Until recently, most conservation efforts, apart from work on forest genetic resources, have concentrated on *ex situ* conservation, particularly seed genebanks. In the 1950s and 1960s, major advances in plant breeding brought about the 'Green Revolution' which resulted in the wide-scale adoption of high-yielding varieties and genetically uniform cultivars of staple crops, particularly wheat and rice. Consequently, global concern about the loss of genetic diversity in these crops increased, as farmers abandoned their locally adapted landraces and traditional varieties, replacing them with improved, yet genetically uniform modern ones. In response to this concern, the International Agricultural Research Centres (IARCs) of the Consultative Group on International Agricultural Research (CGIAR) started to assemble germplasm collections of the major crop species within their respective mandates. It is in this context that the International Board for Plant Genetic Resources (IBPGR) was established in 1974 to coordinate the global effort to systematically collect and conserve the world's threatened plant genetic diversity.

Today, as a result of this global effort, there are over 1300 genebanks and germplasm collections around the world, maintaining an estimated 6,100,000 accessions (FAO, 1996). It is important

(eds J.M.M. Engels, V. Ramanatha Rao, A.H.D. Brown and M.T. Jackson)

to mention that these collecting and conservation activities focused largely on the major food crops, including cereals and some legumes, that is, species that can be conserved easily as seed. This has resulted in over-representation of those species in the world's major genebanks, as well as in the fact that conservation strategies and concepts were biased towards such material. It is only more recently that the establishment of field genebanks, allowing conservation of species for which seed conservation is not appropriate or impossible, as well as the development of new storage technologies, including *in vitro* conservation and cryopreservation, has been given due attention by the international community.

In this chapter, we first briefly review the main achievements made and problems faced with *ex situ* conservation of plant genetic resources. We then present the main areas for research in the development of *ex situ* conservation technologies and strategies to improve conservation efforts. It is important to mention that only technical and managerial aspects of *ex situ* conservation are addressed in this chapter, and that organizational and policy aspects have not been taken into consideration.

Main Achievements of and Difficulties with *ex situ* Conservation of Plant Genetic Resources

Orthodox seed conservation

Many of the world's major food plants produce seeds that undergo maturation drying, and are thus tolerant to extensive desiccation and can be stored dry at low temperature. Seeds of this type are termed 'orthodox' (Roberts, 1973). Storage of such orthodox seeds is the most widely practised method of *ex situ* conservation of plant genetic resources, as indicated in the *Report on the State of the World's Plant Genetic Resources for Food and Agriculture* (FAO, 1996), since 90% of the 6.1 million accessions stored in genebanks are maintained as seed. Over the years, techniques have been devised which allow seeds of many species to be conserved in this way for several decades. These techniques involve drying seeds to low moisture content (3–7% fresh weight basis, depending on the species) and storing them in hermetically sealed containers, at low temperature, preferably at −18°C or cooler (FAO/IPGRI, 1994). All relevant techniques are well established and a series of practical documents have been published which cover the main aspects of seed conservation, including the design of seed storage facilities for genetic conservation (Cromarty *et al.*, 1982); principles of seed testing which need to be understood when monitoring the viability of seed accessions maintained in genebanks (Ellis *et al.*, 1985a); methods for removing dormancy and germinating seeds (Ellis *et al.*, 1985b); as well as suitable methods for processing and handling seeds in genebanks (Hanson, 1985).

Another technical achievement in the area of orthodox seed conservation concerns the development of the so called 'ultra-dry' seed storage technology (Walters, 1998), which is based on the principle that desiccating seeds to much lower moisture contents than those generally used in standard procedures will allow us to store them for an extended period at room temperature, thereby avoiding the requirement for refrigeration facilities. This technology, along with its achievements and limitations, will be discussed further in another section of this chapter.

Seeds are also a convenient form for distributing germplasm to farmers, breeders, scientists and other users. Moreover, since seeds are less likely to carry diseases, in comparison with other plant material, their use for exchange of plant germplasm can facilitate quarantine procedures.

Conservation of non-orthodox seed and vegetatively propagated species

In contrast to orthodox seeds, a considerable number of species, predominantly tropical or subtropical in origin, such as coconut, cacao and many forest and fruit tree species, produce seeds which do not undergo maturation drying and are shed at relatively high moisture content (Chin, 1988). Such seeds are unable to withstand desiccation and are often sensitive to chilling. They cannot be maintained under the conventional seed storage conditions described above, that is, storage at low moisture content and low temperature. Seeds of this type are called 'recalcitrant' and have to be kept in moist, relatively warm conditions to maintain viability (Roberts, 1973; Chin and Roberts, 1980). Even when recalcitrant seeds are stored in an optimal manner, their lifespan is limited to weeks, occasionally months. Of more than 7000 species for which information on seed storage behaviour has been published (Hong *et al.*, 1996), approximately 3% are recorded as recalcitrant and an additional 4% as possibly recalcitrant.

More recent investigations have identified species exhibiting 'intermediate' storage behaviour. While such seeds can tolerate desiccation to fairly low moisture contents, once dried, they become particularly susceptible to injury caused by low temperature (Ellis *et al.*, 1990, 1991). Even though a continuum in desiccation sensitivity is observed within the intermediate seed storage category, from highly desiccation sensitive to relatively tolerant (Berjak and Pammenter, 1994), the storage life of intermediate seeds can be prolonged by further drying, but it remains impossible to achieve the long-term conservation of orthodox seeds. About 1% of the aforementioned 7000 species studied and included in the Compendium on Seed Storage Behaviour are reported as producing intermediate seeds and another 1% have been characterized as possibly intermediate (Hong *et al.*, 1996). Included in this category are some economically important species such as coffee, citrus, rubber, oil palm and many tropical forest tree species.

It should be noted that the percentages of intermediate and recalcitrant seed producing species cited above are likely to be largely underestimated. These figures are based on scientific and technical publications, which, by default, concern mainly temperate species. In addition, it can be expected that a large proportion of the species for which no information is available, which are predominantly from tropical or subtropical origin, exhibit recalcitrant, or to a lesser extent intermediate seed storage behaviour. As an example, it has been estimated that more than 70% of tree species in tropical forest ecosystems have recalcitrant seeds (Ouédraogo *et al.*, 1999).

There are other species for which conservation as seed is problematic. Firstly, there are those that do not produce seeds at all and, consequently, are propagated vegetatively, for example banana and plantain (*Musa* spp.). Secondly, there are crops such as potato (*Solanum tuberosum*), other root and tuber crops such as yams (*Dioscorea* spp.), cassava (*Manihot esculenta*) and sweet potato (*Ipomoea batatas*), and sugarcane (*Saccharum* spp.) that have either some sterile genotypes and/or some that produce orthodox seed. However, these seeds are highly heterozygous and, therefore, of limited utility for the conservation of particular genotypes. These crops are usually propagated vegetatively to maintain genotypes as clones (Simmonds, 1982).

Traditionally, the field genebank has been the *ex situ* storage method of choice for the aforementioned 'problem materials'. According to the *Report on the State of the World's Plant Genetic Resources for Food and Agriculture* (FAO, 1996), around 527,000 accessions are maintained in field genebanks. In some ways, this method offers a satisfactory approach to conservation. The genetic resources under conservation can be readily accessed and observed, thus permitting detailed evaluation. However, there are certain drawbacks that limit its efficiency and threaten its security (Withers and Engels, 1990; Engelmann, 1997). The genetic resources are exposed to pests, diseases and other natural hazards such as drought, weather damage, human error and vandalism. In addition, they are not in a condition that is readily conducive to germplasm exchange because of the great risks of disease transfer through the exchange of vegetative material. Field genebanks are costly to maintain and, as a consequence, are prone to economic decisions that may limit the level of replication of accessions, the quality of maintenance and even their survival in times of economic stringency. Even under the best circumstances, field genebanks require considerable inputs in the form of land (often needing multiple sites to allow for rotation), labour, management and materials and, in addition, their capacity to ensure the maintenance of much diversity is limited.

In vitro *techniques for collection, multiplication and storage*

Tissue culture techniques are of great interest for the collection, multiplication and storage of plant germplasm (Engelmann, 1997). For the collection of species that produce recalcitrant seeds, and of vegetatively propagated material, techniques have been developed which enable a collector to introduce the material *in vitro*, under aseptic conditions, directly in the field (Withers, 1995). This approach will allow germplasm collections to be made in remote areas (e.g. in the case of highly recalcitrant cacao seeds), or when the transport of the collected fruits would become prohibitively expensive (e.g. collecting coconut germplasm). Also in cases where the target species does not have seeds or other storage organs to be collected, or when budwood would quickly lose viability or is highly contaminated, the establishment of aseptic cultures in the field will facilitate collecting and improve its efficiency.

Tissue culture systems allow propagation of plant material with high multiplication rates in an aseptic environment. During the last 30 years, *in vitro* propagation techniques, mainly based on micropropagation and somatic embryogenesis have been extensively developed and applied to well over 1000 different species (George, 1993a, b). Virus-free plants can be obtained through meristem culture in combination with thermotherapy, thus ensuring the production of disease-free stocks and simplifying quarantine procedures for the international exchange of germplasm. The miniaturization of explants allows a reduction in space requirements and consequently labour costs for the maintenance of germplasm collections (Ashmore, 1997).

Different *in vitro* conservation methods are employed, depending on the storage duration required (Engelmann, 1997; Withers and Engelmann, 1998). For short- and medium-term storage, various techniques have been devised that allow reduction of growth and increase the intervals between subcultures. *In vitro* conservation techniques using slow growth storage have been developed for a wide range of species, including temperate woody plants, fruit trees, horticultural species, as well as numerous tropical species. However, despite the availability of such techniques, the *Report on the State of the World's Plant Genetic Resources for Food and Agriculture* (FAO, 1996) indicates that only around 38,000 accessions are conserved *in vitro* worldwide, because many conservation programmes are unable to meet requirements for relatively sophisticated equipment, reliable electricity supply and trained staff. In addition, only a limited amount of genetic diversity can be maintained *in vitro*. Slow growth storage is used routinely in a limited number of national, regional and international germplasm conservation centres with a few species including banana, some root and tuber crops, and temperate fruits (Engelmann, 1999a).

For long-term storage, cryopreservation, that is storage at ultra-low temperature, usually that of liquid nitrogen (−196°C), is employed. At this temperature, all cellular divisions and metabolic processes are stopped. The plant material can thus be stored without alteration or modification for a theoretically unlimited period of time. Moreover, cultures are stored in a small volume, are protected from contamination, and require very limited maintenance. It is essential to recognize that, due to the various problems and limitations encountered with both protected areas and field genebanks (Withers and Engels, 1990; Maxted *et al.*, 1997), cryopreservation currently offers the only safe and cost-effective option for the long-term conservation of genetic resources of problem species.

Concerning *in vitro* cultured material, cryopreservation protocols are now available for cell suspensions, calli, apices, zygotic and somatic embryos of several hundreds of species of temperate and tropical origin (Kartha and Engelmann, 1994; Engelmann, 1997; Engelmann and Dussert, 2000). Most of this work has been performed within the framework of academic studies and has involved only one or a few genotypes. However, due to the development in the last 3–4 years of new cryopreservation procedures for apices and embryos (encapsulation-desiccation, desiccation, pregrowth-desiccation and vitrification), reports involving a larger number of genotypes/varieties are becoming more frequent (Benson, 1999; Engelmann and Takagi, 2000). These new freezing procedures generally lead to satisfactory survival rates with a wide range of genotypes using the same technique. There is an increasing number of cases where techniques can be considered operational on a routine and large-scale basis. However, cryopreservation is significantly more advanced for vegetatively propagated crops than for recalcitrant seed species (Engelmann, 1999b). In general, there is only a limited number of cases where cryopreservation is currently used in a genetic resources conservation context. Examples include *in vitro* shoot tips of potato, cassava, *Musa* and pear (Schäfer-Menuhr *et al.*, 1997; Escobar *et al.*, 2000; Panis *et al.*, 2000; Reed, 2000), seeds of some short-lived or endangered orthodox species (Stanwood, 1985), dormant buds of various tree species (Sakai, 1995), and pollen of some horticultural species (Ganeshan and Rajashekaran, 2000).

Management of germplasm collections

No comprehensive, independent review of genebank facilities has been made to date, except for that performed recently in the genebanks of the centres of the CGIAR (SGRP, 1996). However, it is evident that a limited number of genebanks operate at very high standards, whereas many others are at present only capable of performing the basic conservation role of a genebank to a limited extent (FAO, 1996). The main difficulties faced by those genebanks are due to a number of factors, including inadequacy of existing infrastructures, lack of adequate equipment such as cooling units, seed drying and cleaning devices, unreliability of electricity supply, funding and staffing constraints, and inadequate management practices. As a result, seeds are often stored under sub-optimal conditions and need to be regenerated more frequently, thus imposing additional costs on the, often, already insufficient operating budgets of the genebanks. Very important difficulties faced in particular with the regeneration of the seed collections have been mentioned by many countries in the *Report on the State of the World's Plant Genetic Resources for Food and Agriculture* (FAO, 1996).

In both seed and field genebanks, the technologies for storing germplasm samples are relatively easy to apply under most operational circumstances. The problems relate more to resource constraints that affect the performance of essential operations. This is critical in the case of core activities, of maintaining the viability and genetic integrity of the stored accessions, as well as sufficient stocks, to meet user demands (Engels, 2001). Furthermore, budgets for activities of a public nature such as conservation are permanently shrinking and, consequently, the importance of efficient and cost-effective genebank management has increased over the years and has become a decisive element in the long-term *ex situ* conservation of plant genetic resources.

Other important aspects of genebank operations concern the characterization and documentation of germplasm. Characterization descriptors concern the highly heritable characters that are independent of the environment; for example taxonomic characters, in contrast to evaluation descriptors that relate mainly to traits of agronomic importance that are often highly environment- or method-specific. Characterization of accessions provides essential information for genebank management as well as for plant breeding. The extent to which germplasm collections have been characterized varies widely between genebanks and species (FAO, 1996), but is far from complete in many instances. A study performed in 1987 by Plucknett *et al.* estimated that 80 to 95% of the world collections lacked characterization or evaluation data. This situation has not substantially improved since that time and has led to a limited use of *ex situ* conserved germplasm.

As regards the extent to which *ex situ* accessions are documented, the situation is highly contrasted (FAO, 1996). Some, mainly developed, countries have fully computerized documentation systems and relatively complete accession data, while many others lack information on the accessions in their collections, including the so-called passport data. Another important aspect at the global level is the lack of integrated, compatible systems that would allow for easy exchange of information.

Main Challenges and Priorities for Improving *ex situ* Conservation of Plant Genetic Resources

Conservation research

Priorities for research should focus on improving our understanding of relevant biological mechanisms directly related to conservation as well as on developing improved conservation techniques. Fundamental research includes studies on seed desiccation sensitivity and recalcitrance, on the determination of critical seed moisture content and its interaction with temperature. Technologies for conserving seed, tissue and other plant parts need further development and particular attention should be given to the problem species identified previously, that is, those which produce recalcitrant or intermediate seeds. In addition, several research topics for orthodox seed conservation have been identified and are treated below.

Orthodox seed species

Determining critical seed moisture content

The preferred conditions recommended for long-term seed storage are 3–7% moisture content, depending on the species, at −18°C or lower (FAO and IPGRI, 1994). In the late 1980s, research was initiated to investigate the effects of very low moisture content on seed longevity, based on the assumption that further reduction in moisture content would have a beneficial effect on seed longevity. One of the aims of such research was the prospect of developing 'low-input' alternatives to medium- to long-term cold storage of seed germplasm through its storage at room temperature (i.e. ultra-dry storage). However, it was demonstrated that drying seeds beyond a critical moisture content provided no additional benefit to longevity and may even accelerate seed ageing rates (Ellis *et al.*, 1988; Vertucci and Roos, 1993). Further research on the determination of optimal seed moisture content led to a debate regarding the various parameters involved (Walters and Engels, 1998). In particular, questions were raised with regard to the critical relative humidity, storage temperature and their interaction, as well as the consequences of equilibrating seeds to levels of relative humidity lower than the critical level. The latter was prompted by the finding that the optimal seed moisture content increased as the storage temperature was lowered, suggesting that there may be a danger of actually over-drying seed (Vertucci and Roos, 1993).

A collaborative research project involving the National Seed Storage Laboratory at Fort Collins (USA), the International Crops Research Institute for the Semi-Arid Tropics (ICRISAT) at Hyderabad, India, and the Institute of Crop Germplasm Resources of the Chinese Academy of Agricultural Sciences (CAAS), Beijing, was thus initiated to investigate several aspects related to the interaction between storage temperature and optimal seed moisture content for lettuce seed as a model. Seeds at moisture contents between 0.0 and 0.10 g H_2O g^{-1} dw (i.e. 0 to 9% moisture content, fresh weight basis) were stored at −18°C, 20, 35 and 50°C. After more than 3 years of storage, it appeared that seeds stored at 50°C showed deterioration for all water contents, but ageing was slowest in seeds with water contents lower than 0.018 g H_2O g^{-1} dw (i.e. 1.77% moisture content, fwb) (Rao *et al.*, 1999). This suggests that a critical water content exists at this or a lower water content. Seeds stored at 35°C showed an optimal water content between 0.020 and 0.035 g H_2O g^{-1} dw (i.e. between 1.9 and 3.4% moisture content, fwb). At 20°C, the optimal water content lay between 0.019 and 0.055 g H_2O g^{-1} dw (1.8 and 5.2% moisture content, fwb). From these results, it can be concluded that the optimal water content for storing lettuce seeds at room temperature, that is, around 25°C in tropical countries, lies somewhere between 0.019 and 0.055 g H_2O g^{-1} dw (1.8 and 5.2% moisture content, fwb). The former moisture content is lower than those typically used in conventional conservation practices. These values will be made progressively more accurate as the storage experiment progresses. However, storage at −20°C (at either moisture content) was far superior to storage at 20°C, since no deterioration was observed after more than 3 years of storage.

Developing low-input storage techniques

Various research projects have focused on the development of the ultra-dry seed technology, which would allow the storage of seed germplasm at room temperature, without the need for refrigeration equipment. The current state-of-the-art in this research area has been assessed in a Satellite Symposium on ultra-dry seed storage and longevity organized during the Second International Conference on Seed Science and Technology that took place in Guangzhou, China in 1997. From the papers presented during this symposium, which have been published in 1998 as a Supplement of the journal *Seed Science Research*, it can be concluded that although drying seed to very low moisture content prior to storage seems to have fewer advantages than was initially expected, ultra-dry storage is still considered to be a useful practical, low-cost technique in those circumstances where no adequate refrigeration can be provided. Additional research on various aspects of the ultra-dry seed storage technology, including, notably, drying techniques such as sun/shade drying (Hay and Probert, 2000), vacuum drying or freeze drying (Côme, 1983) and on its applicability to a broader number of species should therefore be continued.

Improving and monitoring viability

The viability of conserved accessions depends on their initial quality and how they have been processed for storage, as well as on the actual storage conditions. In the case of seed, there is evidence that a very small decrease in initial viability can mean a substantial reduction in storage life. This effect needs further investigation and quantification. To date, relatively little attention has been given to the handling of germplasm in the field during regeneration and during subsequent processing stages prior to its arrival at the genebank. Poor practices during these initial stages may be jeopardizing the effort and expense made on providing optimum conditions for the storage of the germplasm. Factors that affect initial viability include, among others, the growing conditions, disease status and time of harvest of the plants. Kameswara Rao and Jackson (1996) demonstrated that seeds of three *japonica* rice cultivars that were produced in cooler environments had a relatively higher potential longevity. Maximum potential seed longevity of these rice cultivars coincided with mass maturity and did not decline afterwards. In contrast, experiments performed with seeds of barley, pepper and tomato indicated that maximum potential longevity was not attained until some time after mass maturity (Ellis *et al.*, 1993). In barley, potential longevity declined after the maximum had been achieved, whereas in pepper and tomato, no substantial decline in potential longevity was detected after the maximum had been attained.

Further research should also be aimed at optimizing the growing conditions and minimizing the detrimental effects of diseases. This requires comparative investigation of the effects of seed-borne diseases and seed treatments on viability and genetic stability during storage as well as on subsequent growing out of plants (Kaiser and Hannan, 1986).

Monitoring the viability of collections during storage necessitates regular germination tests that require a substantial amount of seeds of valuable germplasm. Research into non-destructive methods of measuring viability, explored the use of triphenyl tetrazolium chloride (TTC) to detect viability loss and seed leachate conductivity as an indicator of membrane damage and seed degeneration. Reports indicate that measures of seed vigour might be earlier indicators of seed deterioration than viability (E. Roos, Colorado, personal communication). This is an important aspect for further investigation since retaining the genetic integrity of the stored sample depends on early detection of seed deterioration and adequate regeneration of the sample. For example, undetected death in part of an accession can induce genetic shift.

Non-orthodox seed species

Understanding seed recalcitrance

A number of processes or mechanisms have been suggested to confer, or contribute to, desiccation tolerance (see Pammenter and Berjak, 1999, for a review). Different processes may confer protection against the consequences of loss of water at different hydration levels, and the absence, or ineffective expression, of one or more of these could determine the relative degree of desiccation sensitivity of seeds of individual species. The mechanisms that have been implicated to date include intracellular physical characteristics such as the amount of insoluble reserves accumulated and the degree of vacuolation; intracellular de-differentiation; 'switching-off' of metabolism; presence and efficient operation of antioxidant systems; accumulation of putatively protective molecules such as late embryogenic abundant proteins and various sugars; development of certain amphipathic molecules; an effective peripheral oleosin layer around lipid bodies; and the presence and operation of repair mechanisms during rehydration. It is likely that additional mechanisms will be identified in time. This information has been gathered from a number of independent studies performed on different materials. A comprehensive research programme incorporating all the above elements, focused on a well-identified biological model, such as coffee, which includes species with a wide range of desiccation sensitivities (Dussert *et al.*, 1998), would allow us to make significant progress in the understanding of recalcitrance.

Developing improved conservation techniques

There are various technical options to consider for improving storage of recalcitrant seeds. First of all, especially with species for which no or only a little information is available, it is advisable, before undertaking any 'high tech' research, to examine the development pattern of seeds and to run preliminary experiments to determine their desiccation sensitivity as well as to define germination and storage conditions. The International Plant Genetic Resources Institute (IPGRI), in collaboration with numerous institutions worldwide, has developed a protocol for screening tropical forest tree seeds for their desiccation sensitivity and storage behaviour (IPGRI/DFSC, 1999), which might be applicable to seeds of other species, after required modification and adaptation. Even if such experiments do not allow definition of long-term storage conditions, any increase in the storage duration achieved by conventional means can have beneficial consequences; for example by keeping seeds alive until the next planting season or to transport recalcitrant seeds from the forest to a prepared planting site. Another medium-term option can be to conserve young seedlings arrested in their development by storage at low temperature and/or under low light intensity (Hawkes, 1980) as experimented with *Symphonia globulifera* and *Dryobalanops aromatica* (Corbineau and Côme, 1986; Marzalina *et al.*, 1992).

For long-term storage, it is clear that cryopreservation, often coupled with *in vitro* culture, represents the only option. With some species such as tea, mahogany, neem or coffee, seeds are relatively small and tolerant to desiccation, and can thus be cryopreserved directly after partial desiccation under the laminar flow (Hu *et al.*, 1994; Marzalina, 1995; Berjak and Dumet, 1996; Dussert *et al.*, 1997). With other species which are more desiccation sensitive, very precisely controlled desiccation and cooling conditions may also allow freezing of whole seeds, as demonstrated recently with several coffee species (Dussert *et al.*, 1998).

In cases where seeds are not amenable to cryopreservation, excised embryos or embryonic axes should be used. The desiccation technique has been employed preferentially for freezing such materials and there is scope for various technical improvements in the current cryopreservation protocols. Pre-growth of embryos on media containing cryoprotective substances may confer tissues with increased tolerance to further desiccation and reduce the heterogeneity of the material. Berjak and co-workers have demonstrated that flash drying, followed by ultra-rapid freezing has been very effective for cryopreservation of several species such as *Landolphia kirkii* and tea (Berjak *et al.*, 1989; Wesley-Smith *et al.*, 1992). The hypothesis of these researchers is that very rapid dehydration imposes a stasis on metabolism and precludes the deleterious reactions that would take place under lower desiccation rates and that ultra-rapid freezing induces vitrification of internal solutes or the formation of ice crystals too small to disrupt cellular integrity. Even though some species have proven far too desiccation sensitive to be cryopreserved this way (Pammenter *et al.*, 1993), this is a potentially interesting approach which deserves further research and experimentation with additional species. Other cryopreservation techniques, including pregrowth-desiccation, encapsulation-dehydration and vitrification, which have been seldom employed so far, should be tested (Engelmann, 1997). Finally, it should be emphasized that selecting embryos at the right developmental stage, which will vary from one species to the other, is of critical importance for the success of any cryopreservation experiment (Engelmann *et al.*, 1995). However, in the above cases, basic protocols for disinfection, inoculation *in vitro*, germination of embryos or embryonic axes, plantlet development, and possibly limited propagation will have to be established prior to any cryopreservation experiment.

With species for which attempts to freeze whole embryos or embryonic axes have proven unsuccessful, various authors have suggested using shoot apices sampled from embryos, adventitious buds or somatic embryos induced from embryonic tissues (Pence, 1995; Berjak *et al.*, 1996). This might be the only solution for those species that do not have well-defined embryos but this will require that more sophisticated tissue culture procedures are developed and mastered. In addition to these technical difficulties, using adventitious explants would reduce the range of genetic variability captured (Pence, 1995; Berjak *et al.*, 1996), especially when using somatic embryogenesis, since response to inducing treatments is generally highly genotype-specific and somatic embryo cultures might be obtained from a limited number of genotypes only. In cases where apices are employed, it might be more practical and efficient to sample them on *in vitro* plantlets rather than on embryos to reduce the risks of contamination and to use more homogeneous material. Finally, cryopreservation of pollen may represent an additional option for genetic resource conservation of difficult material (Towill and Walters, 2000).

Materials other than seed

Pollen usually has a relatively short viability when conserved under classical storage conditions (partial desiccation followed by storage at sub-zero temperature), and has therefore been used only to a limited extent in germplasm conservation (Hoekstra, 1995). However, long-term storage of pollen is feasible using cryopreservation (Towill and Walters, 2000). Cryogenic procedures have been established for pollen of a large number of species, including mainly desiccation-tolerant pollen but also several desiccation-sensitive ones (Hanna and Towill, 1995). Cryopreserved collections of pollen have already been established for several crops, including horticultural and fruit tree species (Ganeshan and Rajashekaran, 2000).

Despite these recent successes using pollen for conservation purposes, several disadvantages of pollen storage should be mentioned. The small amount produced by many species; the lack of transmission of organelle genomes via pollen; the loss of sex-linked genes in dioecious species; and the limited plant regeneration capacity are the most important ones. While considering the advantages and disadvantages of pollen conservation, it can be concluded that storing pollen, in addition to other materials, is useful in the framework of the establishment of complementary conservation strategies (Withers, 1991), and is of special interest for species that produce recalcitrant seeds. This is particularly relevant since there is no correlation between seed storage behaviour for a given species and the desiccation sensitivity of its pollen (Hoekstra, 1995). Furthermore, transfer of pests and diseases through pollen is rare (except for some virus diseases), thus allowing the safe movement and exchange of germplasm in the form of pollen.

DNA storage is rapidly increasing in importance. DNA from the nucleus, mitochondrion and chloroplasts are now routinely extracted and immobilized into nitro-cellulose sheets where the DNA can be probed with numerous cloned genes. With the development of PCR one can now routinely amplify specific oligonucleotides or genes from the entire mixture of genomic DNA. These advances have led to the formation of an international network of DNA repositories for the storage of genomic DNA (Adams, 1997). The advantage of this technique is that it is efficient and simple and overcomes physical limitations or constraints. The disadvantage lies in problems with subsequent gene isolation, cloning and transfer (Maxted *et al.*, 1997).

Germplasm management procedures and strategies

The objective of any genebank management procedure is to maintain genetic integrity and viability of the accessions during conservation and to assure their accessibility for use in adequate quantity and quality at the lowest possible cost. As already mentioned in this chapter, genebanks very often face financial constraints that hamper their efficient operation. It is therefore very important to improve genebank management procedures to make them more efficient and cost-effective. Another important problem is that the existing concepts and procedures are biased towards conservation of orthodox seeds, and therefore are often not adapted to the conservation of recalcitrant seed and vegetatively propagated species. Finally, new technologies including *in vitro* culture and conservation, cryopreservation, molecular techniques, data processing software and multimedia communication technology have been developed over recent years, and they should be integrated to the extent possible, when relevant, in genebank management procedures. Special attention should be given to the development of appropriate techniques for genebanks in developing countries where specialized equipment is frequently lacking and resources are usually limited. The following section examines how these new factors might modify some important genebank management procedures and concepts.

Exploration and collection

In vitro collecting procedures, which have been established for a number of recalcitrant and vegetatively propagated species, allow improvement of the efficiency of collecting missions (Withers, 1995). They should be used more systematically for collecting germplasm of relevant species. With this aim, IPGRI is preparing a technical bulletin on the utilization of *in vitro* collecting techniques (Pence *et al.*, 2000). An understanding of the extent and distribution of diversity within a population is essential for effective sampling. The use of molecular techniques in studying genetic diversity has contributed to a better understanding of the genetic diversity of numerous species. Ecogeographic surveys provide information on species distribution as well as intraspecific diversity. IPGRI is preparing a technical bulletin on ecogeographic surveys for wild species (Guarino *et al.*, 2000). Molecular techniques can be applied during such surveys for a proper assessment of the genetic diversity patterns, which would then permit a more effective sampling of a particular region (Rao and Riley, 1994).

Characterization and evaluation

Molecular techniques have very important applications for the characterization and evaluation of plant genetic resources, to complement the morphological and biochemical descriptors used classically. More specifically, these techniques can be used for identifying genotypes, including duplicate accessions; fingerprinting genotypes; analysing genetic diversity in collections; and assembling a core collection. More details on the various techniques available and on their applicability for various purposes can be found in two publications on the use of molecular techniques for plant genetic resources (Ayad *et al.*, 1997; Karp *et al.*, 1997).

Seed regeneration procedures

Regeneration procedures carry risks that may compromise the genetic integrity of the samples. Some of these risks are inherent in the sample selection procedures followed, including the consequences of the size of the population grown out. Other risks include the effects of diseases, pests and abiotic stresses that regenerated samples may be exposed to. Therefore, collections must be managed with the aim of minimizing the frequency of regeneration. Development of appropriate strategies towards this end is recognized as a priority area for continued investigation. In addition, further research is needed in order to evaluate the level and causes of genetic shift and drift during regeneration. Improvement of cultivation procedures is also necessary in order to minimize the risks of such genetic changes, including enhancement of pollination, reduction of plant competition and determination of optimal population sizes. Guidelines aiming at facilitating the development of optimum procedures for regeneration of seed germplasm have been developed (Sackville Hamilton and Chorlton, 1997).

Accession management

The number of accessions in a genebank has a direct impact on operational costs. The number of accessions can be reduced by lumping or discarding accessions. Splitting accessions will increase the number of accessions but may also facilitate their management and improve the quality of genebank operations. However, before taking such decisions, their impact at the genetic, operational and economic levels should be carefully evaluated (van Hintum *et al.*, Chapter 11, this volume).

Management of *in vitro* collections

In the case of *in vitro* collections, the management problems relate to slow growth storage where maintaining the genetic stability of the cultures and minimizing the workload of subculturing, are objectives. The procedures worked out for managing cassava under slow growth (IPGRI/CIAT, 1994) need to be expanded to other collections of cassava and downstreamed for wider application to other species. Towards this aim, IPGRI is currently preparing guidelines that will assist genebank curators in the management of field and *in vitro* germplasm collections (Reed *et al.*, 2000). Some of the important managerial issues addressed in this document concern the need for safety duplication of the collections and for the establishment of linkages between field and *in vitro* collections.

Important considerations for a wider application of *in vitro* culture techniques to conservation of plant genetic resources will be the reproducibility and flexibility of procedures, and experiments should be performed to test these parameters. Some crops, such as banana, plantain and sugarcane, are prone to somaclonal variation, and this phenomenon is accentuated when the material is grown *in vitro*. It is therefore essential to develop tools allowing detection of somaclonal variants in the germplasm collections at the earliest stage possible. With this aim, specific molecular probes may be designed and used to screen the *in vitro* collections, as envisaged for *Musa* (Côte *et al.*, 1993; Damasco *et al.*, 1996).

Enhancing use of collections

Besides efficiency considerations of where to store germplasm accessions and under which conditions, genebank curators might want to compose subsets of accessions from a given collection to promote their utilization. Whether a genebank needs to focus on specific traits rather than on neutral properties will depend on the needs of potential users and their assessment will require adequate interaction. In this context, the establishment of core collections will be a powerful tool for efficient and cost-effective germplasm collection management (Hodgkin *et al.*, 1995; Johnson and Hodgkin, 1999). Other aspects of how collection management can contribute to improved use of germplasm accessions can be found in Engels (2001).

Germplasm health aspects

Germplasm health problems affect the collection, conservation, utilization and distribution of plant genetic resources and are thus among the critical constraints to the management of plant germplasm collections. Although many genebanks have plant protection specialists, their activities are generally linked to plant breeding and crop protection and not necessarily to germplasm management. It is therefore very important to better integrate germplasm health aspects, including new biotechnological tools for detection, indexing and eradication of pathogens in routine genebank operations. IPGRI is currently preparing technical guidelines to facilitate this process (Morales, unpublished).

Documentation

With the increased impact of globalization and of the political importance of plant genetic resources the role and importance of information on germplasm has grown dramatically. In fact, the biological material itself is of limited use without the corresponding information that has, in turn, increased the need for more substantial documentation. Most of the above-mentioned routine genebank operations generate information that is key to the efficient functioning of the genebank and safe and efficient conservation. Many genebanks have computerized documentation systems, which greatly facilitate the storage and maintenance of data, as well as their retrieval. Most systems can also be used for the processing and analysis of the data and, thus, be instrumental in facilitating routine operations. A helpful overview of the various aspects of genebank documentation can be found in the *Guidebook for Genetic Resources Documentation* (Painting *et al.*, 1993). The descriptor lists, developed and produced by IPGRI, are a principal tool to assist curators to better document their collections and to contribute to a better standardization of the resulting information.

Active/base collection concept

When the optimum seed moisture content is known, and the seeds have been packaged, they should be stored at the best available temperature. The Genebank Standards recommends a preferred temperature of −18°C or below for the storage of base collections (FAO and IPGRI, 1994). Temperatures above zero are acceptable for active collections. This two-tiered storage concept of the base collection for long-term storage and active collection for accessions which are frequently used, is largely based on experience with the storage of orthodox cereal seeds, and needs to be adapted to non-orthodox and vegetatively propagated species. It is further based on the assumption that there will be a considerable turnover of germplasm stored in the active collection. These assumptions and conditions might not always apply and/or the genebank might not have two or more storage rooms operating at different temperatures.

A recent analysis of this storage concept concluded that it might be more cost-effective for genebanks to store only those accessions in the active collection which are actually being used by breeders or other users (N.R. Sackville Hamilton, J.M.M. Engels and Th.J.L van Hintum, unpublished). This would justify the maintenance of subsamples of the same accession in a separate room under less stringent conditions. In this same analysis the concept of the 'most original sample' (MOS) has been introduced. An MOS consists of seeds of a given accession that have undergone the lowest number of regenerations since the material was collected or donated to the genebank. It is recommended that the MOS is stored under the best possible conditions, never used for distribution purposes, and stored in the genebank that has accepted national or global responsibilities for its conservation. The minimum sample size of the MOS should be determined by the amount of seed required to regenerate the MOS, the seed required for viability testing, and that needed to regenerate material for distribution. For additional security reasons a second, but smaller sub-sample of the same material (the 'secondary MOS') should be stored at a distant genebank as a back-up. This sample should be maintained under a black-box arrangement and stored under conditions which are at least as good as those used for the primary MOS.

Conclusion

In this chapter, we have presented the main achievements in and problems with the *ex situ* conservation of plant genetic resources, and identified the main priorities to improve the conservation and management of germplasm collections. During recent years, dramatic progress has been made with the development of new conservation techniques for non-orthodox and vegetatively propagated species, and the current *ex situ* conservation concepts should be modified accordingly to accommodate these technological advances. The incorporation of other techniques, such as molecular marker technologies in genebank operational procedures, should have a positive impact on the efficiency, security and cost-effectiveness of *ex situ* conservation and utilization of plant genetic resources.

Considering the fact that the requirements for optimal conservation vary from species to species, as well as the available infrastructural and human resources, it is important to consider all these aspects as well as the wider socio-economic conditions under which a given conservation effort takes place when deciding how to optimize these parameters into the conservation strategy. It is now well recognized that an appropriate conservation strategy for a particular plant genepool requires a holistic approach, combining the different *ex situ* and *in situ* conservation techniques available in a complementary manner. *In situ* and *ex situ* methods, including a range of techniques for the latter, are options available for the different genepool elements (i.e. cultivated species, including landraces and modern varieties, wild relatives, weedy types, etc.). Selection of the appropriate method should be based on a range of criteria, including the biological nature of the species in question, practicality and feasibility of the particular method chosen (which depends on the availability of the necessary infrastructure) as well as the cost-effectiveness and security afforded by its application (Maxted *et al.*, 1997). Considerations of complementarity with respect to the efficiency and cost-effectiveness of the various conservation methods chosen are also important. In many instances, the development of appropriate complementary conservation strategies requires further research to define the criteria, refine the method and test its application for a range of genepools and situations. An important area in this is the linkage between *in situ* and *ex situ* components of the strategy, especially with respect to the dynamic nature of the former and the static, but potentially more secure approach, of the latter.

Such research can be effectively carried out through collaborative studies, involving fundamental and applied research organizations within countries as well as through close cooperation with international institutions concerned with conservation research. In particular, the establishment of research networks can be an efficient way to cope with the enormous range of plant species on which little or no information is available.

References

Adams, R.P. (1997) Conservation of DNA: DNA banking. In: Callow, J.A., Ford-Lloyd, B.V. and Newbury, H.J. (eds.) *Biotechnology and Plant Genetic Resources Conservation and Use.* CAB International, Wallingford, UK, pp. 163–174.

Ashmore, S. (1997) *Status Report on the Development and Application of* in vitro *Techniques for the Conservation and Use of Plant Genetic Resources.* Engelmann, F. (vol. ed.). International Plant Genetic Resources Institute, Rome, Italy.

Ayad, W.G., Hodgkin, T., Jaradat, A. and Rao, V.R. (eds) (1997) *Molecular Genetic Techniques for Plant Genetic Resources. Report of an IPGRI Workshop, 9–11 October 1995, Rome.* International Plant Genetic Resources Institute, Rome, Italy.

Benson, E.E. (ed.) (1999) *Plant Conservation Biotechnology.* Taylor & Francis, London.

Berjak, P. and Dumet, D. (1996) Cryopreservation of seeds and isolated embryonic axes of neem (*Azadirachta indica*). *Cryo-Letters* 17, 99–104.

Berjak, P. and Pammenter, N.W. (1994) Recalcitrance is not an all-or-nothing situation. *Seed Science Research* 4, 263–264.

Berjak, P., Farrant, J.M., Mycock, D.J. and Pammenter, N.W. (1989) Homoiohydrous (recalcitrant) seeds: the enigma of their desiccation sensitivity and the state of water in axes of *Landolphia kirkii* Dyer. *Planta* 186, 249–261.

Berjak, P., Mycock, D., Wesley-Smith, J., Dumet, D. and Watt, P. (1996) Strategies for *in vitro* conservation of hydrated germplasm. In: Normah, M.N., Narimah, M.K. and Clyde, M.M. (eds) In Vitro *Conservation of Plant Genetic Resources.* Plant Biotechnology Laboratory, Faculty of Life Sciences, Universiti Kebangsaan Malaysia, Kuala Lumpur, Malaysia, pp. 19–52.

Chin, H.F. (1988) *Recalcitrant Seeds: a Status Report.* International Plant Genetic Resources Institute, Rome, Italy.

Chin, H.F. and Roberts, E.H. (eds) (1980) *Recalcitrant Crop Seeds.* Tropical Press Sdn. Bhd., Kuala Lumpur, Malaysia.

Côme, D. (1983) Post harvest physiology of seeds as related to quality and germinability. In: Lieberman, M. (ed.) *Post-harvest Physiology and Crop Preservation.* Plenum Press, New York, pp. 165–190.

Corbineau, F. and Côme, D. (1986) Experiments on the storage and seedlings of *Symphonia globulifera* L.f. (Guttiferae). *Seed Science Technology* 14, 585–591.

Côte, F.X., Sandoval, J., Marie, P. and Auboiron, E. (1993) Variations in micropropagated bananas and plantains: literature survey. *Fruits* 48, 15–23.

Cromarty, A.S., Ellis, R.H. and Roberts, E.H. (1982) *The Design of Seed Storage Facilities for Genetic Conservation.* Handbooks for Genebanks No. 1. International Board for Plant Genetic Resources, Rome, Italy.

Damasco, O.P., Graham, G.C., Henry, R.J., Adkins, S.W. and Smith, M.K. (1996) Random amplified polymorphic DNA (RAPD) detection of dwarf off-types in micropropagated Cavendish (*Musa* spp. AAA) bananas. *Plant Cell Reports* 16, 118–123.

Dussert, S., Chabrillange, N., Engelmann, F., Anthony, F. and Hamon, S. (1997) Cryopreservation of coffee (*Coffea arabica* L.) seeds: importance of the precooling temperature. *Cryo-Letters* 18, 269–276.

Dussert, S., Chabrillange, N., Engelmann, F., Anthony, F. and Hamon, S. (1998) Cryopreservation of seeds of four coffee species (*Coffea arabica, C. costatifructa, C. racemosa* and *C. sessiliflora*): importance of seed water content and of freezing rate. *Seed Science Research* 8, 9–15.

Ellis, R.H., Hong, T.D. and Roberts, E.H. (1985a) *Handbook of Seed Technology for Genebanks* Vol. I: *Principles and Methodology.* Handbooks for Genebanks No. 2. International Board for Plant Genetic Resources, Rome, Italy.

Ellis, R.H., Hong, T.D. and Roberts, E.H. (1985b) *Handbook of Seed Technology for Genebanks* Vol. II: *Compendium of Specific Germination Information and Test Recommendations.* Handbooks for Genebanks No. 3. International Board for Plant Genetic Resources, Rome, Italy.

Ellis, R.H., Hong, T.D. and Roberts, E.H. (1988) A low-moisture-content limit to the logarithmic relation between seed moisture and longevity in twelve species. *Annals of Botany* 63, 601–611.

Ellis, R.H., Hong, T.D. and Roberts, E.H. (1990) An intermediate category of seed storage behaviour? I. Coffee. *Journal of Experimental Botany* 41, 1167–1174.

Ellis, R.H., Hong, T., Roberts, E.H. and Soetisna, U. (1991) Seed storage behaviour in *Elaeis guineensis. Seed Science Research* 1, 99–104.

Ellis, R.H., Demir, I. and Pieto Filho, C. (1993) Changes in seed quality during seed development in contrasting crops. In: Côme, D. and Corbineau, F. (eds) *Fourth International Workshop on Seeds, Basic and Applied Aspects of Seed Biology,* Vol. 3. ASFIS, Paris, France, pp. 897–904.

Engelmann, F. (1997) *In vitro* conservation methods. In: Ford-Lloyd, B.V., Newburry J.H. and Callow, J.A. (eds) *Biotechnology and Plant Genetic Resources: Conservation and Use.* CAB International, Wallingford, UK, pp. 119–162.

Engelmann, F. (ed.) (1999a) Management of field and *in vitro* germplasm collections. *Proceedings of a Consultation Meeting, 15–20 January 1996, CIAT, Cali, Colombia.* International Plant Genetic Resources Institute, Rome, Italy.

Engelmann, F. (1999b) Alternative methods for the storage of recalcitrant seeds – an update. In: Marzalina, M., Khoo, K.C., Tsan, F.Y. and Krishnapillay, B. (eds) *IUFRO Seed Symposium 1998 'Recalcitrant Seeds', 12–15 October 1998, Kuala Lumpur, Malaysia.* FRIM (Forest Research Institute Malaysia), Kepong, Malaysia, pp. 159–170.

Engelmann, F. and Dussert, S. (2000) Current development and use of cryopreservation for the conservation of plant genetic resources. *Cahiers Agricultures* 9, 237–245.

Engelmann, F. and Takagi, H. (eds) (2000) *Cryopreservation of Tropical Plant Germplasm – Current Research Progress and Applications.* Japan International Centre for Agricultural Sciences, Tsukuba/International Plant Genetic Resources Institute, Rome, Italy.

Engelmann, F., Dumet, D., Chabrillange, N., Abdelnour-Esquivel, A., Assy-Bah, B., Dereuddre, J. and Duval, Y. (1995) Cryopreservation of zygotic and somatic embryos from recalcitrant and intermediate-seed species. *IPGRI/FAO Plant Genetic Resources Newsletter* 103, 27–31.

Engels, J.M.M. (2001) Genebank management: an essential activity to link conservation and plant breeding. *Plant Genetic Resources Newsletter* 125 (in press).

Engels, J.M.M. and Wood, D. (1999) Conservation of agrobiodiversity. In: Wood, D. and Lenné, J.M. (eds) *Agrobiodiversity: Characterisation, Utilisation and Management.* CAB International, Wallingford, UK, pp. 355–386.

Escobar, R.H., Debouck, D. and Roca, W.M. (2000) Development of cassava cryopreservation. In: Engelmann, F. and Takagi, H. (eds.) *Cryopreservation of Tropical Plant Germplasm – Current Research Progress and Applications.* Japan International Centre for Agricultural Sciences, Tsukuba/International Plant Genetic Resources Institute, Rome, Italy, pp. 222–226.

FAO (1996) *Report on the State of the World's Plant Genetic Resources for Food and Agriculture.* Food and Agriculture Organization of the United Nations, Rome, Italy.

FAO and IPGRI (1994) *Genebank Standards.* Food and Agriculture Organization of the United Nations, Rome, International Plant Genetic Resources Institute, Rome, Italy.

Ganeshan, S. and Rajashekaran, R.K. (2000) Current status of pollen cryopreservation research: relevance to tropical horticulture. In: Engelmann, F. and Takagi, H. (eds) *Cryopreservation of Tropical Plant Germplasm – Current Research Progress and Applications.* Japan International Centre for Agricultural Sciences, Tsukuba/International Plant Genetic Resources Institute, Rome, Italy, pp. 360–365.

George, E.F. (ed.) (1993a) *Plant Propagation by Tissue Culture.* Part 1 *The Technology,* 2nd edn. Exegetics Ltd, Edington, UK.

George, E.F. (ed.) (1993b) *Plant Propagation by Tissue Culture.* Part 2 *In Practice,* 2nd edn. Exegetics Ltd, Edington, UK.

Guarino, L., Maxted, N. and Chiwona, E.A. (eds) (2000) *A Methodological Model for Ecogeographic Surveys of Crops.* IPGRI Technical Bulletins Series. International Plant Genetic Resources Institute, Rome.

Hanna, W.W. and Towill, L.E. (1995) Long-term pollen storage. In: Janick, J.E. (ed.) *Plant Breeding Reviews,* Vol. 13, John Wiley & Sons, New York, pp.179–207.

Hanson, J. (1985) *Procedures for Handling Seeds in Genebanks. Practical Manuals for Genebanks*: No. 1. International Board for Plant Genetic Resources, Rome, Italy.

Hawkes, J.G. (1980) Genetic conservation of 'recalcitrant species' – an overview. In: Withers, L.A. and Williams, J.T. (eds) *Crop Genetic Resources – the Conservation of Difficult Material. Proceedings of an International Workshop held at the University of Reading, UK, 8–11 September 1980.* IUBS Series B42, IUBS/IBPGR/IGF, Paris, France, pp. 83–92.

Hay, F. and Probert, R. (2000) Keeping seeds alive. In: Black, M. and Bewley, J.D. (eds) *Seed Technology and its Biological Basis.* Sheffield Academic Press, Sheffield, UK.

Hodgkin, T., Brown, A.H.D., van Hintum, Th.J.L. and Morales, E.A.V. (eds) (1995) *Core Collections of Plant Genetic Resources.* John Wiley & Sons, Chichester, UK.

Hoekstra, F. (1995) Collecting pollen for genetic resources conservation. In: Guarino, L., Rao, V.R. and Reid, R. (eds) *Collecting Plant Genetic Diversity. Technical Guidelines.* CAB International, Wallingford, UK, pp. 527–550.

Hong, T.D., Linington, S. and Ellis, R.H. (1996) *Seed Storage Behaviour: a Compendium.* Handbooks for Genebanks. No. 4. International Plant Genetic Resources Institute, Rome, Italy.

Hu, J., Guo, C.G. and Shi, S.X. (1994) Partial drying and post-thaw conditioning improve the survival and germination of cryopreserved seeds of tea (*Camellia sinensis*). IPGRI/FAO *Plant Genetic Resources Newsletter,* 98, 25–28.

IPGRI/CIAT (1994) *Establishment and Operation of a Pilot* In Vitro *Active Genebank. Report of a CIAT-IBPGR Collaborative Project Using Cassava* (Manihot esculenta *Crantz*) *as a Model.* A joint publication of IPGRI, Rome and CIAT, Cali, Colombia.

IPGRI/DFSC (1999) Desiccation and storage protocol – March 1999. In: *The Project on Handling and Storage of Recalcitrant and Intermediate Tropical Forest Tree Seeds,* Newsletter 5, April 1999, pp. 23–39.

Johnson, R.C. and Hodgkin, T. (eds) (1999) *Core Collections for Today and Tomorrow.* International Plant Genetic Resources, Rome, Italy.

Kaiser, W.J. and Hannan, R.W. (1986) Incidence of seedborne *Ascochyta lentis* in lentil germplasm. *Phytopathology* 76, 355–360.

Kameswara Rao, N. and Jackson, M.T. (1996) Seed production environment and storage longevity of japonica rices (*Oryza sativa* L.). *Seed Science Research* 6, 17–21.

Karp, A., Kresovich, S., Bhat, K.V., Ayad, W.G. and Hodgkin, T. (1997) *Molecular Tools in Plant Genetic Resources Conservation: a Guide to the Technologies.* IPGRI Technical Bulletin No. 2. International Plant Genetic Resources Institute, Rome, Italy.

Kartha, K.K. and Engelmann, F. (1994) Cryopreservation and germplasm storage. In: Vasil, I.K. and Thorpe, T.A. (eds) *Plant Cell and Tissue Culture.* Kluwer, Dordrecht, The Netherlands, pp. 185–230.

Marzalina, M. (1995) Penyimpanan biji benih mahogani (*Swietenia macrophylla* King). [Storage of mahogany seeds (*Swietenia macrophylla* King]. PhD thesis, Universiti Kebangsaan Malaysia.

Marzalina, M., Yap, S.K. and Krishnapillay, B. (1992) Effect of relative light intensities on *Dryobalanops aromatica* seedlings' growth. In: Ho, H.W., Vidyadran, M.K., Norhani, A., Jainudeen, M.R. and Bahaman, R. (eds) *Proceedings of the National IRPA (Intensification of Research in Priority Areas) Seminar,* Vol. II, *6–11 January 1992, Kuala Lumpur.* Universiti Pertanian Malaysia, Serdang, Selangor, Malaysia, pp. 439–440.

Maxted, N., Ford-Lloyd, B.V. and Hawkes, J.G. (1997) Complementary conservation strategies. In: Maxted, N., Ford-Lloyd, B.V. and Hawkes, J.G. (eds) *Plant Genetic Resources Conservation.* Chapman and Hall, London, pp. 15–39.

Ouédraogo, A.S., Thomsen, K., Engels, J.M.M. and Engelmann, F. (1999) Challenges and opportunities for enhanced use of recalcitrant and intermediate tropical forest tree seeds through improved handling and storage. In: Marzalina, M., Khoo, K.C., Tsan, F.Y. and Krishnapillay, B. (eds) *IUFRO Seed Symposium 1998 'Recalcitrant Seeds', 12–15 October 1998, Kuala Lumpur, Malaysia.* FRIM (Forest Research Institute Malaysia), Kepong, Malaysia, pp. 227–234.

Painting, K.A., Perry, M.C., Denning, R.A. and Ayad, W.G. (1993) *Guidebook for Genetic Resources Documentation.* International Board for Plant Genetic Resources, Rome, Italy.

Pammenter, N.W. and Berjak, P. (1999) A review of recalcitrant seed physiology in relation to desiccation-tolerance mechanisms. *Seed Science Research* 9, 13–37.

Pammenter, N.M., Vertucci, C.W. and Berjak, P. (1993) Responses to dehydration in relation to non-freezable water in desiccation-sensitive and tolerant seeds. In: Côme, D. and Corbineau, F. (eds) *Proceedings of the Fourth International Workshop on Seeds: Basic and Applied Aspects of Seed Biology.* AFSIS, Paris, France, pp. 867–872.

Panis, B., Schoofs, H., Thinh, N.T. and Swennen, R. (2000) Cryopreservation of proliferating meristem cultures of banana. In: Engelmann, F. and Takagi, H. (eds) *Cryopreservation of Tropical Plant Germplasm – Current Research Progress and Applications.* Japan International Centre for Agricultural Sciences, Tsukuba/International Plant Genetic Resources Institute, Rome, Italy, pp. 238–244.

Pence, V.C. (1995) Cryopreservation of recalcitrant seeds. In: Bajaj, Y.P.S. (ed.) *Biotechnology in Agriculture and Forestry* Vol. 32 *Cryopreservation of Plant Germplasm I.* Springer Verlag, Berlin, Germany, pp. 29–52.

Pence, V.C., Engelmann, F., Sandoval, J. and Villalobos, V. (eds) (2000) In Vitro *Germplasm Collecting Techniques.* International Plant Genetic Resources Institute and Food and Agriculture Organization of the United Nations, Rome, Italy.

Plucknett, D.L., Smith, N.J.H., Williams, J.T. and Anishetty, N.M. (eds) (1987) *Gene Banks and the World's Food.* Princeton University Press, Princeton, New Jersey.

Rao, V.R. and Riley, K.W. (1994) The use of biotechnology for conservation and utilization of plant genetic resources. IPGRI/FAO *Plant Genetic Resources Newsletter* 97, 3–20.

Rao, K., Hu, X., Walters, C., Engelmann, F. and Engels, J. (1999) Ultra-dry seed technology and genebanks. In: *Book of Abstracts, VI International Workshop on Seed Biology, Mérida, Mexico, 24–28 January 1999.* Facultad de Quimica, Universidad Nacional Autónoma de México, Mexico City, p. 104.

Reed, B.M. (2000) Genotype considerations in temperate fruit crop cryopreservation. In: Engelmann, F. and Takagi, H. (eds) *Cryopreservation of Tropical Plant Germplasm – Current Research Progress and Applications.* Japan International Centre for Agricultural Sciences, Tsukuba/International Plant Genetic Resources Institute, Rome, Italy, pp. 200–204.

Reed, B.M., Brennan, R.M. and Benson, E.E. (2000) Cryopreservation: an *in vitro* method for conserving Ribes germplasm in international genebanks. In: Engelmann, F. and Takagi, H. (eds) *Cryopreservation of Tropical Plant Germplasm – Current Research Progress and Applications.* Japan International Centre for Agricultural Sciences, Tsukuba/International Plant Genetic Resources Institute, Rome, Italy, pp. 470–472.

Roberts, E.H. (1973) Predicting the viability of seeds. *Seed Science and Technology* 1, 499–514.

Sackville Hamilton, N.R. and Chorlton, K.H. (1997) *Regeneration of Accessions in Seed Collections: a Decision Guide.* Handbooks for Genebanks No. 5. Engels, J.M.M. (vol. ed.) International Plant Genetic Resources Institute, Rome, Italy.

Sakai, A. (1995) Cryopreservation for germplasm collection in woody plants. In: Jain, S., Gupta, P. and Newton, R. (eds) *Somatic Embryogenesis in Woody Plants,* Vol. 1. Kluwer, Dordrecht, The Netherlands, pp. 293–315.

Schäfer-Menuhr, A., Mix-Wagner, G. and Schumacher, H.M. (1997) Cryopreservation of potato cultivars – design of a method for routine application in genebanks. *Acta Horticulturae* 447, 477–482.

SGRP (1996) *Report of the Internally Commissioned External Review of the CGIAR Genebank Operations.* International Plant Genetic Resources Institute, Rome, Italy.

Simmonds, N.W. (1982) The context of the workshop. In: Withers, L.A. and Williams, J.T. (eds) *Crop Genetic Resources – the Conservation of Difficult Material.* IUBS Series B42, IUBS/IBPGR/IGF, Paris, France, pp. 1–3.

Stanwood, P.C. (1985) Cryopreservation of seed germplasm for genetic conservation. In: Kartha, K.K. (ed.) *Cryopreservation of Plant Cells and Organs.* CRC Press, Boca Raton, Florida, pp. 199–226.

Towill, L.E. and Walters, C. (2000) Cryopreservation of pollen. In: Engelmann, F. and Takagi, H. (eds) *Cryopreservation of Tropical Plant Germplasm – Current Research Progress and Applications.* Japan International Centre for Agricultural Sciences, Tsukuba/International Plant Genetic Resources Institute, Rome, Italy, pp. 115–129.

UNCED (1992) *Convention on Biological Diversity.* United Nations Conference on Environment and Development, Geneva.

Vertucci, C. and Roos, E.E. (1993) Theoretical basis of protocols for seed storage. II. The influence of temperature on optimal seed levels. *Seed Science Research* 3, 201–213.

Walters, C. (ed.) (1998) Ultra-dry seed storage. *Seed Science Research* 8, Suppl. 1, 11–14.

Walters, C. and Engels, J. (1998) The effects of storing seeds under extremely dry conditions. *Seed Science Research* 8, Suppl. 1, 3–8.

Wesley-Smith, J., Vertucci, C.W., Berjak, P., Pammenter, N.W. and Crane, J. (1992) Cryopreservation of desiccation-sensitive axes of *Camellia sinensis* in relation to dehydration, freezing rate and the thermal properties of tissue water. *Journal of Plant Physiology* 140, 596–604.

Withers, L.A. (1991) Biotechnology and plant genetic resources conservation. In: Paroda, R.S. and Arora, R.K. (eds) *Plant Genetic Resources Conservation and Management, Concepts and Approaches.* International Board for Plant Genetic Resources, Regional Office for South and Southeast Asia, New Delhi, India, pp. 223–229.

Withers, L.A. (1995) Collecting *in vitro* for genetic resources conservation. In: Guarino, L., Ramanatha Rao, V. and Reid, R. (eds) *Collecting Plant Genetic Diversity.* CAB International, Wallingford, UK, pp. 511–515.

Withers, L.A. and Engelmann, F. (1998) *In vitro* conservation of plant genetic resources. In: Altman, A. (ed.) *Biotechnology in Agriculture.* Marcel Dekker Inc., New York, pp. 57–88.

Withers, L.A. and Engels, J.M.M. (1990) The test tube genebank – a safe alternative to field conservation. *IBPGR Newsletter for Asia and the Pacific* 3, 1–2.

10 The Establishment of a Regional Germplasm Centre in the Pacific Island Region

M. Taylor
TaroGen, Secretariat of the Pacific Community (SPC), Suva, Fiji

Introduction

There are several options available as conservation strategies, which generally fall into two categories, *in situ* and *ex situ*. The Convention on Biological Diversity directs attention to *ex situ* storage in Article 9 of the Convention (Lesser, 1998) to:

- adopt measures for the *ex situ* conservation of components of biological diversity, preferably in countries of origin of such components; and
- establish and maintain facilities for *ex situ* conservation of, and research on plants, animals and microorganisms, preferably in countries of origin of genetic resources.

The Global Plan of Action (GPA) (FAO, 1996) states that conservation could be better achieved by placing more emphasis on a smaller number of quality facilities. This would in turn reduce costs, increase efficiency, and free other facilities and funds for the purpose of developing plant genetic resources for food and agriculture (PGRFA). The GPA recommends strengthening of regional and international facilities, and the drafting of appropriate legal agreements, which would enable countries to place collections within them without compromising their access to, or control and ownership of materials. The Plan emphasizes the need for international and regional cooperation on a wide range of interrelated issues including access, conservation, utilization and benefit sharing. This cooperation is essential as no country is independent in terms of PGRFA needed to sustain and improve its major crops.

The majority of Pacific Island crops are vegetatively propagated and therefore lend themselves to conservation in field genebanks. This has been the preferred method of conservation over the last 20 years. Papua New Guinea has been the focus of many collecting missions because of its diversity (6% of the world's diversity), resulting in field collections of banana, yam, taro and cassava. In many cases, these collections no longer exist, and where collections are still being maintained, accessions have been lost. Losses have been largely due to insufficient resources, a factor aggravated by pests and diseases, and climatic problems. In particular, field genebanks are notorious for their demands on labour and other resources; the secure maintenance of accessions is dependent on constant vigilance and upkeep. In many of the Pacific Island countries, these resources are quite limited to the extent that a member of staff often has multiple responsibilities. Furthermore, funds are often scarce, and conservation is rarely a high priority. On some of the smaller islands suitable land is also a limiting factor.

Pest and disease problems, too, can cause major disasters. In the early 1990s taro leaf blight caused by *Phytophthora colocasiae*, wiped out all of the local varieties of taro in Samoa, and destroyed a very valuable export market. A field collection in the Solomon Islands is currently under severe pressure from a fatal viral disease. Yam collections in the

Pacific Island region commonly suffer from anthracnose disease, which leads to the loss of accessions. Climatic conditions in the Pacific Island region can be extreme; drought, flooding and cyclones are relatively common events, with obvious consequences for the security of field collections. These problems are likely to occur with any national germplasm collection, whether maintained *in vivo* or *in vitro*. Some of the countries do have tissue culture laboratories but again resources are limited and so the most these laboratories can do is to multiply elite material for distribution to growers.

The fragmented nature of the region lends itself, however, to regional strategies, and conservation is one such activity where regional policies and practices would seem to be the best option. If individual countries do not have the resources for germplasm conservation, then it is best carried out on a regional basis. This becomes even more logical when one considers the commonality of the major crops: taro, yam, banana, sweet potato and cassava. Furthermore, many of the varieties are also the same, but exist under different names in different countries. Therefore, when considering the resources within the region, it seemed logical to pool these resources, and establish a regional centre for conservation of the region's germplasm. Regional genebanks have been established in various parts of the world: for example, in Lusaka, Zambia, the SADC Plant Genetic Resources Centre (SPGRC) established under the auspices of the Southern African Development Community (SADC), and the Centro Agronómico Tropical de Investigación y Ensenañza (CATIE) for Central American countries, in Turrialba, Costa Rica. These genebanks have been set up as collaborative ventures between a number of countries in the same geographical region to conserve the germplasm from that region, and to support research.

The need to conserve and the recognition of the importance of regional cooperation in the conservation of plant genetic resources in the Pacific Island region was acknowledged by the Ministers of Agriculture for six Pacific Island countries at a regional meeting held in Fiji in 1996. They endorsed a resolution that recognized the links to equitable sharing of benefits, and the need for regional policies. Prior to the establishment of the Regional Germplasm Centre (RGC) in Fiji there already was a general consensus on the need to conserve genetic resources within a regional framework. The question then arose concerning the preferred method of conservation. Ideally, no one method should exist in isolation, and conservation strategies should be considered as complementary. Whichever strategies are chosen, a clear and well-organized management system must ensure that linkages between the components of the conservation and use programme are maintained. However, the resources available often dictate what the conservation strategy will be.

Regional field genebanks may be cumbersome and suffer from the same problems as national collections. Since seed storage for Pacific Island crops is either impossible or has very limited application, a regional approach using *in vitro* conservation seemed to offer the most sustainable and cost-effective strategy for safeguarding genetic resources. Under a European Union (EU)-funded project (Pacific Regional Agricultural Programme), a regional tissue culture unit was established at the University of the South Pacific in Samoa in the late 1980s. This unit successfully conserved collections of taro, sweet potato, banana, cassava and yams for 10 years, and so experience in the region with *in vitro* conservation was positive. One advantage from *in vitro* conservation is access to pathogen-tested germplasm, which is easily distributed as tissue culture plantlets among countries. Access to virus-indexed germplasm is extremely important in a region that is composed of many islands, each with its own strict quarantine regulations. The existence of a regional centre maintaining pathogen-tested tissue-cultured germplasm therefore supports distribution and utilization, as well as being the best option for conservation.

These were the issues that were considered when the concept of an RGC was under discussion. In attempting to find the solution to all of the problems surrounding plant genetic resources conservation in the Pacific Island region, an RGC was seen as the most practical and viable solution where the region's germplasm would be maintained as pathogen-tested tissue cultures. In addition, such a centre would facilitate crop improvement through enabling distribution of pathogen-tested improved cultivars.

The RGC was officially opened in March 1999. Funding for the establishment of the Centre came from the Australian government through AusAID and Australian Centre for International Agricultural Research (ACIAR) projects, and from the EU through the Pacific Regional Agricultural Programme. The RGC is presently being supported by funds from different projects which operate in the Centre. The EU supports the Plant Protection

Service (PPS) within the Secretariat of the Pacific Community (SPC), and the RGC operates within the PPS. In addition, there is support from AusAID through the Taro Genetic Resources project (TaroGen), and from the EU through the South Pacific Yam Network (SPYN) project.

Conservation

Strategies

The important food crops of the Pacific Island region are taro, yam, sweet potato, cassava and bananas. However, the number of varieties in each country varies, with relatively few in some of the Polynesian countries such as Samoa and Tonga. There are significantly larger numbers in Papua New Guinea and the Solomon Islands. For example, the taro collections recently established by TaroGen in these two countries number 900 and 700, respectively. SPC has 22 member countries and therefore the RGC has to offer genetic resources conservation to all these member countries. As the RGC has a capacity for some 5000 accessions there is an obvious need to rationalize collections and develop appropriate conservation strategies. So although the conservation of genetic resources through a regional centre allows pooling of resources, there is still a need to be economical. Simply maintaining the collections *in vitro* could put a strain on these resources through the management of large *in vitro* collections held under standard conditions. Therefore in conserving the region's genetic resources, both the numbers of samples for conservation and the *in vitro* techniques require consideration. With standard *in vitro* conditions, subculturing occurs at relatively frequent intervals. At each subculture stage, there is the risk of losing material either through contamination or human error or both. There is also the question of genetic integrity, which can be lost when plant material is in culture for long periods of time. With these issues in mind, it was decided that the RGC would utilize slow growth for short- and medium-term storage and, where possible, cryopreservation for long-term storage.

There also was the issue of numbers of samples. With the TaroGen project a total of 2500 taro accessions have been collected. To date no duplicate samples have been identified and none eliminated. But even when this eventually occurs, the total number of accessions in the taro collection will still be high. For this reason it was decided to establish a core collection. Within the TaroGen project procedures for developing core collections have been agreed. Within countries, accessions will be separated on the basis of ecogeographical regions, and cluster analysis applied to morphological and use data (including pest and disease reactions) to distinguish different groups within regions. Molecular marker information will be used to add varieties with unique fingerprints, and to remove duplicates. About 400 taro accessions will be fingerprinted, and about half this number will be conserved at the RGC. Two molecular marker techniques have been developed for taro germplasm, namely microsatellites and inter simple sequence repeats (ISSR). Primer pairs have already been designed and applied to 17 taro (including one *Colocasia esculenta* var. *antiquorum* and a single *Xanthosoma* accession) from Fiji, Federated States of Micronesia (FSM), Hawaii, Niue, Papua New Guinea and Samoa. So far eight primers pairs have been developed, and the goal is to develop 20 for core samples from national collections. Thirty-three ISSR primers were initially screened and four have been chosen. DNA fingerprinting and virus indexing is being carried out at the University of Queensland under an ACIAR-funded project which runs parallel to the TaroGen project.

For security purposes, collections should be duplicated. The TaroGen project will hold a workshop before the end of the project to provide information for the countries on all conservation methodologies for taro. Countries will then be able to determine whether or not they have the resources to maintain some accessions for their immediate use, or as working collections for breeding purposes. Some duplication will occur in this way. However, it is crucial that the core collections that are maintained in the RGC are conserved in full elsewhere. Discussions are ongoing to see if the laboratory that exists at the University of the South Pacific, Samoa, could take on this role.

Methodology

Research carried out in the regional tissue culture unit in Samoa showed that temperature reduction was the most practical method for reducing the growth rate of most of the crops held in the unit. Morphological changes were observed when sweet potatoes were cultured on media containing sugar alcohols such as sorbitol and mannitol. When cultures were grown on an unsupplemented Murashige and Skoog (1962) medium at a temperature of 20°C, the subculture period was extended to 6–8 months. It was noted, however, that if the temperature dropped below 20°C, the plants became stressed and there was significant leaf senescence. In contrast, studies at the International Potato Centre (CIP) on a range of temperatures for slow growth storage of sweet potato found that 15°C was optimal (Lizarraga *et al.*, 1992). Some preliminary experiments were also carried out with taro looking at the effect of combining low temperature, reduced light and osmoticums. The inclusion of mannitol in the culture medium did suppress growth, but some morphological changes were also observed. In addition, a phytotoxic effect was observed when mannitol was used with cultures initiated directly from the field.

There are reports in the literature of taro being stored for more than 8 years with transfer intervals of approximately 3 years at 9°C in total darkness (Bessembinder *et al.*, 1993). However, the experiment was conducted with only one clone, and the report does not state which taro variety was used, var. *esculenta* or var. *antiquorum*. Staritsky *et al.* (1986) also reported that taro could be conserved for 3 years at 9°C and remain viable. Similarly, three species of *Xanthosoma* could be stored in the dark for at least 2 years at 13°C (Zandvoort *et al.*, 1994).

It was generally felt that temperature reduction would be the best option to use for slow growth, because sufficient research has not been carried out on the utilization of growth retardants, and their possible effect on genetic integrity. In fact, the most widely applied slow growth storage technique is temperature reduction, often combined with a decrease in light intensity or culture in the dark (Engelmann, 1997).

Different methods for conserving taro are being investigated as part of TaroGen. Temperature reduction is being evaluated for slow growth storage. The study carried out by the International Centre for Tropical Agriculture (CIAT) on cassava is being used as a model. Local Fijian accessions are maintained at 20°C on a modified Murashige and Skoog (1962) medium. These accessions were also planted in the field by the Ministry of Agriculture. All inputs are being recorded in both genebanks using spreadsheets developed in an ACIAR project ('Economics of preserving genetic diversity in PNG in the context of world agriculture') so that at the end of the culture period direct comparisons can be made between the two systems. Genetic integrity will also be monitored through DNA fingerprinting.

Even though a regional centre might be more economical, and therefore more sustainable than several national collections, there is still a need to be realistic about what can be achieved with the resources available. Core collections of taro and yam of about 400 accessions will be established as part of TaroGen and SPYN using cryopreservation. With the taro project, a vitrification protocol (Thinh, 1997) is being evaluated, since it has already achieved 80–100% survival after cryopreservation with four different cultivars. Similarly, an encapsulation-dehydration method developed at the Institut de Recherche pour le Développement (IRD), Montpellier, France, will be evaluated for the yam collection. With both the taro and yam core collections accessions will also be maintained under slow growth for a period of 12 months while the cryopreservation is being monitored.

With the other crops in the RGC, slow growth using temperature reduction will be implemented. The International Plant Genetic Resources Institute (IPGRI) has provided a grant to carry out some preliminary research on the cryopreservation of sweet potato. This will look at the vitrification method used at CIP (Steponkus *et al.*, 1992; Golmirzaie and Panta, 1997). This technique was relatively successful initially with potato; with sweet potato the survival rate over a range of genotypes has been 30–60%.

Access Issues

Quarantine

The Pacific Islands are very conscious of the importance of quarantine, and each island follows strict rules and regulations concerning the importation of plant material. As a general rule, the relative safety of tissue-cultured material, and even more so with virus-indexed material, is acknowledged by quarantine staff in the various islands. Fiji quarantine is especially sensitive about taro. When taro leaf blight destroyed the taro export market of Samoa, Fiji was quick to take up this market, but cultivated varieties that are susceptible to the fungus. The SPC has yet to develop a policy that enables the RGC to import germplasm as tissue cultures, but not pathogen-tested. Currently, specific requests must be made and these are considered for approval by a special committee. This can take time and can cause significant delays in the progress of a project. A general policy to automatically grant a permit to import germplasm to the RGC and not for distribution within Fiji will facilitate this aspect of the project. A recent example was the importation of coconut embryos to RGC, Fiji, from some of the Pacific Island countries. These embryos came from populations on atolls that were in danger of being lost, and so there was some urgency for their collection. The embryos are merely being held in the RGC until the tissue culture laboratory associated with the international genebank in Medang, Papua New Guinea, is ready to receive the germplasm.

Intellectual Property Rights (IPR)

The Convention on Biological Diversity (CBD) recognizes each country's sovereignty over its plant genetic resources. However, the RGC must have germplasm that is accessible to the countries of the region; otherwise it becomes merely a museum containing resources that cannot be utilized. Within the Pacific Island region, none of the countries yet has any form of national legislation that governs access to germplasm. Some countries such as Fiji are in the process of developing such legislation but it will be some time before it is implemented. For TaroGen, this was an issue that was considered at the beginning of the project. As a result a Code of Conduct was formulated with certain conditions, such as:

- taro germplasm acquired under TaroGen is for research purposes only, and remains the property of the original source country;
- germplasm acquired under TaroGen will be freely exchanged between the participants of the project; and
- any material acquired during the project will not be transferred beyond the project participants without prior informed consent of the original source country, and without the use of Material Transfer Agreements (MTAs).

In October 1999, the SPC presented a draft MTA to govern access to and use of genetic resources at a meeting of the Heads of Agriculture of some of its member countries. The MTA was accepted and this is now being used for all distributions from the RGC. When germplasm is requested from the RGC, an MTA together with an SPC policy statement is sent to the potential recipient of the germplasm, and no germplasm is distributed until a signed MTA is returned.

It is anticipated that the SPC will establish a regional policy on germplasm exchange in which some germplasm may be moved freely without any prior informed consent and that there will be other germplasm for which such consent is required. Currently it seems that prior informed consent will be required for any traditional cultivars that are stored in the RGC, whereas improved material resulting from externally funded regional projects can be freely distributed, within the region at least. The Pacific Island countries have yet to decide on an exchange policy with countries outside the Pacific Island region. The use of an MTA is the first step in working with this IPR issue.

Future Considerations

FAO International Network of Ex Situ *Collections*

One consideration for the future is membership of the FAO International Network of *Ex Situ* Collections. This network provides a legal framework for countries, groups of countries, and institutions to hold material in trust for the international community. It was established with the collections of 12 international agricultural research centres (IARCs) of the Consultative Group on International Agricultural Research (CGIAR), with close to 500,000 germplasm accessions of food crops. The agreements were signed in October 1994. More recently the International Coconut Genebank in India became part of the network. These agreements were developed in accordance with the CBD.

The Global Plan of Action (FAO, 1996), recognized the potential of networks on all levels and stated as its intermediate objectives the following:

- to develop and strengthen national, regional and international networks, including the existing FAO International Network of *Ex Situ* Collections within the FAO Global System and in accordance with policies and strategies set out by the Commission on Genetic Resources for Food and Agriculture;
- to assemble therein sufficient capacity to provide options to countries for the voluntary storage – preferably within each region – of appropriate genetic materials and their duplicates;

- to provide for the transfer and the ongoing conservation of this material under applicable international legal agreements, which ensure the sovereign rights of the countries of origin, and with appropriate technical and financial support.

Being part of this international network can only serve to improve the effectiveness and efficiency of a regional genebank. In the same way that countries are uniting within the region to increase their effectiveness in germplasm conservation, joining with other genebanks within an international network can only serve to improve a regional genebank, and add to the security of the germplasm. Membership would mean that countries that have entrusted their germplasm to that genebank can be assured that conservation standards are maintained at the highest level, and that the distribution of their genetic resources is being 'controlled' and accurately monitored. Being part of this network ensures that a regional genebank would become instantly aware of any new developments in germplasm policy.

One criticism that does occur is that the system does not provide opportunity for monetary benefit. However, this is not strictly true. There is no direct payment mechanism for financial compensation of suppliers of materials designated for open exchange, but through the use of MTAs there is scope to negotiate for some form of compensation should the material provided through the system be commercialized.

Regional networks

Membership of an international network will position the regional genebank to participate more effectively in global discussions, providing access to information, technical support or legal advice. The establishment of a regional network within which the regional genebank operates could also be of significant value in strengthening genetic conservation and utilization in the region. The regional network can work to strengthen national programmes, collaboration, develop conservation technologies, and exchange and disseminate information on plant genetic resources. The regional genebank or germplasm centre can act as the hub in coordinating the network. IARCs are supporting a number of networks in Asia in which a centre acts as the hub and manages or coordinates the network. For example, CIP has been coordinating a network called the Southeast Asian Programme for Potato Research and Development (SAPPRAD) for cooperative testing of new potato and sweet potato varieties in Asian countries.

The Global Plan of Action specifically noted the need for a genetic resources network in the Pacific. This was further endorsed at the GPA Implementation Meeting held in the Philippines in December 1998. At this meeting all representatives from the Pacific Island countries endorsed the importance of stronger collaboration, stressing the need to develop collaborative arrangements for accessing genetic resources and sharing benefits. Other areas requiring attention that could be functions of a network for the Pacific were:

- a need to locate, document and describe Pacific plant germplasm collections;
- improved communication facilities among countries to enhance information exchange about their own plant genetic resources, their needs and research uses; and
- regional training courses for genetic conservation, documentation and use.

There seems to be a need for a network in the region in which a regional genebank could function in a coordinating role. The network would be responsible for collecting all information on Pacific germplasm, disseminating information, advising on conservation strategies, identifying areas or projects or both that require attention, and promoting evaluation of genetic resources. Another function of such a network could be the development and implementation of a plant variety protection (PVP) system that is appropriate to the plants that the genebank or network is responsible for. If a system exists whereby plants can be registered then it can facilitate access and benefit sharing.

Funding

The RGC is largely supported by donor funds. There is obviously a need to look at more long-term, sustainable funding. Funds can come from the budgets of the countries for which the genebank holds accessions, that is, some form of member country contribution. A further source of funds could come from donor agencies almost in the form of a multilateral fund that recognizes the ongoing contribution that germplasm exchange makes to the agricultural community and society. All plant-based projects utilize germplasm but there is very rarely any recognition of this within a proposal; there should be acknowledgement of this through appropriate funding.

Conclusions

It is hoped that the RGC will acquire international status as a germplasm centre, and that it will be of immense benefit to agriculture in the region. Some of the international centres of the CGIAR do conserve some crops important to the Pacific region. For example, the International Institute for Tropical Agriculture (IITA) in Nigeria conserves yams. However, Pacific species and varieties tend not to have high priority and none of the centres has accepted a responsibility to conserve what is the region's most important crop, namely taro. There is, therefore, a need for a germplasm centre in the Pacific. Although initially attention will focus on the major root and tuber crops, there are a number of crops, classified as minor, that are endemic to the region, and important nutritionally such as *Abelmoschus manihot* or bele. Attempts will be made to establish collections of these in the RGC.

References

Bessembinder, J.J.E., Staritsky, G. and Zandvoort, E.A. (1993) Long-term *in vitro* storage of *Colocasia esculenta* under minimal conditions. *Plant Cell, Tissue and Organ Culture* 33, 121–127.

Engelmann, F. (1997) *In vitro* conservation methods. In: Callow, J.A., Ford-Lloyd, B.V. and Newbury, H.J. (eds) *Biotechnology and Plant Genetic Resources: Conservation and Use.* Biotechnology in Agriculture Series 19. CAB International, Wallingford, UK, pp. 119–161.

FAO (1996) *Report of the International Technical Conference on Plant Genetic Resources, Germany, 1996,* ITCPGR/96/REP. FAO, Rome, Italy.

Golmirzaie, A.M. and Panta, A. (1997) Advances in potato cryopreservation by vitrification. In: *CIP Program Report.* International Potato Center, Lima, Peru, pp. 71–76.

Lesser, W.H. (1998) *Sustainable Use of Genetic Resources under the Convention on Biological Diversity.* CAB International, Wallingford, UK.

Lizarraga, L., Panta, A., Espinoza, N. and Dodds, J.H. (1992) Tissue culture of *Ipomoea batatas*: micropropagation and maintenance, *CIP Research Guide* 2. International Potato Center, Lima, Peru.

Murashige, T. and Skoog, F. (1962) A revised medium for rapid growth and bioassays with tobacco tissue cultures. *Physiologia Plantarum* 15, 473–497.

Staritsky, G., Dekkers, A.J., Louwaars, N.P. and Zandvoort, E.A. (1986) *In vitro* conservation of aroid germplasm at reduced temperatures and under osmotic stress. In: Withers, L.A. and Alderson, P.G. (eds) *Plant Tissue Culture and its Agricultural Applications.* Butterworth, London, pp. 277–284.

Steponkus, P.L., Langis, R. and Fujikawa, S. (1992) Cryopreservation of plant tissues by vitrification. In: Steponkus, P.L. (ed.) *Advances in Low Temperature Biology*, Vol. 1. JAI Press Ltd., London, pp. 1–61.

Thinh, N.T. (1997) Cryopreservation of germplasm of vegetatively propagated tropical monocots by vitrification. PhD thesis, Kobe University, Japan.

Zandvoort, E.E., Hulshof, M.J.H. and Staritsky, G. (1994) *In vitro* storage of *Xanthosoma* spp. under minimal growth conditions. *Plant Cell, Tissue and Organ Culture* 36, 309–316.

11 Accession Management Strategies: Splitting and Lumping

Th.J.L. van Hintum,[1] N.R. Sackville Hamilton,[2] J.M.M. Engels[3] and R. van Treuren[1]

[1] Centre for Genetic Resources, The Netherlands (CGN), Plant Research International, Wageningen, The Netherlands; [2] Institute of Grassland and Environmental Research (IGER), Aberystwyth, UK; [3] International Plant Genetic Resources Institute (IPGRI), Rome, Italy

Introduction

Creating collections of plant genetic resources for *ex situ* conservation can be relatively easy: choose a crop, and gather all, or a particular subset, of the available samples. In fact this strategy has been common practice in the past, and has resulted in many collections with huge numbers of accessions. According to the FAO *Report on the State of the World's Plant Genetic Resources for Food and Agriculture* (FAO, 1996), the number of genebank collections has grown to more than 1300 in which more than 6 million accessions are stored and conserved as seed and more than half a million accessions held in field genebanks.

After a first period of collecting, documenting and evaluating accessions, the genebank community has gradually shifted towards a phase of developing strategies for improving the composition and management of collections. Currently, there is a need to streamline conservation activities and increase efficiency. This is due in part to two factors: the pressure of decreasing budgets, and the changing role of governments as privatization increases.

Accession management is not new. Since the first genebanks started conserving germplasm, decisions have been made concerning the composition of accessions. This starts during the collecting phase, and continues during preparation of the samples for inclusion in the genebank collection. A good example is the common practice to separate the different genera or species in a mixed sample. In many genebanks it stops at that point. Other genebanks go further, by separating accessions of selfing crops into different morphotypes (Lehmann and Mansfeld, 1957). Only rarely are accessions being lumped prior to inclusion in the collection.

Controlling the number of accessions is one of the key considerations in managing the genebank budget. Reducing the number of accessions is only possible by discarding or lumping accessions. However, this may directly conflict with the objectives of the genebank, and have major genetic consequences such as loss of alleles, genotypes, or even complete populations. Splitting accessions increases the number of accessions, but can facilitate the management of the accessions and improve the quality of genebank operations. In this chapter we provide an overview of the most important considerations for lumping or splitting accessions.

Genetic markers are used to describe the genetic make-up of genebank accessions. In this way it is possible to formulate expectations, and to monitor the consequences of splitting or lumping. This approach is illustrated using examples of recent and ongoing studies in this area.

Considerations for Splitting or Lumping

Under which circumstances is splitting or lumping of genebank accessions acceptable or even desirable? There are three categories of issues that need to be considered. First, there are the genetic issues, such as the distribution of genetic diversity within and between accessions. The second category consists of operational issues, such as the impact of splitting on regeneration protocols. Finally, there are the economic aspects, which will often be the starting point of the attempts to increase the efficiency of genebank management.

Genetic issues

It is obvious that if two accessions in a collection are identical in all aspects, one of them is redundant and can be removed. However, when a cross-fertilizing crop is considered, two accessions will very rarely be identical. Therefore, the question 'Are the accessions identical?' should be reformulated as 'Are the accessions sufficiently different to consider them distinct?' This is related to another question: 'How is the genetic diversity distributed within and between the accessions?' Similarly, the applicability of splitting accessions is related to how many different genotypes constitute an accession. The distribution of genetic variation has several relevant aspects such as the genetic variation for single traits, the correlation between traits, and the role of the trait concerned in reproduction.

Distribution of genetic variation

If separation of an accession into groups of different genotypes for a specific trait is being considered, the trait must be observable, as should the distribution of its expression, and its genetic basis. These factors are also relevant in decisions to lump accessions. The first factor is obvious. If it is impossible or very difficult to observe the phenotypic expression of the trait, it cannot be used as a criterion for separation or lumping. If the expression of a trait within an accession has a discrete distribution, that is, with a number of distinct states, and if it is possible to distinguish between the different classes, then splitting can be a useful and feasible option. However, if the expression follows a continuous distribution as a result of environmental effects or a large number of genes involved in the expression, splitting the accession into two or more new accessions is less straightforward. Splitting is likely to result in lower within-accession genetic variance as compared to the original, but this will in most cases have only a relatively small impact on the reduction of the probability of drift and selection. This of course assumes that these latter aspects are the reasons for splitting an accession. Fitness traits are an exception to this rule.

The effect of splitting on the reduction of intra-accession diversity is even lower for outbreeding species since they will contain heterozygotes, in which recessive genes may be hidden. Therefore, even if the identification of trait states is perfect, the progeny of the separated fractions may not necessarily conform to the parental state. Another issue for outbreeding species is the possibility of inbreeding depression, which may result from splitting when too few plants are used for a new accession in an attempt to reduce variance.

Conversely, forming one new accession by combining two or more original accessions may be a useful option if the accessions have similar genetic distributions. This will result in a relatively small increase in the probability of drift and selection. Obviously, accessions should not be combined if they are homozygous for different discrete states. Thus, accessions containing a small number of discrete, easily identifiable homozygous genotypes are the best candidates for splitting, while accessions with high continuous variation within accessions and low variation between accessions are the best candidates for lumping.

Correlation between traits

If a particular trait is not correlated with other traits within the same accession, it will not provide a strong basis in supporting decisions for splitting or lumping. Even if a trait is present in a small number of discrete, easily identifiable homozygous states, and not correlated with other traits, separating on this basis will only result in reducing the diversity for that particular trait. Therefore, splitting is not logical except if it concerns an economically very important trait, such as a disease resistance. If this trait occurs at a low frequency it might be lost or no longer noticed if not fixed in a separate accession.

If the variation for traits is correlated within an accession, this could be a candidate for splitting. Correlation between traits may arise from a number of genetic mechanisms. The first is pleiotropy, the phenomenon whereby a single gene affects several traits. Since this correlation is not based on an association between alleles from different genes, it is irrelevant in decisions for splitting. The second mechanism is genetic linkage, in which case separation on the basis of one gene may result in separation of a number of linked genes. Since the number will be limited, this mechanism is also of limited importance for accession management. However, the third mechanism, correlation by descent, is very important. For mixtures of inbred lines or apomicts, this correlation is complete since all genes will be correlated. For example, imagine an accession of a selfing crop that contains a few distinct genotypes. Splitting on the basis of a few well-chosen traits would result in complete separation of all traits that show diversity in the accession.

Also, with respect to lumping accessions, the correlation between traits is an important consideration. If, for example, two homozygous accessions of a selfing crop are thought to be duplicates, only a few discriminating traits are needed for confirmation. In the case of an outbreeder, however, a much more complete description of the genetic variation is needed.

Reproductive traits

Since variation for traits with a direct influence on the fitness of the plant will increase the danger of genetic drift and selection during regeneration, these traits are very relevant for accession management decisions. Variation in flowering date, number of flowers, seed set, cold tolerance, or maturity date, can cause a severe change in the genetic make-up of the accession through selection, which should generally be avoided. Variation in such traits can therefore be a justified reason for splitting, or a reason not to lump. Alternatives that may avoid selection by reducing the selective advantage are generally rather labour intensive, and therefore not appealing. For example, the influence in differences in seed set can be avoided by harvesting equal amounts of seeds from all plants, and the effects of differences in maturity date can be overcome by harvesting at several intervals.

Operational issues

Splitting or lumping accessions has implications for the number of accessions and for the diversity maintained in accessions. These factors will have a major impact on nearly all operations in a genebank: regeneration, characterization, evaluation and documentation, storage, monitoring viability and facilitating use.

Regeneration

Many aspects that influence accession management decisions are related to regeneration. The greater the number of accessions, the less input for regeneration per accession is available for a given genebank capacity. But reducing the number of accessions by lumping usually increases diversity within the accessions; and the more diversity within the accession, the higher the magnitude of shift and selection if not properly managed.

One of the most difficult aspects of collection management in a genebank is maintaining the genetic integrity of accessions during regeneration. This is easy in the case of an accession containing only one genotype of a selfing crop, but very difficult in the case of mixtures of mainly selfing lines, or with populations of largely cross-fertilizing plants. Major threats are drift as a result of random effects (usually sampling too few plants), selection as a result of a higher fitness of some genotypes, and contamination as a result of pollination between accessions. Contamination of seed lots during seed handling before and after growing the plants is also a threat. If there is more diversity within the accession, it will be more difficult to maintain it.

Variation for reproductive traits is especially difficult to maintain in accessions. It is also more difficult to recognize contamination in a heterogeneous accession as compared with a homogeneous accession. Therefore, to maintain the genetic integrity of accessions, a heterogeneous accession will demand more attention than a homogeneous one, and can therefore be more expensive. In some cases splitting one heterogeneous accession to create several homogeneous accessions might prove beneficial for achieving the aims with a given level of resources. However, in the case of outbreeding species the extent of splitting should not be taken too far since it might result in relatively uniform lines that may suffer from inbreeding depression.

Even for an 'easy' crop such as barley, maintaining the genetic integrity of genebank accessions proves very difficult. A recent study revealed that the effective population size in barley regenerations using an estimated 600 plants, was only 4.7 (Parzies *et al.*, 2000). Van Hintum and Visser (1995) showed that duplicate barley accessions had developed into quite different mixtures in different genebanks. Both studies looked at the results of procedures from the past; these procedures might have improved.

In some cases intentional loss of diversity within accessions due to drift might be an option. If it is clear that avoiding selection and contamination or maintaining large numbers of accessions is too expensive, one might consider choosing the diversity between accessions rather than the diversity within, by using relatively few plants per regeneration. By disregarding the effects of inbreeding depression, it is easier to isolate the accessions for regeneration and hence to maintain larger numbers of accessions.

Characterization, evaluation and documentation

Intra-accession variation complicates the description and documentation. This may be solved by simply ignoring the low-frequency genotypes, notwithstanding the fact that these may contribute to the value of the accession, and possibly the collection as a whole.

If intra-accession diversity for a qualitative trait should be scored, obviously this diversity must be assessed. Depending on the frequency and genetic background of the trait, this may mean that relatively large numbers of plants should be observed. For example, if a recessive allele occurs at a frequency of 5% in a panmictic population, 1197 plants must be observed to be 95% sure that the allele is detected.[1] Moreover, once the diversity is observed, the question arises as to how it should be recorded and documented. Several systems have been devised to solve this problem, including a scoring protocol that denotes the presence of diversity without specifying what, the recording of estimates of the frequencies of each score, or intermediate approaches (e.g. Rana *et al.*, 1991; van Hintum, 1993).

In the case of quantitative traits there is another problem connected to intra-accession diversity, namely that environmental variance is entangled with genetic diversity. Large-scale experiments are needed to quantify the genetic component. This is very rarely done, and consequently within-accession variation for such traits will often simply be neglected altogether. Splitting accessions into more homogeneous accessions may help to reduce these problems. In the case of lumping accessions, care should be taken that these problems are not created or aggravated.

Storage and monitoring viability

The next group of issues to consider are related to the storage of the germplasm and monitoring its viability. Again, the key elements for consideration are the number of accessions and the intra-accession diversity. The minimum number of seeds that are stored and distributed is often dictated by the requirement of maintaining the entire diversity of an original sample, and to distributing that same range of diversity to users. Obviously, if the diversity in the sample is eliminated by splitting the separate genotypes into separate accessions, this will have an impact on the number of seeds required.

Intra-accession diversity has implications for monitoring the viability of the samples only if this diversity affects storability. However, since this is typically a trait that is difficult to observe, it is difficult to separate the genotypes in an accession on that basis. More importantly, if lumping occurs, the situation of mixing and subsequently storing seed lots with different seed quality should be avoided.

Facilitating use

Facilitating the use of genebank material is a key element in genebank operation. For the user the quality of a genebank depends to a large extent on the ease of use of the conserved germplasm. Lumping or splitting can have a direct effect on this.

If a user is looking for traits that can be observed from single plants, there might be a preference for highly variable accessions containing much variation. For the curator it might be an option to lump the accessions in the collection into a limited number of genetically similar

[1] The chance of finding a phenotype occurring with the frequency f_p in a sample of n plants is $1-(1-f_p)^n$. The frequency of a phenotype (f_p) corresponding with a homozygous recessive allele occurring with frequency f_g in a panmictic population will be f_g^2.

groups. A user might prefer to screen this limited number of lumped accessions rather than the large number of individual accessions. But also for the curator it might be preferable to distribute only a limited number of lumped accessions rather than large numbers of individual accessions.

In other situations there might be a preference to evaluate larger numbers of less variable accessions, in which case the curator may consider splitting accessions to facilitate use. This applies if it is difficult to distinguish single plants within a dense population, such as in grass swards. It also applies for traits where the contrast between desirable and undesirable genotypes is not visually striking, so that it is difficult to detect elite plants within a population. Finally, it applies where the user's methodology is based on scoring whole plots, and observing individual plants is inconvenient. In all three cases, identifying elite plots from among a large number of relatively uniform accessions may be more convenient than trying to identify elite individuals within a few plots.

Economic issues

Genebank management strategies in general serve an economic purpose: optimizing the use of the limited financial resources to achieve the aims set by the genebank, usually conservation and facilitating use of genetic resources. In the case of accession management decisions this means that the optimum should be sought in terms of number of accessions and intra-accession diversity. Important parameters are, among others, the costs of labour, experimental fields and greenhouses. But there are more difficult factors such as the risk of contamination with a given labour input per accession for regeneration, and the loss of value of the accession in the case of a certain degree of contamination or loss of alleles. Economic theory to support these decisions is not available yet, though some developments are encouraging.

The Role of Genetic Markers

Decisions on lumping or splitting samples are generally made on the basis of experience, practical considerations and population genetic theory. Recently, new technology has become available that allows the formulation of expectations or the monitoring of the genetic impact of accession management decisions. Using genetic marker technology it is now possible to quantify the intra-accession diversity, for example, the change in genetic composition of an accession before and after regeneration. We will describe recent and ongoing studies, involving a number of crops, carried out to support practical decisions that were needed at an operational genebank, in this case the Centre for Genetic Resources, The Netherlands (CGN).

Lumping cabbage and brussels sprouts accessions

The Netherlands has a long history of selection and breeding of *Brassica oleracea*. Breeders and farmers have made their own selections of landraces and old cultivars. The Dutch material in the CGN *B. oleracea* collection consists to a large extent of such selections, so-called 'umbrella varieties'. Sometimes there were up to 16 selections from one landrace combined into one umbrella variety. Since cabbage is an insect pollinated biennial crop, and regeneration is difficult and expensive, it was decided to limit the number of accessions as far as possible (Boukema and van Hintum, 1994).

Material derived from the same parental landrace was planted side by side. Assisted by *B. oleracea* experts involved in commercial plant breeding and variety registration, groups of very similar selections were composed. Other selections were kept as individual accessions. In some cases a number of groups from a single umbrella variety were created on the basis of maturity or other distinctive traits. As a result the number of accessions in the CGN collection of Dutch *B. oleracea* was reduced from 273 to 54, a reduction of 80%.

Subsequently, the process of lumping accessions was validated by an isoenzyme study, using a number of cabbage and brussels sprouts groups. We tested the hypothesis that isoenzyme markers would correctly place a single accession in one of the groups. It appeared that most of the accessions were correctly classified. All misclassifications were within similar groups. In two cases the isoenzyme patterns suggested that the groups could have been even larger. In one of these cases this was a real option since it involved two groups made from the same umbrella variety. In the other case it involved groups with a common genetic background but a distinct identity as defined by morphology and history (van Hintum *et al.*, 1996).

Lumping flax accessions

Sometimes genebanks receive collections of potentially valuable germplasm, but with hardly any documentation. An example is the CGN flax collection. For about 30% of the accessions there was no more than a coded accession name consisting of a few letters and a number, such as 'M 25–341' or '324-Rm', for example. These names allowed grouping of the material in series, such as the M 25 or the Rm series. To investigate the genetic relationships of the accessions within and between the different series, an amplified fragment length polymorphism (AFLP) study was carried out on 29 accessions belonging to three of such series. Subsequently, an analysis of molecular variance (Excoffier *et al.*, 1992) was used to compare the genetic variation observed within and among accessions. Substantial differences in intra-accession variation were observed and accessions that were not significantly different were bulked into groups. As a result, more or less homogeneous accessions remained separate entries while more heterogeneous accessions could often be lumped (Fig. 11.1). It appeared that the 29 accessions could be reduced to 14, reducing the among-population component of variance by only 2.6%. This flax case study showed that it was possible to achieve a considerable reduction in collection size while at the same time maintaining similar levels of variation among accessions (van Treuren *et al.*, 2001).

Diversity in barley accessions

Some genebanks routinely divide a barley accession into the morphotypes that are observed within the accession, while at the same time also maintaining a sample with the original constitution. This can result in a considerable number of accessions arising from one original landrace. However, since morphological variation is often restricted, molecular markers can assist in the identification of the different genotypes. In order to determine the applicability of AFLPs for that purpose, intra-accession variation was studied in two cultivars, two landraces, and two wild populations from the CGN barley collection (Fig. 11.2).

The wild material, *Hordeum vulgare* ssp. *spontaneum*, was surprisingly uniform, probably because of the collecting protocol. Each accession could very well have been derived from a single plant. Despite the genetic purification that is normally practised in the development of new varieties, genotypic variation was detected in the cultivars. Assuming that this level of variation can be considered purely as 'background noise', the only accession exhibiting a substantial level of genotypic variation was the landrace from Nepal. However, the genetic differences observed between the genotypes displayed a gradual pattern rather than the separation of genotypes into a number of distinct classes. In this case, splitting the accession into separate lines could not be done unambiguously (van Treuren and van Hintum, 2001).

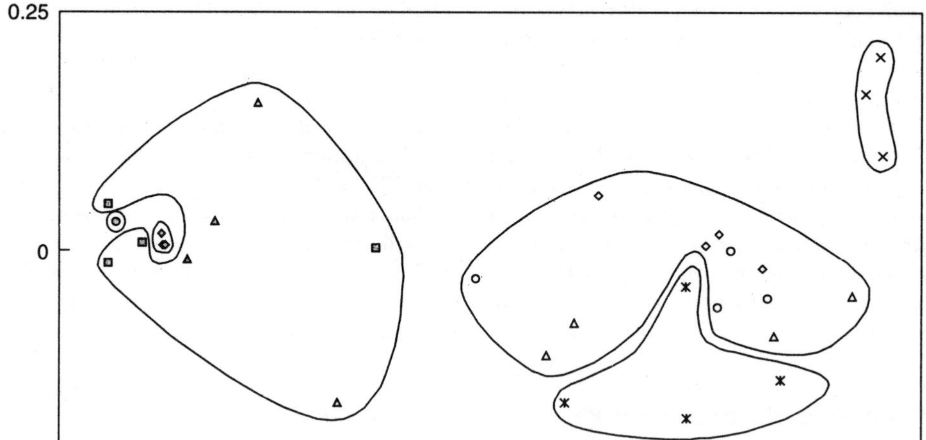

Fig. 11.1. Principal coordinate (PCO) plot of flax plants from the M 25 series. Common symbols are plants from the same accession and the clusters of accessions that qualified for lumping are circled on the plot. The *x*-axis represents 28.0% and the *y*-axis 8.5% of the variation (from van Treuren *et al.* (2001), reproduced with kind permission from Springer-Verlag).

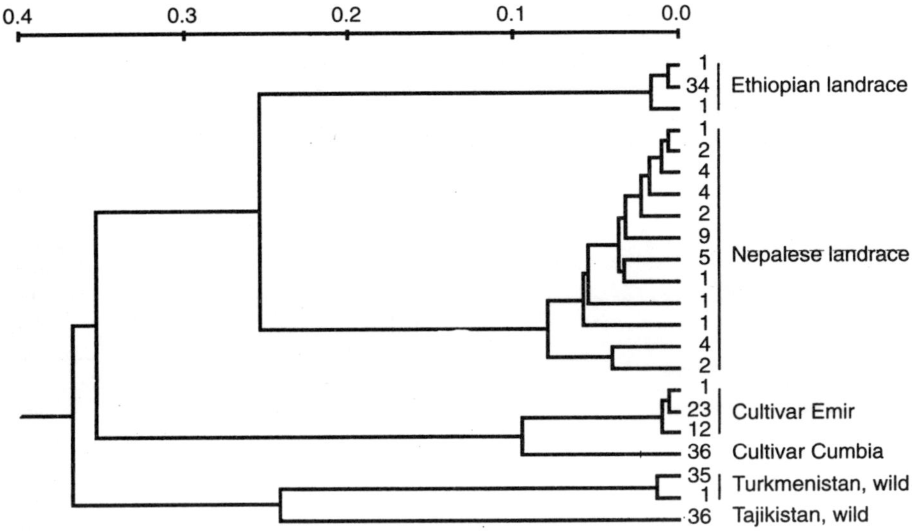

Fig. 11.2. UPGMA cluster analysis of the different barley genotypes based on 104 polymorphic AFLP markers. The numbers at the end of the branches indicate the frequency at which the genotype occurred in the sample (from van Treuren and van Hintum (2001), reproduced with kind permission from Kluwer Academic Publishers).

Conclusions

Accession management decisions concerning the composition of accessions have been made since the first genebanks started operating. However, the pressure to increase efficiency in genebank operations requires a more scientific approach to provide a sound decision-making basis for accession management. This scientific basis of accession management has now become available; marker technology especially can help to monitor genetic consequences of accession management decisions. The only element missing is the economic theory to support these decisions.

References

Boukema, I.W. and van Hintum, Th.J.L. (1994) *Brassica oleracea*, a case of an integral approach to genetic resources conservation. In: *Evaluation and Exploitation of Genetic Resources, Pre-Breeding. Proceedings of the Genetic Resources Section Meeting of EUCARPIA*, pp. 123–129.

Excoffier, L., Smouse, P.E. and Quattro, J.M. (1992) Analysis of molecular variance inferred from metric distances among DNA haplotypes: application to human mitochondrial DNA restriction data. *Genetics* 131, 479–491.

FAO (1996) *FAO Report on the State of the World's Plant Genetic Resources for Food and Agriculture.* Food and Agriculture Organization of the United Nations, Rome, Italy.

van Hintum, Th.J.L. (1993) A computer compatible system for scoring heterogeneous populations. *Genetic Resources and Crop Evolution* 40, 133–136.

van Hintum, Th.J.L. and Visser, D.L. (1995) Duplication within and between germplasm collections. II. Duplication in four European barley collections. *Genetic Resources and Crop Evolution* 42, 135–145.

van Hintum, Th.J.L., Boukema, I.W. and Visser, D.L. (1996) Reduction of duplication in a *Brassica oleracea* germplasm collection. *Genetic Resources and Crop Evolution* 43, 343–349.

Lehmann, C.O. and Mansfeld, R. (1957) Zur Technik der Sortimentserhaltung. *Kulturpflanze* 5, 108–138.

Parzies, H.K., Spoor, W. and Ennos, R.A. (2000) Genetic diversity of barley landrace accessions (*Hordeum vulgare* ssp. *vulgare*) conserved for different lengths of time in *ex situ* genebanks. *Heredity* 84, 476–486.

Rana, R.S., Sapra, R.L., Agrawal, R.C. and Gambhir, R. (1991) *Plant Genetic Resources Documentation and Information Management.* National Bureau of Plant Genetic Resources, New Delhi.

van Treuren, R. and van Hintum, Th.J.L. (2001) Identification of intra-accession genetic diversity in selfing crops using AFLP markers: implications for collection management. *Genetic Resources and Crop Evolution* 48, 287–295.

van Treuren, R., van Soest, L.J.M. and van Hintum, Th.J.L. (2001) Marker-assisted rationalization of genetic resources collections: a case study in flax using AFLPs. *Theoretical and Applied Genetics* 103, 144–152.

12 The Deployment and Management of Genetic Diversity in Agroecosystems

S.D. Sastrapradja[1] and P. Balakhrishna[2]

[1]Indonesian Institute for Sciences, Center for Research in Biotechnology, Raya Bogor, Cibinong, Bogor, Indonesia; [2]Regional Biodiversity Programme, Asia, IUCN-World Conservation Union, Colombo, Sri Lanka

Introduction

The fact that plant genetic resources (PGR) are the very raw materials for crop improvement is generally understood by agriculturists as well as biologists in both developed and developing countries. They are also aware that recently human civilization has undergone rapid change which has had a great impact on PGR existence. Ways and means to conserve PGR for crop improvement have been sought to ensure their continuous availability. Unfortunately, for policy makers in general and in many Vavilovian centres in particular, where commitment to PGR conservation is a determining factor for its implementation, the importance of such resources is usually too abstract and hard to grasp.

As early as 1936 Harlan and Martini noted that crop germplasm in Vavilovian centres was vulnerable to loss due to technological and economic changes. But not until late 1960 did the global efforts to collect and then conserve genetic resources begin in a more concerted way. At the same time the Green Revolution saw the introduction of high yielding varieties (HYVs) of major crops (rice and wheat) in many developing countries. With the introduction of the Green Revolution's products (HYVs), many local varieties of these crops were replaced. A group of concerned scientists assisted the Food and Agriculture Organization (FAO) to take an initiative for a global movement to conserve these crops (Frankel and Bennet, 1970). In 1983 FAO established an intergovernmental forum, the Commission on Plant Genetic Resources, which has now become the Commission on Genetic Resources for Food and Agriculture. This is a permanent intergovernmental body which monitors the implementation of the International Undertaking (IU), a non-binding agreement on PGR which is currently being revised to bring it into harmony with the Convention on Biological Diversity (CBD).

The CBD was signed in 1992 and came into force in 1993. Biological diversity embraces a much larger scope than PGR, in fact, PGR are only a small part of biological diversity (UNEP, 1994). Unlike the FAO International Undertaking, which is a legally non-binding instrument, the CBD is legally binding, hence countries are more serious in implementing the convention compared with the FAO-IU. Moreover, the Conference of the Parties, which is the highest governing body for the CBD, has decided that the Global Environment Facility (GEF) is its temporary financial mechanism allowing the agreed activities on biological diversity to be carried out.

Issues like pollution, depletion of the ozone layer, land degradation, or water scarcity are more easily understood by the public at large than the loss of biological diversity. Although 7 years have passed since the CBD came into force and 5 years since the FAO-Global Plan of Action (GPA) (FAO, 1996) was

 Managing Plant Genetic Diversity (eds J.M.M. Engels, V. Ramanatha Rao, A.H.D. Brown and M.T. Jackson)

launched, the importance of biological diversity, including PGR, for sustainable development has not been mainstreamed into many sectors of governments responsible for implementing development activities. In such a situation it becomes obvious that there is a need to identify key issues within the programme of work of the CBD and the FAO-GPA which can be used as eye-openers to those who should play an active role in national development.

Agrobiodiversity refers primarily to genetic variability in cultivated plants and domesticated animals together with their progenitors and closely related wild species, maintained within agroecosystems and surrounding natural environments. Thus, plants and animals harvested from the wild are also included in this term (Thrupp, 1998). To provide parties of the CBD with a standard term on agrobiodiversity, the Subsidiary Body of Scientific, Technical and Technological Advice (SBSTTA) of the CBD in its fifth meeting (2000) defined the scope of agrobiodiversity, as

> all components of biological diversity of relevance to food and agriculture, and all components of biological diversity that constitute the agroecosystem: the variety and variability of animals, plants and microorganisms, at the genetic, species and ecosystem levels, which are necessary to sustain key function of the agroecosystem, its structure and processes. (UNEP, 2000a)

It should be noted that, within this scope, farmers are identified as managers of the agrobiodiversity. The man-made ecosystems which are intended for agriculture purposes are, therefore, named agroecosystems. The relationship between agrobiodiversity, crop genetic resources, management in agroecosystems, and the complexity of the issues in the context of socio-economy are discussed in this chapter with the emphasis on the Asia Region. Not only does Asia represent several Vavilovian centres, but it is also rich in agroecosystems, agrobiodiversity and population.

This chapter deals with genetic diversity in agroecosystems, the role of agrobiodiversity and genetic diversity in national development, the management of genetic diversity on-farm, challenges to the scientific community and, finally, ways to move forward.

Genetic Diversity in Agroecosystems

Archaeological evidence indicates that humans began domesticating plants and animals approximately 10,000 years ago in several locations of the globe. Through trial and error food crops were identified, selected, cultivated, tested, maintained and improved. This process of domestication continues today. In the emerging agriculture, consideration was not only given to the selection of plants but also to the physical environment in which they grow. Water availability and soil fertility were the determinant factors for cultivating annuals, biennials, perennials or combinations of these. Humans modified natural ecosystems into many types of agroecosystem for the cultivation of plants and the raising of animals. Today's agricultural landscape presents a wide range of practices covering traditional ways of mixed farming to a more intensified monoculture system. In Asia, where most of the population live in rural areas, various types of agroecosystem can be found. In forested areas of South-East Asia, for example, slash and burn agriculture is still going on, while in many parts of the highland areas, simple tools such as those employed in the Stone Age are still being used (Koentjaraningrat, 1994). In general, those farmers practising traditional agriculture are subsistence farmers. In lowland areas where irrigation has been developed, intensified agricultural systems dominate the scene. Unlike those of developed countries, farmers in South-East Asia have small land holdings which, on average, are less than 1 ha (Conway, 1997).

Harwood (1979) described four stages of small farm development: primitive hunting-gathering, subsistence level farming, early consumer, and primary mechanization. The change from hunter-gatherer stage to subsistence level farming allowed a larger human population to live in a given area. Many species of plants were grown together in a semi-cultivated or cultivated way in a given area to satisfy basic human need, that is, food security. In addition to food crops, medicinal plants, fruit trees, and firewood species were also cultivated for their own daily uses. Seeds and other planting materials were routinely gathered from the farm for the next planting season. The same holds true for the need for green manure and organic fertilizer. In this way subsistence farming is non-commercial in nature. Thus, knowledge and practices for managing crops and natural resources were embedded in the local cultures of peoples and households. The species selected for food, medicine and spiritual purposes varied from one place to another. The combination of cultural preference and physical environment induced the formation of so-called landraces or traditional varieties of crops which were unique in each agroecosystem.

Growth of markets and the global spread of a few crops led to the reduction of the varieties of plants grown. Meanwhile, new technologies for farming were introduced which pushed the productivity level of staple food (rice, maize and cassava) still higher. Then came the era of the Green Revolution which is characterized by uniformity and efficiency of the HYVs. Such technology was suitable for arable land where water and soil fertility were secured. Small farming systems which occupied marginal land and less favourable environments, with many traditional crops, were hardly affected by this new technology. HYVs demand arable land, good irrigation and high chemical inputs. Therefore, lowland areas with the above-mentioned conditions are those where HYVs are being cultivated.

The landraces or traditional varieties of major crops which have developed over time in many places in the world are thus being replaced by modern varieties. The displacement of rice landraces between 1970 and 1990 by the International Rice Research Institute (IRRI) through the Green Revolution in South-East Asia (Myanmar, Philippines, Indonesia, Vietnam and Thailand) was documented by Perlas and Vellvé (1997). In 1990 traditional varieties occupied only 45% of the area under rice cultivation in Myanmar, < 20% in the Philippines, < 20% in Indonesia, and 25% in Vietnam. In Thailand, however, modern varieties have not much impact on the existence of traditional varieties, hence they are planted on only less than 10% of the rice growing area. In India, about 70% of the rice area and 95% of the wheat land are now occupied by a handful of HYVs (Jardhari and Kothari, 1996).

New HYVs were not without negative impacts. In Asia, for example, the most populated countries, China, India and Indonesia, welcomed the Green Revolution technology and enjoyed the results of it. However, based on the analysis of the production performance of rice over the past two decades, early indications of unsustainability were revealed ('Rural Asia: beyond the Green Revolution', unpublished Asian Development Bank seminar). The growth of yield per unit area of rice demonstrates a declining trend. The rice producing countries enjoyed an increase in productivity from 1960 to 1990 of, on average, 2.1% a year. This number has dropped to 1% since 1990 (Brown, 1997). One of the explanations is that plant breeders have largely exploited the genetic potential for increasing the share of photosynthate that goes to seed (Brown, 1997). A number of experiments present evidence that microelement deficiencies and toxicities contributed partly to the declining yield (Conway, 1997). Such a trend, of course, should send an alarm signal to policy-makers to find new ways to keep the production on track.

One of the options for securing sufficient food production at the national level is to extend intensive agriculture, which relies on employing HYVs and chemicals in less favourable environments. It should be noted, however, that such areas are usually associated with small farming production systems, many crops and, in general, lack of resources. Moreover, the existing technology associated with HYVs and high chemical inputs mainly deals with single crop farming, Difficult as they are for agricultural intensification, the ADB recommended that less favourable environments be the target areas for agricultural productivity improvement. Should that recommendation be implemented, the traditional agroecosystems and the PGR they contain will go, too.

Meeting the Needs of National Development through Agrobiodiversity

After the Second World War many countries rich in biological diversity, which is predominantly contained in natural forests and oceans, began to exploit this resource for national development. Agriculture in its broadest sense, including forestry and fishery, suddenly became the backbone of their economy. Unfortunately, however, the means to utilize biodiversity is mainly by way of direct exploitation. Timbers and other forest products are harvested from natural forests. The same holds true for fisheries. Although sustainable management, both in forestry and fishery, is theoretically accepted as an optimal solution, such ideal management is hardly followed in practice due to, among other things, lack of appropriate technology for large-scale operation in tropical regions. To extract timber species from natural forest, for example, a technology of selective cutting was developed and followed to ensure the sustainability of the timber stand. But in reality, sustainability of timber production could not be attained because natural regeneration of timber species after cutting was poor, hence the recovery of timber stands did not happen as expected. In the case of cultivated species, such as those of food crops that have been in cultivation for centuries, technology for sustainable production has progressed accordingly.

When the new technology for food production became well known with the introduction of the Green Revolution, every government was eager to welcome it. The yields of the major food crops were indeed doubled or even tripled. At the national level, the miracle of the Green Revolution was praised due to the fact that many countries could avoid importing food. At the community level, however, the Green Revolution seemed to benefit only the resourceful farmers (Conway, 1997). The impact of the Green Revolution on poverty alleviation is subject to interpretation. The poverty level in some countries before and after the Green Revolution was compared and the result showed that the Green Revolution had indeed decreased the number of the poor at the national level (Conway, 1997; Anonymous, 1999). However, in the opinion of many commentators, the technology benefited large farmers at the expense of small ones (Conway, 1997). The challenge to governments thus remains to find ways and means to eradicate poverty and at the same time provide a good quality of life for farmers who seemed to be unaffected by the Green Revolution.

During the Earth Summit meeting in Rio de Janeiro in 1992 countries expressed their commitment to aim at sustainable national development. This means that governments will do their utmost to meet adequately the needs of the citizens today and also those of future generations. In other words, the environment in which we live today should be kept in such a way that coming generations can meet their needs as we do now. Adequate food production, which will secure the availability of food, is the basic element of any national development. The commitment of the international community to ensure food and nutrition security to every child, woman and man on our planet was stated in the Bellagio Declaration in 1989 (Anonymous, 1996) on *Overcoming Hunger in the 1990s*. The FAO (1996) reaffirmed the need to achieve the goal of 'Food for All' because, today, over 10 years after Bellagio, more than 800 million people still suffer from hunger and malnutrition. Many believe that this is not caused by inadequacy of food, but rather its distribution. On the other hand, there is an increasing concern about the Earth's capacity to produce enough food for the growing population (Brown, 1997), despite the fact that science has tremendously improved world food production over the past three decades.

The globalization of the world economy has affected the life of almost everyone on Earth. Farming communities are no exception. The traditional farming community which depends very much on agriculture, is facing pressures to produce more in order to be able to buy various things that are offered in the markets. Unfortunately, however, many cannot rely solely on their farm products to keep pace with such increasing needs. They are compelled to work outside the farm to gather additional income. Those who live near to cities are tempted to abandon farming and migrate into the city. Cities provide them with an opportunity to earn money more easily. As a consequence, most farming communities are no longer isolated. Radios are common and television is also available to some. Both radio and television bring information on modern agriculture as well. New crops are introduced and/or new varieties of such crops, such as melon, tomato, cabbage, chilli pepper, eggplant and bitter gourd, which invade the traditional farming systems. These new crops and varieties are highly valued and the prices they fetch on the market are attractive. Naturally, farmers plant what the market demands despite the fact that new seed often has to be purchased at the beginning of every planting season due to the hybrid nature of those seeds. In this way, traditional varieties of many crops, including those classified as minor crops, are pushed out from the farming systems, which leads to their genetic erosion. Unfortunately, not every country is prepared to take action to collect and conserve these endangered local crops, and the variation within them, in genebanks.

At the global level genebanks of the international agricultural research institutes and several national genebanks conserve germplasm samples of predominantly major crops that are endangered, in *ex situ* facilities. However, not all traditional farmer varieties are represented in these *ex situ* germplasm collections. Therefore, *ex situ* conservation approaches only are not sufficient to conserve all the PGR needed for crop improvement. One possible way to conserve traditional farmers' varieties is through *in situ* conservation on-farm, an option that is gaining more attention nowadays (Brush, 1999). This kind of conservation is conducted in farmers' fields and they are the actors who manage the system within which PGR are just one component.

With regard to *in situ* conservation on-farm, Brush (1999) listed five reasons for promoting it, which are, among others, that agroecosystems will continue to generate new genetic resources, and provide natural laboratories for agricultural research. Brown (1999), describing an attempt to strengthen the scientific basis of *in situ*/on-farm conservation, presents several advantages of *in situ*/on-farm conservation to farmers as well as to management issues including environmental policy. Worede *et al.* (1999) showed that in Ethiopia *in situ*/on-farm conservation is one of the most effective strategies for poor resource farmers to cultivate marginal land with low-input agriculture in a sustainable manner. To support such a farmer-based conservation approach research on genetic, ecological and social dynamic aspects of landraces has now become one of the major programme activities of the International Plant Genetic Resources Institute (IPGRI)(www.ipgri.cgiar.org).

To ensure the genetic integrity of landraces on-farm, Qualset *et al.* (1997) are of the opinion that the agricultural system as a whole needs to be conserved. The question is then how the attitude of farmers, who maintain the system, is expected not to change with time. While the idea proposed is acceptable, from a practical point of view it is unlikely to work. Farmers, like other communities, are socially dynamic and responsive to change. Therefore, according to Pistorius and van Wijk (2000) *in situ*/on-farm conservation should not only be perceived as the possible maintenance of traditional crops but, more importantly, should be linked with rural development programmes. Thus, tangible benefits should be associated with such a conservation effort, if farmers are expected to cooperate.

It will be interesting to learn from the Ethiopian experience whether or not the attitude of farming communities to conserve landraces will change, for instance if poverty is alleviated. In Indonesia, the diversity of crops on individual fields is generally closely linked with subsistence agriculture and poverty (A. Sudjarwo, personal communication, 2000). When government workers advised such farmers to plant more valuable crops for potential markets, farmers did not want to listen. However, when merchants persuaded them to plant certain crops, which were sometimes new species to them, farmers complied readily with the request. The merchants provided them with cash security in advance and farmers did not need to worry whether or not there was a market, an assurance which government officials cannot give. Thus, on-farm conservation is vulnerable to economic incentives provided by other commodities and because of social change. It becomes apparent that incentives for farmers to maintain crop diversity on their farms for national or global needs are indeed necessary.

There is no doubt that *in situ*/on-farm conservation provides an alternative way for countries which lack genebanks to maintain genetic resources and for the global community to complement *ex situ* conservation in genebanks. Attractive as it sounds, in developing countries, conservation of genetic resources is often considered a long-term exercise, whereas politicians would like to see the results of activities for national development during their term of office (E. Salim, personal communication, 2000). Moreover, the site specificity of crops in various agroecosystems means that national policy-makers as well as scientists in general may put less effort into this issue since the impacts, if successful, are felt only locally rather than nationally. Of course, the importance of existing diversity for future crop improvement is well recognized by scientists. However, unless the short-term benefits to farming communities of maintaining diversity are demonstrated, attention from the government cannot be expected.

It is obvious that *in situ* on-farm conservation in countries with many types of agroecosystem, especially those in developing countries, needs government support. The political commitment of governments to enhance food security at the household level has been expressed at the global level (FAO Leipzig Conference, 1996). Such a commitment needs to be integrated into national development plans and translated into the workplans of various sectors of government. Furthermore, in order to reach the household level, a strategy different from that planned at the national level seems to be necessary. After all, the agroecosystems in a country are the building blocks of the national agricultural system. The specificity of agroecosystems and the genetic resources they contain play an important role in guaranteeing local food availability which, in turn, will contribute to national food security. Each agroecosystem offers various sources of carbohydrate, protein and vitamins for local use which, in some cases, differ from those promoted at the national level. If in many countries in South-East Asia rice is considered the main carbohydrate source at the national level, in certain agroecosystems sweet potato, cassava, maize, taro and other minor tuber crops are consumed daily. The same holds true for protein sources. There are a number of legume species, including *Phaseolus* beans, wing bean and velvet bean, which are readily available at the household level. These minor crops are hardly recognized as important food at the national level and hence should be taken into account for targeting household food security.

Managing diversity in agroecosystems

In dealing with the management of biodiversity the CBD stresses the importance of the ecosystem approach (UNEP, 1998). Within the CBD agricultural biodiversity is defined as covering a broad range of elements, including a wide range of ecosystem services. The SBSTTA of the CBD, in its decision V/10 (UNEP, 2000b) elaborates on this ecosystem approach and it is expected that this approach leads to direct as well as indirect values of biological diversity, including that of ecosystem functioning. It considers human dimensions to be an integral part of biodiversity management. In his note on agrobiodiversity, the Executive Secretary of the CBD (UNEP, 2000b) described the application of the ecosystem approach for managing agrobiodiversity, using rice integrated pest management (IPM) as an example. It is shown that farmers who are the managers of their crops, after participating in training on IPM, were able to improve production while pesticide inputs were greatly reduced. He further pointed out that the IPM approach, applied to the rice ecosystem, is consistent with the principles of the ecosystem approach as defined by SBSTTA of the CBD (UNEP, 2000a). However, nowadays in farming rice is mostly cultivated in monoculture while in traditional agricultural practices rice is multiple-cropped with many other species. In Java, the accompanying field crops are maize, cassava, sweet potato and taro, while the legume species include long bean, *Phaseolus* bean and wing bean. They are planted along the borders of the rice field. In addition to rice-based agroecosystems there are several other systems which are based on maize, cassava or other species. In Nusa Tenggara Timur maize is the main crop which is mixed with cassava, upland rice, legumes and water gourd (KEPAS, 1986). Cassava is the main crop planted in drier areas of Gunung Kidul (Central Java), Sukabumi Selatan (West Java) and Pacitan (East Java) and is quite often mixed with biannual legumes such as velvet bean and *Phaseolus* bean.

According to the World Resources Institute (2000), agroecosystems cover more than one-quarter of the global land area. However, almost three-quarters of the land has poor soil fertility. In addition, about one-half of the arable lands occupy steep terrain. To make matters worse, about 40% of agricultural land has been severely degraded. When calculating the size of the human population within various agroecosystems existing in the world and comparing different regions, it is shown that Asia's agroecosystems are the most densely populated. China has a population of 1275 million of which 67% live in rural areas; India has 1000 million of which 72% are in rural areas; and Indonesia has 209.4 million of which 62% live in rural areas. The process of land degradation will continue in the coming years considering that the growth rate in India is 1.9% and in Indonesia 1.6%. Although the Chinese population growth rate is only 1%, the actual increase of the human population in terms of numbers is still high (*Asiaweek*, 2000). Without serious attempts to improve such degraded land, more and more agroecosystems will become wasteland or be converted into other uses. With the steadily growing population in Asia agriculture is the only sector that can absorb a large labour force. Thus, maintaining agricultural land from degradation so that agricultural activities can be sustained will solve in part the problem of job opportunities in rural areas.

A traditional agroecosystem is conceptualized as a web of social relationships between a specific group of people with plants and animals which they keep in a particular space. It is a major repository of PGR. When an ecosystem approach is applied to the management of PGR within an agroecosystem the decentralization of management to the lowest appropriate level is applicable. Farmers are the key player at this level. Their desire to respond to change should be taken into account when designing the *in situ*/on-farm conservation. Moreover, the sustainability of efforts is in the hands of the young generation of farmers. Many of them are no longer interested in farming, which is indeed a hard job and in terms of cash income is not attractive. It is, therefore, necessary to complement efforts to implement *in situ*/on-farm conservation with other economic activities so that the young farmers' generation can choose the best options for their lives.

In Indonesia, for example, the political will of the government to provide enough food at household level has promoted national interest in dealing with agroecosystems and local crop diversity (Badan Urusan Ketahanan Pangan, unpublished). At the moment the country is heavily dependent on rice as the main staple. Indonesia is compelled to import more than 3.5 Mt (Biro Pusat Statistik, 1997) annually since 1993, while in 1984 Indonesia had achieved self-sufficiency in rice production. The increase of rice production between 1969 and 1984 was promoted by the application of HYVs and the technology accompanying it. However, at the beginning of 1993 rice productivity started to decline. The decrease in rice area, especially in Java, coupled with the decline of productivity, contributed to this phenomenon (Purwoto *et al.*, 1998). Another worry is that noodles and bread are now widely accepted as snack food between meals. In this way, wheat, which is not grown in Indonesia, has to be imported in large amounts as well, close to 4 Mt $year^{-1}$. Both rice and wheat imports are heavily subsidized by the government. Therefore, the Department of Agriculture is determined to alter this economically unhealthy situation in the years to come. Since 1972, realizing that the country was heavily dependent on rice as its staple, the government has tried to promote food diversification with the hope that the other carbohydrate producing crops could be cultivated to decrease the demand for rice. Sago palm in eastern Indonesia, sweet potato and taro in the highlands of Irian Jaya, yam in the drier areas of the country, elephant yam and other minor tuber crops may be developed to substitute rice demand in different parts of the country. Together with the minor legumes and fruits which grow in different agroecosystems, they offer a range of alternatives to developing food security not only at local level but also at national level. Unfortunately, however, no serious attempt to create food diversification has materialized.

The future of various agroecosystems and the related diversity of PGR in them is dependent on human culture, particularly that of farmers, whose livelihood is inseparable from their crops. The link between farmers, agroecosystems and PGR as discussed earlier is indeed very complex. The deployment of PGR in an agroecosystem is not a simple matter which can be resolved easily by formal institutions, such as the government, or through informal arrangements, such as non-governmental organizations (NGOs). Like a natural ecosystem, an agroecosystem is a hierarchy of systems from the lowest level of plants, animals and microbes, to farms and the village level. Each lower level of the hierarchy becomes a component of the next higher one. In order to ensure its proper functioning each agroecosystem should maintain the following four characters: productivity, stability, sustainability and equitability. To illustrate these properties, Conway (1997) compares the home garden and the rice field ecosystem. In the home garden system, for example, productivity shows a higher net income, while in the rice system a higher gross income is being generated. With regard to stability, the home garden system allows year-round production while seasonal production with vulnerability to climate and disease variation characterizes the rice field system. Sustainability in home gardens means maintenance of soil fertility and protection from soil erosion; rice fields face heavy pest and disease attack. As far as equitability is concerned, the beneficiaries of home gardens are all households while rice fields benefit the landowner. Here it is shown that the diverse agroecosystem tends to be more sustainable than a comparable field crop system.

As managers of an agroecosystem, farmers should be concerned not only with the crops of interest but also with the totality of the system, including the associated biodiversity and the abiotic components (Almekinders and Struik, 2000). Scientists are in a position to assist farmers in making an agroecosystem work for the four functions mentioned above. Through research the right combination of crops, which will give high productivity, can be made. Moreover, the stability of productivity, which will ensure sustainability in times of stress and protection of soil from erosion, should be attempted. It should be kept in mind that the social and economic aspects which will enhance the livelihood of farmers cannot be separated from the technical aspects of agroecosystems. Therefore, involving farmers in research activities from the very beginning will ensure the adoption of the research results. In this way, equity will be achieved if many farmers are involved. Studies on various agroecosystems in Indonesia were conducted by KEPAS (Research Group on Agroecosystems); on upland agriculture in East Java (KEPAS, 1985a), swampland agriculture of Southern Kalimantan (KEPAS, 1985b) and dryland agriculture in Nusa Tenggara Timur (KEPAS, 1986), developing models which follow the four principles.

The challenge to the scientific community

For more than two decades scientific research on various aspects of *in situ* and *ex situ* conservation of PGR has been conducted all over the world. However, further research is needed in several areas of PGR conservation, especially with regard to the minor crops. In tropical developing countries, where modern improved varieties have not yet been adopted, significant diversity of crops is still managed by farmers in their farming systems. The traditional Dayak community of Long Apo village in Kalimantan (Indonesia), for example, uses 150 food plants of which 67 are wild species. No less than 90 species are cultivated around their compounds (Soedjito, 1999). In other agroecosystems, such as that of 'pekarangan' (home garden) in Java, on average 56 different useful plants are grown around a farmer's house. The size of the land occupied by these plants is only 0.5 ha (Conway, 1997). A case study of Merkal village (Karnataka State) in India shows that villagers are using more than 140 useful plants in their everyday life. Those plants are either cultivated in their fields or taken from the neighbouring forested areas (Bhatta and Bhat, 1997). Considering the enormous amount of diversity, both in terms of farming systems as well as PGR cultivated therein, *in situ* on-farm conservation as an alternative development in germplasm conservation, is gaining growing interest from the scientific community. The challenge remains: how well can scientific information be translated by policy-makers so that it can be integrated into policy actions which, in turn, will enable many actors (scientists, conservationists, seed suppliers and government officials) to work together. The following areas have been proposed as 'entry points' for research.

PGR for food security

Food security is one of the top priorities, which many developing countries are still struggling to achieve. It is true that the application of HYVs since the late 1960s has increased production of major crops significantly in those countries. However, the rate of population growth is keeping pace with production and thus will compel each country to produce more food in the years to come. For that purpose new HYVs which are better than the existing ones need to be developed at the national level. The PGR of the major crops which are needed to assemble such new HYVs are kept in international and national genebanks and, therefore, are readily available. With regard to food security at the local and household level, minor crops will remain the source of daily dietary requirements. Germplasm of these minor crops is not well represented in the genebanks. The importance of the minor crops for household food security was stressed by the Consultative Group on International Agricultural Research (CGIAR) System Review. An international consultation to discuss the role of the International Agricultural Research Centres (IARCs) for crops not specifically covered in their mandates was held in Chennai, India. A number of activities, including public awareness, conservation, processing and marketing, were identified to be appropriate for the work of the CGIAR System (Anonymous, 1999). In order to enhance the role of PGR in local food security, research on the development of improved local varieties which are adapted to a less favourable environment (LFE), on the right combination of crops cultivated, on the level of soil fertility, on the role of soil microbiology, and on the availability of water, needs to be promoted. Improved local PGR will assist farmers to obtain better yields from their lands and, thus, help to secure their need for food. At the same time the diversity of local crops can be maintained within the agroecosystems and, consequently, genetic diversity at large.

Increasing value addition of PGR for income generation

The production of food crops alone is not sufficient to sustain life nowadays. Technology to enhance the added value of crops needs to be developed as well as to increase the monetary gains. Cassava, for example, is produced abundantly by many developing countries as food. The price is low if sold as a food crop. However, flour extracted from cassava, known as tapioca, fetches a good price on the market. By processing it further into other products even higher prices can be obtained. Many crops cultivated in traditional farming systems have not been developed into marketable products. Many species planted or growing in agroecosystems are valued as medicinal or ornamental plants. Research on and development of these crops will promote their value further. Markets to generate monetary income have to be created for new products which are developed from such crops. However, creating such markets for new products is not easy, but once it is done, job opportunities for young farmers may be generated.

Agribusiness development seems to be the answer for boosting the market for local crops. A recent increase in demand for medicinal herbs in Indonesia can be used as an illustration. A number of medicinal plant species are being cultivated by farmers in their backyards or as components of the agroecosystem. The herbal medicine factories rely on farmers for the supply of raw materials. However, they demand quality products and continuity of supplies. Therefore, farmers need to improve the quality of their produce by using good planting materials and, at the same time, also improve their cultivation. The factories process the raw materials into marketable products and distribute the products to their agents. A chain of activities from farmers' fields to consumers through factories and their agents requires many workers and also generates farmers' income. At the same time the development of machinery for processing the medicines is being promoted.

Balancing diversity and environmental quality

Like other components in an ecosystem, agrobiodiversity has a role to play as a valuable ingredient for ecosystem functioning (Thrupp, 1998). The more diverse an agroecosystem is in crop species, the more stable the system is. Diversity of crops, for example, will provide protection from soil erosion by avoiding water runoff. In this way nutrient recycling as well as soil biota functioning are not disrupted. Moreover, crop diversity prevents epidemics of harmful insects and diseases. Thus, application of insecticides is minimized which is environmentally desirable. Moreover, an appropriate combination of crops will require less chemical fertilizer as compared with a monoculture. A mixed cropping system makes the best use of available nutrients, water and sunlight. Each crop matures at a different time so that there is an assurance of adequate ground coverage as protection against wind and water erosion. Quite often, multipurpose tree species are a component of agroecosystems. Fast growing legumes species such as *Leucena leucocephela* and *Sesbania grandiflora* are valuable for firewood. In many less favourable areas in developing countries, however, there is a problem of getting sufficient firewood. Such a demand, if it cannot be fulfilled locally, will affect the safety of the protected areas, because people will get the firewood from nearby forests. This is turn can cause environmental problems such as soil erosion and biological diversity loss. At the national level the issue of firewood for cooking is considered trivial, therefore it has never been included in the discussion of food security. However, at the local level, the problem is real because even if raw food materials are available they cannot be consumed in the absence of firewood. Scientific research on crop diversity and environmental quality may offer technical solutions to these problems.

Traditional farming systems

There is a need to understand the reasons for the threat or disappearance of such traditional farming systems in order to mitigate the loss of genetic diversity. For example, the policies of governments to encourage HYVs, market forces, and the introduction of new crops, may cause the disappearance of traditional farming systems. It is worrying to note that the number of landless farmers tends to increase over time. By keeping traditional farming systems alive, and at the same time putting agribusiness in place, these landless farmers will have an opportunity to remain in the agricultural sector.

Scientific information for policy formulation

In many developing countries, policy in agriculture promotes rapid change which leads eventually to a decrease in agroecosystem types. Monocultures dominate the existing agroecosystems because of their efficiency. Uniformity in crop varieties signifies 'modern' agriculture with its high inputs. Credits are provided so that farmers can readily adopt the system. Such technology is appropriate for certain agroecological conditions, leaving the less favourable areas almost untouched. However, agricultural expansion as a way to produce food by growing more major crops will, sooner rather than later, expand to these areas. To diversify agricultural practices that are suitable for agroecological and socio-economic conditions, new policies such as small credit schemes for village enterprise, marketing facilities for minor crops, and investment in the development of processing equipment, need to be formulated. Scientific information should provide a sound basis for the development of such policies.

Ways Forward

Agrobiodiversity offers a powerful defence against the fatigue of the Green Revolution and the looming food shortages. Intensification of efforts to conserve agrobiodiversity by way of *in situ*/on-farm conservation can help us to decrease the burden on the environment, as well as to ensure the future availability of PGR for plant improvement. Such efforts must include:

1. *Farmers' empowerment.* In many regions of the world farmers who live in marginal lands and have limited resources are still practising traditional agriculture despite the promotion of more efficient technology. If farmers are expected not to abandon their crop diversity as part of an *in situ*/on-farm conservation strategy, farmers need to be empowered to enlarge their food basket with local grains, tuber crops, fruits and vegetables. Not only does such empowerment lead to enhanced livelihood security, but it also allows increased usage of crop varieties that are being neglected (Swaminathan, 1994).
2. *Mobilization of scientists.* Considering the large number of farmers who need to be empowered, scientific criteria for selecting places and areas to be included in an *in situ*/on-farm conservation programme should be a priority. Improved local crops, the betterment of agricultural practices, as well as adding increased value to local agricultural products, and the social implications of the changing economic atmosphere of the country as a whole, are the immediate issues for scientists dealing with *in situ*/on-farm conservation. Both National Agricultural Research Systems (NARS) and centres of the Consultative Group on International Agricultural Research should work together in providing scientific information for policy-makers and practical solutions for farmers.
3. *Government commitment.* Without government commitment to deal with farmers who have limited resources, an *in situ*/on-farm conservation programme will merely be a scientific exercise. Therefore, as suggested by Pistorius and van Wijk (2000), to be beneficial to farmers, efforts towards *in situ*/on-farm conservation should be integrated into rural development programmes. Politically, all governments that are members of FAO expressed such a commitment in 1996. What is needed now is the political action to materialize the commitment by allocating sufficient resources to carry out rural development programmes.
4. *NGO involvement.* Unlike in the formal governance system, the NGOs' way of operating is flexible and less bureaucratic. Moreover, members of NGOs are usually young, energetic and full of dedication and idealism. Donors trust them to work at the grass-roots level with communities. Therefore, the role of NGOs in *in situ*/on-farm conservation activities should be recognized and be included in the overall efforts. A forum of NGOs, scientists, private sector and government representatives at the local level, which meets periodically, may provide a vehicle for working together effectively and in a harmonious way.
5. *Private sector participation.* By nature the private sector is geared towards profit making. Unless they see the dividend of their efforts, they are reluctant to act. Poor farmers are not their business targets. However, there is a possibility that they will join efforts for *in situ*/on-farm conservation if invited. They have to be convinced that they have a significant role to play in an effort to help those who are less fortunate. Consequently, the private sector is not in the group of stakeholders concerned with the conservation of genetic diversity, especially on-farm conservation.
6. *Global community attention.* At the moment, not all local crop genetic diversity which still exists in farmers' fields around the globe, is represented in genebanks. Therefore, if *in situ*/on-farm conservation is considered as an important complement to existing *ex situ* conservation efforts, ways and means to make this complementarity work should be attempted. The global community is in a position to assist those countries which are identified as centres of diversity for specific crops to conserve their local crops' genetic diversity on-farm. There is no simple way of doing this, because on-farm conservation involves not only plant species, and their respective habitats, but also farmers as managers of the plant diversity. Dealing with farmers whose culture is as diverse as plant diversity is certainly a great challenge.

References

Anonymous (1996) *Prioritisation for conservation of agrobiodiversity: Recommendations for GEF.* Report submitted to GEF. M.S. Swaminathan Research Foundation, Chennai, India.

Anonymous (1999) *Enlarging the Basis of Food Security: Role of Underutilized Species.* Proceedings of the International Consultation. M.S. Swaminathan Research Foundation, Chennai, India.

Almekinders, C. and Struik, P.C. (2000) Diversity in different components and at different scales. In: Almekinders, C. and de Boef, W. (eds) *Encouraging Diversity: The Conservation and Development of Plant Genetic Resources.* Intermediate Technology Publications, London.

Asiaweek (2000) Bottom Line. *Asiaweek* 22 September 2000.

Bhatta, G. and Bhat, P. (1997) Conserving biodiversity – what people say. A case study of Merkal Village, Sringeri Taluk, Karnataka State.

Biro Pusat Statistik (1997) *Statistik Indonesia 1979.* Jakarta.

Brown, A.H.D. (1999) The genetic structure of crop landraces and the challenge to conserve them in-situ on farm. In: Brush, S.B. (ed.) *Genes in the Field: On Farm Conservation of Crop Diversity.* Lewis Publishers, Boca Raton/IDRC, Ottawa/IPGRI, Rome, Italy.

Brown, L.R. (1997) *The Agricultural Link: How Environmental Deterioration could Disrupt Economic Progress.* Worldwatch Institute, Washington, DC.

Brush, S.B. (1999) The issues of in-situ conservation of crop genetic resources. In: Brush, S.B. (ed.) *Genes in the Field: On Farm Conservation of Crop Diversity.* Lewis Publishers, Boca Raton/IDRC, Ottawa/IPGRI, Rome, Italy.

Conway, G.R. (1997) *The Doubly Green Revolution.* Penguin Books, Harmondsworth, UK.

FAO (1996) *The Conservation and Sustainable Utilization of Plant Genetic Resources for Food and Agriculture: The Global Plan of Action.* FAO, Rome, Italy.

FAO (1996) *Synthesis of the Technical Background Documents.* World Food Summit, Food and Agriculture Organization of the United Nations, Rome, Italy.

Frankel, O.H. and Bennet, E. (eds) (1970) *Genetic Resources in Plants. International Biological Programme Handbook* No. 11. Blackwell Scientific Publications, Oxford, UK.

Harlan, H.R. and Martini, M.L. (1936) *Problems and Results of Barley Breeding. USDA Yearbook of Agriculture.* US Government Printing Office, Washington, DC.

Harwood, R.R. (1979) *Small Farm Development. Understanding and Improving Farming Systems in the Humid Tropics.* Westview Press, Colorado.

Jardhari, V. and Kothari, A. (1996) *Conserving Agricultural Biodiversity: the Case of Tehri Garthwal and Implications for National Policy. Using Diversity.* Proceedings of a Workshop. International Development Research Centre, New Delhi, India.

KEPAS (1985a). *The Critical Uplands of Eastern Java: an Agroecosystem Analysis.* Badan Penelitian dan Pengembangan Pertanian, Departemen Pertanian, Jakarta.

KEPAS (1985b). *Swampland Agroecosystem of Southern Kalimantan.* Badan Penelitian dan Pengembangan Pertanian, Departemen Pertanian, Jakarta.

KEPAS (1986) *Agro-ekosistem Daerah Kering di Nusa Tenggara Timur.* Badan Penelitian dan Pengembangan Pertanian. Departemen Pertanian, Jakarta.

Koentjaraningrat (1994) *Irian Jaya.* Penerbit Jambatan, Jakarta.

Perlas, N. and Vellvé, R. (1997) *Oryza Nirvana?* A SEARICE Publication, Manila, Philippines.

Pistorius, R. and van Wijk, J. (2000) On farm conservation: a matter of global concern or local survival? In: Almekinders, C. and de Boef, W. (eds) *Encouraging Diversity: The Conservation and Development of Plant Genetic Resources.* Intermediate Technology Publications, London.

Purwoto, A., Sri Hartoyo and Suryana, A. (1998) *Penawaran, Permintaan dan Konsumsi Pangan Nabati Indonesia. Prosiding Wiyakarya Nasional Pangan dan Gizi V!, Jakarta, Serpong, 17–20 February 1998.* Lemabga Ilmu Pengetahuan Indonesia.

Qualset, C.O., Damania, A.B., Zanatta, A.C.A. and Brush, S.B. (1997) Locally based crop plant conservation. In: Maxted, N., Ford-Lloyd, B.V. and Hawkes, J.G. (eds) *Plant Genetic Conservation. The* in situ *Approach.* Chapman and Hall, London, pp. 160–175.

Soedjito, H. (1999) Masyarakat Dayak: Peladang Berpindah dan Pelestari Plasma Nutfah. In: Adimihardja, K. (ed.) *Petani.* Humaniora Utama Press, Bandung.

Swaminathan M.S. (1994) *Ecotechnology and Rural Employment: A Dialogue.* MSSRF, Chennai, India, pp. 366–379.

Thrupp, L.A. (1998) *Cultivating Diversity: Agrobiodiversity and Food Security.* World Resources Institute, Washington, DC.

UNEP (1994) Text and Annexes. UNEP/CBD/94/1, Switzerland. CBD, United Nations Environment Programme (UNEP), Nairobi, Kenya.

UNEP (1998) A program of change. *Decisions from the Fourth Meeting of the Conference of the Parties to the Convention on Biological Diversity*. Secretariat of the Convention on Biological Diversity, Montreal, Canada.

UNEP (2000a) UNEP/CBD/COP/5/3, 25 February 2000. CBD, United Nations Environment Programme (UNEP), Nairobi, Kenya.

UNEP (2000b) UNEP/CBD/COP/5/INF/11, 28 April 2000. CBD, United Nations Environment Programme (UNEP), Nairobi, Kenya.

Worede, M., Tesemma, T. and Feyissa, R. (1999) Keeping diversity alive: an Ethiopian perspective. In: Brush, S.B. (ed.) *Genes in the Field: On-farm Conservation of Crop Diversity.* Lewis Publishers, Boca Raton/IDRC, Ottawa/IPGRI, Rome, Italy.

World Resources Institute (2000) *A Guide to World Resources 2000–2001. People and Ecosystems: The Fraying Web of Life.* World Resources Institute, Washington, DC.

13 Combining Static and Dynamic Management of PGR: a Case Study of *Beta* Genetic Resources

L. Frese
Federal Centre for Breeding Research on Cultivated Plants (BAZ), Gene Bank, Braunschweig, Germany

Introduction

The advantages of maintaining accessions in an *ex situ* collection are many and diverse. Firstly, if managed according to the international guidelines for genebanks the samples are expected to be safe. Secondly, access to the seed or planting material is generally easy. Thirdly, during the course of time much information is generated on genebank accessions and can be accessed with increasing ease through the Internet. To ensure reliable information on the samples now and in future it is essential that the accessions do not change their character. Genebanks must therefore manage static collections of plant genetic resources for food and agriculture (PGRFA) while keeping their genetic integrity.

Evolution and domestication are dynamic processes with a biological, time and space dimension. When we collect seeds, plant parts or plants and conserve them as genebank accessions we sample only a specific segment of the species' evolution or the crops' domestication history. In contrast to *ex situ* collections it is a declared aim of *in situ* and on-farm conservation programmes to allow genetic changes to take place in plant populations. Bretting and Duvick (1997) emphasized the dynamic nature of *in situ* and on-farm conservation and suggested that the term *in situ* conservation, including *in situ* on-farm conservation, should be substituted by 'dynamic conservation' and *ex situ* conservation by 'static conservation'. Taking into account that maintenance of plant genetic resources 'on the farm' always comprises a selection component, the term on-farm management of PGRFA has become popular (Gass *et al.*, 1999) and in fact describes what farmers in traditional farming communities have achieved for centuries: they manage the genetic diversity on their farms as they manage the soil fertility of their farm fields.

The Global Plan of Action (GPA) developed by the 4th International Technical Conference on PGRFA held at Leipzig appeals to countries to 'support on-farm management and improvement of PGRFA' in 'Activity 2' (FAO, 1996). The recommendation aims at the development of national programmes that encourages farmers to use and maintain their crop germplasm. The main target group of the recommendation are countries where farmers still reproduce their own seed material. With the beginning of variety breeding based on scientific knowledge, in industrialized countries, the traditional seed supply systems have changed. Around 1900 especially skilled seed producing farmers set up seed production companies and created the economic basis of today's breeding industry. The development of medium- and large-sized breeding companies led to economically efficient task sharing between the seed producers and the growers. Currently, in Germany only very few farmers develop and maintain their own varieties and these

 Managing Plant Genetic Diversity
(eds J.M.M. Engels, V. Ramanatha Rao, A.H.D. Brown and M.T. Jackson)

farmers belong to the ecological farming sector, mainly. Against the background of this historical development it is difficult to define what on-farm management of PGRFA could mean for farmers in a highly industrialized country like Germany.

The German Ministry of Agriculture is currently developing an integrated national concept for PGRFA whereby *in situ* and on-farm measures shall supplement *ex situ* activities and vice versa. *In situ* and on-farm measures need to be planned warranting a continued adaptation of a wide range of crops and related wild species to environmental changes. This would supplement the already existing extensive *ex situ* conservation and utilization programme. While the objectives of *in situ* conservation projects of wild species can be formulated more straightforwardly, it appears to be more difficult to elaborate on the function of farmers as managers of crop germplasm on their farms. Traditional seed supply systems no longer exist and for conventional farmers the only incentive to produce seeds of modern varieties of self-pollinating cereals on the farm is saving seed costs. However, there is also a growing public interest in growing 'old landraces'. Several to a few hundred accessions of major and minor crops are reproduced by engaged individuals or seed saver associations on a limited acreage today (Maier, 2000).

There are reasons to subsidize farmers for the cultivation of obsolete crops and varieties on their farm because the management of PGRFA can be regarded as an ecological service which has to be rewarded by society. As early as 15 years ago, it was argued that the production and use of a higher diversity of crops will help to reduce environmental problems arising from extremely narrow crop rotation systems (Dambroth, 1985). On-farm management projects are considered as potential founders of regional production initiatives serving specific market niches that may grow in time and may finally significantly contribute to a higher genetic diversity in crop production. Some crops have indeed found market niches as regional products like the old potato variety 'Bamberger Hörnle' or the 'Einkorn' wheat.

The establishment of on-farm management programmes has been called for with the argument that we must promote the continued evolution of crops. It can be assumed that the evolution of a crop on the farm, that is, a measurable net gain in useful genetic variation, will be hard to measure, even if a programme lasted a few decades. What we could prove during the duration of a public funded programme is useful adaptation of crop genetic resources to different ecological conditions and we could test new concepts for the maintenance and more efficient utilization of inherited traits. The key question is: which useful and measurable effects can we expect to achieve with regard to sustainable agriculture and use of PGRFA and how farmers in a highly task-sharing and market-oriented society can support on-farm management and improvement of PGRFA in an economically feasible way.

According to Jana (1999) conservationists and genebank curators should seek for the *best practice* that helps to prevent loss of genetic diversity using either *in situ*, on-farm or *ex situ* methods or a combination thereof. The common objective of *in situ*, on-farm and *ex situ* programmes is to manage genetic resources so as to guarantee their evolution, adaptation and availability for international exchange. Hence, for the sake of clarity, in this chapter procedures and measures that guarantee long-term conservation of the genetic integrity of accessions, easy access to collections and access to reliable information linked with them are called 'static management' of PGRFA. Procedures and measures that sustain the genepool and its evolution and assist in operating the adaptation process of the crop are called 'dynamic management' of PGRFA.

Organizations like the World *Beta* Network (WBN) are challenged to develop a germplasm conservation and management concept integrating static and dynamic measures. Germplasm management concepts that may be useful to supplement the static management of PGRFA in countries with a highly developed seed industry have, for example, been discussed by Schnell (1980), Namkoong (1989), Bretting and Duvick (1997), Jana (1999) and more specifically for the genus *Beta* by the Swedish plant breeder Bosemark (1989). The first operational programme linking static and dynamic management procedures for wild *Beta* was developed by the Turkish genebank. Populations of *Beta* species occur in Gene Management Zones where they are maintained *in situ* along with the target species (Tan *et al.*, 2000). This is the only example of a national programme managing and monitoring *Beta* populations in their natural environment. In the cultivated species the static management of genetic resources is currently supplemented by dynamic management approaches for sugarbeet in France and Germany. A more detailed description of the underlying concepts is provided in the following sections.

Description of the Crop, its Genepool and Breeding History

The sugarbeet is produced on about 7.5 Mha in countries mainly located in the Northern hemisphere. Fodder, garden and leaf beets are of much less economic importance. In addition, the garden beet has some economic significance as a storage vegetable in Eastern Europe. The genus consists of four sections: *Beta* (primary genepool), *Corollinae* and *Nanae* (secondary genepool) as well as *Procumbentes* (tertiary genepool) (Table 13.1) (Buttler, 1977; Letschert, 1993; Lange *et al.*, 1999).

The genus *Beta* is native to Europe and adjacent areas (Fig. 13.1). Sections *Nanae* (Greece) and *Procumbentes* (Canary Islands) have a limited distribution area, while wild species of section *Beta* occur along the coastline from the south of Sweden to Morocco and from the Canary Islands to Iran. Section *Corollinae* has a large distribution area in Turkey and neighbouring countries. The centre of diversity is probably located where the species distribution of sections *Beta* and *Corollinae* overlap (eastern Turkey and the western part of Transcaucasia). The domestication of beets probably started in the Euphrates and Tigris regions and continued in Turkey and Greece from where cultivated beets were introduced to northern Europe (Boughey 1981). Cultivated beets have also occurred in China since the 5th century (Sun Yi Chu, 1994) and in Arabic countries.

One of the youngest cultivated forms, the sugarbeet, has become a cash crop of worldwide importance which has been cultivated on a large scale only since 1806 when Napoleon decreed that beet should be grown for sugar. As the sugarbeet was probably selected from one single cultivated population only, the 'White Silesian', the genetic base of the crop is supposed to be very narrow. The 'White Silesian' beet had a rather low sugar content; however, the German scientist Achard considered the good root shape of this fodder-beet-like type as a favourable trait and started to select within this population on higher sugar content and yield. The sugar content increased from 4% to 16.5% between 1784 and 1981 (Winner, 1981).

Table 13.1. Taxonomy of the genus *Beta*.

Primary genepool	Section *Beta* syn. *vulgaris* Ulbrich
	B. vulgaris L.
	subsp. *vulgaris* (cultivated beets)
	Leaf beet group
	Garden beet group
	Fodder beet group
	Sugar beet group
	subsp. *maritima* (L.) Arcang.
	subsp. *adanensis* (Pamuk.) Ford-Lloyd & Will.
	B. macrocarpa Guss.
	B. patula Ait.
Secondary genepool	Section *Corollinae* Ulbrich
	Base species
	B. corolliflora Zosimovich
	B. macrorhiza Steven
	B. lomatogona Fisch & Meyer
	Hybrid species
	B. intermedia Bunge
	B. trigyna Wald. & Kid.
	Section *Nanae* Ulbrich
	B. nana Boiss. & Heldr.
Tertiary genepool	Section *Procumbentes* Ulbrich syn. *Patellares*
	B. procumbens Smith
	B. webbiana Moq.
	B. patellaris Moq.

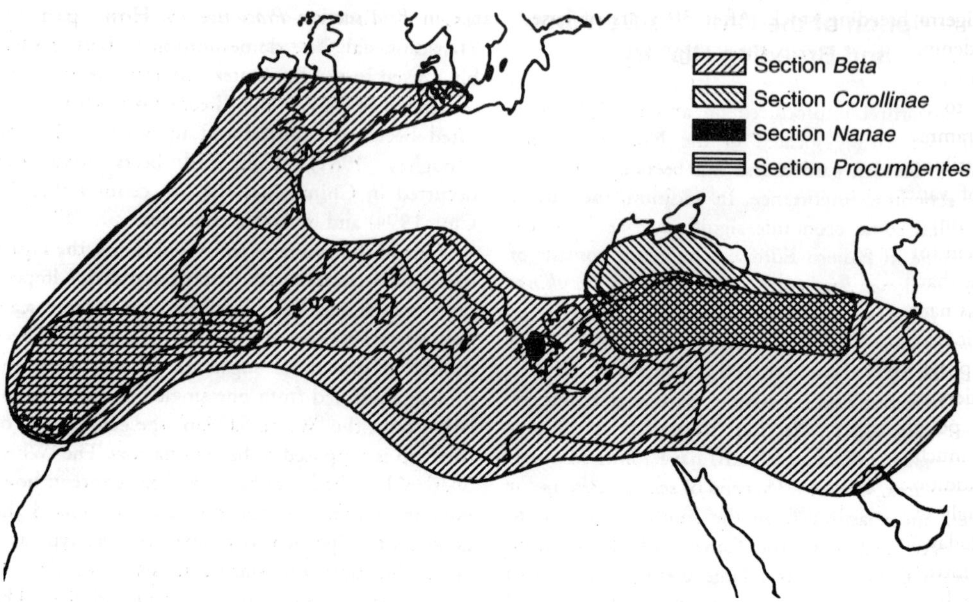

Fig. 13.1. Distribution of wild species of the genus *Beta* grouped by sections.

Before 1960 open pollinated multigerm varieties were developed using family selection methods and the breeding material had a comparatively broad genetic variation (Desprez and Desprez, 1993). Compared with potato, barley and other economically important crops, sugarbeet did not seriously suffer from pest and disease attacks or adverse environmental conditions in the main production areas (Lewellen, 1992). Although almost all beet pests and diseases were already known, the sugarbeet crop was considered as a relatively healthy crop until the 1960s. However, because of the growing acreage, the sugarbeet was increasingly cultivated in short crop rotation, amplifying disease problems. During the 1960s the rising cost of hand labour became an even more pressing problem. The ordinary sugarbeet seed ball contains 3–4 seeds, seedlings emerge in clumps and had to be thinned by hand. Savitsky (1950) found in a seed field of about 1.5 ha one monogerm, homozygous plant. The monogerm genotype occurred in a component of the synthetic 'Michigan Hybrid 18' (Lewellen, 1992). The monogerm character was essentially inherited by a single recessive gene which became of great economic importance in sugarbeet breeding and production. Lines derived from that plant such as SLC101 (Savitsky, 1952) were extensively used in breeding programmes.

In 1942 Owen (1948, 1954) discovered cytoplasmic male sterile (cms) germplasm. After the Second World War, sugarbeet breeders focused their work on the development of monogerm, cms hybrid varieties with a high sugar quality, a high yield of recoverable sugar (Oltmann *et al.*, 1984) and a good level of field resistance to diseases. In the beginning, only one monogerm seed parent line (C562) was resistant to bolting and all of the first monogerm varieties from the Hilleshög company, which reached more than 60% of the European market (cv. 'Monohill' and others), were derived from this line.

For the maintenance of cms lines specific genotypes are required that, when crossed with a cms plant, produce a 100% male sterile progeny. In sugarbeet breeding these maintainer lines are called 'O-types' which were first selected from multigerm material. Hence, monogerm O-types had to be developed by crossing multigerm O-types with monogerm material tracing back to the single plant found by Savitsky. All breeders used this monogerm material to introduce the trait into their elite cms- and O-type elite breeding stock. During this process the female breeding pool passed through a genetic bottleneck. Breeders had to enlarge the female genepool by crossing monogerm germplasm with a wide range of breeding families from the multigerm breeding stock. After 30 years of base broadening, the female genepool has sufficient variability (Frese and Desprez, 1999). Today, breeders wish to broaden the whole basis of the breeding programme as such (female and male pool).

Strikingly, for sugarbeet breeders the collective set of varieties and breeding material still is the most important genetic resource used for the development of improved varieties today. One could argue that the genetic basis of the crop as such is not as narrow as generally assumed. It seems rather that it is a lack of specific traits that hampers breeding progress. Indeed, because commercial plant breeders use a large number of different, heterozygous pollinator populations, hybrid varieties still have much genetic variation (Bosemark, 1979).

Additionally, sugarbeet is a wind-pollinated, strongly out-crossing crop. Therefore, plant breeders today may profit from exotic germplasm that was introgressed in the sugarbeet breeding pool, either by chance or deliberately by breeders. First introductions of wild germplasm into the sugarbeet probably occurred at the beginning of the 20th century, in Russia, the USA and Italy. Cultivated × wild beet crosses were, for example, described by Tjebbes (1933) who used *Beta vulgaris* subsp. *maritima* from the North Sea coast with a sugar content ranging from 15.7% to 17.6%, and Munerati (1932) who crossed a population from the Po estuary with sugarbeets to introduce genetic variation for resistance to *Cercospora beticola*. The Munerati material, in particular, has been widely used in breeding. Due to a negative correlation between leaf spot resistance and exploitable sugar yield, selection of varieties with a high resistance and a high sugar quality and yield proved to be extremely difficult.

Probably because of these early experiences with wild beet crosses, in the 1970s and early 1980s there was a great fear that introgression of undesirable genes of wild or exotic germplasm along with the desired disease resistance trait would destroy the results of costly selection on high sugar quality and bolting resistance. However, the view on potential benefits arising from the utilization of exotic germplasm began to change when soil-borne diseases like the beet cyst nematode (*Heterodera schachtii*) (Hellinga, 1943) or the beet necrotic yellow vein virus (BNYVV) (Grünewald *et al.*, 1983) started to spread and threaten sugarbeet production in the whole northern hemisphere. In 1956 Savitsky (1960) detected a strong *H. schachtii* resistance in *Beta* section *Procumbentes*. However, due to strong crossing barriers between section *Beta* and section *Procumbentes*, the utilization of this source proved to be very difficult and time-consuming. Since the strong nematode resistance in section *Procumbentes* was the only source, scientists from the public and commercial sector struggled through the problems and finally succeeded in producing resistant germplasm used in variety breeding (Uphoff, 1997). It is easy to understand that this experience did little to promote a broader use of exotic material in breeding programmes (Desprez and Desprez, 1996).

In the 1980s, the continued collection and evaluation efforts of the USDA/ARS programme yielded more and more exciting results on new sources of resistances, for example to the BNYVV (Doney and Whitney, 1990) in *B. vulgaris* subsp. *maritima*, which crosses easily with sugarbeet. Since then, the interest in utilization of *Beta* genetic resources collections has been increasing worldwide. Breeders are mainly searching for disease resistance genes in collections to supplement their breeding pool. Nowadays, they keep more than just a single accession with a desired trait in their working collection to capture allelic variants or additional genes just in case the one used is broken by a new pathotype. The introduction of additional genetic variation for sugar content and yield genes from exotic germplasm together with the target trait is thereby welcomed as a positive side-effect that can benefit breeding progress in the long run. Today, breeders wish to broaden the whole basis of the breeding programme as such (female and male pool). The sources for long-term strategic breeding projects are stored in genebanks.

Static Management of Beet Genetic Resources by Genebanks

Characters like the cms and monogermy have had a substantial impact on agricultural practices and the economy of the sugarbeet. cms facilitated breeding of genetical monogerm hybrid varieties. cms and monogermy were found within the sugarbeet breeding material. With a growing demand for pest and disease resistant varieties for sustainable agricultural production, breeders are perceiving that the genetic variation contained in the elite stock will not suffice to match all the breeding aims. *Beta* genetic resources received increased interest in the

1980s when scientists began to report on new sources of resistance in *B. vulgaris* subsp. *maritima* against diseases like rhizomania and *Cercospora beticola*. Collecting activities were intensified to rescue endangered germplasm and to manage this potentially valuable material in static collections. Today, about 9500 accessions of wild and cultivated forms are maintained within the WBN, a network of decentralized collections located in 23 countries. The International Database for *Beta* (IDBB) managed by the BAZ Gene Bank serves as a central link between collections and as a network management tool (Frese, 1992; Germeier and Frese, 2000). In most cases, users contact the national genebanks directly and request information and/or seeds which are provided at no cost. Users looking for a specifically composed set of accessions often request the IDBB manager to search for suitable material, to compile a list of accessions and to provide the addresses of genebanks holding the germplasm. Within the WBN, germplasm curators and other scientists collaborate closely. Task-sharing in all relevant fields, that is, collecting missions, collection management, characterization, evaluation and utilization, is either jointly organized or information on national activities is exchanged.

Together the members of the WBN seek to complete the world *Beta* genetic resources holding. The IDBB is used to detect geographic and taxonomic gaps in the global holding (Hazekamp and Frese, 1992) and is applied as a planning tool to prevent unintentional duplication of collecting missions and germplasm accessions. In the 1980s and 1990s many collecting missions have been implemented to rescue endangered landraces and wild populations or to gather material that could be of use in breeding programmes. A review of the collecting activities was presented by Doney *et al.* (1995). Since then only a few collecting missions have been implemented in Italy and Azerbaijan, mainly to cover existing geographic gaps in the world holding. Decreasing collecting activities indicate that a large proportion of the *Beta* genepool is already stored in genebanks.

All three genepools of *Beta* (see Table 13.1) contain useful traits (Dale *et al.*, 1985; van Geyt *et al.*, 1990; Lewellen, 1992; Paul *et al.*, 1992; Stanescu, 1994; Büttner *et al.*, 1997; Yu, 1997; Panella, 1998). Examples are provided in Table 13.2. Within the framework of the European project 'Evaluation and enhancement of *Beta* collections for extensification of agricultural production – GENRES CT95 42' funded by the Commission of the European Countries a *Beta* core collection has been developed to improve access to the genetic variation summarized in Table 13.2. Since 1996 the core collection is mainly evaluated for disease resistance. This more recent activity is supplementing the continued evaluation work carried out by various institutions under the coordination of the USDA/ARS programme (Doney, 1998) and sporadic evaluation work carried out by breeding companies and research institutes in European, North African and Asian countries. There is an increasing amount of evaluation data available in genebank information systems and breeders are using this information to identify material useful for the introgression of novel genetic variation into their elite breeding pools.

If a trait is purposefully introduced into the breeding pool through crossing and selection, this is called introgression (Simmonds, 1993). The introgression of BNYVV resistance genes from wild sources into the sugarbeet is a recent example. The disease is transmitted by the vector *Polymyxa betae* and caused by the BNYVV. In the case of strong BNYVV resistance, which is possibly inherited by two closely linked major genes (Scholten, 1997), the trait is introgressed into sugarbeet elite breeding material starting with the fixation of the gene in a donor line through selection of a resistant plant in a wild beet accession and the development of a non-segregating, resistant inbred line. Subsequently, the line is crossed with the sugarbeet (Büttner *et al.*, 1997). BNYVV resistances genes can be attributed to resistance donor names like 'Holly-1-4' (gene *Rz*), collection numbers (WB42, gene *Rz2*) (Scholten, 1997) or genebank accession numbers (BGRC54817/RNR870909, gene not yet identified) (Büttner *et al.*, 1997). The original donors are well identified and if the accessions are properly managed in collections the resistance data documented in genebank information systems will remain valid and valuable for future generations of plant breeders. Static management of this specific material is the best possible conservation practice.

From the breeders' point of view static management of genetic resources may not be the most efficient way to maintain traits that are quantitatively inherited. There is the risk of losing minor genes for disease resistance when heterogeneous wild beet accessions are maintained *ex situ* in a region where the disease hardly occurs. In *Beta* little is known about the co-adaptive interaction between diseases and the wild species harbouring resistance genes. The co-adaptive processes resulting in the evolution of a number of valuable resistance genes are far from being understood. Nevertheless, *B. vulgaris* subsp. *maritima* growing in the Po estuary in north-eastern Italy and the leaf spot disease (*C. beticola*) may serve as a hypothetical but realistic example. The region is known for regular and heavy disease epidemics and we can assume co-adaptation between plant and pathogen. A seed sample of *B. vulgaris* subsp.

Table 13.2. Occurrence of useful breeding traits in species of the genus *Beta*.

	Section																		
	I									II					III	IV			
	Taxon code																		
Trait	1	2	3	4	5	6	7	8	9	10	11	12	13	14	15	16	17	18	19
cms	■	■				■	■												
Genic male sterility	■					■			■										
Salt tolerance							■		■		■	■	■	■	■				
Frost tolerance											■	■	■	■	■	■			
Curly top						■					■	■					■	■	■
Yellowing viruses							■						■	■	■		■	■	
Beet mosaic virus							■				■		■		■				■
BNYV virus							■				■						■	■	■
Yellow wilt			■				■												
Peronospora farinosa	■		■		■		■	■			■				■				
Erysiphe betae							■												■
Rhizoctonia solani				■			■												
Cercospora beticola							■				■			■	■		■	■	■
Polymyxa betae							■				■			■	■		■	■	■
Black leg disease				■	■		■												
Erwinia subsp.	■																■	■	■
Heterodera schachtii							■										■	■	■
Heterodera trifolii							■										■	■	■
Meloidogyne hapla							■										■	■	■
Myzus persicae							■				■								
Pegomya subsp.															■				■

Taxon codes used: *B. vulgaris* subsp. (1), *B. vulgaris* subsp. *vulgaris* (2), *B. vulgaris* subsp. *vulgaris* leaf beet group (3), garden beet group (4), fodder beet group (5), sugarbeet group (6), *B. vulgaris* subsp. *maritima* (7), subsp. *adanensis* (8), *B. macrocarpa* (9), *B. patula* (10), *B. corolliflora* (11), *B. macrorhiza* (12), *B. lomatogona* (13), *B. intermedia* (14), *B. trigyna* (15), *B. nana* (16), *B. procumbens* (17), *B. webbiana* (18), *B. patellaris* (19). ■ = variation detected.

maritima collected there will contain genetic variation for leaf spot disease resistance. If we regenerate the sample in an alien environment without any leaf spot disease stress, disease resistant genotypes will no longer have a selective advantage and will gradually decline in frequency with increasing number of seed reproduction cycles. Sylvén (1937, cited in Hill *et al.,* 1998) investigated forage yield in meadow fescue (*Festuca pratensis*) selected for forage production in north Sweden and reported strong changes in yield after only a single generation of seed production in south Sweden. Hill *et al.* (1998) pointed out that selective genetic shift is a particular problem of static management of outcrossing species outside their natural enviroment. If there is a risk of losing genes through genetic shift, dynamic management procedures complementing the static management will play an important role.

Dynamic Management of Beet Genetic Resources

The term 'dynamic' is associated with different concepts of beet genetic resources management. Firstly, wild beet populations are maintained and managed in their natural habitats (Tan *et al.*, 2000). The WBN should, therefore, promote the dynamic management of wild species and should seek to transfer the Turkish programme to other *Beta* distribution areas in Europe. For such a programme more information on the structure of genetic diversity of wild beets is required, facilitating realization of effective management projects for all species. Information on the sea beet (*B. vulgaris* subsp. *maritima*) is already increasing. The species occurs in a dynamic habitat along the west Atlantic and Mediterranean sea shores on shingle beaches in a narrow band between high tide and inland. Mainly during the cold season, tidal movements and sea currents alter the coastline by soil erosion, which sometimes destroys wild beet populations. The same process creates new habitats by washing soil ashore. Seed balls of the sea beet are dispersed by sea currents along the linear habitat. Founder populations arise at suitable places and start to exchange genes with neighbour populations. Currently, methods based on molecular marker techniques are being developed that will allow us to determine at which spatial distance the rate of gene flow between populations drops below a critical value and indicates the beginning of a new, genetically distinct population (Raybould and Clarke, 1999). We have started to understand the genetic structure of this species and the dynamics of gene flow within its habitat. This knowledge mainly stems from investigations on the origin of weedy beets in sugarbeet production fields and from risk assessment studies for genetically modified sugarbeet.

Owing to the large distribution area of the sea beet and closely related wild species (Fig. 13.1, section *Beta*), dynamic management programmes would require the determination and selection of sites with genetically distinct populations and a monitoring scheme to ensure that the sites are safe and populations can continue to exist.

Secondly, the term 'dynamic' stands for utilization of genetic resources through incorporation of sea beet germplasm into the sugarbeet. Thirdly, the term describes a concept of pooling resistance genes contained in genebank accessions. Evaluation of static collections is a prerequisite for this approach. Concentration of genes in dynamic genepools is the prominent objective.

A programme integrating static and dynamic management elements has been developed by the French *Beta* Network. The charter of the French crop specific networks regulating the required contribution of network partners were laid down for cereals by Mitteau (1997) in an exemplary manner. The underlying principle is free access to material held or created within the network for those partners actively contributing to the network activities.

B. vulgaris subsp. *maritima* is distributed along the French Atlantic and Mediterranean sea shores. A geographically stratified sample of 22 native accessions was chosen for the development of 'buffer' populations. Buffer populations link the wild with the cultivated species and purposefully reconstruct the natural gene flow between leaf beets and wild beets. The original wild populations are maintained in their natural environment while plants produced from collected seed samples are allowed to recombine with sugarbeet at various locations 'on the farm', that is, in the fields of sugarbeet breeders or any institution or individual willing to contribute to the common work (Table 13.3). In this system evolution can act on the original wild source as well as on pre-breeding material. The dynamic management of wild beet germplasm in buffer populations has no trait-specific focus. It rather concentrates on the elimination of two undesirable wild characters while keeping as much of

Table 13.3. Organization and development of 'buffer' populations. Based on the principles developed by Doggett and Eberhart (1968).

	Country of the participant(s)						
	BEL	DNK	FRA	GER	GRC	NLD	SWE
	Number of subpopulations						
Year	2	2	6	6	2	2	2
1996	Sugarbeet (Doggett population) × *Beta vulgaris* subsp. *maritima* (No.1 ... 22)						
1997	Production of F_2 seeds						
1998	Production of F_3 seeds						
1999	Mutual exchange of F_3 families						
2000	Intermating of 22 subpopulations at each location to produce a buffer population. The buffer population is segregating for male sterility and male fertility (1:1) and contains 50% of the wild genome						
Next years	Adaptation of the buffer population through mild selection on bolting resistance and root shape. Exploitation of the population at the discretion of each participant						

the wild genome as possible. Wild beet material is often early bolting and develops fibrous or fanged roots. It is known that screening of annual (wild) beets on pest and disease resistance is therefore difficult. Biennial forms that resemble cultivated beets can be tested in the field more reliably than can the original, early bolting wild types. The programme has been set up so as to allow introduction and recombination of a large amount of the wild beet genome in the buffer while imposing a mild selection pressure on essential agronomic traits, that is, bolting resistance and sugarbeet-like roots. Through the specific design of the breeding scheme 50% of the genome of the wild parents can be maintained and can recombine in the buffer populations for as many generations as breeders wish to allow. Half-sib families and inbred lines can be tested and selected at the breeders' discretion without changing the buffer.

A different, trait-specific, collaborative programme for the management of genes has been suggested by a plant breeding company in Germany. The programme is part of a long-term breeding strategy aiming at the improvement of the level of leaf spot resistance in sugarbeets. A comprehensive review on breeding for *C. beticola* resistance was presented by Skaracis and Biancardi (2000) and is the basis of the following descriptions. The first resistant breeding material was developed by the Italian breeder Munerati from a sugarbeet × *B. vulgaris* subsp. *maritima*. The wild species was sampled from the Po river bank where, although endangered by ditch upkeep, it still exists. Around 1935 the most resistant material, in particular line R 581, was distributed to breeders in Europe and the USA where it was widely used in breeding programmes. Although other resistant sources were detected in section *Beta*, and in the more distantly related sections *Corollinae* and *Procumbentes*, the Munerati material is essentially the only source used in current breeding. The inheritance of the *C. beticola* resistance is of polygenic nature with a broad-sense heritability of 60–70% (Smith and Gaskill, 1970) and a narrow sense heritability of 24% (Smith and Ruppel, 1974). The investigations of Koch (1997) confirmed the quantitative nature of the resistance. Using an *in vitro* test system, Koch described three quantitative trait loci (QTL) on linkage groups 2, 3 and 5 explaining almost 50% of the total variation of the segregating F2 mapping population. In contrast to Koch (1997), Schäfer-Pregl *et al.* (1999) tested their F2 and test-cross material under natural disease attack conditions at Monselice (Italy), a location with the highest leaf spot disease pressure known in Europe. Schäfer-Pregl *et al.* (1999) were able to localize QTLs for resistance to *C. beticola* with highly significant likelihood odds ratios (LOD) on linkage groups 2, 6 and 9, if F2 data were considered only. QTLs were also detected on linkage group 4 and 5, and under artificial inoculation another QTL occurs on linkage group 3. Statistically, there are at least three to five chromosomes carrying QTLs which contribute to the resistance expression and these minor genes probably originate to a large extent from the Munerati source. The findings from QTL investigations not only confirm the quantitative nature of the resistance but also provide an idea of how minor genes contributing to the resistance trait expression are distributed over the genome.

Leaf spot resistance qualifies for collaborative conservation activity because: (i) the resistance is inherited by many genes; (ii) promising genebank accessions show only a certain resistance level; (iii) the desired genes are probably scattered all over the natural distribution area of the species *B. vulgaris* subsp. *maritima*; and (iv) many accessions have a strong tendency to early bolting and fanged roots. Because of the difficulty experienced by all sugarbeet breeders to develop highly leaf spot resistant varieties and the economic risks connected with a long-term breeding strategy, a collaboration between companies appeared to be advantageous.

The data required for the choice of accessions with variation for leaf spot resistance provided the GRIN database and the GENRES CT95 42 European Community germplasm screening project. Owing to recent systematic evaluation work wild and cultivated accessions have been detected scoring '3' (= low susceptibility to the leaf spot disease). A map with the collecting sites of the respective wild populations and cultivated accessions is presented in Fig. 13.2. These accessions are maintained as a static collection in genebanks. They may contain minor genes not yet present in the breeding pool. Currently, the accessions are retested in the USA to confirm the earlier evaluation results. If these accessions are combined in *Cercospora* resistance genepools (CRGs) more minor genes can be assembled and can recombine. The German plant breeding company KWS has therefore developed a breeding concept that aims at pooling all available *B. vulgaris* subsp. *maritima* accessions with low disease susceptibility in two different CRGs, that is, a pool built up from wild beet material distributed along the Atlantic coast and a second one created from Mediterranean wild beet populations. As described for the French 'buffer' population programme, basic characters like bolting resistance and a better root shape will be introduced in the pools. For that purpose sugarbeet material will be crossed with the bulked Atlantic respectively Mediterranean wild germplasm. Through mild mass selection for acceptable root shape and bolting resistance biennial material with a sugarbeet-like root shape will be developed that yields more reliable evaluation results than annual types when screened for disease resistance. By backcrossing with the wild beet up to 75% of the wild beet genes can be maintained in such CRGs.

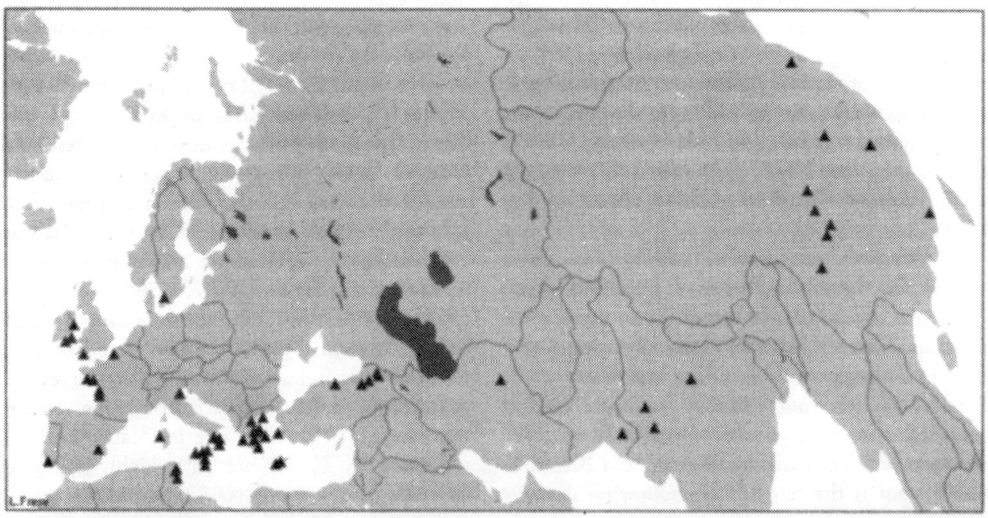

Fig. 13.2. Geographic distribution of the *B. vulgaris* accessions showing variation for resistance to leaf spot disease (*Cercospora beticola*).

The CRGs will not only be subjected to mild selection to re-domesticate the material, but also need to be subjected to artificial (in the breeding garden) or natural selection pressure for disease resistance by growing CRGs on farm fields in areas with a regular and strong disease epidemic like at Monselice in northern Italy. In this way, all available genes contributing to the leaf spot disease could be maintained 'on the farm'.

Discussion and Conclusions

Activity 2 of the GPA contains a recommendation to 'support on farm management and improvement of PGRFA'. But how can this recommendation be realized in the case of *Beta*? The cultivated species *B. vulgaris* subsp. *vulgaris* consists of the leaf, garden, fodder and sugarbeet. In Europe most of the old beet landraces or obsolete varieties are no longer used by farmers and have either been lost or are conserved in genebanks. In a market-oriented society there is no strong economic incentive that stimulates farmers to use old beet varieties; to the contrary, fodder beet production is steadily decreasing in Germany (1950: 566,000 ha, 1997: 17,000 ha) (Keller *et al.*, 1999), the garden beet is grown on about 800 ha only and the use of leaf beet is even more limited. Hence, the reintroduction of beet germplasm from genebanks into agricultural production systems with the objective of managing it 'on the farm', requires either idealism, an interesting market niche in the high price sector and/or subsidies from the government. All three requirements have weak points: idealism may subside with the change of generations and life objectives of people, market demands may be cursory as may be political and administrative support.

In rural areas of the Mediterranean region small farmers and gardeners still produce their own seeds of leaf beets. In a village like Liapades (Kerkira, Greece) population sizes of a few individuals up to 30 seed bearer plants can still be found today. There is no strong spatial isolation between populations which most likely exchange genetic material. In Italy gene flow from *B. vulgaris* subsp. *maritima* into subsp. *vulgaris* and vice versa can be observed (Frese and Burenin, 1994). If it were possible to maintain the local seed production in a village like Liapades or the introgression from wild into cultivated form it would supplement static management of beet germplasm. We can assume continued evolution under such situations which can create new genetic variation in local populations during a period of 10^4 to 10^9 years. However, since land use may suddenly change or gardeners may decide to abandon cultivation of leaf beets, the situation in villages like Liapades may not be stable enough to provide a long-term and meaningful contribution to the static management of PGR. In *Beta*, dynamic management of genetic resources probably needs to be organized by institutions of the public services together with breeding companies or skilled seed-producing farmers. The basic management concepts were outlined in previous chapters.

Buffer populations organizing the gene flow from the wild ancestor of beets to the cultivated material can be described as one element of dynamic management programmes. We can think about additional elements. In the global *Beta* holding distinct groups of leaf, garden and fodder beets exist, like leaf beets with a broad and fleshy petiole and glossy, smooth green leaves. An assemblage of the various accessions could be called a 'type group'. Within the group the individual accessions may be clearly distinct because of different leaf size or petiole length. We could pool all accessions belonging to a group, send seed lots to cooperating breeders or farmers and reproduce the *type specific genepool* on-farm, preferably on different sites in Europe, to allow a maximum adaptation of the type specific genepool. The purposeful management of type specific genepools would reconstruct the selection by farmers which created landraces of beets in the past until the onset of modern plant breeding in Europe. While the individual accessions are still managed by genebanks, the type group as such can participate in the evolutionary process, which is the major advantage arising from dynamic management of PGRFA. While we select on essential agronomic traits (bolting resistance and root shape) in buffer populations only, the management aim for type specific pools would be to maintain a range of typical characters that distinguishes the group from others. In the case of the *trait specific pools* we aim at the maintenance of all genes and alleles contributing to the expression of a specific character. The *Cercospora* disease is only an example trait and can be replaced by others, for example drought tolerance.

The creation of buffer populations, type and trait specific genepools can be seen as components of management programmes that will contribute to the realization of GPA Activity 2 in Europe. The advantage of these components is that they facilitate the definition of objectives and the justification of projects. The objectives are:

- Maintenance and monitoring of wild species populations and populations of cultivated forms in their natural habitat (including home gardens, farm fields).
- Promotion of gene flow from the wild ancestor into the cultivated species and basic adaptation of the source material to agricultural practices (buffer populations).
- Promotion of evolutionary processes within a restricted number of distinct cultivated types (type specific genepools).
- Concentration of valuable, trait specific genes in genepools to avoid unintentional loss of genetic variation by genetic shift and drift through static management (trait specific genepools).

GPA Activity 2 calls, in the first instance, for political support for a technical recommendation that, in 1996, was considered of crucial importance for the sustainable conservation and use of PGRFA. It must be clear that dynamic management programmes based on this recommendation will bind capacities for a long period. The most prominent example of a long-term programme in agricultural science is a series of experiments in microevolution that were conducted at the University of California (Davis) over a period of four decades (Allard, 1988). There are probably only a few more such cases in the world. Political support that finally results in the creation of a budget line required for the technical realization of dynamic management programmes will more likely be provided at the national (Germany) or regional (European Community) level if the objectives of a programme can be categorized so that decision-makers can set priorities. To be meaningful and successful the organization of a cooperative European network dealing with the dynamic management of beet populations cannot be based solely on input in kind. Rather a framework programme is required that guarantees long-term support.

References

Allard, R.W. (1988) Genetic changes associated with the evolution of adaptedness in cultivated plants and their wild progenitors. *Journal of Heredity* 79, 225–238.

Bosemark, N.O. (1979) Genetic poverty of the sugarbeet in Europe. In: *Proceedings of the Conference on Broadening the Genetic Base of Crops, Wageningen, 1978.* Pudoc, Wageningen, The Netherlands, pp. 29–33.

Bosemark, N.O. (1989) Prospect for beet breeding and use of genetic resources. In: *Report of an International Workshop on* Beta *Genetic Resources. International Crop Network* Series 3. International Board for Plant Genetic Resources, Rome, Italy, pp. 89–97.

Boughey, C.L. (1981) Evolutionary and taxonomic studies in wild and cultivated beets. PhD thesis, University of Birmingham, UK.

Bretting, P.K. and Duvick, D.N. (1997) Dynamic conservation of plant genetic resources. *Advances in Agronomy* 61, 1–51.

Buttler, K.P. (1977) Revision von *Beta* Sektion *Corollinae* (*Chenopodiacea*). I. Selbststerile Basisarten. *Mitt. Bot. München* 13, 255–336.

Büttner, G., Frese, L. and Steinrücken, G. (1997) Selektion von Rizomania-Resistenzgenen aus Wildrüben (*Beta vulgaris* L.). *IIRB Beiträge und Poster von IIRB-Kongressen, 19–22 Juni 1995 in Beaune, Frankreich, 13–15 Februar 1996 in Brüssel, Belgien.* IFZ, Göttingen/ IIRB, Brüssel.

Dale, M.F.B., Ford-Lloyd, B.V. and Arnold, M.H. (1985) Variation in some agronomically important characters in a germplasm collection of beet (*B. vulgaris* L.). *Euphytica* 34, 449–455.

Dambroth, M. (1985) Industriepflanzenanbau einzige Alternative zur nachhaltigen Lösung der Agrarmarktprobleme. Sonderdruck aus Agar-Übersicht No. 3, 15 März 1985. Landbuchverlag GmbH, Hannover, Germany.

Desprez, M. and Desprez. B. (1993) Évolution de methods de sélection de la betterave sucrière des origines à nos jours. *Comptes-Rendus de l'Académie d'Agriculture de France* 79(6), 71–84.

Desprez, M. and Desprez, B. (1996) Importance and difficulties of using genetic resources illustrated by the example of beet cyst nematode. Poster presented at the 4th International Technical Conference on Plant Genetic Resources, Leipzig, Germany, 17–23 June 1996.

Doggett, H. and Eberhart, S.A. (1968) Recurrent selection in sorghum. *Crop Science* 8, 199–121.

Doney, D.L. (1998) *Beta* evaluation and sugarbeet enhancement from wild sources. In: Frese, L., Panella, L., Srivastava, H.M. and Lange, W. (eds) *International* Beta *Genetic Resources Network. A Report on the 4th International* Beta

Genetic Resources Workshop and World Beta *Network Conference held at the Aegean Agricultural Research Institute, Izmir, Turkey, 28 February–3 March 1996.* International Crop Network Series 12. International Plant Genetic Resources Institute, Rome, Italy, pp. 73–76.
Doney, D.L. and Whitney, E.D. (1990) Genetic enhancement in *Beta* for disease resistance using wild relatives: a strong case for the value of genetic conservation. *Economic Botany* 44(4), 445–451.
Doney, D.L., Ford-Lloyd, B.V., Frese, L. and Tan, A. (1995) Scientists worldwide rally to rescue the native beet of the Mediterranean. *Diversity* 11 (1&2), 124–125.
FAO (1996) *Global Plan of Action for the Conservation and Sustainable Utilisation of Plant Genetic Resources* and the Leipzig Declaration adopted by the International Technical Conference on Plant Genetic Resources, Leipzig, Germany, 17–23 June 1996. Food and Agricultural Organization of the United Nations, Rome, Italy.
Frese, L., (ed.) (1992) *International* Beta *Genetic Resources Network. A Report on the 2nd International* Beta *Genetic Resources Workshop held at the Institute of Crop Science and Plant Breeding, Braunschweig, Germany, 24–28 June 1991.* International Crop Network Series 7. International Plant Genetic Resources Institute, Rome, Italy.
Frese, L. and Burenin, V.I. (1994) *Sammlung genetischer Ressourcen von* Beta, Lactuca *und* Cichorium *in Mittel- und Norditalien.* Reisebericht für das Bundesministerium für Ernährung, Landwirtschaft und Forsten (BML), Bonn, Germany.
Frese, L. and Desprez, B. (1999) Utilisation of *Beta* genetic resources. In: Gass, T., Frese, L., Begemann, F. and Lipmann, E. (compilers) *Implementation of the Global Plan of Action in Europe – Conservation and Sustainable Utilisation of Plant Genetic Resources for Food and Agriculture. Proceedings of the European Symposium, 30 June–3 July 1998, Braunschweig, Germany.* IPGRI, Rome, Italy, pp. 172–180.
Gass, T., Frese, L., Begemann, F. and Lipmann, E. (compilers) (1999) *Implementation of the Global Plan of Action in Europe – Conservation and Sustainable Utilisation of Plant Genetic Resources for Food and Agriculture. Proceedings of the European Symposium, 30 June–3 July 1998, Braunschweig, Germany.* IPGRI, Rome, Italy.
Germeier, C. and Frese, L. (2000) Access to information on plant genetic resources of beets. (*Beta* subsp.). In: Maggioni, L., Frese, L., Germeier, C. and Lipman, E. (eds) *International* Beta *Genetic Resources Network. Report of a working group on* Beta. *First meeting, Broom's Barn, Bury St Edmunds, UK, 9–10 September 1999.* IPGRI, Rome, Italy, pp. 55–64.
Grünewald, I., Horak, I. and Schlösser, E. (1983) Rizomania. III. Verbreitung im Hessischen Ried und im Raum Worms sowie Beziehungen zum Boden-pH und zur Fruchtfolge. *Zuckerindustrie* 108(7), 650–652.
Hazekamp, T. and Frese, L. (1992) Application of mapping systems for the analysis of the geographical origin of collected material. In: Frese, L. (ed.) *International* Beta *Genetic Resources Network. A Report on the 2nd International* Beta *Genetic Resources Workshop held at the Institute of Crop Science and Plant Breeding, Braunschweig, Germany, 24–28 June 1991.* International Crop Network Series 7. International Plant Genetic Resources Institute, Rome, Italy.
Hellinga, J.J.A. (1943) Verslag over het onderzoek van grondmonsters op bietenaaltjes verricht in samenwerking met de suikerbietenfabriken in 1941 and 1942. *Verslagen van het Institute voor Rationele Suikerproduktie* 13, 47–66.
Hill, J., Becker, H.C. and Tigerstedt, P.M.A. (1998) *Plant Breeding* Series 4. *Quantitative and Ecological Aspects of Plant Breeding.* Chapman and Hall, London, UK.
Jana, S. (1999) Some recent issues on the conservation of crop genetic resources in developing countries. *Genome* 42, 562–569.
Keller, E.R., Hanus, H. and Heyland, K.-U. (eds) (1999) *Handbuch des Pflanzenbaus.* Band 3: *Knollen- und Wurzelfrüchte, Körner- und Futterleguminosen.* Verlag Eugen Ulmer GmbH & Co., Stuttgart, Germany.
Koch, G. (1997) Genetische Untersuchungen zur *Cercospora beticola* Resistenz in Zuckerrüben. *GPZ, Vorträge für Pflanzenzüchtung,* 37, 54–64.
Lange, W., Brandenburg, W.A. and De Bock, Th.S.M. (1999) Taxonomy and cultonomy of beet (*Beta vulgaris* L.) *Botanical Journal of the Linnean Society,* 130, 81–96.
Letschert, J.P.W. (1993) Beta section Beta: biogeographical patterns of variation and taxonomy. Wageningen Agricultural University Papers 93–1.
Lewellen, R.T. (1992) Use of plant introductions to improve populations and hybrids of sugarbeet. In: *Use of Plant Introductions in Cultivar Development,* Part 2, CSSA Special Publication No. 20, pp. 117–136.
Maier, S. (2000) Biosphärenreservate in Deutschland – geeignete Ort für das on-farm Management von Kulturpflanzen? *Save Report* Spring, 7–9.
Mesbah, M. (1997) Characterisation of alien chromosomes in monosomic additions of *Beta.* PhD thesis, Wageningen Agricultural University, Wageningen, The Netherlands.
Mitteau, M. (1997) *Catalogue of Genetic Resources Bread Wheat and Barley.* BRG, Paris, France.
Munerati, O. (1932) Sull' incrocio della barbabietola coltivata con la beta selvaggia della costa adriatica. *L'Industria Saccarifera Italiana* 25, 303–304.

Namkoong, G. (1989) Population genetics and the dynamics of conservation. In: Knutson, L. and Stoner, A.K. (eds) *Biotic Diversity and Germplasm Preservation, Global Imperatives.* Beltsville Symposia in Agricultural Research, 13. Kluwer Academic Publishers, Dordrecht, The Netherlands.

Oltmann, W., Burba, M. and Bolz, G. (1984) *Die Qualität der Zuckerrübe. Bedeutung, Beurteilungskriterien und züchterische Maßnahmen zu ihrer Verbesserung.* Paul Parey, Berlin, Germany.

Owen, F.V. (1948) Utilization of male-sterility in breeding superior-yielding sugarbeets. *Proc. Am. Soc. Sugar Beet Tech.* 5, 156–161.

Owen, F.V. (1954) Hybrid sugarbeets made by utilizing both cytoplasmic and Mendelian male sterility. *Proc. Am. Soc. Sugar Beet Tech.* 8, 64.

Panella, L. (1998) Screening and utilizing *Beta* genetic resources with resistance to *Rhizoctonia* root rot and *Cercospora* leaf spot in a sugar beet breeding programme. In: Frese, L., Panella, L., Srivastava, H.M. and Lange, W. (eds) *International* Beta *Genetic Resources Network. A Report on the 4th International* Beta *Genetic Resources Workshop and World* Beta *Network Conference held at the Aegean Agricultural Research Institute, Izmir, Turkey, 28 February–3 March 1996.* International Crop Network Series 12. International Plant Genetic Resources Institute, Rome, Italy, pp. 62–72.

Paul, H., Henken, B., de Bock, Th.S.M. and Lange, W. (1992) Resistance to *Polymyxa betae* in *Beta* species of the section *Procumbentes*, in hybrids with *B. vulgaris* and in monosomic chromosome additions of *B. procumbens* in *B. vulgaris. Plant Breeding* 109, 265–273.

Raybould, A.F. and Clarke, R.T. (1999) Estimating geneflow among sea beet populations using molecular markers (abstract). In: Maggioni, L., Frese, L., Germeier, C. and Lipmann, E. (eds) *International* Beta *Genetic Resources Network. Report of a working group on* Beta. *First meeting, Broom's Barn, Bury St Edmunds, UK, 9–10 September 1999.* IPGRI, Rome, Italy, p. 73.

Savitsky, H. (1950) Monogerm sugarbeets in the United States. *Proc. Am. Soc. Sugar Beet Tech.* 6, 156–159.

Savitsky, V.F. (1952) Methods and results of breeding work with monogerm beets. *Proc. Am. Soc. Sugar Beet Tech.* 7, 344–350.

Savitsky, H. (1960) Meiosis in an F_1 hybrid between a Turkish wild beet (*Beta vulgaris* ssp. *maritima*) and *Beta procumbens. J. Am. Soc. Sugar Beet Tech.* 11, 49–67.

Schäfer-Pregl, R., Borchardt, D.C., Barzen, E., Glass, C., Mechelke, W., Seitzer, J.F. and Salamini, F. (1999) Localization of QTLs for tolerance to *Cercospora beticola* on sugar beet linkage groups. *Theoretical and Applied Genetics* 99, 829–836.

Schnell, F.W. (1980) Aspekte der genetischen Diversität im Problemkreis der Pflanzenzüchtung. *Göttinger Pflanzenzüchtungs-Seminar* 4, 5–15.

Scholten, O.E. (1997) Characterisation and inheritance of resistance to beet necrotic yellow vein virus in *Beta*. Thesis, Wageningen Agricultural University, Wageningen, The Netherlands.

Simmonds, N.W. (1993) Introgression and incorporation. Strategies for the use of crop genetic resources. *Biological Reviews* 68, 539–562.

Skaracis, G.N. and Biancardi, E. (2000) Breeding for *Cercospora* resistance in sugar beet. In: Cercospora beticola Sacc. *Biology, Agronomic Influences and Control Measures in Sugar Beet.* Advances in Sugar Beet Research, Vol. 2, International Institute for Beet Research, Brussels, Belgium.

Smith, G.A. and Gaskill, J.O. (1970) Inheritance of resistance to *Cercospora* leaf spot in sugar beet, *J. ASSBT* 16, 172–180.

Smith, G.A. and Ruppel, E.G. (1974) Heritability of resistance to *Cercospora* leaf spot in sugar beet. *Crop Science* 14, 113–115.

Stanescu, Z. (1994) Contribution of Romanian scientific research to the activities of the World *Beta* Network. In: Frese, L. and Doney, D.L. (eds) *International* Beta *Genetic Resources Network. A report on the 3rd International* Beta *Genetic Resources Workshop and World* Beta *Network Conference held at the North Dakota State University, Fargo, USA, 4–6 August 1994.* International Crop Network Series 11. International Plant Genetic Resources Institute, Rome, Italy, p. 31.

Sun Yi Chu (1994) *Beta* germplasm collection and its application status in China. In: Frese, L. and Doney, D.L. (eds) *International* Beta *Genetic Resources Network. A report on the 3rd International* Beta *Genetic Resources Workshop and World* Beta *Network Conference held at the North Dakota State University, Fargo, USA, 4–6 August 1994.* International Crop Network Series 11. International Plant Genetic Resources Institute, Rome, Italy, pp. 27–29.

Tan, A., Aykas, L. and Inal, A. (1999) Status of *Beta* genetic resources in Turkey. In: Maggioni, L., Frese, L., Germeier, C. and Lipman E. (eds) *International* Beta *Genetic Resources Network. Report of a working group on* Beta. *First meeting, Broom's Barn, Bury St Edmunds, UK, 9–10 September 1999.* IPGRI, Rome, Italy, pp. 43–47.

Tjebbes, K. (1933) The wild beets of the North Sea region. *Botaniska Notiser* 14, 305–315.

Uphoff, H. (1997) Nematodenresistente Zuckerrüben – Strategien zur Züchtung resistenter Sorten. *Vortr. Pflanzenzücht.* 37, 46–51.

van Geyt, J.P.C., Lange, W., Oleo, M. and de Bock, Th.S.M. (1990) Natural variation within the genus *Beta* and its possible use for breeding sugarbeet. *Euphytica* 49, 57–76.

Winner, C. (1981) *Zuckerrübenbau.* DLG-Verlag, Frankfurt, Germany.

Yu, M.H. (1997) Sugarbeet root-knot nematode resistance and breeding aspects [Abstract]. *60th IIRB Congress, Cambridge, UK, 30 June–4 July 1997.*

14 Rice, Farmers and Genebanks: a Case Study in the Cagayan Valley, Philippines

J.-L. Pham,[1*] S.R. Morin,[1] L.S. Sebastian,[2] G.A. Abrigo,[2] M.A. Calibo,[1] S.M. Quilloy,[1] L. Hipolito[2] and M.T. Jackson[1]

[1] *Genetic Resources Center, International Rice Research Institute (IRRI), Makati City, the Philippines;* [2] *Philippine Rice Research Institute (PhilRice), Maligaya, Muñoz, Nueva Ecija, the Philippines*

Introduction

The current debate on the release of transgenic crops is a useful reminder of how complex the deployment of new varieties is. Not only must the expected benefits be taken into account but also all the potential consequences at very different levels: agroecological, social, economic and political. Long before genetically modified organisms, the release of semi-dwarf high yielding varieties by rice breeders had a tremendous impact on the life of rice farmers and consumers. The first modern rice variety, IR8, was released in 1966. In Asia, rice production increased by 114% between 1966 and 1996, surpassing the 82% population increase during the same period (Hossain and Pingali, 1998). The breeding of high-yielding varieties made this increase possible, but was also associated with dramatic changes in rice farming practices and economy. The release of modern rice varieties has had an impact on which varieties rice farmers grow, why they grow them and how they grow them.

For a long time, only the first point really mattered to genetic resources conservationists, as *ex situ* conservation was receiving most of their attention. Priority was given to the collection of local landraces threatened by the adoption of modern varieties, in order to make them available to all rice genetic resource users. Given the increasing interest in *in situ* conservation on-farm, the conservation of rice genetic resources must include new research activities in agroecosystems that can lead to a better understanding of what this approach means and how it can be implemented (Bellon *et al.*, 1997). This is particularly needed in rapidly changing agroecosystems, where the conservation of agrobiodiversity needs to be supported by providing farmers with the appropriate options. This is a matter of urgency. The risk of genetic erosion is higher in agroecosystems submitted to changes in their cultural, economic or technological environment. While cases of coexistence of local and modern varieties are reported for several crops worldwide (see the examples cited in the next section), it does not mean that such 'equilibrium' situations can be reached in all cases, nor are they durable. Furthermore, developing *in situ* conservation strategies for rapidly changing agroecosystems is an important issue as it may eventually result in the preservation of diversity and its integration in agricultural development policies in much larger areas than those of traditional agroecosystems.

*Present address: Centre IRD, Montpellier, France

(eds J.M.M. Engels, V. Ramanatha Rao, A.H.D. Brown and M.T. Jackson)

In the Philippines, two main research institutions are involved in rice biodiversity conservation, the International Rice Research Institute (IRRI) and the Philippine Rice Research Institute (PhilRice). PhilRice maintains the national collection of rice genetic resources. IRRI holds in the International Rice Genebank the world's largest rice collection (Jackson *et al.*, 1997). In 1996, IRRI and PhilRice implemented collaborative research activities to study rice diversity and its management by farmers in the Cagayan Valley, and identify opportunities to involve farmers in the overall framework of rice genetic resources conservation (Pham *et al.*, 1996).[1]

This chapter presents some of the results of this work. It shows how the survey of the on-farm diversity of rice varieties of the Cagayan Valley allowed the description of diversity in terms of genetic groups and of agronomic and cultural functional groups. Then, the identification of the constraints and threats to this diversity resulted in the definition of testable strategies to sustain it. This research illustrates the role that genebanks and research institutions can play in the maintenance of genetic diversity on-farm.

Study Sites

Despite the intensity of the Green Revolution, rice agriculture intensification has not been uniform all over Asia. As observed in other major crops and countries, for example potato in Peru (Brush *et al.*, 1992), and maize in Mexico (Bellon and Brush, 1994; Louette *et al.*, 1997) and Burkina Faso (Sanou, 1996), agroecosystems can be found where modern varieties coexist with traditional ones.

The Cagayan Valley, located in Northern Luzon, the Philippines, is a good example of these situations. Three different rice ecosystems are present in the Cagayan Valley: rainfed upland, rainfed lowland and irrigated lowland. A differential impact of modern varieties was observed from the upland to the irrigated lowland ecosystem from a survey of 16 households by ecosystem (four villages/ecosystem, four households/village) (Bellon *et al.*, 1998). The ratio of modern to traditional varieties was 24:61, 20:19 and 43:5 in the rainfed upland, rainfed lowland and irrigated ecosystems, respectively. The analysis of the genetic polymorphism at 16 isozyme loci of 149 accessions of traditional and modern varieties showed that a gradient of genetic diversity followed a similar pattern, as the Nei's heterozygosity index demonstrated, that is 0.25 in the upland, 0.21 in the rainfed lowland and 0.15 in the irrigated ecosystem.

The results presented in this chapter mainly come from the study of the rainfed lowland ecosystem. Although not as genetically diverse as the upland ecosystem, we considered this ecosystem represented the main target for maintaining genetic diversity on-farm because of the much larger production areas it represents in Cagayan. Because of the obvious competition between traditional and modern varieties and the changing agroeconomic conditions, in particular the development of irrigation, the rainfed lowland ecosystem offers a unique challenge to study the dynamics of on-farm rice diversity, and to develop strategies to sustain this diversity.

The study sites were located in three municipalities adjacent to each other, in the centre of Cagayan Province. On the western side of the Cagayan river were the municipalities of Solana and western Amulung. On the eastern side were those of Iguig and eastern Amulung. In these municipalities, 15 barangays (villages) and 207 households were selected to represent a diversity of agroecological, economic and ethnic conditions.

The selection of study sites was done with the help of local officials from the Philippines Department of Agriculture Region and the Cagayan Valley Lowland and Marine Research Outreach Station.

Methods

Farmers' classification of varieties

In order to complement the study on farmers' perceptions of varieties presented in Bellon *et al.* (1998), we conducted an analysis to understand the farmers' classification of rice varieties. Given the high number of variety names encountered in the Cagayan Valley, we suspected that farmers have their own classification of the varieties they use, and perceive and manage these varieties as elements of larger groups.

The primary methodology used for identifying variety classes was successive pile-sorting, where an informant is asked to sort a set of items into smaller and smaller piles until each pile is a single item

[1]This project was conducted within the component 'On-farm conservation' of the project 'Safeguarding and Preservation of the Biodiversity of the Rice Genepool' funded by the Swiss Agency for Development and Cooperation.

(Bernard, 1988; Borgatti, 1992). If two varieties are split on the first split, their relationship is one, on the second split two and so on. Higher values (later splits), imply greater perceived similarity between individual varieties. From pile-sorting a similarity matrix can be produced in which pairs of varieties with a higher mean value are considered more similar than those pairs with a lower mean value. The resulting matrix was plotted using multidimensional scaling and only the 20 most commonly known varieties were used.

Farmers were asked to give the reasons for splitting at every split. The earlier splits (especially splits one and two) are indicative of more general criteria for distinguishing all available varieties and the later splits tend to be more precise and variety-based details.

Changes in diversity over time

In 1996 and 1998, two surveys were conducted in the rainfed lowland ecosystem. Among the questions that were put to the households were the name of the varieties that they were planting, and the origin of the seeds.

Microsatellite polymorphism analysis

Two distinct sets of accessions were studied for microsatellite polymorphism: the same set of 149 accessions collected in the three ecosystems previously studied for isozyme analysis, and another set of 205 accessions collected in the rainfed lowland ecosystem.

DNA was extracted from healthy leaves of 4-week-old plants using the CTAB (cetrimide-trimethylammoniumbromide) method. Polymerase chain reaction (PCR) was carried out in 1× PCR buffer (100 mM Tris-HCl, 500 mM KCl, 0.1% gelatin); 1 mM $MgCl_2$; 0.1 mM dNTP mix; 0.2 μM primer-reverse and forward (Research Genetics); 1 unit *Taq* polymerase and 20 μl genomic DNA. Amplification was carried out using an MJ Research thermal cycler with the following profile: initial denaturation for 5 min at 94°C; 35 cycles of 94°C for 1 min, 55°C for 1 min, and 72°C for 2 min, and a final extension of 72°C for 5 min. PCR products were viewed using silver staining after electrophoresis on a 6% polyacrylamide gel.

Eighteen microsatellite primers were used for the first set of 149 accessions: RM 1, 2, 3, 5, 6, 9, 11, 12, 13, 15, 16, 17, 18, 122, 148, 164, 167, 168. For the second set of 230 accessions, the following eighteen primers were used: RM 1, 3, 6, 11, 12, 18, 25, 26, 27, 60, 122, 167, 168, 169, 254, 255, 258 and 261. This set of primers slightly differed from those used for the first set of accessions because of practical and technical reasons. All primers (Panaud *et al.*, 1996; Temnykh *et al.*, in press; Cho *et al.*, in press) were purchased from Research Genetics.

Diversity Assessment

The analysis of farmers' classification of the 20 most frequent varieties in the study villages gave a clear picture. The clusters of varieties could be interpreted as functional groups: varieties were clustered together because farmers recognized them as having similar traits and patterns of use.

Three primary sorting groups came out from the multidimensional scaling (Fig. 14.1), which can be interpreted as follows from the reasons farmers gave for splitting:

- The group of glutinous varieties: Imelda, Imelda Diket, Diket, Bongkitan. These varieties all share the fundamental characteristic of being glutinous varieties. Glutinous variety grains are sticky when cooked. In Cagayan Valley, glutinous varieties are used for special cakes and sweets.
- The group of short growth duration varieties: IR68, C22, IR66, BPRI10, PSBRc12 and PSBRc10. The short duration group includes those varieties that mature in a relatively short period, usually between 90 and 130 days. The members of the short duration cluster are all modern varieties. Short duration is a characteristic that is valued by farmers because it allows for multiple crops per season.
- The long duration group includes the varieties Elonelon, Java, Wagwag pino, Wagwag, Wagwag tawataw, Wagwag bilog and Wagwag red. Long duration varieties are those that mature in more than 130 days. The long duration group is characterized by traditional varieties. It is possible to further classify the long duration group by recognizing that it includes a major subgroup of all the Wagwag types.

The Wagwag Varieties

The Wagwag variety type is prevalent throughout the region and it can be found in 14 of the 15 research villages. Figure 14.2 shows that out of the 72 variety

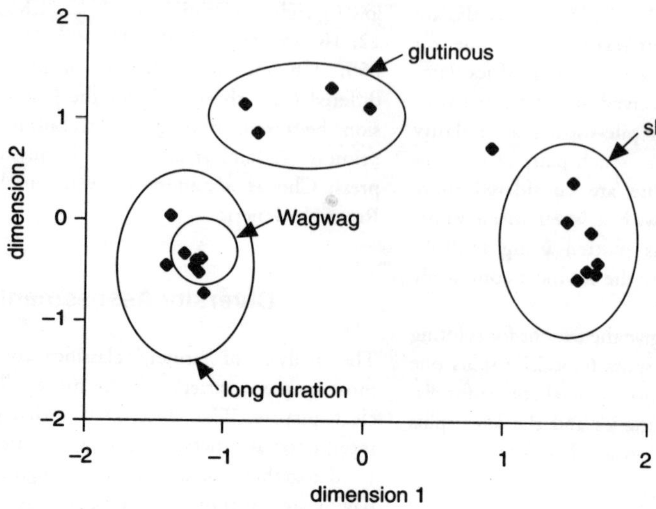

Fig. 14.1. Cultural perception of most frequent rice varieties in the Cagayan Valley. First two axes of a multiscaling analysis of the similarity matrix resulting from the successive pile-sorting.

names that were collected in the rainfed lowland ecosystem in Cagayan, four accounted for more than 50% of the plots. Out of these four varieties, two were modern varieties (IR66 and PSBRc10) and two were Wagwag varieties (Wagwag and Wagwag pino). Wagwag varieties are long duration photosensitive varieties. They are valued for their taste, cooking quality and adaptation to the variable local growing conditions. These varieties generally command a better price at the local and regional markets. Breeders have long recognized the agronomic value of Wagwag varieties but their use in breeding programmes has been hampered by their low combining ability.

The genetic analysis showed that the Wagwag varieties bring an original contribution to the overall rice genetic diversity in the Cagayan. Although not all accessions of Wagwag varieties clustered together – in that respect, the microsatellite analysis tended to blur the image previously obtained from isozyme data (Bellon *et al.*, 1998) – a large group of them were clearly separated from the rest of the accessions (Fig. 14.3). To illustrate what the genetic loss would be if all Wagwag varieties were to disappear, the genetic diversity at each of the microsatellite loci was computed for the second set of samples collected in the rainfed lowland ecosystem, and for this set minus the Wagwag varieties (Fig. 14.4). It shows that a decrease in genetic diversity would be observed at the majority of studied loci.

Threats to Rice Diversity

Double cropping with short duration varieties

The survey on farmers' perceptions conducted in 1996 demonstrated that long duration is the main reason for farmers in the irrigated and rainfed lowland ecosystem to discard varieties (Bellon *et al.*, 1998; Fig. 14.5). The analysis of the cropping calendar shows that the use of long duration varieties by farmers precludes the shift to double cropping (Fig. 14.5). Thus, the development of irrigation, which makes the double cropping of short duration varieties possible, appears to have potential negative consequences on rice diversity in Cagayan. This is an example of how the development of a technology (irrigation) induces perturbation of the agroecological niche occupied by particular varieties.

Threats to genetic diversity revealed by natural catastrophes

In 1997 and 1998, two major weather phenomena affected the Cagayan Valley. In 1997, El Niño caused a severe drought that affected Cagayan Valley and much of the Philippines. The total amount of rain in 1997 was lower than usual and

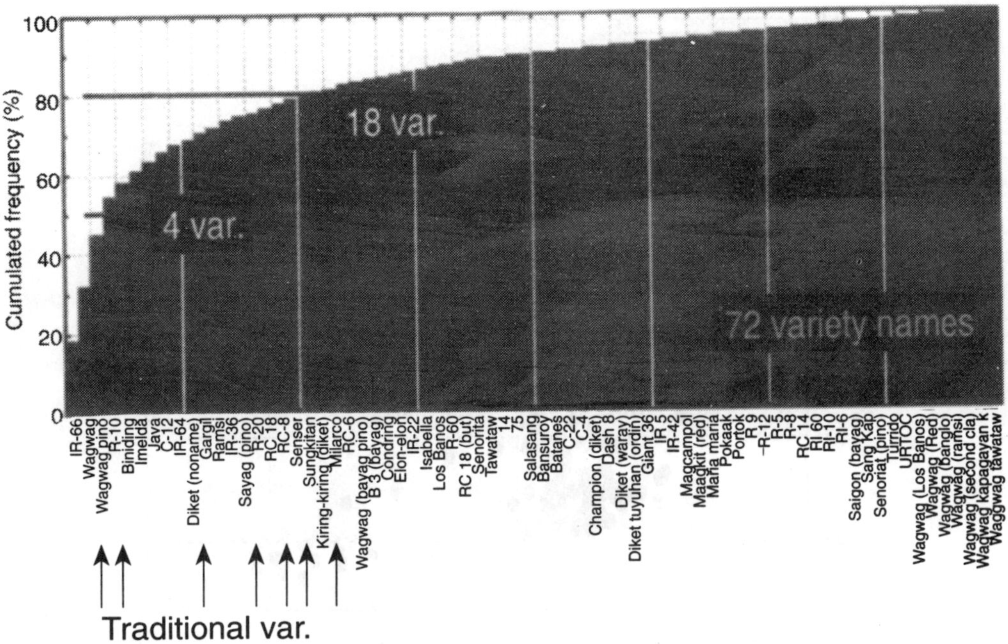

Fig. 14.2. Distribution of varieties in the Cagayan Valley (rainfed lowland ecosystem, wet season 1996, 15 villages, 14 farmers per village). The relative frequency of each variety name was computed.The bars represent the cumulative frequencies. The most frequent varieties are on the right-hand side of the graph.

the timing of rain was not good. The drought came when the rice plants were at the seedling stage, a stage when the tolerance to drought is nearly nil. Some farmers who had decided to wait for more rains were never able to plant.

In September and October 1998, the typhoons Loleng and Iliang hit the valley and caused severe infrastructural damage and early season flooding. The level and intensity of these floods was devastating. Again, rice seedlings were lost and even plants in later growth stages were badly affected.

Our surveys demonstrated that these catastrophes had a major impact on the frequency of traditional and modern varieties in Cagayan. The use of traditional varieties by farmers decreased from roughly 45% in 1996 to about 25% in 1998 (Fig. 14.6). The surveys and discussions with farmers and extension agents provided four main explanations for this rapid change in the varieties grown by Cagayan farmers.

1. *Deficient household seed storage technology.* Due to the humid climate conditions, the normal seed storage conditions in farming households in Cagayan do not permit farmers to maintain the germination ability of seeds for much longer than 6–9 months. This means that farmers cannot 'jump' a production season: if they do not produce seeds for a given variety during a given season, they will have to find an external source to get seeds to be able to plant the variety at the next season. Obviously, another option for them would be not to plant the variety.

2. *Lack of infrastructure for seeds of traditional varieties.* In a situation where seed stocks of most farmers were affected, farmers had to rely on external sources to obtain seeds for the next planting season. The seed stores generally carry only modern varieties, and certified seed growers, part of the Department of Agriculture's system of seed procurement strategy, grow only modern varieties.

3. *Support for the use of modern varieties.* In 1997 and 1998 the Municipal Agriculture Offices sponsored a 'plant now pay later' scheme. In this programme farmers are given seeds at no cost and, on harvest, are expected to pay for them. The seeds given in the scheme are from the certified seed growers and are always modern varieties, and sometimes only the

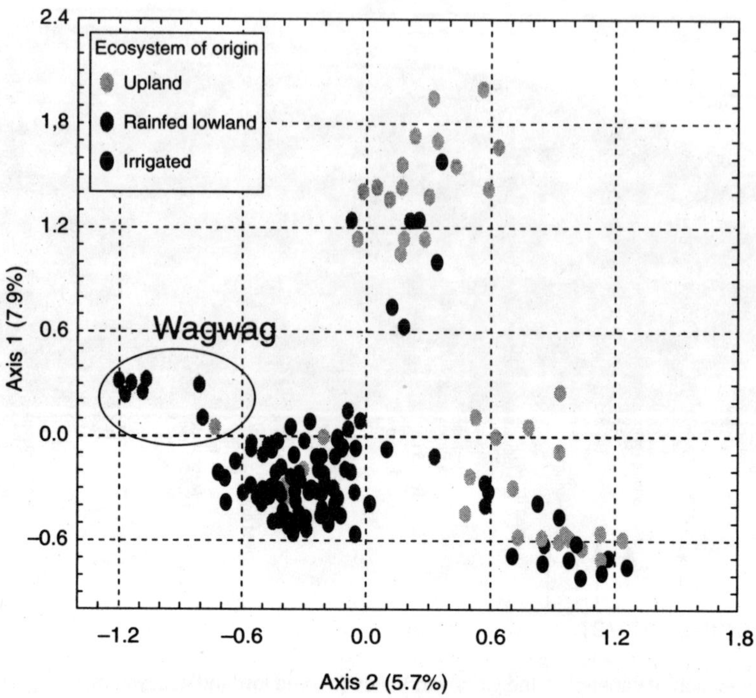

Fig. 14.3. Microsatellite polymorphism of 149 accessions from three rice ecosystems in the Cagayan Valley: first two axes of a correspondence analysis (data from 18 loci).

recommended varieties. The varieties available in 1998 were IR66 and PSBRc28, the former a popular but older modern variety, and the latter a new and currently recommended variety. Traditional varieties are not planted by certified seed growers and were not included in the scheme.

4. *Resilience of irrigated plots.* The varieties that were planted on irrigated plots were obviously less affected by the drought than the varieties planted on rainfed plots. Therefore, irrigation sustains the use of the modern varieties, as farmers plant only modern varieties on irrigated plots (Morin *et al.*, 1998).

Strategies to Sustain On-farm Diversity

The research in Cagayan shows the importance of a continuous monitoring of the rice varieties being grown. While genebank scientists can provide the methodology to do the surveys and initiate the collection of baseline data, they may lack the proximity to the field or the resources to perform the survey at regular intervals. The implementation of 'diversity lighthouses' managed in collaboration with farming communities and local extension offices is needed to provide data on the changes of diversity over time, whether due to particular climatic circumstances or not. These data would also be useful to help farmers obtain seeds from a variety they have lost. Community-based projects in Vietnam and Nepal were successful in developing the use of biodiversity registries in farming communities (Sthapit *et al.*, 2000). In Cagayan Valley, the assessment of the diversity of rice varieties and the analysis of threats to this diversity indicate that strategies need to be identified to sustain the cultivation of long duration varieties in the rainfed lowland agroecosystem.

Making diversity a viable option for farmers

There is a general consensus that farmers are not conservationists by nature but are 'conservationists' through use. In other words, farmers have to be pro-

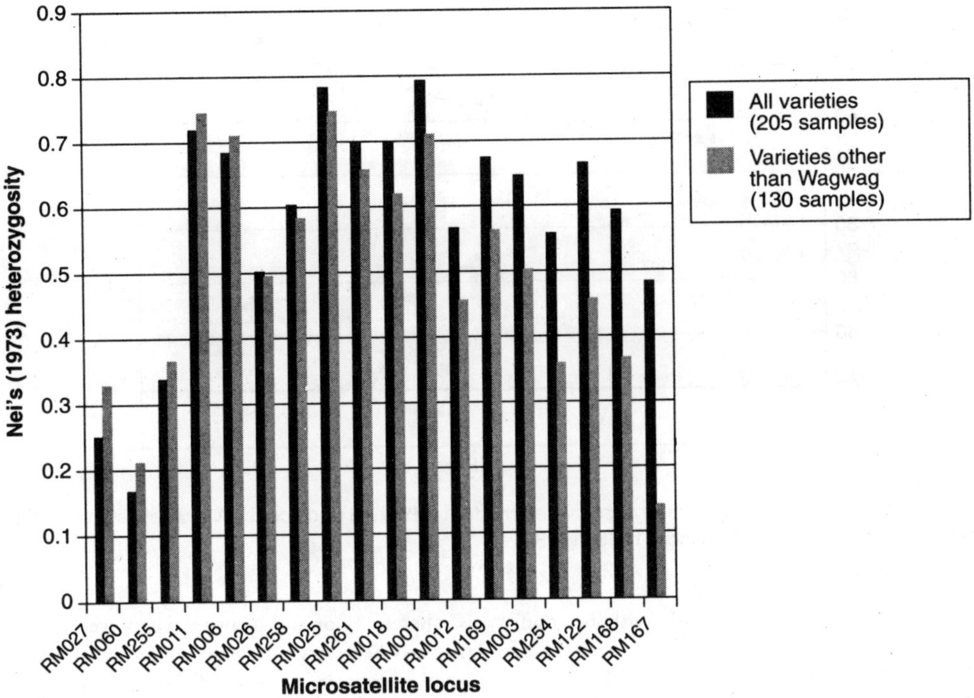

Fig. 14.4. Comparison of the genetic diversity at each of the 18 microsatellites for two sets of accessions: set of 205 accessions from the rainfed lowland ecosystem and subset of 130 accessions other than Wagwag varieties.

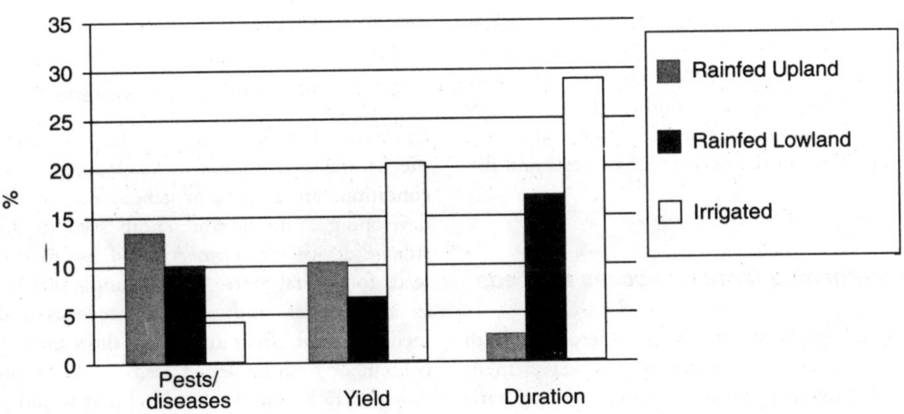

Fig. 14.5. Farmers' reasons for discarding traditional varieties in three rice ecosystems in the Cagayan Valley (from Bellon *et al.*, 1998).

vided with the right technical and economic options, so that they see the advantages in growing the varieties targeted by the conservationists. Cagayan farmers cannot afford to maintain long duration varieties if this results in a loss of income.

The idea of investigating new cropping patterns came from the observation of the practices of a farmer who was planting his traditional varieties in late October, a full 3 months after his neighbours. According to him, this practice posed no risk and

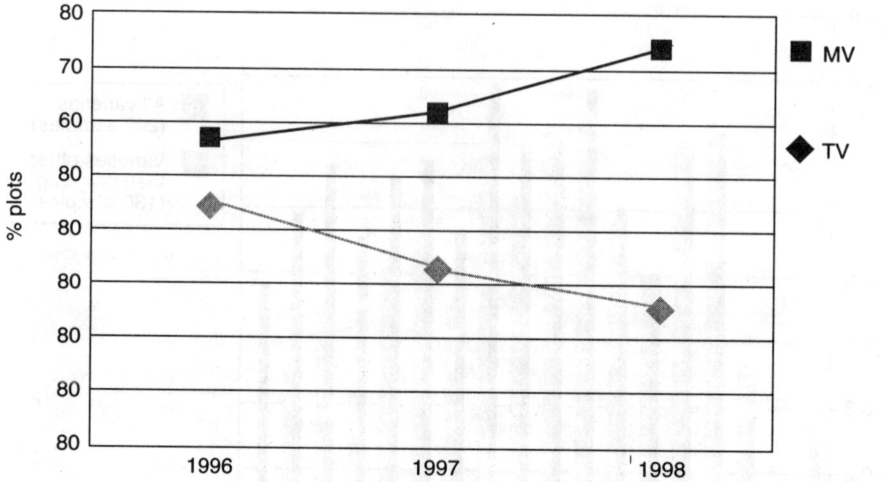

Fig. 14.6. Changes in the proportion of cultivated modern (MV) and traditional (TV) varieties, in selected municipalities in Cagayan Province, 1996–1998.

he felt he achieved higher yields with his traditional varieties than his neighbours. Field trials conducted on the experimental fields of IRRI in Los Baños confirmed these observations. Late planting not only permitted a decrease in maturity of the Wagwag varieties, but there was also an increase in yield (Figs 14.7 and 14.8). Consequently, it is possible to propose a new cropping pattern, which would allow farmers to do double-cropping with both modern and traditional varieties (Fig. 14.9). However, large-scale tests need now to be conducted. It will be extremely important to assess the potential impact of this pattern on the occurrence of pests and diseases.

Strengthening farmers' access to seeds

The seed supply system is considered by both breeders and conservationists to be a key element in the deployment and management of crop varieties in agroecosystems. The development of local genebanks managed by farming communities, and of local seed markets, has been the objective of numerous non-governmental organizations (see for example Salazar, 1992) involved in on-farm conservation activities. More recently, the IPGRI-coordinated project for *in situ* conservation on-farm has also integrated the development of local seed markets into its research agenda. It is, therefore, of no surprise when we conclude that it is necessary to strengthen farmers' access to seeds to sustain on-farm diversity. However, in this chapter, we will not discuss the possible implementation of local genebanks or the implementation of seed flow mechanisms within and between farming communities. We present two activities in which institutional genebanks can play an active role.

Improve on-farm storage

As discussed in the analysis of the consequences of the natural catastrophes in Cagayan, poor storage conditions are a cause of genetic erosion. We are developing a simple and cheap seed drying and storage device that farmers could use to store the seeds for several years. With a simple plastic drum as a container, and old rice seeds as a drying medium agent, preliminary tests show the moisture content of fresh-harvested seeds could be brought down to 12%, that is, at a level that would permit the conservation of seeds in the closed drum for several years. A prototype of the device is currently being tested by a pilot group of farmers.

Linking farmers and genebanks

In November 1998, we went to Cagayan Province to take seeds back to farmers who had participated in our project. The seeds had been collected from

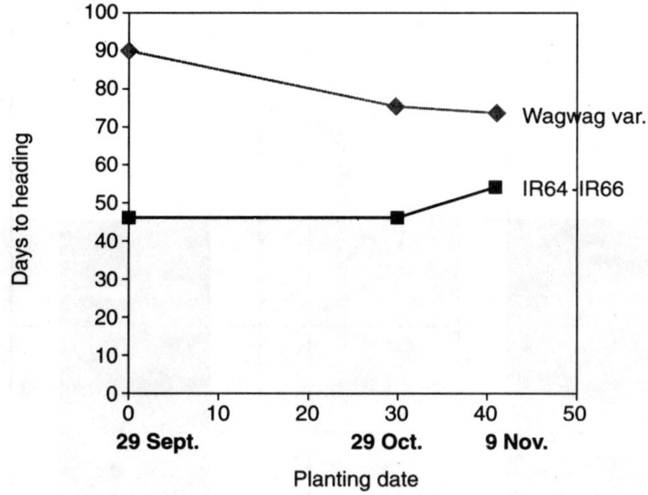

Fig. 14.7. Effect of planting date on the heading date of Wagwag varieties (from Calibo *et al.*, unpublished). Five Wagwag varieties (Wagwag, Wagwag pino, Wagwag red, Wagwag bilog, and Wagwag tawataw) and two IR varieties (IR64 and IR66) were studied in 1998 (4 m × 5 m plots, split-plot design, six repetitions). As no significant differences were observed among Wagwag varieties and among IR varieties, only the mean values for each of these two groups are shown.

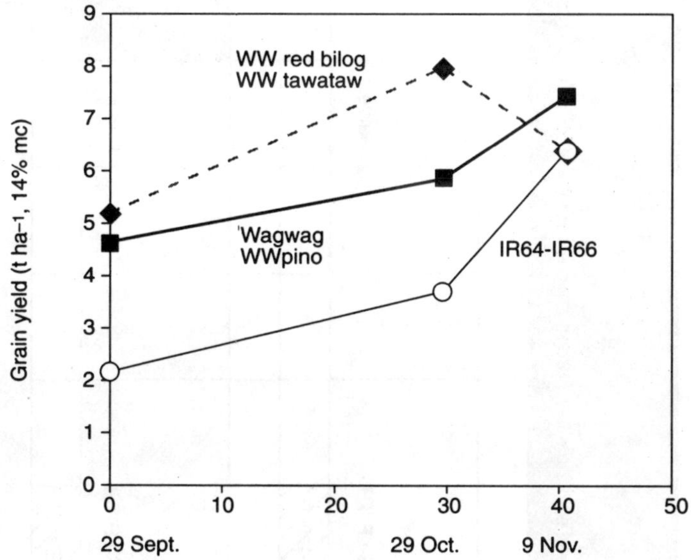

Fig. 14.8. Effect of planting date on the yield of Wagwag varieties (from Calibo *et al.*, unpublished). Same field trial as Fig. 14.7. Mean values are shown for groups of varieties that significantly differ from each other.

farmers in 1996 and planted and characterized at IRRI in 1997. No multiplication had been purposely conducted with the objective of seed distribution, which explains the relatively small amount of seeds distributed for traditional varieties. A total of 28 varieties, including both modern and traditional types, were distributed to farmers in 15 villages. In all, about 1.5 t of seeds were given. A

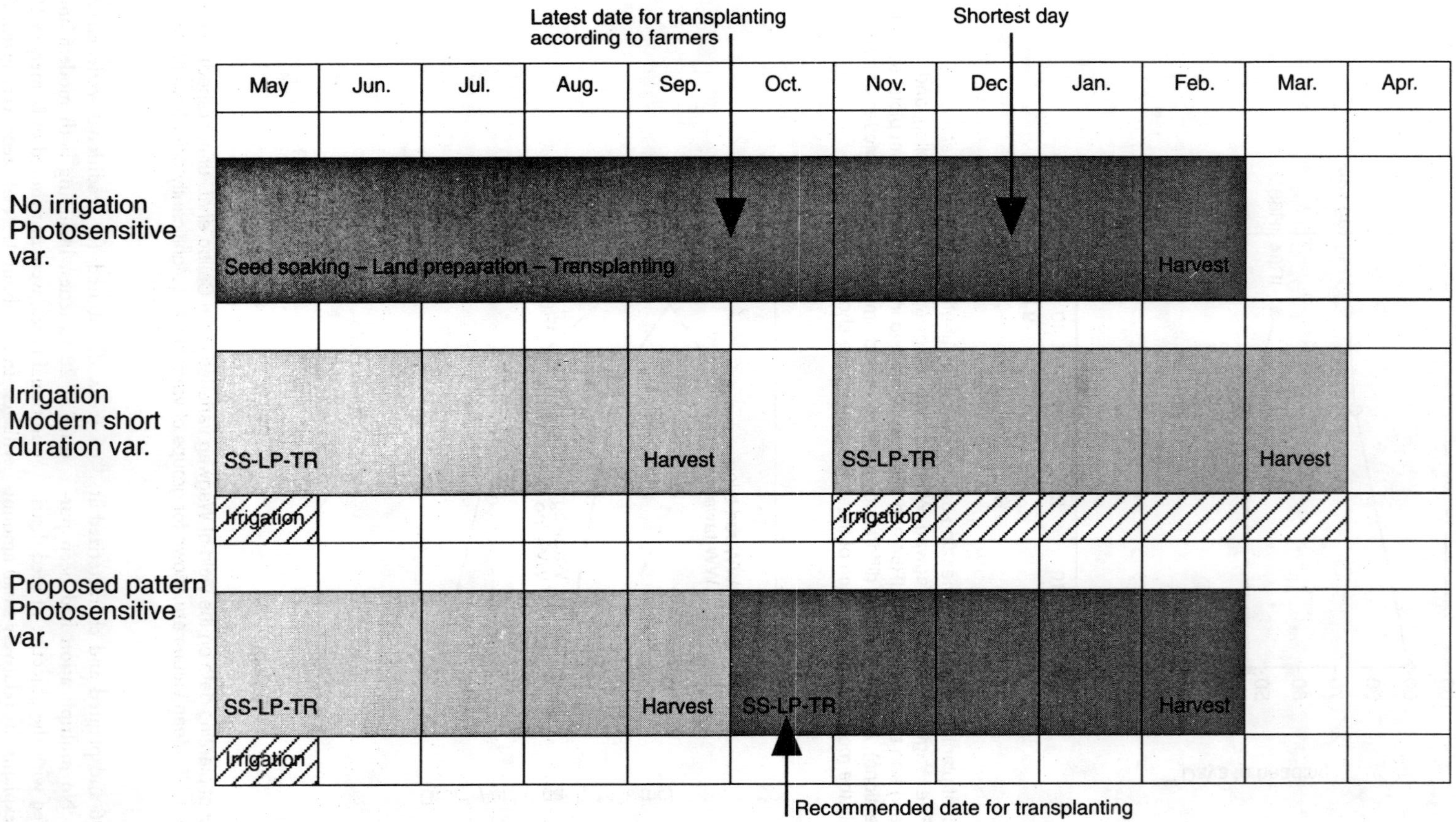

Fig. 14.9. Current and proposed rice cropping patterns in the Cagayan Valley. From the top: (1) current single crop pattern for late maturing varieties without irrigation; (2) current double crop pattern for early maturing varieties; (3) proposed cropping pattern: by delaying the sowing and transplanting of late maturing varieties, these varieties can be used in a double cropping pattern in combination with early maturing varieties.

total of 609 bags of modern variety seeds (2 kg) and 105 bags of traditional variety seeds (1 kg) were distributed.

It appeared 2 years later that the small amount of seeds distributed had limited the efficiency of the distribution. Only 57% and 32% of the bags of modern and traditional varieties respectively, were successfully multiplied. In particular, the small multiplication plots implemented by individual farmers were affected by limited floods, while larger plots, possibly conducted at the community level, would have been more resilient. Nevertheless, the distribution helped to change the ongoing trend of decreasing number of farmers growing traditional varieties (175 in 1996, 110 in 1997, 84 in 1998) as it went up to 148 in 1999.

Although this distribution was not organized in response to the catastrophes in Cagayan, its impact illustrates the need for genebanks to develop an expertise in the restoration of local diversity. One of the activities included in the FAO's Global Plan of Action is the assistance to farmers in disaster situations to restore agricultural systems. The example of the Cagayan Valley shows that disasters do not necessarily happen on a very large scale. The design and logistics for local operation of diversity restoration might have to be very different from that conducted at a national or regional level.

On-farm Conservation: an Additional Role for Genebanks

The increasing interest in *in situ* conservation on-farm makes the institutional genebanks face a new challenge. While being mainly familiar with the closed, controlled environment of cold rooms and multiplication fields and the uninterrupted flow of seed requests, genebanks must now face the open, changing agroecosystems where farmers are not only the end-users but also the decision-makers. On-farm conservation cannot be imposed on farmers. The idea of freezing genetic lanscapes (Iltis, 1974) has long been abandoned. The research activities of the Genetic Resources Centre of IRRI and PhilRice illustrate the role these genebanks and research institutions can play in collaboration with other stakeholders to contribute to the development of the right balance of incentives for farmers to maintain diversity.

The role of genebanks and research institutions is to:

- assess existing *in situ* on-farm diversity and its structure;
- assess the potential benefits of this diversity to both farmers and conservationists;
- identify endangered varieties or varietal groups, and the threats to these varieties.

Their role is then to develop options to make diversity a viable option for farmers:

- to identify technical or policy opportunities for the continued cultivation of these varieties (or to change policies that negatively affect diversity);
- to contribute to the transfer of knowledge/technology to farmers through the appropriate channels.

Finally, genebanks and research institutions have an important role to play to strengthen farmers' access to diversity:

- to understand the impact of seed policies on farmers' access to genetic diversity;
- to understand the technical constraints faced by farmers in conserving genetic resources and to improve or develop seed technologies at the local level;
- to develop channels for the reintroduction of lost varieties when needed, and develop links between farmers and genebanks not only to sustain the diversity on-farm but also to improve its management by farmers.

Understanding the processes that affect the dynamics of diversity in ecosystems goes far beyond the conservation objective. It may provide plant breeders and extension agencies with new options for better deployment and management of diversity in agroecosystems.

Acknowledgements

We gratefully acknowledge the support we received from the local authorities of the study villages and municipalities and from the Municipal Agricultural Offices. We thank all the farmers and their families who gave us their time, seed samples and answered our numerous questions.

References

Bellon, M.R. and Brush, S.B. (1994) Keepers of maize in Chiapas, Mexico. *Economic Botany* 48, 196–209.

Bellon, M.R., Pham, J.L. and Jackson, M.T. (1997) Genetic conservation: a role for rice farmers. In: Maxted, N., Ford-Lloyd, B.V. and Hawkes, J.G. (eds) *Plant Conservation: the* in situ *Approach.* Chapman and Hall, London, pp. 263–289.

Bellon, M.R., Pham, J.L., Sebastian, L.S., Francisco, S.R., Loresto, G.C., Erasga, D., Sanchez, P., Calibo, M., Abrigo, G. and Quilloy, S. (1998) Farmers' perceptions of varietal diversity: implications for on-farm conservation of rice. In: Smale, M. (ed.) *Farmers, Gene Banks and Crop Breeding.* Kluwer Academic Publishing, Dordrecht, The Netherlands, pp. 95–108.

Bernard, R.H. (1988) *Research Methods In Cultural Anthropology.* Sage, Newbury Park, California.

Borgatti, S.P. (1992) A*NTROPAC 4.0 User's Guide.* Analytic Technologies, Columbia, South Carolina.

Brush, S.B., Taylor J.E. and Bellon, M.R. (1992) Biological diversity and technology adoption in Andean potato agriculture. *Journal of Development Economics* 39, 365–387.

Hossain, M. and Pingali, P.L. (1998) Rice research, technological progress, and impact on productivity and poverty: an overview. In: Hossain, M. and Pingali, P.L. (eds) *Impact of Rice Research. Proceedings of the International Conference on the Impact of Rice Research, 3–5 June 1996, Bangkok, Thailand.* Thailand Development Research Institute, Bangkok/International Rice Research Institute, Los Baños, Laguna, Philippines, pp. 1–25.

Iltis, H.H. (1974) Freezing the genetic landscape: the preservation of diversity in cultivated plants as an urgent social responsibility of plant geneticist and plant taxonomist. *Maize Genetics Cooperation Newsletter* 48, 199–200.

Jackson, M.T., Loresto, G.C., Appa Rao, S., Jones, M., Guimaraes, E.P. and Ng, N.Q. (1997) Rice. In: Fucillo, D., Sears, L. and Stapleton, P. (eds) *Biodiversity in Trust: Conservation and Use of Plant Genetic Resources in CGIAR Centres.* Cambridge University Press, Cambridge, UK, pp. 273–291.

Louette, D., Charrier, A. and Berthaud, J. (1997) In situ conservation of maize in Mexico: Genetic diversity and maize seed management in a traditional community. *Economic Botany* 51, 20–38.

Morin, S.R., Pham, J.L., Sebastian, L.S., Abrigo, G., Erasga, D, Bellon, M.R., Calibo, M. and Sanchez, P. (1998) The role of indigenous technical knowledge in on-farm conservation of rice genetic resources in the Cagayan Valley, Philippines. In: *People, Earth and Culture. Readings in Indigenous Knowledge Systems on Biodiversity Management and Utilization.* Book Series No. 165/1998. Philippine Council for Agriculture, Forestry and Natural Resources Research and Development, Department of Science and Technology/National Commission for Culture and the Arts, Los Baños, Laguna, Philippines, pp. 137–150.

Panaud, O., Chen, X. and McCouch, S.R. (1996) Development of microsatellite markers and characterization of simple sequence length polymorphism (SSLP) in rice (*Oryza sativa* L.). *Molecular and General Genetics* 252, 597–607.

Pham, J.L., Bellon, M.R. and Jackson, M.T. (1996) A research program for on-farm conservation of rice genetic resources. *International Rice Research Notes* 21, 10–11.

Salazar, R. (1992) Community plant genetic resource management: experiences in Southeast Asia. In: Cooper, D., Vellvé, R. and Hobbelink, H. (eds) *Growing Diversity: Genetic Resources and Local Food Security.* Intermediate Technology Publications, London, pp. 17–29.

Sanou, J. (1996) Analyse de la variabilité génétique des cultivars locaux de maïs de la zone de savane Ouest africaine en vue de sa gestion et de son utilization. PhD thesis, Ecole Nationale Supérieure Agronomique de Montpellier, France.

Sthapit, B.R., Sajise, P. and Jarvis, D. (2000) Strengthening scientific basis of *in situ* conservation on-farm: Learning participatory experiences from Nepal and Vietnam. Paper prepared for Congress on Cultures and Biodiversity (CUBIC), 20–31 July 2000, Kunming, China.

15 A Study on the On-farm Maintenance of Farmers' Varieties of Sorghum in Malawi

E.A. Chiwona
Malawi Plant Genetic Resources Centre, Chitedze Agricultural Research Station, Lilongwe, Malawi

Introduction

For several years of germplasm collection and conservation of crop genetic resources, *ex situ* has been the main conservation strategy against the loss of genetic diversity in crops. However, it has become evident that views that led to the dismissal of the *in situ* conservation strategy are no longer valid. For example, it has been realized that the *ex situ* facilities (seed storage and field genebanks) cannot accommodate the full range of useful genetic diversity, nor can the facilities conserve the dynamic process of crop evolution and farmers' knowledge of crop selection and maintenance inherent in the development of farmers' varieties (FAO, 1996; Maxted *et al.*, 1997). These shortcomings of an *ex situ* conservation strategy can, however, be addressed by an *in situ* on-farm conservation approach. Therefore, in order to conserve the full range of genetic diversity of crops, there is a need to consider both *ex situ* and *in situ* on-farm conservation methods.

In situ on-farm conservation is a conservation technique which aims at maintaining genetic diversity of crops and their wild relatives in farmers' fields (FAO, 1996; Maxted *et al.*, 1997). However, the technique for implementing on-farm conservation of crop genetic diversity has not been developed; it is still in its initial stages.

A study to understand the on-farm maintenance of farmers' varieties of sorghum was carried out in the Lower Shire Valley, Malawi, with the aim of generating information that could be used to develop a strategy for implementing on-farm conservation in the country. For this study, farmers' varieties are defined as variable plant populations, adapted to local agroecological conditions, which are named, selected and maintained by the farming communities to meet their socioeconomic, cultural and ecological needs. Specifically, the study's objectives were to: (i) assess the status of farmers' varieties of sorghum and its associated crop species in the target area; (ii) identify methods used in maintaining the varieties; (iii) identify key factors influencing farmers' decisions in maintaining or limiting farmers' varieties; and (iv) propose a way forward for the development of an on-farm conservation strategy.

Materials and Methods

The choice of the crop species

Sorghum (*Sorghum bicolor* (L.) Moench) was selected as a target crop species for the study due to its importance in Malawi. While sorghum is the world's fifth largest cereal crop, it is the most

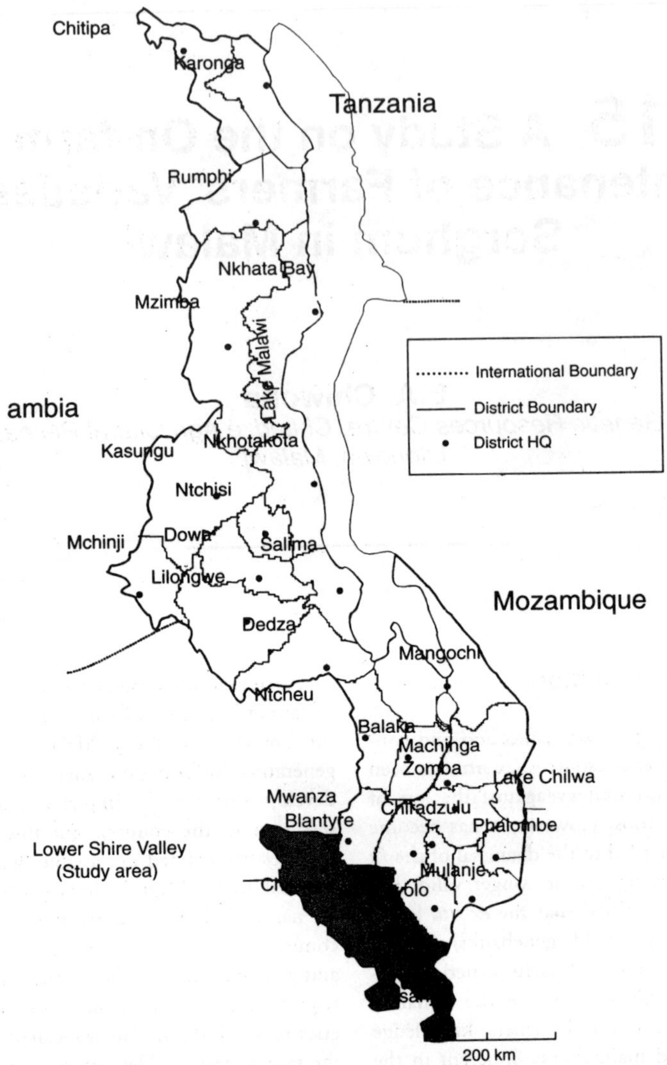

Fig. 15.1. Location of Lower Shire Valley, Malawi.

important indigenous cereal crop in the country. The Lower Shire Valley (Fig. 15.1) is the main sorghum producing area in the country, hence its choice as study site.

The study targeted all farmers' varieties of sorghum. In the absence of more efficient techniques such as molecular markers, different farmers' varieties were distinguished on the basis of local names used in the study area. The problem with this method, however, was the apparent use of different local names for the same variety. This was especially so when one moved from Chikwawa to Nsanje (Fig. 15.1). For example, Gonkho in Chikwawa is a sorghum variety characterized by a goose's neck but the same name in Nsanje means sorghum.

The target area

Due to limited resources, five out of eleven agroecological zones (Fig. 15.2), representing 45% of the zones available to smallholder farming in the Lower Shire Valley were selected as the target area. The selected zones included the Makande Plains (LS 2),

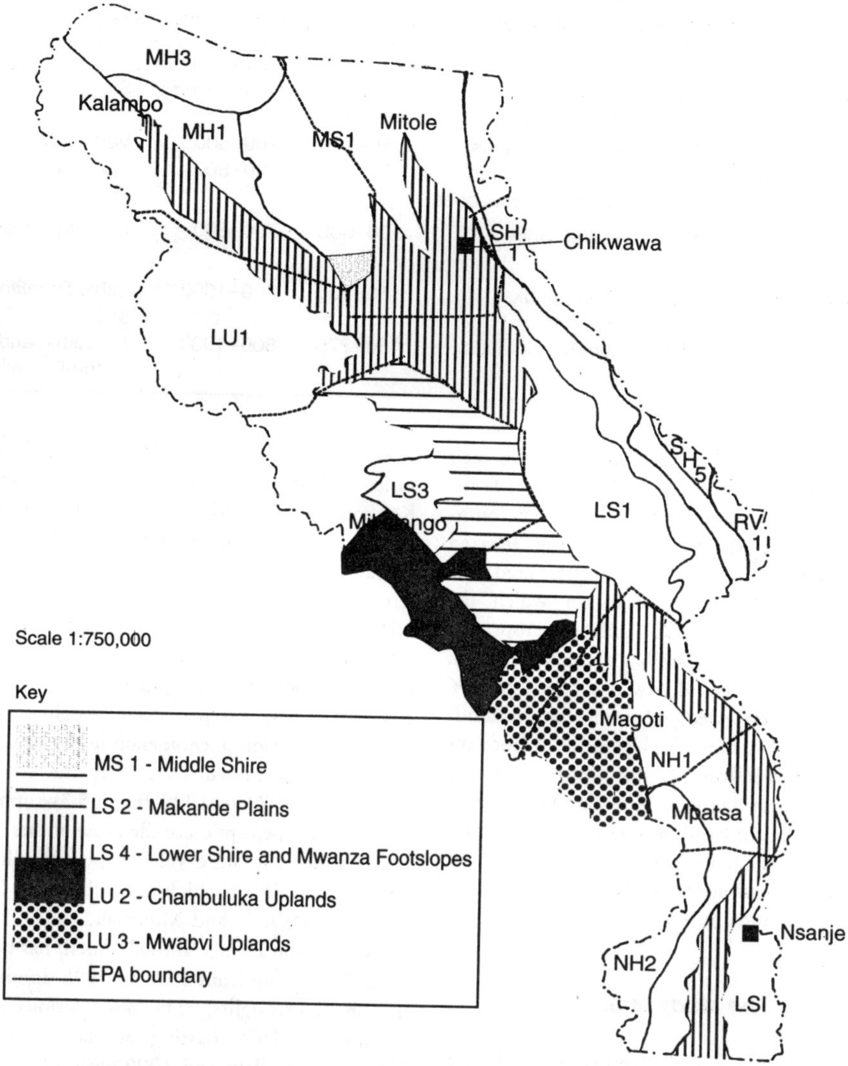

Fig. 15.2. Selected agroecological zones in the Lower Shire Valley, Malawi.

Lower Shire and Mwanza Footslopes (LS 4), Middle Shire (MS 1), Chambuluka Uplands (LU 2) and Mwabvi Uplands (LU 3). The agroecological zones were described mainly on the basis of soil type (Table 15.1).

Village selection, number of farm families and interviews

Again, due to limited resources, a sample of 44 villages representing about 10% of the villages was randomly drawn from 442 villages covering the selected agroecological zones. The selected villages were then marked on 1 : 50,000 maps of the Lower Shire Valley as sites which were then used during field work. The number of farm families targeted in the study was 200. This number was considered adequate given limited resources. Besides, most of the social surveys that have been conducted in the whole country have had sample sizes ranging from 200 to 300.

From each of the selected villages, the number of farm families to be interviewed was calculated by using the following formula:

$$\text{Sample size (N)} = P/TP \times 200$$

Table 15.1. Characteristics of the selected agroecological zones. (From Venema, 1991.)

Natural region	Zone code	Agroecological zone	Altitude (m asl)	Mean annual rainfall (mm)	Soil group
Lower Shire and Mwanza Valley (LS)	LS 2	Makande Plains	50–200	700–800	Vertic soil
	LS 4	Lower Shire and Mwanza Footslopes	60–200	700–800	Fluvic & Eutric-Fersailic soils
Middle Shire Valley (MS)	MS 1	Middle Shire	100–500	700–800	Paralithic soils
Lower Shire Uplands (LU)	LU 2	Chambuluka Uplands	150–400	800–1000	Lithic-Paralithic soils
	LU 3	Mwabvi Uplands	150–275	800–1000	Paralithic and Eutric-Fersailic soils

where P is the number of farming families in each selected village; and TP is the total number of farming families in all the 44 selected villages. The number of farm families for each of the selected villages was drawn from the census register used during the Starter Pack Programme in which the Government provided seed of different crops and fertilizer (free) to farming families in the Shire Valley. From the Lower Shire and Mwanza Footslopes, 96 families were interviewed, 42 from the Makande Plains, 27 from Chambuluka Uplands, 20 from the Middle Shire and 9 from Mwabvi Uplands, giving a total of 194 farm families. The family members interviewed were husband and wife, or either husband or wife alone, depending on who was at home. A participatory rural appraisal method, guided by a structured questionnaire, was used.

The study team

The study team comprised members from the following organizations:

- Malawi Plant Genetic Resources Centre (MPGRC)
- Shire Valley Agricultural Development Division (SVADD)
- Chileka Meteorological Services.

Results and Discussion

Status of farmers' varieties in the target area

Table 15.2 shows that 4 out of 15 farmers' varieties of sorghum have been completely lost in the study area, representing 27% loss. However, among the existing varieties, only Kawaladzuwa, Wayawaya and Katsonte/Kachilawende were widely distributed. According to the families interviewed, Kawaladzuwa, Wayawaya and Katsonte/Kachilawende were preferred due to their early maturing compared with the other varieties. This was a clear demonstration of serious genetic erosion taking place in the farmers' varieties. The major cause of genetic erosion, as reported by the families, was persistent drought, which has become a common phenomenon in the area. Analysis of total annual rainfall received in the Lower Shire Valley from 1979 to 1998 (Fig. 15.3) supports the farm families' perception of the cause of these losses.

Although the Shire Valley experiences low average annual rainfall which varies between 700 and 400 mm (Mkanda and Munthali, 1991), Fig. 15.3 shows that the average annual rainfall has been seriously fluctuating from 1979 to 1998 dominated by dry spells (droughts). The most serious droughts occurred in 1979 (lasting 30 days) followed by 1991 with 26 days and 1990 with 21 days. These fluctuations have had serious impact on crop production in general, and farmers' varieties of sorghum in particular, the result of which has been the selection of early maturing sorghum varieties.

Table 15.2 also shows that while Wayawaya and Kawaladzuwa were among the three varieties that were widely distributed in the area, they dominated different agroecological zones. Wayawaya dominated the Makande plains (LS 2) and Chambuluka Uplands (LU 2), while Kawaladzuwa dominated the Lower Shire and Mwanza Footslopes (LS 4), Middle Shire (MS 1) and Mwabvi Uplands (LU 3). The difference in distribution of the two varieties was attributed to adaptation to different soil types that characterized the agroecological zones.

Table 15.2. Frequency of cultivated and lost farmers' varieties of sorghum reported in various agroecological zones (in percentages).

Sorghum varieties reported	LS 2	LS 4	MS 1	LU 2	LU 3	Average
Wayawaya	36	1	3	45	13	19.6
Katsonte/Kachilawende	12	14	6	13	22	13.4
Kapile	7	5	0	5	0	3.4*
Thengalamanga	8	15	0	15	17	11
Kawaladzuwa	10	33	59	5	44	30.4
Dikwa	2	3	0	3	4	2.4*
Gonhko	3	9	•	3	•	5*
Kalombo/Kalumbu	10	6	3	8	•	7*
Zendembani	3	2	3	0	•	2*
Chunga/Phatamfuli/Misinde	7	1	23	3	0	6.8
Shabalala	2	11	3	0	0	3.8*
Masotong'o	0	0	0	0	•	**
Bandela	•	0	•	•	0	**
Kapsyabanda	0	0	•	•	0	**
Mchirawanyumbu	0	0	•	•	0	**

Note: 0, the variety used to exist but has now been lost; •, the variety never existed in the agroecological zone; *, rare; **, lost/disappeared.

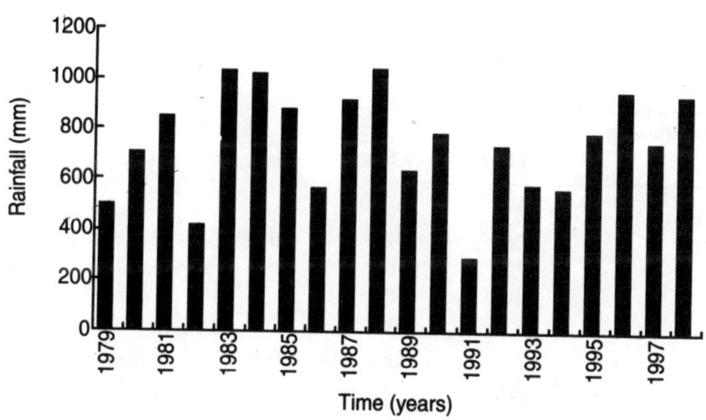

Fig. 15.3. Total annual rainfall in the Lower Shire Valley, Malawi, 1979–1998.

Table 15.1 shows that the Makande Plains, in which Wayawaya was most common, are mostly dominated by vertisols which are characterized by high clay percentage (> 30%), whereas the Lower Shire and Mwanza Footslopes, in which Kawaladzuwa was most common, are dominated by fluvic soils characterized by continuous soil rejuvenation through the deposition of sediments on the surface by floodwater. In the case of Wayawaya also being most common in the Chambuluka Uplands, the possible explanation could be its close proximity to the Makande Plains (Fig. 15.2). Agroecological zones that are close to each other are likely to have stronger seed exchange than agroecological zones far apart. A similar explanation also applies to Kawaladzuwa being most common in the Middle Shire and Mwabvi Uplands.

Methods used in the management of farmers' varieties

Seed selection

Seed selection is an important process in the maintenance process of varieties (whether improved or not). This ensures that high quality seeds are planted every season and that purity of the varieties is maintained.

It was therefore not surprising that all the families carried out seed selection of some sort and were well versed as to why they did it as well as what characteristics make good and bad seeds. Differences, however, were found in the time of selection both within and between agroecological zones (Table 15.3).

In Table 15.3, it can be seen that although different families carried out seed selection at different times in each agroecological zone, the majority of the families selected their seed between harvest and threshing periods. The reasons given for seed selection at this time were that it was the best time to select good seed for safe-keeping against weevils and other pests and that it was easier to separate seeds of different varieties. However, a few families in the Lower Shire and Mwanza Footslopes and Chambuluka Uplands selected their seed at planting time. According to those families that selected during planting time, it was the best time for seed selection since one of the characteristics they look for is the ability of seeds to survive pests and disease attack.

Gender involvement in seed selection

Table 15.3 shows that women were predominantly involved in seed selection in all the agroecological zones. The major reason for this was that, in the Shire Valley, women are traditionally responsible for food crops of which farmers' varieties of sorghum played a major role. In the case of men, it was reported that their main interest was in cash crops such as cotton and improved crop varieties as these too are commonly for sale due to their poor storage.

Storage techniques

SEED DRESSING. It was found that a number of techniques were used to control pests in the store (Table 15.3). These included application of Actellic powder (Pirimiphos-methyl), Sevin (Carbarly 85 S), 'Good hands', use of ash dust, smoke, Neem and ultra sun drying. Use of Actellic powder, which is the current recommended pesticide, was most

Table 15.3. Frequency of methods for maintaining farmers' varieties of sorghum in various agroecological zones (in percentages).

Methods used in maintaining farmers' varieties	LS 2	LS 4	MS 1	LU 2	LU 3	Average
Time of seed selection						
Harvest-threshing	69	67	93	62	73	72.8
After threshing	17	0	7	0	0	4.8
Planting time	.14	33		38	27	22.4
Gender involvement						
Women	86	62	93	94	67	80.4
Men	14	5	7	6	33	13
Both		33				6.6
Seed dressing						
Actellic powder	34	40	14	68	27	36.6
Sevin powder	10	13	5	5	20	10.6
Nothing/good hands	23	32	57	11	47	34.0
Use of ash	10	6	14	5	0	7.0
Smoking over fireplace	14	7	5	11	6	8.6
Neem	7	1	5	0	0	2.5
Ultra sun drying	2	0	0	0	0	0.4
Type of storage						
Bags	88	70	56	100	80	78.8
Clay pots	6	13	11	0	0	6.0
Chikwa	6	2	0	0	13	4.2
Bench inside house	0	15	0	0	7	4.4
Granary	0	0	33	0	0	6.6

Note: 0, not in use or not reported.

common in all the agroecological zones except in the Middle Shire Valley (MS 1) and Mwabvi Uplands (LU 3). The use of 'Good hands' came second. This method is interesting because the scientific basis for it is not known. In the villages where the method of 'Good hands' was used, it was reported that when certain people of either sex and of any age implement seed processing, that is from harvest to threshing to storage, there is less pest attack on the seed. However, the families using the method reported that it was not as effective as use of Actellic powder.

Another interesting point on how families control pest attack on stored seed, was the use of Sevin (Carbarly 85 S). Sevin is a recommended cotton pesticide for controlling elegant grasshoppers, red and spiny bollworms and other pests. Cotton is the main cash crop for smallholder farmers in the Shire Valley, hence Sevin's easy availability in the area. It has not, however, been recommended for use in food crops. It seemed that the families were not aware of the dangers of this chemical to human health. While it was apparent that the traditional methods such as smoking the seeds and use of ash were disappearing (except 'Good hands') (Table 15.3), use of ash in cow peas was still common.

SEED STORAGE TYPES. Table 15.3 shows that several storage types were being used across the target area. These included the use of bags (made out of jute), clay pots, woven baskets called Chikwa, and platform/benches in houses. Of these types, the use of bags was most common. According to information gathered during the study, the use of bags is quite recent. Before the 1970s, the common storage types were the use of Chikwa for sorghum and pearl millet, pots for cow peas and granaries for maize. These old types have almost disappeared as a result of theft and the introduction of handy jute bags.

Key factors influencing farmers' decisions on the maintenance of sorghum and its associated crop species

One of the objectives of the study was to find out key factors that influence farming families in either continuing or discontinuing with the maintenance process of the farmers' varieties. The approach chosen was first to find out if the family would be interested in continuing with the maintenance process. Depending on the type of answer, the family was asked to justify it. The answers and their justifications are summarized in Table 15.4. The majority of the families (84%) across the target area were interested in continuing with the maintenance process. The reasons given were: (i) farmers' varieties produce hard grains that make storage handling easier and are easier to pound; (ii) food security: by growing both farmers' varieties and improved varieties of different crop species, families felt more food secure; this was because of recurrent droughts in the target area; (iii) inability to purchase seed of improved crop varieties every season; and (iv) keeping farmers' varieties was part of their tradition.

One of the most serious problems facing the acceptance of improved varieties of sorghum and pearl millet in Malawi is related to pest problems owing to the softness of the grains. Weevils attack the improved varieties more easily compared with the hard grains of farmers' varieties. The second reason, that food security was important, was due to the unpredictability of rainfall. Since it is difficult to predict how much rainfall will be received, especially in the Lower Shire Valley, families grow both improved and farmers' varieties so as to be more food secure. On the other hand, the group of families not interested in continuing with the maintenance process gave the following reasons: (i) low yield of farmers' varieties compared with improved varieties; (ii) late maturity and hence inability to match with current rains; (iii) bird damage; and (iv) lack of drought resistance.

Overall, low yield of the farmers' varieties, followed by late maturity, are the major factors that influence the families to discontinue with the maintenance process of farmers' varieties and to switch to modern varieties.

Conclusions

The findings of the study led to the following conclusions:

- Serious genetic erosion has been threatening the continued existence of farmers' varieties of sorghum in the Lower Shire Valley. The dominance of Wayawaya and Kawaladzuwa, both of which are early maturing varieties, is a clear indication of genetic erosion which has occurred in

Table 15.4. Frequency of factors influencing farmers' decisions in the maintenance of farmers' varieties in various agroecological zones (in percentages).

Key factors affecting the maintenance process	LS 2	LS4	MS 1	LU 2	LU 3	Average
Percentage of farmers interested in continuing the maintenance of varieties	78	88	95	76	81	83.6
Hard grains	18	28	35	11	31	24.6
Food security	52	40	54	50	44	48
Unable to purchase improved seeds every year	8	20	6	15	6	11
Percentage of farmers not interested in continuing the maintenance of varieties	22	12	5	24	19	16.4
Low yielding	8	6	1	12	10	7.4
Late maturing	8	5	2	6	0	4.2
Bird damage	3	1	1	0	9	2.8
Drought	3	0	1	6	0	2.0

Note: 0, not reported.

sorghum because of the replacement of late maturing varieties by the early maturing varieties. Persistent drought was found to be the major contributing factor to genetic erosion of farmers' varieties in sorghum.

- Use of local names of farmers' varieties was confusing as one moved from one place to another; local names, used for particular sorghum varieties in Chikwawa were found to be different from those names used in Nsanje for the same varieties.
- Most farming families select the seed of farmers' varieties between harvest and threshing period. However a few families carried out seed selection at planting time.
- Women were found to be the custodians of farmers' varieties in the Shire Valley.
- Just like the findings of a previous survey (Chiwona, 1998), the majority of farming families showed interest in keeping farmers' varieties mainly for food security and grain hardiness.

Way Forward for an On-farm Conservation Strategy

Based on the study results, personal experience gained during field work, as well as on the outcome of the 1998 survey, the following way forward for the development of an on-farm conservation strategy for Malawi is presented.

Verification of farmers' varieties

Local names of farmers' varieties of sorghum were used for differentiating the varieties in the study. It seemed, however, that different local names were in some cases used for similar or the same varieties. In order to be certain whether the study dealt with different sorghum varieties, variety verification of the collected samples has to be conducted.

Conducting similar studies on other crops and in other parts of the country

Similar studies should be conducted on different crops and in different agroecological zones of the country in order to be able to compare the results.

Conservation through use

Conservation through use of farmers' varieties is a concept which is based on the principle that farming families will only keep farmers' varieties that are useful to them (Worede, 1999). Less favoured sorghum varieties need to be made more attractive in order to ensure that more farm families will be interested in their maintenance. One proposed way to do this would be by improving the farmers' varieties through participatory plant breeding.

The study results indicated that farmers' varieties in general and particularly those that are less dominant, are low yielding compared with modern crop varieties. However, it is possible that, through participatory plant breeding, yields of farmers' varieties can be improved, while maintaining diversity in such materials (Worede *et al.*, 2000). Therefore, by enhancing the performance of farmers' varieties this can result in increased utilization and in turn this will improve their conservation on-farm. However, the effect of participatory plant breeding on genetic diversity will have to be investigated. If an effect exists, it will be important to establish levels of genetic loss that can be tolerated.

Acknowledgements

Without the funding received from the SADC Plant Genetic Resources Centre (SPGRC), the study would not have been undertaken. I would, therefore, like to thank most sincerely SPGRC for providing funding to undertake the study as well as for the guidance received from the SPGRC *In situ* and Under-utilised Plants Regional Crop Working Group Committee. My profound gratitude also goes to all the farm families as well as the extension workers from the SVADD for giving their precious time in providing the requested information on various issues of the study.

My thanks also go to the Chief Meteorological Officer and his staff for providing weather data for the study area, the Government of Malawi through the Director of Agricultural Research and Technical Services for approving the study, Mr W. Kumwenda, Mr A. Kanyika of the Department of Agricultural Research and Technical Services (Malawi) and the SAT 21 Editorial Committee, particularly Dr Jan Engels and Dr Devra Jarvis for their constructive comments on the paper and Mr M. Komwa for assisting with the production of maps. Furthermore, I acknowledge the tremendous contributions of the MPGRC staff, in particular, Mr K.F. Kapila, Mr R.D. Chitezi, Mr F.E. Kambadya and Mr E. Kamwela for their active participation in field data collection.

References

Chiwona, E.A. (1998) On-farm conservation survey of sorghum landraces in Malawi. A paper presented at the 7th *In situ* and Under-utilised Plant RCWG meeting held in Lusaka, Zambia, 26–27 October 1998.

FAO (Food and Agriculture Organization of the United Nations) (1996) *Global Plan of Action for the Conservation and Sustainable Utilisation of Plant Genetic Resources for Food and Agriculture.* Division of Plant Production and Protection, FAO, Rome, Italy.

Maxted, N., Ford-Lloyd, B.V. and Hawkes, J.G. (1997) *Plant Genetic Conservation.* Chapman and Hall, London.

Mkanda, F.X. and Munthali, S.M. (1991) Causes of mortality of nyala (*Tragelaphus angusi* Gay) in Lengwe National Park, Malawi. *African Journal of Ecology* 29, 28–36.

Venema, J.H. (1991) *Land Resource Appraisal of the Ngabu Agricultural Development Division.* Malawi Government, Ministry of Agriculture Land Husbandry Branch, Lilongwe, Malawi.

Worede, M. (1999) Working to raise productivity while keeping diversity alive. In: *Unitarian Service Committee of Canada (USC) Sparks Newsletter*, pp. 4–5.

Worede, M., Tesemma, T. and Feyissa, R. (2000) Keeping diversity alive: an Ethiopian perspective. In: Brush, S.B. (ed.) *Genes in the Field: On-Farm Conservation of Crop Diversity.* Lewis Publishers, USA, pp. 143–161.

16 The Role of Bioinformatics in Germplasm Conservation and Use

B.W.S. Sobral*
National Center for Genome Resources, Santa Fe, New Mexico, USA

Introduction

Biology is at the threshold of a major paradigm shift, or revolution (www.ncgr.org/bioinformatics/paper/ and www.agbio.cabweb.org; Sobral *et al.*, 2001). A collective leap in knowledge is about to come from investments in the biological sciences. Until now experimentation has driven biological information. Increasingly, it will be information that will drive biological experimentation. Biology is moving from primarily a descriptive, experimental discipline to a predictive, mathematical discipline. This trend began in the 1960s with systems ecology and the rise of modelling to predict the behaviour of ecosystems. It failed and lost favour because of the difficulty in choosing parameter values but has resurfaced in the form of agent-based models. German scientists may have first used the word bioinformatics about 100 years ago, with the meaning of producing a mathematical model of living organisms. However, the current use of bioinformatics is much less defined than mathematical modelling and the word seems to have lost an agreed meaning. One driving force for the paradigm shift is the shift from 'cottage industry' laboratories to large-scale factories or 'data farms'. As a result of engineering of laboratories, changes in the resolution at which biological questions can be posed (from molecules to ecosystems) are now envisioned. To provide the capability for such questions, a data management infrastructure is required by biologists, much like that which has been built to support other information-intensive communities.

The focus of most of biology in the 20th century was to reduce biological phenomena and their understanding to the level of molecules and their behaviour. However, it is increasingly recognized that many, if not most, biological functions come from the interactions among many components. One thing this suggests is that biology will start to need to incorporate the 'systems thought' that is typical of engineering. To achieve some of the benefits possible from an integrative approach to biological questions and information, it is necessary to invest in biological information resources (databases and associated tools) and to create standards for interpretation and comparisons of the data and their transformation into information and, hopefully, knowledge. Also, it is necessary to integrate information at various levels and with an environmental context. To do so requires:

- engineering of integrated software systems;
- high-performance hardware systems to support distributed collaborations and complex data analyses;
- high-speed connectivity to the Internet.

Biology has also become 'big business'. Private investment in biological research and development has been growing (NSF S&T report, 1998). However, to realize the potential that is offered by paradigm shifts and biological revolutions will

*Present address: Virginia Bioinformatics Institute, Virginia Tech, Blacksburg, Virginia 24061, USA

(eds J.M.M. Engels, V. Ramanatha Rao, A.H.D. Brown and M.T. Jackson)

require a different type of research organization, with a team focus. This is caused by the necessity of multidisciplinary approaches to tackle the opportunities and challenges posed by biology as an information-driven science. Biological information is doubling roughly every 6 months, based only on DNA sequence data. This is faster than the exponential rate of increase of computing power, as suggested by Moore's Law (an empirical observation made long ago that has held until today: the doubling of processor power every 12 months). In the past decade, more scientific information has been created than in all of previous human history (Kaku, 1997).

Leaps in information technology (IT) are well known and of equal importance now to biology and other fields, precisely because of their impact on biology as it becomes an information-driven science. The direction appears to be one of distributed computing eventually allowing high-end virtual reality, modelling and visualization to become routine over the Internet (Butler, 1999). Some argue that the world's largest supercomputer will eventually be the Internet itself.

Supercomputing power at commodity prices will help scientists interested in biology to tackle complex systems, thus prompting a profound change in science itself. Ruzena Bajcsy (assistant director of the National Science Foundation's directorate for computer and information science engineering) believes that it will end the Descartian chapter of reductionist research and lead to modelling of complex systems from organisms to the whole Earth (according to Butler, 1999). Thus, science that has been previously divided into experimental and theoretical will increasingly include a new form: virtual reality (VR) or cyberscience, providing the ability to simulate complex systems (Kaku, 1997).

One can define terms in a variety of ways, especially in a rapidly changing scientific and technological environment. However, there are critical areas that form the foundation for the biological revolution being described here. These areas will be key focal and organizational points in the next 5 years and beyond. The areas are defined below for the purposes of this chapter:

- *Bioinformatics*: the IT infrastructure required to support biological information storage, acquisition and retrieval, along with the networking needs to enable connectivity to distributed research groups worldwide. An example is the design, construction and maintenance of databases. This is typically an engineering domain.
- *Computational biology*: development of new algorithms for data analysis and interpretation. An example would be a new mathematical approach toward extracting knowledge from a gene expression dataset. This is typically a mathematical or computer science domain, although some biologists that are well-versed in mathematics have been getting involved here as well.
- *Bioengineering*: the development of new, smaller, cheaper laboratory instrumentation for high-throughput biological data production. Typically an engineering domain and very related to and involved with nanotechnology.

Organismal Germplasm Information Resources

Scientists or managers working with germplasm have different needs and objectives, though broadly speaking these relate to conservation or utilization. Whatever their perspectives, it appears that all those involved in germplasm collection, conservation and utilization would benefit from professionally built, deployed and maintained information resources for their favourite organisms.

To engineer laboratory information management systems (LIMS), laboratory managers and scientists needed to step back and think about the process of laboratory data generation and acquisition. As a result, various LIMS have been built over the years, with varying success. Integration of molecular information is important but it is even more important (and challenging) to integrate molecular and organismal data, the latter typically provided through germplasm repositories and mutant stock centres. This need is clear because organisms are more than the sum of their parts. In addition, it is within this organismal context that data can be transformed into useful information and, perhaps, knowledge. Unfortunately, there has been less development of germplasm information management systems (GIMS) when compared with LIMS; thus the basic informatics infrastructure is missing. However, an interesting prototype for such a GIMS has been studied and implemented by members of the Consultative Group on International Agricultural Research (CGIAR, www.cgiar.org/); it is known as the International Crop Information

System (ICIS www.cgiar.org/icis/homepagetext.htm). Preliminary implementations have occurred for wheat and maize at the International Maize and Wheat Improvement Centre (CIMMYT) in Mexico and for rice at the International Rice Research Institute (IRRI) in the Philippines.

Not only are the types of information resource changing from libraries to electronic repositories, but the very nature of scientific publishing in biology may be undergoing a change. For example, when the genome of *Arabidopsis thaliana* was completed in 2000, there were publications describing the results. However, those papers contained very little of the raw data from the genome and focused on highlights that the authors considered to be of relevance. It is simply impossible to think of a manuscript that would contain all the data from a genomic sequencing effort. Thus the only public repository that contains almost all of the relevant data and thus is the most useful to a broad range of scientists interested in *A. thaliana* or other organisms, is The *Arabidopsis thaliana* Information Resource (TAIR www.arabidopsis.org).

Despite these wonderfully rich organismal information resources, typically supported by the communities they serve, there is a need for their integration at a higher level to allow any investigator with Internet access to have a complete view of the model system without needing to navigate many sites and integrate the information manually. This has been the experience, for example, of the human genome, yeast genome genome-www.stanford.edu/Saccharomyces/ and *Arabidopsis* genome www.arabidopsis.org communities. It makes sense therefore for the germplasm communities worldwide to learn from the lessons of those that have earlier undergone the genomic shift.

Once integration of molecular information with organismal (phenotypic) information is achieved in a robust manner, then it is very likely that the resulting system will modify the way we think about germplasm conservation and breeding. Typically, plant breeding is done by crossing and selecting from progeny. With the opportunity to make predictions concerning the outcomes, and to explicitly model *in silico* (i.e. model in the computer) the desired genotype × environment × breeding scheme combinations to optimize for specific traits, breeding shifts in its character. One of the changes is that breeders start to become model testers themselves and model systems (or other information rich systems) become useful tools for model building. Once models of phenotype × genotype × environment are verified through explicit breeding experiments, the task becomes one of moving the models themselves around through breeding in different organisms. One very interesting effort that is pursuing this type of paradigm shift is the Quantitative Genetics (QU-Gene www.pig.ag.uq.edu.au/qu-gene) system. QU-Gene is a simulation platform for quantitative analysis of genetic models.

Finally, once integration from the molecular to the organismal levels has occurred, the next frontier becomes integration of ecosystems information (environment and interactions among organisms and populations) as the concluding step in bringing together molecules to ecosystems in a useful manner (because ecosystems are more than the sum of their component organisms). There are various efforts tackling how to bring together environmental information to the organismal level. One such effort is represented by FloraMap www.floramap-ciat.org, developed by the International Centre for Tropical Agriculture (CIAT). FloraMap is a computer tool for predicting the distribution of plants and other organisms in the wild. Approaches like FloraMap could enable scientists to link disparate studies based on environmental characteristics, for example.

The Role of Molecular Information Resources Useful to Germplasm Scientists

The main reason to support the new era of industrialized biological research, from the perspective of germplasm scientists, is that as a result of integration of various levels of information, we should be able to ask more effectively what the relationship between genotype and phenotype might be for our favourite biological processes, traits or organisms. This has been the driving question underlying genetics since Mendel and Darwin.

Industrial-scale biology has the capability to offer the needed data to address the question of the relationship between genotype and phenotype in various manners, organisms and traits. This is because the industrialized laboratory can generate data to test the relevant hypotheses. Industrial laboratories are exquisitely capable of grinding up organisms and separating and measuring the behaviour of their parts in novel ways. As a result, such projects are capable of generating information about the proteins expressed by genes, the role of those proteins in phenotypes and the variation in gene expression patterns in populations. When such data are integrated effectively, predictive models can be forwarded and comparisons across organisms become possible.

It is necessary to provide integration for at least the following types of molecular data:

- structural genomics: DNA sequences (to complete genomes) and maps (genetic, physical or cytological);
- gene expression: mRNA profiling, and single gene profiles (Northerns);
- biochemistry: pathways (metabolic and signalling), metabolites, proteomics.

Of course, then it must be possible to provide a context for these types of molecular data, and it is the germplasm (or organismic) data that provide such a context. Germplasm data might include natural variants and induced mutant collections with their respective phenotypic characterization, closely tied to environmental and geographical data as appropriate.

Structural genomics

The 'data types' described above progress from a molecular starting point ever closer to phenotype. For each of these types of data it will be necessary to provide the means for acquisition, storage, querying, analysis and visualization. This is best handled through generalized 'components' as it is easier to solve the informatics needs for each data type and then use the general solution across multiple organisms to enable the power of comparisons. This is most famously illustrated through the use of similarity searches of DNA sequence databases using tools such as BLAST. DNA sequence databases such as GENBANK are warehouses that contain all public DNA sequences from all organisms. These can be thought of as being one level of 'component' or subsystem within a whole that integrates the different components through some architecture. That a warehouse may be updated from various distributed specialized databases is particularly encouraging, especially if communication standards pave the way.

As a genome is studied using industrialized laboratories, maps become useful viewing paradigms for genome-scale information, especially for genetic applications. In addition, the capability to compare maps across species boundaries, much as is possible in DNA sequences in GENBANK, is of paramount interest to those working in less well funded crop species.

Gene expression

Methods for studying large-scale gene expression, while unparalleled in power, also promise to increase data flow by orders of magnitude, unleashing a deluge of gene expression data. The increase in production of gene expression data has not been matched by corresponding development in software tools and databases to analyse and exploit the results of such experiments. Techniques for examining overall gene expression are so new that even the technology vendors do not yet understand all the complexities involved, so many of the datasets that have been generated are not being analysed correctly or completely. Gene expression datasets tend to be larger than sequence entries and only a fraction of data produced is of direct interest to the group generating the data. Thus the extraction of information tends to be sparse. Other groups should be able to glean this under-utilized information by querying public gene expression systems.

There is an inclination to treat gene expression data in terms of biological sequence, as a much larger set of data, but approachable in the same way with the same basic set of tools. This is incorrect. While some of the tools and regular expressions have applications to gene expression, the data are not only different in kind but much more complex. In a sequence database, the core data are strings of nucleic (or amino) acid residues. There may be variation of a given sequence across a population, but within an individual, the sequence is invariant. In contrast, gene expression is highly dynamic and it is this variation in the expression patterns which researchers will use to attribute relationships between genes. Another difference between gene expression data and other biological data is that they contain little inherent information. The value or meaning of gene expression data comes from the context of the experiment. What were the taxonomy, sex and developmental stages of the organism?

What were its growth conditions? From what organ and tissue was the sample extracted? What protocols were used in sample preparation?

Biochemistry

The availability of complete genomes, starting with prokaryotes, has also allowed metabolic reconstruction based on identified genes, enzymes and known pathways. This is currently achieved through use of annotation (metadata) of DNA sequence data from a given genome. With the imminent completion of the *A. thaliana* genome (scheduled for completion in 2000) metabolic reconstruction of *A. thaliana* based on this approach is now possible, as shown in a preliminary manner at www.intl-pag.org/pag/8/abstracts/pag8906.html.

Moving away from the structural DNA sequence data towards information that is more related to the state of a living organism, such as information on mRNA populations (gene expression), metabolites and biochemical pathways that are active in a given environment, starts to permit integration of information aimed at functional knowledge about organismal behaviour or performance. Thus biology starts to move away from reductionism and into whole organism performance and organism × environment interactions. Biochemistry is a crucial step towards understanding the function of genes and biological processes.

Efforts to provide the information infrastructure for biochemical bioinformatics are already developed. However, only recently have efforts been focused on integrating biochemical knowledge with modelling and experimentation. An example of data acquisition and integration within this context is the PathDB effort www.ncgr.org/software/pathdb/, which is positioned to acquire through literature scanning and curation the combined biochemical knowledge of the last 70 or so years of biochemical research, in an organism prioritized manner. By combining this approach for data acquisition with tools for modelling and predicting the behaviour of biochemical pathways, such as GEPASI www.ncgr.org/software/gepasi/, biochemistry is becoming integrated with genomics, in a path that can move towards functional knowledge of living organisms. Next steps are poised at supporting experiments that include data acquisition and analysis of metabolites produced by cells, organs or tissues.

Integration of Information Resources

For all types of biological data, it is necessary to build information resources that are capable of data acquisition, storage, analysis and visualization. The need for integration across data types is explicit if we need to make predictions that integrate data types. This is required for predictions concerning organismal performance or phenotypes. Various approaches to deliver this requirement to scientists can be taken and some have already yielded interesting results. Some of the most prominent efforts have come in the form of systems to enable querying across heterogeneous databases, and in the promotion of component-based software development. In addition, a few groups have built integrated systems, depending on which definition of 'integrated' is in mind.

Work to enable cross-database querying addresses the problem of having to query databases independently and at the exclusion of problems of analysis and visualization. Such work has been performed by various groups, with results published (for example, Chung and Wong, 1999; Baker *et al.*, 1999). All of these efforts take unmediated multi-database or federated-database strategies (Karp, 1996), choosing to query source databases 'on the fly', rather than to replicate their data in a data warehouse. While these systems are potentially useful for software systems that enable user-friendly browsing, visualization and analysis of genomic data, they do not in themselves constitute 'integrated systems' if one thinks of seamless access to data of different types and from different organisms.

One common and simple approach to data integration is to use hypertext-link networks of HTML pages via the World Wide Web. This represents a simplified type of component-based design. One web page points to another by way of a URL. The URL is a component interface because it represents a contract to show some information or provide some service, without any promise of how such responsibilities will be met. Systems of HTML pages can be built from the bottom-up and are forgiving of heterogeneity. The 'interfaces' between HTML pages, however, are limited in the richness of their semantics and after one 'component' (i.e. web page) invokes another, HTML and browser technology (in its current state) does not really allow ongoing interactions between them. In addition, browser-based user interfaces are discontinuous, requiring an expensive server call to process each set of user inputs, and consequently can be unresponsive to use. HTML browser-based approaches can be most cost-effective but they are limited and in the long run may be destined for displacement by systems that take better advantages of the powerful computers on our desktops.

The ACeDB system www.sanger.ac.uk/Software/Acedb/ represents an early, ambitious attempt to provide biologically meaningful visualization, along with browsing and querying capabilities, for data of a variety of different types and sometimes spanning multiple species. It is widely used, especially in the agricultural community, and as an open-source project, has an enthusiastic and public-spirited development community.

Ritter (1994) proposed an 'Integrated Genomic Database (IGD)'. This was perhaps the earliest reference to an attempt to develop an integrated system for genomics. IGD approached both the problems of integrating data of different types that comes from multiple, heterogeneous repositories, and the problems of providing an intuitive, graphical interface to allow browsing, querying and visualization of that data. It used a data warehousing approach and employed ACeDB tools for visualization and browsing. IGD included map, sequence, phenotype and population data, and focused on the human genome. Although it was implemented successfully as a prototype, development seems to have been discontinued.

Various groups have also experimented with 'software bus' architectures for achieving integration of distributed, heterogeneous information resources. For example, one integrated system (ISYS, Siepel *et. al.*, 2000 and www.ncgr.org/research/isys/) prototype is being proposed as a method to allow users to seamlessly access germplasm and molecular resources by way of a software bus on the client side, written in Java.

Finally, several companies have built 'next-generation' integrated systems for their own commercial use or for sale to others, using promising new technologies such as Java and CORBA to achieve platform-neutral, distributed deployments, and using component-based architectures for maximum flexibility, extensibility and re-usability. Unfortunately, the commercial systems tend to be prohibitively expensive, and inaccessible to most not-for-profit researchers.

The 'core' subsystems (or components) of what needs to be developed and integrated to be effective for germplasm conservation have been described by Larry Smarr www.ncsa.uiuc.edu/People/ls/talks/. Components should support acquisition, storage and analysis of information on genomes, proteins, biochemical pathways, cellular systems, organismic models, ecological systems, geographic biodiversity and environmental interactions. In addition, the system should adapt to emerging information infrastructure (bioinformatics), such as: scalable computing; distributed sensors, data, people, computers; web- and object-oriented software architecture; and decentralization of content and software authoring.

Unresolved Issues

Data curation

Data population and curation should be as painless as information technology and evolutionary (iterative) software development can make them. Processes should be set up and maintained by germplasm researchers. Ongoing data curation and editing must be seriously considered and resources allocated to support these activities at the outset. It is not necessary to place editors near software developers or research scientists (although more than one of these attributes may be found in single people on occasion), though communication channels need to exist and methods need to be determined, tested, applied and evolved continuously. With DNA sequences, for example, it would be optimal to provide methods for semi-automated data acquisition and preliminary analysis. One example of such a system is the *Phytophthora* Genome Initiative (Waugh *et al.*, 1999, and www.ncgr.org/research/pgi/), where simple tasks are performed automatically for geographically distributed researchers collaborating on the study of a biological process. As a result of these collaborations and application of a genomic approach, new scientific knowledge about *Phytophthora* and its interactions with host plants has rapidly developed (for example, see Kamoun *et al.*, 1999; Qutob *et al.*, 2000).

Additionally, gene families can be of particular interest to biologists and yield important basic and practical information through their study. While large warehouses provide a place for all public data to reside, it is important that highly curated datasets are developed through analysis by experts in the field, and the results provided freely to the public research community. An example of such a dataset is the Plant Disease Resistance Genes

Database www.ncgr.org/research/rgenes/). Similar datasets could be produced for key plant gene families, or plant gene families as compared with *Oryza sativa*, *Medicago truncatula* and *A. thaliana*, thus providing insight into the three main groups of flowering plants, from a social and economic perspective.

Restricted vocabularies

To enable efficient searching across the wide range of experimental conditions in expression databases, it will be important to develop restricted vocabularies to replace the use of free text. However, the problems associated with establishing satisfactory nomenclatures, especially across species, are substantial. Even in taxonomy, where there has been an established formalism for centuries, there is disagreement on vocabulary. The situation is much less settled for such recent and dynamic domains as developmental biology, plant and animal pathologies, gene and protein naming conventions, metabolic relationships or laboratory protocols, all of which will play a major role in determining the utility of gene expression systems. There is a need for communities to collaborate and share restricted vocabularies. An example of successful collaboration in this respect is the development of the System-wide Information Network for Genetic Resources (SINGER). SINGER is a common gateway to the genebank databases managed in 12 centres of CGIAR. In developing SINGER, it was necessary to adopt a common structure based on agreed taxonomy and other descriptors while retaining the identity and independence of the individual databases contributing to SINGER in terms of their software and hardware platforms and structure.

Conclusions

The industrialization of biological research laboratories has created an unprecedented deluge of biological data. Consequently, there is a huge need in biological research for professional data management infrastructure to support the biological revolution promised for the 21st century. Fortunately, information technologies have been developed and are evolving elsewhere to support the needs of biological researchers worldwide. Funding models for developing, deploying and maintaining a bioinformation technology infrastructure are underdeveloped, much like the mechanisms to acquire, describe, understand and store humanity's plant and animal biodiversity. Thus, the needs (and limitations) of bioinformatics databases and germplasm bank managers are similar in many respects. The requirements, development and evolution of the information system infrastructure to support 21st century biology need to be elucidated through collaborations involving biologists, computer scientists, mathematicians and engineers. The curation (editorialization) of biological data in these systems must be done by those with appropriate biological knowledge in ways that are supportive rather than restrictive of diverse opinions, while providing means of standardization of terminology and methodology. In the age of the Internet, it is highly repetitive and non-productive to develop information resources for single or small groups of users; thus infrastructure should be built supported by the community and for the community. When public systems are developed and deployed, they should be developed openly for maximum leverage of research efforts and investments.

Germplasm is composed of populations of living organisms. Whole organism characteristics are precisely those that are most needed to provide an integrative framework for thinking about molecular data. Thus, germplasm collections, if adequately characterized to enable comparative queries using molecular datasets, provide the means to integrate molecular data meaningfully to diverse stakeholders, from molecular biologists, to breeders to conservation geneticists (see pages 21–23 in Serageldin and Persley, 2000). A spirit of scientific enquiry and dedication, focused on the linking of laboratory, field and virtual experimentation and modelling is the desired outcome. With the right mix of skills, people and organizations it is a bright possibility that comes with our possible future.

Acknowledgements

I want to thank Toby Hodgkin (IPGRI) for fruitful conversations and shared interests, Jennifer Weller (NCGR) and Harry Mangalam (GATC) for thoughts on gene expression, Adam Siepel (NCGR) for thoughts on systems integration, and Pedro Mendes (NCGR) for thoughts on biochemistry. In addition, I want to thank Paul Fox (CIMMYT), Graham McLaren (IRRI), Joe Tohme (CIAT), Dapeng Zeng (CIP), Michael Jackson (IRRI) and Dave Hoisington (CIMMYT) for shared efforts on understanding what it will take to make germplasm resources a part of the biological revolution.

References

Baker, P.G., Goble, C., Bechhofer, S., Patton, N.W., Stevens, R. and Brass, A. (1999) An ontology for bioinformatics applications. *Bioinformatics* 15, 510–520.

Butler, D. (1999) *Nature* 402, c67–c70.

Chung, S.Y. and Wong, L. (1999) Kleisli: a new tool for data integration in biology. *Trends in Biotechnology* 17, 351–355.

Kaku, M. (1997) *Visions: How Science Will Revolutionize the 21st Century*. Anchor Books, New York.

Kamoun, S., Hraber, P.T., Sobral, B.W.S., Nuss, D. and Govers, F. (1999) Initial assessment of gene diversity for the oomycete pathogen *Phytophthora infestans* based on expressed sequences. *Fungal Genetics and Biology* 28, 94–106.

Karp, P. (1996) A strategy for database interoperation. *Journal of Computational Biology* 2, 573–586.

National Science Board, Science and Engineering Indicators—1998. National Science Foundation (NSB98–1), Arlington, Virginia.

Qutob, D., Hraber, P.T., Sobral, B.W.S. and Gijzen, M. (2000) Expressed sequences from *Phytophthora sojae*. *Plant Physiology* 123, 243–254.

Ritter, O. (1994) The integrated genomic database (IGD). In: Suhai, S. (ed.) *Computational Methods in Genome Research*. Plenum Press, New York, pp. 57–73.

Serageldin, I. and Persley, G.J. (2000) *Promethean Science: Agricultural Biotechnology, the Environment, and the Poor*. Consultative Group on International Agricultural Research, Washington, DC.

Siepel, A., Farmer, A., Tolopko, A., Zhuang, M., Mendes, P., Beavis, W. and Sobral, B.W.S. (2000) ISYS: a decentralized, component-based approach to the integration of heterogeneous bioinformatics resources. *Bioinformatics* (submitted).

Sobral, B.W.S., Waugh, M. and Beavis, W. (2001) Information systems approaches to support discovery in agricultural genomics. In: Phillips, R.L. and Vasil, I.K. (eds) *Advances in Cellular and Molecular Biology of Plants*, Volume I: *DNA-Based Markers in Plants*.

Waugh, M., Hraber, P.T., Weller, J., Chen, G., Farmer, A., Inman, J., Kiphardt, D., Wu, Y. and Sobral, B.W.S. (1999) The *Phytophthora* genome initiative database: informatics and analysis for distributed pathogenomic research. *Nucleic Acids Research* 28(1), 87–90.

17 Distributed Databases Retrieval Systems in Germany as a National Approach in an International Context

S. Harrer, F. Begemann, J.D. Jiménez Krause and S. Roscher
German Centre for Documentation and Information in Agriculture (ZADI), Information Centre Genetic Resources (IGR), Bonn, Germany

Introduction

International context

Information requirements for genetic resources are included in the Convention on Biological Diversity (CBD) as well as in the Food and Agriculture Organization's (FAO's) Global Plan of Action (GPA) for the Conservation and Sustainable Utilization of Plant Genetic Resources for Food and Agriculture. The CBD covers the conservation and sustainable use of biodiversity as two of its three objectives, along with the fair and equitable sharing of benefits arising from the use of genetic resources. Its Article 17 (UNEP, 1992) stresses that the exchange of information should be facilitated, especially information relevant to the conservation and sustainable use of biological diversity. For the case of plant genetic resources for food and agriculture (PGRFA) the same objectives are specified in the GPA. Its Activity 17 (FAO, 1996) postulates the construction of comprehensive information systems for plant genetic resources to facilitate access to them and their management and utilization through the assembly, exchange and provision of useful information. Thus, the challenge to establish information systems on genetic resources has been recognized internationally. Their implementation includes both national and international mandates and activities.

Information requirements on genetic resources

Biodiversity in the sense of the CBD and the GPA includes the genetic diversity at the infraspecific level (e.g. varieties and cultivars), at the species level, and the diversity of habitats and ecosystems, including human-managed ecosystems for domesticated and cultivated species. Conservation of biodiversity has to target genetic resources *in situ* as well as *ex situ*, that is, in their natural or human-managed habitats, and in collections. Information systems on genetic resources have to cover this range of diversity levels and reference sites. For the purpose of food production or other uses such information systems should ideally contain factual data on properties and performance of the genetic material. For the purpose of conservation and sustainable exploitation georeferenced data on spatial patterns of biodiversity (species distribution, infraspecific or ecosystem variability in space) are desirable.

National Approach to Establish an Information System on PGR in Germany: the Federal Information System Genetic Resources

Aims and objectives

In Germany, the CBD and GPA will be implemented *inter alia* by establishing a set of related internet databases: The Federal Information System Genetic Resources (BIG) integrates databases on the wild flora of Germany, collections of botanical gardens, accessions of the largest German genebank, as well as other relevant databases.

The objective of BIG is to integrate the available data on genetic resources provided by agencies of the federal ministries, by universities and other research institutions. BIG aims at providing easy access to these data for decision-makers in public institutions (e.g. nature conservation services), for scientists in and outside universities, and for the interested public such as non-governmental organizations. At the same time, BIG will be important for private industry, in particular breeding enterprises or companies dealing with natural plant substances.

Taking advantage of synergism, BIG (www.big-flora.de) aims at developing an integrated information system on plant genetic resources (PGR) that covers a wide range of taxonomic, genetic, biological, ecological, economic and geographical information. It will permit complex searches in heterogeneous, decentralized databases, and thus facilitate access to actual germplasm, both *in situ* and *ex situ*.

Institutional collaboration

Four German institutions agreed to pool their extensive databases on wild and cultivated plants and their expertise in database management to establish BIG as an online information system on the Internet:

- The German Federal Agency for Nature Conservation (BfN) at Bonn holds databases on the wild flora (*in situ*), species distribution and ecology, as well as databases on the protection status of plant species according to national and European Union (EU) legislation. It acts as the German Scientific Authority to the Convention on International Trade in Endangered Species of Wild Fauna and Flora (CITES).
- On behalf of the Association of Botanical Gardens, the Department for Special Botany and the Botanical Garden of the Ruhr University Bochum (RUB) documents collections (*ex situ*) of botanical gardens in Germany.
- The Institute of Plant Genetics and Crop Plant Research (IPK) at Gatersleben performs basic research on cultivated plant taxonomy. It maintains the database of nearly 100,000 accessions of cultivated plants (*ex situ*), and is developing a database on *Mansfeld's World Manual of Agricultural and Horticultural Crops*.
- The Information Centre for Genetic Resources (IGR) within the German Centre for Agricultural Documentation and Information (ZADI) at Bonn has many years of experience providing information systems on the Internet. Besides a number of crop specific databases, it manages the central German database on plant genetic resources for food and agriculture on the Internet.

Technical realization of BIG as a multi-database retrieval system

Information systems on genetic resources have to cover a wide range of taxonomic, genetic, biological, ecological, economic and geographical data. Therefore technology is needed that can support an open, multipurpose, multiparticipant system. Technically this will be realized by a central view for accessing the online information system of various interlinked, but independent and heterogeneous databases, running under different hardware and software environments.

A user-friendly online interface allows queries of these different databases by a simple and common syntax. This user interface is part of the BIG World Wide Web-server (see Fig. 17.1) and can be used by any ordinary browser software. A BIG repository containing a reflection of different taxonomies including synonyms and a thesaurus will provide keywords like plant names or attributes (e.g. breeding traits) or other factual data such as geographical references for searching the system. The WWW-server will send the query as an http-request to the so-called BIG-Kernel (search agent). In the BIG-Kernel only one query will be composed using Extensible Markup Language (XML) as a general exchange format and sent to all relevant databases of BIG. Interfaces at each database will translate the

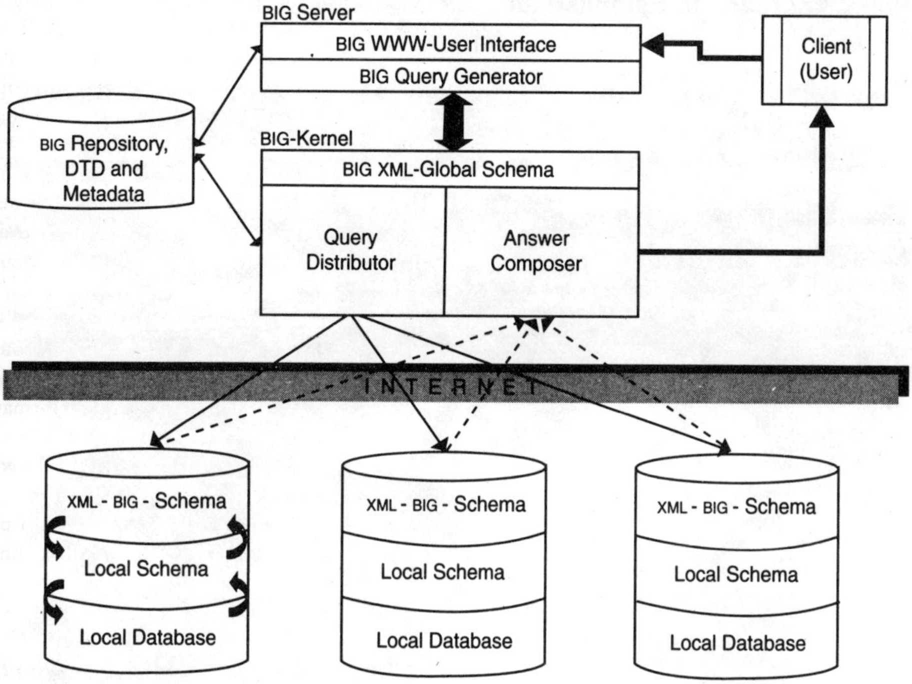

Fig. 17.1. BIG's distributed database architecture.

XML-queries in local database syntax and local responses in XML-answers. The BIG-Kernel finally compiles one answer out of all XML-answers and sends this back to the user.

Integration of geographic information systems (GIS)

As spatial analysis gets more and more important in the field of managing genetic resources, the need for geographic information systems (GIS) is growing tremendously. But visualization and exploration of spatial data must be easy and intuitive. Therefore a user-friendly GIS interface is integrated into BIG.

Interactive maps allow the user to define a particular area of interest. The GIS will transform the geometry of this area into geographic names, or vice versa. This is needed because the decentralized databases store different kinds of geo-objects. Therefore, the user interface has to compile different queries for each database. The GIS is also used for visualization and exploration of spatial data included in the result set. For example, the result set includes distribution maps of a taxon on a global scale as well as on a national scale.

Figure 17.2 shows how GIS can be used to compare and to plan future genebank collecting strategies. In Germany there are two main genebanks, at Braunschweig and Gatersleben. Figure 17.2 also shows the number of *Hordeum* samples at both genebanks, distinguished by country of collection. Thus the regional focus of the two genebanks can be compared. Additionally this helps to coordinate and plan future collecting missions and to identify gaps in the collections.

Added value through integration of associated databases

The Online-Information System for Evaluation Data or EVA (www.genres.de/eva) provides specific information on the agronomic and breeding characteristics of germplasm in Germany. Beyond passport data of genebank accessions, breeders and researchers can find reliable characterization and evaluation information such as yield, agronomic

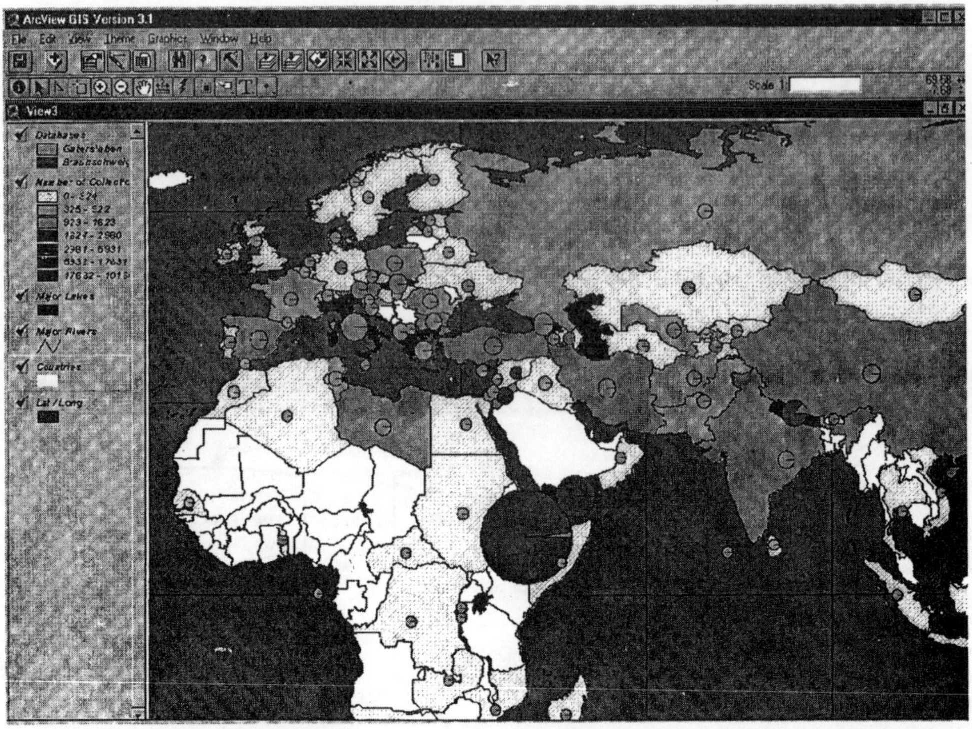

Fig. 17.2. Comparison of the number of *Hordeum* samples of both German genebanks, distinguished by country. The size of pies and the intensity of shading of countries refer to the total number of collected *Hordeum* samples by both genebanks. The shading of pie segments refers to the proportion that was collected either by the genebank in Gatersleben (pale) or by the genebank in Braunschweig (dark).

performance, resistance and quality from different sources (Harrer, 1999). Currently, the EVA prototype database consists of evaluation data for barley (11,500 accessions), potato (2000 varieties) and fruit crops (mainly apple, 1000 varieties). Once fully established, the two online information systems BIG and EVA will be integrated, thereby covering the full range of genetic resources *in situ* and *ex situ*, from evaluation data of agricultural crops to data on the distribution of endangered and protected wild species.

International Perspectives

The concept of BIG is generally applicable to taxon-related information. The methodology and tools that are under development for BIG can thus be easily transferred to other databases. The whole system has a modular design and the communication language XML is internationally agreed. In this way, BIG will provide new opportunities for regional or international information networking such as the information exchange within the framework of the FAO International Undertaking and World Information and Early Warning System (WIEWS). The FAO/IPGRI Multicrop Passport Descriptors could be used as the common fields for XML-linked national databases in order to develop a proposed World Information Networking in PGRFA (WIN/PGRFA) (Fig. 17.3).

A first step has been initiated between Germany (ZADI) and the USA (Germplasm Resources Information Network, GRIN) to develop a bilateral retrieval system over both national online databases. Another project named EPGRIS (European Plant Genetic Resources Information Infrastructure) and coordinated by the Centre for Genetic Resources, The Netherlands (CGN) may complement this activity by establishing a European Information Network

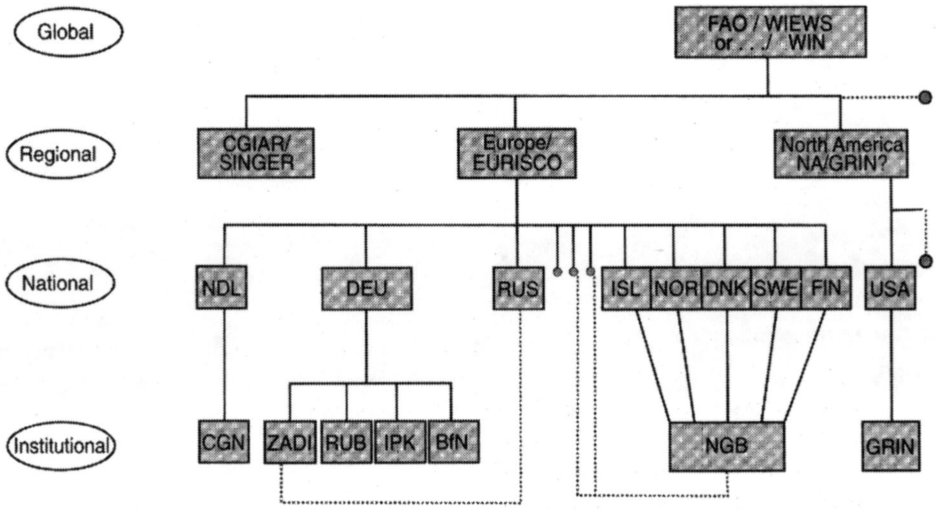

Fig. 17.3. Model of the proposed World Information Networking in PGRFA (WIN/PGRFA).

with a central European Search Catalogue (EURISCO). Other participating institutions of EPGRIS will be Bureau des Ressources Genetiques (BRG) from France, IPGRI, Instituto Nacional de Investigação Agrária (INIA) from Portugal, Nordic Genebank (NGB) from Sweden, Czech Research Institute of Crop Production (RICP) from the Czech Republic and ZADI.

References

FAO (1996) *Global Plan of Action for the Conservation and Sustainable Utilization of Plant Genetic Resources for Food and Agriculture.* FAO, Rome, Italy.

Harrer, S. (1999) EVA – Aufbau eines Informationssystems für Evaluierungsdaten pflanzengenetischer Ressourcen in der Bundesrepublik Deutschland. In: Begemann, F., Harrer, S. and Jiménez Krause, J.D. (eds) *Dokumentation und Informationssysteme im Bereich pflanzengenetischer Ressourcen in Deutschland. Schriften zu Genetischen Ressourcen.* Band 12. Informationszentrums für Genetische Ressourcen (IGR) der Zentralstelle für Agrardokumentation und -information (ZADI), Bonn, Germany.

UNEP (United Nations Environment Programme) (1992) *Text of the Convention on Biological Diversity.* www.biodiv.org/chm/conv/default.htm

[illegible] Model of the components of the Information Network on [illegible]

[illegible]

References

[illegible]

18 The Potential Role of Passport Data in the Conservation and Use of Plant Genetic Resources

Th. Hazekamp

International Plant Genetic Resources Institute (IPGRI), Rome, Italy

Introduction

In the study and management of biodiversity, genebanks draw upon a large number of data sources and data types. Passport data are just one of these, but they provide collection curators and users alike with convenient ways to group collection material into logical and more manageable units. In addition, passport data in particular have the ability to provide the linkage between genebank data and other (external) data sources. For example, taxonomic passport data can link genebank data to a large variety of biological data sources, while geographical passport descriptors provide a link to many spatial data sources. The combination of genebank data with external data sources adds considerable depth to biodiversity studies upon which genebanks and their user community base strategies for conservation and use. This chapter assesses the current use of passport data by genebanks, the constraints encountered, and the potential for improved use of passport data. Besides the extensive use of scientific literature, this chapter includes the results of a survey that was sent to 100 medium to large genebanks worldwide. The survey investigated how specific types of passport data are used by genebanks, which additional data sources are used, and how the use of passport data could be improved.

Current Use of Passport Data

Passport data are used by genebanks for a variety of tasks. For the purpose of this chapter the genebank activities are divided into: (i) germplasm collection and acquisition; (ii) germplasm management; and (iii) germplasm use. Examples of the use of passport data are discussed below for each of these areas.

Germplasm collection and acquisition

Collecting germplasm is done for a number of reasons: to obtain material for biosystematic research or for genetic diversity studies, for conservation, and for immediate use in breeding programmes (von Bothmer and Seberg, 1995). In the context of genebanks, collecting efforts will generally be part of an integrated conservation strategy for the taxon in question. An ecogeographic survey is commonly used as the basis for a conservation strategy. Maxted *et al.* (1995) provide comprehensive details on the components of an ecogeographic survey. Passport data of existing collections, both genebanks and herbaria, are an important source of information since they help to establish the historical occurrence of the species in particular areas and the sampling it has been subjected to in the past. One example is a study of the *Vicia narbonensis* complex (Bennett and Maxted, 1997). A detailed knowledge of the ecogeographic distribution of the target taxon helps to identify potential collecting sites. Jones *et al.* (1997) have used geographical information systems (GIS) to link the latitude, longitude and altitude of the location of origin of wild *Phaseolus vulgaris* with climatological datasets to locate 'bean-favouring' climates. Such techniques, used in conjunction with molecular marker techniques, can make significant contributions to the development of a more effective collecting procedure as part of the overall conservation strategy. Ferguson *et al.* (1998) studied the geographical distribution of diversity of four wild relatives of cultivated lentil using geographical passport data from ICARDA's (International Centre for Agricultural Research in the Dry Areas) wild lentil collection in conjunction with randomly amplified polymorphic DNA (RAPD) data. In another type of study, Greene *et al.* (1999) showed that as part of the evaluation of newly collected germplasm, the passport data in combination with GIS data were successfully used to help curators with post-collection decisions, particularly which of the newly acquired samples to include in the genebank collection.

Currently, worldwide *ex situ* germplasm collections are estimated to number around 6 million accessions (FAO, 1998). Duplication of germplasm among collections is substantial. Lyman (1984) estimated that at least 50% of the germplasm held consists of duplicated accessions. Eliminating this type of duplication has often been suggested as a way to reduce the costs associated with the operation of genebanks. Methodologies based on passport data have been used to assist in the identification of replicated germplasm collections. The work by Knüpffer (1989) on the European Barley Database provides a good example. Van Hintum and Knüpffer (1995) further discussed the complexities of identifying duplicates. They concluded that passport data were useful in giving a first indication of probable duplicates, but in very few cases would lead to a direct and definitive identification of duplicates. In most cases definitive verification requires additional data from field observations and genetic markers, as demonstrated in subsequent research by van Hintum and Visser (1995) on barley.

Germplasm management

The germplasm management cycle starts after the decision to include a specific sample in a genebank collection. The complexities of managing and maintaining genebanks accessions are substantial, due to the very diverse nature of the material. A vast body of biological information related to phenology, reproductive biology and storage behaviour is necessary to develop adequate maintenance programmes. The use of these was confirmed by the genebanks responding to the International Plant Genetic Resources Institute (IPGRI) survey on use of passport data by genebanks (see Box 18.1). Many of these sources of biological or agroecological information that are important to genebanks will not be generated by the genebanks, but will be developed by scientists in related disciplines. The use of standard reference systems for scientific names or geographical coordinates permits genebanks to use these 'external' information resources together with their own data and the germplasm it relates to. In particular passport data with descriptors such as species name, geographical origin and ecological descriptors enable genebanks to make this linkage to data sources that originate from other biological or ecogeographical disciplines.

Germplasm use

Genebanks have been well aware of the need to promote the use of the collections they maintain. This is clearly demonstrated in the explicit linkage between conservation and use in the *Global Plan of Action for the Conservation and Sustainable Utilization of Plant Genetic Resources* (FAO, 1996). Plant breeders have traditionally been an important user group for genebank collections. Most genebanks responding to the IPGRI survey on the use of passport data (see Box 18.1) specifically listed the provision of source material to breeding programmes as one of their main objectives. Many factors can influence the utilization of germplasm by users. Engels (2001) provides some examples of how the management of germplasm can be tailored to facilitate the use of germplasm by specific user groups. The relative lack of evaluation data has been frequently cited as one of the main restrictions to the use of genebank collections by breeding programmes (FAO, 1998). To a certain degree this is due to the extensive collecting that has taken place in the past, especially during the 1970s and 1980s. Although this has helped to safeguard diversity that would otherwise have been lost, many collections have grown to a size where the resources to properly characterize and evaluate the material in the collections are simply not available (Spagnoletti Zeuli and Qualset, 1993). This means that users will often have to use more indirect criteria to select germplasm from genebank collections.

The geographic origin of the accession is regarded as a very useful criterion. This is based on the understanding that plant populations are adapted to local conditions and that accessions from a similar geographic origin probably share a larger part of their genetic makeup than more distant accessions (Peeters and Martinelli, 1989). However, some care should be taken in using ecogeographical factors as the predictors of patterns of diversity. Habitats free of significant human interference, even the habitats of some semi-wild or wild species, will be increasingly difficult to encounter. Various authors have pointed out that humans can have a significant influence on the development of patterns of diversity (Spagnoletti Zeuli *et al.*, 1984; Smith *et al.*, 1994; Pickersgill, 1998). It is likely that anthropogenic factors could increasingly interfere with an analysis of diversity on pure ecogeographical grounds. In this context there is much interest among genebanks to use ethnobotanical data, as indicated by the responses to the IPGRI survey on the use of passport data (see Box 18.1). The use of this kind of data should increase our understanding of the distribution of crop diversity.

Another development that needs mentioning in relation to enhancing the use of germplasm collections is the development of core collections. The core collection aims to define a small subset of accessions that represents a large part of the diversity found in the entire collection (Frankel and Brown, 1984). Such a small subset, often around 5–10% of the size of the total collection, can undergo more in-depth evaluation and as such provide a good starting point for users looking for particular traits. The initial core collection concept and its further promotion (Brown, 1989; Hodgkin, 1990) have led to the development of methodologies to designate core subsets (Hodgkin *et al.*, 1995a). By the mid-1990s, a substantial number of core collections had already been established (Diwan *et al.*, 1995) and their number is growing steadily.

Different methodologies exist to designate accessions to a core collection. In general, a mixture of passport, characterization and evaluation data are used to determine whether an accession should be part of the core set. Nevertheless passport data, and in particular the geographical origin, are usually one of the first criteria used to determine selection (Prasada Rao and Rao, 1995; Tohme *et al.*, 1995; Clark *et al.*, 1997). However, the methodologies to designate core collections are still relatively new and experience with the use of core collections is limited. Although passport data are currently seen as quite suitable to provide much of the data to identify the distinct groups of material, further experience with core collections is necessary to determine whether this actually leads to optimally configured groups or whether the balance with the other types of data should be changed (Hodgkin *et al.*, 1995b).

Constraints

Despite the fact that passport data are generally considered important, there are several very serious constraints associated with their use. Incompleteness and errors are often-cited deficiencies when referring to genebank data, and passport data are no exception (Williams, 1989; Beuselinck and Steiner, 1992; van Hintum and Knüpffer, 1995). Plucknett *et al.* (1987) estimated that at least 65% of all accessions in the world's genebanks are lacking passport data. Although this situation seems to have improved somewhat, the latest report on the *State of the World's Plant Genetic Resources for Food and Agriculture* indicates that serious deficiencies remain (FAO, 1998).

Besides the lack of good quality data, there are various data exchange issues. Even for relatively simple descriptors such as country codes, different implementations can be found (van Hintum, 1995). Lack of data standardization makes widespread use of genebank data difficult. When considering the data quality aspects of passport data, it is important to take into account that systematic germplasm conservation efforts have been going on for more than a century. In the US for example, plant introduction PI1 of the National Plant Germplasm System Collections is a *Brassica oleracea* accession that was received in February 1898 (www.ars-grin.gov/cgi-bin/npgs/html/taxon.pl?7668). Over the years the insights on how collections should be managed and documented have undergone many changes. For quite some time, the emphasis has been on the acquisition of new material rather than proper documentation of the germplasm and its natural habitat (Steiner and Greene, 1996). However, the initial selection of

Box 18.1. IPGRI survey on use of passport data by genebanks (April/May 2000).

This survey was sent to 100 institutions around the world that are holding medium to large *ex situ* germplasm collections. The purpose of this survey was to investigate:

1. How specific types of passport data are used by genebanks;
2. Which external data sources are used by genebanks; and
3. Where and how the use of passport data could be improved.

The survey on use concerned four categories of passport data, defined as follows:

- Taxonomic: genus, species, subspecies, variety name, pedigree and biological status of sample.
- Geographical: latitude, longitude, elevation, country of origin, state/province and location.
- Ecological: site environment, soil characteristics and vegetation type.
- Ethnobotanical: local use, cultivation methods and processing of plant parts.

By 1 June 2000, 46 questionnaires (46% response) had been received.

Use of passport data generated by genebanks

A) For decision-making related to *germplasm collection/acquisition activities* (i.e. especially the planning of germplasm collection/acquisition), please indicate, according to your own experience, how important the various types of passport data are:

Passport data type		(n)	% Irrelevant	% Helpful	% Essential
Ecological data	Own genebank	43	5	51	44
	Other genebanks	42	3	71	26
Ethnobotanical data	Own genebank	44	16	75	9
	Other genebanks	44	16	75	9
Geographical data	Own genebank	46	4	11	85
	Other genebanks	44	3	36	61
Taxonomic data	Own genebank	46	0	13	87
	Other genebanks	44	0	32	68

Notes: the most cited additional data sources are evaluation and characterization data, floras, taxonomic monographs and spatial data of geophysical and climatological origin.

B) For decision-making related to *germplasm maintenance/management activities* (i.e. the planning phase of maintenance activities) please indicate, according to your own experience, how important the various types of passport data are:

Passport data type		(n)	% Irrelevant	% Helpful	% Essential
Ecological data	Own genebank	41	14	49	37
	Other genebanks	40	17	60	23
Ethnobotanical data	Own genebank	38	28	61	11
	Other genebanks	38	29	71	0
Geographical data	Own genebank	44	2	41	57
	Other genebanks	42	7	50	43
Taxonomic data	Own genebank	45	4	16	80
	Other genebanks	43	7	26	67

Notes: the majority of additional sources of information deal with basic biological data such as phenological, agronomic and seed storage.

Box 18.1. Continued

C) For decision-making related to *germplasm use activities* (i.e. the targeting of specific germplasm including collection rationalization through core collections) please indicate, according to your own experience, how important the various types of passport data are:

Passport data type		(n)	% Irrelevant	% Helpful	% Essential
Ecological data	Own genebank	43	3	60	37
	Other genebanks	43	5	72	23
Ethnobotanical data	Own genebank	43	18	56	26
	Other genebanks	41	19	66	15
Geographical data	Own genebank	45	2	18	80
	Other genebanks	43	2	33	65
Taxonomic data	Own genebank	46	0	11	89
	Other genebanks	44	0	23	77

Notes: the most cited additional data sources are evaluation and characterization data, ethnobotanical data and spatial data of geophysical and climatological origin.

Overall notes:

- Taxonomic and geographical passport data are consistently receiving the highest scores (essential).
- Ethnobotanical data and to a certain extent ecological data seem relatively 'undervalued' in genebanks, but much sought after from other sources.

Potential for novel or improved use of passport data:

1. Using GIS to link morphological or molecular traits with ecogeographical parameters.
2. Data standardization issues that hinder data exchange or networking or both (in particular taxonomic treatments).

Constraints:

- Available resources (funding, GIS skills), low quality passport data.

germplasm is frequently made on passport data. As a result germplasm with incomplete passport data are routinely excluded from further studies (Tolbert *et al.*, 1979; Bennett and Maxted, 1997).

Data completeness and validation

In some cases, additional passport data will be available, but have not been computer-encoded yet due to other priorities. In addition, data quality standards tend to change over time. What were acceptable standards 15 years ago might be totally unacceptable today. Origin data at the country level might have been acceptable in the absence of PC-based GIS. Nowadays the absence of latitude and longitude data would be considered a major obstacle towards a better use of genebank data.

Sometimes no additional original data are available. In some circumstances retro-classification can be considered as a way to improve the data, and Steiner and Greene (1996) provide an example of such an approach. They rightly point out that retro-classified data are not equivalent to original data, and need careful interpretation. For example, retro-classifying longitude, latitude and elevation from a location description will most likely yield approximate values. This would make this kind of data questionable for analysis of microhabitats at the local level, especially when the landscape is ecologically very diverse. However it might be perfectly suitable for a study on a more global level, and as such make a meaningful addition to the original data. Tools such as the Getty Thesaurus of Geographic Names (www.shiva.pub.getty.edu/tgn_browser/) could be helpful in the retro-classification of data. The thesaurus is a structured vocabulary containing around 1 million names and other information about places. It includes all continents and nations of the modern political world, as well as historical place names, physical features, and administrative entities, such as cities and nations. Other tools, such as FLORAMAP (Jones and Gladkov, 1998), can be used to retro-classify accessions for climatic variables.

Data validation is an essential step in any data entry process. Without proper validation, the quality of the dataset is compromised from the outset. A very disciplined and systematic approach to data validation is necessary, especially in genebanks where collections are built up over a long period of time and data processing can be stretched over equally long periods.

Accumulation of data across collections

Whenever the data quality of individual accession records has been brought up to an acceptable level, the usefulness of the data can be further improved by accumulating data for a particular genepool across collections. The accumulation of data across collections results in access to a single dataset that gives the widest possible coverage of the genepool in terms of distribution and genetic diversity. As such it will facilitate more comprehensive diversity studies. In fairly recent times, the establishment of a large number of European central crop databases (Gass *et al.*, 1997) and also the accumulation of Consultative Group on International Agricultural Research (CGIAR) genebank data in the System-wide Information Network for Genetic Resources (SINGER, www.singer.cgiar.org/) are some examples. Regional, subregional or crop-oriented genetic resources networks often prove a very good vehicle to organize such collaborative data activities. In some cases, such as the European Collaborative Programme on Genetic Resources, a specific network of germplasm documentation specialists was formed to give this kind of work a strong foundation.

Data standardization

Of course the accumulation of data from different collections makes issues related to data standardization more prominent. Genetic resources documentation systems have been developed in many different places and using many different approaches. With the increase of international collaboration among genebanks, different approaches made it increasingly difficult to exchange information and the need to formulate international formats for the documentation of crop germplasm became apparent. To assist in this task, IPGRI started the production of crop descriptor lists in 1979, and more than 80 descriptor lists have been produced. In 1996, IPGRI went one step further and defined a set of multi-crop passport descriptors (Hazekamp *et al.*, 1997). These descriptors aim to provide consistent coding schemes for a set of common passport descriptors of all crops and therefore will facilitate the accumulation of passport data into multicrop information systems.

Several genebanks mentioned data standardization as obstacles to a fuller use of passport data (see Box 18.1). In particular, differences in taxonomic treatments were mentioned. Indeed the lack of standardization in taxonomic treatments seriously hampers the exchange of basic biological data. This affects all aspects of genebank work. Not only are genebanks reliant on a wealth of basic biological data to fulfil their collection acquisition and maintenance tasks, access of users to genebank collections is equally hampered by inconsistent taxonomies. Scientific names are notoriously inconsistent among the collection databases of different organizations (Hulden, 1997). Not only do taxonomic treatments frequently change as a result of new research, but also because conflicting treatments coexist for some species. This creates major problems when trying to exchange basic biological information.

The Species 2000 project (www.sp2000.org/) aims to create a taxonomic database with an entry for every species in the world. The prototype IV, February 2000 version of the database already contains 196,000 species, 114,000 synonyms, 122,000 common names and 13,000 references from five contributing databases. This database is a valuable asset and will provide a unified structure through which a wealth of basic biological information can be linked. It is essential for genebanks to adhere to such taxonomic standards, not only to facilitate data exchange between genebanks, but also to ensure an effective linkage with related biological disciplines. While initiatives such as the Species 2000 project will go a long way towards solving the use of different taxonomic treatments, there is still a need for adequate taxonomic capacity within national programmes to apply such taxonomic standards correctly.

Novel or Improved Use of Passport Data

When asked about potential areas for novel or improved use of passport data, the IPGRI survey respondents most frequently mentioned GIS related topics (see Box 18.1), in particular using GIS to link morphological or molecular traits with ecogeographic parameters. Indeed, GIS provides an intuitive framework for analysing the way biodiversity is dispersed in space and time. Many aspects of genebank work, from collecting through maintenance to use, contain elements where a spatial representation of data can broaden our insight into the structure of the biodiversity of concern.

The main constraints seen by the responding genebanks are the low quality of passport data, and resource issues. In most cases, the geographical origin of accessions is documented at the country level, but the meaningful application of GIS does require precise longitude, latitude and, to a lesser extent, elevation data. Some approaches concerning data quality were outlined earlier in the chapter. Individually, or as part of a network, genebanks will have to determine how much scope for improvement there is vis-à-vis the resources required for data improvement work and developing a GIS capability. In this context the threshold for initiating GIS studies is declining as a result of decreasing prices for the necessary hardware and software. Even though the learning curve is becoming less steep, as a result of more user-friendly software packages, basic training in GIS concepts will always be required. However networks can provide an ideal environment for national genebanks to share costs and the benefits of GIS.

Conclusions

In general there is no misunderstanding about the importance of passport data. Passport data connect all genebank data. Through species name, geographical and ecological descriptors, they have the potential to link to many biological and georeferenced external data sources. This expands the utility of the genebank data enormously.

It is also clear that the quality of the passport data is far from optimal. To a certain extent this cannot be changed. Some data are irretrievably lost or were simply never collected. However several genebanks indicated that data are available, but resource limitations have prevented full computerization. Furthermore, in some cases, data could be improved by retro-classification. Individually, or as part of a network activity, genebanks should assess this potential and set their priorities accordingly.

Quality passport data not only allow genebanks to link to other data sources and make a fuller use of these, but will also allow users to gain fuller access to the collections. Users will often start their selection of genebank material based on passport data, in particular taxonomic and geographic data. This will provide a first global grouping of the material. Further refinement of a selection will include characterization and evaluation data. Accessions with incomplete passport data will have a high chance of being eliminated from the selection process at the outset. Examples of this kind are not difficult to find in the scientific literature. Early elimination from the selection of an accession with poor passport data will also mean that its related characterization and evaluation data will have reduced chances of ever being used. Poor passport data therefore have a knock-on effect that may result in lower utilization of genebank data overall. It should be realized that the criteria that define acceptable data quality change over time as a result of new ways of using and accessing data. The maturation of GIS technology has made its wider use in genetic diversity studies feasible. However, it requires good quality latitude and longitude data to work properly. Data quality standards for georeferenced genebank data need to be set at a level where such applications can be used.

Furthermore, the way data on germplasm collections are accessed is changing. Wide Area Networks such as the Internet have made it possible to make data available to large audiences at very reduced costs or even at virtually no cost at all. Genebanks also make their data available via the Internet. Users that have direct access to germplasm databases will route their requests less frequently through collection curators. This effectively eliminates the possibility for curators to use their knowledge to compensate for incomplete data in the germplasm databases. Therefore online data will require high data quality standards.

It is important that genebanks have a good understanding of the search patterns applied by users and optimize their data to facilitate the location of appropriate material. These search patterns might vary depending on the type of material. For wild material geographical and ecological descriptors might be extremely important, while for cultivated material descriptors such as variety name and pedigree are more significant. The optimalization of data for use hinges not only on complete and correct data. Data standardization is another factor determining transparent access to and use of collections. The absence of consistent documentation will cause fragmentation of data and will make them difficult to use. Although work on documentation standards has been going on for some time, there are still important areas to address. Consistent scientific nomenclature is one such area. Often crop genetic resources networks are very useful to discuss and agree on a specific taxonomic treatment to represent joint data. Furthermore, genebanks may have ecogeographical and ethnobotanical data that are currently in a narrative form and therefore difficult to use. Current standards need to be considered for suitability and possibly undergo revision to capture these data. Biochemical and molecular data are being generated in large quantities by genebanks and related biological sciences. Currently ways to study a better linkage between these data and the 'traditional' genebank data are under investigation.

In many cases passport data will not only link internal genebank activities, they will also act as the primary gateway between genebanks and external interest groups. A sub-optimal gateway will restrict the use and sharing of resources. Once genebanks have optimized their passport data and linkages to external data sources are facilitated, genebanks and their collections will become more visible to the external world through reciprocal linkages. Without doubt this will not only have a positive effect on the use of germplasm collections, but also create opportunities for genebanks to establish scientific collaboration in new areas.

Acknowledgements

The author would like to thank Dr Jan Engels and Dr Luigi Guarino for their scientific inputs during the writing of this chapter.

References

Bennett, S.J. and Maxted, N. (1997) An ecogeographic analysis of the *Vicia narbonensis* complex. *Genetic Resources and Crop Evolution* 44(5), 411–428.

Beuselinck, P.R. and Steiner, J.J. (1992) A proposed framework for identifying, quantifying and utilizing plant germplasm resources. *Field Crops Research* 29, 261–272.

von Bothmer, R. and Seberg, O. (1995) Strategies for the collecting of wild species. In: Guarino, L., Ramanatha Rao, V. and Reid, R. (eds) *Collecting Plant Genetic Diversity: Technical Guidelines.* CAB International, Wallingford, UK, pp. 93–111.

Brown, A.H.D. (1989) The case for core collections. In: Brown, A.D.H., Frankel, O.H., Marshall, D.R. and Williams, J.T. (eds) *The Use of Plant Genetic Resources.* Cambridge University Press, Cambridge, UK, pp. 136–156.

Clark, R., Shands, L.H.L., Brettling, P.K. and Eberhart, S.A. (1997) Germplasm regeneration: Developments in population genetics and their implications; managing large diverse germplasm collections. *Crop Science* 37, 1–6.

Diwan, N., McIntosh, M.S. and Bauchan, G.R. (1995) Methods of developing a core collection of annual *Medicago* species. *Theoretical and Applied Genetics* 90(6), 755–761.

Engels, J.M.M. (2001) Genebank management: an essential activity to link conservation and plant breeding. *Plant Genetic Resources Newsletter* 125.

FAO (1996) *The Global Plan of Action for the Conservation and Sustainable Utilization of Plant Genetic Resources for Food and Agriculture.* Food and Agriculture Organization of the United Nations, Rome, Italy.

FAO (1998) *The State of the World's Plant Genetic Resources for Food and Agriculture.* Food and Agriculture Organization of the United Nations, Rome, Italy.

Ferguson, M.E., Ford-Lloyd, B.V., Robertson, L.D., Maxted, N. and Newbury, H.J. (1998) Mapping the geographical distribution of genetic variation in the genus *Lens* for enhanced conservation of plant genetic diversity. *Molecular Ecology* 7, 1743–1755.

Frankel, O.H. and Brown, A.H.D. (1984) Current plant genetic resources: a critical appraisal. In: *Genetics: New Frontiers,* Vol. IV. Oxford and IBH Publ. Co., New Delhi, India, pp. 1–11.

Gass, T., Lipmann, E. and Maggioni, L. (1997) The role of central crop databases in the European Cooperative Programme for Crop Genetic Resources Networks. In: Lipmann, E., Jongen, M.W.M., van Hintum, Th.J.L., Gass, T. and Maggioni, L. (compilers) *Central Crop Databases: Tools for Plant Genetic Resources Management. Report of a Workshop, 13–16 October 1996, Budapest, Hungary.* International Plant Genetic Resources Institute, Rome, Italy/CGN, Wageningen, The Netherlands, pp. 20–25.

Greene, S.L., Hart, Th.C. and Afonin, A. (1999) Using geographic information to acquire wild crop germplasm for *ex situ* collections: II Post-collection analysis. *Crop Science* 39, 843–849.

Hazekamp, Th., Serwinski, J. and Alercia, A. (1997) Multi-crop passport descriptors. In: Lipmann, E., Jongen, M.W.M., van Hintum, Th.J.L., Gass, T. and Maggioni, L. (compilers) *Central Crop Databases: Tools for Plant Genetic Resources Management. Report of a Workshop, 13–16 October 1996, Budapest, Hungary.* International Plant Genetic Resources Institute, Rome, Italy/CGN, Wageningen, The Netherlands, pp. 35–39.

van Hintum, Th.J.L (1995) Standardization in plant genetic resources documentation. III: Country and region codes. In: van Hintum, Th.J.L., Jongen, M.W.M. and Hazekamp, Th. (eds) *Standardization in Plant Genetic Resources Documentation. Report of the Second Technical Meeting of Focal Points for Documentation in East European Genebanks, Plant Breeding and Acclimatization Institute, Radzikow, Poland, 10–14 October 1995.* Centre for Genetic Resources, Wageningen, The Netherlands, pp. 89–96.

van Hintum, Th.J.L. and Knüpffer, H. (1995) Duplication within and between germplasm collections I: Identifying duplication on the basis of passport data. *Genetic Resources and Crop Evolution* 42(2), 127–133.

van Hintum, Th.J.L. and Visser, D.L. (1995) Duplication within and between germplasm collections II: Duplication in four European barley collections. *Genetic Resources and Crop Evolution* 42(2), 135–145.

Hodgkin, T. (1990) The core collection concept. In: van Hintum, Th.J.L., Frese, L. and Perret, P.M. (eds) *Crop Networks: Searching for New Concepts for Collaborative Genetic Resources Management.* IBPGR, Rome, Italy, pp. 43–48.

Hodgkin, T., Brown, A.H.D., van Hintum, Th.J.L. and Morales, E.A.V. (eds) (1995a) *Core Collections of Plant Genetic Resources.* John Wiley & Sons, Chichester, UK.

Hodgkin, T., Brown, A.H.D., van Hintum, Th.J.L. and Morales, E.A.V. (1995b) Future directions. In: Hodgkin, T., Brown, A.H.D., van Hintum, Th.J.L. and Morales, E.A.V. (eds) *Core Collections of Plant Genetic Resources.* John Wiley & Sons, Chichester, UK, pp. 253–260.

Hulden, M. (1997) Standardization of central crop databases. In: Lipmann, E., Jongen, M.W.M., van Hintum, Th.J.L., Gass, T. and Maggioni, L. (compilers) *Central Crop Databases: Tools for Plant Genetic Resources Management. Report of a Workshop, 13–16 October 1996, Budapest, Hungary.* International Plant Genetic Resources Institute, Rome, Italy/CGN, Wageningen, The Netherlands, pp. 26–34.

Jones, P. and Gladkov, A. (1998) *FloraMap: a Computer Tool for the Distribution of Plants and Other Organisms in the Wild.* CIAT, Cali, Colombia.

Jones, P.G., Beebe, S.E., Tohme, J. and Galwey, N.W. (1997) The use of geographical information systems in biodiversity exploration and conservation. *Biodiversity and Conservation* 6(7), 947–958.

Knüpffer, H. (1989) Identification of duplicates in the European Barley Database. In: *Report of a Working Group on Barley (Third Meeting), April 1989, Gatersleben, Germany.* IBPGR, Rome, Italy, pp. 22–43.

Lyman, J.M. (1984) Progress and planning for germplasm conservation of major food crops. *Plant Genetic Resources Newsletter* 60, 3–21.

Maxted, N., van Slageren, M.W. and Rihan, J.R. (1995) Ecogeographical surveys. In: Guarino, L., Ramanatha Rao, V. and Reid, R. (eds) *Collecting Plant Genetic Diversity: Technical Guidelines.* CAB International, Wallingford, UK, pp. 255–286.

Peeters, J.P. and Martinelli, J.A. (1989) Hierarchical cluster analysis as a tool to manage variation in germplasm collections. *Theoretical and Applied Genetics* 78, 42–48.

Pickersgill, B. (1998) Crop introductions and the development of secondary areas of diversity. In: Prendergast, H.D.V., Etkin, N.L., Harris, D.R. and Houghton, P.J. (eds) *Plants for Food and Medicine.* Royal Botanic Gardens, Kew, UK, pp. 93–105.

Plucknett, D.L., Smith, N.J.H., Williams, J.T. and Anishetty, N.M. (1987) *Genebanks and the World's Food.* Princeton University Press, New Jersey.

Prasada Rao, K.E. and Rao, V.R. (1995) The use of characterization data in developing a core collection for sorghum. In: Hodgkin, T., Brown, A.H.D., van Hintum, Th.J.L. and Morales, E.A.V. (eds) *Core Collections of Plant Genetic Resources.* John Wiley & Sons, Chichester, UK, pp. 109–116.

Smith, S.E., Johnson, D.W., Conta, D.M. and Hotchkiss, J.R. (1994) Using climatological, geographical, and taxonomic information to identify sources of mature-plant salt tolerance in alfalfa. *Crop Science* 34(3), 690–694.

Spagnoletti Zeuli, P.L. and Qualset, C.O. (1993) Evaluation of five strategies for obtaining a core subset from a large genetic resource collection of durum wheat. *Theoretical and Applied Genetics* 87, 295–304.

Spagnoletti Zeuli, P.L., De Pace, C. and Porceddu, E. (1984) Variation in durum wheat populations from three geographical origins. I. Material and spike characteristics. *Euphytica* 33, 563–575.

Steiner, J.J and Greene, S.L. (1996) Proposed ecological descriptors and their utility for plant germplasm collections. *Crop Science* 36, 439–451.

Tohme, J., Jones, P., Beebe, S. and Iwanaga, M. (1995) The combined use of agroecological and characterization data to establish the CIAT *Phaseolus vulgaris* core collection. In: Hodgkin, T., Brown, A.H.D., van Hintum, Th.J.L. and Morales, E.A.V. (eds.) *Core Collections of Plant Genetic Resources*. John Wiley & Sons, Chichester, UK, pp. 95–107.

Tolbert, D.M., Qualset, C.O., Jain, S.K. and Craddock, J.C. (1979) A diversity analysis of a world collection of barley. *Crop Science* 19, 789–794.

Williams, J.T. (1989) Practical considerations relevant to effective evaluation. In: Brown, A.H.D., Frankel, O.H., Marshall, D.R. and Williams, J.T. (eds) *The Use of Plant Genetic Resources*. Cambridge University Press, Cambridge, UK, pp. 235–244.

19 *In situ* Conservation of Wild Species Related to Crop Plants: the Case of Turkey

A. Tan and A.S. Tan
Aegean Agricultural Research Institute, Menemen, Izmir, Turkey

Introduction

The *in situ* conservation of wild species is one activity of the Global Plan of Action, and can be defined as the maintenance of viable diverse populations of species in their natural surroundings or ecosystems. The conservation of wild species in genetic reserves, termed gene management zones (GMZs), involves the target species, location, designation, management and monitoring of genetic diversity in a particular natural location. This technique allows the evolution of the species in its ecosystem. The first step of *in situ* conservation is the selection of target wild species. Then should follow the survey of potential sites, the designation of genetic reserves or GMZs in accordance to genetic variation within that target species, and the management plan for monitoring genetic diversity.

The project '*In situ* Conservation of Genetic Diversity in Turkey' aims to maintain the genetic diversity of wild crop relatives and forest species in their natural habitats. The project involves both woody and herbaceous species and is integrated in a multiple-species and multiple-site approach. The project initiates and develops a mechanism to foster the ongoing National Plant Genetic Resources Research Programme for identifying, designating and managing the areas for *in situ* conservation of nationally and globally significant wild crop relatives and woody species. The project has been implemented in cooperation with both the formal sector (Ministry of Agriculture and Rural Affairs, Ministry of Forestry, Ministry of Environment) and the informal sector (universities, non-governmental organizations (NGOs)). The pilot part of the project was supported by Global Environment Facility (GEF). The project components have been set up according to the principles of the *in situ* technique for wild species (Firat and Tan, 1995; Firat and Tan, 1997; Anonymous, 1998a). The activities of the *In situ* Conservation of Genetic Diversity project of Turkey are shown in Box 19.1.

The selection of target species is an important step for an *in situ* conservation programme. Therefore, selection of target species or taxa should be objective, based on logical, scientific, economic principles related to the significance of species (Anonymous, 1996; Maxted and Hawkes, 1997). Factors that influence the conservation value of target species are listed in Box 19.2. The target species should have global or national importance for potential economic use. The highest priority can be given to species in the primary genepool proposed by Harlan and de Wet (1971). Various threat factors, such as over-grazing, forest decline, disease and pest epidemics, and changes in farming system, should be taken into account. When selecting the species to conserve *in situ* the highest emphasis should be given to the most genetically distinct groups of taxa (the widespread species with a wide range of adaptation). For example, *Aegilops speltoides* and *Lens orientalis* are distributed in Turkey throughout the northern belt of the Fertile Crescent. These species share different

Box 19.1. *In situ* project components.

- Ecogeographical surveys and inventory
- Estimation of genetic diversity
- Designation of GMZs
- Management and monitoring of GMZs
- Data management
- Public awareness
- National plan for *in situ* conservation

Box 19.2. Factors influencing conservation value.

- Potential to economic use
- Threat of genetic erosion
- Genetic distinction
- Ecogeographical distinction
- Conservation agency priorities
- Biologically important species
- Relative cost of conservation

types of habitat in different ecosystems, for example forest and marginal ecosystems, and they have a wide range of adaptability. The priorities of conservation agencies can differ from those of genetic resource conservation, so the selection of target species may vary depending on their conservation policies. If a biologically important keystone species has priority, its associated species can be maintained in the same reserve. The cultural importance of the species should also be taken into account. Sustainability of conservation and its relative cost are also important factors for the selection of target species. The globally and nationally important wild legumes (wild *Lens* species, wild *Vicia sativa sensu lato, Vicia johannes*, wild *Pisum sativum sensu lato*), wild cereals (*Triticum boeoticum, Triticum dicoccoides, Triticum tauschii, A. speltoides* var. *aucheri* and var. *lugistica*), wild fruits (*Castanea sativa, Prunus divericata*) and some forest species (*Abies equitrojona, Abies cilicica, Pinus nigra, Pinus brutia*) have been selected as target species, in three different locations: Kaz Mountain (North of Aegean Region), Central Taurus Mountains (Bolkar and Aladag Mountains), and Ceylanpinar (in south-east Anatolia). The target areas are shown schematically in Fig. 19.1. While selecting the target species to be conserved *in situ*, the annual species (*Triticum* spp., *Lens* spp., *V. sativa sensu lato, P. sativum sensu lato*), perennial and vegetatively propagated fruit trees (*C. sativa, P. divaricata*), and forest trees have been taken into account, including the status and potential for *in situ* conservation of those taxa. The pilot study sites were also selected from different types of ecosystem, such as the forest ecosystem (Kaz Mountain and Central Taurus Mountains), and the marginal ecosystem (Ceylanpinar). Ceylanpinar and Central Taurus Mountains are located in the Northern belt of the Fertile Crescent and the centre of diversity of various crop species such as cereals and legumes. The rich Kaz Mountain location has elements of three phytogeographic regions; namely, the Mediterranean, Irano-Turanian and Euro-Siberian Regions, and includes endemic tree and herb species. Table 19.1 shows the species targeted to conserve *in situ* at the selected locations. The selection of different regions and locations for a species allows a comparison of the genetic variation in populations in the different ecosystems.

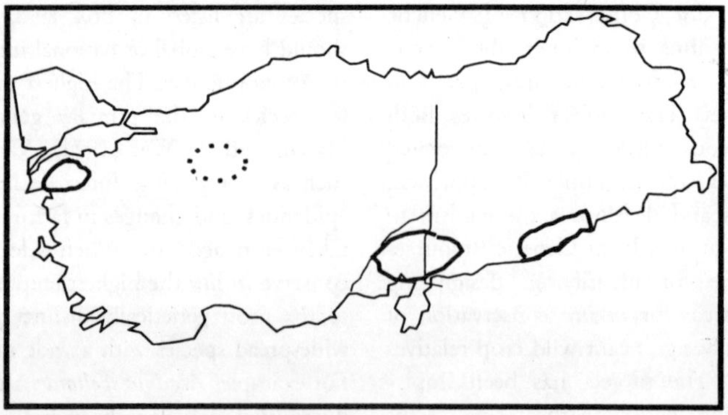

Fig. 19.1. *In situ* conservation sites.

Table 19.1. The target species at three selected locations.

	Selected study locations for *in situ* conservation of wild species		
Target species	Kaz Mountain	Central Taurus Mountains	Ceylanpinar
Aegilops speltoides		+	+
Triticum tauschii			+
Triticum boeoticum			+
Triticium dicoccoides			+
Lens ervoides		+	
Lens orientalis		+	+
Pisum sativum sensu lato		+	
Vicia sativa sensu lato		+	
Vicia johannes		+	
Castanea sativa	+		
Prunus divaricata	+		
Abies equitrojona	+		
Abies cilicica		+	
Pinus brutia	+	+	
Pinus nigra	+		

Surveys and Inventories

Surveys and inventories in the selected locations provide the initial assessment for identifying the suitable sites for selected species to be maintained. Ecogeographical surveys involve the detailed analysis and interpretation of data. Sites in the selected areas should be visited in different seasons and years in order to acquire data and information on the existence and abundance of target species, and frequency of associated species. The inventories of target taxa and associated species can be made on several of the transects or quadrats that are set for proper representation of the study site. The number and direction of transects depends on size of the sites, visual variability of topography, vegetation of the study site and distribution and variation of target species. Transects are also used as the sampling unit of target species for genetic analysis.

Surveys and inventories have been conducted for the target species of wild legumes, wild cereals, wild fruits and forest species, in the three selected locations (Anonymous, 1998a; Karagoz, 1998; Kucuk *et al.*, 1998; Sabanci *et al.*, 1998). The modified Braun-Blanquet method was used for the inventory, with several transects set in each study site at three selected locations. At the GMZs of those locations the abundance of target and associated species was estimated. The inventory in each study site was obtained in order to complete the selection of the GMZs of target species (Firat and Tan, 1997; Anonymous, 1998a; Karagoz, 1998; Kucuk *et al.*, 1998; Sabanci *et al.*, 1998; Tan, 1998; Kitiki and Tan, 1998). Herbarium specimens were also collected for the identification of associated species in the study sites. The physical and ecological description of study sites (Box 19.3) were achieved by the surveys (Anonymous, 1998). During the initial surveys the candidate study sites were selected in each location and for each target species. Then the collecting data were interpreted with selection criteria given in Boxes 19.2 and 19.3. For example, after the initial survey in Kaz Mountain, 16 chestnut sites and nine plum sites were selected as candidate study sites. But the candidate study sites were reduced to five for chestnut and four for plum as candidate GMZs for those species. At these sites four to six transects were set up and the inventory study was initiated. At the chestnut sites, an average of 70 species per site (minimum 52, maximum 110) was found. At the plum sites the average was 130 species (minimum 114 and maximum 144). It may be assumed that the ecological factors bearing on local demographic and selection processes within the target species population also affect the local distribution of many other species in the same area.

To support a complementary approach between *in situ* and *ex situ* conservation, samples of some selected target and associated species at the study sites of each location, have been collected for *ex situ* conservation.

Box 19.3. *In situ* survey outputs.

- Physical description of area surveyed: size, topography, ownership, designation, ownership of adjacent land
- Ecological description of area surveyed: detailed description of plant community types currently present at the sites
- Other descriptions: geology, soil, and climate information; fire history, logging, grazing, other impacts and uses.

Estimation of Genetic Diversity of Target Species

Genetic diversity is affected by genetic processes and life history characteristics, such as mutation rate, selection, gene flow, genetic drift and mating system. *In situ* conservation efforts should be guided by levels of patterns of genetic diversity, and gene flow of target species. Therefore, genetic diversity measurements are needed to designate the appropriate genetic reserve of target species with high variation. Diversity measurements are also needed to monitor any change in the pattern of diversity in the *in situ* GMZs. Morphological measurements are the classical but indirect method to achieve or estimate diversity. Nevertheless, morphological characters play a central role in the analysis of variation of wild relatives of crops. Molecular techniques are more accurate for assessing the genetic diversity (Newbury and Ford-Lloyd, 1997).

The target species are sampled along the transects at study sites in each location and genetic diversity is measured by using isozyme and molecular markers. Seed storage proteins (gliadins and glutenins in wheat) and two isozymes (endopeptidase-1 and aminopeptidase-2) were used to determine the level of genetic variation at seven loci of *T. tauschii* populations from Ceylanpinar sites and to select the GMZs of this species for *in situ* conservation (Eser *et al.*, 1998) Seven black pine populations from Kaz Mountain GMZs were assayed for 15 enzyme systems, and 22 allozyme loci were identified, 12 of which are supported by evidence based on segregation of band-pattern variants in megagametophytes of heterozygous mothers. Linkage was examined in 41 pairs of loci, but significant (< 0.05) nine joint segregation was detected in only three pairs, and in only one (6Pgd1-6Pgd2) is relatively strong linkage suggested (r = 0.13). (Dogan *et al.*, 1998). The four populations of *Abies equitrojani* from Kaz Mountain sites were studied. Nineteen loci coding allozymes in 12 enzyme systems were described. Ten of the 19 loci were polymorphic in one or more of the populations. The populations differ significantly (< 0.05) in allele frequencies at six loci. They also appear to differ in levels of expected heterozygosity and fixation index. Although this endemic fir species has a narrow distribution, considerable genetic variation was observed in the populations (Gulbaba *et al.*, 1998). Morphological variation is also obtained from the sampled populations. The fruit morphology of targeted wild plum and chestnut populations from the Kaz Mountain GMZs was studied. Significant variation between and within populations of these species was observed (Onal *et al.*, 1998). This information assisted in the choice of GMZs of target species to maintain *in situ*.

Genetic Reserves or Gene Management Zones

Designation of reserves

GMZs can be defined as long-term monitoring sites that contain one or more diverse populations of various target species to be conserved *in situ*. A series of genetic reserves should be designated, in order to represent the ecological ranges of target species so as to support sufficient environmental heterogeneity of selected species. Each genetic reserve is an open laboratory, permits the conservation of several taxa in a single reserve and allows continued evolution of the component species. Genetic reserves should be easily accessible, relatively isolated from exotic gene flow, and include a wide range of biological diversity and of the genetic diversity of target species (Anonymous, 1996; Tan, 1996; Tan and Ulubelde, 1998). The selection criteria (Box 19.4) of genetic reserves or GMZs should be applied to all reserves to establish the appropriate set of reserves for each target species. Reserves should consist of core and buffer zones, with corridors if necessary. The core zones are the central areas with stable habitat of target species surrounded by a buffer zone, which allows research applications. If multiple small genetic reserves are selected, a corridor can be set up between those in order to consider one

Box 19.4. Considerations in the selection of GMZs (Anonymous,1996).

- Target species must have the primary consideration
- GMZs should capture as much genetic variation as possible and should have species richness
- GMZ sites under consideration should be accessible, sustainable and suitable for efficient populations management. Their numbers and size may vary according to the availability of resources
- Appropriate size and number of target species should be determined in terms of evolutionary potential, genetic integrity and protection values
- The establishment of a GMZ can be in either natural or semi-natural environments

single reserve for easy management. The size and number of genetic reserves of target species may vary according to the availability of resources and depending on the species; whether woody or non-woody, annual or perennial (Anonymous, 1996; Tan and Ulubelde, 1998).

In the Turkish project for *in situ* conservation of genetic diversity for wild species, the candidate GMZs have been selected in accordance with the selection criteria. Workshops were organized in each location and for each target species. The information about the habitats of target species, and the data on morphological and genetic diversity have been considered and more than one GMZ of each target species has been designated. Of the five candidate GMZs of chestnut at Kaz Mountain, three were designated as GMZs, and among four candidate GMZs of wild plum, three were chosen. One of the chestnut GMZs consisted of two small populations linked with a corridor. The other GMZ for chestnut neighboured a black pine site so those two were considered as one GMZ. All GMZs of target species consisted of core and buffer zones. The core zone will be kept as the *in situ* reserve of original populations, while the buffer zones will be open laboratories for further studies, such as testing grazing pressure. The sizes of the plum and chestnut GMZs, including buffer zones, vary from 150 ha to 479 ha, and from 176 ha to 886 ha, respectively (Anonymous, 1998a).

The management plan of genetic reserves

The designated genetic reserves are and will continue to be subject to the pressure of various impacts. Therefore, a management plan for each genetic reserve should be developed in order to regulate human intervention to ensure the maintenance of genetic diversity of the target species populations. Monitoring of genetic reserves is also needed to ensure that management intervention remains appropriate. Box 19.5 lists the many factors that need consideration during the preparation of management plans (Tan and Ulubelde, 1998).

The objectives of *in situ* conservation of the target species should be carefully defined in the management plan in order to ensure their implementation. The components of a management plan are given in Box 19.6. A management plan is prepared for each target species in Turkey. They include detailed information on purpose, genetic structure of the target species (degree of homogeneity, molecular variation, mode of reproduction, size of target populations and their distribution); environments (geographical information of each site, inventory of associated species and their distribution, land ownership, local demographics, past and current history including potential threats); management action required; schedule; action plan; and appropriate assignments of responsible scientific and technical groups.

Intervention may be applied to the buffer zones or outside the core zones in conducting the case studies. According to the result of case studies, practices such as harvesting in the chestnut GMZs or grazing at the plum GMZs, may be required to maintain the target populations. When necessary, additional local seeding may be required to maintain or increase the target populations of plum and chestnut. This activity may be necessary for plum GMZs, since these populations are small. Monitoring is an essential step to observe the changes of populations and associated species at

Box 19.5. Considerations for the management plan.

- Maintenance of genetic diversity is the most important objective of the management plan. Maintenance of the ecosystems for target species is also essential
- The management plan should recognize the possible contrast that may arise between species richness and diversity within the target species
- Manipulations of various kinds will be necessary within GMZs (e.g. limited grazing, selective logging, fire, pruning)
- Since best practices are still largely unknown, various alternative treatments will need to be tested and compared with the management regime
- The plan should address the amount of interference that would be acceptable in the face of severe potential damage to the target species
- Buffer zones should be a component of all GMZs and their management included in the plan
- Community participation and support will be vital for effective maintenance of GMZs
- GMZs must be supported by complementary *ex situ* conservation activities
- GMZs should be accessible to scientists; both within and outside the country, including collaborative projects with the international scientific community
- Management plans are likely to vary for woody and non-woody species (wild crop relatives and forestry species)

Box 19.6. Components of the management plan.

- Articulation of the objective and purpose
- Biological and genetic structure of target species
- Description of the physical and biological environment of GMZs
- Description of any conflict and problems associated with GMZs
- Description of the management practices required to achieve the objectives
- Monitoring of the community dynamics within the GMZs to assess management effectiveness
- Link with national conservation plans and policies
- Community participation

each GMZ of plum and chestnut. Monitoring will take place in core zones and in the permanent plots (transects) of each GMZ. Primary monitoring is required at 5 yearly intervals, especially for the vegetation (both target species and associated species) and ecological characteristics. Before implementing management procedures, case studies on intervention will be made with the overall aim of preserving a large proportion of variability while limiting the level of intervention, grazing, harvesting and so on, and to define the changes in the composition of the vegetation communities at the sites. In the GMZs, permanent plots should be selected as controls and the sites where no intervention will take place should be monitored so as to have a measure of the effects of particular management practices. In wild plum and chestnut GMZs the transects will be the permanent plots.

It is more difficult to give specific practical advice for the management of the wild plum and chestnut than for other forest tree species. Common cultural practices cannot be applied to those species found in forest areas. Regular inventories can be carried out every 2 years, for both the target species and for the associated species at the permanent transects of the GMZ sites. Previous records of the inventory should be compared with present records. For wild plum, the removal of mature overstorey trees may be necessary after a period of time, to allow development of seedlings or young trees which are characteristically found at forest margins or as understorey components where light intensity is high. Low fruit set of wild plum populations was observed in GMZs for various reasons. If this situation resulted from low pollination in the population, the introduction of managed apiaries may be attempted. Since the chestnut is a wind-pollinated species, orchards of cultivated improved varieties or local types from other regions should not be located near the GMZ. Pruning or thinning should not be applied in mature trees. If necessary such practices can be applied to young trees. Thinning should be at random in the GMZs for plum and chestnut.

A GMZ for wild fruit species probably requires larger sites than required for annual herbaceous species. The planting of trees from seed sources outside GMZs would not be permitted so as to prevent pollution of the genepool from exotic sources. Grazing would not be permitted in core areas of wild plum and chestnut to encourage development of seedlings, associated species and the forest understorey. Some level of grazing may be permitted in the plum GMZs especially in the sites where grazing is already practised. That level of grazing, however, should be monitored in a case study in buffer zones. In some cases, immediate intervention may be required, such as fencing, especially in the heavily grazed plum GMZ at Yukariçavus, to control but not eliminate grazing. Grazing control is essential in these GMZs to allow the regeneration of associated species. To overcome community disaffection, community and public awareness are important. Grafting or other forms of propagation should be avoided. If the restoration or regeneration of the population is necessary, then only 'local' natural (from the same sites or similar sites) seeds should be used. The GMZs for wild plum and chestnut were selected by considering many factors which may affect directly or indirectly the success of long-term *in situ* conservation of those target species in the selected sites, such as existing level of morphological variation of target species, level of biological and ecological diversity, presence of more than one target species, and sustainability of the sites for conservation. Chestnut fruit harvesting is common at Kazdag. It may be very difficult to restrict the harvesting by local communities from designated chestnut GMZs. At least, however, selective harvesting can be restricted.

For annual wild species, such as wild legumes and cereals, grazing pressure is a key impact factor. Therefore, in the GMZs of the Central Taurus Mountains and in Ceylanpinar this impact will be tested carefully and research will be undertaken in buffer zones.

Because of the importance of community participation to the implementation of the management plans of GMZs, various workshops were organized for public and community awareness and to encourage their participation.

Data Management

For *in situ* conservation of wild crop relatives, a comprehensive database should be built up (Tan and Tan, 1998). The *in situ* database should also be linked with existing *ex situ* conservation databases and should be interfaced with geographical information systems (GIS). Information related to *in situ* conservation accrues from a range of sources. Thus, for effective use of *in situ* databases, the acquisition of information from different sources should be carefully planned. Information on *in situ* conservation consists of the project activity components such as target taxa at site, habitat data, geographic parameters, genetic diversity (morphological variation, isozyme variation, molecular variation), management plans of GMZs and monitoring records. The database should be suitable for data processing, verification and exchange. GIS contributes effectively to assess all data from *in situ* conservation in order to integrate and manage the spatial type of information.

The data management system for *in situ* conservation of wild species in Turkey has been integrated with *ex situ* databases, and GIS was used for the documentation of spatial diversity (Tan and Tan, 1998; Anonymous, 1998a,b). The database for the project comprises digitized map sheets and drawings, linking to external databases, including *ex situ* databases, integrating tabular data and converting existing digital geographic data. Using this database a couple of different analyses for planning the *in situ* conservation of wild species have been carried out. The results of these analyses together with the underlying database, the spatially grouped maps of GMZs and the potential sites for GMZs for designated priority species were produced and 'Project Atlas' has been produced and published (Anonymous, 1998b).

National Plan for *in situ* Conservation

A National Plan for *in situ* Conservation can be prepared to provide the basis for review of the project to date, facilitate the coordination and cooperation for the management of GMZs into other nature preservation strategies in the country, and outline the implementation of the National Plan. The National Plan for *in situ* Conservation provides a mechanism for the country to set priorities and present a plan of action that ensures the protection of wild crop relatives and forest genetic resources in their natural habitat beyond designated *in situ* reserves. For this reason, a National Plan for *in situ* Conservation of genetic diversity in Turkey has been prepared (Tan, 1998; Kaya *et al.*, 1998). The objective of the plan is to determine the priorities and strategies for effective management and conservation, sustainable utilization and monitoring of genetic diversity. This plan is expected to help expand *in situ* conservation activities into a network of GMZs that includes other areas of Turkey and incorporates additional important wild crop relatives, forest species and especially endangered species beyond those targeted species in the project. The plan also includes the needs and opportunities for landraces as part of the *in situ* genetic resources conservation spectrum. In the first three sections of the National Plan, the importance of *in situ* conservation of genetic diversity and the present status of plant genetic resources, their conservation and related biodiversity conservation legislation in Turkey are spelt out. In the fourth section of the National Plan, the objectives, priorities and strategies for *in situ* conservation of wild species, forest tree species as well as landraces are presented. The National Plan, in the fifth section, strongly recommends the following actions for the successful implementation of the plan:

1. The preparation of GMZ framework regulations and legislation;
2. The formulation of a Regional Board for *In situ* Conservation and an Implementation Board for *In situ* Conservation;
3. Sustained and effective coordination of activities among the implementing Ministries (MARA, MOF, ME) and NGOs;
4. A clear statement of the responsibilities of partner institutions;
5. The identification of sources of funding;
6. A strong public awareness programme.

Institutional arrangements and priority research topics are included in the plan.

The National Plan for *in situ* Conservation is also consistent and complementary with other national, regional or global efforts to conserve wild plant genetic resources. The plan is compatible with and supports the National Plant Genetic Resources Research Programme of MARA, and contributes to, but does not replace, the more comprehensive National Biodiversity Conservation Action Plan of Turkey. The plan also leads the way for the continuation and replication of the pilot part of the *in situ* conservation project, which can provide a supportive platform for continued assistance.

Conclusions

The basic essential steps for the *in situ* conservation of wild species, and the monitoring of existing genetic diversity of their target populations are summarized below (Anonymous, 1996; Tan and Ulubelde, 1998):

1. Identify target taxa (decide which species need to be included in the *in situ* conservation).
2. Review the conservation status of target taxa.
3. Undertake an ecogeographical survey.
4. Select potential sites to sample maximum genetic diversity (select a site or a range of sites to cover as much as possible of the range of diversity of the chosen species).
5. Review ecological requirements for the continuing viability of populations.
6. Maintain at least two GMZs in each ecogeographic zone on a regional basis (more than one site should be chosen for each species group or ecosystem to cover climatic and edaphic diversity).
7. Ensure that GMZs encompass at least the minimum viable population sizes of the target taxa.
8. Monitor populations within each GMZ over appropriate time periods (1–10 years) using transects and study plots within the GMZ.
9. Install exclusion fencing and buffer zones when shown to be necessary.

10. Prepare a 5-year management plan for target taxon.
11. Encourage public participation, through workshops, supporting interpretation and media programmes.
12. Promote the conserved material through the use of published reports and catalogues for appropriate utilization.
13. Ensure that each site should include (whenever possible) several management practices (a range of grazing practices, tree-felling, pruning, grafting, burning, etc.).
14. Make sites large enough to contain at least 1000 individuals of each species.
15. Seek cooperation with nature conservation organizations and agencies, when establishing the *in situ* reserves and for advice on management practices.
16. Maintain duplicate samples of the species of crop plant relatives in the reserves in *ex situ* genebanks (at seed or *in vitro* banks) whenever possible.

The *in situ* conservation project of Turkey targets the conservation of wild relatives of crop species and other selected plants with significant genetic variability. This project is the first attempt to integrate the various components of *in situ* conservation for multiple species (woody and non-woody) at multiple sites. By complementing the ongoing *ex situ* programme, it will develop the institutional and technical mechanisms for a comprehensive strategy for plant genetic resources conservation in Turkey. Success in these endeavours could provide a model applicable in other parts of the world. Although other countries and agencies have undertaken *in situ* projects, these have targeted relatively few species or a single location, and have not focused on genetic reserve management. In Mexico, for instance, *in situ* conservation of the wild endemic perennial *Zea diploperennis* was prescribed in the Sierra de Manantlan Biosphere reserve (Benz, 1988). In Israel the wild tetraploid wheat (*Triticum dicoccoides*) was conserved at Ammiad, Eastern Galilee, along with wild barley and wild oats (Anikster and Noy-Meir, 1991). This study was among the first on *in situ* conservation of viable diverse populations of wild species related to crops. The study has guided the Turkish project in setting up the methodology and strategy for wild relatives of cereals. The *T. dicoccoides* populations at Ammiad and Ceylanpinar can be compared in a collaborative study in future as a regional *in situ* conservation effort.

In situ conservation is different from the concept of protected areas. In this respect, the significance of *in situ* conservation is that it targets the conservation of genetic diversity of specific plant species, and not necessarily the protection of entire ecosystems. The integration of *in situ* with *ex situ* strategies for the conservation of genetic resources conservation in this project is also innovative. It uses participatory approaches between the formal (the Ministries and institutions) and informal sectors in Turkey for the setting, design and maintenance of *in situ* conservation areas. In this respect, the project is the first of its kind, and is a model for other countries that are rich in genetic diversity and aim to implement the internationally agreed Global Plan of Action.

References

Anikster, Y. and Noy-Meir, I. (1991) The wild wheat field laboratory at Ammiad. *Israel Journal of Botany* 40, 351–362.

Anonymous (1996) *Genetik Cesitliligin* In situ, *Yerinde, Muhafazası Projesi: Gen Muhafaza Zonları. Workshop, 28–29 Mart. 1996. Menemen Izmir.* [*In situ* Conservation of Genetic Diversity Project Workshop on Gene Management Zones, 28–29 March 1996, Menemen Izmir] ETAE (AARI) No. 92, Menemen, Turkey.

Anonymous (1998a) *Final Report of* In situ *Conservation of Genetic Diversity Project in Turkey*. MARA, Turkey.

Anonymous (1998b) *Information Technology Management Plan (ITMP), Database Design and GIS Applications Development Project.* Project Atlas, MARA, Ankara, Turkey.

Benz, B.F. (1988) *In situ* conservation of the Genus *Zea* in the Sierra de Manatlan Biosphere Reserve. In: Russell, N. and Listman, G.M. (eds) *Recent Advances in the Conservation and Utilisation of Genetic Resources: Proceedings of Global Maize Germplasm Workshop.* CIMMYT, Mexico, pp. 59–69.

Dogan, B., Ozer, A.S., Gulbaba, A.G., Velioglu, E., Doerksen, A.H. and Adams, W.T. (1998) Inheritance and linkage of allozymes in black pine (*Pinus nigra* Arnold.) from Turkey. In: Zencirci, N., Kaya, Z., Anikster, Y. and Adams, W.T. (eds) *Proceedings of International Symposium on* In situ *Conservation of Plant Genetic Diversity.* CRIFC, Ankara, Turkey, pp. 249–257.

Eser, V., Gocmen, B., Erisen, S., Baran, I., Donmez, E., Barut, A.A. and Karagoz, A. (1998) Determination of biochemical variation in an *Aegilops tauschii* population collected from Ceylanpınar. In: Zencirci, N., Kaya, Z., Anikster, Y. and Adams, W.T. (eds) *Proceedings of International Symposium on* In situ *Conservation of Plant Genetic Diversity.* CRIFC, Ankara, Turkey, pp. 93–99.

Firat, A.E. and Tan, A. (1997) *In situ* conservation of genetic diversity in Turkey. In: Maxted, N., Ford-Lloyd, B.V. and Hawkes, J.G. (eds) *Plant Genetic Conservation, The* In situ *Approach.* Chapman and Hall, London, pp. 254–262.

Firat, A.E. and Tan, A. (1995) Turkey maintains pivotal role in global genetic resources. *Diversity* 11, 61–63.

Gulbaba, A.G., Velioglu, E., Ozer, A.S., Dogan, B., Doerksen, A.H. and Adams, W.T. (1998) Population genetic structure of Kazdağ Fir (*Abies equitorjani* Ashers. Et Sint), a narrow endemic to Turkey. Implication for *in situ* conservation. In: Zencirci, N., Kaya, Z., Anikster, Y. and Adams, W.T. (eds) *Proceedings of International Symposium on* In situ *Conservation of Plant Genetic Diversity.* CRIFC, Ankara, Turkey, pp. 271–281.

Harlan, J.R. and de Wet, J.M.J. (1971) Towards a rational classification of cultivated plants. *Taxon* 20, 509–517.

Karagoz, A. (1998) *In situ* conservation of plant genetic resources in the Ceylanpınar State Farm. In: Zencirci, N., Kaya, Z., Anikster, Y. and Adams, W.T. (eds) *Proceedings of International Symposium on* In situ *Conservation of Plant Genetic Diversity.* CRIFC, Ankara, Turkey, pp. 87–93.

Kaya, Z., Kün, E. and Güner, A. (1998) National plan for *in situ* conservation of plant genetic diversity in Turkey. In: Zencirci, N., Kaya, Z., Anikster, Y. and Adams, W.T. (eds) *Proceedings of International Symposium on* In situ *Conservation of Plant Genetic Diversity.* CRIFC, Ankara, Turkey, pp. 33–47.

Kitiki, A. and Tan, A. (1998) Vegetation survey in the southern part of the Anatolian Diagonal and possible gene management zones for wild crop relatives. In: Zencirci, N., Kaya, Z., Anikster, Y. and Adams, W.T. (eds). *Proceedings of International Symposium on* In situ *Conservation of Plant Genetic Diversity.* CRIFC, Ankara, Turkey, pp. 129–135.

Kucuk, S.A., Tan, A.S., Sabancı, C.O., Cinsoy, A.S., Onal, K. and Kostak, S. (1998) Ecogeographical and floristic differentiation of chestnut gene management zones in Kazdağ. In: Zencirci, N., Kaya, Z., Anikster, Y. and Adams, W.T. (eds) *Proceedings of International Symposium on* In situ *Conservation of Plant Genetic Diversity.* CRIFC, Ankara, Turkey, pp. 135–149.

Maxted, N. and Hawkes, J.G. (1997) Selection of target species. In: Maxted, N., Ford-Lloyd, B.V. and Hawkes, J.G. (eds) *Plant Genetic Conservation, The* In situ *Approach.* Chapman and Hall, London, UK, pp. 43–68.

Newbury, H.J. and Ford-Lloyd, B.V. (1997) The estimation of genetic diversity. In: Maxted, N., Ford-Lloyd, B.V. and Hawkes, J.G. (eds) *Plant Genetic Conservation, The* In situ *Approach.* Chapman and Hall, London, UK, pp. 192–206.

Onal, M.K., Sabancı, C.O., Kucuk, S.A. and Cinsoy, A.S. (1998) The pomological variation patterns of wild plum (*Prunus divaricata* Ledeb.) and chestnut (*Castanea sativa* Miller) *in* Kazdağ (Mt Ida) area of Turkey. In: Zencirci, N., Kaya, Z., Anikster, Y. and Adams, W.T. (eds) *Proceedings of International Symposium on* In situ *Conservation of Plant Genetic Diversity.* CRIFC, Ankara, Turkey, pp. 149–155.

Sabancı, C.O., Onal, K., Tan, A.S., Kucuk, S.A., Kostak, S. and Cinsoy, A.S. (1998) Ecogeographical and floristic differentiation at plum gene management zones at Kazdağ. In: Zencirci, N., Kaya, Z., Anikster, Y. and Adams, W.T. (eds) *Proceedings of International Symposium on* In situ *Conservation of Plant Genetic Diversity.* CRIFC, Ankara, Turkey, pp. 155–163.

Tan, A. (1996) Establishment of gene management zone (GMZ) for an *In situ* Conservation Programme in Turkey. *Report of the Working Group on* Prunus. *Fifth Meeting, 1–3 February 1996. Menemen, Izmir-Turkey.* IPGRI, ECP/GR, Rome, Italy.

Tan, A. (1998) Current status of plant genetic resources conservation in Turkey. In: Zencirci, N., Kaya, Z., Anikster, Y. and Adams, W.T. (eds) *Proceedings of International Symposium on* In situ *Conservation of Plant Genetic Diversity.* CRIFC, Ankara, Turkey, pp. 5–17.

Tan, A. and Tan, A.S. (1998) Database management systems for conservation of genetic diversity in Turkey. In: Zencirci, N., Kaya, Z., Anikster, Y. and Adams, W.T. (eds) *Proceedings of International Symposium on* In situ *Conservation of Plant Genetic Diversity.* CRIFC, Ankara, Turkey, pp. 309–323.

Tan A.S. and Ulubelde, M. (1998) Selection criteria and planning of Gene Management Zones (GMZs) for *in situ* conservation. In: Zencirci, N., Kaya, Z., Anikster, Y. and Adams, W.T. (eds) *Proceedings of International Symposium on* In situ *Conservation of Plant Genetic Diversity.* CRIFC, Ankara, Turkey, pp. 363–373.

20 Metapopulation Dynamics of Lima Bean (*Phaseolus lunatus* L.) in the Central Valley of Costa Rica

O.J. Rocha,[1] J. Degreef,[2] D. Barrantes,[1] E. Castro,[1] G. Macaya[3] and L. Guarino[4]

[1]*Escuela de Biología, Universidad de Costa Rica, Ciudad Universitaria 'Rodrigo Facio', San Pedro de Montes de Oca, San José, Costa Rica;* [2]*National Botanic Garden of Belgium, Domaine de Bouchout, Belgium;* [3]*Centro de Investigación en Biología Celular y Molecular, Universidad de Costa Rica, Ciudad Universitaria 'Rodrigo Facio', San Pedro de Montes de Oca, San José, Costa Rica;* [4]*International Plant Genetic Resources Institute, Regional Office for the Americas, IPGRI c/o CIAT, Cali, Colombia*

Introduction

Metapopulation studies provide a useful new approach to studying the ecology and genetics of populations at the landscape level. This is particularly true when populations function as a product of local dynamics and the regional processes of migration, extinction and colonization (Hanski and Gilpin, 1991). In most cases, metapopulations of plants occur in habitats that represent temporal niches along successional series (Poschold, 1996). They may also occur in spatial niches of the natural landscape with small and isolated areas, such as in treefall or wind-fall areas in forests (Valverde and Silverton, 1997; Gillman, 1997) and areas that experience fire occasionally (Menges *et al.*, 1998). Human activity also creates metapopulations, for example by intensive land use, changes in agricultural practices and urban development, that may result in fragmentation of the landscape and temporal degradation of suitable habitats (Rocha *et al.*, 1997). Dispersal processes, which play a key role in the survival of metapopulations, have changed markedly since humans started to cultivate crops and during the development of the man-made landscape (Poschold, 1996).

Metapopulation structure may break down under high disturbance rates (Husband and Barrett, 1996; Valverde and Silverton, 1997; Ouborg *et al.*, 1999). Because plants are relatively immobile, present strong spatial structure and restricted dispersal, habitat fragmentation and patch isolation may lessen the probabilities of recolonization events (Giles and Goudet, 1997; Ouborg *et al.*, 1999). Thus, isolation by distance may result in a disruption of the structure of the metapopulation (Giles and Goudet, 1997). In contrast, it has been proposed that for *Primula vulgaris*, local populations may never become extinct even under high disturbance rates and, thus, seed dispersal may not play a major role in local population re-establishment (Husband and Barrett, 1996). In this scenario, the structure of the metapopulation would also be disrupted.

Metapopulation dynamics also affect the genetic structure of populations, their diversity and differentiation (Giles and Goudet, 1997). Gene flow is an important factor in genetic structure at the landscape level. Other authors have shown that gene flow appears to be high for tropical plants (Hamrick *et al.*, 1991; Nason and Hamrick, 1997). But for most cases, ascertaining the implications on species or nature conservation is only partially possible because of the lack in knowledge about interactions between populations. Therefore, it is more useful at the moment to study some important aspects of the metapopulation concept such as dispersability and dispersal processes to give this theoretical concept a more practical basis in order to use it for nature conservation purposes.

Wild populations of Lima bean (*Phaseolus lunatus*) can be found throughout the Central Valley of Costa Rica (Standley, 1937; Rocha *et al.*, 1997). They are usually found in open areas, disturbed areas with grass and scattered trees (Standley, 1937; Debouck, 1987). Moreover, Lima bean is more frequently found where coffee is grown under shade (traditional coffee plantations), as well as in wastelands around these plantations. This habitat has a forest-like structure with a diverse mixture of shade trees, and several varieties of bananas (*Musa* spp.). Typically, agricultural activities are less intense in this agroecosystem, and do not rely on heavy use of herbicides for the elimination of weeds (Rocha *et al.*, 1997). However, it has been demonstrated that because of changes in agricultural practices and changes in land use due to urban development, the populations of Lima bean in the Central Valley are fragmented and undergo local extinction and recolonization (Rocha *et al.*, 1997). In this study, we describe the metapopulation structure of wild populations of Lima bean in the Central Valley of Costa Rica. In addition, we describe the metapopulation dynamics using a Markovian model that incorporates the different stages through which local populations pass. The transition probabilities among these stages are calculated and the extinction and recolonization rate estimated.

Materials and Methods

Study organism

Wild Lima beans are short-lived perennial vines. Typically, they show indeterminate growth during the rainy season, but cease growth during the dry season, when most plants flower and later lose their leaves. The flowers are borne on inflorescences and require the visitation of pollinators to transfer pollen from the anthers to the stigma (Rocha, unpublished). Lima beans are self-compatible and are predominantly self-pollinated, but with significant variation in the outcrossing rate (Maquet *et al.*, 1996, Bi, 1999).

Study area

This study was conducted in the Central Valley of Costa Rica. This region is located in the geographic centre of the country and comprises two intermontane valleys separated by the low Ochomogo pass at the continental divide: the Central Valley of the Pacific slope and the much smaller Valle del Guarco on the Caribbean side. However, most of this study was conducted in the upper valley and watershed of the Río Grande de Tárcoles (latitude range 9° 54′ N – 10° 07′ N, longitude range 83° 50′ W – 84° 28′ W). We also included in this study a portion of the Valle del Guarco and the small valley of Acosta. As a whole, the study area is approximately 2100 km^2.

Ecogeographical surveys

In early November 1992, we started a large-scale survey of populations of *P. lunatus* in the study area. This survey was conducted with the collaboration of six advanced students of the Escuela de Biología, Universidad de Costa Rica. When we surveyed a given area, we examined all roads and trails looking for plants. We also interviewed local farmers in order to find out whether or not this species was present in the area, and asked for permission to visit their farms. Where *P. lunatus* was found, we defined as one population any group of plants separated (regardless of its size) from other plants by at least 500 m. For each population we collected a voucher specimen, fruit and seed samples and made a description of the site, including parameters such as the size of the population, topography, associated vegetation and geographical coordinates using a Garmin 48 GPS unit. Similar surveys were conducted again in 1993, 1994 and 1999. This information was used to determine the distribution of distances between populations.

Metapopulation dynamics

In order to study the metapopulation dynamics we established six sampling transects along main roads in the study area (Fig. 20.1). In 1994, we found 103 populations along these sampling transects. The initial number of populations on each transect varied between 14 and 23. All populations have been visited every 2 weeks since January 1995. During each visit, we recorded the phenological status of each population by counting the number of individual plants that had foliage and the number of individuals bearing flower bud, flowers, immature fruits and/or mature fruits with seeds. In addition, we recorded any disturbances experienced by each population, such as fire, weeding (manually or with herbicides) and habitat destruction due to urban development.

Metapopulation dynamics were studied, by determining the transition probabilities between five possible population stages, namely: (i) populations that remain vegetative during the year; (ii) populations that flowered during the year; (iii) populations that produced seeds during the year; (iv) populations that became extinct; and (v) populations that recolonized an empty site (Fig. 20.2). Lefkovich matrices were used to describe the dynamics of these populations using the UNIFIED LIFE MODELS software program (Legendre and Clobert, 1995)

Populations were considered extinct if all plants previously present had disappeared. Moreover, in this model we did not distinguish between recolonization events occurring from seeds present in the seed bank, or recolonization due to the arrival of seeds from adjacent populations. Years were defined on the basis of the phenology of the species, and were thus considered to begin in May, at the beginning of the rainy season, which is accompanied by the flush of new leaves. The end of the dry season, when plants are leafless and seeds have already been dispersed, defined the end of the year.

Results

Spatial structure

The study of spatial distribution of the population of Lima bean in the Central Valley of Costa Rica has shown that, for any given population, the mean distance to the nearest populations is 1768.7 m

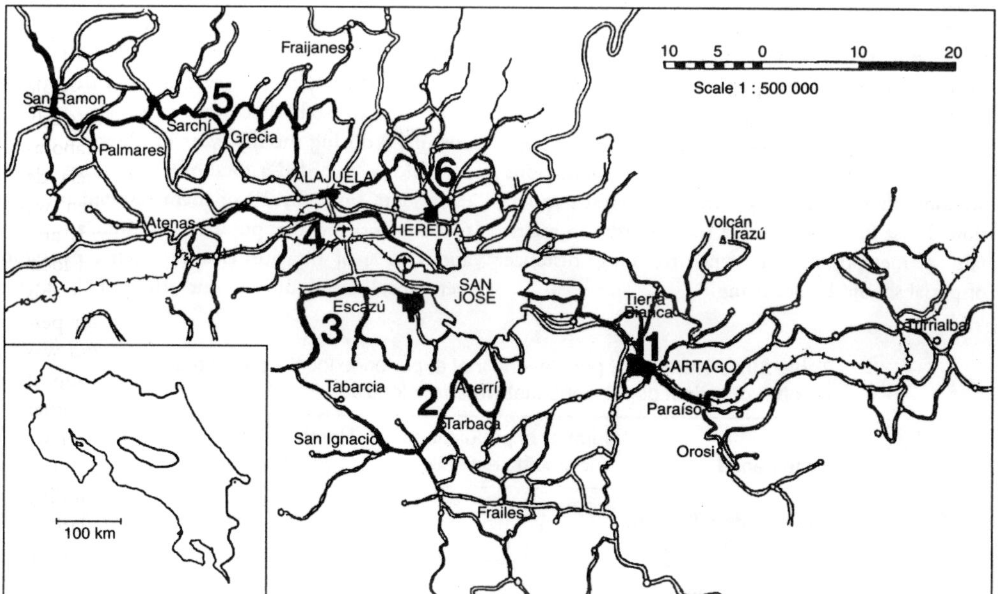

Fig. 20.1. Location of the study area showing the six transects defined to conduct this study. Number of populations in each transect: transect 1 = 17, transect 2 = 14, transect 3 = 14, transect 4 = 15, transect 5 = 20 and transect 6 = 23.

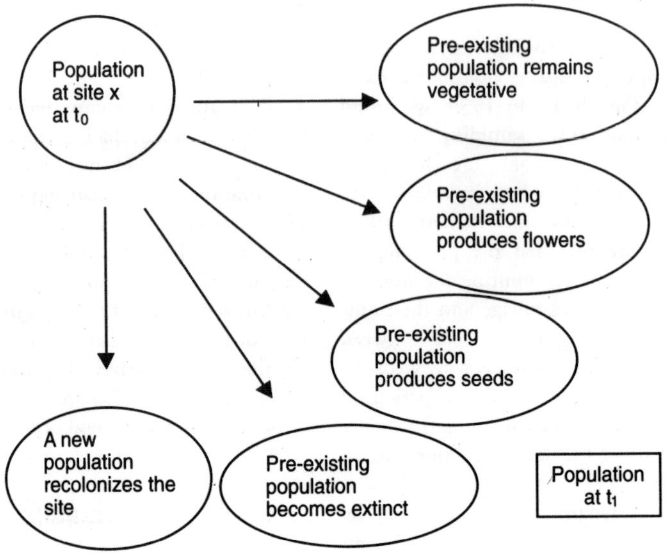

Fig. 20.2. Possible transitions for the metapopulation dynamics.

(n = 497 populations). However, the range for the distance to the nearest population is between 165 m and 3697 m. Distances smaller than 500 m might result from errors in the reading obtained from the GPS unit. In addition, such short distances might also indicate new populations found in areas where older populations have become locally extinct. Overall, our data show that 58% of the populations are within 1 km of the nearest population. However, about 8.6% of the populations are more than 3 km away from the nearest population. These findings indicate that in order to maintain a metapopulation structure, long distance dispersal should be occurring.

Metapopulation dynamics

Our results show that the occurrence of local population extinction is common in wild populations of Lima bean (Table 20.1). We found that as many as 68% of the populations experienced the elimination of all the adults. However, most of the populations were re-established within the same year (51 out of 70 during the first year of observations). In general, we recorded a total of 91 episodes of local extinction, but only 19 of them resulted in a permanent loss of the populations, indicating that about 79% of the sites that experience extinction were recolonized during our study. This finding

Table 20.1. Number of populations that experience local extinction and number of extinct populations that were recolonized.

Time period	Number of populations undergoing extinction	Number of recolonized populations
October 94 – April 95	70	51
May 95 – April 96	0	0
May 96 – April 97	15	0
May 97 – April 98	0	10
May 98 – April 99	2	1
May 99 – April 2000	4	0

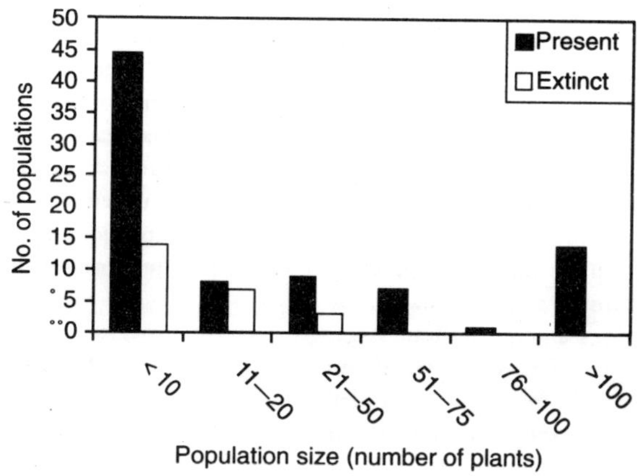

Fig. 20.3. Distribution of populations with plants (sizes in 1999) and extinct populations (sizes in 1994) according to population size categories.

Table 20.2. Number of populations that are found in each of the possible stages at the end of each year. There were 103 populations initially found in the six transects.

	Population stage			
Time period	Remained vegetative	Only flowered	Produced seeds	Populations remaining
October 94 – April 95	7	23	3	33
May 95 – April 96	14	22	48	84
May 96 – April 97	13	7	49	69
May 97 – April 98	14	7	58	79
May 98 – April 99	12	5	61	78
May 99 – April 2000	44	0	30	74

indicates that, on average, we expect about 18 populations to become locally extinct each year in the Central Valley. Our data also reveal that recolonization of empty sites is also common; a total of 62 sites were recolonized in the 5 year period considered here.

Our data also revealed that small populations are more likely to experience local extinction. Based on the census conducted in 1999, we found that most of the populations currently present are very small (Fig. 20.3). Moreover, the 1994 sizes of the populations that have become extinct during the study period also indicate that 58% of them were very small (Fig. 20.3). We have not recorded local extinction of populations with more than 50 plants.

Not all populations are equally likely to produce seeds (Table 20.2). Nearly one-fifth of the populations (18) in the study area failed to produce flowers every year. In addition, only 49% of the populations produced seeds each year. This finding suggests that populations differ in their ability to maintain a seed reservoir in the soil. Moreover, populations that produce seeds in a given year have a high probability of producing seeds in future years (probability of remaining in same group = 0.824), while those that become extinct also have a great probability of staying extinct (probability of remaining extinct = 0.787) (Fig. 20.4).

Metapopulation dynamics were studied using matrix analyses. The UNIFIED LIFE MODEL software program was used to determine the demographic parameters, in particular, metapopulation growth rate (lambdaM, the rate at which the number of populations in the metapopulation increases or decreases). This analysis revealed that in spite of the frequent extinctions recorded in this study, the metapopulation growth rate calculated from the matrix of average transition probabilities is close to one (lambda = 0.990). However, the trajectory of the number of populations in the metapopulation varies according to the initial scenario, that is, the distribution of populations among the four stages considered in this study. For example, if we consider only the 33 active populations recorded in 1995 as our initial metapopulation size, the number of populations will decline at a rapid rate (Fig. 20.5). However, if we considered that all recorded populations were vegetative at the beginning of the study (initial metapopulation size = 103), the number of populations will decline for a few years but then become rather stable (Fig. 20.6). Moreover, there is another possible scenario, using the mean number of active populations in each category as the initial metapopulation size. Under this scenario, the number of populations also shows a decline, but eventually the number of populations becomes stable (see Fig. 20.7). Overall, matrix analyses show that wild populations of Lima beans in the Central Valley are not under a clear risk of extinction. This finding is somewhat counterintuitive given the fact that only 24% of the locations where populations experience extinction are colonized again.

Discussion

The objective of this study was to describe the metapopulation dynamics of wild populations of Lima bean in the Central Valley of Costa Rica as a basis for predicting the probability of persistence. We found that most of the populations are within 1 km of the nearest other population. However, some populations can be more than 3 km away from other populations. Our findings support the notion that wild populations of Lima beans in the Central Valley of Costa Rica

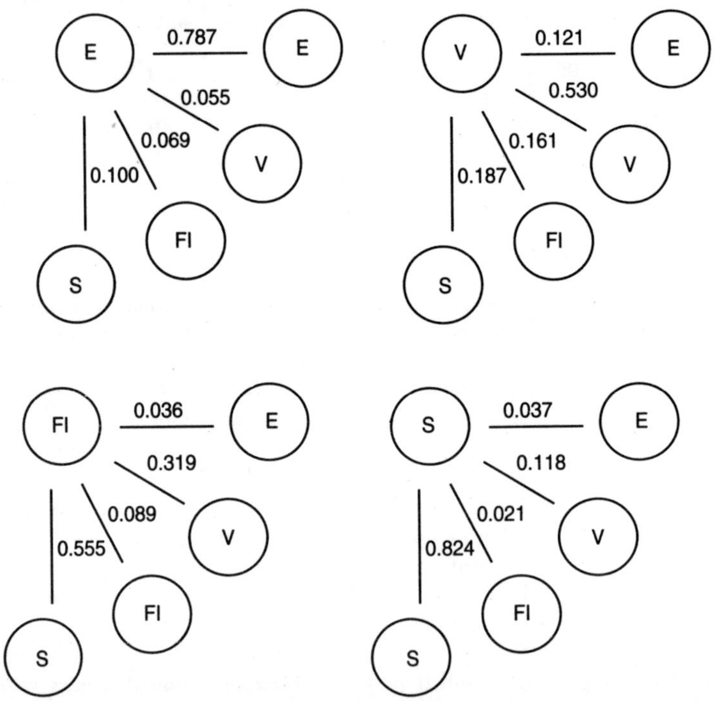

Fig. 20.4. Preliminary estimate of all possible transitions among stages (E, extinct; V, vegetative; FI, flowering; S, seeds).

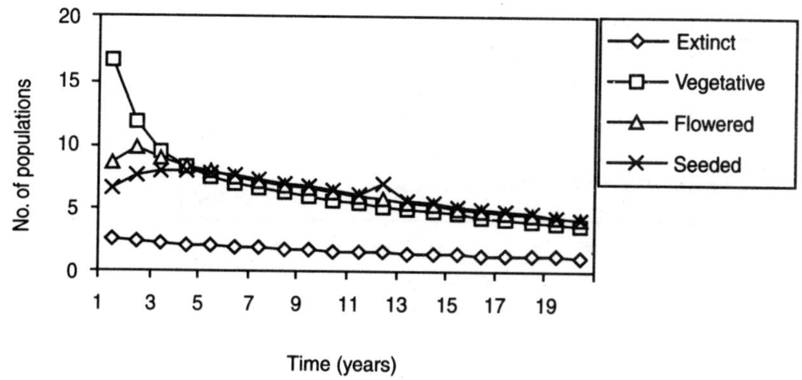

Fig. 20.5. Simulation of metapopulation dynamics considering the 33 populations that were active in 1995 as the initial metapopulation. Lines indicate the number of populations in each of the four stages considered here.

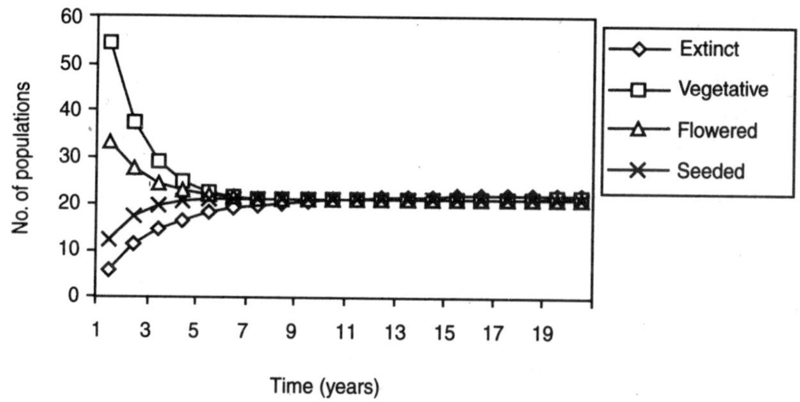

Fig. 20.6. Simulation of metapopulation dynamics under the assumption that all 103 initial populations were vegetative. Lines indicate the number of populations in each of the four stages considered here.

experience events of local extinction and recolonization. Overall, we found that at the metapopulation level, the growth rate is 0.990, indicating that the number of populations in the area is likely to remain fairly constant once a stable population size is established.

Genetic structure

Other authors have studied the ecological and genetic consequences of metapopulation dynamics of plants (Martinez-Ramos and Alvarez-Buylla, 1995; McCaulley *et al.*, 1995; Quintana-Ascencio and Menges, 1996; Giles and Goudet, 1997). Arens *et al.* (1998) studied the remaining black poplar trees on the bank of the Dutch branches of the Rhine river. They found that new small populations of young trees derived from seeds were the main form of recolonzation, while vegetative propagation is a very local strategy. However, they question whether the genetic diversity in these black poplar populations is sufficient for recolonization of the river bank and survival of the metapopulation. In

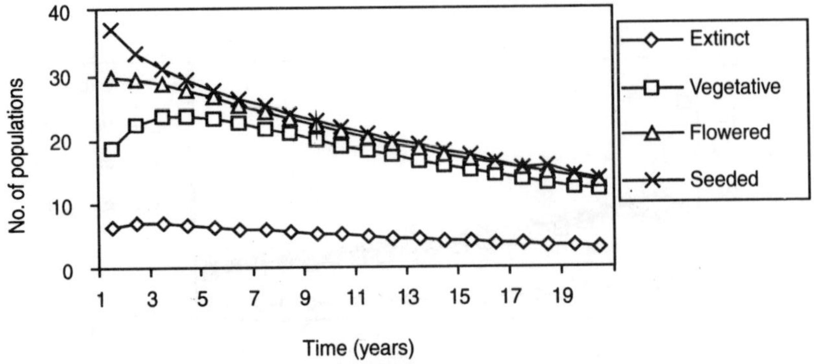

Fig. 20.7. Simulation of metapopulation dynamics considering the mean number of populations in each category as the initial metapopulation. Lines indicate the number of populations in each of the four stages considered here.

another study, Giles and Goudet (1997) studied the levels of population differentiation in *Silene dioica* metapopulations in a successional gradient in the Skeppsvik archipelago. They found that newly founded populations were more differentiated than those of intermediate age, which suggests that colonization dynamics increase genetic variance between populations. Other studies have shown that local founding events are important determinants of the genetic structure of the metapopulation (Giles and Goudet, 1997; Gillman *et al.*, 1997; Ouborg *et al.*, 1999). For example, McCaulley *et al.* (1995) showed that long-distance dispersal events are biologically very important for plants because they affect colonization probabilities, probabilities of population persistence in a fragmented habitat and metapopulation structure. Similarly, Giles and Goudet (1997) found that isolation by distance was the best model to explain colonization events of *S. dioica* in the Skeppsvik archipelago, which in turns helps to determine population differentiation.

Rocha *et al.* (1997) proposed that wild populations of Lima bean in the Central Valley of Costa Rica should be regarded as making up a metapopulation. They reported that between 29% and 45% of the populations disappeared during 3 years of monitoring. In addition, they also reported that new populations were found in the area, and proposed that these new populations represented the reappearance of previously extinct populations. Our data confirm that there is variation in the extinction rates between years. Similarly, our data also show that many extinct populations are re-established within the next year. These findings strongly suggest that the seed bank plays an important role in the dynamics of these populations, as suggested by Degreef (1998).

Here, we report that small populations are more likely to experience extinction. The genetic consequences of such extinction can be quite variable. First, small populations might also have small seed banks, and hence limited capacity for recolonization. Second, over-representation of one or a few individuals in the seed bank can result in changes in the genetic constitution of the new plants.

Soil seed bank

Others have studied the role of the soil seed bank in the dynamics of metapopulations (Husband and Barrett, 1996). It has been pointed out that seed dormancy needs to be incorporated into models considering the dynamics of plant metapopulations (Husband and Barrett, 1996). Washitani *et al.* (1997) studied the role of the ability of *Aster kantoensis* to form a persistent soil seed bank on the dynamic of metapopulations. They proposed that a persistent seed bank is of crucial importance for the survival of metapopulations composed of ephemeral populations. In contrast, they pointed out that in the absence of seed banks, dispersal to sporadically occurring safe-sites is the key to the long-term persistence of the metapopulation.

Degreef (1998) has shown that seeds of wild Lima bean can establish a persistent seed bank. Overall, seeds can remain dormant in the soil for about 3 years, but most seeds germinate within the first year. These findings indicate that the seed bank may play a major role in the metapopulation dynamics of wild Lima bean in the Central Valley of Costa Rica, suggesting that most recolonization events come from dormant seeds in the soil seed bank. However, a significant fraction of extinct populations is likely to remain extinct for longer than 1 year, suggesting that long distance seed dispersal must also be important in maintaining the structure of the metapopulation. However, our data show that there are populations that are very isolated from other populations, suggesting that the persistence of these populations may be in doubt in the long term.

Our data show that most of the populations that remain vegetative during one year are likely to fail to produce seeds the next year. Degreef (1998) showed that populations found in sites with a high degree of habitat degradation also show lower probabilities of seedling establishment. These findings suggest that the absence of seed production and low levels of seedling establishment would result in a loss of the soil seed bank, which, in turn, is a threat to the ability of local populations to buffer the negative effects of a severe disturbance. In contrast, our data also show that seed producing populations are also likely to maintain their seed production in subsequent years and, hence, are more likely to recover after a local disturbance that causes local extinction. In the latter case, only severe man-made disturbance would result in permanent loss of the population.

Other studies by our team have revealed that genetic variation among wild populations of Lima beans is highly structured (Maquet *et al.*, 1996; Zoro, 1999; Vargas *et al.*, 2000, 2001). These studies have shown that most of the variation is found between populations and that there is little variation within populations. The findings described support the notion that local extinction and recolonization may play an important role in the metapopulation structure, as proposed by Rocha *et al.* (1997). The impact of bottlenecks as severe as those reported here may be reinforced by the high rate of inbreeding that these populations will experience. Bottlenecks result from the loss of living plants, as well as non-random contribution of all individuals in the population to the soil seed bank. This is particularly true if only a small proportion of the individuals in the population contribute seeds to the soil bank, resulting in the loss of alleles.

In more severe situations, recolonization might result from the arrival of seeds from adjacent populations. This is particularly true in those populations that fail to produce seeds for many years. Degreef (1998) showed that seeds are likely to remain dormant in the soil for little over 3 years. For this reason there are likely to be many populations that lack soil seed banks. Under these circumstances, recolonization is likely to result from one or two seeds dispersed from a nearby population, and the new population will reflect the genetic variation of the new founders (see McCaulley *et al.*, 1995).

Overall, we found that the number of wild populations in the study area is declining. This is the result of habitat fragmentation and degradation. Habitat fragmentation may lead to isolation by distance, which may disrupt the ability of dispersal processes to maintain the structure of the metapopulation. Habitat degradation due to changes in land use and agricultural practices may also result in the breakdown of the metapopulation. If dispersal processes are limited, it is reasonable to propose that there are several metapopulation units in the study area, each including a group of populations that are linked by dispersal processes (pollen dispersal and seed dispersal). Therefore, it is important to study dispersability and dispersal processes to give a better background for the development of more practical strategies for conservation. In addition, more attention should be given to determine the impact of local extinction and the origin of the seeds that re-establish a given population (soil seed bank or newly dispersed seeds) on the genetic structure of the metapopulation. Indeed a metapopulation framework strictly requires that populations experience a finite rate of immigration of seeds from neighbouring populations, with or without the migration of pollen, to replace local extinction. Otherwise the ultimate fate of each component population is essentially independent of its neighbours.

References

Arens, P., Coops, H., Janzen, J. and Vosman, B. (1998) Molecular genetics analysis of black poplar (*Populus nigra* L.) along Dutch rivers. *Molecular Ecology* 7, 11–18.

Debouck, D. (1987) Phaseolus Germoplasm. International Board for Plant Genetic Resources, Rome, Italy. Mimeograph.

Degreef, J. (1998) Développement d'un modèle démographique et applications à la conservation *in situ* de populations sauvages de haricot de Lima (*Phaseolus lunatus* L.) dans la vallée centrale du Costa Rica. Thèse de doctorat. Faculté Universitaire des Sciences Agronomiques de Gembloux, Belgique.

Giles, B.E. and Goudet, J. (1997) Genetic differentiation in *Silene dioica* metapopulations: estimation of spatiotemporal effects in a successional plant species. *American Naturalist* 149, 507–526.

Gillman, M., Maxted, N., Ford-Lloyd, B.V. and Hawkes, J.G. (1996) In: Maxted, N., Ford-Lloyd, B.V. and Hawkes, J.G. (eds) *Plant Genetic Conservation: the* in situ *Approach*. Chapman and Hall, London, pp. 14–131.

Gliddon, C. and Goudet, J. (1994) The genetic structure of metapopulations and conservation biology. In: Loeschke, V., Tomiuk, J. and Jain, S.K. *Conservation Genetics*. Birkhauser Verlag, Germany, pp. 107–114.

Hamrick, J.L., Godt, M.J.W., Murawski, D.A. and Loveless, M.D. (1991) Correlation between species traits and allozyme diversity: implications for conservation biology. In: Falk, D.A. and Holsinger, K.E. (eds) *Genetics and Conservation of Rare Plants*. Oxford University Press, New York, pp. 75–86.

Hanski, I. and Gilpin, M. (1991) Metapopulation dynamics: brief history and conceptual domain. *Biological Journal of the Linnean Society* 42, 3–16.

Husband, B.C. and Barrett, S.C.H. (1996) A metapopulation perspective in plant population biology. *Journal of Ecology* 84, 461–469.

Legendre, S. and Clover, J. (1995) ULM, a software for conservation and evolutionary biologist. *Journal of Applied Statistics* 22, 817–834.

McCaulley, D.E., Raveill, J. and Antonovics, J. (1995) Local founding events as determinants of genetic structure in plant metapopulations. *Heredity* 75, 630–636.

Maquet, A., Zoro, I., Rocha, O. and Baudoin, J. (1996) Case studies on breeding systems and its consequences for germoplasm conservation; 1. Isozyme diversity in wild Lima beans populations in Central Costa Rica. *Genetic Resources Crop Evolution* 43, 309–318.

Martinez-Ramos, M. and Alvarez-Buylla, E.R. (1995) Seed dispersal and patch dynamics in tropical rain forest: a demographic approach. *Ecoscience* 2, 223–229.

Menges, E.S., Hawkes, C.V., Platt, W.J. and Peet, R.K. (1998) Interactive effects of fire and microhabitat on plants of Florida scrub. *Ecological Applications* 8, 935- 946.

Murillo, C., Macaya, G. and Rocha, O.J. In preparation. Genetic variation of wild populations of *Phaseolus lunatus* in the Central Valley of Costa Rica using AFLP markers. To be submitted to Conservation Biology.

Nason, J.D. and Hamrick, J.L. (1997) Reproductive and genetic consequences of forest fragmentation: two case studies of neotropical canopy trees. *Journal of Heredity* 88, 264–276.

Ouborg, N.J., Piquot, J., van Groenendael, J.M. and van Groenendael, J.M. (1999) Population genetics, molecular markers and the study of dispersal in plants. *Journal of Ecology* 87, 551–568.

Poschold, P. (1996) The metapopulation concept – a consideration from the plant ecological viewpoint. *Ziet. Okol. Natur.* 5, 161–185.

Quintana-Ascencio, P.F. and Menges, E.S. (1996) Inferring metapopulation dynamics from patch-level incidence of Florida scrub plants. *Conservation Biology* 10, 1210–1219.

Rocha, O., Macaya, G. and Baudoin, J.P. (1997) Causes of local extinction and recolonization, determined by 3 years of monitoring wild populations of *Phaseolus lunatus* L. in the Central Valley of Costa Rica. *Plant Genetic Resources Newsletter* 112, 44–48.

Standley, P. (1937) *Flora of Costa Rica*. Field Museum of Natural History, Chicago, USA.

Valverde, T. and Silverton, J. (1997) A metapopulation model for *Primula vulgare*, a temperate forest understorey herb. *Journal of Ecology* 85, 193–210.

Vargas E.M., Macaya, G., Baudoin, J.P. and Rocha, O.J. (2000) Variation in the content of phaseolin in thirty-seven wild populations of lima beans (*Phaseolus lunatus* L.) in the Central Valley of Costa Rica. *Plant Genetic Resources Newsletter* 121, 53–58.

Vargas, E.M., Macaya, G., Baudoin, J.P. and Rocha, O.J. (2001) Case studies on breeding systems and its consequences for germplasm conservation: 3. Electrophoretic mobility of phaseolins in wild populations of Lima beans (*Phaseolus lunatus* L.) in the Central Valley of Costa Rica. *Genetic Resources and Crop Evolution* 48, 109–120.

Washitani, I., Takenaka, A., Kuramoto, N. and Inoue, K. (1997) *Aster kantoensis* Kitam., an endangered flood plain endemic plant in Japan: its ability to form persistent seed banks. *Biological Conservation* 82, 67–72.
Wolf, A., Brodmann, P.A. and Harrison, S. (1999) Distribution of the rare serpentine sunflower, *Helianthus exilis* (Asteraceae): the role of habitat availability, dispersal limitation and species interactions. *Oikos* 84, 69–76.
Zoro, B.I. (1999) Variabilité génétique des populations sauvages de *Phaseolus lunatus* L. dans la vallée centrale du Costa Rica et ses implications dans la mise au point d'une strategie de conservation *in situ*. Thèse de doctorat. Faculté Universitaire des Sciences Agronomiques de Gembloux, Belgique.

21 Inventories for *in situ* Conservation of Broadleaved Forest Genetic Resources in South-eastern Europe

I. Blada,[1] A.H. Alexandrov,[2] G. Postolache[3], J. Turok[4] and N. Donita[1]

[1]*Forest Research and Management Institute, Bucharest, Romania;* [2]*Forest Research Institute, Sofia, Bulgaria;* [3]*Institute of Botany, Chisinău, Moldova;* [4]*International Plant Genetic Resources Institute (IPGRI), Rome, Italy*

Introduction

South-eastern Europe, in particular the Balkan Peninsula, is a region characterized by indigenous broadleaved forest resources that are valued for their quality and natural diversity. This diversity is a result of adaptation to the heterogeneous ecological conditions associated with the effects of several phytogeographic zones, and the mountainous nature of the region with its great variety of elevations, soils and exposures. Several forest tree species are widely occurring, economically important European broadleaves, such as European beech (*Fagus sylvatica* L.), sessile oak (*Quercus petraea* (Matt.) Liebl.) and pedunculate oak (*Quercus robur* L.), while other species of these genera are rare (*Fagus orientalis* Lipsky, *Quercus macrolepis* Kotschy, *Quercus pubescens* Willd., *Quercus trojana* Webb), even endemic to the region (*Quercus dalechampii* Ten.). Many tree species with scattered distribution patterns, for example, 'Noble Hardwoods' such as maple and ash (*Acer campestre* L., *Acer platanoides* L., *Acer pseudoplatanus* L.; *Fraxinus excelsior* L., *Fraxinus ornus* L. and *Fraxinus angustifolia* Vahl), are particularly valued for their ecological role in the species-mixed ecosystem and for the high-quality timber and other forest products that they provide.

Because of their ecological and economic importance, the broadleaved tree species in south-eastern Europe have received attention in conservation, tree breeding and afforestation programmes in Bulgaria (Dobrinov *et al.*, 1982; Anonymous, 1992; Garelkov *et al.*, 1995; Marinov *et al.*, 1995), Moldova (Tashkevici, 1984; Postolache, 1995) and Romania (Benea and Stanescu, 1981; Enescu *et al.*, 1997).

Owing to the impacts of industrial air pollution, repeated extreme summer drought and subsequent attack by pests and diseases, broadleaved forest genetic resources in the region have been described as threatened during recent years (Blada, 1998a,b; Postolache, 1998; Alexandrov *et al.*, 2000). Low fructification and the absence of, or sporadic occurrence of, natural regeneration are particular concerns in oak species and several Noble Hardwoods.

In situ conservation is generally applied as the principal approach for conserving the genetic diversity of long-lived, largely outbreeding and undomesticated forest tree species (Ziehe *et al.*, 1989; Eriksson *et al.*, 1993). The key characteristic of *in situ* conservation is its dynamic nature, allowing for continued evolution of gene resource populations. *Ex situ* conservation measures complement this approach. In order to design genetic conservation strategies, reliable information is needed about the levels, patterns and processes of genetic variation.

 Managing Plant Genetic Diversity (eds J.M.M. Engels, V. Ramanatha Rao, A.H.D. Brown and M.T. Jackson)

While the genetic information obtained from field trials and genetic marker studies is very limited, substantial research work has been done in the three countries in the past with regard to the ecogeographic distribution of resources, from which patterns of genetic variation can be predicted.

Since 1997 scientists from Bulgaria, Moldova and Romania have jointly undertaken to develop genetic conservation strategies for 13 selected broadleaved species of the genera beech (*Fagus*), oak (*Quercus*), maple (*Acer*) and ash (*Fraxinus*). The objective of this chapter is to present the outcomes to date of the joint project aimed at inventories of resources, which have led to the development of distribution maps and establishment of databases of gene conservation units after 2 years of work.

Methods

The work was based on existing information from various sources: data from the latest national forest management inventories, botanical and geobotanical literature, old maps and registers of selected seed sources. Using the existing data and additional information from the field observations, maps of distribution areas of the 13 species were constructed, digitized and produced in a standard geographical information system (GIS) (scale 1:600,000).

Gene conservation units including seed stands, forest reserves and other categories were described according to common passport data. This inventory was conducted in all three countries during the years 1997–1999. The majority of seed stands had been identified during previous inventories at national level (Enescu, 1982; Dobrinov *et al.*, 1982; Anonymous, 1992). The following criteria were used for selecting seed stands: age, indigenous origin, health state, adaptability, stem quality and productivity. The passport data for each gene conservation unit were compiled and stored in a common database; they include:

- number
- state forestry unit
- compartment
- sub-compartment
- region of provenance (for seed stands)
- latitude, longitude
- altitude (m)
- area (ha)
- species composition
- age
- height (m)
- diameter ($D_{1.3}$) at breast height (cm)
- growing stock (m^3)
- grade (site quality level).

Results

Natural distribution of species

In Romania and Bulgaria, European beech occurs from low hills to mountain regions, but occasionally it can be found in lowlands. In Moldova, the species' occurrence is limited to a very small hilly area in the central part of the country (Fig. 21.1). This area is called Codri and represents the eastern limit of the distribution range of the species in Europe (Borza, 1937; Soceava and Lipatova, 1952; Gheideman, 1969; Tashkevici, 1984; Postolache, 1995). Along with sessile oak, beech is a dominant species and it occupies very steep slopes at altitudes between 200 and 400 m.

The natural vertical span of beech ranges from 200 to 1800 m in Bulgaria's Balkan Mountains, although tree groups and single trees are found outside these limits. The lowermost occurrences were found at 150 m in Bozhuritsa (Vidin), and at 200–300 m in Ludogorie. The natural distribution range of beech in Romania covers a large part of the mountain territory of the Carpathians, between 600 and 1300 m in the north of the country and between 700 and 1450 m in the south, and in large parts of the inner Carpathian area between 300 and 700 m. Beech can reach the upper tree limit in the sub-alpine belt, at altitudes of 1550 m in the Parâng Mountains and 1600 m in the Vâlcan Mountains (Muica, 1995). The species may be found at low altitudes (50–100 m) in the Danube's Strait and in Oltenia. Within the present range there are a few large, compact forest areas: the eastern slopes of the Eastern Carpathians between Bistrita Valley and Uzului Valley or in the Banat Mountains. The remaining area is strongly fragmented, with individual forest stands of different size occurring with greater density in some areas.

Beech occurs mainly in natural pure, but also in species-mixed stands. Large pure stands mainly occupy northern slopes of the mountain belt with moist, nutrient-rich soils. In the upper part of the

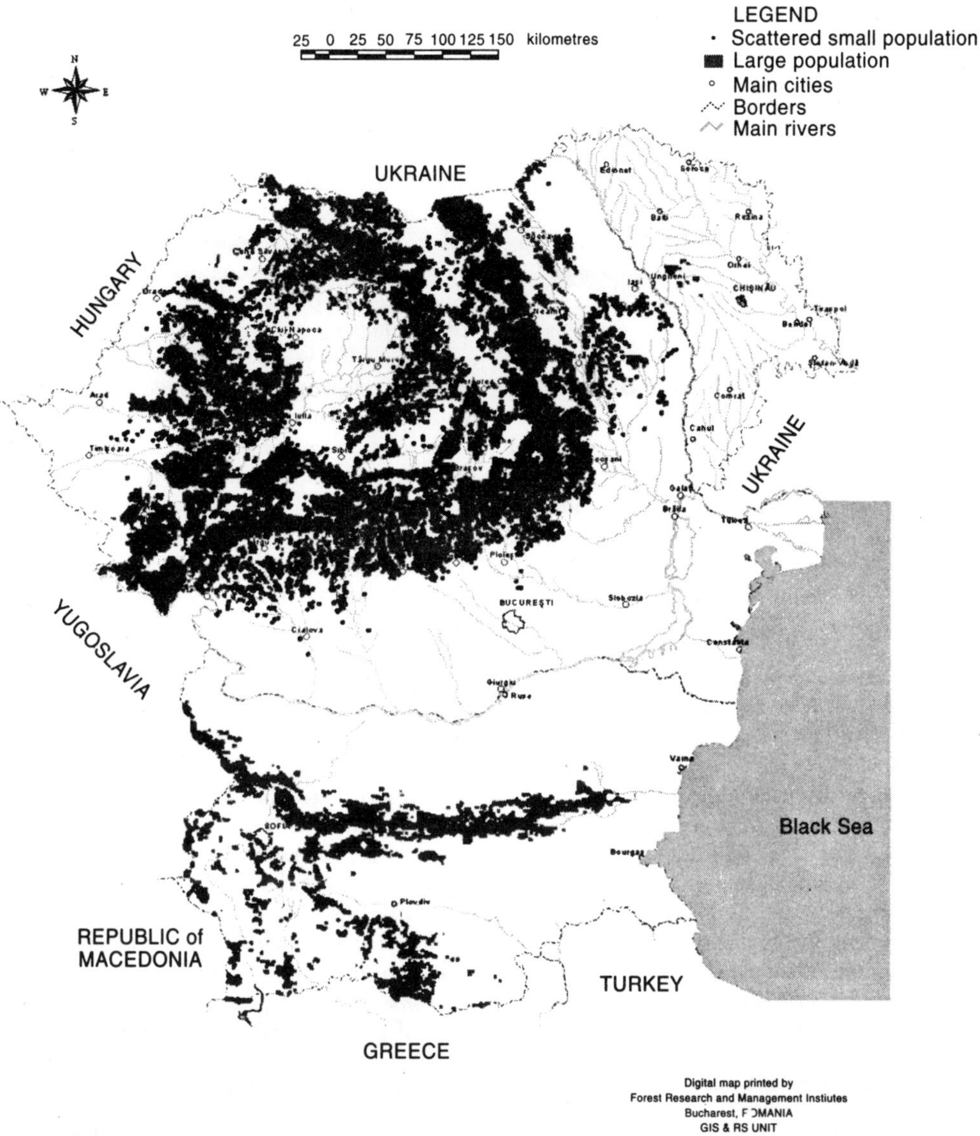

Fig. 21.1. Eastern limit of the distribution of European beech (*Fagus sylvatica*) in Bulgaria, Moldova and Romania, where it occurs in pure and species-mixed stands in hilly areas.

vertical range it is frequently associated with *Picea abies* (L.) Karst., *Abies alba* Mill., *Ulmus montana* With., *Betula pendula* Roth, *A. pseudoplatanus* L., *Populus tremula* L. and *Sorbus aucuparia* L. In the lower part, it is mixed with *Q. petraea* Liebl, *Carpinus betulus* L., *Tilia tomentosa* Moench. and *Prunus avium* L.

The distribution range of Oriental beech (*F. orientalis* Lipsky) includes the eastern part of the Balkan Peninsula. A contact zone between the natural ranges of European beech and Oriental beech runs through northern Greece and Bulgaria (Delkov, 1988). Isolated occurrences of *F. orientalis* outside its natural range were recorded in Dobrudja and central Bulgaria (Czeczott, 1932), in eastern Serbia (Glisic, 1973), in Macedonia, in

Banat and Moldova. The occurrence in Moldova was not confirmed in the present inventory. In Romania, Oriental beech is extremely rare and can be found as scattered trees in several places. As a Pontic species it is one of the principal species in the Strandja Mountains, Bulgaria, where it comprises pure and mixed formations situated in deep ravines and on moist, north-facing sites. Oriental beech grows in mixed stands mainly with *C. betulus* and, to a lesser extent, *A. pseudoplatanus* and *T. tomentosa*.

Inventory of seed stands

A summary of the results of the inventory of seed stands in two species of beech, five of oak and three of maple is presented in Table 21.1. During the inventory, a total of 4764 seed stands covering 71,778 ha were characterized in all three countries.

The seed stands represent mature, indigenous populations with superior phenotypic performance, selected in various site conditions throughout the distribution range in the three countries. Although the selection of seed stands followed the purpose of seed production for afforestation programmes and for tree breeding, they are considered as an adequate sample of the existing genetic variation for genetic conservation. Seed stands are widely recognized as a category that links the conservation of genetic resources and their use (e.g. Kleinschmit, 1995). Most seed stands were identified for sessile oak and European beech, the two principal, widely distributed and economically important broadleaved species (see Table 21.1).

A total of 1655 seed stands of sessile oak (27,155 ha) were described, among them 1010 in Romania. The vertical range was between 115 m (Mehedinti) to 1020 m in Harghita. The average age of individual seed stands varied between 50 and 170 years. In Bulgaria, seed stands of sessile oak were also designated in all vertical zones (100–1500 m) and the average age was 96 years.

Compared with sessile oak, the distribution of pedunculate oak (*Q. robur*) is fragmented and uneven in the three countries (Fig. 21.2). The occurrence of pedunculate oak was drastically reduced due to human activity in the lowlands during the past 100 years (Cuza, 1994; Marinov *et al.*, 1995). It usually grows in altitudes up to 400 m, but its upper limit in Romania's Arges, Brasov and Bihor counties ranges between 520 and 660 m. The species usually forms small stands or single trees; large populations are exceptional. This overall situation is reflected in the number, distribution and size of the selected seed stands (see Table 21.1). Although the average area of the stands (13.4 ha) is comparable with the other species, approximately one-fifth of the seed stands were very small, not exceeding 2 ha.

A relatively high proportion of seed stands of pedunculate oak was selected in Moldova, where this species plays an essential role, both economic and ecological (Borza, 1937; Postolache, 1995, 1998). In addition to seed stands, two different categories of gene conservation units were established. 'Optimal' forest genetic conservation stands were defined on the basis of ecological site conditions (nutrient rich soils, humidity). Conservation stands with high diversity were identified, particularly in the contact zone of pedunculate oak with sessile oak and pubescent oak. An overview of the three

Table 21.1. Overview of seed stands of broadleaved species in Bulgaria, Moldova and Romania.

	Bulgaria		Moldova		Romania		Total	
Species	No.	ha	No.	ha	No.	ha	No.	ha
Quercus petraea	522	7,101	123	5,315	1,010	14,739	1,655	27,155
Quercus robur	41	440	81	1,828	400	4,758	522	7,026
Quercus cerris	145	2,143	—	—	60	732	205	2,875
Quercus frainetto	348	4,433	—	—	90	1,007	438	5,440
Quercus pubescens	4	41	—	—	5	12	9	53
Fagus sylvatica	898	11,824	21	229	344	7,628	1,263	19,681
Fagus orientalis	171	1,863	—	—	—	—	171	1,863
Acer campestre	161	2,101	75	1,528	—	—	236	3,629
Acer platanoides	49	836	51	1,171	—	—	100	2,007
Acer pseudoplatanus	154	1,959	6	77	5	13	165	2,049
Total	2,493	32,741	357	10,148	1,914	28,889	4,764	71,778

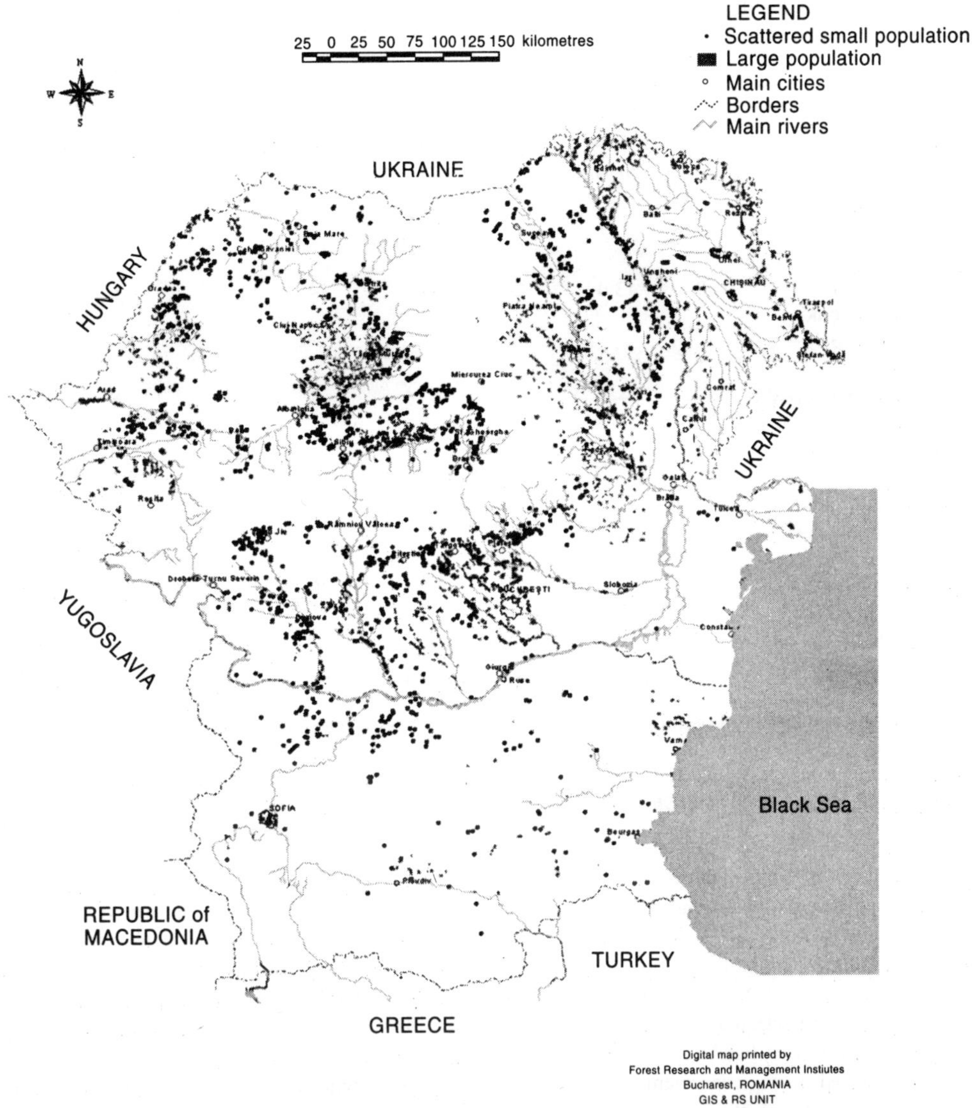

Fig. 21.2. The distribution of pedunculate oak (*Quercus robur*) in south-eastern Europe is uneven and fragmented, owing to intensive human activity in the lowlands.

categories is provided in Table 21.2. *Q. cerris* and *Q. frainetto* do not occur in Moldova.

From a conservation point of view, particular attention should be paid to pubescent oak. Owing to the dramatic reduction of its area in the past (Marinov *et al.*, 1995; Stanescu *et al.*, 1997), only nine suitable seed stands with total area of 53 ha could be selected in Bulgaria and Romania. The health status of the resources was also assessed during the inventory. Similar to *Q. cerris* and *Q. frainetto*, pubescent oak showed a strong decline in health over the past decades owing to repeated drought periods, pests and diseases and lack of fructification. Pedunculate oak and especially sessile oak have been less affected by the negative factors (see also Blada, 1998b; Alexandrov *et al.*, 2000).

European beech was represented by 1263 seed

Table 21.2. Gene conservation units of pedunculate oak (*Quercus robur*) in Moldova.

Forest type	'Optimal' conservation stands		Seed stands		High diversity stands		Total	
	No.	Area (ha)	No.	Area (ha)	No.	Area (ha)	No.	Area (ha)
Forests of pedunculate oak with cherry, North Moldova	5	75.7	18	460.1	9	277.6[a]	32	813.4
Sessile oak and pedunculate oak forests, Nistru Plateau	8	43.4	20	374.5	14	143.2[a]	42	561.1
Pedunculate oak forests/ black locust, left bank of Nistru river	—	—	2	48.5	—	—	2	48.5
Sessile oak, pedunculate oak and beech forests, central Moldova	15	270	35	817.4	12	307.8[a]	62	1395.2
Pubescent oak forests, southern Moldova	6	159.2	6	68.6	20	464.4[b]	32	692.2
Nature-protected areas	9	181.7	2	59.3	2	6.8	13	247.8
Total	43	730.0	81	1828.4	57	1199.8	181	3758.2

[a]Forest genetic resources of pedunculate oak (*Q. robur*) with sessile oak (*Q. petraea*) delimited in the contact zone.
[b]Forest genetic resources of pedunculate oak (*Q. robur*) with pubescent oak (*Q. pubescens*) including 96.5 ha in the contact zone of three oak species (*Q. robur*, *Q. petraea* and *Q. pubescens*).

stands covering 19,681 ha (Table 21.1). The highest number of stands was designated in Bulgaria (898, total area 11,824 ha). Their average size was 13.2 ha ranging from 2 ha to 54.7 ha. The average age was 103 years (from 40 to 210 years). The total growing stock in the selected seed stands amounted to 3,483,540 m^3 or 294.6 m^3 ha^{-1}. The stand quality index, assessed during the inventory, confirmed the superior quality of the selected stands. The number of seed stands is considered to be sufficient and representative of the diversity of beech in Bulgaria, not only for meeting the country's demand for reproductive material, but also for conservation purposes. However, the irregular seed crop requires the use of effective methods for beech mast conservation (*ex situ*) in the years of abundant seeding. In spite of the fact that the total area of natural beech forests in Romania is twice that of Bulgaria, the number of selected seed stands was only 344. This was due to the practice of almost exclusive natural regeneration of beech – thus very little demand for reproductive material – in Romania's forestry management. The average size of the seed stands was 22.2 ha; the size of 24 populations was larger than 200 ha. The average age of individual seed stands ranged from 55 to 175 years. Most seed stands in Romania and in Bulgaria were identified in the vertical range between 400 and 800 m, although emphasis was put on selecting seed stands representative for the entire range, that is, between 100 and 1800 m.

As beech in Moldova is situated at the eastern border of the distribution area, the genetic resources of this species have a special importance, attributed to their adaptation in marginal environmental conditions. Therefore, three different categories of gene conservation units were designated for beech, covering a total area of 421 ha (including 21 seed stands on 229 ha).

The seed stands of Oriental beech covered 1863 ha distributed within 171 units selected in Bulgaria. One of the problems with the management of these stands is that overmature stands currently prevail and these need to be regenerated. In some of them, where the undergrowth is formed by *Rhododendron ponticum* L., the regeneration process is hampered by the occurrence of a thick layer of undecomposed leaves.

A number of seed stands of the three maple (*Acer*) species were identified in Bulgaria (364 stands with 4896 ha) and Moldova (132; 2776 ha). In Bulgaria, the effective area covered by the maple species in these species-mixed seed stands varied greatly: for field maple (*A. campestre* L.) between 0.4 and 46 ha, Norway maple (*A. platanoides* L.) from 1 to 55 ha and sycamore (*A. pseudoplatanus* L.) from 0.1 to 41 ha. Twelve seed stands were identified to best represent the different site conditions and growth potential of the species in the country (Table 21.3). These populations could become part of a wider, European effort towards the conservation of the species' genetic resources in its entire distribution area (Rusanen, 1998). The distribution of sycamore is concentrated in western and north-western Bulgaria, from 150 to 1600 m altitude (Fig. 21.3). The location and size of the seed stands correspond to the geographic distribution of resources and seed stands can be found in the entire natural range in these two countries. The average age of the seed stands was 107 years. Although all three maple species have a wide distribution in Romania, only five stands of sycamore (13 ha) were identified. The work on selecting and describing seed stands of maple in Romania will be continued.

Inventory of other units for the conservation of genetic resources

Genetic conservation of forest tree populations takes place in both unmanaged and managed forests. The unmanaged category concerns different types of nature-protected areas, such as national parks, nature reserves or forest reserves. Besides their conventional functions including ecosystem and habitat preservation, protected areas contribute to maintaining and preserving the genetic diversity of forest tree species (Koski *et al.*, 1997). Seventeen national parks with a total area of around 1.3 Mha were previously established in Romania. Seventy-five nature-protected areas in Bulgaria cover 1.4 Mha. All of them include the forest communities with beech, oak and maple species. The main limitation of nature-protected areas with regard to genetic conservation is related to their distribution; they often aim at protecting very particular landscapes or environments rather than being representative of the existing forest types.

Forest reserves are strictly protected areas where the natural conditions allow population genetic processes to continue. A total of 318 forest reserves covering 75,477 ha with significant proportions of beech, oak and maple species were designated in Bulgaria and Moldova. These were characterized during the present inventory according to the common passport data and stored in the database of *in situ* gene conservation units. There are still large surfaces with virgin or quasi-virgin forest areas in south-eastern Europe (Cuza, 1994; Stoiculescu, 1994). Forest reserves often concentrate on these areas, which are characterized by exceptional composition, structural diversity and high stability. They also represent true *in situ* 'laboratories' for studying the structure, organization and functioning of the forest ecosystem, including genetic structures and processes.

Ex situ conservation units, including seed orchards and clone collections were established for the main economically important broadleaved species in Bulgaria and Romania in the past. Although these units were not included in the present inventory, information about them can be obtained from various sources (Dobrinov *et al.*, 1982; Postolache, 1995; Enescu *et al.*, 1997; Postolache, 1998; Blada, 1998a,b; Alexandrov *et al.*, 2000).

The inventory of resources and characterization of *in situ* gene conservation units for ash (*Fraxinus*) species will be completed in the year 2000.

Developing Integrated Genetic Conservation Strategies

The knowledge obtained through the present inventory provides a solid basis for developing and implementing integrated genetic conservation strategies by the national programmes in the countries involved. It helped to identify priorities for the conservation of diversity both within and between species, particularly for the rare and threatened broadleaved species, which have often been neglected in forest management and tree breeding surveys. The data related to risks and value of resources will need to be integrated into the decision-making framework, which can be used for genetic conservation management in

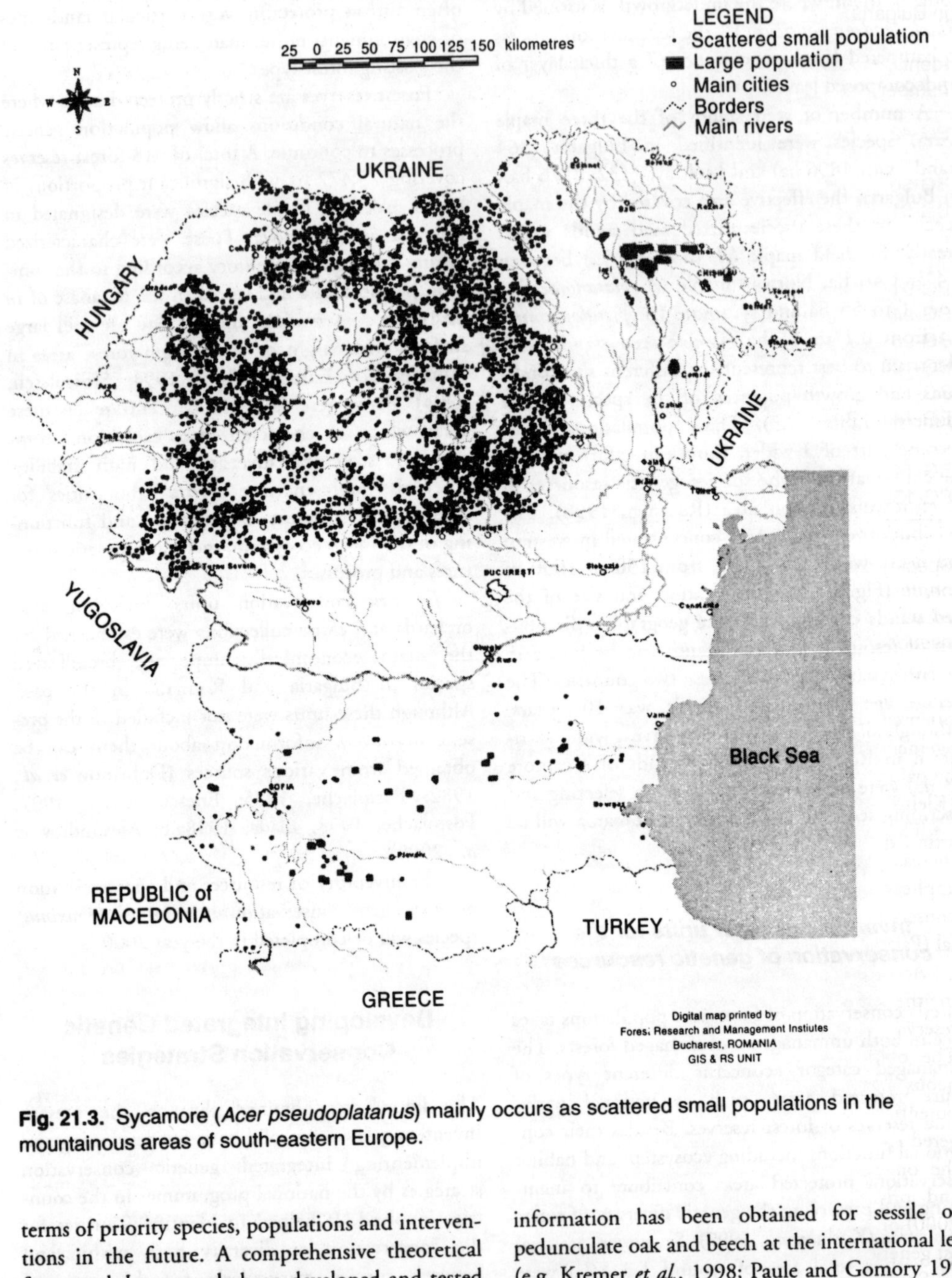

Fig. 21.3. Sycamore (*Acer pseudoplatanus*) mainly occurs as scattered small populations in the mountainous areas of south-eastern Europe.

terms of priority species, populations and interventions in the future. A comprehensive theoretical framework has recently been developed and tested in a field study in Brazil (Koshy *et al.*, Chapter 25, this volume).

The existing knowledge of the patterns of ecogeographic variation in the region needs to be combined with genetic information from field trials or genetic marker studies. While significant genetic information has been obtained for sessile oak, pedunculate oak and beech at the international level (e.g. Kremer *et al.*, 1998; Paule and Gomory 1998; Von Wuehlisch *et al.*, 1998), knowledge of the processes maintaining genetic variation in rare species is hardly available. In fact, a pilot study was recently started using microsatellite markers on common ash (*F. excelsior*) in Bulgaria (Hausman, 1999). The general trend in Europe towards an ecologically

Table 21.3. Seed stands identified as priority gene conservation units of sycamore (*A. pseudoplatanus*) in Bulgaria.

Ident. number	Forestry unit	Compartment	Altitude (m)	Composition (1–10)	Area (ha)	Age	V (m^3)	Site grade
BG-42	Borovets	303a	1200	Aba 6, Pia 1, Fas 2, **Aps 1**; Sac, Lad, Soa	2.7	100	43	2
BG-50	Borovets	578v	1300	**Aps 8**, Pia 1, Pis 1	0.5	40	70	2
BG-51	Borovets	578zh	1300	**Aps 10**	0.7	40	120	2
BG-71	Elena	271e	850	**Aps 4**, Fas 2, Apl 2, Fre 2; Pia	9.4	110	1310	1
BG-91	Kazanlak	192m	1000	Pis 6, **Aps 4**	5.3	35	410	1
BG-94	Koprivshtitsa	57zh	1100	**Aps 10**	0.3	35	60	1
BG-96	Kotel	31	700	Fas 4, **Aps 3**, Apl 2, Tip 1	3.0	140	330	2
BG-108	Novi Pazar	123v	250	Fre 8, Rop 1, **Aps 1**	1.5	35	30	2
BG-137	Sliven	1a	1000	Fas 8, **Aps 1**	5.5	140	320	2
BG-140	Staro Oryahovo	3371	200	Qur 9, **Aps 1**	3.5	40	130	1
BG-147	Varna	305k	150	Plo 4, **Aps 3**, Qur 2, Fro 1	0.8	70	90	1
BG-158	Vratsa	607d	400	Bep 8, Tit 1, **Aps 1**	5.3	20	40	1

V, growing stock.
Species composition: Aba, *Abies alba*; Apl, *Acer platanoides*; **Aps**, ***Acer pseudoplatanus***; Bep, *Betula pendula*; Fre, *Fraxinus excelsior*; Fro, *Fraxinus oxycarpa*; Fas, *Fagus sylvatica*; Lad, *Larix decidua*; Pia, *Picea abies*; Plo, *Platanus orientalis*; Pis, *Pinus sylvestris*; Qur, *Quercus robur*; Rop, *Robinia pseudoacacia*; Soa, *Sorbus aucuparia*; Sac, *Salix caprea*; Tip, *Tilia platyphyllos*; Tit, *Tilia tomentosa*.

oriented, uneven-aged, close-to-nature forestry management suggests that increased attention needs to be paid to Noble Hardwoods and other rare species (Kleinschmit, 1995; Rusanen, 1998).

The conservation of these species also requires increased efforts towards the development and application of complementary methods for *ex situ* conservation and propagation of the genetic material (Palada-Nicolau and Hausman, 1999).

According to the national laws and regulations in the three countries, seed stands and forest reserves are recognized as *in situ* conservation units. The objectives of their protection are heterogeneous and include genetic variation, timber yield potential, ecological services, and rare and endangered species. The consequences and challenges of the ongoing processes towards partial restitution and privatization of forest lands (e.g. Staddon, 2000) on the conservation and management of forest genetic resources will need to be evaluated.

Seed stands and forest reserves currently represent around 2–3% of the total area covered by the respective species. *In situ* conservation is, however, a dynamic process that needs to be integrated into regular forest management, which is applied in the remaining forest area. Common principles for silvicultural measures that contribute to the sustainable, close-to-nature and multifunctional management of forests for dynamic gene conservation need to be implemented and adapted to local needs (Rotach, 1999). This is usually feasible, with a few main principles to be taken into consideration:

- natural regeneration, which supports the biological and economic stability and continuity of forest stands, should be promoted where possible;
- for artificial regeneration, planting stock should originate from mating within the same population;
- silvicultural management should give due attention to preserving the adaptability of the principal as well as important associated species;
- due attention should also be given to evaluating and identifying new potential gene conservation units;
- reproductive material should be collected from seed stands and used according to national rules and legislation.

We conclude that south-eastern Europe is home to very important forest genetic resources in a wide range of broadleaved species. The practical collaboration among countries has shown a stimulating effect for the development of joint genetic conservation strategies, and their implementation by each country. The genetic resources in south-eastern Europe represent a valuable source of genes not only for the countries involved but also for the whole international community.

Acknowledgements

The activities carried out within this project were supported by the Forest Research and Management Institute, Bucharest, Romania, the Forest Research Institute, Sofia, Bulgaria and the Institute of Botany of the National Academy of Sciences in Chisinău, Moldova. The financial contribution provided by the Government of Luxembourg is acknowledged.

References

Alexandrov, A.H., Popov, E., Genov, K. and Hinkov, G. (2000) Social broadleaves genetic resources in Bulgaria. In: *EUFORGEN Social Broadleaves Network. Second meeting, 3–6 June 1999, Birmensdorf, Switzerland.* International Plant Genetic Resources Institute, Rome, Italy, pp. 41–52.

Anonymous (1992) *Guidelines for the Establishment, Management and Use of the Forest Seed Production Base.* Forestry Committee, Sofia (in Bulgarian).

Benea, V. and Stanescu, V. (1981) Resurse genetice vegetale ale padurilor [Plant genetic resources in forests]. In: Padurile Romaniei, Bucuresti (in Romanian), pp. 125–141.

Blada, I. (1998a) Beech and oak genetic resources in Romania. In: *EUFORGEN Social Broadleaves Network. First meeting, 23–25 October 1997, Bordeaux, France.* International Plant Genetic Resources Institute, Rome, Italy, pp. 5–10.

Blada, I. (1998b) Conservation of forest genetic resources in Romania with special reference to Noble Hardwoods. In: *EUFORGEN Noble Hardwoods Network. Second meeting, 22–25 March 1997, Lourizan, Spain.* International Plant Genetic Resources Institute, Rome, Italy, pp. 6–16.

Borza, A. (1937) *Cercetări fitosociologice asupra pădurilor Basarabiei.* [Phytosociological research in the forests of Bessarabia]. Cluj, Romania, (in Romanian).

Cuza, P. (1994) *Structura populationala a stejarului pedunculat* (Quercus robur *L.*) *din Republica Moldova* [Population structure of pedunculate oak in the Republic of Moldova]. Chisinău, Moldova (in Romanian).

Czeczott, H. (1932) Distribution of *Fagus orientalis* Lipsky. In: *Die Buchenwälder Europas* (Rübel, E. (ed.). Verlag Hans Huber, Bern, pp. 362–387.

Delkov, N. (1988). *Dendrology.* Agricultural Publishing House, Sofia.

Dobrinov, I., Doikov, G. and Gagov, V. (1982) *Forest Genetic Fund in Bulgaria.* Zemizdat, Sofia (in Bulgarian).

Enescu, V. (1982) *Producerea semintelor forestiere genetic ameliorate* [Production of improved forest seed materials]. Ceres, Bucuresti (in Romanian).

Enescu, V., Chereches, D. and Bandiu, C. (1997) *Conservarea Biodiversitatii si a Resurselor Genetice Forestiere* [Conservation of biodiversity and forest genetic resources]. S.C. Agris, Bucuresti (in Romanian).

Eriksson, G., Namkoong, G. and Roberds, J.H. (1993) Dynamic gene conservation for uncertain futures. *Forest Ecology and Management* 62, 15–37.

Garelkov, D., Stiptsov, V., Kalinkov, V., Turlakov, P., Bojinov, Ch., Bouzov, B., Nedelin, G. and Bobev, R. (1995) *The Beech Forests in Bulgaria.* Zemizdat, Sofia (in Bulgarian).

Gheideman, T.S. (1969) *Bukovaja dubrava v Moldavii* [Beech-oak forest in Moldova]. Chisinău, Moldova (in Russian).

Glisic, M.V. (1973) Prilog poznavanju varilijabiteta balkanske bukve (*Fagus moesiaca* Domin/ Maly/ Czeczott). [A contribution to the knowledge of variability in Balkan beech]. *Institut za šumarstvo i drvnu industriju, Zbornik radova (Beograd)* 12, 5–25 (in Serbian).

Hausman, J.F. (1999) Genetic conservation of broadleaved tree species. *IPGRI Newsletter for Europe* 16, 3.

Kleinschmit, J. (1995) Practical implications of the forest genetic resources conservation program in Germany. *Silvae Genetica* 44 (5–6), 269–274.

Koski, V., Skrøppa, T., Paule, L., Wolf, H. and Turok, J. (1997) *Technical Guidelines for Genetic Conservation of Norway Spruce* (Picea abies (L.) *Karst*). International Plant Genetic Resources Institute, Rome, Italy.

Kremer, A., Petit, R.J. and Ducousso, A. (1998) Structure of gene diversity, geneflow and gene conservation in *Quercus petraea.* In: *EUFORGEN Social Broadleaves Network. First meeting, 23–25 October 1997, Bordeaux, France.* International Plant Genetic Resources Institute, Rome, Italy, pp. 133–144.

Marinov, M., Kostadinov, K., Popov, G., Stiptsov, V., Bojinov, Ch., Denev, D. and Horozov, S. (1995) *The Oak Forests in Bulgaria.* Zemizdat, Sofia (in Bulgarian).

Muica, C. (1995) *Muntii Valcanului: Structura si evolutia peisajului* [Muntii Valcanului: Structure and landscape evolution]. Editura Academiei, Bucuresti (in Romanian).

Palada-Nicolau, M. and Hausman, J.F. (1999) Oak somatic embryogenesis: carbohydrate accumulation during somatic versus zygotic embryo development. *Abstracts of the International Congress on Application of Biotechnology to Forest Genetics, Vitoria-Gasteiz, Spain, September 1999.*

Paule, L. and Gomory, D. (1998) Genetic diversity of beech populations in Europe. In: *EUFORGEN Social Broadleaves Network. First meeting, 23–25 October 1997, Bordeaux, France.* International Plant Genetic Resources Institute, Rome, Italy, pp. 152–163.

Postolache, Gh. (1995) *Vegetatia Republicii Moldova* [Vegetation of the Republic of Moldova]. Stiinta, Chisinău, Moldova (in Romanian).

Postolache, Gh. (1998) Present status of the conservation and use of broadleaved forest genetic resources in Moldova. In: *EUFORGEN Social Broadleaves Network. First meeting, 23–25 October 1997, Bordeaux, France.* International Plant Genetic Resources Institute, Rome, Italy, pp. 11–12.

Rotach, P. (1999) *In situ* conservation and promotion of Noble Hardwoods: silvicultural management strategies. In: *EUFORGEN Noble Hardwoods Network. Third meeting, 13–16 June 1998, Sagadi, Estonia.* International Plant Genetic Resources Institute, Rome, Italy, pp. 39–50.

Rusanen, M. (1998) European gene conservation strategy for Noble Hardwoods: Norway maple and sycamore. In: *EUFORGEN Noble Hardwoods Network. Second meeting, 22–25 March 1997, Lourizan, Spain.* International Plant Genetic Resources Institute, Rome, Italy, pp. 40–43.

Soceava, V. and Lipatova, V. (1952) Rasprostranenie buka v lesach Moldavii [Beech distribution in the forests of Moldova]. In: *Trudy Botaniceskogo Instituta im. V.L. Komarova*, ser. III, vyp. 8. Moskva – Leningrad (in Russian), pp. 259–288.

Staddon, C. (2000) Restitution of forest property in post-communist Bulgaria. *Natural Resources Forum* 24, 237–246.

Stanescu, V., Sofletea, N. and Popescu, O. (1997) *Flora forestiera lemnoasa a Romaniei* [Forest tree flora in Romania]. Editura Ceres, Brasov, Romania (in Romanian).

Stoiculescu, C.D. (1994) *Problema ariilor forestiere protejate din Romania* [Protected forest areas in Romania]. Prosit No. 2, Timisoara, Romania (in Romanian).

Tashkevici, G.L. (1984) *Ochrana i vosstanovlenie bukovych lesov* [Protection and regeneration of beech forests]. Stiinta, Chisinău, Moldova (in Russian).

Von Wuehlisch, G., Liesebach, M., Muhs, H.-J. and Stephan, R. (1998) A network of international beech provenance trials. In: *EUFORGEN Social Broadleaves Network. First meeting, 23–25 October 1997, Bordeaux, France.* International Plant Genetic Resources Institute, Rome, Italy, pp. 164–172.

Ziehe, M., Gregorius, H.-R., Glock, H., Hattemer, H.H. and Herzog, S. (1989) Gene resources and gene conservation in forest trees: general concepts. In: Scholz, F., Gregorius, H.-R. and Rudin, D. (eds) *Genetic Effects of Air Pollutants in Forest Tree Populations.* Springer Verlag, Berlin, pp. 173–185.

22 Forest Genebanks: a New Approach to Conserving Forest Tree Genetic Resources

R. Uma Shaanker,[1,3] K.N. Ganeshaiah,[2,3] M. Nageswara Rao,[1,4] and G. Ravikanth[1,4]

[1]*Departments of Crop Physiology and [2]Genetics and Plant Breeding, University of Agricultural Sciences, Bangalore, India; [3]Jawaharlal Nehru Centre for Advanced Scientific Research, Jakkur, Bangalore, India; [4]Ashoka Trust for Research in Ecology and the Environment, Hebbal, Bangalore, India*

Introduction

Several measures that aim to conserve the genetic resources of crop plants and their wild relatives have been implemented worldwide (Wilkes, 1995). In addition to seed genebanks, these include *ex situ* gardens, field genebanks and *in situ* methods of conservation. Several programmes to conserve the genetic resources of crop plants *in situ* have also been initiated wherein the resources are conserved in their native habitats where these resources originated. In addition, some 12,000 to 15,000 threatened plant species are being cultivated in *ex situ* gardens such as in botanical gardens, genebanks and arboreta (Scarascia-Mugnozza and Perrino, 2000).

While these strategies are meeting the needs of crop species, there is concern that they might be less suited for the conservation of forest tree species. This might be particularly the case for forest trees that are either endangered or threatened and for species with recalcitrant seeds. *Ex situ* gardens cannot accommodate large populations of tree species and consequently their genetic base is narrow. Further, they should be deliberately managed to minimize drift and selection for adaptation to the garden environment (Frankel *et al.*, 1995). *In situ* conservation can complement such *ex situ* approaches and allow for adaptive evolution. Ideally *in situ* programmes require information on where to conserve coupled with long-term support and commitment, and this has limited the adoption of this method (Uma Shaanker and Ganeshaiah, 1997). Obtaining spatially explicit information on the distribution of intraspecific genetic variability and the identification of 'hotspots' of genetic variability is very tedious and for tropical species has only recently been attempted (Uma Shaanker and Ganeshaiah, 1997). In some species, only a few sites can be set aside for conservation, as contending pressure on land use restricts the proportion of the total genetic variability in a given species that can be set aside. Genetic diversity of populations may erode over time in *in situ* sites, especially if the populations are small and fragmented or disjunct. A decrease in genetic diversity parameters from adult to progeny population has been reported in several forest tree species whose populations are sparse or isolated with little or no gene flow among the populations or patches (Doligez and Joly, 1997).

We suggest an additional approach, referred to as forest genebanks, for the conservation of forest genetic resources. The forest genebank combines benefits from several conventional conservation protocols and is dynamic and evolutionary. We discuss the various steps involved in the establishment of the forest genebanks with particular reference to the conservation of the genetic resources of *Phyllanthus emblica*, an important medicinal plant species of India.

Forest Genebank: an Alternate Strategy for Conservation of Forest Genetic Resources

The forest genebank is an *in situ* site that serves as a repository of genes from as many diverse populations of a species as possible to represent the widest possible spectrum of genetic variability. The bank functions as a 'sink' into which genepools from various 'source' sites are introduced and maintained (Uma Shaanker and Ganeshaiah, 1997). In this way, the forest genebank is a modification of the existing protocols of *in situ* conservation with provision for substantial gene flow into it. In other words, the forest genebank could be viewed as facilitating 'gene corridors' among populations to convey to a central 'sink' the elements of each of the individual populations.

By virtue of such gene introductions, the forest genebanks facilitate the maintenance of the full allelic set of the species. Further, because of a continuous interaction between and among the different allelic sets maintained at the banks, genetic diversity would evolve in response to local selection pressures. Thus the forest genebank incorporates aspects of both *in situ* conservation and field genebanks. Forest genebanks are of particular relevance in the conservation of genetic resources of long-lived forest tree species. They could potentially be very useful in incorporating the genes or genepools from threatened, endangered and fragmented populations. In fact, in such populations, forest genebanks could help avert the loss of diversity through drift, population extinction and other processes.

Establishment of Forest Genebanks: Processes

The salient steps involved in the establishment of the forest genebanks are:

1. *Mapping the geographic distribution of the species.* A primary requirement in the establishment of forest genebanks is to identify the spatial distribution of the species and then to identify where populations of the species are reasonably large. Detailed information on the sites with reference to their threats and protection status would further help to prioritize them for potential conservation.
2. *Mapping the genetic diversity of populations.* From among the sites identified from (1), the genetic diversity of the populations needs to be determined and the hot-spots of genetic diversity identified. Further, based on the genetic diversity parameters, assessment needs to be made on the genetic differentiation of the populations to identify those that might constitute candidates for the conservation of genetic resources.
3. *Identification of donor and recipient populations.* Sites that can serve as donor (from where the genepool can be imported) and recipient (a genetic hot-spot and a 'sink' to receive the genepool from outside) have to be identified. The choice of the donor and recipient can be guided among several criteria by the extent of genetic variation in the population, the presence of rare alleles, population size, local heterogeneity, threat to the population and so on. Ideally, a recipient could be that which is allele-rich and has a broad genetic base with long-term demographic and genetic stability. For example, populations in protected areas such as wildlife sanctuaries and national parks could be important candidates for the recipient population. On the other hand, the donor populations could be among those which: (i) have certain rare or unique alleles; (ii) have distinct genetic configurations; and (iii) are threatened or endangered and are likely to be lost if not rescued.
4. *Process of gene introduction from 'source' to 'sink' population.* The actual process of genetic donation can be effected through the introduction of either seeds or pollen grains from the donors to the recipient populations. The interval at which such gene donations can occur would depend on the longevity of the species and its breeding system.
5. *Monitoring of genetic diversity in the forest genebanks.* An effective forest genebank should serve as a repository of the widest possible array of genetic variation. Periodic injection of the genetic elements from previously identified populations could assist this purpose. However, it might be necessary to monitor the levels of genetic diversity from time to time to test for any significant erosion of diversity.

Forest Genebank for *Phyllanthus emblica*: a Case Study

P. emblica is an important native, medicinal, forest plant in India. Uma Shaanker and Ganeshaiah (1997) proposed the establishment of a forest genebank for the conservation of this non-timber species. The fruit is a rich source of vitamin C and forms an integral component of indigenous (ayurvedic) medicine systems. The fruits are extracted from natural populations and are reported to contribute substantially to the household economy of a number of forest fringe communities (Uma Shankar *et al.*, 1996). Lately, the resources have been seriously threatened with loss of the genetic diversity of the populations because of heavy demand for the fruits and the liberal trade of the products both within and outside the country. In fact, studies have demonstrated that human activities including harvesting of the fruits have had a detrimental effect on the fitness, regeneration and genetic diversity of the population of the species at two forest sites in south India (Wickneswari and Boyle, 2000; Padmini *et al.*, 2001).

Mapping the geographic and genetic diversity distribution of P. emblica *in south India*

As a part of the Medicinal Plant Conservation Program, 30 Medicinal Plant Conservation Areas (MPCAs) in three southern states of India have been established for the *in situ* conservation of medicinal plants by the Foundation for Revitalization of Local Health Traditions, Bangalore. Populations of *P. emblica* occur in these sites but only four of the seven sites where the species occurred (BRT Hills (Karnataka), Thenmalai, Kollihills and Petchiparai (in Tamil Nadu)) had sizeable populations. Using isozyme electrophoresis, we mapped the genetic diversity of the four populations to prioritize the sites and to identify the hot-spots for the conservation of genetic diversity of the species at the regional level.

Based on several genetic parameters including allele richness, genetic diversity and rare alleles, our study showed that the population at Petchiparai is the most diverse and rich, followed by those at BRT Hills, Kollihills and Thenmalai (Fig. 22.1). Furthermore, the Petchiparai population was found to be most representative of the populations. It had the highest mean frequency of alleles with a coefficient of variability of their occurrence nearly equal to that of the entire population. The squared Euclidean distance of the Petchiparai population with all other populations was the least. This is also reflected in the principal component analysis of the populations; while there appears to be clear genetic differentiation among the populations, the Petchiparai population members span almost the entire spectrum of genetic variability (Fig. 22.2).

What conservation strategy to adopt?

Conventionally, two strategies might be proposed for the conservation of the genetic resources of *Phyllanthus* among the populations examined. First, based on several of the genetic diversity parameters discussed, the Petchiparai population could be targeted for conservation. Second, and in the event that the first strategy is not possible, the BRT and Thenmalai populations could be conserved together to represent the entire array of variability. However, either of the strategies might not afford the maximum conservation requirements of the species. A certain proportion of the space between the principal component axes is bound to go unrepresented. This might be especially critical if the population(s) in question is faced with threats and with the possibility of losing the genetic variability contained in it.

Forest genebank for P. emblica

Uma Shaanker and Ganeshaiah (1997) proposed that under these circumstances, the existing protocol of *in situ* conservation could be modified and a forest genebank established. In their proposal, one of the sites, designated as the 'sink' site, could serve as a repository of genetic material from other 'source' sites. Such a bank would then contain the full set of alleles of the species and would provide a continuous turnover of genetic material within and among populations. They proposed that the forest genebank for *Phyllanthus* could be located at BRT Hills considering the fact that the population is protected (Fig. 22.3). The genepool from other populations, which incidentally are not protected, including that from Petchiparai, can be incorporated into the forest genebank by introducing either pollen grains or seed material. The extent of introduction could be made in proportion to the rich-

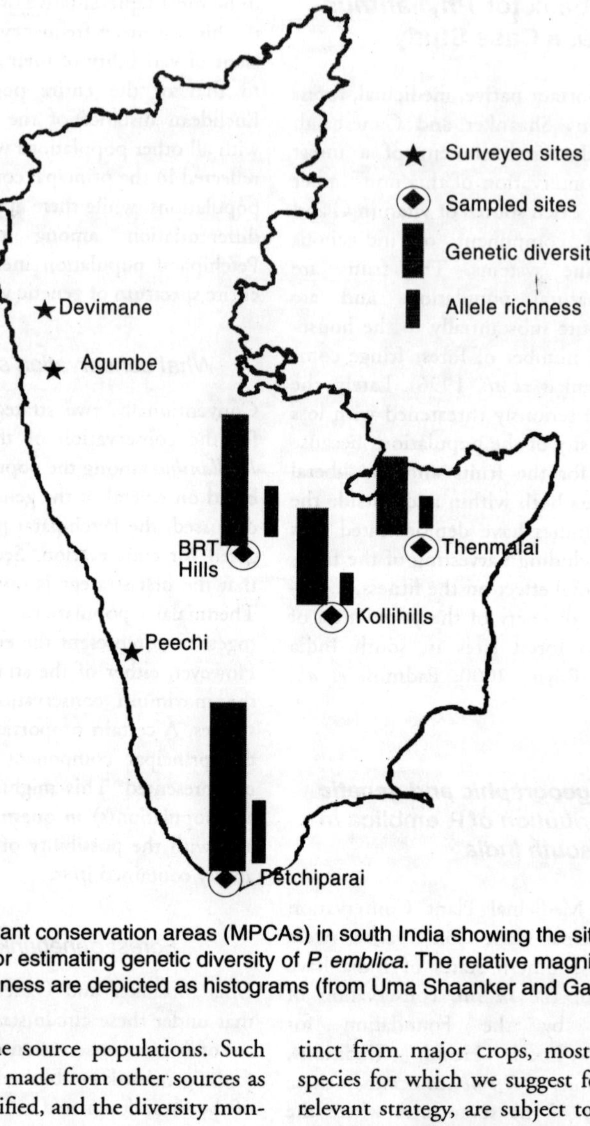

Fig. 22.1. Medicinal plant conservation areas (MPCAs) in south India showing the sites surveyed and the origin of samples for estimating genetic diversity of *P. emblica*. The relative magnitudes of the genetic diversity and allele richness are depicted as histograms (from Uma Shaanker and Ganeshaiah, 1997).

ness or diversity of the source populations. Such introductions could be made from other sources as well, if and when identified, and the diversity monitored periodically.

Caveats and Rejoinders

The bulking and growing of genetic diversity away from its site of origin, albeit as components added to a natural population, has elements in common with Simmonds' (1962) proposal to use mass reservoirs as a conservation method in crops. The proposal has been extensively debated ever since (see Frankel *et al.* (1995) for a critical review). As distinct from major crops, most populations of the species for which we suggest forest genebanks as a relevant strategy, are subject to heavy use and vulnerable to extinction. In such situations, focusing the conservation effort on a few forest genebanks could be the only alternative to the loss of major portions of the species variation.

Conservation biologists generally view critically the introduction of propagules from one population into another. This is because: (i) there is potential for outbreeding depression and the loss of alleles due to competition (Campbell and Waser, 1987); (ii) gene flow from one population to the other may prevent or disrupt local adaptive differentiation (Antonovics and Bradshaw, 1970;

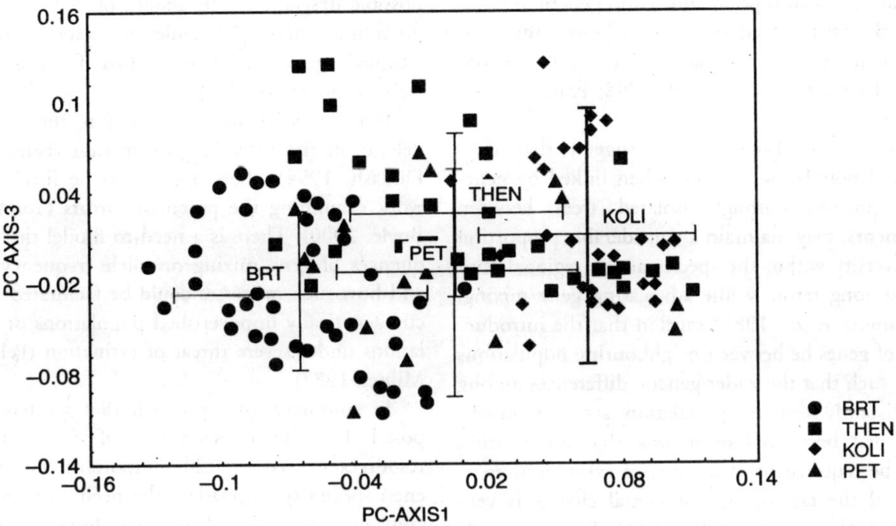

Fig. 22.2. Principal component analysis of the populations of *P. emblica*. The standard deviations from the mean of the two axes for each of the populations are depicted (BRT, BRT Hills; THEN, Thenmalai; KOLI, Kollihills; PET, Petchiparai). Figure modified from Uma Shaanker and Ganeshaiah (1997).

Simberloff, 1988; Adams and Burczyk, 2000); (iii) if populations are sufficiently well differentiated genetically, the species is better conserved by conserving as many small units of population there are than mixing the populations (Dole and Sun, 1992); and finally (iv) it is feared that introduction of genes will tend to cause an overall erosion of the genetic variability, most of which would otherwise have been distributed among populations.

On the other hand, there are a number of proponents of the gene mixing or introduction school. For example, several workers have argued in favour of artificial dispersal of seeds among populations to increase gene flow (Price and Waser, 1979; Waser, 1993). In fact the practice of transporting individuals from long distances for breeding purposes has been undertaken to minimize inbreeding and maximize genetic variability (Ralls *et al.*, 1988). Gene

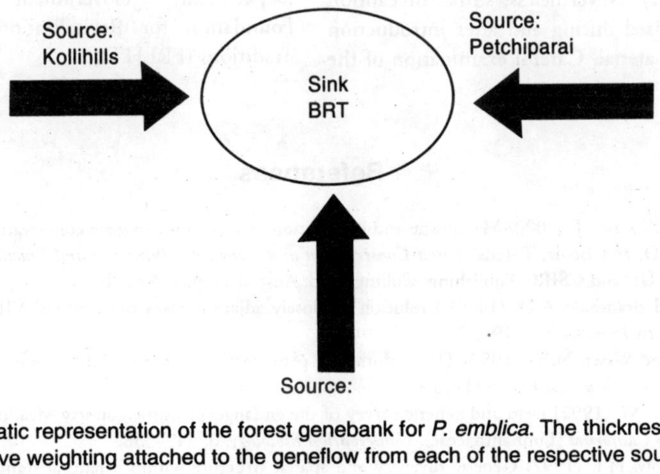

Fig. 22.3. Schematic representation of the forest genebank for *P. emblica*. The thickness of each arrow indicates the relative weighting attached to the geneflow from each of the respective source sites.

introductions into populations and genetic reserves have also been considered beneficial especially if the immigrant gene is well adapted to local environmental conditions (Ellstrand, 1992; Frankel *et al.*, 1995).

Young and Boyle (2000) suggest that fragmented populations, if and when linked by geneflow possibly through isolated trees between fragments, may maintain a considerable proportion of diversity within the species at the regional level in the long term. While advocating gene mixing, Vrijenhoek *et al.* (1985) caution that the introduction of genes be between neighbouring populations only, such that the wider genetic differences among the distantly separated populations are maintained.

It has been well recognized that intervention may be required to avoid species extinction, especially if the rate of environmental change is very rapid (Ledig and Kitzmiller, 1992; Eriksson *et al.*, 1993). Further, under circumstances of environmental disturbances when the genetic diversity of the population is threatened, genetic rescue operations using *ex situ* reserves and artificial migrations may be necessary. A classic example is offered by the conservation concerns over the highly endangered black rhinoceros (*Diceros bicornis*) and white rhinoceros (*Certotherium simum*), both of which have been decimated by very high levels of poaching (Merenlender *et al.*, 1989).

The apprehensions regarding gene mixing and the consequent lowering of fitness can be allayed by the fact that immigration of ill-adapted genes into a population is likely to be eliminated through the process of natural selection (Ledig, 1986; Millar and Libby, 1991). Nevertheless, sufficient caution should be exercised during and after introduction of the genetic material. Careful examination of the possible dysgenic effects should precede any introduction of foreign propagules into a local genepool, weighed against the risks of loss if the genepool were not to be incorporated.

It is ironic to note that despite the recurrent debate on the issue of gene mixing (Fenster and Dudash, 1994), there appears to be little critical work examining the potential threats (Young and Boyle, 2000). There is a need to model the consequences of gene mixing on allele frequency shifts and how such processes could be facilitated to rescue genetically impoverished populations or populations under severe threat of extinction (Pella and Milner, 1987).

In summary, the approach that we have proposed for the conservation of forest genetic resources of economically important and threatened species re-emphasizes the need to re-evaluate conventional approaches. The forest genebank approach is not so much aimed at preserving the genetic diversity of populations of trees in the genetic resource management units (Ledig, 1988; Millar and Libby, 1991); rather it is aimed at enriching the total genetic diversity contained at a designated *in situ* site with the possibility of evolving diversity.

Acknowledgements

This work was supported by grants from the International Plant Genetic Resources Institute (IPGRI), Center for International Forestry Research (CIFOR), Department of Forest, Ecology and the Environment and Karnataka Forest Department, Government of Karnataka, Foundation for Revitalization of Local Health Traditions (FRLHT).

References

Adams, W.T. and Burczyk, J. (2000) Magnitude and implications of gene flow in gene conservation reserves. In: Young, A., Boshier, D. and Boyle, T. (eds) *Forest Conservation and Genetics – Principles and Practice.* CAB International, Wallingford, UK and CSIRO Publishing, Collingwood, Australia, pp. 215–224.

Antonovics, J. and Bradshaw, A.D. (1970) Evolution in closely adjacent plant populations. VIII. Clinal patterns at a mine boundary. *Heredity* 25, 349–362.

Campbell, D.R. and Waser, N.M. (1987) The evolution of plant mating systems: multilocus simulations of pollen dispersal. *American Naturalist* 129, 593–609.

Dole, J.A. and Sun, M. (1992) Field and genetic survey of the endangered Buttle country Meadow-foam – *Limnanthes floccosa* subsp *California* (Limnanthaceae). *Conservation Biolology*, 6, 549–558.

Doligez, A. and Joly, H.I. (1997) Genetic diversity and spatial structure within a natural stand of a tropical species, *Carapa procera* (Meliaceae) in French Guyana. *Heredity* 79, 72–82.

Ellstrand, N.C. (1992). Gene flow through pollen: Implications for conservation. *Oikos* 63, 77–86.

Eriksson, G., Namkoong, G. and Roberds, J.H. (1993) Dynamic gene conservation for uncertain futures. *Forest Ecology and Management* 62, 15–37.

Fenster C.B. and Dudash, M.R. (1994) Genetic considerations for plant population restoration and conservation. In: Bowles, M.L. and Whelan, C.J. (eds) *Restoration of Endangered Species: Conceptual Issues, Planning, and Implementation.* Cambridge University Press, Cambridge, UK, pp. 34–62.

Frankel, O.H., Brown, A.H.D. and Burdon, J.J. (1995) *The Conservation of Plant Biodiversity.* Cambridge University Press, Cambridge, UK.

Ledig, F.T. (1986) Conservation strategies for forest genetic resources. *Forest Ecology and Management* 14, 77–90.

Ledig, F.T. (1988) The conservation of diversity in forest trees: why and how should genes be conserved? *Bioscience* 38, 471–479.

Ledig, F.T. and Kitzmiller, J.H. (1992) Genetic strategies for reforestation in the face of global climate change. *Forest Ecology and Management* 50, 153–169.

Merenlender, A.M., Woodruff, D.S., Ryder, O.A., Kock, R. and Vahala, J. (1989) Allozyme variation and differentiation in African and Indian Rhinoceros. *Journal of Heredity* 80, 377–382.

Millar, C.I. and Libby, W.J. (1991) Strategies for conserving clinal, ecotypic and disjunct population diversity in widespread species. In: Falk, D.A. and Holsinger, K.E. (eds) *Genetics and Conservation of Rare Plants.* Oxford University Press, New York, pp. 149–170.

Padmini, S., Nageswara Rao, M., Ganeshaiah, K.N. and Uma Shaanker, R. (2001) Genetic diversity of *Phyllanthus emblica* in tropical forests of south India: Impact of anthropogenic pressures. *Journal of Tropical Forest Science* 13(2), 297–310.

Pella, J.L. and Milner, G.B. (1987) Use of genetic markers in stock composition analysis In: Ryman, N. and Utter, F. (eds) *Population Genetics and Fishery Management.* University of Washington Press, Seattle, pp. 247–276.

Price, M.V. and Waser, N.M. (1979) Pollen dispersal and optimal outcrossing in *Delphinium nelsoni. Nature* 277, 294–297.

Ralls, K., Ballou, J.D. and Templeton, A. (1988) Estimation of lethal equivalents and the cost of inbreeding in mammals. *Conservation Biology* 2, 185–193.

Scarascia-Mugnozza, G.T. and Perrino, P. (2000) The history of *ex situ* germplasm conservation and use of genetic resources. In: *International Conference on Science and Technology for Managing Plant Genetic Diversity in the 21st Century, Kuala Lumpur, Malaysia, 12–16 June 2000.*

Simberloff, D. (1988) The contribution of population and community biology to conservation science. *Annual Review of Ecology and Systematics* 19, 473–511.

Simmonds, N.W. (1962) Variability in crop plants, its use and conservation. *Biological Reviews of the Cambridge Philosophical Society* 37, 442–465.

Uma Shaanker, R. and Ganeshaiah, K.N. (1997) Mapping genetic diversity of *Phyllanthus emblica*: Forest gene banks as a new approach for *in situ* conservation of genetic resources. *Current Science* 73, 163–168.

Uma Shankar, Murali, K.S., Uma Shaanker, R., Ganeshaiah, K.N. and Bawa, K.S. (1996) Extraction of non-timber forest products in the forests of Biligiri Rangan Hills, India. 3 Productivity, extraction and prospects of sustainable harvest of Amla, *Phyllanthus emblica* (Euphorbiaceae). *Economic Botany* 50, 270–279.

Vrijenhoek, R.C., Douglus, M.E. and Meffe, G.M. (1985) Conservation genetics of endangered fish populations in Arizona. *Science* 229, 400–402.

Waser, N.M., (1993) Population structure, optimal out breeding and associative mating in angiosperms. In: Thornhill, N.W (ed.) *The Natural History of Inbreeding and Outbreeding.* The University of Chicago Press, Chicago, pp. 173–199.

Wickneswari, R. and Boyle, T.J. (2000) Effects of logging and other forms of harvesting on genetic diversity in humid tropical forests. In: Young, A., Boshier D. and Boyle, T. (eds) *Forest Conservation and Genetics – Principles and Practice.* CAB International, Wallingford, UK and CSIRO Publishing, Collingwood, Australia, pp. 115–122.

Wilkes, G. (1995) Gene banks. In: Nierenberg, W.A. (ed.) *Encyclopedia of Environmental Biology*, Vol. 2. Academic Press, San Diego, pp. 181–190.

Young, A.G and Boyle, T.J. (2000) Forest fragmentation. In: Young, A., Boshier, D. and Boyle T., (eds) *Forest Conservation and Genetics – Principles and Practice.* CAB International, Wallingford, UK and CSIRO Publishing, Collingwood, Australia, pp. 123–134.

23 Human Impacts on the *Coffea arabica* Genepool in Ethiopia and the Need for its *in situ* Conservation

Tadesse Woldermariam Gole,[1] M. Denich,[1] Demel Teketay[2] and P.L.G. Vlek[1]

[1]*Centre for Development Research (ZEF), University of Bonn, Bonn, Germany;*
[2]*Ethiopian Agricultural Research Organization, Addis Ababa, Ethiopia*

Introduction

Conservation and sustainable use of plant genetic resources have long been focal points of national and international agenda (FAO, 1974, 1989, 1998). There is a growing awareness of conservation measures by national and international organizations (FAO, 1989, 1998). There have been tremendous successes in developing methods for the conservation of genetic resources *ex situ* (Maxted *et al.*, 1997c; Dulloo *et al.*, 1998). On the other hand, despite its overall advantage and promotion by the international community, *in situ* conservation is still inadequate (Williams, 1997). This is because its implementation requires the appropriate socio-economic, policy and political conditions as well as scientific understanding of the natural environment and biological characteristics of the species. Such issues are often local and require localized solutions.

The beauty of *in situ* conservation lies in that it keeps the genetic structure of a population intact in a dynamic process, while allowing the evolutionary processes to continue as plants adapt to changes in environmental conditions (Eriksson *et al.*, 1993). This involves conservation at the ecosystem level, the highest and most complex level of biodiversity. Ecosystem conservation unites the abiotic and biotic worlds, including their processes and entities (Noss, 1996).

Coffea arabica is commercially one of the most important crops and, at the same time, one of the most neglected crops in the world, with regard to genetic conservation (Tewolde and Egziabher, 1990). For decades it has been and still is a very valuable commodity in international trade, second only to oil (Pendergrast, 1999). It is the most important foreign currency earner for more than 80 developing countries, and was responsible for the transfer of over US$13 billion from developed countries to developing countries in 1983 (Cannell, 1983), and US$18 billion in 1994 (Raina *et al.*, 1998). For Ethiopia, it is 4–5% of the gross domestic product (GDP), 20% of the government revenue, 60% of the total foreign exchange earnings and a livelihood for more than 25% of its population (Tafesse, 1996).

It is difficult to conserve coffee germplasm *ex situ* in genebanks, since its seeds do not stay viable for a long time. *In situ* conservation in the forest ecosystem housing its genetic resources is the best and most reliable option. But the forests housing much of the coffee genepools are being lost at an alarming rate. Hence, there is an urgent need to save the wild coffee genepool (IBPGR, 1980; Tewolde and Egziabher, 1990; Dulloo *et al.*, 1998). In this chapter, we discuss the need for the conservation of Arabica coffee in the forests of south-western Ethiopia. Beginning with the geographic and genetic origin of coffee, we go on to describe

 Managing Plant Genetic Diversity
(eds J.M.M. Engels, V. Ramanatha Rao, A.H.D. Brown and M.T. Jackson)

the diversity of the genepools, human impacts on the forest landscape, and the current conservation status of coffee. Finally, the strategies for *in situ* genetic reserves, the challenging constraints for conservation efforts, and some recommendations and actions are presented.

Origin and Distribution

All cultivated species of coffee have their origin in Africa. *C. arabica* is geographically isolated and genetically distinct from the rest of the species in genus *Coffea*. It is confined to two isolated mountain forests on the western and eastern sides of the Great Rift Valley (GRV) in southern Ethiopia, while the distribution of other coffee species overlaps elsewhere in the central and western parts of Africa (FAO, 1968; Mesfin and Lisanework, 1996). *C. arabica* is the only species occurring in Ethiopia.

In its place of origin, coffee has been used as a stimulant and a special kind of food since time immemorial. It was known to the rest of the world from Arabia, hence the specific name. The Arabs introduced coffee from Ethiopia to Yemen during the 13th century (Haarer, 1962), where the habit of drinking coffee was developed in the 15th century. This habit gradually spread to the rest of the world, leading to the increased interest of some countries to produce coffee as a commodity on a large scale. The Dutch first introduced the coffee plantation to Java in 1690 (Ferwerda, 1976). A coffee plant from Java was taken to Amsterdam, then to Paris and the rest of the Dutch and French colonies in the tropics, especially Latin America (Fig. 23.1). Today, Latin American countries, led by Brazil, are the major producers of Arabica coffee.

Genetically, *C. arabica* is the only tetraploid and self-fertile species of *Coffea*, with chromosome number $2n = 4x = 44$, while others are diploid ($2n = 2x = 22$) and self-infertile (Lashermes *et al.*, 1999, 2000). Recent molecular characterization of *C. arabica* indicated its possible origin as an allotetraploid (Raina *et al.*, 1998; Lashermes *et al.*, 1999, 2000). According to Raina *et al.* (1998), the diploid wild ancestors of *C. arabica* are *Coffea eugenioides* and *Coffea congensis*, while Lashermes *et al.* (1999) found them to be *C. eugenioides* and *Coffea canephora*. The differences in these results could be due to the inadequacy of the analysis techniques; hence further investigation is needed.

Diversity and Significance of the Ethiopian Coffee Genepool

Ethiopia possesses enormous genetic variability of Arabica coffee. Its genepool is largely found in the

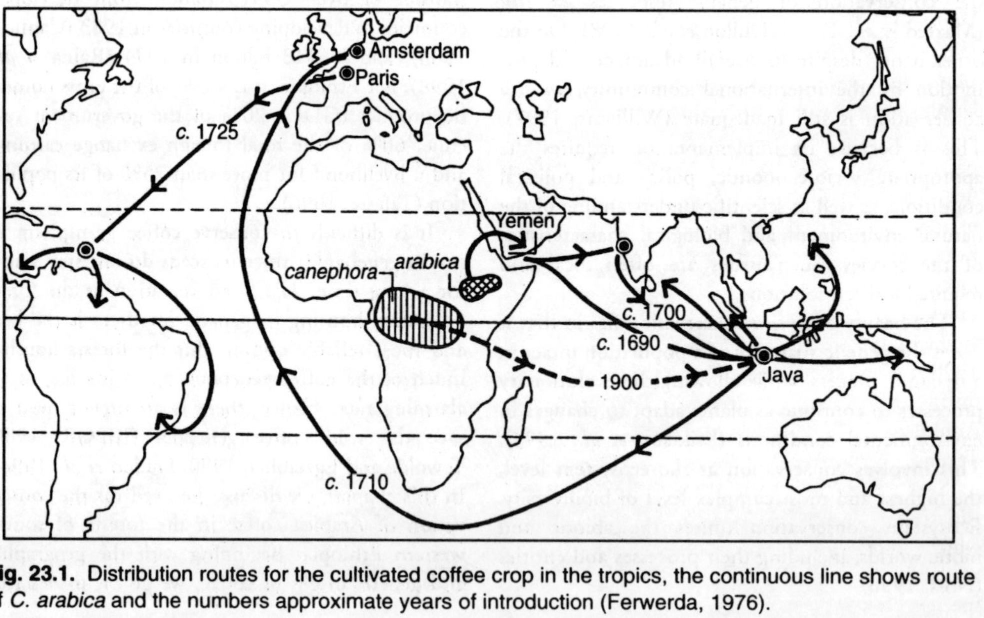

Fig. 23.1. Distribution routes for the cultivated coffee crop in the tropics, the continuous line shows route of *C. arabica* and the numbers approximate years of introduction (Ferwerda, 1976).

mountain rainforests of the south-west part of the country (Sylvian, 1958; FAO, 1968; Tewolde and Egziabher, 1990; Paulos and Demel, 1999), south-eastern forests (Mesfin and Lisanework, 1996) and as cultivated landraces in home gardens (Melaku, 1984; MCTD, 1989).

Recent studies on the genetic diversity of the species indicated east–west differentiation of the centres of diversification of *C. arabica* (Lashermes *et al.*, 1996; Montagnon and Bouharmont, 1996). Lashermes *et al.* (1996) analysed the genetic diversity among cultivated and sub-spontaneous accessions of *C. arabica* and detected genetic variation among the accessions from west of the GRV and east of the GRV in Ethiopia and the cultivated plants outside Ethiopia. The phenotypic diversity analysis of the accessions of coffee plants from similar origin by Montagnon and Bouharmont (1996) also confirmed the existence of similar variation. Accessions from the eastern part of the GRV are more similar to the coffee cultivars in other parts of the world, indicating this area to be a different centre of domestication from the western part and the source of the distribution of coffee to the outside world. On the other hand, accessions from the western part of the GRV are different genetically and morphologically from other populations and also have higher genetic variability within a population.

The majority of the coffee genepool east of the GRV is in home gardens and on farms as cultivated landraces, with some wild populations in the forests (Mesfin and Lisanework, 1996). The western portion consists mainly of semi-forest and forest populations that represent the partially domesticated and wild populations of the species. Surveys in different parts of the country found over 130 farmer-identified cultivated coffee landraces both east and west of the GRV (Admasu *et al.*, 1989). Through the Ethiopian National Coffee Collection Programme, more than 600 coffee types were collected and documented from 1966 to 1984 (IAR, 1986). Further studies may well discover more new coffee types. In 1998 a new coffee type, which is morphologically distinct from previous records, was found in the Anfilo forest of West Wollega Zone (Demel Teketay, personal observation).

Sylvian (1958) witnessed the existence of a great variation among the wild coffee plants in Ethiopia, which constitute the best sources of germplasm available for programmes of improvement of the species. The fact that the spread of Arabica coffee around the world was based on only a few trees suggests the importance of the Ethiopian germplasm for the world coffee industry as a whole (Wrigley, 1988). There may be great danger for the coffee industry and the people depending on it, if the production of coffee fails due to unforeseen natural calamities such as the outbreak of disease or pests. For instance, the occurrence of coffee leaf rust in Sri Lanka in 1869 forced that country to abandon coffee production and shift to tea (Demel, 1999). In Ethiopia, economic coffee production is still possible though leaf rust is endemic to the country. Even in the case of an outbreak of a new disease called coffee berry disease (CBD) in 1971, coffee production in Ethiopia was not significantly affected (Tewolde and Egziabher, 1990). It took a very short period of time to release CBD resistant varieties in Ethiopia as compared with other East African countries with similar problems. This is attributed to the availability of genetic diversity that is large enough to withstand diseases and pests (Demel, 1999).

The Forest Coffee Ecosystem

The forest coffee ecosystem (FCE) is part of the forest ecosystem in which Arabica coffee occurs as a natural member of the plant community. The FCE is the major source of coffee produced in Ethiopia. There are four major coffee production systems in Ethiopia, namely: forest, semi-forest, garden and plantations. The first two production systems are part of the FCE. In the forest coffee production system, coffee is harvested directly from wild coffee plants growing in Afromontane rainforests of west and south-western Ethiopia by subsistence farmers. Currently, the forest coffee production system represents 9% of the total land covered by coffee, and contributes about 5–6% of the total coffee production in the country. Semi-forest coffee represents the production system in which forest coffee is manipulated through thinning of overstorey trees, removal of ground vegetation and enrichment of empty spaces in the forests by transplanting naturally regenerated or raised seedlings. In some instances, naturally regenerated or raised coffee seedlings are also planted under naturally growing or planted trees. This system represents about 24% of the total land covered by coffee, and contributes about 20% of the total coffee production in the country (Paulos and Demel, 1999). In total, the FCE accounts for 33% of the land covered by coffee and 25% of the coffee produced in Ethiopia. Local farmers use traditional practices to produce the coffee in the FCE, which serves as the means of livelihood for millions of people.

Human Impact on *C. arabica* Genepool

According to climatic climax vegetation cover of Ethiopia, the whole plateau of the south-western part could have been covered by forest vegetation (Anonymous, 1988). Even in the recent past, the area was considered to be remote and inaccessible, and botanists and explorers did not visit the forest until the mid-1930s. In the late 1940s and 1950s, coffee production enterprises began to flourish in the area (Sylvian, 1958). This has resulted in an increase in the human population due to increased immigration from other parts of the country.

Reusing (1998) assessed the changes in forest vegetation of Ethiopia, especially of the south-west, using aerial photos of the 1970s and 1996/97, and satellite images of the 1990s. He found that between 1971 and 1975 approximately 40% of the highland plateau of south-west Ethiopia was covered by closed high forest. The closed high forest declined to only *c.* 18% by 1997 (Table 23.1), which is a loss of *c.* 60% in less than 30 years. Deforestation due to conversion to other kinds of land use is a serious threat. Around 235,400 ha of closed and slightly disturbed forest were deforested between 1971 and 1997, a loss of 10,000 ha of forest every year.

The pattern of deforestation is always centred on human settlement areas and progresses into the forest interior. Reusing (1998) indicated that the areas severely damaged by deforestation are Mizan Teferi, Tepi, Gore and Bonga. Although the exact range of distribution and area covered by FCE was not yet mapped and quantified, most human settlement areas coincide with such areas, since the main source of income for the local community of the region is coffee production from forest and semi-forest areas. FCE is located at mid-altitudes that have suitable agroclimatic conditions for the production of several cash crops including coffee and food crops, and of course for human habitation. Deforestation as result of the expansion of coffee, tea and rubber tree plantations is also becoming common (Table 23.2). Recently, around 30 small enterprises were licensed to start coffee plantations in the Keffa Zone.

Current Conservation Status

Several individuals and organizations have been involved in the efforts to collect, conserve and utilize coffee germplasm since the early 1960s (Meyer, 1965; FAO, 1968; IAR, 1986; Berthaud and Charrier, 1988). The FAO Coffee Mission collected germplasm from most of the coffee growing areas, while Meyer's collections were only from the south-west and central parts of Ethiopia. The Institut Français de recherché scientifique pour le développement en coopération (ORSTOM) also collected coffee samples from wild populations in the south-western forest of Ethiopia (Berthaud and Charrier, 1988), following the FAO mission. The Ethiopian National Coffee Collection Programme continued to collect coffee germplasm from all over the country, including areas that were not covered by other germplasm collection missions (IAR, 1986). Accessions of germplasm collected to date are found in different field genebanks around the world and within Ethiopia (Table 23.3).

Currently, the Institute of Biodiversity Conservation and Research (IBCR) maintains more than 4000 accessions at the Chochie Biodiversity Unit in Jimma on 115 ha (Paulos and Demel, 1999). In addition, the Jimma Branch of the Institute of Agricultural Research currently maintains about 600 coffee types and 700 random selections of lines that show varying resistance to CBD (IAR, 1986; Demel *et al.*, 1998).

Efforts to conserve *C. arabica in situ* have not gone beyond the proposal stage. The Plant Genetic Resources Centre of Ethiopia (now IBCR) proposed the establishment of a coffee genetic

Table 23.1. Changes in forest vegetation cover in south-west Ethiopia (Reusing, 1998).

	1971–1975		1996–1997	
Forest cover class	Area (ha)	% Area	Area (ha)	% Area
Closed high forest	1,158,300	38.42	556,700	18.47
Slightly disturbed forest	90,900	3.02	23,600	0.783
Highly disturbed forest	Not determined	Not determined	667,400	22.14

Table 23.2. Some of the forest areas converted to plantations in south-western Ethiopia (source: Feyera Senbeta, Addis Ababa, 2000, personal communication).

Plantation	Deforested area (ha)	Status, February 2000
Bebeka Coffee (Berhan forest)	5,000	Deforested
Tepi Coffee (Part of Giz Meret forest)	6,000	Deforested
Tepi Palm (Meti forest)	1,000	Deforested
Midrock Coffee plantation	3,000	In progress
Bonga Tea Plantation	3,000	Deforested
East African Plc. Tea plantation	3,000	In progress
Rubber plantation	25,000	Planned

resources conservation programme (Melaku and Hailu, no date, unpublished proposal). The proposal was to launch a programme for further study, selection and establishment of conservation sites, though there was no further development beyond the writing of the proposal.

In 1998, the Coffee Improvement Project of Ethiopia proposed the establishment of three *in situ* conservation reserves in the south-western forest (Demel *et al.*, 1998). The sites selected were Kontir-Berhan (*c.* 20,000 ha), Boginda-Yeba (*c.* 5500 ha) and Geba-Dogi River (18,600 ha) forests. Complementary *ex situ* conservation in different agroecological zones was also proposed. The project was submitted to the Coffee and Tea Authority but implementation has not yet started because of financial constraints.

Approaches for *in situ* Conservation

In this section technical strategies for effective planning and implementation of coffee genetic reserves are discussed. We will address basic issues such as: habitat characterization and ecological studies, mapping the distribution of the range of wild coffee populations, assessment of population genetic structure of different isolated populations, and design and management plans of the reserves. The objective of setting up coffee genetic reserves is clear: to conserve the genetic diversity of coffee in its natural environment, without disrupting the rights of the local community for traditional and sustainable use of resources. In short, it is to optimize conservation and utilization on a sustainable basis.

Habitat characterization/ecological studies

Sound scientific understanding of the biophysical conditions of the ecosystem, ecogeographic range, and fundamental biological and niche adaptations of the target species is essential (Maxted *et al.*,

Table 23.3. Major *C. arabica* field genebank collections.

Country	Institute	Total nos of accessions
Brazil[a]	Centro Nacional de Recursos Geneticos	275
Colombia[a]	Centro Nationale de Investigaciones de Café Pedro Uribe Mejia	886
Costa Rica[a]	Centro Agronomico Tropicale de Investigacion y Enseñanza	1498
Côte d'Ivoire[a]	ORSTOM-Institut Français de recherché scientifique pour le développement en coopération	1787
Ethiopia[b]	Institute of Biodiversity Conservation and Research	4000
Ethiopia[b]	Jimma Research Station	679
India[a]	Central Coffee Research Institute, Kamataka	329
Kenya[a]	Coffee Research Foundation	592
Madagascar[a]	Recherche Agricole a Madagascar	329
Tanzania[a]	Tanzanian Agricultural Research Organization	42
USA[a]	US Department of Agriculture	292

[a]Dulloo *et al.*, 1998. [b]Paulos and Demel, 1999.

1997a, b, c) to identify and select representative sites for conservation. In ecological studies, human beings are often neglected as an active part of the ecosystem. Actually, there is no ecosystem where human impact has not altered the natural conditions. The FCE in this case is highly influenced and modified by the people living in it, since they depend on it for coffee production, honey and food crops. It is essential to understand how human beings alter the natural ecosystem. In order to determine the management regimes and monitor a reserve, it is also necessary to characterize the ecological requirements of coffee, such as floristic composition of the forest, edaphic conditions, climate, topography and site conditions, and how these conditions respond to different perturbations. For FCE of south-west Ethiopia, this aspect has not been investigated. Hence, a study to assess ecological dynamics of the Geba-Dogi Rivers FCE has started with a research grant from the Centre for Development Research (ZEF) of the University of Bonn. The aims of this project are: to assess the trends and patterns of ecological dynamics of the forest ecosystem along anthropogenic and environmental gradients; to assess the floristic composition and distribution patterns of the plant communities of the forest ecosystem; to assess the patterns of distribution and population structure of coffee in different forest cover types in the landscape; and to identify priority sites for *in situ* conservation of wild coffee populations. Keystone ecosystems for conserving coffee genetic resources will be identified and used to design and prepare management plans of coffee genetic reserves.

A keystone ecosystem, as defined by Stohlgren *et al.* (1997), is a portion of a landscape that is particularly important for a given ecological or management question. It can be an ecosystem that contains high plant richness, distinctive species compositions, or distinctive ecological processes that benefit many other species and ecosystems. Such distinct ecosystems with wild coffee populations in the forest landscape of south-west Ethiopia are, therefore, important keystone ecosystems for the conservation of *C. arabica*.

Mapping of the distribution of wild coffee populations

FCEs in south-west Ethiopia are distributed in different parts of the large forest landscape, mainly due to topographic variations. Identification and mapping of important ecosystems for conservation within the forest landscape is essential to help managers and development practitioners to carry out management activities. The keystone ecosystems, when mapped and presented to managers and development planners, will provide useful information about what has to be done and where. It helps planners by providing basic information before they make large-scale investment decisions that may cause the total conversion of the forest area to other land uses like plantations or further settlement.

Assessment of population genetics

A study on the population genetic structure of coffee in Ethiopia has not even been attempted, except for a few studies by breeders. To ensure the conservation of the maximum genetic variability and allelic diversity, analysis of the population genetics of coffee plants collected from different forest areas and coffee growing regions is essential. Newbury and Ford-Lloyd (1997) emphasized the need for measuring plant genetic diversity with regard to conservation and making decisions about whether an *in situ* reserve is actually needed; selecting appropriate sites for *in situ* conservation; and monitoring any change in the pattern of diversity within an *in situ* reserve. Conserving only representatives of some populations of a species is insufficient unless the whole range of genetic diversity is represented. To conserve the whole range of genetic diversity of a taxon, it is essential to know the population genetic structure. Higher genetic diversity will enable more resistance to any kind of environmental impact and increase adaptation and fitness in the face of environmental change. Hence, this should be one of the principal strategies in the conservation efforts for a species.

Design and management of genetic reserves

The FCE is not only important for genetic conservation, it is also the basis of livelihood for millions of people in the region. The design and management plan of coffee genetic reserves in the area should consider a mechanism that maximizes the conservation effort without jeopardizing the production system of the local communities. In this regard, the Man and Biosphere (MAB) programme of the United Nations Educational, Scientific and Cultural Organization (UNESCO) is a good strategy since it combines conservation and sustainable management for human use (Hawkes *et al.*, 1997). MAB reserve design is considered as an optimal reserve strategy (Hawkes *et al.*, 1997) since it has a special spatial arrangement in which human settlement and the use of resources by traditional patterns of land use is allowed in some zones. A biosphere reserve should be a representative environment, internationally recognized for its conservation value and provision of scientific knowledge, skills and human values to support sustainable development (UNESCO, 1987).

A reserve in an FCE should be as large as possible. Large reserves have the advantage of including landscapes with diverse ecogeography and minimizing edge effects (Hawkes *et al.*, 1997). It is also advantageous in cases of catastrophes that might destroy everything if the size of the reserve is small. Ethiopia experienced a catastrophic event in which over 70,000 ha of forest in the south-eastern part was burned down by accidental fires in February–March 2000. This forest also housed wild coffee populations. Therefore, we recommend as large an area as possible for coffee genetic reserves.

Biosphere reserves normally consist of three components: one or several core zones, a buffer zone and a transition zone, each with different characteristic conservation roles. A conceptual reserve design for coffee in the FCE of south-west Ethiopia is shown in Fig. 23.2. The core zone is a strictly protected reserve area where all kinds of human interaction should be avoided, except for research and monitoring of the natural vegetation dynamics. The areas to be selected for such zones should also be more or less undisturbed high forest with positive wild coffee population structure and other plants in the community.

Two buffer zones around the core zone are recommended. In Buffer Zone I, that is, a slightly disturbed forest zone, collection of non-timber forest products like honey and other traditional uses are allowed. Collection of coffee from wild plants is also allowed. But, the users are not allowed to manipulate the canopy or ground vegetation to enhance coffee production or other forms of agricultural production. Monitoring and research activities are also

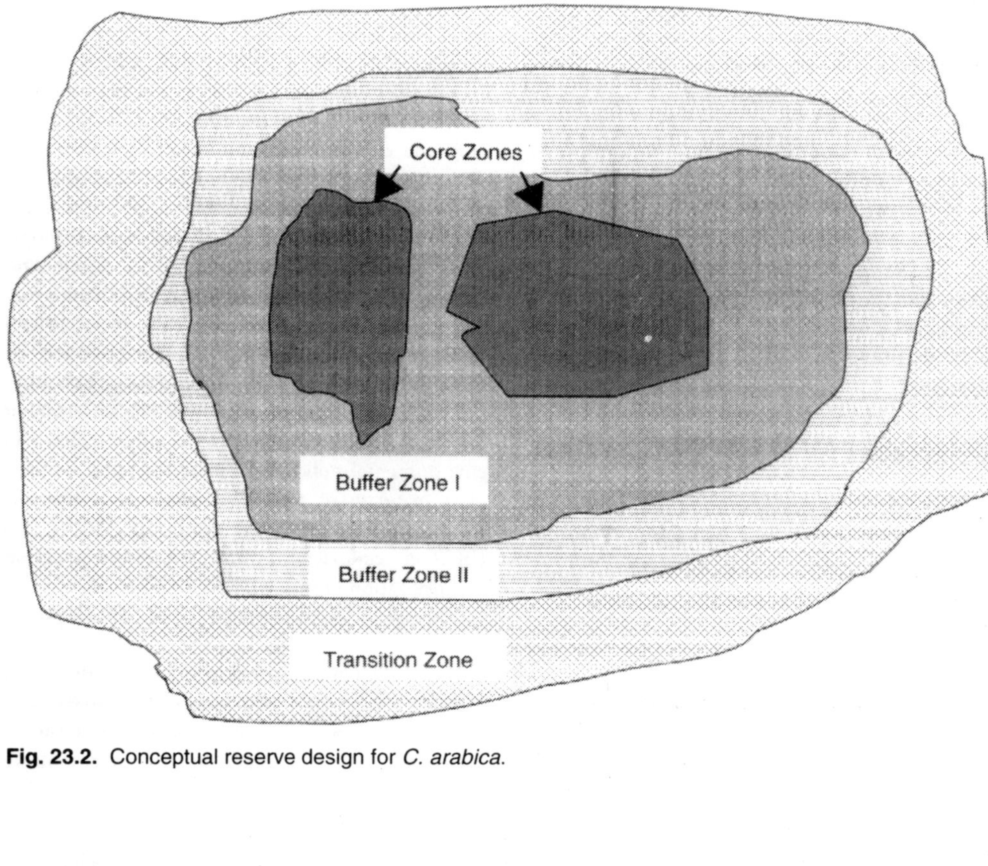

Fig. 23.2. Conceptual reserve design for *C. arabica*.

part of the management in this zone. In Buffer Zone II (socio-buffering), the traditional system of coffee production from the forest is allowed. Farmers will be allowed to modify the shrub layer vegetation to some extent, to improve the coffee production from the natural system. But no enrichment planting or conversion of the land to other agricultural use is allowed. Research and monitoring continue to assess the management regime and impacts on the core zone. Use right is restricted to those who are already living within the area set aside for the reserve. No further settlement is allowed.

In the transition zone, the traditional forest coffee production system, garden coffee and some agricultural practices are allowed. Some habitat restoration and research are carried out on those areas that were highly degraded. The farming practice should involve the traditional agroforestry system, without total clearing of the forest vegetation. Local communities are also allowed to manipulate the coffee populations in areas under their possession, using local landraces. Incentives of some kind, depending on local need should also be provided, to reward their effort in assisting conservation. This can include basic public services like schools, clinics or health centres, access roads and the like.

This is only a conceptual design and management plan, considering the strong dependency of a high number of people on the forest ecosystem and the immediate need for conservation measures to save the genetic resources of coffee. The actual intervention mechanism has to be based on sound scientific knowledge after detailed studies. The idea here is to save the resources first, then study and use them sustainably and equitably. In addition to the genetic reserve, complementary *ex situ* conservation sites such as field genebanks in different agroecological zones around the reserves should be established.

Constraints for *in situ* Conservation

There are several constraints associated with establishing *in situ* reserve sites for coffee. The constraints can be grouped into three major categories: technical, political and economic constraints.

Technical or scientific constraints

Despite its paramount social and economic importance, little is known about the biology and ecological requirements of *C. arabica* in its wild populations. Identifying important priority areas for conservation is difficult based on the current knowledge of the forest ecosystem. To set up priority areas for conserving genetic resources, sound scientific knowledge about geographical distribution, population genetics, population ecology and the dynamics and characteristics of the keystone ecosystem for the species is required. The ongoing project attempts to characterize and identify important keystone ecosystems for coffee genepool conservation and will contribute to filling the knowledge gap in population ecology and habitat characteristics of the species.

Further research is needed on population genetics, overall distribution patterns and exact locations of distinct populations. Knowledge of distribution and population genetic structure is one of the major challenges for coffee research in Ethiopia, since it is expensive and needs up-to-date laboratory equipment and highly skilled personnel, which Ethiopia cannot afford.

Political or policy constraints

For the first time, Ethiopia now has an environmental policy and a conservation strategy (EPA, 1997). Genetic, species and ecosystem diversity conservation and management are core points in both the environmental policy and the conservation strategy. To implement the policies through further legislation and legal orders, an institutional framework is in place at the federal level. Ethiopia has nine federal states and two special administrative regions. The current federal government structure of the country gives the right to all federal states to use and develop their respective regions independent of central government influence. However, the absence of respective government organs to implement the national conservation and environmental policy at regional level and the lack of a mandate of national organizations like the Environmental Protection Authority (EPA) and IBCR at regional level make the implementation and realization of such a policy impractical.

Ethiopian administrative and development institutions suffer from frequent restructuring and changes of mandate. Sometimes, an institution at ministry level merges with another ministry or else is totally avoided. The responsibilities of one institution are given to another, in which case policies are also liable to change. Corrective measures are required from the government to extend all national level organizations to regions and to minimize institutional changes.

Economic constraints

Conservation of plant genetic resources is expensive, since it is a long-term venture and needs long-term investment for monitoring and management of the genetic estate (Hawtin and Hodgkin, 1997). The initial and maintenance cost of genetic conservation in Ethiopia will be even more expensive due to poor infrastructure and lack of adequate scientific information about plant species and ecosystems. This requires high investment in training, personnel and research facilities.

It is an unattainable dream to expect Ethiopia to invest in such costly conservation of genetic resources because of its poor economy. Ethiopia remains the poorest of the poor countries in the world, where small-scale, age-old traditional agriculture is the mainstay of the economy. This situation is a threat to conservation since poor farmers are forced to deforest and convert natural ecosystems into agricultural land to earn their livelihood. Moreover, the poor Ethiopian economy has less to offer as incentives or alternative livelihoods to poor farmers in order to protect the FCE from deforestation. This is the main economic constraint concerning FCE conservation.

Support from international communities, either from bilateral donation or multilateral donor agencies is needed to save the last remaining forest ecosystems that house the genepool of *C. arabica*, since conserved genetic resources are public goods at local, national and international levels (Hawtin and Hodgkin, 1997). Williams (1997) suggested two steps to assist developing countries, like Ethiopia, in conserving genetic resources: first, outside funding for the initial planning and, second, the provision of blueprints and organizational needs to national governments by an international organization, and the adoption of these as part of the national plan. Generous support from different national and international organizations in the past helped to conserve crop germplasm in Ethiopia and to provide institutional capacity for the IBCR, enabling Ethiopia to establish a genebank.

There are efforts on the side of the government to conserve important genetic resources. National institutional frameworks and policies have been formulated to implement conservation of genetic resources. But there is still a need for financial resources and coordination of the relevant government institutions and regional government bodies to initiate conservation activity.

Actions to be Taken

The first and most basic step towards a conservation programme should be strengthening the capacity of the responsible institutions. The responsible organization, in collaboration with other organizations that have direct or indirect stakes in forest resources management need to set up a mechanism for conservation of priority areas. The following actions are recommended for the success of the coffee biosphere reserve:

- Designate the FCEs as UNESCO biosphere reserves and ensure sustainable use and conservation by legislation and legal orders.
- Design a mechanism of financial resources generation from different sources to sustainably manage the biosphere reserve.
- Research on the FCE to bridge the knowledge gap for further conservation and utilization decisions. On this aspect, the national responsible body should be able to collaborate with international organizations like the International Plant Genetic Resources Institute (IPGRI) and local research institutions in order to use available human and material resources.
- Regular monitoring and modification of the management regime as appropriate for conservation.

Acknowledgements

We would like to thank Drs Assefa Admassie, T. Borsch, D. Virchow, M. Andreini, Ms M. Plunkett and Dr T. Brown for proofreading and useful comments. Mr Feyera Senbeta kindly provided us with some data on deforestation. The Gottlieb Daimler and Karl Benz Foundation, ZEF and the Catholic Academic Exchange Service (KAAD) financed the research stay of the first author in Germany.

References

Admasu, S., Masresha, F., Mehari, E. and Tefsetewolde, B. (1989) *Coffee Area Specialization.* Ministry of Coffee and Tea Development, Addis Ababa, Ethiopia.

Anonymous (1988) *National Atlas of Ethiopia.* Ethiopian Mapping Agency, Addis Ababa, Ethiopia.

Berthuad, J. and Charrier, A. (1988) Genetic resources of *Coffea.* In: Clarke, R.J. and Macrae, R. (eds) *Coffee,* Vol. 4. Elsevier Applied Science, London, pp. 1–42.

Cannell, M.G.R. (1983) Coffee. *Biologist* 30, 257–263.

Charrier, A. and Berthaud, J. (1985) Botanical classification of coffee. In: Clifford, M.N. and Willson, K.C. (eds) *Coffee: Botany, Biochemistry and Production of Beans and Beverages.* Croom Helm, London, pp. 13–47.

Demel, T. (1999) History, botany, and ecological requirements of coffee. *Walia* 20, 28–50.

Demel, T., Ababu, A., Getahun, M. and Mehari, E. (1998) *Study on Forest Coffee Conservation.* Coffee Improvement Project, Addis Ababa, Ethiopia.

Dulloo, M.E., Guarino, L., Engelmann, F., Maxted, N., Newbury, J.H., Attere, F. and Ford-Lloyd, B.V. (1998) Complementary conservation strategies for the genus *Coffea*: a case study of Mascarene *Coffea* species. *Genetic Resources and Crop Evolution* 45, 565–579.

EPA (1997) *Environmental Policy of Ethiopia.* Environmental Protection Authority (EPA), Addis Ababa, Ethiopia.

Eriksson, G., Namkoong, G. and Roberds, J.H. (1993) Dynamic gene conservation for uncertain futures. *Forest Ecology and Management* 62, 15–37.

FAO (1968) FAO *Coffee Mission to Ethiopia 1964–65.* FAO, Rome, Italy.

FAO (1974) *FAO Technical Conference on Crop Genetic Resources, 12–16 March 1973, Rome.* FAO, Rome, Italy.

FAO (1989) *Plant Genetic Resources, their Conservation* in situ *for Human Use.* FAO, Rome, Italy.

FAO (1998) *The State of the World's Plant Genetic Resources for Food and Agriculture.* FAO, Rome, Italy

Ferwerda, F.P. (1976) Coffees *Coffea* spp. (Rubiaceae). In: Simmonds, N.W. (ed.) *Evolution of Crop Plants.* Longman, London, pp. 257–260.

Haarer, A.E. (1962) *Modern Coffee Production,* 2nd edn. Leonard Hill, London.

Hawkes, J.G., Maxted, N. and Zohary, D. (1997) Reserve design. In: Maxted, N., Ford-Lloyd, B.V. and Hawkes, J.G. (eds) *Plant Genetic Resources, the* in situ *Approach.* Chapman and Hall, London, pp. 132–143.

Hawtin, G.C. and Hodgkin, T. (1997) Towards the future. In: Maxted, N., Ford-Lloyd, B.V. and Hawkes, J.G. (eds) *Plant Genetic Resources, the* in situ *Approach.* Chapman and Hall, London, pp. 368–383.

IAR (1986) *Department of Coffee Progress Report 1983/84.* Institute of Agricultural Research, Addis Ababa, Ethiopia.

IBPGR (1980) Coffee genetic resources. *IBPGR Working Group on Genetic Resources of* Coffea arabica, *11–13 December 1979, Rome, Italy.*

Lashermes, P., Trouslot, P., Anthony, F., Combes, M.C. and Charrier, A. (1996) Genetic diversity for RAPD markers between cultivated and wild accessions of *Coffea arabica. Euphytica* 87, 59–64.

Lashermes, P., Combes, M.C., Robert, J., Trouslot, P., D'Hont, A., Anthony, F. and Charrier, A. (1999) Molecular characterization and origin of the *Coffea arabica* L. genome. *Theoretical and Applied Genetics* 99, 259–266.

Lashermes, P., Andrzejewski, S., Bertrand, B., Combes, M.C., Dussert, S., Graziosi, G., Trouslot, P. and Anthony, F. (2000) Molecular analysis of introgressive breeding in coffee (*Coffea arabica* L.). *Theoretical and Applied Genetics* 100, 139–146.

Maxted, N., Hawkes, J.G., Guarino, L. and Sawkins, M. (1997a) Towards the selection of taxa for plant genetic conservation. *Genetic Resources and Crop Evolution* 44, 337–348.

Maxted, M., Guarino, L. and Dulloo, M.E. (1997b) Management and monitoring. In: Maxted, N., Ford-Lloyd, B.V. and Hawkes, J.G. (eds) *Plant Genetic Resources, the* in situ *Approach.* Chapman and Hall, London, pp. 144–159.

Maxted, M., Ford-Lloyd, B.V. and Hawkes, J.G. (1997c) Complementary conservation strategies. In: Maxted, N., Ford-Lloyd, B.V. and Hawkes, J.G. (eds) *Plant Genetic Resources, the* in situ *Approach.* Chapman and Hall, London, pp. 15–39.

MCTD (1989) *Report on Coffee Yield Assessment for 1990.* Ministry of Coffee and Tea Development, Addis Ababa, Ethiopia.

Melaku Worede (1984) Coffee genetic resources in Ethiopia, conservation and utilization with particular reference to CBD resistance. In: *Proceedings First Regional Workshop on Coffee Berry Disease, 19–24 July 1982.* AAASA, Addis Ababa, Ethiopia, pp. 203–211.

Mesfin Tadesse and Lisanework Nigatu (1996) An ecological and ethnobotanical study of wild or spontaneous coffee, *Coffea arabica* in Ethiopia. In: van der Maesen, L.J.G., van der Burgt, X.M. and van Medenbach de Rooy, J.M. (eds) *The Biodiversity of African Plants, Proceedings XIVth AETFAT Congress, 22–27 August 1994, Wageningen, The Netherlands.* Kluwer Academic Publishers, Dordrecht, The Netherlands, pp. 277–294.

Meyer, F. (1965) Notes on wild *Coffea arabica* from southwestern Ethiopia, with some historical considerations. *Economic Botany* 19, 136–151.

Montagnon, C. and Bouharmont, P. (1996) Multivariate analysis of phenotypic diversity of *Coffea arabica*. *Genetic Resources and Crop Evolution* 43, 221–227.

Newbury, H.J. and Ford-Lloyd, B.V. (1997) Estimation of genetic diversity. In: Maxted, N., Ford-Lloyd, B.V. and Hawkes, J.G. (eds) *Plant Genetic Resources, the* in situ *Approach*. Chapman and Hall, London, pp. 192–206.

Noss, R.F. (1996) Ecosystems as conservation targets. *TREE* 11, 351.

Paulos Dubale and Demel Teketay (1999) The need for forest coffee germplasm conservation in Ethiopia and its significance in the control of coffee diseases. In: EARO (ed.) *Proceedings of Coffee Berry Disease Workshop*. Ethiopian Agricultural Research Organization (EARO), Addis Ababa, Ethiopia, pp. 125–135.

Pendergrast, M. (1999) *Uncommon Grounds: the History of Coffee and how it Transformed our World*. Basic Books, New York, USA.

Raina, S.N., Mukai, Y. and Yamanoto, M. (1998) In situ hybridization identifies the diploid progenitor species of *Coffea arabica* (Rubiaceae). *Theoretical and Applied Genetics* 97, 1204–1209.

Reusing, M. (1998) *Monitoring of Natural High Forests in Ethiopia*. Ministry of Agriculture and GTZ, Addis Ababa, Ethiopia.

Stohlgren, T.J., Coughenour, M.B., Chong, G.W., Binkley, D. and Kalhan, A. (1997) Landscape analysis of plant diversity. *Landscape Ecology* 12, 155–170.

Sylvian, P.G. (1958) Ethiopian Coffee–its significance for the world coffee problems. *Economic Botany* 12, 111–139.

Tafesse, A. (1996) Agroecological zones of southwest Ethiopia. *Matreialien Zur Ostafrica-Forschung*, 13.

Tewolde, B. and Egziabher, G. (1990) The importance of Ethiopian forests in the conservation of Arabica coffee genepools. *Mitt Inst. Allg. Bot. Hamburg* 23a, 65–72.

UNESCO (1987) *Practical Guide to the Man and Biosphere Programme (MAB)*. UNESCO, Paris.

Williams, J.T. (1997) Technical and political factors constraining reserve placements. In: Maxted, N., Ford-Lloyd, B.V. and Hawkes, J.G. (eds) *Plant Genetic Resources, the* in situ *Approach*. Chapman and Hall, London, pp. 88–98.

Wrigley, G. (1988) *Coffee*, Tropical Agriculture Series. Longman Scientific and John Wiley & Sons, New York.

24 Indicators for Sustainable Management of Plant Genetic Resources: How Well are we Doing?

A.H.D. Brown and C.L. Brubaker
Centre for Plant Biodiversity Research, CSIRO Plant Industry, Canberra, Australia

Introduction

The 20th century has seen an enormous transition in our appreciation of plant genetic diversity. The century began with the rediscovery of Mendel's laws of inheritance, and through Johanssen the dawning of modern plant breeding. The basis and method for scientific plant improvement were to hand. In addition, the post-Darwinian world saw no short supply of genetic diversity within crop species, nor was such diversity viewed as threatened.

Then through the century awareness has grown that genetic resources are indeed limited. With the spread of agricultural development worldwide, local varieties and minor species have given way to modern introduced varieties of major crops. By the mid-1970s, concern for plant genetic resources (PGR) was widespread and efforts to conserve them underway. PGR were generally divided into domesticated populations (landraces, obsolete cultivars, breeders' populations, modern varieties of crops, genetic stocks) and wild populations (wild plant species especially those related to crops). The dominant strategies were to conserve domesticated germplasm in genebanks *ex situ*, and wild populations in nature reserves *in situ* (Frankel and Soulé, 1981).

Since 1980, several issues, concerns and forces have emerged to question claims of progress toward adequate PGR conservation (Marshall, 1989). The major ones are:

- National awareness of the value of plant genetic resources has grown, particularly in economically poor but 'gene rich' countries in the centres of crop diversity. Since the Convention on Biological Diversity such countries have assumed greater responsibility for the conservation of their own indigenous genetic resources and expect to receive their fair share of benefits from their use.
- The number of accessions in *ex situ* collections has continued to grow, while their use and maintenance of their viability remain problematic.
- Several hundred crop or wild plant species, previously used by humans, are now classed as Underutilized or Neglected. They are poorly represented in *ex situ* collections and without increased attention stand to be the main arena of genetic erosion.
- Botanical gardens organizations have sought to take an increasing role in the conservation *ex situ* of wild species. How their efforts in seed banking of wild species are to be integrated into other PGR conservation measures is an open issue.
- Despite the increase in areas protected for nature conservation, nature reserves generally hold a limited and biased slice of agricultural biodiversity. Many factors (biological, political, economic, amenity) other than concern for wild crop relatives, or even of the utility of wild species determine where such reserves are set

aside. Just how biased the coverage is is not known. There are very few areas expressly reserved to conserve wild relatives of crops.

- In many countries, farmers, particularly those in marginal areas, continue to grow their own local varieties (Brush, 1995). Yet the stability of such on-farm systems as a conservation strategy in the face of quickening development is not known.
- New genetic technologies, especially DNA sequencing on a larger scale than thought possible, has meant the flowering of molecular systematics and new ways to measure genetic similarity and divergence of plant species and populations.

How well are we managing plant genetic resources for sustainable human benefit? This simple question requires objective pointers or indicators as to whether and where the task is being addressed well and where it is not. Governments, industry, scientists and the public require means of measuring progress in this task, and for sounding early warnings of emergencies.

Indicators

Purpose and scope

Indicators are tools to monitor progress and point to emerging problems. Saunders *et al.* (1998) define an indicator as a significant physical, chemical, biological, social or economic variable that is measurable in a defined way for management purposes. Indicators that measure the quality of the physical environment, for example, atmospheric CO_2 concentration, are relatively straightforward and interpretable. However, suitable indicators for tracking the state of biological resources locally and globally are more difficult to agree upon.

In a broad sense, national and international statistics on cropping areas, yields, droughts, plant disease outbreaks, plagues, farm income, agricultural production and trade, even famine, are indirect measures of human wellbeing and indirect indicators of sustainable use of PGR. However, these are only the broadest pointers to policy decisions. Here we limit the discussion to indicators closer to the biodiversity at stake, technical questions and managerial decisions at the level of institutions and researchers.

The developers of indicators for sustainable management of PGR face what seems to be a fundamental and apparently insoluble dilemma. On the one hand, we are aware that the system is a highly complex one in which a great variety of *forces* (or pressures) can potentially act to threaten PGR. It includes the actions of individual farmers, scientists, breeders, communities, industry, conservation agencies and governments. Consequently, an equally large array of parameters of varied complexity, scale and cost can be nominated as important *variables* to monitor. On the other hand, while an innumerable array of variables might seem necessary to explain 'the system', such a suite will attract no support from users as indicators. There are too many to follow, to interpret and to act upon, and a reductionist approach is inescapable.

Deciding on a list of indicators is only the first step. Interpretation of data presents challenges that need prior thought. One intent is to have absolute standards (e.g. a specific number of varieties should underpin crop production in a certain area, or the germinability of genebank accessions should remain above 85%). Alternatively, the intent may be to monitor change over time with specific rates of change deemed desirable or acceptable (a certain number of accessions regenerated per year).

Benchmarks or threshold values need to be determined scientifically. In contrast, targets are tools of policy and are specified levels or ranges for the indicators set jointly by groups of stakeholders. Further detailed discussion on the policy interpretation of each indicator lies outside the brief of this chapter.

Biodiversity management

As well as for environmental quality, there is an upsurge in interest in developing indicators for conserving biodiversity at the ecosystem, species and gene level. As governments seek to implement the Convention on Biological Diversity, they require indicators to measure progress, to monitor the main human impacts on biodiversity (sustainable development, conservation and use) and to give early warning of any irreplaceable loss.

Despite the difficulty of the task, it is clear that experts and conservation agencies in several countries are proposing indicators to encourage and improve the management of biodiversity in nature. Foresters have perhaps led the way in calling for indicators to monitor the genetic effects of direct exploitation of forests (Ledig, 1992; McKenney *et al.*, 1994; Boyle and Sayer, 1995; Namkoong *et al.*, 1996; Boyle, 2000). Using population genetics theory, Savolainen (2000) considered the basic evolutionary processes generating and maintaining diversity, and discussed heterozygosity of marker genes as a useful indicator of historical effects. A classic example is the contrast between the extremely low values of heterozygosity in *Pinus resinosa* and the high values in *Pinus sylvestris*, indicating the lingering effects of bottlenecks in population size of *P. resinosa* in the distant past.

As part of a wider project seeking indicators for national reporting on the 'State of the Australian Environment' (the term included both the physical environment and the biodiversity it supports), Brown *et al.* (1997b) proposed seven indicators to monitor natural genetic diversity in selected indicative taxa. These were:

1. Number of sub-specific taxa;
2. Population size, number and physical location;
3. Environmental amplitude of populations;
4. Genetic diversity at marker loci within individuals and populations;
5. Quantitative genetic variation;
6. Interpopulation genetic structure; and
7. Mating system.

Of these, Saunders *et al.* (1998) integrated the first four into the overall proposal for national biodiversity indicators at all levels.

Desirable properties of indicators

In attempting to develop indicators for Canadian forests, McKenney *et al.* (1994) gave as guidelines for the selection of indicators the following criteria.

> [Ideal indicators] are easy to implement, are based on good experimental design and analysis, do not disturb the system, avoid fads, indicate both process and flows as well as states and stocks, give early warnings, contain some dramatic 'flagship' species and other 'umbrella' species, target all scales, are participatory for all stakeholders, and have clear specific objectives.

Table 24.1 is a longer wish list of properties of the ideal indicator from the perspectives of the operator of indicators, the manager and the user of PGR, and the wider community. However, it is easier to draw up such a list of ideal properties, and to lengthen it still further, than it is to find indicators that satisfy such a list. We consider that the first three items, namely, validity, clarity of interpretation and simplicity in measuring genetic diversity, must be given top billing when assessing indicators.

Potential Indicators According to Genepool and Conservation Strategy

McKenney *et al.* (1994) divided indicators into those that are 'species-based', and those that are 'system-based'. Ferris and Humphrey (1999) called these structural and compositional indicators. Others have used the division into Pressure, Condition or State, and Response. We use the partition mentioned above as two axes of classification. The first axis is the kind of genepool (primary or cultivated versus secondary and tertiary or wild relatives), and the second refers to the conservation strategy (*in situ* versus *ex situ*) (Table 24.2).

Primary genepools in situ

The state of crop plant diversity actually under cultivation on farms, in pastures or in gardens has been the original focus of the genetic resources movement. It is therefore a natural subject for the development of indicators, particularly to watch for the process of 'genetic erosion'. The Food and Agriculture Organization (FAO) documentation for the Leipzig technical conference (FAO, 1998) contained a list of possible indicators of genetic erosion of landraces (Table 24.3). These indicators are clearly candidates for our task here, because genetic erosion, or the steady loss of genetic diversity in on-farm agriculture, is perhaps the key 'pressure' on the sustainable management of domesticated PGR. Brown (1999) took a different tack and considered variables for the study of genetic structure of landraces that were being conserved on-farm. The approach was based on testing five putative advantages of *in situ* conservation, and the resulting indicators were numerous and wide-ranging.

In Table 24.2 we reduce these and other potential variables into a set of four. Topping the list is the number and frequency of occurrence of distinct landraces of a crop (in a given area). Of course this begs the question of how reliable is the classification and recognition of landraces, and how widespread is the common use of particular names. In some crops, such as sorghum in Ethiopia, a single

Table 24.1. Checklist of desirable properties for indicators for managing genetic resources (adapted from Saunders *et al.*, 1998).

Indicator intrinsic properties
Scientifically valid and credible
Easy to understand, unambiguous and robust
Simple and cheap to assay
Use accepted and well-documented methods
Cost-effective (i.e. an expensive indicator should yield greater information)
Adaptable for use at a range of spatial scales
Aggregative (capable of meaningful summation over items and scales)
Capable of being monitored easily to show trends over time
Managers of PGR
Relevant to management objectives and fit in a policy framework
Part of the management cycle and not an end in itself
Focus on the use of information rather than the gaining of it
Render progress evident
Kept under review and refined when necessary
Users of PGR
Developed with all people involved: stakeholders, monitors
Reflect an essential, fundamental and highly valued element of the object being monitored
Provide early warning of emerging issues or problems
Other essential elements for indicator development and use
Partnerships between communities, governments, companies and research agencies setting up and running the process and sharing information
The provision of adequate resources (time, expertise, funds)
A commitment to collect new data
Continuing research and development for improving indicators and determining cause and effect

field might be planted to as many as 20 landraces, while in others, such as lucerne in Morocco, a farmer's variety might not bear a distinctive name, but just be known as 'local variety'. Breeding system and plant morphology affect the way in which such naming interacts with farmer management of this diversity.

Several studies have been made of the reliability of farmer names in reflecting diversity consistently. Remarkable reliability has been found, for example, in farmer recognition of sorghum landraces in Ethiopia (Teshome *et al.*, 1997). Such studies are important in underpinning reliance on names for estimating indicators. Further, if the studies include assay of divergence and distinctiveness for genetic markers for a sample of such landrace populations, we can get estimates of the fundamental levels of polymorphism involved.

However, an important point is that names are signals to adaptive or yielding attributes and through them to farmer management strategies. Named populations will be treated in particular ways, almost irrespective of the genetic differences. As such, names, which probably evolve themselves, affect, if not control, the immediate future of these populations. A variable landrace, bearing a certain name, may be reputed to be flooding tolerant, and planted accordingly, thus setting up selection pressure to emphasize that attribute.

Another factor in the equation is the environment where crops are grown. Hence the second area for indicators to address, is a measure of the range of environments that the landraces of a crop species currently occupy. This aspect is not explicit in the FAO list (Table 24.3), but is listed second in Table 24.2. The notion here is that the wider the span of environments that a crop occupies, the greater the adaptive genetic diversity that crop might house. Using survey data on crop occurrence, geographical information systems and multivariate methods may offer ways in which environments can be classified and typified. This would lead to measures of the extent of occupied environments and such measures could be compared over time for changes. Eyzaguirre (unpublished) has suggested this kind of indicator for taro (*Colocasia esculenta*) in China, a species that is important for food security and has income generating potential.

Table 24.2. Indicators for sustainable management of plant genetic resources.

Genepool	*In situ*	*Ex situ*
Primary	Number and frequency of landraces, and area planted to them Environmental amplitude of crop area Durability and evolution of farmer management and farmer selection criteria Security of traditional knowledge	Number of accessions in the genebank Country distribution of genebanks Coverage in collection of crop diversity Extent of usage and representation in core collections Collection health, accession viability Documentation and evaluation of collection Backup duplication provisions
Secondary and tertiary	Distribution in protected areas, that cover the species' environmental range Population numbers and size, particularly of rare wild crop relatives Gene diversity, divergence and distribution	Number and frequency of accessions *used* Coverage of species range Evolutionary relationships and taxonomic clarity 'Prebreeding' activities, including evaluation DNA banks

In measuring sustainable management of crop diversity on-farm, we must recognize that farmer decisions lie at the root of what diversity will be planted. Hence we need to obtain measures of how durable are farmers' selection criteria and management practices and how they are evolving. For example, populations selected for multiple purposes such as for ease of growth, grain, fuel or fodder straw production, may be more diverse than those cropped solely for sale to the market, particularly when there is market stress on uniformity. Multiple use is also the basis of an indicator for taro where different genotypes and different organs of the plant are put to separate uses (Eyzaguirre, unpublished).

The last indicator is to meet concerns about loss of knowledge as to how and why certain species and landraces are grown or used. It also addresses the new demands of benefit sharing. Sustainable management of resources existing in local varieties calls for the creation of mechanisms to secure traditional knowledge. In a sense this activity is parallel to the systems of plant breeders rights that exist in developed agriculture. Monitoring the existence and scope of such systems of protection will be indicative of forces to help encourage the sustaining of diversity on-farm.

We therefore focus on four areas, the crops themselves, the environment they occur in, the farmer management practices and the information. Each relates back to the different landrace populations as entities. For this reason, the number of landraces, coupled with their frequency and area occupied by each landrace forms the fundamental variable for the minimum indicator.

Primary genepools ex situ

Since the 1960s, the growth of institutional genebanks that hold accessions as dried seed, in tissue culture, or as living plants in the field has been prodigious. Along with such an investment have come considerable thought, research and experience into how to judge progress in genebanking. Table 24.2 includes some well-established variables (such as viability, documentation, duplication) used in evaluating collections.

However, the primary datum is the number of accessions for each crop held in each of the large institutional genebanks. The 1998 report on the state of the world's genetic resources contains data collected from worldwide surveys. The collections are further divided into whether the material is held in long-term conservation, medium-term accessibility, or indeterminate storage. In several of the larger collections, accessions have been counted twice because they are in both conditions. Indeed, the same report estimates that of the ± 6 million accessions stored worldwide, between 1 and 2 million are 'unique'.

These figures are perhaps the 'best' examples we have of indicators of genetic resources management. They exemplify both the benefits and the problems or pitfalls of making inferences based on

Table 24.3. Indicators of genetic erosion of landraces as proposed by FAO (1998).

Occurrence of landraces
Number of landraces
Vernacular names of landraces
Main distinguishing morphological or agronomic traits
Changes in area devoted to landrace cultivation
Changes in agronomic practices
Indigenous and local knowledge

Population diversity of landraces
Phenotypic diversity
Population size
Extent of distribution in a target region

Modern varieties
Ratio of crop area growing modern varieties to that growing landraces
Seed source and distribution mechanisms
Promotion of new agronomic practices
Reasons for introduction of new varieties

Note that the 1999 FAO meeting in Prague on methodology of the World Information and Early Warning System on Plant Genetic Resources defined genetic erosion as 'A permanent reduction in the number, evenness and distinctness of alleles, or combinations of alleles, of actual or potential agricultural importance in a defined geographical area.'

them. To illustrate this we summarize in Table 24.4 the FAO (1998) estimates of the total number of accessions for the eight main cereal crops and compare these with those give by Holden (1984) in his overview of the situation a decade earlier. The picture is similar for legumes and forages.

In nearly all cases there is a doubling of the numbers in collections. If this is true we could certainly infer that the numbers in store cannot keep growing at this rate. It is likely that part of the growth represents collections or samples that were in hand in 1984 but not included in Holden's total (e.g. for oat). Likewise there could be material that escaped the latest census. The point is, however, that there has been at least a doubling in the size of the formal commitment of the PGR community in numbers of conserved units. Part of this growth may represent the deliberate duplication of material from one country to another (or indeed duplication within a collection in different storage conditions, or *in vitro* and in the field). This has come about from policies of insurance backup storage, repatriation, or introduction of large collections to new areas.

Table 24.4. Estimates of the total number of accessions stored in genebanks worldwide.

	Holden (1984)	FAO (1998)
Cereals		
Wheat	401,500	788,654
Barley	280,300	486,724
Rice	212,200	420,341
Maize	99,700	261,684
Oat	37,000	155,049
Sorghum	91,000	168,550
Millet	56,000	145,476
Rye	18,300	27,132
Total cereals	1,200,000	2,454,000
Grain legumes	185,140	871,577
Forage legumes	84,200	168,530
Forage grasses	127,900	240,978

Whatever the degree of planned duplication and of inadvertent redundancy, each accession is a managed unit, kept and recorded as distinct. For example, it is of interest to know that a collection has a total of 80,000 accessions made up of 40,000 in long-term store and the same 40,000 in medium term. The basic figure is the total number of accessions stored. This statistic can be judged and modified with values for the remaining indicators in this group.

The first of these asks how well the material in genebanks covers the existing crop diversity. This may be hard to assess, because it requires knowledge of what is in hand and what is not. Yet the filling of gaps has been stated often as a need, and various methods of identifying gaps are available to do this.

Next is the question of usage, which might be measured by the number of samples dispatched per year. Again this figure can be misleading and likely to overestimate usage because of requests or responses in excess of needs. Equally, however, the number of appearances in the pedigrees of cultivars is a vast underestimate of usage. It ignores the base screening population from which particular selections were made, or the extent of sampling on which biological research was conducted. The proportion of a collection that forms the basis of a core collection is another indicator of sampling usage. In a recent worldwide survey of core collections, Brown and Spillane (1999) found that the core collection procedures have been adopted for a wide range of crop types. However, the adoption rate was apparently greater in medium-sized germplasm collections, and the larger collections have so far not been subject to such sampling principles.

The other suggested indicators are straightforward and familiar, and deal with the state of the germplasm collection. Here we can build on the international experience in the monitoring of genebanking. The International Plant Genetic Resources Institute (IPGRI) and FAO (FAO and IPGRI, 1994) have recently published standards for genebank management that provide benchmarks for indicators. Holden *et al.* (1993) have mapped out in detail how such variables could be combined into a 'score' to attach to each accession. The average over the whole collection could function as an indicator. It is interesting to consider why such proposals were not widely adopted as a way to lobby funding bodies for more resources to redress problem areas. Perhaps the risk was too great of unfair public comparisons between genebanks that ignored radically different levels of resources. This history shows how important partnerships will be in formulating and using indicators that will enjoy wide support (Table 24.1).

Monitoring the health of collections, particularly of field genebanks, and assessment of accession viability are essential housekeeping functions. The FAO (1998) overview inevitably found a diverse picture on these parameters, and particularly stressed the growing backlog of samples for regeneration. Documentation and evaluation also inevitably but perhaps less critically, were also variable across the spectrum of collections. These activities may never be totally complete and perhaps a more helpful indicator is a measure of the annual effort being devoted to them, rather than an estimate of the proportion of the task not done.

Once again the overall indicator that captures most of the genebanking effort is the total number of accessions of a crop conserved in a genebank. This could primarily reflect the number to which an institution is committed to conserve. The secondary indicators come into play in assisting in improving the growth and management of that collection, and hence its value and sustainability.

Secondary genepools in situ

From one of the more traditional modes of conserving agricultural plant genetic resources, we turn to the other and polar opposite, namely the conservation of wild related species in nature reserves. The first question to consider is why emphasize wild species that either are used directly, or are wild relatives of crops? After all, modern molecular technology gives access to the 'quaternary' genepool, that is without limit. Should our thinking be broader to encompass other forms of biodiversity than higher plants, such as fungi, at the level of ecosystems?

However, we argue that the PGR community has a special concern for and focus on the wild relatives of domesticated plants. Their roles as alternative hosts in coevolutionary relationships, and their long history as proven sources of useful genes assure them of that focus. Indeed they are natural 'flagship' species, and have proved important in achieving political support for conservation activity in the cases of *Zea diploperennis* in Mexico and *Triticum dicoccoides* in Israel (Frankel *et al.*, 1995), and the Antalya Gene Management Zone in Turkey (see Tan and Tan, Chapter 19, this volume). But to judge progress and give early warnings we need a broader view than the few celebrated cases.

One problem is the number of species to consider, presumably at least an order of magnitude higher than their related crops. As well, there are two attitudes to the diversity in wild relatives. On the one hand some (e.g. Marshall, 1989) have pointed out that for many crops, many areas of the world are colonized by certain wild relatives of crops (e.g. wild *Avena* species), and that such cases would not merit monitoring. On the other hand, others have argued that highly specific ecotypes or populations have proven value (e.g. the sources of certain crown rust resistances in *Avena* spp. in Israel), and these sometimes weedy populations collected beside farmers fields are endangered and should be monitored.

With the emergence of IUCN red book lists, it becomes possible to get an overview of the situation at least at the species level. The task at the national level is to consider the indigenous species of genera that have domesticated or economically useful species, decide which species are rare and threatened, and for those obtain estimates of population number and sizes. A second approach is to consider which crop-related wild species are present in nature reserves and other protected areas and which are not. Of those omitted, the question to check is whether their population distribution, number and sizes are such that they need no conservation action.

Table 24.5 is an example of what can be done with such lists. It lists the genera related to economic plants that are included in Briggs and Leigh's (1996) authoritative survey of the conservation status of rare or threatened plants of Australia (ROTAP), and the number of native species that belong to each category. Of the crop genera listed, appreciable proportions of wild species are of conservation concern. Over half of these were too 'poorly known' to classify as to their endangerment status. While some are known to be in reserves, only about 20% of species related to field and vegetable crops and 'at or likely to be at risk' could be checked off as adequately reserved.

The situation for eucalypts is perhaps most indicative. They are big plants, easy to see and well

Table 24.5. The number of Australian native species congeneric with or closely related to crops, and the number of those species in various conservation categories.

Genus	Australian	Extinct	Endangered	Vulnerable	Rare	Poorly known	In reserves
Field Crops							
Glycine	*c.* 25			1	1	4	3
Gossypium	17				2	4	3
Sorghum	17				1	1	
Nicotiana	20				1	3	1
Amaranthus	27					1	
Cajanus	11		1			1	
Corchorus	*c.* 40		1		1	5	2
Vegetables							
Solanum	*c.* 100			3	4	6	5
Dioscorea	5					1	
Ipomoea	50			1	2	2	1
Apium	4			1			
Oils, fruits, nuts, spices							
Eucalyptus	*c.* 800		13	68	98	80	151
Melaleuca	*c.* 250			2	9	26	16
Macadamia	7		1	4	2		5
Ficus	40					5	
Syzygium	*c.* 80			4	13	2	12
Cinnamomum	5				2		2
Piper	8				1		1
Musa	3	1[a]			1		1

Summarized ROTAP definitions (following Briggs and Leigh, 1996):
Extinct, not collected or otherwise verified over the past 50 years despite thorough searching.
Endangered, in serious risk of disappearing from the wild within 10–20 years if present land use and other threats continue to operate. Includes populations too small to be assured survival even if conserved.
Vulnerable, not presently endangered but at risk over a longer period (20–50 years) under current conditions or because it occurs on land whose future use will place it at risk.
Rare, taxon that is rare in Australia but does not have an identifiable threat. Small number of large populations in a restricted area or small populations over large areas, or combinations of the two preceding conditions.
Poorly known, suspected to belong to one of the above categories.
Crop genera not listed: *Oryza, Linum, Vigna, Chenopodium, Cucumis, Citrullis, Abelmoschus, Alocasia, Citrus, Prunus, Olea, Rubus, Myristica.*
[a]*Musa fitzalanii* F. Muell. Daintree's River banana, presently known only from the type specimen.

studied, yet 10% of species are probably of conservation concern and their vulnerability too poorly known to classify. Of the 259 species of conservation interest, only 151 are adequately represented in reserves. At the other extreme the situation for crop genera for which no species are listed, might imply that their species are indeed at no risk (e.g. *Oryza*), or that they have escaped attention.

Our suggested indicators (Table 24.2) exploit the general tendency for genetic diversity to increase with increasing population size (Young *et al.*, 1996). Area occupied could be a surrogate for size. Another contender is population density, because density may often be easier to measure than population size to which it is probably related. However, as Gram and Sork (1999) found, density can also be an indicator of habitat quality. In their study, populations with small densities had different genotypes than those with high densities, and thus they recommend choosing populations to represent a range of densities. Another possible approach borrows from that of ecosystem conservationists who have used geographic environmental data to partition areas and select areas for reservation that represent environmental heterogeneity for the species concerned (e.g. Belbin, 1993).

In assessing collecting priorities among alleles, the 2 × 2 classification of alleles based on their *population frequency* (common versus rare) and distribution among populations of a species (widespread versus localized) has proved useful. In a parallel manner, Murray *et al.* (1999) have classified species based on their *population sizes* (abundant versus sparse) and geographic distribution in a given area (everywhere versus somewhere, or sometime). They sought life-history characteristics that would be indicators of different kinds of rarity among species, and hence vulnerability to loss. In our case, the number of wild related species for which such classifications are possible from existing data indicates our progress in predicting vulnerable genepools.

As before, interpretation of basic data on occurrence, distribution and reservation in protected areas would be much more meaningful with genetic and systematic studies of the population genetic structure of these species. Evidence of cryptic genomic diversity, subspecies or unrecognized species such as have come to light in perennial *Glycine* species in Australia, or evidence of highly non-uniform levels of diversity in geographic patterns (see Schoen and Brown, 1991) are important pointers to improving reservation strategies.

Secondary genepools *ex situ*

The final cell of Table 24.2 concerns indicators for the sustainable management of accessions of wild related species in *ex situ* collections. The primary, long-established reason for such samples is to have them at hand for study and use in breeding (Frankel and Soulé, 1981). Hence, except for the indicator dealing with coverage, the indicators we suggest have a strong flavour of measuring actual use. The primary indicator is the number of different accessions (of each of the relevant wild species in a collection) actually used in research or breeding, and the frequency of such use in a given time interval. These could be computed from statistics on requests, or from surveys of the literature reporting materials used in research as done by Dudnik *et al.* (2001).

As a numerical example, Table 24.6 contains some measures of the pattern of use for the CSIRO collection of indigenous relatives of *Gossypium* over the last few years. Although the numbers are very small, they illustrate ways to track the use of diversity with indicators. The data are for dispatches from the genebank in answer to requests, for accessions used in first hybrids with either diploid or tetraploid cottons, and used in research crosses among the wild species themselves. Since indicators are meant to be comparative, the intent here is to compare usage of the three genome groups of species: the subtropical C-genome (two species), the tropical G- (three species) and the highly interesting K-group (12 species) that come from the Kimberley region of Western Australia. The latter are rare and very difficult to handle in a genebank. Simply getting seed is a challenge in many cases. From the basic frequency distribution of number of accessions used once, twice, three times, etc., we compute the total number of different accessions used (*richness*); and the coefficient of variation (CV) of the frequency of use (*evenness*). To make the figures comparable, the richness values in the table are adjusted to a total usage of 30 by resampling procedures. The evenness figures are transformed to a negative exponential. This means that complete or maximum evenness, which would be identically the same frequency for all accessions used gives a CV of zero and a value in the table of 1.0. Evenness values on this scale range from zero to one, with higher values being more desirable, other things being equal.

The data point to the effect of practical limits on using the diversity in the K-group in dispatches.

Table 24.6. Numerical example: Patterns of usage in germplasm shipments and in hybridization research of the CSIRO collection of indigenous *Gossypium* species.

	C-genome	G-genome	K-genome
Number of species in the genome group	2	3	12
Number of accessions in the collection (June 2000)	154	228	81
Number in long-term store	56	112	15
Dispatches			
Number of samples sent on request	45	59	64
Number of different accessions sent	31	45	27
Richness (for a total of 30 samples)	18.4	21.2	16.4
Evenness (exp[-CV of use])	0.63	0.66	0.54
SW Information Index (SE)	3.34 (0.07)	3.73 (0.05)	3.12 (0.07)
Hybridization with cotton (4X or 2X)			
Number of cross combinations attempted	32	32	100
Number of accessions used as parents	12	8	33
Richness (for a total of 30 crosses)	10.4	7.6	18.4
Evenness (exp[-CV of use])	0.66	0.69	0.59
SW Information Index (SE)	2.4 (0.07)	2.0 (0.06)	3.4 (0.05)
Hybridization among wild accessions			
Number of cross combinations attempted	68	33	72
Number of accessions used as parents	25	11	81
Richness (for a total of 30 crosses)	15,8	9.5	17.8
Evenness (exp[-CV of use])	0.55	0.58	0.46
SW Information Index (SE)	3.05 (0.07)	2.25 (0.09)	3.31 (0.08)

Richness, the number of different accessions in the total effort, standardized to a given size by geometric resampling procedures (Brown, 1989).
Evenness, coefficient of variation of the frequency of accession use.
SW Information Index, Shannon-Weaver Information Index, combining both richness and evenness (Hutcheson, 1970).

On the other hand, a richer sample of K-accessions has been used in crosses while the effort in the G-genome has been highly concentrated. There are sound reasons for these trends, especially the scientific novelty and importance of the K-crosses, but the point to make here is that statistics to track these patterns are useful in crosses as indicators. Other factors that affect usage in crosses in this case of *Gossypium* research include rate of success, coverage of the species range, presence of useful traits, ease of growing and abundance of flowering and seeding.

How can we measure the adequacy of coverage of diversity in nature by the samples in hand? What intensity and how many samples per species in total should be the target? Sampling benchmarks are already much debated in the literature on genetic resources. They vary widely depending on the kind of species (crop, wild herb, forest tree), the abundance of the material (scarce or plentiful), the purpose of the sample (locally common versus species rare alleles) and the assumptions of the authors. Here we simply want to stress that a sampling of the complete geographic range of each wild species is a major goal.

An important area that commands much current research attention is the study of species relationships and development of molecular phylogenies. Appraising and counting such studies, although potentially involving heated debate, could indicate how well we are coming to know the heritage of crop genetic diversity, and evolution under domestication.

The other main area of endeavour for wild relatives concerns 'prebreeding' activities. These are research attempts to make the genetic resources in genepools distant from breeders' populations, more available to breeders. Examples are the development of near-isogenic lines from wild species (e.g. barley) or the development of hybrid populations for molecular mapping of desirable traits and linked markers for manipulating traits in breeding programmes. Summaries of the number of lines, mapping populations and traits developed for each crop would provide an instructive overview of the penetration of wild genetic resources into crop improvement.

A tool of potential importance is the development of collections of DNA samples *in glacie* from wild species (see Frankel *et al.*, 1995, for discussion). If readily available, this kind of material could significantly expedite research, and test hypotheses. Such research may lead to more efficient use of living samples, even if it is not the direct source of useful genes. As for any germplasm collection, the size, composition, quality of accessions and documentation, and actual usage of DNA banks all need monitoring. Presently DNA samples *in glacie* form an adjunct method, rather than replacing other conservation methods and, as such, may be themselves evidence of the usage of *ex situ* wild collections, to which they are linked.

Role of Newer Molecular Techniques in the Development of Indicators

A primary aim of this volume is to map out the future role that new technologies can take in managing PGR. In concert with this aim, we here ask what established and emerging molecular techniques might offer in devising, implementing or improving indicators for sustainable management of PGR.

Molecular techniques give the power to monitor genetic variation right at the elemental level of DNA sequences. Starting with the introduction of isozymes, there has been an explosion in the technologies available for directly assaying genetic differences among organisms. It is now possible to compare organisms from the genome level (using for example fluorescent *in situ* hybridization (FISH) and genomic *in situ* hybridization (GISH)), down to the level of single nucleotides (DNA sequencing and single nucleotide polymorphisms (SNPs)). Because the genome is assayed directly, these new technologies circumvent the often poor correspondence between morphological and genetic diversity in crop species. Thus, in theory they could increase the validity and credibility of indicators by providing markers that offer greater clarity of interpretation.

The immediate and obvious benefit is the flexibility and precision by which genetic diversity can be assayed. Marker systems can be tailored to specific organisms to accommodate differences in breeding systems and relative levels of genetic diversity, and can be scaled depending on the number of accessions to be screened, how many loci are needed, and which sequences in the genome are to be sampled. Furthermore many anonymous DNA markers (e.g. randomly amplified polymorphic DNA (RAPDs), restricted fragment length polymorphisms (RFLPs), amplified fragment length polymorphisms (AFLPs)) can be drawn from genomic maps guaranteeing that genetic diversity assays evenly sample the genome. With sequence tagged sites (STSs) developed from expressed sequence tags (ESTs) it is even possible to use expressed genes specific to life history stages, rather than anonymous sequence differences to assay genetic differences among accessions. Because database comparisons can often identify the functional product of an EST, the genebank manager not only gets an indicator of genetic diversity and relationships among accessions but an increase in the informational content of the sampled accession.

There are clear benefits to the greater use of these more precise measures of genetic variation. Equally clearly, they are costly in human and financial resources. They can only be employed in a limited number of collections. Therefore, the selection of which species and which samples is crucial. Since the aim is to obtain the maximum amount of useful information from a limited sample, the use of core collections is an obvious approach. Core collections are designated using all the data available, to make their entries representative of genetic diversity. The basic procedure is to recognize groups of related or similar accessions within the collection, and sample from each group. DNA sequence analysis provides the opportunity to measure how different these empirically derived groups are, and test for relationships between them.

Here is a fundamental gain in genetic knowledge, not only to prove that two individuals or gene copies differ, but to be able to place them in a phylogenetic hierarchy of relationships, based on recency of a shared ancestor. Once this is done, the phylogenetic diversity of the collection can be estimated (Crozier, 1997). Calculating the phylogenetic diversity of a collection allows PGR managers to extend and improve core collections of secondary and tertiary genepools. Maintenance of these wild related species is often problematic relative to primary genepool accessions (Brown *et al.*, 1997a). In speciose groups, maintaining large unchecked collections of every related species is simply not feasible. Phylogenetic diversity measures can be used to identify a subset of related wild species that maximizes the genetic information content of the collection (Crozier, 1997).

A case in point would be the wild Australian cotton (*Gossypium*) species (Brown and Brubaker, 2000). Molecular phylogenetic analyses identified three main lineages among the 17 species; lineages that are now recognized as distinct genomes. The recognition of these three evolutionary lineages is a key variable in three areas: (i) sources of agronomically useful traits; (ii) ability to be incorporated into cotton breeding programmes; and (iii) risk assessment of transgene escape from genetically modified cotton cultivars (Brubaker *et al.*, 1999). In all three areas, the phylogenetic topology has proved to have predictive value and allows the identification of indicator species for each lineage. Equally important, the phylogenetic topology allows germplasm managers to identify high priority species for germplasm further collection and maintenance in what is a difficult group to manage.

Towards both Richness and Balance

Up until now, the pursuit of diversity 'richness' might be seen as typifying the prevailing philosophy of PGR. Sampling and conservation strategies have sought to maximize the diversity of collections mainly in terms of the total numbers of different genetic entities or types, be they alleles, genotypes or clones. Richness has been the guiding concept of diversity that has shaped much of the thinking. Such a view of diversity is straightforward. The problem with 'richness' from a managerial point of view is that richness is a function of numbers. Larger collections and populations will have more genotypes. There is no evaluation of effort, no setting of priorities competitively. If richness is the dominating indicator, there is no check on growth, nor a check against bias or imbalance in a collection. Large numbers of individuals close at hand can readily be sampled to bolster the number of genotypes, which necessarily must increase merely by enlarged numbers, although inefficiently.

There are many examples of this problem in wild species. Our collection of perennial *Glycine* species began as moderate numbers of accessions of the species near to hand in south-eastern Australia. In later phases it has become clear that the high species diversity is in the monsoonal areas of northern Australia and particularly the remote Kimberley district. Because of ease of access, the collection is biased in numbers to populations near at hand. The same story holds for the wild Australian *Gossypium* species, which also have striking diversity in the remote north-west. Ideally our collections should be more balanced.

Table 24.7 gives a comparison of the kind of questions that are either dominated by the richness

Table 24.7. Comparison of typical questions and perspectives from the previous period of emphasis on 'richness' with those in the new era of 'balance'.

	Richness	Balance
Genetic erosion	Does loss of an allele or clone have to be complete to constitute genetic erosion? Should we be concerned with the loss of *any* allele from a species?	How can we increase the evenness in frequency of use of genotypes, varieties and crops?
Germplasm collections	Have we collected and conserved *all* alleles and genotypes and maintained *all* accessions? Is richness maximized?	Are collections organized for ease of use? Has collection *evenness* improved, by discarding redundancies, gap filling, etc.?
Clonal species	Are *all* clonal genotypes growing in a secure field genebank and *in vitro*?	Have field genebanks been brought to manageable size and duplicated to ensure they are not lost due to catastrophes?
Wild relatives	Are *all* wild related species collected and conserved?	Are structured representative samples of wild relatives on hand for study and use?
In situ conservation	Has a full description of all the genotypes and all the forces been made to ensure conservation of all?	Are programmes of participatory plant breeding and improvement in place?

or the balance in approach. We contend that what is now needed is the other concept of point diversity, namely the evenness of frequencies of types. The concept here is that when two items are drawn from a collection or from an assembly of populations they will differ with appreciable probability.

The discussion of Table 24.2 gave primacy to data that have both numbers and frequencies of types. These frequency distributions provide the basis on which both richness and balance can be computed. One possibility for a combined index, admittedly somewhat abstruse, is to compute the Shannon-Weaver Information Index, which brings together both richness and evenness. The relative unfamiliarity of such an index and its dependence on the kinds and extent of the basic data, however, work against its adoption, but we believe this is the direction in which we should head.

The next challenge for improving balance is to incorporate measures of divergence such as phylogenetic distance, as outlined above. Ideally we wish to measure progress by the sustained retention of a large number of types (richness), all in appreciable frequencies (evenness) and covering a wide spectrum of diversity (phylogenetic divergence). Such measures still do not address any parameters of differential value of types; which are useful alleles and which are not, etc. Should we hand the search for indicators over to the economists? There may be arguments for doing that, as economic value is likely to be a principal factor in determining which fraction of biodiversity will survive. However, we are challenged to sustain PGR despite the pressure of market forces, and indicators that focus on the diversity itself are essential in this process.

Conclusions

The 20th century leaves us with a challenging and essential task: the sustainable management of our physical and biological resources. An indicator is a significant physical, chemical, biological, social or economic variable that is measurable in a defined way for management purposes. For plant biodiversity, indicators are needed at the levels of communities, species and genes. This last is the most difficult, and yet the most important. Many measures are plausible, ranging from the number of varieties of a crop species and the area patterns of their planting *in situ*, or the number of distinct accessions in genebanks *ex situ*, through to measures of DNA sequence and genome diversity among populations. They differ in cost, precision and scale of application. Modern technologies have roles in the devising of structured sampling strategies, and the more accurate measuring of population and species relationships. The phylogenetic framework will be an important tool in assessing priorities for action. These approaches will help to make both the collections of PGR, and their conservation and use, richer and more balanced in the coming century.

Acknowledgements

This paper was completed while AHDB was a visiting research fellow at the Department of Systematic Botany, University of Osnabrueck and honorary research fellow of IPGRI in Rome. AHDB thanks in particular Drs Toby Hodgkin, Jan Engels, Devra Jarvis, Pablo Eyzaguirre and Herbert Hurka for comments on the manuscript.

References

Belbin, L. (1993) Environmental representativeness: Regional partitioning and reserve selection. *Biological Conservation* 66, 223–230.

Boyle, T.J. (2000) Criteria and indicators for the conservation of genetic diversity. In: Young, A., Boshier, D. and Boyle, T. (eds) *Forest Conservation and Genetics – Principles and Practice*. CAB International, Wallingford, UK, and CSIRO Publishing, Collingwood, Australia, pp. 239–251.

Boyle, T.J.B. and Sayer, J.A. (1995) Measuring, monitoring and conserving biodiversity in managed tropical forests. *Commonwealth Forestry Review* 74, 20–25.

Briggs, J.H. and Leigh, J.L. (1996) Rare or threatened plants of Australia. CSIRO, Collingwood, Australia.

Brown, A.H.D. (1989) Core collections: a practical approach to genetic resources management. *Genome* 31, 818–824.

Brown, A.H.D. (1999) The genetic structure of crop landraces and the challenge to conserve them *in situ* on farms. In: Brush, S.B. (ed.) *Genes in the Field: On-Farm Conservation of Crop Diversity*. Lewis Publishers, Boca Raton, Florida, pp. 29–48.

Brown, A.H.D. and Brubaker, C.L. (2000) Genetics and the conservation of Australian wild relatives of crops. *Australian Journal of Botany* 48, 297–303.

Brown, A.H.D. and Spillane, C. (1999) Implementing core collections – principles, procedures, progress, problems and promise. In: *Core Collections for Today and Tomorrow*. IPGRI, Rome, Italy, pp. 1–9.

Brown, A.H.D, Brubaker, C.L. and Grace, J.P. (1997a) The regeneration of germplasm samples. Wild versus cultivated species. *Crop Science* 37, 7–13.

Brown, A., Young, A., Burdon, J., Christidis, L., Clarke, G., Coates, D. and Sherwin, W. (1997b) *Genetic indicators for state of the environment reporting*, Australia: State of the Environment Technical Paper Series (environmental indicators). Department of the Environment, Sport and Territories, Canberra, Australia, 29pp.

Brubaker, C.L., Brown, A.H.D., Stewart, J.McD., Kilby, M.J. and Grace, J.P. (1999) Production of fertile hybrid germplasm with diploid Australian *Gossypium* species for cotton improvement. *Euphytica* 108, 199–213.

Brush, S.B. (1995) *In situ* conservation of landraces in centers of crop diversity. *Crop Science* 35, 346–354.

Crozier, R.H. (1997) Preserving the information content of species: genetic diversity, phylogeny and conservation worth. *Annual Review of Ecology and Systematics* 28, 243–268.

Dudnik, N.S., Thormann, I. and Hodgkin, T. (2001) The extent and use of plant genetic resources in research – a literature survey. *Crop Science* 41, 6–10.

FAO (Food and Agriculture Organization of the United Nations) (1998) *The State of the World's Plant Genetic Resources for Food and Agriculture*. FAO, Rome, Italy.

FAO and IPGRI (1994) *Genebank Standards*. FAO, Rome and International Plant Genetic Resources Institute, Rome, Italy.

Ferris, R. and Humphrey, J.W. (1999) A review of potential biodiversity indicators for application in British forests. *Forestry* 72, 313–328.

Frankel, O.H. and Soulé, M.E. (1981) *Conservation and Evolution*. Cambridge University Press, Cambridge, UK.

Frankel, O.H., Brown, A.H.D. and Burdon, J.J. (1995) *The Conservation of Plant Biodiversity*. Cambridge University Press, Cambridge, UK.

Gram, W.K. and Sork, V.L. (1999) Population density as a predictor of genetic variation for woody plant species. *Conservation Biology* 13, 1079–1087.

Holden, J.H.W. (1984) The second ten years. In: Holden, J.H.W. and Williams, J.T. (eds) *Crop Genetic Resources: Conservation and Evaluation*. Allen and Unwin, London, pp. 227–285.

Holden, J.H.W., Williams, J.T. and Peacock, W.J. (1993) *Genes, Crops and the Environment*. Cambridge University Press, Cambridge, UK.

Hutcheson, K. (1970) A test for comparing diversities based on the Shannon formula. *Journal of Theoretical Biology* 29, 151–154.

Ledig, F.T. (1992) Human impacts on genetic diversity in forest ecosystems. *Oikos* 63, 87–108.

McKenney, D.W., Sims, R.A., Soule, M.E., Mackey, B.G. and Campbell, K.L. (1994) *Towards a set of biodiversity indicators for Canadian forests: Proceedings of a forest biodiversity indicators workshop*. National Resources Canada, Canadian Forest Service. Sault Ste. Marie, Ontario, Canada.

Marshall, D.R. (1989) Crop genetic resources: Current and emerging issues. In: Brown, A.H.D., Clegg, M.T., Kahler, A.L. and Weir, B.S. (eds) *Plant Population Genetics, Breeding and Genetic Resources*. Sinauer Associates, Sunderland, Massachusetts, pp. 367–388.

Murray, B.R., Rice, B.L., Keith, K.A., Myerscough, P.J., Howell, J., Floyd, A.G., Mills, K. and Westoby, M. (1999) Species in the tail of rank-abundance curves. *Ecology* 80, 1806–1816.

Namkoong, G., Boyle, T., Gregorius, H.-R., Joly, H., Savolainen, O., Ratnam, W. and Young, A. (1996) *Testing Criteria and Indicators for Assessing the Sustainability of Forest Management: Genetic Criteria and Indicators*. CIFOR, Bogor, Indonesia.

Saunders, D., Margules, C. and Hill, B. (1998) *Environmental Indicators for National State of the Environment Reporting – Biodiversity*, Australia: State of the Environment (Environmental indicator reports). Department of the Environment, Canberra, Australia.

Savolainen, O. (2000) Criteria and indicators for genetically sustainable forestry. In: Turek, J. and Geburek, Th. (eds) *International Collaboration on Forest Genetic Resources: the Role of Europe. Proceedings of 2nd EUFORGEN Steering Committee, 1998 Vienna, Austria*. IPGRI, Rome, Italy, pp. 26–29.

Schoen, D.J. and Brown, A.H.D. (1991) Intraspecific variation in population gene diversity and effective population size correlates with the mating system in plants. *Proceedings of the National Academy of Sciences USA* 88, 4494–4497.

Teshome, A., Baum, B.R., Fahrig, L., Torrance, J.K., Arnason, J.T. and Lambert, J.D. (1997) Sorghum (*Sorghum bicolor* (L.) Moesch) landrace variation and classification in north Shewa and south Welo regions of Ethiopia. *Euphytica* 97, 255–263.

Young, A., Boyle, T. and Brown, A. (1996) The population genetic consequences of habitat fragmentation for plants. *TREE* 11, 413–418.

25 Decision-making Strategies for Conservation and Use of Forest Genetic Resources

M.P. Koshy,[1] G. Namkoong,[1] P. Kageyama,[2] A. Stella,[2] F. Gandara[2] and W.A. Neves do Amaral[3]

[1]Department of Forest Sciences, University of British Columbia, Vancouver, British Columbia, Canada; [2]Instituto de Pesquisas e Estudos Florestais, Piracicaba, Brazil; [3]International Plant Genetic Resources Institute (IPGRI), Rome, Italy

Introduction

The concept of forest genetic resources refers to the environmental, social, economic, cultural and scientific values of the heritable materials contained within and among tree species. Conservation of forest genetic resources is regarded as constituting the actions and policies that assure the continued existence, evolution and availability of these resources in the future. A central problem is striking a balance between resource use and the effectiveness of conservation management such that high levels of efficiency can be expected. The problem related to the conservation and use of genetic resources is particularly challenging since it is not only the genes that must be conserved, but also the underlying evolutionary processes (Namkoong and Koshy, 2000).

While the goal of conserving forest genetic resources can be simply stated, its implementation can be very complex. With thousands of woody species distributed among several local populations, which are intermating groups of individuals, each with thousands of variable genetic loci, and with relatively little data on most, the temptation is to assume that anything that is done is of equal value and to use de facto priority by convenience or familiarity. However, some genetic resources may be resilient and require only benign neglect, while others may experience degradation that portends the extinction of genes or species, unless conservation efforts are targeted to them. It is important to understand usefulness of the kinds of data on the status and dynamics of genetic resources that can affect gene conservation efficiency. Also, the kinds of species that are of significance to conserve and their vulnerabilities are important information to collate with the kinds of *in situ* and *ex situ* management techniques that can efficiently assuage the threats (Namkoong, 1998).

National governments need methods for establishing priorities for conservation that consider the large potential number of species for which they may be responsible. Sometimes, the focus may be on species that funding agencies may be particularly familiar with for their charismatic appeal. As an alternative, we propose to focus on perceived threats, on perceived value for economic or ecological flagship reasons and on species that have peculiar and possibly susceptible stages of life history such as non-traditional pollination or particular seed germination mechanisms. With limited funding, some way is needed to choose more rationally species or populations to work with that are most at risk, most important, and for which the means we have available for conservation can be effective.

This requires some knowledge of the status and dynamics of genetic diversity, a rating of its potential value, an evaluation of the threats, and the potential for conservation management. The product can be a priority ranking for management or a classification of species into priority groups. The concept of triage (Myers, 1979) is perhaps applicable here. Given the goal of conserving the gene resource, tree species could be classified into three classes: those for which whatever measures undertaken would not help conservation, those species that will survive without any management, and those which will survive if management means within reach is carried out. It is important that the third category of species be identified and resources allocated to them (Vane-Wright, 1996).

In an ideal case, a listing and mapping of species' occurrences would be available, the kind and extent of threats to populations would be known, and methods of conservation and management would be well established. It would then be possible to evaluate risks, the costs, and certainties of how management would reduce those risks, and by evaluating those resources for economic or ecological payoffs, priorities could be established. Such detailed knowledge, however, is rarely complete.

In most conservation programmes, one person usually has to make decisions on what management actions to take but may also have limited access to research that can help to make those decisions as objective as possible. It seems obvious that the two problems are related but since each kind of question is complicated, it is often difficult to sort out how information is best used to come to rational decisions. As a manager, the conservationist must consider many factors in setting priorities for actions. These include the importance of the species or population to the economy or the ecology of the region, the risk level to which the resources are exposed, and the expected return those alternative management practices may have. In some cases the conservationist can delay decisions until more scientifically based information becomes available. In those cases, the conservationist must decide what features of the biology of the organisms involved are least known but if studied, are most likely to make a difference in how the organisms are managed. In this kind of applied research, we are not so much interested in descriptive biology as we are in the biology of genetically and ecologically critical functions (Purvis and Hector, 2000).

In this chapter, we suggest a way to organize thinking on how we make informed decisions about conservation and how we can construct priorities for conservation management actions. We assume that species and populations differ in their levels of susceptibility to severe loss or extinction. Those that have low population sizes or are widely dispersed, and those that have bottlenecks in their reproductive processes would have obvious susceptibilities. We also assume that different threats may exist such as being exposed to harvesting, or to general environmental degradation. We further assume that it is the combination of susceptibility and threat that induces the risks of loss that should be cause for alarm. That is, targeting action on a species that may have a susceptible life history stage but for which no threat exists, is assumed to be a waste of valuable resources in the short run.

We also assume that we wish to target efforts to those cases in which management can affect either the susceptibility, such as by maintaining pollinators, or the threat, by reducing harvesting and thereby reducing the risk of loss. We suggest a method to estimate the value of reducing risk and to make a classification and prioritization of effective actions so that management decisions can be made as to which actions are most beneficial. We further assume that there are cases where the manager feels that 'if only I knew more' better decisions could be made, but that the costs of obtaining information in time or money also vary substantially and require another level of prioritization.

In a decision tree approach, we try to separate the kinds of information that conservationists may have available and to construct a formal way to set priorities. We are often faced with the practical reality that all the desired information is not available. While general information is available, the ways in which it can best be integrated and deployed for decision-making has rarely been considered in any detail. An integrated decision-making framework provides procedures that can be used by managers for choosing the activities needed. It would allow selection of species, sites, populations and conservation techniques to be considered in an integrated manner to ensure effective conservation of species and the maintenance of their intraspecific diversity. The goal of this study is to produce such a framework and develop an initial step-by-step guide for prioritizing genetic conservation and use programmes to test its applicability in a specific site in Brazil. The development of such an integrated decision support mechanism to establish priorities for a large number of potentially threatened species would provide a valuable tool for conservation decision-makers. This decision-making process assumes that goals of the programme are set and harmonized at different administrative levels of organization.

Factors Affecting Conservation Management Decisions

Risk and value, in general, largely determine conservation management decisions. Threats and susceptibilities are factors that affect risk. In biological conservation, inherent value is in the context of sustainable productivity. This productivity is defined in terms of goals set by the stakeholders. In a broader biological sense, this value has economic and ecological components. While estimating the value, it is important to ascertain comparative value for economic and ecological reasons, though there may be interdependencies. While economic value may be defined in terms of monetary, social or spiritual returns, ecological value will be based on interdependencies of the species for sustainability. To harmonize objectives, it is important to get the judgement of different stakeholders as the values are bound to differ with expectations (Margules and Pressey, 2000; McCann, 2000).

Another major factor that will affect management decisions is the attitude towards risk (Pukkala and Kangas, 1995). For example, in the business world, a small company may not be able to take a large monetary risk, while a larger company may be able to buffer a larger loss. However, in the case of biological diversity, it is very difficult to quantify attitudes to risk. While it may be acceptable to lose a few species at the national or geographical range level, local ecosystems may be very sensitive to loss of even one species. Such sensitivities make it difficult to make decisions, when they span various levels of management. It may be possible to compensate the lower level by the upper level stakeholders, when a management decision is made to suit higher level objectives as a compromise. For example, local communities can be compensated by provincial or federal government. We do not address these issues of valuation but note that some approaches to their resolution are available.

Uncertainty is also a major factor affecting management decisions (Ronen, 1988). Estimates of susceptibilities and threats are often based on vague generalizations with little supporting data. This will affect the precision of the estimates of actual risk. This brings into consideration the importance of additional information and its value (Bunn, 1984). When resources are limited or the management is obligated to maximize returns for the funds committed, it is often a challenge to decide whether to go for further information, which will give a more precise estimate of the susceptibilities and risk or invest in management operations. If such investment is guaranteed to bring better returns, it will be justifiable. A decision analysis will be able to give the manager appropriate values for additional information to make a decision on whether or not to collect more information.

Decision Tree Approach

Decision analysis is the systematic evaluation of alternative actions as a basis for choice among them (Brown *et al.*, 1974). Decision-makers in business, industry, government and other organizations are applying this discipline to their everyday problems.

Decision trees and influence diagrams are generally used in the decision-making process. Decision trees are comprehensive tools for modelling all possible decision options. While influence diagrams produce a compact summary of a problem, decision trees can show the problem in greater detail. Decision trees also describe the events in a chronological order. Once the decision tree is structured in terms of possible actions and events, which are probable consequences of these actions, it can be evaluated in terms of costs and probabilities. The implication of these evaluations for the courses of action can be figured out. This process is called ‘folding back’. The goal of folding back is to calculate what each of the initial acts is worth. Once these values and their probabilities are known, we will be able to calculate the distribution of values. This distribution will help to find the expected value and variance. The expected value is a probability weighted average in that each of the profits or losses is weighed by its probability of occurrence. This expected value is one measure of the value of an act and it is often used as a measure of suitability of actions by maximizing expected value. The variance of the return will give a measure of risk (Bunn, 1984). For example, in the decision tree given in Fig. 25.1, we can estimate the expected value of managing based on collecting genetic data. Assuming that we have collected demographic (d info) and reproduction (ng info) data, and now also have genetic data (g info), the expected value of managing depends on whether or not management is really needed. We have obtained better information and have a higher probability of making the correct decision but may still make a mistake. If we manage and it turns out to be needed, the benefit is \$1,000,000 but it cost (\$1,000 + \$3,000 + \$10,000 + \$100,000) \$114,000 for a net benefit of \$886,000. The probability of making that decision correctly is 0.85. However, if we manage and it is not needed, then subtracting the additional opportunity cost of \$100,000 leaves a net benefit of \$786,000 and the probability of that event is 0.15. The expected value is then \$886,000 × 0.85 + \$786,000 × 0.15 = \$871,000.

When additional information is fed into the tree, the expected value may change and this difference is defined as the value for obtaining information. The cost and benefit of each factor can be evaluated using sensitivity analysis.

Sensitivity Analysis

Sensitivity analysis helps us to understand which variables matter most in the decision. It measures the impact of changing uncertain variables to extreme values, while keeping other variables constant. By examining the impact of reasonable changes in base-case assumptions, sensitivity analysis determines which variable has little impact on the outcome and can be treated as determinist. Sensitivity analysis does not give an explicit answer to the problem, but will give a better understanding of the model. For example, in the decision tree given in Fig. 25.1, sensitivity can be tested for cost of management and cost of collecting information. If the cost of collecting genetic data could be only half of that listed if we had a local facility, we might find that the value of obtaining genetic data and then deciding on management is often of value. If it is sufficiently better then we might want to get data more often and then consider investing in the development of local facilities. However, the final evaluation of options may be insensitive to such cost savings and data costs are not important for management decisions.

This study tries to adapt such a decision tree approach to make conservation decisions and also ascertain the value of additional information. A study in Brazil has been undertaken to test this method. Currently, no data are available to demonstrate sensitivity analysis.

Field Study in Brazil

Area of study

This study was undertaken at two sites in Sao Paulo State in Brazil. The first site is at Caetetus Ecological Station in Galia. It is a 2000 ha semi-deciduous tropical forest with 267 tree species. It is a forest reserve under state care for 15 years. But before that it was a hunting reserve. The second site is in the Morro do Diabo State Park at Teodoro Sampaio. It covers over 35,000 ha of semi-deciduous tropical forest. There are approximately 300 tree species in this area.

Two steps were set up in the prioritization process. In the first step, species enumeration and evaluation of their potential threats and values were carried out. Then, species were ranked for further management considerations. In the second step, use of additional information and its cost were investigated. The field data were collected and collated under the supervision of Brazilian scientists in permanent sample plots and on sampled adult trees in the two sites during 1999 and 2000.

Evaluation of threats and values

A contingent ranking, which is a direct ranking, was adopted for threat and value evaluation. Species were evaluated for major threats such as exploitation intensity, harvesting, fragmentation, fire and fauna. A score ranging from 1 to 5 was assigned by stakeholders for each species. Also, certainty of the score was expressed as a probability. A relative weighting for each of these factors was also made. Total score for each species was calculated as follows:

$$T_k = \sum_{i=1}^{n} s_i w_i p_i$$

where T_k is the threat score for species *k*, and *s*, *w*, and *p* are score, weight and probability for threat *i*.

A method of scoring species value was also adopted that included a combination of economic as well as biological factors. Economic factors could

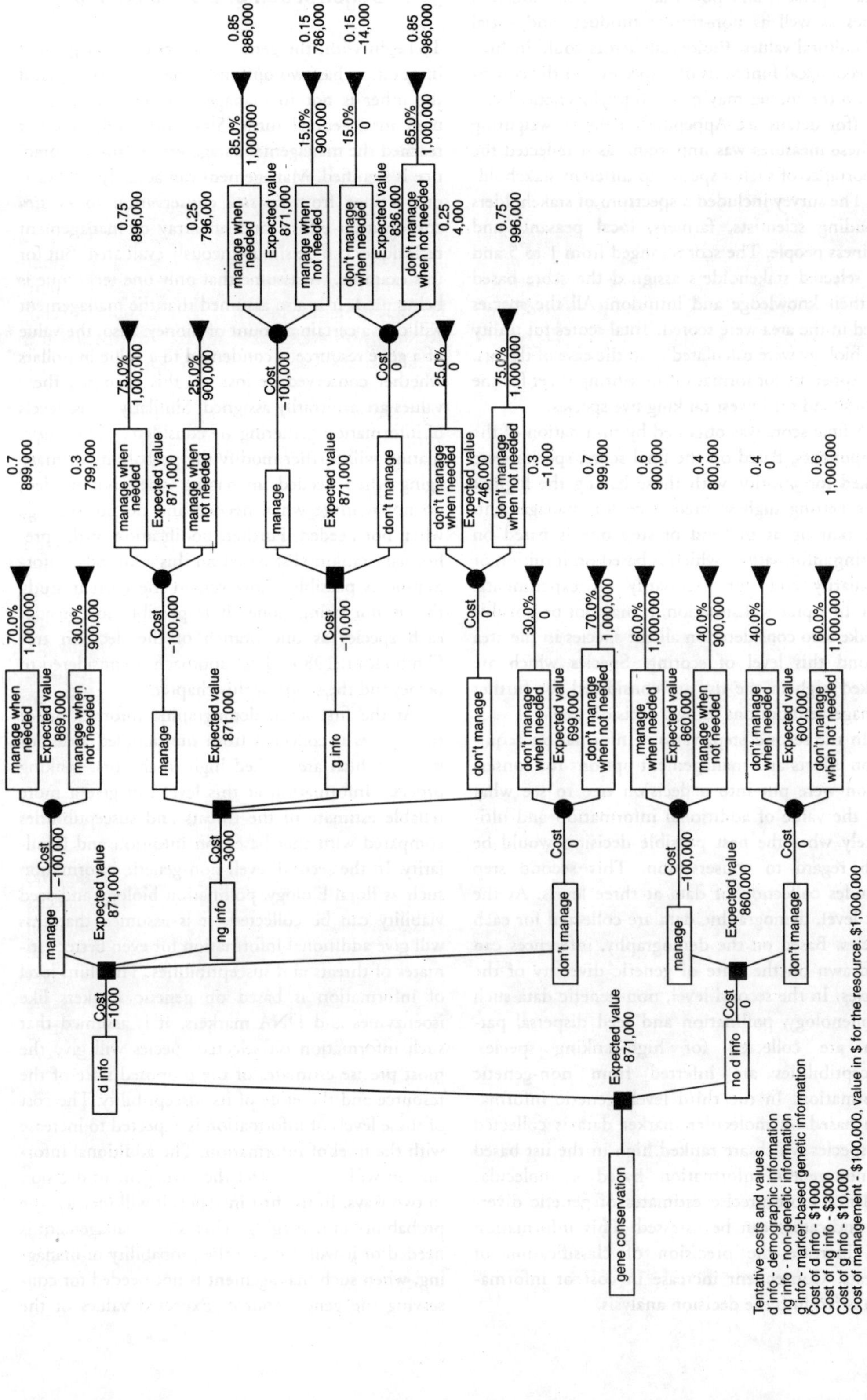

Fig. 25.1. Decision tree.

include present and potential wood and industrial values as well as non-timber products and social and cultural values. Biological factors could include the ecological functions of a species and the contribution the species may make to phylogenetic diversity (for details see Appendix). Relative weighting of these measures was important, as it reflected the importance of such a species to different stakeholders. The survey included a spectrum of stakeholders including scientists, farmers, local peasants and business people. The scores ranged from 1 to 5 and the selected stakeholders assigned the score based on their knowledge and intuition. All the species listed in the area were scored. Total scores for utility and biology were calculated as in the case of threats. See Appendix for format of the scoring sheet for the highest and the lowest ranking five species.

A final score was obtained by summation of the components. Based on the final score, species were ranked for priority, with those having the highest score getting highest preference for management. The ranking at the end of step one is based on existing information, which is based on intuition or familiarity and not necessarily on experimental data. In a practical situation, it may not be possible to take into consideration all the species in the area beyond this level of scoring. Species which are ranked high in the list are considered for further management options using a decision tree.

In the second step, various information acquisition efforts and management options for conservation were put into a decision tree to see what was the value of additional information and ultimately what the best possible decision would be with regard to conservation. This second step includes collection of data at three levels. At the first level, demographic data are collected for each species. Based on the demography, inferences can be drawn on the state of genetic diversity of the species. In the second level, non-genetic data such as phenology, pollination and seed dispersal pattern are collected for high-ranking species. Susceptibilities are inferred from non-genetic information. In the third level, genetic information based on molecular marker data is collected for species which are ranked high in the list based on non-genetic information. Based on molecular evidence more precise estimates of genetic diversity indicators can be assessed. This information will increase the precision of classification of threats. Consequent increase in cost of information is fed into the decision analysis.

Construction of a decision tree

To begin with, the gene conservation management in question has two options. One is to manage and the other is not to manage. The term manage is used in a general form. Since the study has not reached the management stage yet, no specific practice is signified. Management can actually take various actions from *in situ* conservation to *ex situ* conservation efforts and an array of management techniques can be simultaneously evaluated. But for this example, we assume that only one technique is being used. It is also assumed that the management will cost a certain amount of money. Also, the value of a gene resource is condensed to a value in dollars whether conserved or lost. In this example, these values are arbitrarily assigned. Similarly, three levels of information gathering are considered. This information will further modify the probability of managing when needed, managing when not needed, do not manage when needed and do not manage when not needed. Further modification of the preliminary ranking list based on this additional information is possible. However, in the current study this is not being done. It is possible to consider each species as one branch of the decision tree (Thibodeau, 1983). That approach is considered to be beyond the scope of this chapter.

At the first level, demographic information of the species is collected from the sampled areas for species which are ranked high in the first ranking process. Information at this level can give a more reliable estimate of the threats and susceptibilities compared with that based on intuition and familiarity. In the second level, non-genetic information such as floral biology, pollination biology and seed viability can be collected. It is assumed that this will give additional information for even better estimates of threats and susceptibilities. The third level of information is based on genetic markers like isoenzymes and DNA markers. It is assumed that such information on selected species will give the most precise estimates of the proposed state of the resource and therefore of its susceptibility. The cost of these levels of information is expected to increase with the level of information. The additional information will mainly affect the management decision in two ways. In the first instance, it will increase the probability of managing, when such management is needed or it will decrease the probability of managing, when such management is not needed for conserving the gene resource. Expected values of the different options will depend on the mean and distribution of the probabilities. More precise information will reduce the spread of the distribution of such probabilities. This will reduce the variation of the return, thereby reducing the probability of misclassification.

The decision tree used for this analysis is shown in Fig. 25.1. A statistical report of the decision based on expected value and standard deviation is given in Tables 25.1 and 25.2. The values are given for the decision point listed on the left for the scenarios listed across the top. The mean values refer to the expected values for all possible outcomes. The minimum values refer to the worst case scenarios and the maximum values refer to the best possible outcomes. Table 25.1 gives expected values when the cost of collecting genetic information is considered to be $10,000. Table 25.2 gives the same values when the cost of collecting genetic information is considered to be $5000. At each level of information there are different options to choose from.

Tables 25.1 and 25.2 give these options and their expected values. The first decision is whether or not to collect demographic information. In the case where no demographic information is collected, we could make a decision to manage or not to manage. If we decide to manage without demographic information, let us assume that the probability of managing when needed is 60% and managing when not needed is 40%. Similarly, if we decide not to manage, the probability of not managing when needed and not managing when not needed are set as 40% and 60%, respectively. As the field study is not yet completed, these probabilities are arbitrarily set for demonstration and are not based on hard evidence. If we assume that the cost of collection of demographic information is $1000 per species, the decision tree (Fig. 25.1) shows that the decision to collect demographic data is better compared with not collecting the demographic data. The expected value for the demographic data collection branch is $871,000 compared with $860,000 for the no demographic information option (see Table 25.1).

If we consider collecting demographic information, we may follow one of three courses of action: we may decide immediately to go ahead with a management system without further data; we may delay the management decision until genetic data are in hand; or we may decide to not manage without any further data. If we have an experienced manager, decisions may be made with a higher

Table 25.1. Decision tree statistics report (cost of genetic information collection set at $10,000). Expected values are given for different decision options. See Fig. 25.1 for structure of the decision tree.

Demographic information decision	1. Demographic information	2. No demographic information	
Mean	871,000	860,000	
Minimum	796,000	800,000	
Maximum	896,000	900,000	
Mode	896,000	900,000	
Standard deviation	43,301	48,989	
Non-genetic information decision	1. Manage	2. Non-genetic information	3. Do not manage
Mean	869,000	871,000	699,000
Minimum	799,000	796,000	–1,000
Maximum	899,000	896,000	999,000
Mode	899,000	896,000	999,000
Standard deviation	45,825	43,301	458,257
Genetic information decision	1. Manage	2. Genetic information	3. Do not manage
Mean	871,000	871,000	746,000
Minimum	796,000	786,000	–4,000
Maximum	896,000	886,000	996,000
Mode	896,000	886,000	996,000
Standard deviation	43,301	35,707	433,012

Table 25.2. Decision tree statistics report (cost of genetic information collection set at $5000). Expected values are given for different decision options. See Fig. 25.1 for structure of the decision tree.

Demographic information decision	1. Demographic information	2. No demographic information	
Mean	876,000	860,000	
Minimum	791,000	800,000	
Maximum	891,000	900,000	
Mode	891,000	900,000	
Standard deviation	35,707	48,989	
Non-genetic information decision	1. Manage	2. Non-genetic information	3. Do not manage
Mean	869,000	876,000	699,000
Minimum	799,000	791,000	−1,000
Maximum	899,000	891,000	999,000
Mode	899,000	891,000	999,000
Standard deviation	45,825	35,707	458,257
Genetic information decision	1. Manage	2. Genetic information	3. Do not manage
Mean	871,000	876,000	746,000
Minimum	796,000	791,000	−4,000
Maximum	896,000	891,000	996,000
Mode	896,000	891,000	996,000
Standard deviation	43,301	35,707	433,012

degree of accuracy than for the above case. Then, the probability of managing when it is needed may increase to 70% and the probability of managing when not needed may decrease to 30%. It may also be that the decision not to manage when it is indeed not needed, similarly rises to 70%, and not to manage when it is really needed falls to 30%. If we choose the option to get more information at an additional cost, in this case non-genetic information, the cost increases by $3000, the probability of managing when needed further improves and for not managing when needed decreases. Given that collection of non-genetic data costs $3000, the optimal decision is to collect the non-genetic data.

Based on the non-genetic information, there are three options: to manage, not to manage or to collect still more genetic information based on molecular markers at a higher cost. Assuming that the cost for such information is $10,000, it is shown that the optimal decision is not to collect the information (Table 25.1). Though the expected values are the same (see Table 25.1), the maximum and minimum returns are higher for the option of managing without collecting the genetic information. However, if the cost of genetic information collection is only $5000, it becomes the optimal choice (Table 25.2) due to higher expected return, $876,000, compared with $871,000.

As the study is in progress, hypothetical numbers are assigned for the value, cost, return and probability, until the second phase of the study is completed. However, it will demonstrate the capability of the system in making the optimal decisions. Values and probabilities in the decision tree can be assigned as fixed or as a probability density function of varying types with a mean and standard deviation. The structure of the decision tree is shown in Fig. 25.1.

Conclusion

The example cited above illustrates the usefulness of a decision tree approach to make rational management decisions in gene conservation management. In the first step, species were ranked based on intuitive information from multiple stakeholders. This step can also be approached in the same framework as used in the second step with each species constituting one branch of the decision tree. For example, information on any of the susceptibilities or values in the first step can be based on supporting data. This additional information will improve the probability that the value assigned intuitively will be close to the real value. Cost to collect information on such factors can be weighed against its usefulness in improving expected returns. Though the principle laid out in this chapter is generally valid, the example is simplistic. But it can be adapted to include all the factors involved in management. The main challenge will be to estimate population parameters, which will enhance the probability of making the right management decisions. The dollar values for the resources given in the example are subjectively assigned. Estimating appropriate dollar value for species and genetic resources will be a fundamental problem. However, if population parameters for indicators are estimable and values can be assigned with reasonable accuracy, the decision tree will provide an effective tool for rational decision-making for gene conservation. More precise methods for predicting extinction probabilities based on population parameters may become available, as the field of evolutionary genetics evolves in future. Rapid progress in the field of socio-economics of species conservation may also outline methods to effectively estimate dollar values for gene resources.

References

Brown, R.V., Kahr, S.A. and Peterson, C. (1974) *Decision Analysis – An overview.* Holt, Rinehart and Winston, New York.

Bunn, D.W. (1984) *Applied Decision Analysis.* McGraw-Hill, New York.

Margules, C.R. and Pressey, R.L. (2000) Systematic conservation planning. *Nature* 405, 243–253.

McCann, K.S. (2000) The diversity-stability debate. *Nature* 405, 228–233.

Myers, N. (1979) *The Sinking Ark. A New Look at the Problem of Disappearing Species.* Pergamon Press, Oxford, UK.

Namkoong, G. (1998) Genetic diversity for forest policy and management. In: Bunnel, F.L. and Johnson, J.F. (eds) *The Living Dance – Policy and Practices for Biodiversity in Managed Forests.* UBC Press, Vancouver, Canada, pp. 30–44.

Namkoong, G. and Koshy, M.P. (2000) Genetics and speciation in the world's forests. In: Evans, J. (ed.) *Forests Handbook* Vol. I. Blackwell Science, Oxford, UK, pp. 69–82.

Pukkala, T. and Kangas, J. (1995) A method for integrating risk and attitude to risk into forest planning. *Forest Science* 42, 198–205.

Purvis, A. and Hector, A. (2000) Getting the measure of biodiversity. *Nature* 405, 212–219.

Ronen, Y. (1988) *Uncertainty Analysis.* CRC Press, Boca Raton, Florida.

Thibodeau, F.R. (1983) Endangered species: Deciding which species to save. *Environmental Management* 7, 101–107.

Vane-Wright, R.I. (1996) Identifying priorities for the conservation of biodiversity: systematic biological criteria within a socio-political framework. In: Gaston, K.J. (ed.) *Biodiversity.* Blackwell Science, Victoria, Australia, pp. 309–344.

Appendix

1. Utility scores for the five highest scoring and five lowest scoring species. *w* is the weighting for the value, and *Pr* indicates the intuitive probability that the score given is the true score.

		Site		Economic												Social			Other			
				Wood			Pharmaceutical			Food or industrial			Ornamental			Cultural value			Ease of cultivation			Utility
No.	Species	Caetetus	Morro	score	*w*	*Pr*	score	*w*	*Pr*	score	*w*	*Pr*	score	*w*	*Pr*	score	*w*	*Pr*	score	*w*	*Pr*	score
13	*Myroxylon peruiferum*	x	x	5	2	100	5	1	100	1	1	100	3	2	100	4	1	80	5	3	80	37.2
19	*Euterpe edulis*	x		3	2	80	1	1	100	5	1	100	5	2	100	5	1	100	4	3	80	35.4
3	*Hymenaea courbaril*	x	x	5	2	100	3	1	100	2	1	100	5	2	100	5	1	100	5	3	100	45
10	*Jacaratia spinosa*	x	x	1	2	100	2	1	100	5	1	100	5	2	100	5	1	100	5	3	100	39
59	*Maclura tinctoria*	x	x	4	2	100	1	1	100	4	1	80	3	2	100	2	1	80	3	3	80	27
74	*Senna pendula*	x	x	1	2	100	1	1	100	1	1	100	4	2	100	2	1	80	5	3	80	25.6
144	*Aegiphila sellowiana*	x	x	4	2	100	1	1	100	1	1	100	1	2	100	1	1	100	3	3	80	20.2
263	*Piptocarpha sellowii*	x		1	2	100	1	1	100	1	1	100	1	2	100	1	1	100	1	3	0	7
81	*Croton floribundus*	x	x	3	2	80	3	1	100	1	1	100	2	2	80	1	1	100	5	3	80	25
251	*Senna biflora*	x		2	2	50	1	1	100	1	1	100	2	2	80	1	1	100	1	3	0	8.2

2. Ecological scores for the five highest scoring and five lowest scoring species. *w* is the weighting for the value, and *Pr* indicates the intuitive probability that the score given is the true score.

	Site		Ecological															Phylogenetic			
			Rarity			Successional group			Geographic range			Ecological plasticity			Resources for the fauna			No. of species/ genera			Eco-
Species	Caetetus	Morro	score	*w*	*Pr*	score	*w*	*Pr*	score	*w*	*Pr*	score	*w*	*Pr*	score	*w*	*Pr*	score	*w*	*Pr*	score
Myroxylon peruiferum	x	x	5	5	90	5	4	100	1	3	100	3	2	90	1	4	100	5	3	100	69.9
Euterpe edulis	x		1	5	100	3	4	100	4	3	100	5	2	100	5	4	100	5	3	100	74
Hymenaea courbaril	x	x	3	5	90	5	4	100	1	3	100	3	2	90	4	4	100	4	3	100	69.9
Jacaratia spinosa	x	x	4	5	90	3	4	100	2	3	100	3	2	90	4	4	100	5	3	100	72.4
Maclura tinctoria	x	x	5	5	90	4	4	100	2	3	100	3	2	90	4	4	100	5	3	100	80.9
Senna pendula	x	x	3	5	90	2	4	100	1	3	100	1	2	90	1	4	100	1	3	100	33.3
Aegiphila sellowiana	x	x	2	5	90	2	4	100	1	3	100	3	2	90	2	4	90	1	3	100	35.6
Piptocarpha sellowii	x		4	5	90	2	4	100	1	3	90	3	2	90	1	4	100	1	3	100	41.1
Croton floribundus	x	x	1	5	80	2	4	100	1	3	100	3	2	90	1	4	100	1	3	100	27.4
Senna biflora	x		3	5	90	2	4	100	2	3	50	3	2	70	1	4	100	1	3	100	35.7

3. Threat scores for the five highest scoring and five lowest scoring species. *w* is the weighting for the value, and *Pr* indicates the intuitive probability that the score given is the true score.

No.	Species	Site		Exploitation intensity			Harvesting			Fragmentation effect			Fire			Fauna			Threat score
		Caetetus	Morro	score	*w*	*Pr*	score	*w*	*Pr*	Score	*w*	*Pr*	score	*w*	*Pr*	Score	*w*	*Pr*	
1	*Myroxylon peruiferum*	x	x	5	5	90	5	2	100	5	3	80	3	3	80	1	2	80	53.3
2	*Euterpe edulis*	x		4	5	90	5	2	100	5	3	80	2	3	80	3	2	80	49.6
3	*Hymenaea courbaril*	x	x	4	5	90	5	2	100	3	3	80	3	3	80	4	2	80	48.8
8	*Jacaratia spinosa*	x	x	2	5	90	3	2	80	5	3	80	5	3	80	4	2	80	44.2
26	*Maclura tinctoria*	x	x	4	5	90	2	2	80	3	3	80	3	3	80	3	2	80	40.4
258	*Senna pendula*	x	x	1	5	90	1	2	100	1	3	100	1	3	80	2	2	80	15.1
264	*Aegiphila sellowiana*	x	x	1	5	90	1	2	100	1	3	100	1	3	80	1	2	80	13.5
267	*Piptocarpha sellowii*	x		1	5	90	1	2	100	1	3	100	1	3	80	1	2	80	13.5
257	*Croton floribundus*	x	x	1	5	90	1	2	100	1	3	100	1	3	80	2	2	80	15.1
263	*Senna biflora*	x		1	5	90	1	2	100	1	3	80	1	3	80	2	2	80	14.5

4. Total scores for the five highest scoring and five lowest ranking species.

No.	Species	Site		Utility		Ecological		Threat		Final score
		Caetetus	Morro	weight	score	weight	score	weight	score	
1	*Myroxylon peruiferum*	x	x	2	37.2	4	69.9	5	53.3	620.5
2	*Euterpe edulis*	x		2	35.4	4	74	5	49.6	614.8
3	*Hymenaea courbaril*	x	x	2	45	4	69.9	5	48.8	613.6
4	*Jacaratia spinosa*	x	x	2	39	4	72.4	5	44.2	588.6
5	*Maclura tinctoria*	x	x	2	27	4	80.9	5	40.4	579.6
263	*Senna pendula*	x	x	2	25.6	4	33.3	5	15.1	259.9
264	*Aegiphila sellowiana*	x	x	2	20.2	4	35.6	5	13.5	250.3
265	*Piptocarpha sellowii*	x		2	7	4	41.1	5	13.5	245.9
266	*Croton floribundus*	x	x	2	25	4	27.4	5	15.1	235.1
267	*Senna biflora*	x		2	8.2	4	35.7	5	14.5	231.7

26 Germplasm Enhancement to Sustain Genetic Gains in Crop Improvement

R. Ortiz
International Crops Research Institute for the Semi-Arid Tropics (ICRISAT), Patancheru, Andhra Pradesh, India

Germplasm Enhancement or 'Prebreeding'

The genetic composition of a crop population is affected by mutation, migration and recombination, which widen genetic variation. In contrast, the genetic variation is reduced by natural and artificial selection and random genetic drift. Changes in the frequency of favourable alleles that enhance crop adaptation to the human-made environment for agriculture are the driving force during crop evolution and account for current crop genetic diversity (Ortiz, 1999a). Likewise, isolating mechanisms such as natural pollinator(s), geographical barriers, breeding systems, self-incompatibility, sterility and vegetative propagation have influenced crop evolution and domestication.

The term 'prebreeding' refers to the transfer or introgression of genes and gene combinations from non-adapted sources into breeding materials (FAO, 1996a). In the author's view, this strategy for genetic enhancement of crops does not differ significantly from the general paradigm of plant breeding. In his recently revised edition of *Principles of Plant Breeding*, Allard (1999) does not use the term 'prebreeding', but instead focuses much of his revised text on two distinct components of Darwinian evolution: 'the facts of evolution (descent with modification), and selection as the chief agent of evolutionary change'. Indeed, the manipulation of crop evolution (plant breeding) for enhanced economic yield under artificial growing environments (modern agriculture) consists of two phases: collection and generation of variation and reproductive potential followed by selection of the most productive surviving variants. The qualities that determine survival in plant breeders' populations are the same as those observed in nature with additional selective forces relating to perceived value when grown in environments reflective of farmers' fields (Allard, 1999). On this basis, the term 'prebreeding' will be replaced with that of 'germplasm enhancement' reflecting the early component of sustainable plant breeding that deals with identifying a useful character, 'capturing' its genetic diversity, and putting those genes into a 'usable' form (Peloquin *et al.*, 1989; Ortiz, 1999b). Perhaps, a more general and appropriate use and

Dedication
In memory of Dirk R. Vuylsteke (1958–2000), plantain and banana breeder and biotechnologist, whose sudden, untimely and tragic death has orphaned the genetic enhancement of these crops, while the international agricultural research community has lost one of its most outstanding scientists. We will miss his innovative ideas, open-minded style, hard work and commitment to the African small landholder, the *Musa* crop, and research-for-development. However, Dirk will always be a source of inspiration to his colleagues and the new generation of scientists involved in germplasm enhancement of research-neglected tropical crops.

(eds J.M.M. Engels, V. Ramanatha Rao, A.H.D. Brown and M.T. Jackson)

understanding of this term, 'germplasm enhancement', will attract the attention of some professional plant breeders who remain reluctant to use wild, landrace or exotic germplasm in crop improvement.

Germplasm Enhancement with Non-adapted (Exotic) and Landrace Germplasm

There are two approaches for using wild species, landraces and exotic germplasm in plant breeding: introgression and incorporation (Simmonds, 1993). Introgression indicates the transfer of one or a few alleles from exotic stocks to adapted breeding populations that lack the allele(s) controlling a specific characteristic. Often, introgression of useful genes requires its further characterization and introduction into locally adapted or elite material for use in breeding programmes. Incorporation refers to a large-scale programme aiming to develop locally adapted populations using exotic germplasm, which will broaden the genetic base of new breeding materials. As indicated in the *Global Plan of Action for the Conservation and Sustainable Utilization of Plant Genetic Resources for Food and Agriculture* (FAO, 1996b), the broadening of the genetic base of crops will contribute to increasing crop stability and performance.

Germplasm enhancement should be regarded as a long-term activity because exotic germplasm seldom has an immediate use without selection for local adaptation and enhanced yield potential. Public rather than private breeders have been more proactive in using germplasm enhancement because many generations are needed to achieve substantial progress without any guarantee of success. Thus, germplasm enhancement programmes are independent of the local crop genetic base until they become sources of parental material for the conventional breeding pool. It was not surprising therefore that in the *Report on the State of the World's Plant Genetic Resources for Food and Agriculture* (FAO, 1996a) it was indicated in the single paragraph written for this early component of sustainable crop improvement: 'very few country reports mentioned prebreeding or genetic enhancement as a national breeding activity, though a number [of them] called the attention to the importance of such work'. Individual or private plant breeders sometimes consider that the costs of incorporating exotic germplasm into already adapted material outweigh the benefits they could gain. Furthermore, such approaches may have more benefits to society in general (including other plant breeders) rather than direct individual corporate gain. Hence, germplasm enhancement to sustain long-term breeding gains requires collaboration among plant breeders from both public and private sectors.

Crop uniformity and associated epidemics have spurred the development of germplasm enhancement methods, particularly since the 1970s. Analysis of breeding pools led to the belief that new sources of genes for pest and disease resistance were needed to avoid crop vulnerability as observed with vast monocultures of maize devastated by southern corn blight in the USA (Fowler and Mooney, 1990). However, the prediction of major problems associated with the narrow genetic diversity of crops has been challenged by the analysis of genetic gains for yield or by marker-aided analysis of breeding materials (Donini *et al.*, 2000; Petersen, 2000). For example, there has been no levelling-off of progress in genetic improvement of productivity in several major crops (MacKey, 1994). In particular new kinds of gene action are being advocated to explain the continual recovery of transgressive segregants for enhanced productivity from populations even with a narrowing genetic base (Rasmusson and Phillips, 1997). Thus, in this new paradigm of plant breeding, manipulating gene interactions (epistasis) becomes as important as manipulating actual loci.

Furthermore, it seems that selection in breeding populations does not depend solely on the existing variation occurring in the parental stocks (and the permutations of those existing genes) but also depends on intragenic recombination, unequal crossing over, transposable elements, DNA methylation, paramutation and gene amplification. And in turn, elevated epistasis may be achieved through new gene interaction with that *de novo* variation (Rasmusson and Phillips, 1997). In disomic polyploid species alleles at distinct homeologous loci may interact with each other, as they become homozygotes in the respective genomes (MacKey, 1970). Thus in disomic polyploid inbred species heterosis can occur via fixed 'overdominance' between homeologous loci (MacKey, 1987). These kinds of gene action may account for the observed genetic gains in so-called narrow base populations derived from elite breeding stocks (Ortiz *et al.*, 1998a).

Germplasm Enhancement for Broadening the Genetic Base of New Cultivars

Germplasm enhancement also serves to widen the genetic base of breeding materials, in specific genome regions (introgression) or more broadly (incorporation). However germplasm enhancement methods depend on the specific crop biology (Simmonds, 1993). This will be illustrated by a discussion of germplasm enhancement methods for two crops: barley (an inbreeding species), and maize (an outcrossing species).

Inbred crops: barley

One of the most successful public–private research partnerships for introgression of *Hordeum vulgare* subsp. *spontaneum* into barley breeding was achieved by a cooperative programme between Swedish researchers (Lehmann *et al.*, 1998). Private breeders carried out the disease screening while the public researchers did the crosses, multiplication and documentation. A backcross strategy was applied using cultivated barley lines as the maternal recurrent parent. The second disease screening was carried out in the BC_1F_1 to select the parents of the BC_2, and plants of the BC_1F_2 were screened to select the resistant material for the BC_3, which was then selfed for one generation (BC_3F_2) before providing this advance generation to the private plant breeders. As a result of this cooperative project, several hundred BC_3F_2 plants were developed from 89 accessions of *H. vulgare* subsp. *spontaneum* with many unknown resistance genes to powdery mildew. These resistance genes are broadening the genetic base for host plant resistance to a disease in which the recessive *mlo* gene accounts for 30% of the barley grown area in Europe (Jørgensen, 1992). Progress was also made in the transfer of resistance to spot blotch, net blotch, leaf rust and scald from *H. vulgare* subsp. *spontaneum*. However, sterility barriers and chromosome pairing in hybrids have restricted the access to genes from other wild *Hordeum* species (von Bothmer and Jacobsen, 1991).

Another scheme for adding new gene complexes into an elite barley population, which serves as sources of new cultivars, was developed in Canada (Kannenberg and Falk, 1995). This recurrent introgressive population enrichment (RIPE) breeding system uses male sterility facilitated recurrent selection and has four sequential steps with distinct hierarchical levels (base, intermediate, high and elite) following the introduction of exotic germplasm into the system. Selected lines are crossed with male steriles to enhance the amount of elite (E) germplasm through recurrent backcrossing, that is, 50% E in base level, 75% E in intermediate level, 87.5% E in high level, and 93.5% E in elite level. New lines incorporated into the elite level do not exceed 6% to avoid significantly diluting the high-performing, highly adapted E population, which is maintained by a standard 2 years per cycle recurrent selection procedure. The parents of the next cycle of this RIPE system are the highest yielding F_4 lines selected in each cycle. RIPE reduces unfavourable linkages at each step through genetic recombination, while promoting at the same time attractive recombinants to the E level. By intermating of, and selecting within the E population unique and agronomically superior genotypes are obtained through RIPE.

Using exotic genetic resources Nordic breeders have developed experimental populations of barley after six generations, which promoted genetic recombination by breaking linkage groups (Vetelainen *et al.*, 1996a). These results showed that useful variation for agronomic characteristics was incorporated into the locally adapted breeding materials, though further recombination may still be needed to enlarge the range of agronomic variation, and new methods should be developed to assess the breeding value of exotic germplasm because parental phenotypes are poor predictors of progeny performance (Vetelainen *et al.*, 1996b).

Outcrossing species: maize

Plant introductions have not been widely incorporated in maize breeding materials in the USA, and until recently only a few open-pollinated cultivars have been used by maize breeding programmes in this country (Goodman, 1990). Perhaps the early open-pollinated cultivars, available to farmers and plant breeders, had significant variation that accounts for the breeding gains in grain yield reported in this crop (Troyer, 1990). The development of the Illinois High Oil and High Protein strains provides an example of long-term selection in a maize breeding pool derived from 24 ears, which were selected from an original stock of 163 ears harvested from the open-pollinated cultivar

'Burr's white' by G.C. Hopkins in 1896 (Dudley and Lambert, 1992). Genetic analysis in the advanced selection cycles suggests that further breeding progress may be expected owing to the polygenic control of both characteristics in maize. Likewise, this long-term experiment shows that the maize genome is more plastic and amenable to selection than earlier thought.

The United States Department of Agriculture, Pioneer Hi-Bred International and 11 cooperative programmes from Latin America and the USA launched the Latin American Maize Program (LAMP) to assess national germplasm and exchange genetic resources in the region (Sevilla and Holle, 1995). This germplasm was tested for agronomic characteristics from sea level to 3300 m, and from 41°N to 34°S across 32 locations in phase I (14,847 accessions), and in 64 locations (two per region) in phase II (using the best 20% of selected accessions for each location in phase I). The best selections of each country were tested across all locations in phase III. These locations were clustered according to five homologous areas: lowland tropics, temperate and three altitudes. Selected accessions (268) were tested for their combining ability using a tester in at least two locations within each region. The best cross combinations (for hybrids) and heterotic pools were also determined by LAMP. Maize breeders now have access to the most promising accessions identified by LAMP to expand the genetic base of the crop.

The hierarchical open-ended population enrichment (HOPE) breeding system was developed by Kannenberg and Falk (1995) for broadening genetic variation while maintaining germplasm quality in order to obtain promising parental inbred lines for production of commercial hybrids. HOPE gains depend on two complementary heterotic sets, each with four pools according to their agronomic performance (from low to elite). Exotic accessions are added to the lowest pool while selected gene(s) and gene complexes move upwards according to a hierarchy in a progressively more stringent selection procedure throughout the process. In the lowest pool (e.g. open-pollinated exotic germplasm) breeders apply mass-grid selection to maximize genetic recombination while keeping genetic variation. Replicated progeny tests are applied through half-sib family selection to identify about 15% superior genotypes in the intermediate pool for further crossing. Selfed progeny recurrent selection is employed in the high pool to enhance the variation among families and unmask deleterious recessive alleles at this stage. Reciprocal recurrent selection is practised for simultaneous improvement of the elite heterotic pools because this procedure enhances the combining ability between both populations. In this way, population crosses as well as crosses between inbred lines from each heterotic pool improve in performance after each cycle of selection. Two inbreds (CG31 and CG32) were developed in the mid-1990s, using HOPE (Kannenberg and Falk, 1995).

Evolutionary Crop Breeding

Experts assembled by the Food and Agriculture Organization of the United Nations (FAO) in a 1998 workshop on 'Broadening the Genetic Base of Crop Production' recommended that new cultivars should have an enhanced and expanded genetic base to improve the deployment of crop diversity in farmers' fields. They also recommended evolutionary plant breeding on composite (or synthetic) populations derived from a broad germplasm base across diverse environments to ensure wide genetic variation in long-term (10 years) crop improvement programmes. Such distinct populations may be incorporated into regular breeding programmes or even adopted by progressive farmers. Hence, farmers are not only seen as partners for *in situ* or on-farm conservation of plant genetic resources but also in the development of new, locally adapted cultivars. .

Evolutionary crop breeding may lead to a dynamic conservation of plant genetic resources because locally adapted, improved populations will be preserved in distinct environments by plant breeders or farmers. In this way, several genetically distinct populations will be developed, thereby protecting crop diversity at the meta-population level. Furthermore, this approach may also permit the dynamic interaction and co-evolution between four partners: crop and wild relatives, and their respective pathogen populations (Frankel *et al.*, 1995), and the competition of crop populations with other indigenous plants in their permanent or original niches. In this way, *in situ* conservation and on-farm management provide a means to search and test alleles for potential durable resistance to specific crop pests and diseases (Ortiz, 1999a). Hybridization between distinct populations may allow the accumulation of the best genetic combinations in new breeding populations.

Potato: the model crop species for genetic enhancement of polysomic polyploids

Potato is a tetrasomic polyploid crop that was domesticated by ancient Peruvians in the Andes. There are many wild tuber-bearing *Solanum* species but only a few cultivated species. This wild and cultivated germplasm consists of diploid, triploid, tetraploid, pentaploid and hexaploid species. Some researchers have suggested that there is a narrow genetic base of cultivated potato in the northern hemisphere (Mendoza and Haynes, 1974; Hawkes, 1979), and consequently germplasm enhancement methods were advocated to remedy this problem. For example, potato breeders developed long-day adapted populations derived from Andean tetraploid and diploid potato landraces through mass selection (Glendinning, 1979; Plaisted, 1982). These long-day adapted tetraploid populations have been termed as Neotuberosum, and some potato breeders have incorporated this improved germplasm in their programmes.

An alternative approach was to incorporate diploid wild germplasm into breeding stocks using ploidy manipulations, that is, scaling up and down the chromosome number. Most of the pioneering work was developed by Professor Stanley J. Peloquin and his co-workers at the University of Wisconsin-Madison (Peloquin *et al.*, 1989). Chromosome sets of wild and cultivated germplasm have been manipulated using haploids, $2n$ gametes and the endosperm balance number (EBN). Maternal haploids (or sporophytes with the gametic chromosome number) are easily obtained through parthenogenesis from tetraploid landraces and cultivars (Peloquin *et al.*, 1996). Haploids are then crossed with diploid (wild or primitive cultivated) species to develop haploid species hybrids. Most of these haploid species hybrids are vigorous and show improved tuberization in long-day environments. Often, most wild species do not have tubers in these environments. Chromosomes pair and recombination occurs between wild and cultivated tuber-bearing *Solanum* species because of minimal chromosome differentiation among these taxa. The success of a cross between two tuber-bearing *Solanum* species also depends on the EBN. This endosperm dosage system, common to other angiosperm genera, requires a correct proportion of 2:1 maternal to paternal contributions for proper seed development (Ehlenfeldt and Ortiz, 1995). In this breeding strategy wild species and diploid landraces are the source of genetic diversity. The haploids derived from adapted tetraploid cultivars are able to 'capture' this genetic diversity in crosses with the diploid germplasm. These haploid-species hybrids transmit this genetic diversity to the adapted tetraploid breeding pool via $2n$ gametes (or gametes with the sporophytic chromosome number).

Important characteristics such as pest and disease resistance, tolerance to abiotic stresses, and tuber quality have been transferred using this method particularly by scientists of the Centro Internacional de la Papa (CIP) (Ortiz, 1998a). Furthermore, the diploid germplasm provides allelic variation that may enhance yield because non-additive intra- and inter-locus interactions are important for this characteristic in tetraploid potato (Ortiz, in press). A few cultivars and advanced breeding materials have been released following this method in North America. Ploidy manipulations for the genetic enhancement of potato have also been included in breeding programmes across America, Asia and Europe. This breeding method developed for potato may be extended for improving other vegetatively propagated polysomic crops.

Potatoes are clonally propagated from tubers. However, botanical true seed may be an alternative, especially in the developing world. Potato farmers in Bangladesh, China, Egypt, India, Indonesia, Nicaragua, Peru, the Philippines and Vietnam are planting commercial fields using direct seeding or propagules derived from true seed. New potato cultivars for production from true seed should have plant characteristics for true seed production, and for tuber production from true seed propagules. Such a cultivar could be a synthetic tetraploid hybrid from bilateral ($2x \times 2x$) or unilateral sexual polyploidization ($4x \times 2x$) using either seedling or tuberlet propagules derived from true seed (Ortiz, 1997a). Also, synthetic breeding populations derived from polycrosses may be developed for further adaptive testing and local selection of the most promising genotypes towards cultivar release.

A core collection of *Solanum* species will improve the utilization of *Solanum* genetic resources in potato breeding (Ortiz, 2001). This core collection must contain chosen wild and cultivated accessions representing with minimum redundancy the genetic variability of the whole tuber-bearing *Solanum* germplasm and closely related non-tuber bearing *Solanum* species (Ortiz, 1998a). CIP has been developing a core collection of Latin American landraces using hierarchical classification based on taxonomy, ecogeographical origin, morphological characteristics, molecular markers, and reaction to diseases and pests. Recently, CIP scientists have defined a core subset of *Solanum tuberosum* ssp. *andigena*, which contains 306 landraces (Huaman *et al.*, 1999). The number of accessions included in the core was determined primarily by the square root of the number of *andigena* accessions collected (Huaman *et al.*, 2000a). A phenogram was constructed from the descriptor data using a matching coefficient and the unweighted pair group method using arithmetic averages. Accessions were retained in each cluster considering data on resistance to diseases and pests, dry matter content, and number of duplicate accessions identified in the original collection. Allozyme frequencies validated the sampling procedure for this core collection (Huaman *et al.*, 2000b). The most frequent allozymes in the core collection were those observed in the entire collection, and their frequency distributions were homogeneous for most loci.

Breeding for tuber yield of a vegetatively propagated crop such as potato capitalizes on heterosis, that is, identifying the right hybrid with high specific combining ability owing to non-additive intra- and inter-allelic interactions (Ortiz, 1998a). The core collection should be seen as a preliminary source of testing materials for combining ability in a genepool that does not have well-defined heterotic genepools (Huaman *et al.*, 2000b). In this way, broadening the genetic base of potato with locally adapted, pest and disease resistant germplasm will be ensured for the sustainable and environment-friendly production of this crop. However, as demonstrated by allozyme surveys (Ortiz and Huaman, in press), the need for broadening the genetic base in potato may only be needed for specific chromosomes or regions within chromosome arms.

Plantain and banana: extensive progress in understanding leading to rapid success in genetic enhancement

Triploid plantains (*Musa* × *paradisiaca* L.) are interspecific hybrids of the diploid species *Musa acuminata* Colla. and *Musa balbisiana* Colla. Although the *Musa* crop was domesticated in South-east Asia, the humid lowland tropics of West and Central Africa are the secondary centres of plantain diversity. After Columbus's voyages, domesticated *Musa* moved to tropical America and the Caribbean, where triploid dessert bananas (intraspecific hybrids of *M. acuminata*) became an important export commodity, though most cultivars are still farmers' selections from somatic mutants.

Plantain landraces show a wide range of morphological variation (Ortiz, 1997b). However, there is low level of gene flow via pollen among triploid plantains, as determined by Wright's ϕ_{ST} (Ortiz *et al.*, 1998b). This finding was expected because this vegetatively propagated crop has very low male fertility. Hence, the variation observed in plantains arose from mutations accumulated throughout the history of cultivation of this crop (Crouch *et al.*, 2000). Consequently, plantains should be regarded as an old domesticated species. Since the 1980s, plantain breeders have been widening the genetic base of this crop through the use of wild and cultivated diploid germplasm (Vuylsteke *et al.*, 1997). Tetraploid hybrids are obtained by hybridizing $2n$ eggs from triploid plantains and haploid (n) pollen from diploid accessions. On average, it takes 1000 seeds, produced from hand-pollination of 200 plants (0.12 ha), to obtain one selected tetraploid hybrid per year. Tetraploid hybrids combining heavy and stable yield were identified after multilocational testing across sub-Saharan Africa (Ortiz, 1998b) and shared with local breeders across sub-Saharan Africa and around the world, who are testing this breeding material developed by the International Institute of Tropical Agriculture (IITA) for further regional release to farmers (Ortiz *et al.*, 1997).

Based on this early success, an evolutionary crop breeding method was proposed for further improvement of the plantain genome (Ortiz, 1997c). This method ensued from the genetic knowledge accumulated during the conventional cross breeding of plantains at IITA in Nigeria (Ortiz and Vuylsteke, 1996). Heterozygous triploid landraces are the source of adapted allelic diversity, which would be released after these plantain landraces are crossed with a diploid accession showing the desired characteristic (particularly resistance to abiotic stress). High-yielding tetraploid hybrids are selected on the basis of specific combining ability in the segregating population. The selected primary tetraploid hybrids are crossed with selected diploids to obtain improved secondary triploid hybrids (Ortiz *et al.*,

1998c). These hybrids may result from artificial hand-pollination or through polycrosses among selected tetraploid and diploid parents (Ortiz and Crouch, 1997). The parental genotypes are chosen for the isolated polycrosses according to their specific combining ability as determined earlier in artificial tetraploid–diploid crosses. The synthetic populations derived from these polycrosses may be tested in other locations to identify promising offspring for cultivar development. Local selections are needed because the genotype-by-environment interaction affects the performance of *Musa* hybrids across environments (Ortiz, 1998b).

As Vuylsteke highlighted in one of his last writings (2000):

> A broad-based, improved *Musa* germplasm with pest and disease resistance will be a major component to achieve sustainable production of this vegetatively propagated, perennial crop. Such germplasm can be produced through conventional cross breeding, enhanced by the utilization of innovative methods for the introduction of additional genetic variation. Also, the increased use of molecular markers will accelerate the process of recurrent selection of improved *Musa* germplasm and, hence, facilitate the development of new hybrids. Indeed, the prospects of banana and plantain breeding are unlimited and increased efforts will at once initiate a new phase of *Musa* evolution.

Dynamic genepools for crop genetic enhancement

Goldringer *et al.* (1994) claimed that a dynamic management of genetic resources supplements the static conservation of seeds in cold storages of genebanks. They further advocated that dynamic management of genetic resources should be recommended for adapting exotic germplasm to specific local environments. Likewise, dynamic management of genetic resources may preserve genetic diversity between different populations evolving in distinct environments. For example, results from two-dimensional protein electrophoresis revealed that natural selection acted strongly on 11 wheat populations evolving since 1984 in multi-site experiments grown in the major French cultivation areas (David *et al.*, 1997). Plant height, male fertility, earliness, and resistance to powdery mildew improved in the wheat populations included in this multi-site experimental network (Le Boulc'h *et al.*, 1994). Furthermore, variability was maintained for genes enhancing adaptedness to the specific environment, including those providing host resistance to powdery mildew in wheat (Goldringer *et al.*, 1998).

Artificial populations, involving wild, interspecific and cultivated genotypes, have been developed in southern Europe for the dynamic management of sunflower genetic resources (Belhassen *et al.*, 1994). Similarly, nine breeding pools of perennial ryegrass were developed from 42 selected populations (out of 547 natural populations) as a method to preserve genetic diversity as well as to provide material that may be used directly by plant breeders (Balfourier and Charmet, 1994).

Farmer-participatory pearl millet improvement in India and Namibia

Plant breeders may obtain important information from farmers regarding why they select and keep specific landraces in their fields. This indigenous knowledge and farmers' participation in testing and selecting new genotypes will enhance the development and adoption of new breeding materials, and may also ensure the on-farm conservation of agrobiodiversity. The International Crops Research Institute for the Semi-Arid Tropics (ICRISAT) in collaboration with other research partners started participatory approaches for pearl millet breeding in Rajasthan (India) to bridge the gap that sometimes exists between conventional breeders and farmers, and to work together with farmers to assess constraints and find solutions (Weltzien-Rattunde *et al.*, 1998). The appropriate understanding of farmers' needs provides a means for targeting breeding research while participatory research gives farmers the opportunity for choosing, improving and adapting a range of technology options.

As a result of this participatory approach in pearl millet breeding at Rajasthan six key conclusions were made (E. Weltzien-Rattunde, ICRISAT, 2000, personal communication): (i) farmers' knowledge is crucial for the success of participatory breeding; (ii) a dialogue between farmers and breeders enriches both groups' knowledge in crop improvement; (iii) an agreement was reached between the new research partners regarding goals and priority setting for both on-farm conservation and genetic enhancement of pearl millet; (iv) comparative advantages and roles among these new partners may be easily determined for enhanced cooperation; (v) participatory breeding should be regarded as a flexible process in which research procedures and interventions must be adjusted to suit specific physical and socio-economic environments; and (vi) owing to the above, participatory plant breeding is situation specific.

ICRISAT scientists and their partners benefited from this participatory farmer approach by re-targeting pearl millet breeding, and changing the source breeding material as well as the selection methods to better suit the harsh arid environments of Rajasthan, which suffers from low rainfall and poor soils. Likewise, this participatory breeding methodology was adopted by pearl millet breeders in other locations, for example in Namibia (Bidinger, 1998). Okashana-1 is a cultivar grown on almost 50% of the national pearl millet area in this country. The source germplasm was originally developed by ICRISAT breeders working in India by improving African collections. Farmers were involved early in the selection process according to their preferences. The Government of Namibia actively participated in the rapid multiplication and dissemination of high-quality seed. Such a partnership reduced research costs and speeded adoption of Okashana-1. The internal return to public investments in the development and dissemination of this cultivar was 50%, about US$11 million in 1998 (Rohrbach *et al.*, 1999). Two new pearl millet cultivars were also released in 1998 by this partnership between native farmers, Namibian breeders and ICRISAT.

Simultaneously to this farmer participation in variety testing and selection in Namibia, local farmers who were growing Okashana-1 mixed it with their own landraces, and selected outstanding plants resulting from natural crosses. A breeding composite, named after the farmer (Maria Kaharero) from whose field the plants were selected, was formed by random mating. After interaction with farmers a participatory breeding composite (also named Maria Kaharero) was developed. Materials selected by farmers from a diversity nursery were crossed to the original Maria Kaharero. The number of outstanding early, high yielding lines with medium to large seeds from this participatory composite Maria Kaharero exceeded that of another composite bred by conventional breeding procedures (i.e. without farmer participation), though the average performance of the two composites was similar (Bidinger, 1998). It was clear from this participatory breeding that farmers are important research partners to assess and select materials in their own cropping systems, especially in a cross-pollinated crop such as pearl millet, in which hybrid populations may be obtained easily in their own fields for further farmer selection.

Participatory pearl millet breeding may also be enhanced by providing landrace-based topcross hybrids (Mahalakshmi *et al.*, 1992) to farmers for their further testing and selection, especially in arid or semiarid environments. High yielding adapted topcross hybrids have been obtained by crossing a uniform inbred line as the female seed parent with a genetically heterogeneous landrace as the male parent (Bidinger *et al.*, 1994; Yadav *et al.*, 2000a). Furthermore, topcross hybrids were either as stable, or more so, than their pollinators (Mahalakshmi *et al.*, 1992) or more responsive to improved environmental conditions than their pollinators (Bidinger *et al.*, 1995). Recent results indicate that environment affects the combining ability of landrace pollinators (Yadav *et al.*, 2000b), thus specific pollinators should be selected for a particular zone. These topcross hybrids provide a unique prospect to exploit and preserve indigenous agrobiodiversity in an enhanced genotype.

Wild species for enhancement of crops from the semiarid tropics

The most widespread application of this approach has been in resistance breeding with genetic resources of wild species (Lenné and Wood, 1991). The levels of resistance to pests and diseases available in the primary genepool are sometimes low or only a limited number of resistance sources have been incorporated into elite materials by plant breeders. Furthermore, selection pressure on pathogen populations due to the widespread use of homogeneous host plant resistance may result in more virulent strains that may overcome that resistance. Therefore, discovery and incorporation of new genes for resistance from wild species provide a means for sustaining crop improvement. Although durability of resistance cannot be predicted (Johnson, 1992), genetic diversity through pre-emptive breeding may provide buffering against crop losses arising due to changes in the pathogen population (McIntosh, 1992). In this way, new breeding stocks could be generated to combat more virulent pathotypes when they become important.

Germplasm enhancement with wild species has seldom resulted in direct cultivar release, but many parents with 'wild' genes have become available.

Below I provide details from research at ICRISAT and elsewhere concerning germplasm enhancement of dryland crops with wild species. ICRISAT has about 2500 accessions of wild species and weedy relatives of sorghum, pearl millet, groundnut, chickpea and pigeonpea (Table 26.1). Screening of this germplasm has identified several sources of resistance to important pests and diseases (Table 26.2). Backcross followed by selection has been the most common method for gene introgression from wild germplasm to breeding materials. Transfer of new cytoplasmic male sterility to pigeonpea and pearl millet plus the development of chickpea with enhanced yield or pigeonpea with high protein were achieved through this breeding method at ICRISAT. However, some problems still remain for genetic enhancement with wild species: linkage drag, sterility, small sample size of interspecific hybrid population, and restricted genetic recombination in the hybrid germplasm.

Sorghum

Diseases such as downy mildew, anthracnose and grain mould affect this crop. Sources of resistance are available in cultivated germplasm but pathogens may break down this resistance; therefore, additional sources of resistance are needed. Immune host response to these diseases appears to be available in *Sorghum sudanense* and species of other sections. *S. sudanense* belongs to the primary genepool and easily crosses with sorghum but not the species from the other sections that are in the tertiary genepool. It seems that pre-fertilization crossing-barriers exist, owing to either lack of pollen germination or very slow and irregular pollen tube growth (Huelgas *et al.*, 1996). High levels of resistance to insect pests, especially stem borer and shoot fly, are available in five accessions of section *Parasorghum*. Similarly, no oviposition of sorghum midge was observed on 11

Table 26.1. Wild species of crops from the semiarid tropics held at ICRISAT genebank.

Species/section or genus	Number of species	Number of accessions
Sorghum (sect.)		
Sorghastrum	1	8
Parasorghum	5	36
Chaeto sorghum	1	3
Hetero sorghum	1	12
Stipo sorghum	10	106
Sorghum	7	362
Unknown	–	17
Pearl millet (gen.)		
Pennisetum	24	739
Cenchrus	1	11
Chickpea (gen.)		
Cicer	8 annual	58
	10 perennial	77
Pigeonpea (gen.)		
Cajanus	20	213
Rhyncosia	35	303
Paracalyx	1	2
Dunbaria	2	12
Eriosema	4	7
Flemingia	8	18
Groundnut (sect.)		
Rhizomatosae	4	98
Arachis	20	192
Erectoides	6	21
Extranervosae	2	5
Trisemenatae	1	5
Heteranthae	4	57
Caulorhizae	2	31
Procumbentes	8	30

Table 26.2. Promising wild species accessions with disease or pest resistance and quality traits at ICRISAT genebank. (Number screened/selected for each pest or disease in brackets.)

Crop	Pest or disease
Sorghum	Grain mould (308/29), downy mildew (565/94), shoot fly (268/13[a]), stem borer (50/15[a]), midge (20/15)
Pearl millet	Downy mildew (534/220[b] with 100% resistance)
Chickpea	*Aschochyta* blight (54/29), *Botrytis* grey mould (35/2)
Pigeonpea	Sterility mosaic (72/44)
Groundnut	Leaf miner (20/5), *Spodoptera* (10/5), root-knot nematodes (55/9), protein (55/6[c] > 60%), oil (55/2[d] > 58%)

[a]*Parasorghum* are immune.
[b]All *Pennisetum pedicellatum* and 66 out of 68 *Pennisetum polystachyon* accessions are immune.
[c]*Arachis duranensis.*
[d]*Arachis magna* and *Arachis sylvestris.*

Australian sorghum species. An attempt was made to cross maternal sorghum plants with pollen from these accessions, but all the hybrids resembled the female parent (Huelgas *et al.*, 1996). It seems that these Australian sorghum species may be an important source of resistance, e.g. *Sorghum timorense* shows immunity to downy mildew, shoot fly, stem borer and midge, while *Sorghum australians* exhibits high resistance to downy mildew, stem borer and shoot fly. However, crossing barriers are still impeding an easy gene transfer to the breeding pool.

Pearl millet

The four major diseases of this crop are downy mildew, smut, ergot and rust. Though sources of resistance to these diseases are available in the cultivated species, the resistances break at high inoculum or at early inoculation. The inbred line 'Tift 85' was developed by transferring dominant resistance genes to rust and leaf spot from *Pennisetum glaucum* subsp. *monodii* (Hanna, 1987). This inbred line also led to a new source of cytoplasmic male sterility in pearl millet hybrid breeding. Napier grass (*Pennisetum purpureum*), which belongs to the secondary genepool and whose progeny crosses with pearl millet is often sterile hybrid, possesses resistance to many pests plus an outstanding forage yield potential. Genes for earliness, long inflorescence, leaf size and male fertility restoration were transferred from this wild species to pearl millet. Apomixis has been reported in the tertiary genepool of pearl millet, but crosses with species of this genepool are difficult. Besides apomixis, other interesting characteristics available in this genepool are perennial growth habit, drought and cold tolerance, pest resistance and new cytoplasm sources. Some researchers (Dujardin and Hanna, 1989; Hanna, 1992) have developed interspecific hybrids of this genepool with pearl millet through ploidy manipulations. Pollen from species such as *Pennisetum ramosum* or *Pennisetum mezianum* may produce shrivelled and immature seeds, and thus they do not germinate. Embryo rescue may be an alternative path to obtain interspecific hybrids in crosses between the tertiary genepool and pearl millet.

Chickpea

At least 13 *Cicer* species are bearers of useful characteristics such as resistance to wilt, soil borne fungi, grey mould, blight, cyst nematode, leaf miner, bruchid beetle, tolerance to cold, or possess high protein, multi-seeds or twin pods (N. Kameswara Rao, ICRISAT, unpublished data). ICRISAT has cooperated with the International Centre for Agricultural Research in the Dry Areas (ICARDA) to transfer genes for some of these wild species to chickpea, because such species are the only known source of resistance to pests such as seed beetle and cyst nematode. However, some of these species are perennial and not easy to propagate. Among the annual species, *Cicer reticulatum* belongs to the primary genepool and fertile hybrids were obtained in crosses with chickpea. Likewise, high yielding lines were derived from *Cicer echinospermum* from the secondary genepool, and hybrids between the blight resistant *Cicer pinnatifidum* and chickpea were obtained using embryo rescue (Mallikarjuna, 1999). The other *Cicer* annual species and all perennial species belong to the tertiary genepool of chickpea (Singh *et al.*, 1998).

Pigeonpea

Resistance to pod borer, pod-fly, bruchid beetle, wilt, blight, sterility mosaic, leaf spot and nematodes are inadequate in cultivated pigeonpea. Resistance to these biotic factors as well as early flowering, photo-insensitivity, tolerance to drought plus salinity, and high seed protein are available in nine wild species (N. Kameswara Rao, ICRISAT, unpublished data). For example, trichomes, which provide an insect resistance mechanism, are available in wild *Cajanus* species. The resistance to pod borer observed in *Cajanus scarabaeoides* appears to be associated with high density of non-glandular trichomes (Romeis *et al.*, 1999). This species and another five Indian wild species (*Cajanus cajanifolius, Cajanus trinervius, Cajanus albicans, Cajanus lineatus* and *Cajanus sericeus*) are able to cross with pigeonpea, thus, gene transfer may be feasible through recombination. Hybrids between pigeonpea and *Cajanus platycarpus*, a species that possesses genes for photo-insensitivity, earliness plus resistance to wilt and blight, and tolerance to salinity, were obtained through embryo rescue (Mallikarjuna and Moss, 1995).

Groundnut

Resistances to rust, leaf spots, viruses, thrips, aphids, mites and jassids are available in annual and perennial *Arachis* species (N. Kameswara Rao, ICRISAT, 2000, personal communication). Some of these resistances, such as the resistance to leaf spots, are not available in cultivated groundnut. Ploidy barriers and genomic incompatibility make cumbersome the transfer of genes from wild *Arachis* to groundnut. Post-zygotic failure of embryo development seems to be the major barrier for successful gene introgression. Ploidy manipulations offer a means to overcome this problem by chromosome doubling of triploid hybrids and reducing these hexaploids to tetraploids through recurrent backcrossing. For example, the diploid species *Arachis cardenasii* was crossed with groundnut and the ensuing triploid hybrid was colchicine-doubled to restore fertility. The selfing and backcrossing of the hexaploids resulted in chromosome loss and consequently tetraploid genotypes were obtained. Following this approach, several germplasm lines (ICGV 86699 and ICGV 87165) with multiple disease and insect resistance were developed at ICRISAT (Reddy *et al.*, 1996; Moss *et al.*, 1997), and one derived-cultivar (ICGV-SM 86715) was released in Mauritius (Moss *et al.*, 1998). In the USA, the cultivars Spancross and Tamnut 74 were developed using *Arachis monticola* (Hammons, 1970; Simpson and Smith, 1974). Ploidy manipulation or embryo rescue and other tissue culture methods were attempted to transfer resistance to groundnut rosette virus from *Arachis diogoi* (= *Arachis chacoense*, section *Arachis*) and resistance to early leaf spot from *Arachis paraguariensis* (section *Erectoides*) and *Arachis appressipila* (section *Procumbentes*) to groundnut (Moss *et al.*, 1993; Legume Program, 1994; ICRISAT, 1995). However, embryo rescue has not been successful with species outside section *Arachis* owing to genomic incompatibility. A new approach involving gene location plus isolation and transfer through transgenics may be needed to access genes from the other *Arachis* genepools.

Biotechnology and Germplasm Enhancement

Plant breeders determine their genetic response to selection (*R*) by the following equation:

$$R = i\,h^2\sigma_P$$

where *i* is the selection intensity, expressed in standardized units, h^2 is the narrow-sense heritability in the reference population, and σ_P is the phenotypic standard deviation of the selected characteristic. *In vitro* screening may provide a means for achieving a higher intensity of selection. Molecular markers are being used to tag specific chromosome segments bearing the desired gene(s) to be transferred (or incorporated) into the breeding lines (or populations). In this way, indirect selection with co-dominant molecular markers tightly linked to the gene(s) controlling the characteristic(s) of interest improves *R*, because co-dominant markers have an h^2 equal to 1 (Ortiz, 1999c). Transgenic plants offer new genes to the breeding pool (Ortiz, 1998c), thus enhancing the phenotypic standard deviation of the population.

Transgenics

In recent years transgenic plants have been used as parents in US breeding programmes for crops such as maize, soybean and oilseed rape. In the late

1990s, North American farmers grew 4.4 Mha of transgenic maize, 5 Mha of transgenic soybean, and 1.6 Mha of transgenic oilseed rape (*The Wall Street Journal Europe* 5–6/6/1998). It is not surprising that such advances were rapidly achieved and widely accepted in the USA, as most of the food crops of this country are derived from exotic germplasm introductions. The characteristics most commonly incorporated into breeding lines through genetic engineering include resistance to herbicides, insects and viruses.

Molecular markers

Molecular markers are descriptors that offer reproducible results for characterizing genotypes. Similarly, applied plant genomics also improves the understanding of crop genepools, which are being enlarged by including transgenes, and 'native' genepools that are becoming available through comparative analysis of plant biological repertoires (Lee, 1998). Furthermore, finding new genes adds value to traditional agricultural products. Genetic resources available in genebanks are still the best source for a routine gene discovery but this work will be facilitated by gene databases assembled with the aid of applied plant genomics, which can also accelerate the utilization of available genes through transformation or meiotic-based breeding methods (Ortiz, 1998c).

There are many ongoing applications of DNA markers in crop improvement and research (Crouch and Ortiz, 2001). For example, molecular-aided analysis of cultivars released through different decades provides a means to determine the extent of diversity among improved crop germplasm (Donini *et al.*, 2000; Petersen, 2000), and the need to incorporate exotic (or non-adapted) germplasm into the respective breeding pool. Knowledge concerning DNA fingerprints at key genomic regions for adapted and exotic germplasm may also facilitate the introgression of new genes from the non-adapted germplasm. Geneticists may monitor changes in segments of chromosomes through generations of cultivar release (Ortiz, 1999d). Assessing the genotypic composition of released cultivars, generation by generation, provides a means to determine which allelic combinations are consistently selected. Hence, monitoring chromosomal changes caused by breeders' selection at successive cycles of improvement may provide a genetic 'ideotype' for future marker-assisted selection. Selecting for these optimal marker alleles linked to loci controlling agronomic or quality characteristics may facilitate the incorporation of useful genetic variation into the adapted cultivar genepool (Tanksley and McCouch, 1997). In this way, desired alleles may be 'pyramided' in enhanced lines (or potential new cultivars). Likewise, such investigation regarding changes in allelic frequency, aided by molecular marker analysis, may help to establish the amount of genetic diversity still available for further crop improvement.

Positions of useful genes in less investigated target crops can be inferred by cross-referencing to the maps of model crop species, ultimately leading to the possibility of inter-generic transfers (Lee, 1998). For example, owing to gene synteny and comparative mapping, the extreme drought resistance of pearl millet may contribute some day to the development of more hardy and water efficient sorghum, maize, rice and wheat. In the long term, this approach should lead to the isolation and characterization of candidate genes for drought tolerance because the ordering of DNA loci between chromosomes of cereals corresponds well. This conservation of gene order between two genomes offers the potential for using the map location of a gene of interest in one crop to lead to map-based isolation of the homologous gene in another crop. Comparative mapping may also be an important tool for gene location between cultivated and associated wild species. For example, the identification of DNA markers associated with characteristics of interest may be facilitated by comparative mapping, and these markers may facilitate gene transfer plus help to avoid linkage drag. For example, gene cloning from wild *Arachis* that are completely cross-incompatible with cultivated groundnut, may facilitate such a transfer.

Conclusion

Domestication and further breeding of crop species led to an apparent narrow genetic base in some crop genepools. Plant evolution in a human-made environment, however, has not exhausted useful genetic variation in the most important crop species. Assessment of genetic gains in different crop pools does not show that plant breeders have reached a plateau in their populations. However, as indicated by the International Technical

Conference on Plant Genetic Resources (Leipzig, 1996), plant breeders should include wild, landrace, or exotic germplasm in their populations to broaden the genetic base of crop production.

Germplasm enhancement has become an important tool for the genetic improvement of breeding populations by gene introgression or incorporation of wild and landrace genetic resources into respective crop breeding pools, that is, putting genes into a usable form in conventional breeding programmes (Ortiz, 1999b). Recent advances in gene technology also provide a means to broaden the genetic base of crops, and to monitor the incorporation or introgression of new genes into the breeding population(s). In recent years transgenic plants have been incorporated as parents of hybrids in US breeding programmes for crops such as maize and oilseed rape. Molecular markers are being used to tag specific chromosome segments bearing the desired gene(s) to be transferred (or incorporated) into the breeding materials. Last but not least, it is important that international crop breeder networks assess diversity in crop genepools, promote long-term evolutionary breeding programmes, and encourage farmer participatory management of plant genetic resources.

Acknowledgements

To N. Kameswara Rao (ICRISAT) for providing information and tabulated data regarding the utilization of wild species for improving crops of the semiarid tropics, and to him and Dr. Jonathan H. Crouch (ICRISAT), for the review of an early draft of this chapter.

References

Allard, R.W. (1999) *Principles of Plant Breeding*, 2nd edn. John Wiley & Sons, New York.

Balfourier, F. and Charmet, G. (1994) Étude méthodologique de la conservation de ressources génétiques de ray-grass anglais (graminée fourragère) par multiplication en pools de populations naturelles. *Genetics Selection Evolution* 26 (Suppl. 1), 203s–218s.

Belhassen, E., Auge, G., Ji, J., Billot, C., Fernández-Martínez, J., Ruso, J. and Varès, D. (1994) Dynamic management of genetic resources: first generation analysis of sunflower artificial populations. *Genetics Selection Evolution* 26 (Suppl. 1), 241s–253s.

Bidinger, F.R. (1998) Farmer participation in pearl millet research in Namibia. In: *Participatory Plant Improvement. Proceedings of the Workshop, Chennai, India, 27–28 October 1998.* M.S. Swaminathan Research Foundation – ICRISAT, Chennai, India, pp. 21–30.

Bidinger, F.R., Weltzien, R., Mahalakshmi, V., Singh, S.D. and Rao, K.P. (1994) Evaluation of topcross hybrids of pearl millet for arid zone environments. *Euphytica* 76, 215–226.

Bidinger, F.R., Mahalakshmi, V., Talukdar, B.S. and Sharma, R.K. (1995) Improvement of landrace cultivars of pearl millet for arid and semi-arid environments. *Annals of Arid Zone* 34, 105–110.

Crouch, J.H. and Ortiz, R. (2001) DNA markers in the improvement of research-neglected tropical food species. *The Journal of Genetic Engineering and Biotechnology: Practices and Ethics* 1, in press.

Crouch, H.K., Crouch, J.H., Madsen, S., Vuylsteke, D. and Ortiz, R. (2000) Comparative analysis of phenotypic and genotypic diversity among plantain landraces (*Musa* spp., AAB group). *Theoretical and Applied Genetics* 101, 1056–1065.

David, J.L., Zivy, M., Cardin, M.L. and Brabant, P. (1997) Protein evolution in dynamically managed populations of wheat: adaptive responses to macro-environmental conditions. *Theoretical and Applied Genetics* 95, 932–941.

Donini, P., Law, J.R., Koebner, R.M.D., Reeves, J.C. and Cooke, R.J. (2000) Temporal trends in the diversity of UK wheat. *Theoretical and Applied Genetics* 100, 912–917.

Dudley, J.W. and Lambert, R.J. (1992) Ninety generations of selection for oil and protein in maize. *Maydica* 37, 81–87.

Dujardin, M. and Hanna, W.W. (1989) Crossability of pearl millet with wild *Pennisetum* species. *Crop Science* 29, 77–80.

Ehlenfeldt, M.K., and Ortiz, R. (1995) Evidence on the nature and origins of endosperm dosage requirements in *Solanum* and other angiosperm genera. *Sexual Plant Reproduction* 8, 189–196.

FAO (Food and Agriculture Organization of the United Nations) (1996a) *Report on the State of the World's Plant Genetic Resources for Food and Agriculture.* FAO, Rome, Italy.

FAO (1996b) *Global Plan of Action for the Conservation and Sustainable Utilization of Plant Genetic Resources for Food and Agriculture.* FAO, Rome, Italy.

FAO (1998) *Edited contents summary of an informal preparative workshop on Broadening the Genetic Bases of Crop Production, Rome, Italy, 17–20 September 1997.* FAO, Rome, Italy.

Fowler, C. and Mooney, P.R. (1990) Genetic erosion: losing diversity. In: Fowler, C. and Mooney, P.R. (eds) *Shattering: Food, Politics, and the Loss of Genetic Diversity.* University of Arizona Press, Tucson, Arizona, pp. 81–86.

Frankel, O.H., Brown, A.H.D. and Burdon, J.J. (1995) *The Conservation of Plant Biodiversity.* Cambridge University Press, Cambridge, UK.

Glendinning, D.R. (1979) Enriching the potato gene-pool using primitive cultivars. In: Zeven, A.C. and Durzan, D.J. (eds) *Broadening the Genetic Base of Crops.* Pudoc (Centre for Agricultural Publishing and Documentation), Wageningen, The Netherlands, pp. 39–45.

Goldringer, I., Pham, J.L., David, J.L., Brabant, P. and Gallais, A. (1994) Is dynamic management of genetic resources a way of pre-breeding? In: Balfourier, F. and Perretant, M.R. (eds) *Proceedings of the Genetic Resources Section Meeting of EUCARPIA on Evaluation of Genetic Resources: Pre-breeding, Clermont-Ferrand, France, 15–18 March 1994.* INRA, Versailles, France, pp. 163–170.

Goldringer, I., Paillard, S., Enjalbert, J., David, J.L. and Brabant, P. (1998) Divergent evolution of wheat populations conducted under recurrent selection and dynamic management. *Agronomie* 18, 413–425.

Goodman, M.M. (1990) Genetic and germ plasm stocks worth conserving. *Journal of Heredity* 81, 11–16.

Hammons, R.O. (1970) Registration of Spancross peanuts. *Crop Science* 10, 459.

Hanna, W.W. (1987) Utilization of wild relatives of pearl millet. In: *Proceedings of International Pearl Millet Workshop. ICRISAT, Patancheru, 7–21 April 1986.* ICRISAT, Patancheru, India, pp. 33–42.

Hanna, W.W. (1992) Utilization of germplasm from wild species. In: Chapman, G.P. (ed.) *Desertified Grass Lands: their Biology and Management.* The Linnean Society, London, pp. 251–257.

Hawkes, J.G. (1979) Genetic poverty in Europe. In: Zeven, A.C. and Durzan, D.J. (eds) *Broadening the Genetic Base of Crops.* Pudoc (Centre for Agricultural Publishing and Documentation), Wageningen, The Netherlands, pp. 19–27.

Huaman, Z., Ortiz, R. and Gomez, R. (1999) A proposed *Solanum tuberosum* subsp. *andigena* core collection. In: *Impact on a Changing World: Program Report 1997–1998.* International Potato Center, Lima, Peru, pp. 185–194.

Huaman, Z., Ortiz, R. and Gomez, R. (2000a) Selecting a *Solanum tuberosum* subsp. *andigena* core collection using morphological, geographical, disease and pest descriptors. *American Potato Journal* 77 183–190.

Huaman, Z., Ortiz, R., Zhang, D. and Rodriguez, F. (2000b) Isozyme analysis of entire and core collection of *Solanum tuberosum* subsp. *andigena* potato cultivars. *Crop Science* 40, 273–276.

Huelgas, V.C., Lawrence, P., Adkins, S.W., Mufti, M.U. and Gowdin, I.D. (1996) Utilization of the Australian species for sorghum improvement. In: Foale, M.A., Henzell, R.G. and Kneipp, J.F. (eds) *Proc. 3rd Australian Sorghum Conference, Tamworth, 20–22 February 1996.* Australian Institute of Agricultural Sciences, Melbourne, Australia, pp. 369–375.

ICRISAT (International Crops Research Institute for the Semi-Arid Tropics) (1995) Wide hybridisation. In: *ICRISAT Report 1994.* ICRISAT, Patancheru, India, pp. 11–13.

Johnson, R. (1992) Past, present and future opportunities in breeding for disease resistance, with examples from wheat. *Euphytica* 63, 3–22.

Jørgensen, J.H. (1992) Discovery, characterization and exploitation of Mlo powdery mildew resistance in barley. *Euphytica* 63, 141–152.

Kannenberg, L.W. and Falk, D.E. (1995) Models for activation of plant genetic resources for crop breeding programs. *Canadian Journal of Plant Science* 75, 45–53.

Le Boulc'h, V., David, J.L., Brabant, P. and de Vallavielle-Pope, C. (1994) Dynamic conservation of variability: responses of wheat populations to different selective forces including powdery mildew. *Genetics Selection Evolution* 26 (Suppl. 1), 221s–240s.

Lee, M. (1998) Genome projects and genepools: new germplasm for plant breeding? *Proceedings of the National Academy of Sciences USA* 95, 2001–2004.

Legume Program, ICRISAT (1994) Transfer of desirable alleles from the wild relatives of groundnuts. In: *Annual Report 1993.* ICRISAT, Patancheru, Andhra Pradesh, India, pp. 154–155.

Lehmann, L.C., Jönsson, R. and Gustafsson, M. (1998) Identification of resistance genes to powdery mildew isolated from *Hordeum vulgare* spp. *spontaneum* and land races of barley. *Sveriges Utsädesförenings Tidskrift* 108, 94–101.

Lenné, J.M. and Wood, D. (1991) Plant diseases and the use of wild germplasm. *Annual Review of Phytopathology* 29, 35–61.

MacKey, J. (1970) Significance of mating systems for chromosomes and gametes in polyploids. *Hereditas* 66, 165–176.

MacKey, J. (1987) Implications of polyploidy breeding. *Biologisches Zentralblatt bl.* 106, 257–266.

MacKey, J. (1994) The history of cereal yield increase. *Melhoramento* 33, 37–54.

Mahalakshmi, V., Bidinger, F.R., Rao, K.P and Raju, D.S. (1992) Performance and stability of pearl millet topcross hybrids and their variety pollinators. *Crop Science* 32, 928–932.

Mallikarjuna, N. (1999) Ovule and embryo culture to obtain hybrids from interspecific incompatible pollinations in chickpea. *Euphytica* 110, 1–6.

Mallikarjuna, N. and Moss, J.P. (1995) Production of hybrids between *Cajanus platycarpus* and *Cajanus cajan. Euphytica* 83, 43–46.

McIntosh, R.A. (1992) Pre-emptive breeding to control wheat rusts. *Euphytica* 63, 103–113.

Mendoza, H.A. and Haynes, F.L. (1974) Genetic relationship among potato cultivars grown in the United States. *HortScience* 9, 328–330.

Moss, J.P., Singh, A.K., Subrahmanyam, P., Hildebrand, G.L. and Murant, A.F. (1993) Transfer of resistance from a wild *Arachis* species into cultivated groundnut. *International* Arachis *Newsletter* 13, 22–23.

Moss, J.P., Singh, A.K., Reddy, L.J., Nigam, S.N., Subrahmanyam, P., McDonald, D. and Reddy, A.G.S. (1997) Registration of ICGV 87165 peanut germplasm with multiple resistance. *Crop Science* 37, 1028.

Moss, J.P., Singh, A.K., Nigam, S.N., Hildebrand, G.L., Goviden, N., Ismael, F.M., Subrahmanyam, P. and Reddy, L.J. (1998) Registration of ICGV-SM 87615 peanut germplasm. *Crop Science* 38, 572.

Ortiz, R. (1997a) Breeding for potato production from true seed. *Plant Breeding Abstracts* 67, 1355–1360.

Ortiz, R. (1997b) Morphological variation in *Musa* germplasm. *Genetic Resources and Crop Evolution* 44, 393–404.

Ortiz, R. (1997c) Secondary polyploids, heterosis and evolutionary crop breeding for further improvement of the plantain and banana genome. *Theoretical and Applied Genetics* 94, 1113–1120.

Ortiz, R. (1998a) Potato breeding via ploidy manipulations. *Plant Breeding Reviews* 16, 15–86.

Ortiz, R. (1998b) AMMI and stability analyses of bunch mass in multilocational testing of *Musa* germplasm in sub-Saharan Africa. *Journal of American Society for Horticultural Science* 123, 623–627.

Ortiz, R. (1998c) Critical role of plant biotechnology for the genetic improvement of food crops: perspectives for the next millenium. *Electronic Journal of Biotechnology* 1(3), 1–8. www.ejb.org/content/vol1/issue3/full/7/

Ortiz, R. (1999a) Genetic diversity of cultivated crops and in situ conservation of genetic resources. *Botanica Lithuanica* Suppl. 2, 15–30.

Ortiz, R. (1999b) Genetic enhancement and base broadening efforts. In: Gass, T., Frese, L., Begemann, F. and Lipmann, E. (eds) *Conservation and Sustainable Utilization of Plant Genetic Resources for Food and Agriculture – Implementation of the Global Plan of Action in Europe.* International Plant Genetic Resources Institute (IPGRI), Rome, Italy, pp. 191–203.

Ortiz, R. (1999c) Indirect and multitrait selection with genetic markers. In: Crouch, J.H. and Tenkouano, A. (eds) *DNA Markers in Improvement of African Staple Crops.* International Institute of Tropical Agriculture, Ibadan, Nigeria – CTA, Wageningen, The Netherlands, pp. 43–48.

Ortiz, R. (1999d) Not just seed repositories: a more proactive role for genebanks. In: *Nordic Gene Bank 1979–1999*, pp. 45–49.

Ortiz, R. (2001) The state of use of potato genetic diversity. In: Cooper, D., Spillane, C. and Hodgkin, T. (eds) *Broadening the Genetic Bases of Crop Production.* Food and Agriculture Organization of the United Nations (FAO), IPGRI, Rome, Italy, CAB International, Wallingford, UK, pp. 181–200.

Ortiz, R. and Crouch, J.H. (1997) The efficiency of natural and artificial pollinators in plantain (*Musa* spp. AAB group) hybridisation and seed production. *Annals of Botany* 80, 693–695.

Ortiz, R. and Huaman, Z. (2001) Allozyme polymorphism in tetraploid potato gene pools and the effect of human selection. *Theoretical and Applied Genetics* (in press).

Ortiz, R. and Vuylsteke, D. (1996) Recent advances in *Musa* genetics, breeding and biotechnology. *Plant Breeding Abstracts* 66, 1355–1363.

Ortiz, R., Vuylsteke, D., Ferris, S., Okoro, J., N'Guessan, A., Hemeng, O.B., Yeboah, D.K., Afreh-Nuamah, K., Ahiekpor, E.K.S., Fouré, E., Adelaja, B.A., Ayodele, M., Arene, O.B., Ikiediugwu, F.E.O., Agbor, A.N., Nwogu, A.N., Okoro, E., Kayode, G., Ipinmoye, I.K., Akele, S. and Lawrence, A. (1997) Developing new plantain varieties for Africa. *Plant Varieties and Seeds* 10, 39–57.

Ortiz, R., Madsen, S. and Andersen, S.B. (1998a) Diversity in Nordic spring wheat cultivars (1901–1993). *Acta Agriculturae Scandinavica Section B Soil and Plant Science* 48, 229–238.

Ortiz, R., Madsen, S. and Vuylsteke, D. (1998b) Classification of African plantain landraces and banana cultivars using a phenotypic distance index of quantitative descriptors. *Theoretical and Applied Genetics* 96, 904–911.

Ortiz, R., Vuylsteke, D., Crouch, H.B. and Crouch, J.H. (1998c) TM3x: triploid black sigatoka resistant *Musa* hybrid germplasm. *HortScience* 33, 362–365.

Peloquin, S.J., Yerk, G.L., Werner, J.E. and Darmo, E. (1989) Potato breeding with haploids and 2n gametes. *Genome* 31, 1000–1004.

Peloquin, S.J., Gabert, A.C. and Ortiz, R. (1996) Nature of 'pollinator' effect in potato (*Solanum tuberosum* L.) haploid production. *Annals of Botany* 77, 539–542.

Petersen, M. (2000) Genetic diversity of bread wheat during the 20th century. MSc thesis, The Royal Veterinary and Agricultural University, Frederiksberg, Denmark.

Plaisted, R.L. (1982) Progress and future plans for the use of neotuberosum populations. In: *Utilization of the Genetic Resources of the Potato* III. International Potato Center, Lima, Perú, pp. 65–79.

Rasmusson, D.C. and Phillips, R.L. (1997) Plant breeding and genetic diversity from de novo variation and elevated epistasis. *Crop Science* 37, 303–310.

Reddy, L.J., Nigam, S.N., Moss, J.P., Singh, A.K., Subrahmanyam, P., McDonald, D. and Reddy, A.G.S. (1996) Registration of ICGV 86669 peanut germplasm line with multiple disease and insect resistance. *Crop Science* 36, 821.

Rohrbach, D.D., Lechner, W.R., Ipinge, W.R. and Monyo, E.S. (1999) Impact from investments in crop breeding: the case of Okashana 1 in Namibia. *ICRISAT Impact Series* Vol. 4. ICRISAT, Patancheru, Andhra Pradesh, India.

Romeis, J., Shanower, T.G. and Peter, A.J. (1999) Trichomes on pigeonpea [*Cajanus cajan* (L.) Millsp.] and two wild *Cajanus* spp. *Crop Science* 39, 564–569.

Sevilla, R. and Holle, M. (1995) *Recursos genéticos vegetáles.* La Molina, Lima, Perú.

Simmonds, N.W. (1993) Introgression and incorporation: strategies for the use of crop genetic resources. *Biological Reviews* 68, 539–562.

Simpson, C.E. and Smith, O.D. (1974) Tamnut 74. *Texas Agricultural Experiment Station Bulletin* L. 1348.

Singh, K.B., Ocampo, B. and Robertson, L.D. (1998) Diversity for abiotic and biotic stress resistance in the wild annual *Cicer* species. *Genetic Resourses and Crop Evolution* 45, 9–17.

Tanksley, S.D. and McCouch, S.R. (1997) Seed banks and molecular maps: unlocking genetic potential from the wild. *Science* 277, 1063–1066.

Troyer, A.F. (1990) A retrospective view of corn genetic resources. *Journal of Heredity* 81, 17–24.

Vetelainen, M., Nissila, E., Tigerstedt, P.M.A. and von Bothmer, R. (1996a) Utilization of exotic germplasm in Nordic barley breeding and its consequence for adaptation. *Euphytica* 92, 267–273.

Vetelainen, M., Suominen, M. and Nissila, E. (1996b) Agronomic performance of crosses between Nordic and exotic barleys. *Euphytica* 93, 239–248.

von Bothmer, R. and Jacobsen, N. (1991) Interspecific hybrids within the genus Hordeum. In: *Chromosome Engineering in Plants: Genetics, Breeding, Evolution.* Part A. Elsevier Science Publishers, Amsterdam, The Netherlands, pp. 411–431.

Vuylsteke, D. (2000) Breeding bananas and plantains: From intractability to feasibility. *Acta Horticulturae* 540, 149–156.

Vuylsteke, D., Ortiz, R., Ferris, R.S.B. and Crouch, J.H. (1997) Plantain improvement. *Plant Breeding Reviews* 14, 267–320.

Weltzien-Rattunde, E., Whitaker, M.L., Rattunde, H.F.W., Dhamotharan, M. and Anders, M.M. (1998) Participatory approaches in pearl millet breeding. In: Witcombe, J.R., Virk, D.S. and Farrington, J. (eds) *Seeds of Choice: Making the Most of New Varieties for Small Farmers.* Oxford & IBH Publishing Co., New Delhi, India, pp. 143–170.

Yadav, O.P., Weltzien-Rattunde, E., Bidinger, F.R. and Mahalakshmi, V. (2000a) Heterosis in landrace-based topcross hybrids of pearl millet across arid environments. *Euphytica* 112, 285–295.

Yadav, O.P., Weltzien-Rattunde, E., Mahalakshmi, V. and Bidinger, F.R. (2000b) Combining ability of pearl millet landraces originating from arid areas of Rajasthan. *Indian Journal of Genetics* 60, 45–53.

27 Genetic Base Broadening in Autogamous Crops: *Lycopersicon esculentum* Mill. as a Model

G. Saavedra[1,2] and W. Spoor[2]
[1]*Instituto de Investigaciones Agropecuarias, CRI La Platina, Santiago, Chile;* [2]*Scottish Agricultural College, The King's Buildings, Agriculture Building, Edinburgh, UK*

Introduction

In most crop species, exploited populations have been reduced dramatically as a result of domestication and subsequent breeding/selection processes. The lower overall genetic diversity of modern cultivars of autogamous species may also reflect the genetic 'bottlenecks' to which these species have been subject either due to natural phenomenon such as polyploidy, mating and dispersion systems, and geographical barriers, or during their introduction to new regions from the centre of origin. Only a limited number of seeds/propagules (or accessions) were carried back by explorers, which serve as the basis of today's modern cultivars. In addition, many genotypes have been lost as a result of the replacement and disappearance of old varieties and landraces by new, more productive, modern varieties. These modern replacement varieties are apparently better adapted to biotic and abiotic stresses of these localities. Later, commonly used breeding techniques such as backcrosses and recurrent backcrosses to an existing top-variety were effective in terms of producing new varieties with highly prescribed characteristics, but this has been obtained at the further expense of genetic diversity (Sneep, 1979; Rick, 1987; Miller and Tanksley, 1990).

Since the development of scientific breeding, exploitation of the available genetic diversity has been poor; for years plant breeders have confined their programmes to a relatively small part of the overall genetic resource. However, over a similar time frame, a large amount of germplasm such as wild relatives, old varieties, landraces, and other breeding material has been stored in genebanks; this is a valuable, but relatively unexploited, source of genetic variability. Only a small amount of this variability has been introgressed to crop species, usually aiming to solve a specific problem (frequently disease resistance) involving a few genes.

Genetic base broadening is one approach which has been suggested as a means of providing a viable sustainable genetic base from which varieties can be selected either directly or following hybridization with the existing and currently exploited genetic base of a crop species. Genetic base broadening has been defined as 'composite crosses' (Suneson, 1956), 'incorporation' (Simmonds, 1993) and 're-synthesis' (Becker *et al.*, 1995). However, regardless of the term used, the definition *per se* has been the same. Genetic base broadening is the incorporation and re-synthesis of populations from wild relatives, landraces and/or old varieties into relatively new varieties or accessions. Strictly, base broadening should be without preconceived aims, partly because we do not know what is required in the future, and partly because such aims may influence the construction of the initial populations.

 Managing Plant Genetic Diversity
(eds J.M.M. Engels, V. Ramanatha Rao, A.H.D. Brown and M.T. Jackson)

However, the intention must be to create populations that have enhanced ability to respond to biotic and abiotic stresses in future breeding generations. Genetic base broadening involves the systematic utilization of an arrangement of genetic variability in a manner likely to generate a mass of newly adapted genetic stocks to be made available as parents in breeding programmes (Simmonds, 1993). These populations, selected for local adaptation, would contribute to the sustainability of agricultural systems and be a ready source of variability if there should be unexpected environmental change.

FAO's (1996) Global Plan of Action proposed to increase genetic enhancement and base-broadening efforts. There have already been two expert workshops supported by FAO and the International Plant Genetic Resources Institute (IPGRI) to discuss broadening the genetic base of crops; the first was held in Rome (Italy) in 1997 and the second in Edinburgh (UK) in 1999. Therefore, FAO has given considerable support to these actions, because this methodology has the potential to be one of the most environmentally benign of agricultural technologies. The search for genetic resistance or tolerance to biotic and abiotic stresses could decrease the use of the many contaminant and pollutant products used in modern agriculture and yet be more sustainable.

Some Examples of Base Broadening Efforts

Examples of deliberate base broadening activities, although relatively rare, can be found in a range of crops. In outbreeding species such as maize (*Zea mays* L.), tropical germplasm was adapted to conditions in southern maize-growing regions in the USA (Goodman, 1985); Salhuana *et al.* (1993) reported a national project for Germplasm Enhancement of Maize (GEM) in the USA; Kannenberg and Falk (1995) designed a breeding system for maize called HOPE (hierarchical open-ended population enrichment). In sorghum (*Sorghum bicolour* (L.) Moench), Ethiopian and Sudanese landrace germplasm was successfully incorporated into adapted Indian cultivars (Mengesha and Rao, 1982). In clonal crops, normally outbreeders, the narrow genepool of potato (*Solanum tuberosum* L.) has been enhanced by the creation of *Neotuberosum* populations from wild relatives from the *Andigena* Group (Glendinning, 1979; Plaisted, 1982; Mendoza, 1989; Simmonds, 1993). In cassava (*Manihot esculenta* Crantz) (Nassar, 1989) and sugarcane (*Saccharum officinarum* L.) (Chave, 1991; Simmonds, 1993) there is also reported research in this subject. In the case of inbreeding crop species such as barley (*Hordeum vulgare* L.) there have also been approaches for broadening the genetic base, such as incorporation of exotic germplasm into barley breeding pools in the Nordic region (Vetelainen *et al.*, 1996) and the recurrent introgressive population enrichment (RIPE) in Canada (Kannenberg and Falk, 1995). Becker *et al.* (1995) reported re-synthesis research in oil seed rape (*Brassica napus* L.). Other autogamous crops have also been studied for genetic base broadening such as oat (*Avena sativa* L.) (Frey, 1994), soybean (*Glycine max* (L.) Merr.) (Wang, 1994), rice (*Oryza sativa* L.) (Li *et al.*, 1994), and Arabica coffee (*Coffea arabica* L.) (Agwanda and Owuor, 1989). Several other researchers have suggested the use of incorporation in other genetically narrow crops using wild relatives and landraces as sources of genetic variability. For example, Ahmad *et al.* (1995) proposed an incorporation programme in cultivated lentils (*Lens culinaris* ssp. *culinaris*) because of the scarce genetic variation in this crop. Also it has been suggested in tomato (*Lycopersicon esculentum* Mill.) by Garanko (1991) and Rick and Chetelat (1995); and in common bean (*Phaseolus vulgaris* L.) by Welsh *et al.* (1995).

Tomato is one of the many autogamous species which has been affected by the processes of genepool depletion. Poor genetic diversity observed in modern cultivars may reflect the genetic bottlenecks (Rick, 1976; Miller and Tanksley, 1990; Williams and St Clair, 1993) during the domestication and later introduction from Latin America to Europe in the 16th century, and the initial genetic variability of the ancestral form, which may have already been at a lower level (Rick and Chetelat, 1995). Within *L. esculentum* accessions, little genetic diversity can be found (Miller and Tanksley, 1990; De Verna and Paterson, 1991; Breto *et al.*, 1993; Williams and St Clair, 1993), except from induced variation and variability resulting from the introgression of traits from wild species (Rick, 1979). Closely related cultivars presented genetic variation at the interspecific level as detected by randomly amplified polymorphic DNA (RAPD) (Klein-Lankhorst *et al.*, 1991; Nienhuis and dos Santos, 1994; Paran *et al.*, 1995) and microsatellites

(Vosman *et al.*, 1992; Rus-Kortekaas *et al.*, 1994; Arens *et al.*, 1995; Smulders *et al.*, 1997), and also at the intraspecific level (Miller and Tanksley, 1990).

Wild relative tomatoes are a rich source of agronomic and quality characters of use in breeding. Some of them are valuable sources of resistance/tolerance to diseases and arthropod pests; tolerance to cold, drought and salinity; shelf life; content of vitamins C and A, sugars and other secondary compounds useful in the human diet (Taylor, 1986).

As noted earlier, the benefits of base broadening have been more apparent in the case of outbreeding crop species, particularly potato, sugarcane, maize and others. With outbreeding species continuous reassortment of alleles over generations, generates the variability for future selection. However, inbreeding species may present difficulties in crossing with other individuals (particularly with outbreeding progenitor species), and then there is the tendency to inbreed rapidly leading to homozygosity and fixation; this may not be too problematic if such species had low rates of outcrossing, which introduces some degree of genetic diversity.

Thus, at the outset of a project aimed at broadening the genetic base of an autogamous crop species, a number of questions have to be considered:

- How large should be the scale of operation?
- How many parental lines should be utilized?
- What range of parents or how wide should be the choice of parental material?
- Can the population be large enough to generate variability for many years even at low rates of natural outcrossing?
- Will there be a need for continued hybridization?
- How to maintain the created variability over time?
- How to minimize selection particularly during the initiation phase?
- Should the material be exploited at one or many different sites?
- Does the selection of different sites lead to maintenance of variability overall?

Although it is not possible to undertake an entire base broadening approach for any crop within 3 years, or to answer all these questions, this project has chosen tomato as a model because it is a typical autogamous crop with a narrow genetic base, an easy plant to grow and cross; there are large numbers of wild relatives and landraces available in genebanks; and it is possible to obtain at least two generations per year under controlled conditions.

The objectives of this chapter are to analyse progress to date particularly in relation to the architecture of F_2 populations created from interspecific and intraspecific hybridizations of *Lycopersicon* spp. through co-dominant (microsatellites) and dominant (RAPD) molecular markers, and morphological characters, and to speculate about possible strategies for conserving the created genetic variability in such an autogamous crop.

Diversity in Tomato and its Relatives

A random sample of five accessions of *L. esculentum* and five belonging to *Lycopersicon* spp. (known to be involved in the evolutionary history of the cultivated tomato) were selected as parents from different genebanks sources (Table 27.1). The accessions were hybridized in all combinations by hand. Two populations were created, one with the interspecific crosses, and the other within the *L. esculentum* crosses. In total 40 successful F_1s were selfed to produce F_2 generation, of which 36 surviving families were grown under similar agroclimatic conditions and selfed.

DNA was extracted from individual plants in each F_2 accession, then PCR based microsatellites and RAPD were carried out. A molecular markers statistical analysis was performed showing the following indices as diversity measures in the created populations:

Percentage of polymorphic loci

Percentage of polymorphic loci defines a locus as polymorphic when the frequency of the most common allele is no greater than 0.95 (Ayala, 1982). Although considered a very arbitrary and imprecise measurement of diversity, because it depends on how many individuals are examined, the values obtained in the case of microsatellite molecular markers showed a variation within individual accessions in interspecific cross populations of between 62.5% and 87.5%, in comparison with intraspecific crosses within *L. esculentum* accessions of only 25.0% to 37.5%. The difference in percentage of polymorphic loci between both groups was highly significant ($P < 0.01$). As an overall comparison, the *esculentum* populations presented just 37.5% of polymorphic loci, while the populations of interspecific crosses showed 100% (Table 27.2).

Table 27.1. Species and accessions utilized in the experiment.

Species	Accession	Cultivar	Donator	Origin
Lycopersicon esculentum		Limachino	INIA-Chile	Chile
Lycopersicon esculentum	LA 0516	Ace	TGRC-Davis-USA	USA
Lycopersicon esculentum	LA 0534	Lukullus	TGRC-Davis-USA	UK
Lycopersicon esculentum	LA 0502	Marglobe	TGRC-Davis-USA	USA
Lycopersicon esculentum	LA 0180	San Marzano	TGRC-Davis-USA	USA
Lycopersicon esculentum var. *cerasiforme*	LA 1673		TGRC-Davis-USA	Peru
Lycopersicon hirsutum f.sp. *glabratum*	G 29255		USDA-ARS-USA	Ecuador
Lycopersicon parviflorum	T 1264/94		IPK-Germany	Peru
Lycopersicon pennellii var. *puberulum*	LA 1926		TGRC-Davis-USA	Peru
Lycopersicon pimpinellifolium	PI 270449		USDA-ARS-USA	Mexico

The variation shown in RAPD markers between individual crosses grouped in intraspecific *L. esculentum* and interspecific was significant ($P < 0.05$). The range of variation in interspecific crosses was from 14.0% to 48.0%, while in the *esculentum* group it was between 14.0% and 28.0% of polymorphic loci. Overall for each population, the relative values varied from 82.0% for interspecific to 52.0% for intraspecific crosses.

The results obtained with this index are not the best comparison between populations, because they are biased by the selection of the most polymorphic primers in both molecular marker systems utilized in this experiment. However, they give a robust indication of the differences between the created populations.

Observed (n_a) and effective (n_e) number of alleles

These indices are the inverse of homozygosity, and the higher the value of n_a and n_e the less homozygous are the alleles studied. Microsatellite analysis showed that n_e for individual accessions on interspecific crosses had values between 1.418 up to 1.837 and n_a from 1.625 to 1.875, in comparison with crosses between accessions belonging to *L. esculentum* that presented smaller values for n_e ranging from 1.225 to 1.375 and for n_a from 1.250 to 1.375 (Table 27.2). The differences between groups for n_a and n_e were highly significant ($P < 0.01$). In summary, n_e was 24% and n_a was 31% higher for interspecific crosses than in the *esculentum* populations.

RAPD results for both populations are also presented in Table 27.2. As a group, interspecific crosses population showed an n_a of 16.4% and an n_e 10.5% higher than the *esculentum* cross population. The difference between n_a and n_e was 0.378 and 0.230 for interspecific and *esculentum* populations, respectively. Within individual accessions the differences between n_a and n_e showed values from 0.061 to 0.215 for interspecific entries and from 0.046 to 0.128 for *esculentum* crosses. Significant differences ($P < 0.05$) were found within members of individual populations.

Table 27.2. Genetic diversity indices means and standard deviation in F_2 populations of *Lycopersicon* spp. interspecific crosses and intraspecific crosses within *L. esculentum*.

	Microsatellites		RAPD	
	Intraspecific *L. esculentum* F_2 population	Interspecific crosses F_2 population	Intraspecific *L. esculentum* F_2 population	Interspecific crosses F_2 population
Observed number of alleles (n_a)	1.375 ± 0.518	2.000 ± 0.000	1.520 ± 0.505	1.820 ± 0.388
Effective number of alleles (n_e)	1.369 ± 0.510	1.800 ± 0.250	1.290 ± 0.368	1.442 ± 0.362
Gene diversity (H_t)	0.186 ± 0.257	0.431 ± 0.106	0.171 ± 0.197	0.261 ± 0.186
Average gene diversity (H_s)	0.150 ± 0.211	0.359 ± 0.127		
Shannon's information index (I)	0.259 ± 0.357	0.617 ± 0.118	0.258 ± 0.282	0.396 ± 0.252
Percentage of polymorphic loci	37.5	100	52	82
Fixation index (F_{st})	0.180	0.185	0.289	0.327

The closer the difference between n_a and n_e, the higher the similarity of allele frequencies between populations; therefore it is likely that less diversity exists among the accessions analysed. This can be expected on predominantly self-pollinated species because of their tendency to homozygosity. Both molecular markers showed a clear difference between intraspecific *esculentum* and interspecific crosses; this is expected since the difference reflects the lower genetic diversity present in edible tomato accessions and which increases when hybridized to accessions with more variation in their genetic background, such as wild relatives.

Gene diversity (H_t), average gene diversity (H_s) and Shannon's information index (I)

These measures are the most commonly used to estimate diversity. In theory these values should range from 0 to 1 (homozygosity to full heterozygosity), although for dominant markers the maximum level is 0.5 and co-dominant markers never reach the maximum value of 1 for self-pollinating species. For autogamous species, H_t and I are more useful because H_s does not reflect well the amount of genetic variation in such organisms. There will be more homozygotes in a population in which crosses between relatives is common, even though different individuals can carry different alleles if the locus is variable in the population. There will also be more homozygotes in a population in which mating between relatives is common than in a population where it does not occur, even when the allelic frequencies are identical in both populations (Ayala, 1982). Examining microsatellite data of individual crosses, H_s was similar, 0.150 and 0.359, in all individual *L. esculentum* and interspecific crosses, respectively. However, H_t varied from 0.241 to 0.427 in interspecific crosses and from 0.118 to 0.183 in the intraspecific *esculentum* crosses; while I presented a range between 0.354 and 0.596, and 0.166 and 0.255 for interspecific and *esculentum* crosses. For both groups of populations H_t and I showed significant differences ($P < 0.01$). In general, interspecific populations groups showed values nearly twofold higher than *esculentum* populations for all indices (Table 27.2).

With RAPD (Table 27.2) there were no huge differences between both created populations for H_t and I; however the interspecific group presented increases of 34.5% for H_t and 34.8% for I with respect to *esculentum* intraspecific crosses. For individuals, H_t for interspecific entries ranged between 0.060 and 0.181, while I ranged from 0.089 to 0.265. But in *esculentum* crosses accessions H_t ranged from 0.055 to 0.098, and for I from 0.081 to 0.143. The difference in H_t and I values between both populations was found to be significant at the 0.05 level. The gene diversity (H_t) and Shannon's information index (I) results also demonstrate the low levels of diversity present in *L. esculentum* accessions but indicate that there is still variability present within landraces and old varieties. This may be useful for breeding purposes when incorporated into appropriate populations.

Fixation index (F_{st})

This index is usually utilized to analyse the differences in genetic diversity among populations. In this programme the selected *esculentum* parents were homozygous and no differences were found within the accessions, but between them there were differences. Thus for parental groups the average gene diversity (H_s) was 0, but the total gene diversity (H_t) was 1 or 100%, when F_{st} was 1, then the genetic variation found was only between populations.

In the F_2 populations analysed with microsatellite markers, populations of interspecific crosses and *esculentum* crosses showed very close F_{st} value of 0.185 and 0.180, respectively (Table 27.2). Thus about 18% of the genetic variation between populations belonging to each group can be explained as differences between accessions; however 82% of the genetic diversity lies in the differences within the population.

For RAPD markers, the values obtained reflect the high differentiation of genetic diversity among created populations, 28.9% for *esculentum* and 32.7% for the interspecific group. However, there was higher genetic variability within populations; 71.1% and 67.3% of the diversity for *esculentum* and interspecific crosses, respectively.

These results indicate that a sizeable portion of the genetic diversity in created populations lies within accessions, but also the diversity between is not insubstantial; this is one of the objectives in a base broadening approach. All of the indices used help to provide an insight into the population structure of the different genepools.

Relationship between microsatellites and RAPD

The correlation coefficient (*r*) between the different diversity indices utilized was calculated. Effective number of alleles (n_e), gene diversity (H_t) and Shannon's information index (*I*) showed a moderate positive correlation as presented in Table 27.3.

Table 27.3. Correlation coefficient (*r*) for diversity indices between microsatellites and RAPD.

	Correlation coefficient (*r*)
Effective number of alleles (n_e)	0.707**
Gene diversity (H_t)	0.679**
Shannon's information index (*I*)	0.690**

**Significance ($P < 0.01$).

Continuous Characters

Six continuous characters were scored after fruits of the second inflorescence were completely ripe: fruit fresh weight (g), weight of 1000 seeds (g), fruit diameter (cm), fruit length (cm), fruit shape as a ratio between length/diameter, and soluble solid content (°Brix).

The six morphological traits measured are all quantitative, and expression can be strongly influenced by the environment. These traits were chosen as representative of the type of characters that might be of interest to plant breeders utilizing germplasm collections. Indeed these characters may also be used when undertaking a base broadening approach to rapidly improve a population with respect to agronomically important characters.

Fruit weight showed great divergence among the four groups analysed, as shown in Table 27.4. The *L. esculentum* accessions utilized as parents were almost 50 times bigger than *Lycopersicon* spp. parents. The F_2 interspecific crosses produced fruits with a tendency toward small fruits, though their average was higher than the wild relatives.

Fruit diameter and length also showed great differences between the four groups (Table 27.4). The mean in *L. esculentum* parental accessions was almost three times higher than the wild relative parents for fruit diameter and length. However, this mean was only twice as large as the interspecific cross populations, and just slightly larger than *L. esculentum* crosses. The differences within populations were significant at 0.01 level and the interspecific F_2 population exhibited much greater mean performance for both of these characters than the original *Lycopersicon* spp. population.

The fruit ratio, as shown in Table 27.4, was very close for all four groups, with a clear predominant tendency toward round fruits with a mean closer to 1. Of the five *L. esculentum* accessions utilized as parents, two presented flattened or oblate fruits, two were round type and one was elongated shape. All three wild relative parents were round shaped fruits. Within the interspecific population, the accessions presented round shape, slightly elongated and flattened. But in the case of *L. esculentum* crosses the accessions were rounded, flattened and elongated shape. It is known that at least three genes affect fruit shape and normal round shape is dominant.

Soluble solid content is a character highly influenced by environmental factors and is controlled by several genes. Significant differences ($P < 0.05$) were observed between the four populations (Table 27.4). This variation showed that wild relatives had higher content of soluble solid than *L. esculentum*, and this difference was almost the average between both groups of parents in the interspecific group,

Table 27.4. Average and standard deviation for continuous characters in different parental and F_2 populations.

	Interspecific F_2 population	*L. esculentum* F_2 population	*Lycopersicon* spp. parents population	*L. esculentum* parents population
Fruit weight (g)	8.93 ± 6.61	30.97 ± 14.77	1.82 ± 1.23	51.47 ± 45.22
Fruit diameter (cm)	2.41 ± 0.60	3.85 ± 0.77	1.35 ± 0.35	4.51 ± 1.46
Fruit length (cm)	2.18 ± 0.59	3.51 ± 0.54	1.24 ± 0.35	3.90 ± 0.91
Fruit ratio (D/L)	1.12 ± 0.14	1.10 ± 0.18	1.10 ± 0.07	1.17 ± 0.29
Soluble solid content (°Brix)	7.01 ± 1.83	5.95 ± 1.22	9.41 ± 2.52	5.70 ± 0.99
1000 seeds weight	2.34 ± 0.26	3.17 ± 0.67	1.21 ± 0.23	2.91 ± 0.59

but it was found to be similar or very close in the case of *L. esculentum* parents and crosses.

In the case of seed weight and size (Table 27.4), seeds of the *esculentum* parents group were heavier than wild relative parents. However, the interspecific and *esculentum* crosses presented means closer to *L. esculentum* parents. These results suggest that there is a tendency to produce heavier and bigger seeds within the populations influenced by the *L. esculentum* group.

Further Steps

Most domesticated crops need genetic base broadening, some earlier than others to overcome barriers for future crop improvement. The situation of autogamous crop species is especially sensitive where the main bottleneck is the mating system, in practice this leads to a loss in genetic variability in each generation, because of the inevitable selection that occurs. Nevertheless for most inbreeding species a low percentage of outcrossing occurs which can produce a level of variability each time.

Wild and unadapted germplasm represent a rich source of variation. Though exotic germplasm can present problems of adaptation and characters not desirable in a breeding programme, it can help to increase in response to selection as a result of the improvement of genetic diversity (Ortiz, 1998). In this context, adapted germplasm such as old varieties and landraces are also an interesting source of genetic variability for use in base broadening programmes.

The results obtained from the F_2 populations created will, along with data from future generations, allow an appraisal of the different methodologies to be used when developing strategies for base broadening with autogamous species. Traditional approaches by breeders for germplasm evaluation have centred on the use of morphological characters, not surprisingly given the requirements and objectives of current breeding programmes. However, those characters tend to be quantitatively inherited, and should therefore be assessed in a number of environments involving a large number of plants and at considerable cost.

In order to make recommendations on strategies for base broadening, it is essential that appropriate descriptors of the variability be developed, and how they might change with different management approaches, be established. The results to date demonstrate the robust nature of the information from microsatellites and DNA analysis, uninfluenced by environmental factors, and this will be tested further looking at additional generation material.

For the future, it is hoped to use the experience gained with *esculentum* utilizing molecular markers in order to answer some of the questions posed in the introduction. These operational questions need to be examined in order to remove the empiricism that has, by and large, dominated previous base broadening efforts. Such methodologies on their own will not answer all the questions. Some questions will be very much species specific, others are more a matter of resources and finance. Nevertheless, the lessons learnt from pursuing some of the questions, in such an amenable species, will have messages for other autogamous crops in other environments.

Conclusions

Finally, broadening the genetic base of a crop species can take many forms: by creating diverse populations utilizing a wide range of parental material (landraces through to progenitor species); by encouraging exploitation of genotypes in space and time (diversification schemes); by utilizing deliberate genotype mixtures or designing improved landraces (exploiting agronomic combining ability); by developing farmer participatory selection programmes (allowing farm-based adaptation), to name but a few. All of these approaches have merits for different agricultural systems and all need to be considered in order to avoid an unsustainable dependency on a few genotypes.

References

Agwanda, C.O. and Owuor, J.B.O. (1989) Clonal comparative trials in Arabica Coffee (*Coffea arabica* L.). I: The effect of broadening the genetic base on the stability of yield in Kenya. *Kenya Coffee* 54, 639–643 (Abstract).

Ahmad, M., McNeil, D.L., Fautrier, A.G., Armstrong, K.F. and Paterson, A.M. (1995) Genetic relationships in *Lens* species and parentage determination of their interspecific hybrids using RAPD markers. *Theoretical and Applied Genetics* 92, 1091–1098.

Arens, P., Bredemeijer, G., Smulders, M.J.M. and Vosman, B. (1995) Identification of tomato cultivars using microsatellites. *Acta Horticulturae* 412, 49–57.

Ayala, F.J. (1982) *Population and Evolutionary Genetics. A Primer.* The Benjamin/Cummings Publishing Company, Menlo Park, California.

Becker, H.C., Engqvist, G.M. and Karlsson, B. (1995) Comparison of rapeseed cultivars and resynthesized lines based on allozymes and RFLP markers. *Theoretial and Applied Genetics* 91, 62–67.

Breto, M.P., Asins, M.J. and Carbonell, E.A. (1993) Genetic variability in *Lycopersicon* species and their genetic relationship. *Theoretical and Applied Genetics* 86, 113–120.

Chave, J.W. (1991) Sugar cane breeding at the W.I. Central Sugar Cane Breeding Station with special reference to the base broadening programme. In: *Proceedings Ninth Annual Conference Barbados Society of Technologists in Agriculture, Rockley Resort, Barbados, 13–15 November 1991* (Abstract), pp. 63–70.

De Verna, J.W. and Paterson, A.H. (1991) Genetics of *Lycopersicon.* In: Kalloo, G. (ed.) *Genetic Improvement of Tomato.* Springer-Verlag, Berlin, Germany, pp. 22–38.

FAO (1996) *Global Plan of Action for the Conservation and Sustainable Utilization of Plant Genetic Resources for Food and Agriculture, and the Leipzig Declaration.* FAO, Rome, Italy.

Frey, K.J. (1994) Remaking a crop genepool: the case history of *Avena. Proceedings of the National Science Council, Republic of China, part B, Life Sciences* 18, 85–93 (Abstract).

Garanko, I.B. (1991) Germplasm resources in *Lycopersicon.* In: Kalloo, G. (ed.) *Genetic Improvement of Tomato.* Springer-Verlag, Berlin, Germany, pp. 51–57.

Glendinning, D.R. (1979) Enriching the potato genepool using primitive cultivars. In: *Proceedings of the Conference: Broadening the Genetic Base of Crops, Wageningen, The Netherlands,* pp. 39–45.

Goodman, M.M. (1985) Exotic maize germplasm: status, prospects and remedies. *Iowa State Journal of Research* 59, 497–527.

Kannenberg, L.W. and Falk, D.E. (1995) Models for activation of plant genetic resources for crop breeding programs. *Canadian Journal of Plant Science* 75, 45–53.

Klein-Lankhorst, R.M., Vermunt, A., Weide, R., Liharska, T. and Zabel, P. (1991) Isolation of molecular markers for tomato (*L. esculentum*) using random amplified polymorphic DNA (RAPD). *Theoretical and Applied Genetics* 83, 108–114.

Li, C.P., Chen, Y.H., Chern, C.G., Tseng, T.H., Chen, C.C., Lai, M.S. and Kuo, Y.C. (1994) Broadening genetic diversity of cultivated rice through interspecific hybridization within genus *Oryza*.1. *Journal of Agricultural Research of China* 43, 261–274 (Abstract).

Mendoza, H.A. (1989) Population breeding as a tool for germplasm enhancement. *American Potato Journal* 66, 639–653.

Mengesha, M.H. and Prasada-Rao, K.E. (1982) Current situation and future of sorghum germplasm. In: *Sorghum in the Eighties, Proceedings of an International Symposium on Sorghum.* ICRISAT, India, pp. 323–335.

Miller, J.C. and Tanksley, S.D. (1990) RFLP analysis of phylogenetic relationships and genetic variation in the genus *Lycopersicon. Theoretical and Applied Genetics* 80, 437–448.

Nassar, N.M.A. (1989) Broadening the genetic base of cassava, *Manihot esculenta* Krantz, by interspecific hybridisation. *Canadian Journal of Plant Science* 69, 1071–1073.

Nienhuis, J. and dos Santos, J.B. (1994) Genetic distance among tomato cultivars as measured by RAPD molecular markers. *Acta Horticulturae* 376, 59–66.

Ortiz, R. (1998) Genetic enhancement and base broadening efforts. In: Part I, *Utilization of plant genetic resources, Implementation of GPA in Europe – Braunschweig Proceedings, Braunschweig, Germany,* pp. 191–203.

Paran, I., Horowitz, M., Zamir, D. and Wolf, S. (1995) Random amplified polymorphic DNA markers are useful for purity determination of tomato hybrids. *HortScience* 30, 377.

Plaisted, R.L. (1982) Progress and future plans for the use of Neotuberosum populations. In: *Utilisation of the genetic resources of the potato* III. International Potato Center, Lima, Peru, pp. 65–79.

Rick, C.M. (1976) Tomato. In: Simmonds, N.W. (ed.) *Evolution of Crop Plants.* Longman, London, pp. 262–273.

Rick, C.M. (1979) Potential improvement of tomatoes by controlled introgression of genes from wild species. In: *Proceedings of the Conference: Broadening the genetic base of crops, Wageningen, The Netherlands,* pp. 167–173.

Rick, C.M. (1987) Genetic resources in *Lycopersicon.* In: Nevins, D.J. and Jones, R.A. (eds) *Tomato Biotechnology.* Alan R. Liss, New York, pp. 17–27.

Rick, C.M. and Chetelat, R.T. (1995) Utilisation of related wild species for tomato improvement. *Acta Horticulturae* 412, 21–38 (Abstract).

Rus-Kortekaas, W., Smulders, M.J.M., Arens, P. and Vosman, B. (1994) Direct comparison of levels of genetic variation in tomato detected by a GACA-containing microsatellite probe and by random amplified polymorphic DNA. *Genome* 37, 375–381.

Salhuana, W., Pollak, L. and Tiffany, D. (1993) Public/private collaboration proposed to strengthen quality and production of U.S. corn through maize germplasm enhancement. *Diversity* 9–10, 77–79.

Simmonds, N.W. (1993) Introgression and incorporation. Strategies for the use of crop genetic resources. *Biological Reviews* 68, 539–562.

Smulders, M.J.M., Bredemeijer, G., Rus-Kortekaas, W., Arens, P. and Vosman, B. (1997) Use of short microsatellites from database sequences to generate polymorphisms among *Lycopersicon esculentum* cultivars and accessions of other *Lycopersicon* species. *Theoretical and Applied Genetics* 97, 264–272.

Sneep, J. (1979) The breeder's point of view on broadening the genetic base. In: *Proceedings of the Conference: Broadening the genetic base of crops, Wageningen, The Netherlands*, pp. 13–18.

Suneson, C.A. (1956) An evolutionary plant breeding method. *Agronomy Journal* 56, 188–191.

Taylor, I.B. (1986) Biosystematics of the tomato. In: Atherton, J.G. and Rudich, J. (eds) *The Tomato Crop. A Scientific Basis for Improvement.* Chapman and Hall, Cambridge, UK, pp. 1–34.

Vetelainen, M., Nissila, E., Tigerstedt, P.M.A. and von Bothmer, R. (1996) Utilisation of exotic germplasm in Nordic barley breeding and its consequence for adaptation. *Euphytica* 92, 267–273.

Vosman, B., Arens, P., Rus-Kortekaas, W. and Smulders, M.J.M. (1992) Identification of highly polymorphic DNA regions in tomato. *Theoretical and Applied Genetics* 85, 239–244.

Wang, J.L. (1994) Broadening soybean genetic basis in the northeast of China. *Journal of Northeast Agricultural University* 1, 1–8 (Abstract).

Welsh, W., Bushuk, W., Roca, W. and Singh, S.P. (1995) Characterisation of agronomic traits and markers of recombinant inbred lines from intra- and interracial populations of *Phaseolus vulgaris* L. *Theoretical and Applied Genetics* 91, 169–177.

Williams, C.E. and St Clair, D.A. (1993) Phenetic relationship and levels of variability detected by restriction fragment length polymorphism and random amplified polymorphic DNA analysis of cultivated and wild accessions of *Lycopersicon esculentum. Genome* 36, 619–630.

28 An Enhancement Strategy for Rice Germplasm: DNA Marker-assisted Identification of Beneficial QTL for Resistance to Rice Blast

K. Okuno* and S. Fukuoka
National Institute of Agrobiological Resources (NIAR), Tsukuba, Japan

Introduction

In 1991, the National Institute of Agrobiological Resources (NIAR) initiated the Rice Genome Research Programme (RGP) to analyse the molecular basis of the rice genome organization and function. The first phase of this programme from 1991 to 1997 consisted of three major projects: construction of a genetic linkage map using DNA markers; construction of a physical map of rice chromosomes; and large-scale sequencing and cataloguing of cDNA from rice plants.

Construction of a genetic linkage map

McCouch *et al.* (1988) and Saito *et al.* (1991) separately constructed the first linkage map involving restriction fragment length polymorphism (RFLP) markers. During the first phase of the RGP, researchers emphasized the construction of a high-density and high-resolution linkage map using the F_2 population from a cross between japonica and indica rice varieties. A linkage map on all 12 chromosomes was produced with about 2300 DNA markers, comprising cDNA fragments derived from callus, root and shoot, genomic DNA fragments, and randomly amplified polymorphic DNA (RAPD) (Kurata *et al.*, 1994b; Harushima *et al.*, 1998). These DNA markers cover a genetic distance of 1521 cM. It is noteworthy that all the cDNA fragments and most of the genomic fragments were partially sequenced to convert them into STS (sequence tagged site) markers (Inoue *et al.*, 1994). These molecular markers are useful for tagging genes which are responsible for biological and agronomic traits.

Genomic DNA fragments of wheat have been mapped on to the rice linkage map. The synteny analysis of these plants revealed a high conservation in the order of DNA markers (Kurata *et al.*, 1994a). The synteny map and the mapped rice probes are available for the molecular analysis of the corresponding chromosomal regions in other cereals.

Construction of a physical map

Densely distributed DNA markers on the linkage map enable construction of a physical map of the rice genome. A YAC library of large DNA fragments has been constructed, and 2400 YAC clones have been ordered along the rice linkage map. The physical map that was constructed is estimated to cover about 70% of the whole genome. These clones can be used for map-based cloning of genes with agronomic value.

*Present address: National Agricultural Research Center for Hokkaido Region, Sapporo, Japan

Large-scale cDNA analysis

Large-scale cDNA analysis aims to catalogue all the expressed genes in rice (Sasaki *et al.*, 1994). About 15,000 cDNA clones have been partially sequenced. The function of sequenced cDNA clones was deduced by a similarity search to a proteins database. About 25% of the cDNA clones analysed showed a similarity to known proteins; this suggests that most of the expressed genes encode for unknown proteins.

New Approaches to Functional Analysis of the Rice Genome

The results obtained in the first phase of the RGP are now being used as important tools for rice functional genomics. These tools have not only facilitated tasks such as marker-assisted selection of desirable traits and map-based cloning of agronomically significant genes, but also enhanced better understanding of the structure and function of the rice genome. Since 1997, the RGP has mapped a large number of EST (expressed sequence tags) on to YAC clones, and mapped more than 2700 random EST clones to the rice physical map. The EST map is useful for the analysis of the rice genome by producing new STS markers for rice genetic analysis and linking new YAC clones to the physical map to eliminate YAC contig gaps. The results from the EST mapping will provide us with some fundamental information for rice genome sequencing during the second phase of rice genome analysis.

Since 1998, the RGP has entered a new phase focusing on large-scale sequencing of the rice genome, QTL (quantitative trait loci) analysis of agronomically important traits, cloning and characterization of genes, and marker-assisted selection in rice breeding. The programme also includes the development of mutant lines, which are tagged by retrotransposons of rice as a strategy for gene isolation, and functional analysis of genome by microarrays with rice cDNA clones. Large-scale sequencing of the rice genome has been undertaken by international collaboration including Japan, USA, UK, France, Canada, China, Korea, Taiwan, India and Thailand (Sasaki *et al.*, 1999).

Rice Germplasm Enhancement Using DNA Markers

The RGP has had a great impact on both rice genetics and breeding, resulting in a broadened use of the rice genepool. High-density linkage maps and DNA markers, which are distributed over 12 rice chromosomes, are very powerful tools for germplasm enhancement. One of the uses of DNA markers in the enhancement of rice germplasm is the analysis of the linkage relationship between DNA markers and genes for agronomic traits. DNA markers have played the principal role in mapping beneficial QTL for key quantitative traits of agronomic importance, whose chromosomal location is difficult to identify by conventional methods. Using DNA markers, these QTL can be efficiently transferred from unadapted germplasm into elite lines by backcrossing (Tanksley and Nelson, 1996). The progress of DNA marker-assisted studies on the identification of valuable QTL of wild rice relatives can enhance the potential for their use in improving key quantitative traits in elite rice varieties (Xiao *et al.*, 1998). Yano *et al.* (1997) reported that tools supplied by the RGP contributed to the genetic dissection of the QTL that were conditioning the heading time of japonica and indica rice varieties. In addition to DNA marker-assisted selection of agronomic traits in the breeding programme, each of the QTL responsible for agronomically important traits can be tagged by DNA markers and can be isolated by positional cloning. This research on the rice genome will also contribute significantly to the enhancement of the rice germplasm collections.

QTL Analysis for Field Resistance to Rice Blast and Enhancement of Disease Resistance Using DNA Markers

Field resistance and Japanese upland rice

Rice blast, caused by *Pyricularia grisea* Sacc. is a major destructive disease of rice in most of the rice-growing areas worldwide. Breeding efforts are ongoing to transfer resistance genes into a desirable genetic background. Two types of resistance, namely true and field resistance, to rice blast have been described. True resistance is a hypersensitive resistance that is often complete, characterized by a resistant infection type and usually under monogenic control. At least 20 loci for true resistance have been mapped on rice chromosomes and seem to be clustered or occur as multiple alleles (McCouch *et al.*, 1994). Although true resistance is often very effective against rice blast, it may occasionally be broken down by compatible races of the pathogen. For instance, a Japanese blast-resistant variety, Kusabue, was developed by introducing the resistant gene *Pik*, from Chinese germplasm into a Japanese elite variety in 1960. However, the resistance in the new variety was overcome after only 2 years. Several other exotic germplasm resources have also been used as sources of true resistance in rice breeding in Japan, but in all these cases the resistance has not held up in Japanese conditions. To overcome this genetic vulnerability, multilines have been produced, each with different major genes for blast resistance.

Field resistance is characterized by infection that is incomplete, and usually under polygenic control. This type of resistance allows effective control of the parasite under natural conditions and is considered to be durable when exposed to new races of the pathogen. Japanese upland rice is in the primary genepool of Asian cultivated rice and a source of the genes that are responsible for field resistance to rice blast. However, rice breeders in Japan have never succeeded in introducing resistance into elite rice varieties, possibly due to the close linkage between the genes for resistance and certain undesirable characteristics. Since the 1970s, genetic studies have been conducted to detect genes or loci for field resistance in Japanese upland rice. Some studies reported that many genes with additive effects were responsible for the expression of field resistance (Higashi and Kushibuchi, 1978; Higashi and Saito, 1985). These results suggested that the genetic behaviour of this field resistance was so complicated that conventional genetics could not identify the number of genes, and their chromosomal locations.

Detection and mapping of QTL for field resistance to rice blast

Recent progress in QTL analysis using DNA markers provides a better understanding of the genetics of resistance to rice blast. Several newly found resistance genes were mapped on molecular genetic linkage maps by Yu *et al.* (1996), Inukai *et al.* (1996) and Chen *et al.* (1999). Wang *et al.* (1994) reported two complete resistance genes and ten chromosomal regions involving QTL for partial resistance derived from the West African variety Moroberekan. We have worked to detect the chromosomal regions responsible for field resistance to blast in the Japanese upland rice Owarihatamochi. Genetic studies were further conducted in advanced backcrossed lines for DNA marker-assisted selection and map-based cloning of the gene involved in this resistance.

A total of 146 F_4 lines were developed from the cross between the Japanese lowland rice variety, Nipponbare, and the upland variety, Owarihatamochi. Analyses using seven differential races of *P. grisea* showed that these two varieties have no true (complete) resistance genes. Nipponbare was susceptible and Owarihatamochi highly resistant to rice blast in field tests. BC_1F_3 lines from F_3 plants of a cross between both varieties were developed as the mapping population. Aichiasahi, a highly susceptible lowland rice variety, was used as the recurrent parent.

Field resistance of all the materials was tested in an upland nursery bed with two replications. The severity of infection on 40–50 day-old plants was scored from 0 (highly resistant; no symptom) to 10 (highly susceptible; leaves totally dead) based on the diseased leaf area. Total DNA was extracted, digested with nine restriction enzymes and blotted on to a positively charged nylon membrane after electrophoresis. Southern hybridization and signal detection were carried out using an ECL direct nucleic acid labelling and detection kit (Amersham). A total of 299 RFLP probes from RFLP linkage maps and RFLP landmarkers were surveyed for a probe–enzyme combination showing polymorphism between parental varieties. Some 131 selected RFLP markers on 12 rice chromosomes and five SSR (simple sequence repeat) markers on chromosomes 9 and 12 were used for linkage analysis.

The program MAPMAKER/EXP 3.0 (Lander *et al.*, 1987) was used to establish an RFLP linkage map. The PROC GLM in the Statistical Analysis Systems (SAS) was used to determine the relationship between RFLP markers and field resistance. MAPMAKER/QTL (Lander and Botstein, 1989) was used to calculate LOD scores for significant marker loci and also used to obtain estimates of the percentage of phenotypic variance explained by each QTL. Putative QTL for field resistance to rice blast were identified using the LOD threshold of 2.0 in MAPMAKER/QTL.

The analysis of the different rice lines revealed several interesting results:

- The frequency distribution of field resistance in F_4 lines was continuous, ranging from scores of 1 to 10. The scores of Owarihatamochi (resistant) and Nipponbare (susceptible) were 2.8 and 8.0, respectively.
- RFLP markers on chromosomes 2, 4, 8, 9, 11 and 12 were significant at the 5% level in SAS/GLM. The analysis by MAPMAKER/QTL showed four QTL with LOD scores more than 2.0 located on three chromosomes. A Nipponbare allele is resistant in the QTL on chromosome 9, while an Owarihatamochi allele is resistant in the QTL on chromosomes 4 and 12. Two QTL close to RFLP markers G271 and Y8026L on chromosome 4 explained 45.7% and 29.4% of phenotypic variation, respectively. QTL on chromosomes 9 and 12 explained 7.9% and 13.7%, respectively. A multiple QTL model estimated that these four QTL explained 66% of the total phenotypic variation.
- Two BC_1F_1 plants had the Owarihatamochi allele at the RFLP marker locus G271 on chromosome 4. In these plants, four QTL except for the largest one were replaced by the allele from the highly susceptible variety, Aichiasahi. One plant (96BC131) possesses Owarihatamochi alleles in the chromosomal region from G271 to XNpb237. The other plant (96BC122) possesses the chromosomal segment from C513 to G271. Sixteen and 11% of alleles at the RFLP marker loci tested were from Owarihatamochi in 96BC131 and 96BC122, respectively. Eighty-two and 76 BC_1F_3 lines derived from 96BC131 and 96BC122 were subjected to a field resistance test. The scores for Owarihatamochi and Aichiasahi were 0.7 and 9.2. The scores for BC_1F_3 lines derived from 96BC131 varied from 2.5 to 9.2 and showed a bimodal distribution, while those of BC_1F_3 lines from 96BC122 varied from 7.5 to 9.7.
- The BC_1F_3 lines derived from 96BC131 and analysed for mapping were divided into three groups according to the field resistance test: group 1 with scores of less than 6.0; and groups 2 and 3 with scores of more than 6.0. Groups 2 and 3 were identified by the presence or absence of segregation of diseased leaf area among plants within a line. With respect to the genotype of the field resistance gene, group 1 was expected to be homozygous for the resistant allele and group 2 without segregation was expected to be homozygous for the susceptible allele. Group 3 with segregation was expected to be heterozygous for resistant and susceptible alleles. The average score among the lines was 4.20±0.67 (group 1), 8.21±0.66 (group 2) and 7.5±0.51 (group 3). The map position of the locus conferring field resistance, designated *pi21*, was flanked by RFLP markers G271 and G317 at a distance of 5.0 cM and 8.8 cM, respectively.

Japanese upland rice is a potential gene donor for improving field resistance to rice blast. We identified the chromosomal regions for field resistance to rice blast in Japanese upland rice using DNA markers mapped on rice chromosomes. DNA markers and plant materials will be useful for DNA marker-assisted evaluation and enhancement of rice germplasm, marker-assisted selection in rice breeding and positional cloning of the genes involved in this resistance.

In this study, disease severity ranged from 1 to 10 among F_4 lines. We observed lines with higher resistance than Owarihatamochi, and also lower than Nipponbare. This observation suggests that multigenic inheritance is involved in field resistance and it is difficult to determine the number of genes responsible for the resistance by conventional genetic analysis. QTL analysis using DNA markers confirmed three chromosomal regions of field resistance derived from Japanese upland rice including a single recessive gene, *pi21*. Two major QTL for field resistance are located on chromosome 4, supporting the previous results obtained by conventional genetic analysis (Maruyama *et al.*, 1983; Higashi and Saito, 1985). These studies indicate that Japanese upland rice has genes on chromosome 4 that play a major role in the expression of field resistance to rice blast.

The present study provides DNA markers linked to QTL for field resistance. One of them, designated as *pi21*, was successfully located on the RFLP genetic linkage map using backcrossed progeny lines. Although Japanese upland rice was extensively used as a potential gene donor for field resistance in rice breeding, most of the genes for the resistance were not efficiently selected due to their multigenic nature and large environmental variation. Undesirable characteristics in upland rice also limit the introduction of multiple genes into elite rice varieties. DNA markers identified in this study have provided detailed genetic information that will be useful for marker-assisted germplasm enhancement and selection in rice breeding.

For detailed genetic dissection of chromosomal regions and for positional cloning, the frequency of polymorphism between parental varieties is a limiting factor. The frequency of polymorphism between Japanese lowland and upland rice varieties was 38% in this study and is not as high as for the indica/japonica crosses. Use of a substitution line with the chromosomal segment harbouring target genes from indica rice varieties as a parent of a cross is a promising strategy to obtain higher mapping resolution. Highly polymorphic mapping populations are now being developed for map-based cloning of *pi21*.

Finally, we are interested in resolving whether each of the QTL for field resistance against rice blast reacts to specific races of blast fungus. This information will be important to answer the question whether field resistance is durable or vulnerable. The highlight of future research is to manifest the genetic differentiation of disease resistance during the evolution of crop–pathogen interactions and the cascade expression of genes for plant defence responses.

Rice genome research has recently had a great impact on progress in both rice genetics and breeding worldwide. Large-scale sequencing of the rice genome will be completed in the near future, and will accelerate better understanding of the structure and function of the rice genome. All the information and genetic materials provided by rice genome research are also useful and powerful tools for the study of diversity in plant genetic resources. In particular, a DNA marker-assisted enhancement strategy for plant genetic resources can contribute to efficient and effective use of genepools.

References

Chen, D.H., dela Vina, M., Inukai, T., Mackill, D.J., Ronald, P.C. and Nelson, R.J. (1999) Molecular mapping of the blast resistance gene, *Pi44* (t), in a line derived from a durably resistant rice cultivar. *Theoretical and Applied Genetics* 99, 1046–1053.

Harushima, Y., Yano, M., Shomura, A., Sato, M., Shimano, T., Kuboki, Y., Yamamoto, T., Lin, S.Y., Antonio, B.A., Parco, A., Kajiya, H., Huang, N., Yamamoto, K., Nagamura, Y., Kurata, N., Khush, G.S. and Sasaki, T. (1998) A high-density rice genetic linkage map with 2275 markers using a single F_2 population. *Genetics* 148, 479–494.

Higashi, T. and Kushibuchi, K. (1978) Genetic analysis of field resistance to leaf blast (*Pyricularia oryzae*) in Japan. *Japanese Journal of Breeding* 28, 277–286.

Higashi, T. and Saito, S. (1985) Linkage group of field resistance genes of upland rice variety Sensho to leaf blast caused by *Pyricularia oryzae* CAV. *Japanese Journal of Breeding* 35, 438–448.

Inoue, T., Zhong, H.S., Miyao, A., Ashikawa, I., Monna, L., Fukuoka, S., Miyadera, N., Nagamura, Y., Kurata, N., Sasaki, T. and Minobe, Y. (1994) Sequence-tagged sites (STSs) as standard landmarkers in the rice genome. *Theoretical and Applied Genetics* 89, 728–734.

Inukai, T., Zeigler, R.S., Sarkarung, S., Bronson, M., Dung, L.V., Kinoshita, T. and Nelson, R.J. (1996) Development of pre-isogenic lines for rice blast-resistance by marker-aided selection from a recombinant inbred population. *Theoretical and Applied Genetics* 93, 560–567.

Kurata, N., Moore, Y., Nagamura, T., Foote, T., Yano, M., Minobe, Y. and Gale, M. (1994a) Conservation of genome structure between rice and wheat. *Bio/technology* 12, 276–278.

Kurata, N., Nagamura, Y., Yamamoto, K., Harushima, Y., Sue, N., Wu, J., Antonio, B.A., Shomura, A., Shimizu, T., Lin, S.Y., Inoue, T., Fukuda, A., Shimano, T., Kuboki, Y., Toyama, T., Miyamoto, Y., Kirihata, T., Hayasaka, K., Miyao, A., Monna, L., Zhong, H.S., Tamura, Y., Wang, Z.X., Momma, T., Umehara, Y., Sasaki, T. and Minobe, Y. (1994b) A 330 kilobase interval genetic map of rice including 883 expressed sequences. *Nature Genetics* 8, 365–372.

Lander, E.S. and Botstein, D. (1989) Mapping Mendelian factor underlying quantitative traits using RFLP maps. *Genetics* 121, 185–199.

Lander, E.S., Green, P., Abrahamson, J., Barlow, A., Daly, M.J., Lincoln, S.E. and Newburg, L. (1987) MAPMAKER An interactive computer package for constructing primary genetic linkage maps of experimental and natural populations. *Genomics* 1, 174–181.

McCouch, S.R., Kochert, G., Yu, Z.H., Wang, Z.Y., Khush, G.S., Coffman, W.R. and Tanksley, S.D. (1988) Molecular mapping of rice chromosome. *Theoretical and Applied Genetics* 76, 815–829.

McCouch, S.R., Nelson, R.J., Tohme, J. and Zeigler, R.S. (1994) Mapping of blast resistance gene in rice. In: Zeigler, R.S., Leong, S.A. and Teng, P.S. (eds.) *Rice Blast Disease*. IRRI, Manila, Philippines, pp.167–186.

Maruyama, K., Kikuchi, K. and Yokoo, M. (1983) Gene analysis of field resistance to rice blast (*Pyricularia oryzae*) in Rikuto Norin Mochi 4 and its use for breeding. *Bulletin of the National Institute of Agricultural Sciences* D35, 1–31.

Saito, A., Yano, M., Kishimoto, N., Nakagahra, M., Yoshimura, A., Saito, K., Kuhara, S., Ukai, Y., Kawase, M., Nagamine, T., Yoshimura, S., Ideta, O., Ohsawa, R., Hayano, Y., Iwata, N. and Sugiura, M. (1991) Linkage map of restriction fragment length polymorphism loci in rice. *Japanese Journal of Breeding* 41, 665–670.

Sasaki, T., Song, J., Koga-Ban, Y., Matsui, E., Fang, F., Higo, H., Nagasaki, H., Hori, M., Miya, M., Murayama-Kayano, E., Takiguchi, T., Takasuga, A., Niki, T., Ishimaru, K., Ikeda, H., Yamamoto, Y., Mukai, Y., Ohta, I., Miyadera, N., Havukkala, I. and Minobe, Y. (1994) Toward cataloging all rice genes: large-scale sequencing of randomly chosen rice cDNAs from a callus cDNA library. *Plant Journal* 6, 615–624.

Sasaki, T., Bevan, M., Burr, B., Eun, M.Y., Hong, G. and Messing, J. (1999) The international rice genome project. In: Rutger, J.N., Robinson, J.F. and Dilday, R.H. (eds) *Proceedings of the International Symposium on Rice Germplasm Evaluation and Enhancement.* Arkansas Agricultural Experiment Station, Fayetteville, pp.115–118.

Tanksley, S.D. and Nelson, J.C. (1996) Advanced backcross QTL analysis: a method for the simultaneous discovery and transfer of valuable QTLs from unadapted germplasm into elite breeding lines. *Theoretical and Applied Genetics* 92, 191–203.

Wang, G.L., Mackill, D.J., Bonman, M., McCouch, S.R., Champoux, M.C. and Nelson, R.J. (1994) RFLP mapping of genes conferring complete and partial resistance to blast in a durably resistant rice cultivar. *Genetics* 136, 1421–1434.

Xiao, J., Li, J., Grandillo, S., Ahn, S.N., Yuan, L., Tanksley, S.D. and McCouch, S.R. (1998) Identification of trait-improving quantitative trait loci alleles from a wild rice relative, *Oryza rufipogon*. *Genetics* 150, 899–909.

Yano, M., Harushima, Y., Nagamura, Y., Kurata, N., Minobe, Y. and Sasaki, T. (1997) Identification of quantitative trait loci controlling heading date in rice using a high-density linkage map. *Theoretical and Applied Genetics* 95, 1025–1032.

Yu, Z.H., Mackill, D.J., Bonman, M., McCouch, S.R., Guiderdoni, E., Notteghem, J.L. and Tanksley, S.D. (1996) Molecular mapping of genes for resistance to rice blast (*Pyricularia grisea* Sacc.). *Theoretical and Applied Genetics* 93, 859–863s.

29 Prebreeding in Sugarcane with an Emphasis on the Programme of the Mauritius Sugar Industry Research Institute

K. Ramdoyal and G.H. Badaloo
Mauritius Sugar Industry Research Institute, Reduit, Republic of Mauritius

Introduction

Sugar cane (*Saccharum* spp.) is one of the crops for which interspecific hybridization has provided a major breakthrough in its improvement. Modern commercial sugarcane varieties (*Saccharum* hybrids, $2n$=100–130) are derived from interspecific hybridization (Price, 1963a) pioneered in Java (Stevenson, 1965). Before the advent of interspecific crosses, the improvement of sugarcane relied on the selection of naturally occurring variants of *Saccharum officinarum* obtained by collecting expeditions to its centre of origin in New Guinea. Cane improvement during the 19th century was one of variety substitution, mainly due to susceptibility to diseases such as sereh, mosaic, gumming and the root-disease complex that caused extensive losses, low yields or restricted adaptation. The discovery of the fertility of sugarcane reported in Java from Soltwedel's work and in Barbados by Harrison and Bovell in 1888 (Stevenson, 1965), led to the noble improvement programme. Although this programme was restricted to improving disease resistance, it culminated in the production of better noble varieties and added 'new material' for future work. The sugarcane breeding programme underwent radical changes with the utilization of wild *Saccharum spontaneum* in the late 1800s and the early 1900s for incorporating the much sought disease resistance into cultivars, which marked the end of the noble cane era.

The need to broaden the genetic base, given the spectre of yield stagnation, and the need to introgress specific characters from wild and associated genera, raised the interest in prebreeding. Prebreeding is the early phase of any breeding programme utilizing exotic germplasm (Frankel, 1989). Several authors have noted the need for prebreeding to provide high yielding, widely adapted germplasm possessing resistance to major biotic and abiotic stresses and emphasized biological and economic constraints (Stalker, 1980; Frankel, 1989; Ladizinsky, 1989).

Breeding in Mauritius started in 1891 although the history of exchange and utilization of germplasm dates back to the Dutch era (1635–1710), the French (1715–1820) and the British rule (1810–1968) [see reviews by Rouillard (1990) and Ramdoyal and Domaingue (1993)]. The utilization of germplasm is geared towards the development of varieties with improved cane yield and sucrose content, tolerance to biotic and abiotic stresses and good ratooning potential. This chapter reviews the most salient features of prebreeding and germplasm enhancement in sugarcane and reports on the main features of the Mauritius Sugar Industry Research Institute (MSIRI) programme.

 Managing Plant Genetic Diversity
(eds J.M.M. Engels, V. Ramanatha Rao, A.H.D. Brown and M.T. Jackson)

The Genus *Saccharum* and the '*Saccharum* Complex'

Sugar cane belongs to the family Gramineae in the genus *Saccharum*, a member of the tribe *Andropogoneae* abundant in tropical and subtropical regions. The term '*Saccharum* complex' was used by Mukherjee (1957) to denote that the genera *Saccharum*, *Erianthus* (sect. *Ripidium*), *Sclerostachya* and *Narenga* constituted a closely related inter-breeding group involved in the origin of sugarcane, to which Daniels *et al.* (1975) added the genus *Miscanthus* sect. *Diandra* Keng. The taxonomy, evolution, distribution and characteristics of genera in the *Saccharum* complex and the species of the genus *Saccharum* (hereafter referred to as basic species or basic germplasm) have been reviewed by Daniels and Roach (1987). Salient features of the *Saccharum* species and of the allied genera relevant to their use in prebreeding are outlined.

S. officinarum L., an octaploid, $2n = 80$ (Sreenivasan *et al.*, 1987; D'Hont *et al.*, 1995), indigenous to New Guinea, has provided the genetic background and the sucrose genes of modern sugar-cane hybrids. Clones of *S. officinarum* have thick stalks, high juice purity, low fibre and starch content, low vigour and adaptability to environmental stresses, and they are susceptible to diseases. *S. spontaneum* L., $2n = 40–128$ (Panje and Babu, 1960), is distributed widely from New Guinea, the Mediterranean to Africa (Brandes *et al.*, 1939; Mukherjee, 1950). Clones of *S. spontaneum* are highly polymorphic from bushy types to tall stalks with low sucrose and high fibre content. They played an important role in providing disease resistance and vigour to modern hybrids. *Saccharum robustum* Brandes and Jesweit ex Grassl., indigenous to New Guinea, has stalks of up to 10 m high which are hard, woody and pithy with little juice (Stevenson, 1965). Five groups are recognized with two cytotypes $2n = 60$ and $2n = 80$ (Price, 1965). This species contributed towards the evolution of some Hawaiian varieties.

Saccharum barberi Jesw. and *Saccharum sinense* Roxb., $2n = 81–124$ (Bremer, 1966; Price, 1968), presumably evolved in Northern India and China, are characterized by thin to medium stalks, low sucrose, high fibre and tolerance to stress conditions. They are of limited fertility and their use in breeding is restricted. Molecular studies have confirmed that both *S. barberi* and *S. sinense* possess *S. officinarum* and *S. spontaneum* genomes (A. D'Hont, France, 2000, personal communication).

The genus *Erianthus* Michx Sect. *Ripidium* Henrard (Daniels and Roach, 1987), $2n = 20–60$, is distributed in India, South-east Asia to Japan, Indonesia and New Guinea. Seven species are described. Clones of *Erianthus* are highly vigorous, tall with slender stalks of good diameter and display disease resistance, excellent ratooning ability and drought tolerance (BSES, 1990). The genus *Miscanthus* Anderss., $2n = 38–114$, is distributed from Tahiti through Eastern Indonesia, Indo-China to northern China, Siberia and Japan. The species vary from small wiry-leafed types to taller ones, occurring from sea level in Indonesia (Berding and Koike, 1980) to 3300 m in Taiwan (Lo *et al.*, 1978). Four sections have been described with prominent species.

Nobilization

Nobilization refers to the crossing of the wild cane, *S. spontaneum*, to the noble cane *S. officinarum*, and further backcrossing of progenies to the latter (Stevenson, 1965), and includes the planned introgression of the other *Saccharum* species and related genera into the noble cane. Although the first interspecific crosses were carried out by Soltwedel in 1887, the discovery of a highly vigorous, disease-free, naturally occurring F_1 hybrid, Kassoer, led to the creation of the 'wonder cane' POJ 2878 (BC2) of the 'POJ' series in 1921. This clone occupied 90% of the area under cane in Java within 8 years. Concurrently, nobilization of wild *Saccharum* species was conducted in India in 1912 and the use of a Coimbatore form, $2n = 64$, led to another source of varieties, particularly Co 205. In addition to introducing disease resistance in the noble background, nobilization produced unexpected gains in general vigour, increased cane and sugar yields, adaptability to stress conditions and ratooning ability. The early success of interspecific hybridization led to the intercrossing of other species to produce tri-species hybrids that proved very successful in subtropical areas in India (Roach and Daniels, 1987).

Intergeneric hybridization has been gaining importance as a means to broaden the genetic base, to obtain commercially useful characteristics and to increase hybrid vigour (Tai, 1989). Downy mildew (*Peronosclerospora sacchari*) resistance genes have successfully been transferred from *Miscanthus* to sugar cane (Chen and Lo, 1989).

Chromosome transmission during nobilization

Cytological studies showed that *S. officinarum* transmits its whole somatic chromosome complement ($2n$) when crossed with *S. spontaneum* which in turn transmits its haploid (n) chromosome number in the normal way (Roach, 1968), that is, a $2n + n$ transmission process. Hybrids of $n + n$ type were relatively low but may predominate in certain specific crosses (Roach, 1968; D'Hont *et al.*, 1996). Similarly, BC_1 progenies from crosses between *S. officinarum* as the maternal parent and *S. officinarum* × *S. spontaneum* F_1's, showed $2n + n$ chromosome transmission. Chromosome increase ceased in further backcrosses (Price, 1963b; Roach, 1971) such that chromosome numbers within the range of $2n = 100–130$ were reported as from the third nobilization. The $2n + n$ mode of transmission explains the rapid drop in vigour in the progenies from F_1 to BC_1 generations, which may be halted to some extent if the F_1 hybrid is backcrossed to another noble clone or a commercial type hybrid.

The process of $2n + n$ transmission is not clearly understood and a number of hypotheses, including formation of unreduced egg cells, chromosome doubling through endoduplication, selective fertilization of $2n$ eggs and selective survival of $2n + n$ zygotes, postmeiotic endomitosis in the egg, have been put forward by Price (1961) and Bremer (1961a, b, 1962). The implication of these observations is that genetic variability observed is a result of segregation and recombination of whole chromosomes within species (Roach, 1968). Optimum *S. officinarum*:*S. spontaneum* chromosome balance may take several generations to achieve. However, comparative genomic DNA *in situ* hybridization demonstrated that about 10% of the chromosomes of the commercial cultivar R 570 ($2n = 107–115$), originated from *S. spontaneum* and about 10% were identified as recombinant chromosomes between *S. officinarum* and *S. spontaneum* (D'Hont *et al.*, 1996).

Chromosome transmission in crosses of *S. officinarum* × *S. robustum* are usually of $n + n$ type (Price, 1961) whereas $2n + n$ chromosome transmission was reported in crosses between *S. officinarum* and *S. sinense* (Price, 1957). Various intergeneric hybrids between *S. officinarum* and *Erianthus*, *Sclerostachya*, *Miscanthus*, *Erianthus*, *Sorghum*, *Imperata* and *Zea* have been reported (Sreenivasan *et al.*, 1987) and between commercial hybrids and *Erianthus*, *Miscanthus*, *Miscanthidium* (Tai and Miller, 1988; Tai *et al.*, 1991). The range of hybrid types reported includes $n + n$, $n + 2n$, $2n + n$ and $2n + 2n$ (Sreenivasan *et al.*, 1987). Li *et al.* (1951) observed two types of hybrids from a cross between sugarcane and *Miscanthus*. One was the female parent type, with a $2n + n$ mode of chromosome transmission, with taller and thicker stalks, wider leaf margins and higher sugar content. The other was the male type, with $n + n$ mode of chromosome transmission, which was closer to its *Miscanthus* parent with slender stalks, narrower leaves and lower sucrose content.

Chromosome transmission in commercial hybrids × *S. spontaneum* clones in F_1, BC_1 to BC_3 generations is of strictly $n + n$ type, although meiotic irregularity, aneuploidy and chromosomal mosaics resulted in aneuploid gametes (Burner and Legendre, 1993). Chromosomes paired primarily as bivalents and variable numbers of univalents and multivalents were observed.

Genetic base of modern sugar cane varieties

Only a few clones of some basic species have been used to develop modern varieties compared with the wide diversity that is available in world collections (Table 29.1). Arceneaux (1965) reported on the use of 19 *S. officinarum* clones in the production of commercial hybrids of which three of the clones account for 57% of the genealogy of the varieties studied. Two *S. spontaneum* sources, the Java form of *S. spontaneum* ($2n = 112$) introduced mainly through Kassoer and the Coimbatore form of Indian *S. spontaneum* ($2n = 64$), and one *S. sinense* clone, introduced mainly through Chunnee, were extensively used in the basic interspecific hybridization work (Roach, 1971). Random amplified polymorphic DNA (RAPD) analyses revealed little variation between the South African commercial varieties and those at advanced stages of selection (Harvey *et al.*, 1995) and tend to confirm little or no added variation between generations.

In Mauritius, breeders have used the same genetic stock produced by the early hybridization work from Java and India mainly through the BC_2 series of POJ varieties and the derived clones of the basic Indian series (Ramdoyal and Domaingue, 1993). The Mauritian 'miracle' variety M 134/32 bred in 1932 from POJ 2878 × D109 (noble)

Table 29.1. Status of *Saccharum* species and allied genera in world germplasm collections in USA and India, and the MSIRI collection.

Germplasm	United States[a] (Miami)	India[b] (Coimbatore/Kannur)	Mauritius (Réduit)
Saccharum spp.			
S. officinarum	305	764	136
S. robustum	70	145	10
S. spontaneum	237	661	27
S. barberi	55	43	—
S. sinense	37	29	6
S. edule	10	—	—
Allied genera		318	
Erianthus spp.	87		12
Miscanthus spp.	18		
Total	819	1960	191

[a]Comstock *et al.*, 1996; P.V.P. Tai, Canal Point, 2000, personal communication.
[b]T.V. Sreenivasan, Coimbatore, 2000, personal communication.

which occupied 92% of the area under cane in 1952, appears in the lineage of most varieties bred locally (Anon, 1990). Mauritian breeders have used another source of genetic stock through Uba Marot ($2n$ = 112), a naturally occurring vigorous seedling, derived presumably from an introduced Indian 64-chromosome, *S. spontaneum* and a noble cane (Stevenson, 1940). The lineage of Mauritian commercial hybrids released to date, which trace back to the M 134/32, Uba Marot and E 1/37 is summarized in Fig. 29.1 and reveals little or no new addition to their genetic background.

Broadening the Genetic Base

Recognizing the narrowness of the genetic base and the possible stagnation of yields (Walker, 1962), interest in interspecific work grew in a number of countries in the 1960s (Berding and Roach, 1987). In Barbados, an intra noble improvement programme through recurrent selection was designed to produce better noble varieties for interspecific programmes (Walker, 1974). A prebreeding programme included clones from a wide range of latitudes and ecological conditions. Several problems in the programme were noted, such as the lack of male-sterile nobles as the maternal parent and early flowering in *S. spontaneum* from the more northerly latitudes under Barbados conditions, which necessitated photoperiodic treatments. A nobilization programme with *S. spontaneum* where the main criteria were high dry matter yield, good ratooning, high sugar and low fibre showed good progress as a number of clones were promoted to final phase trials (Berding and Roach, 1987).

Nobilization in Hawaii initially emphasized the use of *S. robustum* clones (Heinz, 1967), which contributed towards the evolution of some Hawaiian varieties. Heinz (1980) reported on the use of *S. spontaneum* and observed that the expression of BC_1 clones showed high interaction with elevation. In Australia, nobilization was centred on cytological aspects (Roach, 1969, 1971, 1978) and was later intensified, with several clones reaching the stage of on-farm trials, but none was released commercially (Berding and Roach, 1987). Characterization of clones received more attention to target characters for improvement and use was made of *Erianthus arundinaceus* as a source of genes to combat the poor root syndrome affecting cane in Northern Queensland (Berding and Roach, 1987).

In Taiwan, the use of *Miscanthus* spp. was emphasized as a source of downy mildew and smut resistances (Chen *et al.*, 1982, 1986; Chen and Lo, 1989). F_1 hybrids between commercial hybrid × *Miscanthus* spp. showed improved hybrid vigour for stalk length and some clones showed tolerance to drought and salinity (Chen *et al.*, 1986). Sugar content in a *Saccharum* × *Miscanthus* hybrid showed a steady increase in successive nobilized generations (Lo and Chen, 1988). Clones selected from F_1 *Saccharum* × *Miscanthus* hybrids were productive

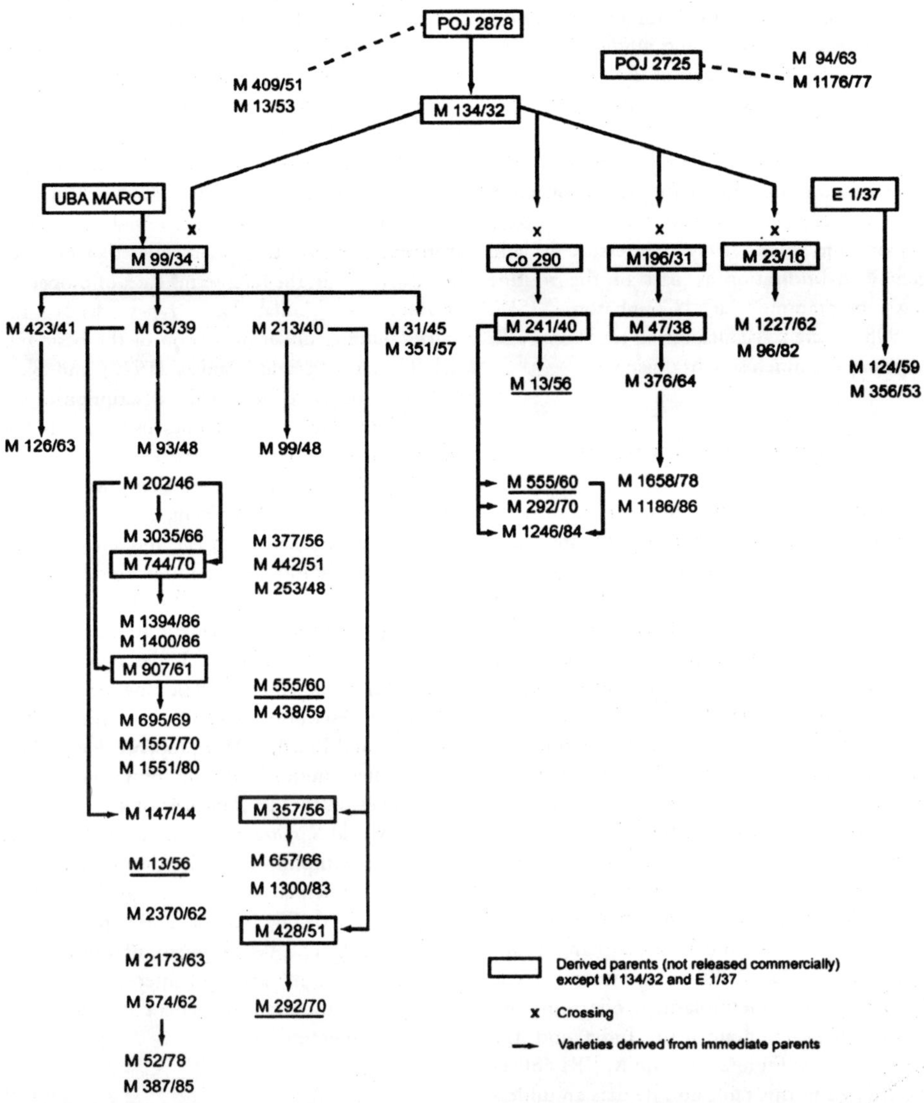

Fig. 29.1. Lineage of sugar cane commercial varieties bred at MSIRI which trace back to M 134/32, Uba Marot and E 1/37.

in fibre yields, with better pulp sheet as compared to a commercial control. Some clones from BC_1 crosses involving sugarcane and sorghum had sucrose content equivalent to commercial controls and proved to be early maturing (TSRI, 1995–96). Sun-Yuan Hsu *et al.* (1988) reported that the use of native *S. spontaneum* helped in generating promising interspecific hybrids as parental stocks.

In the USA, interest in interspecific hybridization focused on resistances to mosaic and borer, cold tolerance and ratooning vigour (Dunkelman and Breaux, 1972). Resistance to mosaic was successfully transferred to BC_2 progenies and a number of clones with gametes from *S. spontaneum* reached agronomic trials. Little progress for cold tolerance was reported as a result of screening difficulties. Tai and Miller (1988) evaluated F_1 crosses between commercial hybrids and *Miscanthus* and *Erianthus* (as male parents) and indicated that thin stalk and low sucrose content were strongly influenced by *Miscanthus* and *Erianthus*. However, mean sucrose content and purity of F_2 and BC_1 seedlings of crosses between commercial hybrids and *Miscanthus* spp., *Miscanthidium* spp. and *Erianthus* spp. were markedly improved over that of F_1 hybrids, but mean stalk diameter was still very small (Tai *et al.*, 1991).

In India, extensive efforts were made to characterize basic *Saccharum* spp. of the world collection (Kandasami *et al.*, 1983; Ramana Rao *et al.*, 1985; Sreenivasan and Nair, 1991). Interspecific and intergeneric hybridization is part of the routine crossing programme at Coimbatore (SBI, 1995–1998) where resistance to red rot (*Glomerella tucamanensis*) is a much sought objective.

Maintenance of germplasm

International concern for the conservation of sugar cane genetic resources grew considerably for many reasons including the safeguarding of genetic material of the *Saccharum* complex from diverse geographical origins. Organized collecting of the *Saccharum* complex was undertaken by the International Society of Sugar Cane Technologists (ISSCT) and International Board of Plant Genetic Resources (IBPGR) (Anon, 1982). The germplasm collected is maintained by the US Department of Agriculture, Agricultural Research Service (USDA/ARS) in Florida at Miami and by the Sugar Cane Breeding Institute at Kannur and Coimbatore, in India. Characteristics of the two world collections in India and the USA have been outlined by Alexander and Viswanathan (1996) and Comstock *et al.* (1996). The status of the World Sugar Cane Germplasm Collection and associated genera maintained in India and the USA are shown in Table 29.1. The MSIRI collection is included in this table and depicts an under-representation of most germplasm groups except for the noble group, of which a sizeable collection is maintained.

Walker (1980) advocated genetic conservation of *S. officinarum* seeds; accordingly, true seeds of 235 accessions of *S. spontaneum* (*c.* 66% of the world collection of *S. spontaneum* in Miami) are conserved at the National Seed Storage Laboratory (NSSL) at Colorado (Tai *et al.*, 1999). The collection represents nearly all the major regions of geographical distribution of the species. Seeds from both selfed clones and polycrosses are stored to minimize genetic drift. In Mauritius, selfed seeds of *S. spontaneum* clones are stored at −20°C.

Characterization of basic germplasm

Characterization of germplasm is important to differentiate accessions and to describe the variability in characters of interest. Breeders have emphasized characterization for direct utilization in commercial breeding programmes rather than evaluating and cataloguing the basic clones. Limited resources are a constraint to investing in activities that may generate dividends in the long term. Lack of proper information on 'unfamiliar' types leads to a corresponding under-utilization of the basic material. Hutchinson and Daniels (1972) and Skinner (1972) gave guidelines for the description of sugarcane clones. Several catalogues using a number of descriptors have been produced for various groups of the World *Saccharum* Collection in India (Kandasami *et al.*, 1983; Ramana Rao *et al.*, 1985; Sreenivasan and Nair, 1991) and characterization data for the World Collection at Florida were provided by Tai *et al.* (1994, 1995). The IBPGR Working Group on the Genetic Resources of Sugar Cane had proposed a list of 32 descriptors and their descriptor states (Anon, 1982). However, a universal system of description has not been widely adopted (Berding and Roach, 1987) probably due to the lack of dedicated curators and time constraints.

Studies on the development of a core collection for the World *S. spontaneum* clones in Florida based on various sampling strategies, using random cluster and principal component analyses, are in progress (P.V.P. Tai, Florida, 2000, personal communication). The development of one or several cores based on characters of interest could help to better access the available genetic diversity for prebreeding programmes.

Genetics of interspecific crosses

Studies on the genetics of traits are important for increasing the efficiency of prebreeding programmes. Quantitative genetics studies in interspecific hybridization have been reported by several authors although many of the underlying assumptions, such as diploid inheritance and no linkage, may not hold true (Hogarth, 1968). Roach (1968) found highly significant male × female interaction effects for yield and sucrose in *S. officinarum* × *S. spontaneum* crosses. Heterosis was observed for early growth, stalk height, cane yield, flowering, pollen production and level of reducing sugars.

Badaloo and Ramdoyal (unpublished data) studied 14 crosses between commercial hybrids (female) and *S. spontaneum* (male) and found families to differ significantly for stalk number, stalk diameter and cane quality characters. Narrow-sense heritability estimated from the partitioning of variances between and within family ranged from relatively low for dry matter % cane (Brix % cane + fibre % cane) (0.17) to moderately low for stalk number and fibre % cane (0.39 and 0.38, respectively) and moderately high for cane quality characters (0.67–0.90) and stalk diameter (0.75). These estimates were generally higher than those obtained from commercial × commercial crosses (Lawrence and Sunil, 1997; Ramdoyal and Badaloo, 1998). The distribution of Brix, sucrose content, juice purity and fibre content in crosses between commercial hybrids and *S. spontaneum* clones were continuous and controlled by polygenes, and additive genetic variance was more important than dominance genetic variance for all four characters (Tai *et al.*, 1992).

For crosses which involved commercial hybrids and *S. robustum*, Rao and Rao (1977) obtained moderate to high narrow sense heritability for stalk weight and height (0.47), stalk number (0.52), Brix (0.58) and stalk diameter (0.60). Genetic analysis of F_1 populations derived from *S. officinarum* × *S. barberi*, *S. robustum*, *S. spontaneum*, commercial hybrids, indicated highest heritability values in the *S. spontaneum* group (Bakshi Ram *et al.*, 1992). Genetic coefficient of variation (GCV) of BC_1 populations was generally greater in crosses involving *S. officinarum* as the recurrent parent relative to the commercial hybrids (Bakshi Ram *et al.*, 1995). However, the mean performance of families and heritability estimates were higher for commercial × F_1 progenies than for the *S. officinarum* × F_1 groups. The extent of genetic variation and heritability estimates indicates that selection and progress would be effective in interspecific hybrids particularly from the *S. spontaneum* source.

Prebreeding at MSIRI

Sugar cane is not native to Mauritius; *Saccharum* species and allied genera are introduced primarily from the Sugarcane World Germplasm Collections and undergo 2 years of quarantine in glasshouses before multiplication in the field. At the beginning of the 20th century, intercrossing of noble clones produced new varieties, some of which reached commercial status (Stevenson, 1965; Rouillard, 1990). The interspecific programme initiated in 1931 utilized mainly the *S. spontaneum* Glagah canes, notably G. Tabongo, G. Kletak and G. Djatiroto in a series of nobilizations and to some extent the *S. sinense*, Uba. Extensive use was made of the BC_2 series (POJ 2364) and BC3 series (POJ 2725, POJ 2878) as well as Uba Marot that were systematically backcrossed to noble varieties. However, no other breeding stock was as successful as M 134/32 and Uba Marot in producing commercial varieties (Fig. 29.1).

In the early 1980s, the interspecific hybridization programme got a new impetus following government policy to use bagasse, the fibrous residue left after cane extraction, as a renewable source of energy. Bagasse is currently used to generate electricity to power sugar mills during the milling season and excess electricity is sold to the commercial grid. The bagasse left after milling contains around 48% moisture, 50% fibre and 1–2% sugar, with a calorific value of 9700 kJ kg^{-1} and has 30–40% of the heat content of coal (Baguant, 1984). Breeding efforts that focused on increasing sugar yield per unit area resulted in selection against fibre (Brown, 1965; Walker, 1974), which averages 13% for commercial varieties bred in Mauritius. Increasing fibre content of commercial varieties but still maintaining an acceptable sugar content was thought to be an attractive economic option that could cut down on imported coal and also reduce CO_2 emissions. Deepchand (2000) reviewed the use of bagasse in cogeneration in the Mauritian Sugar Industry.

Evaluation of *Saccharum* and allied species

During 1983 to 1985, some work was initiated on the evaluation of *S. spontaneum* (20), *S. robustum* (7), *S. sinense* (6), *S. officinarum* (86), commercial hybrids (87) and *Erianthus* spp. (6) for cane quality components, Brix % cane (total solids of juice), pol % cane (apparent sucrose), fibre % cane and dry matter % cane, by the method described by Saint Antoine and Froberville (1964). A ratio of pol : fibre (PF) was derived from the pol % dry matter and fibre % dry matter. Juice purity was also derived from the ratio of Brix % cane and pol % cane. These studies revealed high levels of fibre and dry matter content in *S. spontaneum* clones as compared with *S. officinarum*. Commercial hybrids ranked very close to *S. officinarum*, and the *Erianthus* group ranked very close to *S. robustum*. Brix % cane and pol % cane were the highest in *S. officinarum* and commercial hybrids whereas sucrose content of the *Erianthus* group was even lower than that of *S. spontaneum*. The PF ratio showed equal partitioning of dry matter between pol and fibre in the *S. officinarum* and commercial clones. These investigations have given a good indication of the range and variation of the quality characters in the MSIRI germplasm collection and showed limitations for increasing fibre content in commercial hybrids.

The noble collection comprising 126 clones was further characterized in 1997 for 22 agronomic, morphological and quality characters and disease reaction to *Puccinia melanocephala* and *Mycovellosiella koepkei*. Furthermore, clones were screened for bacilliform virus (Autrey *et al.*, 1990). Principal component analysis (PCA) and discriminant function analysis are being undertaken to characterize clones further and to rationalize the collection.

Evaluation of interspecific crosses

In Mauritius, the wild *S. spontaneum* clones start flowering towards the end of April/May whereas most of the *S. officinarum* and commercial clones flower from mid-May to July. Crossing between noble clones and *S. spontaneum* and the early nobilized clones requires that either the flowering time in noble clones be advanced or that of the wild *S. spontaneum* be delayed. In Mauritius, 4 h night break treatments are applied between 8.00 pm and midnight to delay flowering (Julien, 1971; Julien and Soopramanien, 1975). A delay of 1 to 2 days was realized with each day of treatment (Badaloo *et al.*, 1995) which permitted a better flowering synchrony among clones for targeted crosses. In addition, trials are in progress to induce and advance flowering in noble and commercial clones.

The interspecific hybridization programme of the MSIRI for base-broadening involves the production of F_1, BC_1 and BC_2 seedlings where the recurrent parent is either the noble cane or commercial-type hybrids. F_1 seedlings are selected based on stalk number, stalk diameter and cane quality components and are multiplied in breeding plots for backcrossing. Similarly, BC_1 and BC_2 seedlings are selected and evaluated for plot yield, field refractometric Brix, cane quality and kilo Brix (plot yield × field refractometric Brix). An average to low sucrose commercial variety is planted as a control in the trial. A clone is selected based on a probability of kilo Brix exceeding the standard, giving due consideration to pol % cane, fibre % cane and morphological attributes. Interesting BC_1 and BC_2 clones are multiplied in breeding plots for further backcrossing and promoted to the next stage of selection (second clonal stage) with a pool of 3–5 commercial standard varieties. Selected clones, based on the same criteria as in the previous stage, are multiplied and included in variety trials in major agro-climatic environments for evaluation over a period of 5 years. In general, the production of commercial seedling populations for selection from prebreeding material adds another 3–5 years to the already lengthy selection programme, which ranges from 11 to 15 years. Nevertheless, encouraging results in this field are being obtained as more interspecific/intergeneric derived clones are reaching on-farm trials. Many breeding stations maintain a small prebreeding programme as a means to generate new parents.

Progeny testing

Concurrent to the above selection stream, 30–50 progenies from BC_1 and BC_2 families are evaluated in a progeny trial, with individual seedlings planted as one stool (the whole plant, which has grown from a single seedling or a vegetative cutting emanating from a single location in a field trial) and a standard variety planted at fixed locations as control. Progenies and controls are sampled for cane quality analyses and a number of morphological characters are recorded. Statistical analysis is carried out for each family. Interesting crosses may be repeated until an appropriate population size has been tested per family, which is around 600 progenies, whereas parents that are known to transmit good characteristics to their progenies are further used in crosses.

Table 29.2 summarizes data from samples of crosses for the first, second and third nobilization series for characters of interest and also shows the progress that can be achieved with successive generations of nobilization. Generally, the F_1 crosses between *S. officinarum* and *S. spontaneum*, *S. officinarum* × *S. robustum* and commercials × *S. spon-*

taneum are closer to the wild types, exceeding the standard variety for stalk number and fibre % cane and are inferior for stalk diameter, pol % cane, with a preferential partitioning of dry matter towards fibre. Intergeneric crosses between *S. officinarum* × *Erianthus arundinaceus* and commercial hybrids × *E. arundinaceus* display very good potential for all characters from a commercial angle. The clones of *E. arundinaceus* present in the local collection show tolerance to drought. However, seed setting in these crosses is generally low. The *Erianthus* clones flower sparsely under local conditions with low pollen fertility. Progenies from crosses between commercial hybrids × *S. spontaneum* are generally more vigorous and attractive than the basic nobilization crosses. Commercial hybrids constitute a selected and adapted genetic stock that is easier to hybridize and provide quicker scope for progress. However, their use would not contribute appreciably to genetic base broadening as compared with the original *S. officinarum* clones.

The phenotypic correlations among the various characters measured in commercial × *S. spontaneum* families depict the difficulty in associating all desirable characters in prebreeding material (Table 29.3). High stalk number is associated with low stalk diameter, low sugar content and high fibre content. Jackson (1994) reported similar associations in F_1 crosses where the relationships among attributes corresponded to the association with the original genetic sources. This points to the difficulty of breaking down linkage groups particularly when homologous pairing between the chromosomes of the same species is generally observed (Grivet *et al.*, 1996) and intrachromosomal interspecific recombination represents infrequent events (D'Hont *et al.*, 1996).

For the second (BC_1) and third (BC_2) generation backcrosses, which involved mostly elite commercial parents, a return to the commercial types or noble clones was evident but with better scope for increased stalk number (Table 29.2). Although an increase in pol % cane and purity was observed, it was difficult at this level to identify good segregants, which combined good stalk diameter and high fibre with acceptable sucrose levels. However, ignoring higher fibre content, few clones from the BC_1 and BC_2 generations were promoted to the final phase trials.

New orientations

In light of the difficulties encountered in associating fibre with pol and acceptable stalk diameter, it was necessary to restructure the programme with the main emphasis on genetic base broadening for commercial attributes with an added gain for high fibre clones. Furthermore, prebreeding populations would be evaluated in the marginal dry (< 1500 mm rainfall) and the superhumid zones (> 2500–3600 mm) where better returns are expected from the introgression of exotic genomes from *S. spontaneum* and allied genera which possess the genes for adaptation to diverse stressful environments.

The progeny testing as used above has been further improved by the use of replicated cross prediction trials which have been implemented as from 1999 for the prebreeding programme, in parallel with cross prediction trials for commercial crosses (Badaloo, 1997). Univariate cross prediction methodologies, namely family mean (MEAN), the predicted proportion of genotypes, within interspecific populations, that transgress set targets for each

Table 29.2. Mean of samples of crosses from different nobilized groups for morpho-agronomic and cane quality characters.

Nobilization groups	Stalk number	Stalk diameter	Pol % cane	Fibre % cane	Dry matter % cane	Pol : Fibre index
S. officinarum × *S. spontaneum*	21.5 (192)[a]	19.7 (72)	7.7 (62)	17.0 (130)	27.4 (100)	0.49 (57)
S. officinarum × *S. robustum*	13.6 (113)	22.9 (84)	6.4 (68)	18.4 (153)	27.9 (117)	0.36 (46)
Commercial × *S. spontaneum*	21.7 (181)	18.2 (67)	7.2 (77)	18.6 (155)	29.4 (123)	0.44 (56)
Commercial × *Erianthus*	14.3 (110)	22.7 (84)	10.1 (106)	12.9 (119)	26.0 (111)	0.80 (92)
BC 1[b]	14.0 (94)	24.0 (92)	9.3 (137)	11.0 (113)	23.5 (118)	0.80 (116)
BC 2[b]	14.0 (117)	25.0 (89)	10.1 (140)	11.9 (125)	25.1 (121)	0.87 (113)

[a]Figures in parentheses represent % of standard variety (M 555/60).
[b]Commercial type varieties as recurrent parent.

Table 29.3. Phenotypic correlation coefficients, based on family means, among agronomic and quality characters for commercial × *S. spontaneum* crosses.

	Stalk number (SNO)	Stalk diameter (SD)	Brix % cane (BC)	Pol % cane (PC)	Fibre % cane (FC)	Dry matter % cane (DM)	Pol : Fibre index (PF)	Purity (PUR)
SNO	1.00							
SD	−0.62*	1.00						
BC	−0.50	0.74**	1.00					
PC	−0.48	0.63*	0.96**	1.00				
FC	0.57*	−0.68**	−0.77**	−0.70**	1.00			
DM	0.19	−0.01	0.18	0.23	0.49	1.00		
PF	−0.51	0.73**	0.95**	0.92**	−0.87**	−0.04	1.00	
PUR	−0.47	0.66**	0.91**	0.93**	−0.61*	0.30	0.87**	1.00

*Significant at 5%.
**Significant at 1%.

character (PROB) and the observed proportion of genotypes that transgress set targets (OBS), are being used based on univariate probability integrals of Jinks and Pooni (1976) for a range of agronomic and morphological characters. Bivariate and multivariate cross prediction statistics will also be used for two or more characters simultaneously, based on sum of ranks (RANK) and the observed frequency (FREQ) of genotypes transgressing set targets for two or more characters simultaneously (Badaloo, 1997).

Proposed IT System for Prebreeding and Germplasm Enhancement at MSIRI

The interrelationships between databases and crossing and selection records at the MSIRI for the commercial sugarcane programme were described by Ramdoyal *et al.* (1999). The prebreeding programme operates in parallel to the main sugar cane variety development programme. Most of the prebreeding data, breeding performance derived from cross ratios, cross prediction data, cane quality analyses and morpho-agronomic data are computerized and would be converted into a database.

The information on amount and location of all available germplasm is stored in a relational database known as the current germplasm database, which includes information on all prebreeding parents. Detailed codes for describing *Saccharum* species and allied genera, to indicate the various levels of nobilization, have been established. As opposed to the commercial breeding programme that relies on a computer-aided crossing approach, most of the prebreeding crosses are sorted out manually. Further development is envisaged of a cohesive system of relational databases from which programmes using decision-making rules would be applied to fulfil breeding criteria for the various nobilized groups.

Barriers to Prebreeding and Prospects from Applications of Molecular Biology

Although progress in prebreeding has been recorded at various levels in different programmes, it has not produced the expected gains. Breeders are less inclined to use exotic, unfamiliar or alien material which may yield results in the long term while the pressure to show short-term gains compels them to use adapted commercial types as parents. Nevertheless, the narrowness of the genetic base of modern sugarcanes and the need to introgress specific genes are still of concern and justify the pursuance of prebreeding work. Interspecific/intergeneric hybridization is fraught with difficulties arising from: asynchronous and erratic flowering in some species, low fertility of hybrids, genetic complexity resulting from high ploidy levels, peculiar chromosome transmission and meiotic irregularities, low intergenomic recombination, unfavourable linkage groups, authenticity of true hybrids and long generation interval. Furthermore, the breeding value of the basic parental material for use in prebreeding cannot be appraised from its phenotypic characteristics only. It is generally recognized that only a limited proportion of the total available basic germplasm has been utilized in prebreeding work and scope for progress is substantial. Successful exploitation of the genetic resources of sugarcane in the future will require careful characterization and evaluation of the available germplasm, using different methods including molecular tools, setting up clear objectives before hybridization and an appreciation of factors affecting progress (Berding and Roach, 1987).

Some of the barriers have been overcome to various extents such as controlled flowering (Miller and Li, 1995; Nuss and Berding, 1999), low temperature storage of pollen of *S. spontaneum* and F_1 interspecific hybrids (Tai, 1988, 1993). Molecular techniques now offer new avenues in genome analysis to accelerate benefits arising from the exploitation of new germplasm. A genetic linkage map of *S. spontaneum* 'SES 208' based on RFLP- and PCR-based markers has been developed and polysomic segregation, a genetic behaviour typical of autopolyploid species, has been demonstrated (Da Silva *et al.*, 1995). Genetic maps composed of single-dose DNA markers have been constructed for the genomes of *S. officinarum*, 'La purple $2n = 80$' and *S. robustum*, 'MOL 5829 $2n = 80$' (Sobral, 1997) and it has been shown that a single-dose DNA marker framework map could be used to identify genomic regions controlling quantitative traits of polysomic polyploids, in a manner similar to diploids. Several species-specific RAPD and PCR markers have been identified for *Sorghum giganteum* and *Erianthus* spp. that can assist in the selection of interspecific sugarcane hybrids (Pan *et al.*, 2000).

The feasibility of using comparative genomic *in situ* hybridization (GISH) and fluorescent *in situ* hybridization (FISH) to identify the parental genomes in an interspecific hybrid between *S. officinarum* and *E. arundinaceus* (D'Hont *et al.*, 1995), and a commercial hybrid, R 570, has been shown (D'Hont *et al.*, 1996). With further developments, microsatellite markers could soon become a tool for genetic fingerprinting, genetic mapping and to assist in selecting specific, genetically diverse parents for use in introgression (Cordeiro and Henry, 1999). The development of marker-assisted selection would have a major potential use in introgressing a single quantitative trait locus (QTL) or multiple QTL from an ancestral species into commercial cane (Kearsey and Pooni, 1996). The search for molecular markers for disease resistance (MSIRI, 1999), and sucrose genes (N. Jannoo, MSIRI, 2000, personal communication) to assist in breeding and selection work at the MSIRI may be extended to prebreeding material and could reduce the number of backcrosses necessary to recover the recipient genotype near to the donor target genes.

Acknowledgements

The authors are grateful to Dr J.C. Autrey, Director, and Dr G.C. Soopramanien, Deputy Director (Biology), MSIRI, for reviewing the paper and their interest in publishing this work. The contribution of Mr R. Domaingue and the staff of the Plant Breeding Department are acknowledged.

References

Alexander, K.C. and Viswanathan, R. (1996) Conservation of sugarcane germplasm in India given the occurrence of new viral diseases. In: Croft, B.J., Piggin, C.M., Wallis, E.S. and Hogarth, D.M. (eds) *Sugar Cane Germplasm Conservation and Exchange*. ACIAR Proceedings No. 67, Australian Centre for International Agricultural Research, Queensland, Australia, pp. 19–21.

Anon (1982) Genetic resources of sugarcane. *International Board for Plant Genetic Resources Working Group on the Genetic Resources of Sugar Cane*. IBPGR Secretariat, Rome, Italy, pp. 1–19.

Anon (1990) *Sugar Cane Varieties in Mauritius. A Botanical Description*. Mauritius Sugar Industry Research Institute, Reduit, Mauritius.

Arceneaux, G. (1965) Cultivated sugar cane of the world and their botanical derivation. *Proceedings of the International Society of Sugar Cane Technologists* 12, 844–854.

Autrey, L.J.C., Saumtally, S., Dookun, A. and Boolell, S. (1990) Occurrence of sugarcane bacilliform virus in Mauritius. *Proceedings of the South African Sugar Technologists Association* June, 34–39.

Badaloo, G.H. (1997) Quantitative genetics of sugar cane: cross evaluation for major agronomic and morphological traits. PhD thesis, University of Birmingham, UK.

Badaloo, G.H., Ramdoyal, K., Mangar, M. and Domaingue, R. (1995) Investigation on the efficiency of different light sources and duration of treatment on flower delay. ISSCT, *Sugarcane Breeder's Newsletter*, Feb, 11–12.

Baguant, J. (1984) Electricity production from the biomass of the sugarcane industry in Mauritius. *Biomass* 5, 283–297.

Bakshi Ram, Hemaprabha, G. and Ram, B. (1992) Genetic variability in interspecific progenies in sugar cane (*Saccharum* spp.). *Indian Journal of Genetics and Plant Breeding* 52(2), 192–198.

Bakshi Ram, Hemaprabha, G. and Ram, B. (1995) Influence of noble and commercial hybrid clones on economic traits in nobilization of *Saccharum* species. *Indian Journal of Genetics and Plant Breeding* 55(2), 166–169.

Berding, N. and Koike, H. (1980) Germplasm conservation of the *Saccharum* complex: A collection from the Indonesian Archipelago. *Hawaii Plant Records* 59, 87–176.

Berding, N. and Roach, B.T. (1987) Germplasm conservation, collection, maintenance, and use. In: Heinz, D.J. (ed.) *Sugar Cane Improvement through Breeding*. Elsevier, Amsterdam, pp. 143–210.

Brandes, E.W., Sartoris, G.B. and Grasssl, C.O. (1939) Assembling and evaluating wild forms of sugar cane and related plants. *Proceedings of the International Society of Sugar Cane Technologists* 6, 128–153.

Bremer, G. (1961a) Problems in breeding and cytology of sugar cane. II. The sugar cane breeding from a cytological viewpoint. *Euphytica* 10, 121–133.

Bremer, G. (1961b) Problems in breeding and cytology of sugar cane. III. The cytological crossing research of sugar cane. *Euphytica*, 10, 229–243.

Bremer, G. (1962) Problems in breeding and cytology of sugar cane. V. Chromosome increase in *Saccharum* hybrids in relation to interspecific and intergeneric hybrids in other genera. *Euphytica* 11, 65–80.

Bremer, G. (1966) The origin of North Indian sugar canes. *Genetica* 37, 345–363.

Brown A.D.H. (1965) Correlation between Brix in juice and fibre in commercial hybrid sugar cane populations. *Proceedings of the International Society of Sugar Cane Technologists* 12, 754–759.

BSES (1990) *Annual Report 1990*. Bureau of Sugar Experiment Stations, Australia.

Burner, D.M. and Legendre, B.L. (1993) Chromosome transmission and meiotic stability of sugarcane (*Saccharum* spp.) hybrid derivation. *Crop Science* 33, 600–606.

Chen, W.H., Huang, Y.J. and Shen, I.S. (1982) Utilisation of *Miscanthus* germplasm. *Annual Report 1981–1982.* Taiwan Sugar Research Institute, p. 7.

Chen, W.H., Huan, Y.J., Shen, I.S. and Shih, S.C. (1986) Utilisation of *Miscanthus* germplasm in sugar cane breeding in Taiwan. *Proceedings of the International Society of Sugar Cane Technologists*, 18, 641–648.

Chen, Y.H. and Lo, C.C. (1989) Disease resistance and sugar content in *Saccharum Miscanthus* hybrids. *Taiwan Sugar* 36(3), 9–12.

Comstock, J.C., Schnell, R.J. and Miller, J.D. (1996) Current status of the world sugar cane germplasm collection in Florida. In: Croft, B.J., Piggin, C.M., Wallis, E.S. and Hogarth, D.M. (eds) *Sugar Cane Germplasm Conservation and Exchange*. ACIAR Proceedings No. 67. Australian Centre for International Agricultural Research, Queensland, Australia, pp. 17–18.

Cordeiro, G.M. and Henry, R.J. (1999) Microsatellite markers as an important tool in the genetic analysis of sugarcane (*Saccharum* spp.) genotypes. *Plant and Animal Genome VII Conference, San Diego, Calfornia.*

Da Silva, J., Honeycutt., R.J., Burnquist., W., Al-Janabi, S.M., Sorrells, M.E., Tanksley, S.D. and Sobral, B.W.S. (1995) *Saccharum spontaneum* L. 'SES 208' genetic linkage map combining RFLP- and PCR- based markers. *Molecular Breeding* 1, 165–179.

Daniels, J. and Roach, T. (1987) Taxonomy and evolution. In: Heinz, D.J. (ed.) *Sugarcane Improvement through Breeding*. Elsevier, Amsterdam, pp. 7–84.

Daniels, J., Smith, P., Paton, N. and Williams, C.A. (1975) The origin of the genus *Saccharum*. *Sugarcane Breeders Newsletter*, 36, 24–39.

Deepchand, K. (2000) Bagasse energy development – the Mauritian experience. *International Sugar Journal* 102 (1215),127–138.

D'Hont, A., Rao, P.S., Feldmann, P., Grivet, L., Islam-Faridi, N., Berding, N. and Glaszmann, J.C. (1995) Identification and characterization of intergeneric hybrids, *Saccharum officinarum* x *Erianthus arundinaceus*, with molecular markers and *in situ* hybridisation. *Theoretical and Applied Genetics* 91, 320–326.

D'Hont, A., Grivet, L., Feldmann, P., Rao, P.S., Berding, N. and Glaszmann, J.C. (1996) Characterization of the double genome structure of modern sugar cane cultivars (*Saccharum* spp.) by molecular cytogenetics. *Molecular and General Genetics* 250, 405–413.

Dunkelman, P.H. and Breaux, R.D. (1972) Breeding sugar cane varieties for Louisiana with new germplasm. *Proceedings of the International Society of Sugar Cane Technologists* 14, 233–239.

Frankel, O.H. (1989) Principles and strategies of evaluation. In: Brown, A.D.H., Frankel, O.H., Marshall, D.R. and Williams, J.T. (eds) *The Use of Plant Genetic Resources*. Cambridge University Press, Cambridge, UK, pp. 245–260.

Grivet, L., D'Hont, A., Roques, D., Feldmann, P., Lanaud, C. and Glaszmann, J.C. (1996) RFLP mapping in cultivated sugarcane (*Saccharum* spp.). Genome organization in a high polyploid and aneuploid interspecific hybrid. *Genetics* 142, 987–1000.

Harvey, M., Huckett, B. and Botha, F.C. (1995) Genetic diversity within the South African sugarcane germplasm. *Plant Genome III Conference, San Diego, California.*

Heinz, D.J. (1967) Wild *Saccharum* species for breeding in Hawaii. *Proceedings of the International Society of Sugar Cane Technologists* 12, 1037–1043.

Heinz, D.J. (1980) Thailand *S. spontaneum* hybrid progeny as a new germplasm source in Hawaii. *Proceedings of the International Society of Sugar Cane Technologists* 17, 1347–1356.

Hogarth, D.M. (1968) A review of quantitative genetics in plant breeding with particular reference to sugar cane. *J. AST. Inst. Agric. Sci.* 34, 108–119.

Hutchinson, P.B. and Daniels, J. (1972) A rating scale for sugarcane characterisation. *Proceedings of the International Society of Sugar Cane Technologists* 14, 128–131.

Jackson, P. (1994) Genetic relationships between attributes in sugar cane clones closely related to *Saccharum spontaneum. Euphytica* 79, 101–108.

Jinks, J.L. and Pooni, H.S. (1976) Predicting the properties of recombinant inbred lines derived by single seed descent. *Heredity* 36, 253–266.

Julien, M.H.R. (1971) Photoperiodic control of flowering in sugar cane. PhD thesis, University of Reading, UK.

Julien, M.H.R. and Soopramanien, G.C. (1975) Effects of nights breaks on floral initiation and development in *Saccharum. Crop Science* 15, 625–629.

Kandasami, P.A., Sreenivassan, T.V., Ramana Rao, T.C., Palanichami, K., Natarajan, B.V., Alexander, K.C., Madhusudana Rao, M. and Mohan Raj, D. (1983) *Catalogue on Sugarcane Genetic Resources* 1. Saccharum spontaneum *L.* Sugarcane Breeding Institute, ICAR, Coimbatore.

Kearsey, M.J. and Pooni, H.S. (1996) Applications. In: Kearsey, M.J. and Pooni, H.S. (eds) *The Genetical Analysis of Quantitative Traits.* Chapman and Hall, London, pp. 302–334.

Ladizinsky, G. (1989) Ecology and genetic considerations in collecting and using wild relatives. In: Brown, A.D.H., Frankel, O.H., Marshall, D.R. and Williams, J.T. (eds) *The Use of Plant Genetic Resources.* Cambridge University Press, Cambridge, UK, pp. 297–305.

Lawrence, M.J. and Sunil, H.K. (1997) Quantitative genetics of sugar cane. II. The inheritance of eleven agronomically important characters. *Sugar Cane* 1, 15–22.

Li, H.W., Cheng, C.F. and Leung, T.C. (1951) Genetical analysis of the hybrids obtained in crossing POJ 2725 and *Miscanthus japonicus. Proceedings of the International Society of Sugar Cane Technologists* 7, 266–276.

Lo, C.C., Chia, Y.H., Chen, W.H., Shang, K.C., Shen, I.S. and Shih, S.C. (1978) Collecting *Miscanthus* germplasm in Taiwan. *Proceedings of the International Society of Sugar Cane Technologists* 16, 59–69.

Lo, C.C. and Chen, Y.H. (1988) Disease resistance and sucrose content in *Saccharum Miscanthus* hybrids. *Report of Taiwan Sugar Research Institute* 122, 1–8.

Miller, J.D. and Li, Q.W. (1995) Effects of photoperiod treatments on initiation, emergence and flowering date of elite and exotic clones. *Sugar Cane* 6, 4–11.

MSIRI (1999) *Annual Report 1998.* Mauritius Sugar Industry Research Institute, Reduit, Mauritius.

Mukherjee, S.K. (1950) Search for wild relatives of sugar cane in India. *International Sugar Journal* 52, 261–262.

Mukherjee, S.K. (1957) Origin and distribution of *Saccharum. Botanical Gazette* 119, 55–61.

Nuss, K.J. and Berding, N. (1999) Planned recombination in sugar cane breeding: artificial initiation of flowering in sugar cane in sub-tropical and tropical conditions. *Proceedings of the International Society of Sugar Cane Technologists* 23(2), 504–508.

Pan, Y.B., Burner, D.M. and Legendre, B.L. (2000) Search for DNA markers to assist sugar cane breeding. *Plant and Animal Genome Conference. San Diego, California* (Abstract).

Panje, R.R. and Babu, C.N. (1960) Studies in *Saccharum spontaneum.* Distribution and geographical association of chromosome numbers. *Cytologia* 25(2), 152–172.

Price, S. (1957) Cytological studies in *Saccharum* and allied genera. III. Chromosome numbers in interspecific hybrids. *Botanical Gazette* 118, 146–159.

Price, S. (1961) Cytological studies in *Saccharum* and allied genera. VII. Maternal chromosome transmission by *S. officinarum* in intra-and interspecific crosses. *Botanical Gazette* 122, 298–305.

Price, S. (1963a) Cytogenetics of modern sugar cane. *Economic Botany* 17, 97–106.

Price, S. (1963b) Cytological studies in *Saccharum* and allied genera. VIII. F2 and BC1 progenies from 112 and 136-chromosome *S. officinarum* × *S. spontaneum* hybrids. *Botanical Gazette* 124, 186–190.

Price, S. (1965) Cytology of *Saccharum robustum* and related sympatric species and natural hybrids. *U.S. Dep. Agric. Res. Serv., Tech. Bull.* 1337.

Price, S. (1968) Cytology of Chinese and North Indian sugar canes. *Economic Botany* 22(2), 155–164.

Ramana Rao, T.C., Sreenivasan, T.V. and Palanichami, K. (1985) *Catalogue on Sugar Cane Genetic Resources – II Sugar Cane Breeding Institute*. ICAR, Coimbatore, India.

Ramdoyal, K. and Badaloo, G.H. (1998) Inheritance of agronomic traits in commercial hybrid sugar cane populations in contrasting environments and in different crop cycles. *Journal of Genetics and Breeding* 52, 361–368.

Ramdoyal, K. and Domaingue, R. (1993) Conservation, exchange, evaluation and utilisation of sugar cane germplasm in Mauritius. In: Dulloo, M.E. and Dulymamode, R. (eds) *Plant Genetic Resources in Mauritius: towards a National Strategy. Proceedings of a National Workshop on Plant Genetic Resources in Mauritius, University of Mauritius, 1993*. National Parks and Conservation Service, pp. 65–78.

Ramdoyal, K., Mamet, L.D., Rivet, L., Badaloo, G.H. and Domaingue, R. (1999) Sugar cane hybridisation procedures and computer-aided crossing at the Mauritius Sugar Industry Research Institute. *Sugar Cane* Sept, 5–10.

Rao, T.C.R. and Rao, J.T. (1977) Analysis of quantitative variation in sugar cane hybrid populations involving *Saccharum robustum*. *Proceedings of the International Society of Sugar Cane Technologists* 16 (1), 239–244.

Roach, B.T. (1968) Quantitative effects of hybridisation in *Saccharum officinarum* × *Saccharum spontaneum* crosses. *Proceedings of the International Society of Sugar Cane Technologists* 13, 939–954.

Roach, B.T. (1969) Cytological studies in *Saccharum*. Chromosome transmission in interspecific and intergeneric crosses. *Proceedings of the International Society of Sugar Cane Technologists* 13, 901–920.

Roach, B.T. (1971) Nobilisation of sugar cane. *Proceedings of the International Society of Sugar Cane Technologists* 14, 206–216.

Roach, B.T. (1978) Utilisation of *Saccharum spontaneum* in sugar cane breeding. *Proceedings of the International Society of Sugar Cane Technologists* 16, 43–58.

Roach, B.T. and Daniels, J. (1987) A review of the origin and improvement of sugar cane In: *Copersucar International Sugar Cane Breeding Workshop, Copersucar Technology Center, Piracicaba-SP, Brasil*, pp. 1–31.

Rouillard, G. (1990) Historique de la canne à sucre à l'île Maurice, 1639–1989.

Saint Antoine, J.D. de R. and de Froberville, R. (1964) The direct determination of fibre in cane. *Annual Report of the Mauritius Sugar Industry Research Institute*, 142–146.

SBI (1995–1998) *Annual Report of the Sugarcane Breeding Institute*. Coimbatore, India.

Skinner, J.C. (1972) Description of sugarcane clones: 3. Botanical description. *Proceedings of the International Society of Sugar Cane Technologists* 14, 124–127.

Sobral, B.W.S. (1997) Genetic maps, quantitative traits and genomic comparisons: *Saccharum* and its relation to other Andropogonae. *Plant and Animal V Conference*, pp. 12–16.

Sreenivasan, T.V. and Nair, N.V. (1991) *Catalogue on Sugarcane Genetic Resources* III. Saccharum officinarum *L.* Sugarcane Breeding Institute, ICAR, Coimbatore, India.

Sreenivasan, T.V., Ahloowalia, B.S. and Heinz, D.J. (1987) Cytogenetics. In: Heinz, D.J. (ed.) *Sugar Cane Improvement through Breeding*. Elsevier, Amsterdam, pp. 211–253.

Stalker, H.T. (1980) Utilisation of wild species for crop improvement. *Advances in Agronomy*, 33, 111–147.

Stevenson, G.C. (1940) An investigation into the origin of sugar cane variety Uba Marot. *Department of Agriculture Mauritius Sugar Cane Research Station Bulletin* 17.

Stevenson, G.C. (1965) *Genetics and Breeding of Sugar Cane*. Longman, London, UK.

Sun-Yuan Hsu, Lo, C.C. and Shih, S.C. (1988) Use of *Saccharum spontaneum* derivative in sugar cane crossing. *Report of the Taiwan Sugar Research Institute*, No. 122, December, pp. 1–7.

Tai, P.Y.P. (1988) Long-term storage of *Saccharum spontaneum* L. pollen at low temperature. *Sugarcane* (spring supplement), 12–16.

Tai, P.Y.P. (1989) Progress and problems of intergeneric hybridisation in sugar cane breeding. *Proceedings of the International American Sugar Cane Seminar. Sugar Cane: Meeting the Challenge of the 1990s, September 1989*, pp. 391–395.

Tai, P.Y.P. (1993) Low temperature preservation of F_1 pollen in crosses between noble or commercial sugarcane and *Saccharum spontaneum* L. *Sugar Cane* 5, 8–11.

Tai, P.Y.P. and Miller, J.D. (1988) Phenotypic characteristics of the hybrids of sugar cane × related grasses. *Journal of the American Society of Sugar Cane Technologists* 8, 5–11.

Tai, P.Y.P., Gan, H., Heng He and Miller, J.D. (1991) Phenotypic characteristics of F_1 and BC_1 progenies from sugar cane intergeneric crosses. *Journal of the American Society of Sugar Cane Technologists* 11, 27–38.

Tai, P.Y.P., Hong, H., Gan, H.P. and Miller, J.D. (1992) Variation for juice quality and fibre content in crosses between commercial sugarcane and *Saccharum spontaneum*. *Journal of the American Society of Sugar Cane Technologists* 12, 47–57.

Tai, P.Y.P., Miller, J.D. and Legendre, B.L. (1994) Preservation of *Saccharum spontaneum* germplasm through storage of true seed. *Sugar Cane* 6, 3–8.

Tai, P.Y.P., Miller, J.D. and Legendre, B.L. (1995) Evaluation of World Collection of *Saccharum spontaneum* L. *Proceedings of the International Society of Sugar Cane Technologists* 21, 250–260.

Tai, P.Y.P., Miller, J.D. and Legendre, B.L. (1999) Preservation of *Saccharum spontaneum* germplasm in world collection of sugarcane and related grasses through storage of true seed. *Sugar Cane* 3, 4–10.

TSRI (1995–96) *Annual Report Taiwan Sugar Research Institute.*

Walker, D.I.T. (1962) Family performance at early selection stages as a guide to the breeding programme. *Proceedings of the International Society of Sugar Cane Technologists* 13, 939–954.

Walker, D.I.T. (1974) Utilisation of noble and *S. spontaneum* germplasm in the West Indies. *Proceedings of the International Society of Sugar Cane Technologists* 14, 224–232.

Walker, D.I.T. (1980) Seed storage for genetic conservation of noble canes. *Sugarcane Breeders Newsletter* 43, 10–11.

30 Underutilized Crops: Trends, Challenges and Opportunities in the 21st Century

S. Padulosi,[1] T. Hodgkin,[1] J.T. Williams[2] and N. Haq[2]
[1]International Plant Genetic Resources Institute (IPGRI), Rome, Italy; [2]International Centre for Underutilized Crops (ICUC), Southampton, UK

Introduction

Plant biodiversity represents the primary source for food, feed, shelter, medicines and many other products and means that make life on Earth possible and enjoyable (WCMC, 1992; UNEP, 1995). The number of plant species used by humans around the world (Table 30.1) is only one-third of the number of species which generations of diverse cultures around the world have drawn upon to develop crops that would meet specific needs. The centres of diversification of most common cultivated species are known today (Zeven and de Wet, 1982), but for many other species of local importance, knowledge of the distribution of their genetic diversity and use patterns is still largely limited. Increased reliance on major food crops has been accompanied by a shrinking of the food basket which humankind has been relying upon for generations (Prescott-Allen and Prescott-Allen, 1990). This *nutritional paradox* (Ogle and Grivetti, 1995) has its roots in the agricultural 'simplification', a process that favoured some crops instead of others on the basis of their comparative advantages for growing in a wider range of habitats, their simple cultivation requirements, easier processing and storability, nutritional properties, taste and so on.

Though the simplification process lowered food quality, it increased the chances of successful harvests, which in turn allowed survival through narrow but abundant sustenance (Collins and Hawtin, 1999). However, the shrinking of agricultural biodiversity has reduced both the intra- and interspecific diversity of crops, increasing the level of vulnerability among users, particularly the poorer sections, for whom diversity in crops is a necessity for survival rather than a choice. Extensive literature documents the dramatic effects of genetic erosion in staple crops (Fowler and Mooney, 1990 and references therein), comparably less has been published, however, on the effects of the narrowing of the food basket leading to reduced quality of life.

A change in attitude has been noticeable over the last 5–10 years among policy-makers and the public with regard to the quality of life as related to the quality of food as well as diverse sources of food. Vitamins and other micronutrients are, for instance, being searched for in crops and plant species with greater emphasis than in the past in recognition of their role in combating diet imbalances. Although 'hidden hunger' affects mainly developing countries, particularly children and older people (FAO, 1997), it is increasingly being recorded also among the more vulnerable social groups in developed nations.

If the 20th century witnessed the undertaking of systematic collecting to rescue the genetic resources of staple crops (Pistorius, 1997), the 21st century has started with the awareness of the need to rescue and improve the use of those crops left

Table 30.1. Estimates of number of plant species used around the world.

Part A: Global assessments

Author	Year	No. of species	Uses
Heywood	1991	100,000	Used plants
Paroda and Mal	1993	80,000	Explored by humans since dawn of civilization
Myers	1983	75,000	Edible
Wilson	1992	30,000	Edible
Kunkel	1984	12,650	Edible
Uphof	1968	9,500	Economic uses
Wilson	1992	7,000	Source of food (wild/cultivated)
Terrell *et al.*	1977	3,000	Vascular species of economic importance
Zeven and de Wet	1982	2,489	Cultivated species excluding ornamentals, timber crops and lower plants
Rehm	1994	2,454	Agronomic plants

Part B: Regional or crop-oriented assessments.

Country/region	Reference	Year	No. spp.	Type of species (w, wild; c, cultivated; wd, weedy)
Worldwide				
	Simons	1996	2,500	Tree agroforestry species
	Wickens	1995	542	Edible nuts
	Vietmeyer	1990	1,500	Edible nuts (w/c)
	Heywood	1999	25,000	Medicinal
	Vietmeyer	1990	60,000	Medicinal (w/c)
	Vietmeyer	1990	2,000	Pesticide (w/c)
	Vietmeyer	1990	3,000	Contraceptive (w/c)
Africa				
	Jardin	1967	1,410	Indigenous wild species: 60 grass seeds, 50 legumes, 60 oil seeds, >550 fruits/nuts, >600 spices
West Africa	Burkill	1985	>4,600	Useful
Botswana	Campbell	1986	150	Edible
Ghana	Abbiw	1990	2,500	Useful
Kenya	Juma	1989	>100	Vegetables and fruits (w/c) (Bungoma District, West Kenya)
Kenya	Maundu *et al.*	1999	800	Indigenous food
Morocco	Hmamouchi	1999	340	Medicinal (w/c)
Sahara	Harlan	1989	60	Edible grasses (w)
Sahel	Becker	1984	800	Edible
Southern Africa	Heywood	1999	900	Medicinal
Swaziland	Ogle and Grivetti	1985	>200	Edible (w)
Zambezi and S. Zaire	Malaisse and Parent	1985	184	Edible vegetables (w)
Americas				
	Brucher	1989	170	Economic
Mexico	Alvarez-Buylla *et al.*	1989	338	Home gardens (w/c)
Canada	Bates	1985	30,000	Nursery trade
USA	Yarnell	1964	130	Edible of the Upper Great Lakes region
USA	Yarnell	1964	400	Useful (excluded food) of the Upper Great Lakes region
Peru	Padoch *et al.*	1991	168	Useful in home gardens (w)

Table 30.1. Part B: Regional or crop-oriented assessments (*Continued*).

Country/region	Reference	Year	No. spp.	Type of species (w, wild; c, cultivated; wd, weedy)
Europe				
Portugal	Anonymous	1996	500	Medicinal and aromatic
Ukraine	WCMC	1992	300	Medicinal and aromatic
Russia	Chikov	1973	2,000	Potentially useful in medicine
Asia				
Indian Himalaya	Samant and Dhar	1997	675	Edible (w)
South-east	Jansen *et al.*	1991	1,462	Timber trees
South-east	Jansen *et al.*	1991	285	Feed species
South-east	Jansen *et al.*	1991	228	Vegetables
South-east	Jansen *et al.*	1991	110	Spices and condiments
South-east	Jansen *et al.*	1991	72	Dyeing plants
South-east	Heywood	1999	6,000	Used so far
India	Arora and Nayar	1984	320	Economically important (w/wd)
India	de Padua *et al.*	1999	1,100–1,500	Medicinal
India (Uttar Pradesh)	Heywood	1999	480	Edible (w)
Pakistan	de Padua *et al.*	1999	300	Medicinal
Sri Lanka	de Padua *et al.*	1999	550	Medicinal (flowering plants)
New Guinea	de Padua *et al.*	1999	>600	Medicinal
Bangladesh	Ashraful *et al.*	1999	500	Medicinal
Vietnam	de Padua *et al.*	1999	1,800	Medicinal
Indonesia	de Padua *et al.*	1999	1,000	Medicinal
Philippines	de Padua *et al.*	1999	850	Medicinal (w - used in 'jamus')
Philippines	Madulid	1979	200	Fruits (trees-shrubs)
India	Paroda and Mal	1989	536	Economic (w/wd/c)
India	Arora and Pandey	1996	1,000	Edible (w)
Malaysia	Saw *et al.*	1991	820	Edible fruit trees (diameter >1 cm)
Mediterranean region				
	Bianco	1989	137	Vegetables indigenous
	Bianco	1992	50	Vegetables indigenous
Italy	Hammer *et al.*	1992	522	Cultivated (indigenous and introduced)
Italy	Corbetta	1991	90	Edible (salad, condiment, soups, etc.)
Jordan	Al-Eiswi and Takruri	1989	142	Edible
Tropics				
	Martin *et al.*	1987	2,800	Edible fruits
	Vietmeyer	1990	3,000	Fruits (w/c)
	Heywood	1999	18,000–25,000	Used so far

aside by research, technology, marketing systems as well as conservation efforts. These underutilized crops (referred to also by other terms such as minor, orphan, neglected, underexploited, underdeveloped, lost, new, novel, promising, alternative, local, traditional, niche crops) have been included in worldwide plans of action after having successfully raised the interest of decision-makers. Leading international research organizations such as the Consultative Group on International Agricultural Research (CGIAR), are also among those taking a keen interest in strengthening the work on these species (Swaminathan, 1999).

This global 'opening' towards underutilized species is the result of a gradual change of attitude towards biodiversity and plant genetic resources by many countries. Instrumental in this awareness raising process have been the 1992 Convention on Biological Diversity and the FAO IV International Technical Conference on Plant Genetic Resources for Food and Agriculture held in Germany in 1996 (cfr. Activity 12: 'Promoting development and commercialization of underutilized crops and species') (UNEP, 1992; FAO, 1996a). The Global Forum on Agricultural Research (GFAR) in 1999 also emphasized the role of underutilized species in raising income of the rural poor (Frison *et al.*, unpublished).

This chapter addresses some aspects related to improving the conservation and use of neglected and underutilized crops through the experiences of the International Plant Genetic Resources Institute (IPGRI) (Padulosi *et al.*, 1999) and the International Center for Underutilized Crops (ICUC) (Smith, 1997), both of which have been directly involved with these species since their establishment.

Underutilized Crops: Is there an Agreement on What they Are?

Perhaps no agriculture term has raised more discussions among workers than the word underutilized. Underutilized is commonly applied to refer to species whose potential has not been fully realized. The term itself does not provide any information as to geographical (underutilized where?), social (underutilized by whom?) and economic (underutilized to what degree?) implications. It is thus not surprising that whenever underutilized species are being addressed in national or international fora there is inevitably a call for a clarification over the exact meaning of such a term. The following are some examples to explain the source of confusion on this term. With regard to the *geographical distribution*, often a species could be underutilized in some regions but not in others. The cowpea (*Vigna unguiculata*) is, for instance, a staple crop for millions of people in sub-Saharan Africa, but is considered as an underutilized crop in some Mediterranean countries where it was once widely used and now is grown in some restricted areas (Padulosi *et al.*, 1987). Similarly chickpea (*Cicer arietinum*), considered by many Italian scientists as an underutilized species in their country, is a main pulse crop in Syria and other countries in West Asia. Regarding the *socio-economic implication* of the term, many species represent an important component of the daily diet of millions of people (such as leafy vegetables in sub-Saharan Africa) (Guarino, 1997) but their poor marketing conditions make them largely underutilized in economic terms. With regard to the *time factor*, the degree of underuse of a crop may be subject to a sudden improvement due to dynamic marketing systems present in some countries while the same crop may continue to be poorly marketed and managed by researchers in others. This is the case, for instance, for the vegetable rocket (collective name for the species *Eruca sativa*, *Diplotaxis tenuifolia* and *Diplotaxis muralis*). Rocket has become a highly priced vegetable in Europe through innovative cultivation and commercial practices (Pimpini and Enzo, 1997), while it is among the cheapest vegetables in Egypt and a rich source of micronutrients for the poorer classes (Mohamedien, 1995). The importance of this crop for millions of people in Egypt has, however, not been enough to convince local policy-makers to give it the attention that it deserves. A similar case is that of hulled wheats, a collective name for *Triticum monococcum*, *Triticum dicoccum* and *Triticum spelta*. This is an important speciality crop in Italy and other European countries where *ex situ* and *in situ* conservation are being supported along with considerable research efforts (Padulosi *et al.*, 1996b). However, it is a typical 'life support' crop in remote areas of Turkey with the very last relic populations of *T. monococcum* being cultivated in poor subsistence farming systems (Karagöz, 1996).

Underutilized crops are often being presented as 'new crops' (Vietmeyer, 1990) because commercial companies/researchers have only recently been working on them. In reality, local populations over generations have used these species. Yet the loss of local knowledge and thus the increasing ignorance of new generations of the traditional uses of these crops also contribute to portraying such a misleading image. Of course, a crop can be completely new to an area simply because it has been introduced there recently from a distant country as in the case of the kiwi fruit, unknown outside China or New Zealand until recently (Ferguson, 1999), and of the *Annona* fruit (custard apple) introduced into Lebanon in the recent past.

Because of recurrent confusion among workers on the distinction between the term underutilized and neglected, it is worthwhile reporting here IPGRI's definitions (Eyzaguirre *et al.*, 1999) for these two categories of crops:

- Many *underutilized crops* were once more widely grown but are today falling into disuse for a variety of agronomic, genetic, economic and cultural factors. Farmers and consumers are using these crops less because they are in some way not competitive with other crop species in the same agricultural environment. The general decline of these crops may erode the genetic base and prevent the use of distinctive useful traits in crop adaptation and improvement.

- *Neglected crops* are those grown primarily in their centres of origin or centres of diversity by traditional farmers, where they are still important for the subsistence of local communities. Some species may be globally distributed, but tend to occupy special niches in the local ecology and in production and consumption systems. While these crops continue to be maintained by sociocultural preferences and use practices, they remain inadequately characterized and neglected by research and conservation.

The understanding of the causes behind the low level of use and/or neglect of a crop is, however, what is ultimately needed in order to design an appropriate strategy to address its improvement. In the rest of this chapter, in the interests of readability, we use underutilized in its 'broad sense' to refer to both underutilized and neglected species, unless specified in the text.

Challenges in the Promotion of Underutilized Species

'United commodities stand! Divided new crops fail!' This remark made in a paper addressing the prospects of new crops in the USA at the 1990 International Conference on Advances in New Crops (Duke, 1990) captures the feeling of many workers engaged in the promotion of underutilized crops. Though it is true that the funding for addressing research and hence improvement of these species around the world is still extremely small, other important factors, similarly determinant in the promotion process, should not be underestimated. For example, during a participatory conference organized by IPGRI in 1998 in Aleppo, Syria, on 'Priority Setting for Underutilized and Neglected Plant Species of the Mediterranean region' (Padulosi, 1999a), ten major categories of constraints were identified as most relevant for the promotion of underutilized species. As Table 30.2 shows, some of the constraints were not linked to lack of funds as their solution lies in the strategic mobilization of existing resources to address the real problems hampering the full utilization of the species.

Most challenges in the promotion of underutilized species are now well understood and appreciated and have been addressed in numerous works (Sankary, 1977; Hawkes, unpublished; Lazaroff, 1989; von Maydell, 1989; De Groot and Haq, 1995; FAO, 1996b; Bhag Mal *et al.*, 1997; Maxted *et al.*, 1997; Monti, 1997; Heywood, 1999; Padulosi, 1999a). However, an increasingly key issue in improving the use of underutilized species is the globalization of the agricultural market. This is part of a much wider phenomenon that is changing our life dramatically and includes urbanization, loss of small commercial activities replaced by large enterprises, the homogenization of local cultures as a result of the spreading of a few models and cultures and improved communication. With specific reference to agriculture, globalization is bringing about a further specialization in favour of a few crops that might best serve the 'global village'. Obviously, this trend will lead to a much narrower agricultural basket, estimated to hold only some 150 widely commercialized crops (Prescott-Allen and Prescott-Allen, 1990), in favour of some commodities which will enter newer areas, displacing local ones.

For many supporters of the globalization process, this is the great opportunity for economic growth, but for many more it is a dealer of illusions, carrying prospects of a wealthy life that cannot be met and hiding the high cost to pay in terms of loss of sustainability, traditions, culture and local economies. Without entering the debate over the issue of globalization, we would like to respond to the question that is raised on many occasions by those who see no future for underutilized crops in the global village: why should underutilized species be promoted in an increased global world if these are characterized by regional, national or local importance? The answer is very simple: securing the resource base of underutilized species, particularly in developing countries, is crucial to maintain the 'safety net' of diversified food and natural products that has provided options to address food needs in a sustainable way (Eyzaguirre *et al.*, 1999). Diversification in agricultural systems is indeed an important asset for those fragile social groups who may never be able to afford certain commodities. To these groups, the more diversified the portfolio the greater their self-sustainability and self-reliance in the difficult areas they live, often characterized by conditions not suitable for the cultivation of improved varieties of commodity crops.

Table 30.2. Assessment of constraints in the promotion of underutilized species in the Mediterranean region (higher number corresponds to a greater constraint).

Constraints	MAP	Forest trees	Fruit trees and nuts	Vegetables	Forages/ browses	Industrial	Ornamental	Legumes	Cereals
Low competitiveness	3	3	3	3	3	3	3	3	3
Lack of knowledge on uses	3	3	3	3	3	3	3	3	3
Lack of research on GD assessment and use	3	3	3	3	3	3	3	2	2
Policy and legislation	3	3	3	3	3	3	1	1	1
Loss of traditional knowledge	3	3	2	3	2	2	2	1	1
Lack of market/poor commercialization	3	2	2	2	3	3	3	1	1
Low income	2	3	2	2	3	3	1	2	2
Lack of propagation techniques	3	3	2	1	2	2	3	1	1
Scarce knowledge on cultural practices	3	2	2	2	2	1	3	2	1
Lack of attractive traits	1	2	3	3	1	2	1	1	1

Generally speaking, plants are a fundamental component of the natural resource available to the poor and the role played by underutilized species is indeed central to reducing poverty and empowering the poor so as to allow poor rural communities to pursue resources-based rather than commodity-based development (Burgess, 1994; Blench, 1997). But the people who benefit from underutilized species in a globalized world are not just the poor. The benefits in terms of more balanced diets, diversified income to farmers as well as related sectors of society, better maintenance of agroecosystems and greater use of marginal lands along with enhanced preservation of cultural identity (Padulosi, 1999b and references therein) can be shared by all humankind.

Choice of the Species

In developing work on underutilized crops, research and development agencies face substantial problems. The most obvious is that of deciding which crops to work on. Picking the right species from a broad group of potential candidates is a necessary step to make the best use of limited (as is always the case in this area) resources. Selecting species that can be used as a 'case study' may be possible and will also help to build up a knowledge base in the promotion process and thus enhance future efforts on other underutilized crops.

The selection of the species should be based on the analysis of its contribution to meet the goal that has been set out. The ultimate goal may, however, vary among stakeholders. For organizations like IPGRI and ICUC, the goals in the work on underutilized species include the improvement of food security; enhancement of nutritional balances, sustainability in agriculture and alleviation of poverty through income generation. The contribution of the species will, therefore, be assessed against their present contribution to meet these goals as well their potential contribution to be realized after intervention. Needless to say, the longer the list of criteria we set out for the selection, the more difficult the selection process will be. The 12 criteria recommended for the selection of underutilized food producing trees developed by von Maydell (1989) given in Table 30.3 would certainly be useful to guide the process, but it is unlikely that we would find a single species that would meet all of them.

The complexity of the selection process is made even more challenging by the limited and poor quality of information available on these species. Ultimately, the key factor for a successful selection is the involvement of direct users in the selection process. Adopting this approach will ensure that the species is favoured by the ultimate recipient of our efforts, farmers, rural and forest dwellers. Such an approach was followed by IPGRI in its project on leafy vegetables supported by the Dutch Government, which focused on leafy vegetables in five sub-Saharan African countries. During the first phase of this project, the surveys carried out together with farmers have yielded enough elements to select a number of priority species (Table 30.4) that can be used in the second phase of the project, which will focus on crop and market improvement.

The views of farmers and other stakeholders (universities, research centres, private groups, etc.) should be seen as dynamic and subject to shift in emphasis with regard to type of crop, depending on changes in opportunities and needs. Today, for instance, an increased interest is recorded among national and

Table 30.3. List of criteria for the selection of food producing trees and shrubs in semiarid regions (von Maydell, 1989).

They should meet demands
They should solve problems
They should be accepted by people
There should be no legal restriction
They should have low risk
They should be free from negative properties or effects
They should be adapted to site conditions
They should be easy and safe to establish, with low inputs
They should have fast growth
They should produce high yields
The quality of products should be good
The crops should be compatible with other land use

Table 30.4. Priority species of leafy vegetables for five African countries in the IPGRI project on germplasm management of African leafy vegetables for food security in sub-Saharan Africa.

Priority species	Botswana	Cameroon	Kenya	Senegal	Zimbabwe
Vigna unguiculata	√		√	√	√
Amaranthus spp.	√	√	√ (*A. dubius*)	√	
Cleome gynandra	√				√
Cucurbita spp.	√	√ (*C. maxima* and *C. moschata*)			√ (*C. maxima*)
Solanum nigrum		√			
Corchorus olitorius		√	√		√
Vernonia spp.		√			
Hibiscus sabdariffa		√		√	
Moringa oleifera				√	
Adansonia digitata				√	
Abelmoschus esculentus					√
Brassica juncea					√

international research organizations towards medicinal and aromatic species in view of their role in improving the health of poor and their contribution to combat poverty through income generation (Leaman *et al.*, 1999). Attention to underutilized species may also originate from considerations that are not directly related to food security or poverty alleviation, but to the need to safeguard artistic, landscape and cultural values of these species. This was, for instance, the objective of the international workshop organized by the Italian National Research Council (CNR) in 1997 in Naples (Monti, 1997).

Securing the Resource Base of Underutilized Species

Securing the resource base of underutilized species is a key component of the whole promotion process and is central to IPGRI's concerns. The maintenance of genetic diversity through a complementary *ex situ* and *in situ* approach can ensure that all users, including farmers and breeders, will have the access to material to carry out their activities. In order to understand the relationship between conservation and the other elements in the promotion process, it is useful to see how conservation fits within the whole process. Table 30.5 lists some of the major problem areas that can be identified for underutilized crops and the ways in which they might be addressed. Access to appropriate plant material can be an important factor in addressing most of the areas identified and is particularly relevant to tackle the first two, namely securing genetic resources and documenting and using traditional knowledge.

The provision of genetic diversity to allow crop improvement has been a central element of the Green Revolution (CGIAR, 1994). One of the most important elements of successful work on underutilized crops will be the full recognition of the importance of the work to safeguard their genetic diversity and to ensure its fullest use. The establishment of germplasm collections for underutilized species has been advocated in the Global Plan of Action of FAO (FAO, 1996a) and is one key element of the promotion process pursued by several other international organizations, including IPGRI and ICUC. However, the evidence suggests that *ex situ* conservation of many underutilized species is highly inadequate. More than 6 million accessions of plant genetic resources for food and agriculture are conserved today in some 1300 germplasm collections around the world. Although the number of these collections is an impressive one, an analysis of the type of crop plants they contain is disappointing: about 80% of these belong to major crops and their close relatives. The remaining 20% are other crops, including underutilized crops, which are very poorly represented (less than eight accessions per species) (Padulosi, 1999b). Figure 30.1 shows in detail the statistics related to such a situation.

We must concede that there will never be sufficient resources for large-scale, formal conservation activities. Hence, in order to achieve sustainability of

Table 30.5. Major problems for research and development work on underutilized crops.

Problems	Outputs required	Relevant activities
1. Lack of genetic material	Improved availability of seed and other planting materials Crop improvement programmes Improved planting materials derived from traditional varieties	Set up local germplasm supply systems among rural communities Initiate participatory and other improvement programmes to obtain clean planting materials and improved varieties
2. Loss of germplasm and traditional knowledge	Resource base of selected species secured through *ex situ* and on-farm conservation Appropriate traditional knowledge documented and shared among stakeholders	Assess distribution of species and genetic erosion threats Sample germplasm for *ex situ* maintenance and use Implement on-farm conservation through community-based actions Identify and collate traditional knowledge using participatory procedures based on informed consent (including e.g. recipes on uses)
3. Lack of knowledge on uses, constraints and opportunities	Enhanced information on production levels, use constraints and opportunities Knowledge of gender and other socially significant factors identified	Participatory surveys on uses, constraints and opportunities with communities and other levels of the 'filieres' Analysis of survey data for gender and other socially significant factors
4. Limited income generation	Strategies for adding value and increasing rural incomes using target crops Enhanced competitiveness of selected crops	Development of value adding strategies (through processing, marketing, commercialization etc.) Investigate and identify improved agronomic and production procedures
5. Market, commercialization and demand limitations	Enhanced working alliances among stakeholders in 'filieres' Improved processing and marketing opportunities identified Improved capacities of marketing associations and producer groups	Strengthen operational links in the 'filieres' between seed supply system, processing and distribution stakeholders Develop improved low-cost processing techniques Analyse and identify market opportunities
6. Lack of research and development activities and weak national capacities	Enhanced national capacities to work with neglected and under-utilized crops Enhanced information and knowledge on the selected neglected and underutilized crops Methods to improve nutritional values developed and documented	Carry out short training courses for researchers Develop and undertake community-based participatory courses Characterize crops for agronomic, nutritional and market-related traits Study formal and informal classification systems Investigate methods of maintaining and enhancing nutritional value Investigate new areas of crop growth
7. Lack of links across conservation and production to con-sumption 'filieres'	'Filieres' established or strengthened Participatory networking procedures established	Hold planning workshops for all stakeholders Establish and strengthen operational links between stakeholders
8. Inappropriate (inadequate) policy and legal frameworks	Raised awareness among policy-makers of issues and options for improved policy and legal frameworks Links to existing rural and economic development projects enhanced	Identify inappropriate policy/legal elements Undertake public awareness actions among policy-makers Establish close partnerships with extension workers and others involved in agricultural development

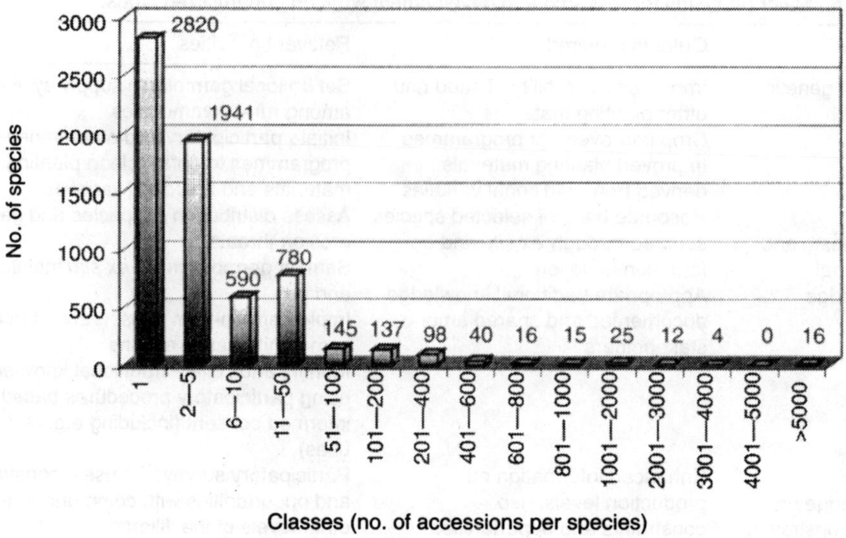

Fig. 30.1. Representation of minor crops in *ex situ* genebanks (see text for explanation).

the process, conservation and use must be closely linked; thus 'conservation through use' becomes important. Most of the conservation will in practice be through the continued use of these species in production systems. For example, in the case of the African leafy vegetables, the understanding of the users' needs (species with larger leaf area per plant and delayed flowering) is crucial to strengthen the on-farm maintenance of those crops. Another crucial strategic aspect in the sustainable promotion of underutilized species is adding value to existing crops. *Cleome gynandra*, for instance, a leafy vegetable with insect repellent properties can be very valuable in intercropping systems. Interestingly, many underutilized species have multiple uses and do not belong to any one specific category of crops (food, medicinal, ornamental, etc.). This means that high levels of diversity will be needed in production systems to meet different production environments, user needs and uses. The material in production provides one element of an integrated conservation strategy where the *ex situ* collections remain small and are developed through sharper analyses of crop distribution patters, genetic diversity, and so on. On the other hand, greater efforts for the gathering of information on the distribution, use and traditional knowledge of these crops will need to be pursued as such data will be very valuable to improve future access to material by researchers and other users. Securing material in production as well as the information on material that is already in production therefore becomes an important element in the strategy for conserving the diversity of underutilized species.

Since needs and use are often primarily local, so the actions will be local and community-based. Local mechanisms that support the deployment of useful diversity will need to be strengthened. For example, household 'filieres', largely run by women, built in rural and forest areas typically around multiple uses of the same crop, should be improved or established anew if no longer present. These chains, linking farmers up to final end-users, play a critical role in securing revenues to rural communities and thus fuelling the very mechanism that will maintain the diversity of these species in the field (see also following chapter on the filiere issue).

Securing and Strengthening the Work on Underutilized Crops

Developing and monitoring the work

Efforts so far have been directed to raise the awareness of the need to work on underutilized species and to start redressing their status of neglect. An area that has not been investigated yet is the analysis of when *a species will evolve* from its status of underuse and neglect *to become a well-utilized crop*. This is a very relevant point if we want to make sure our efforts will lead ultimately to the 'full promotion stage' beyond which the crop will no longer be considered underutilized. It is also very relevant to ensure that efforts to promote a specific crop have the desired effects with respect to the interests and needs of poor farmers, the maintenance of diversity and strengthening sustainability of production. It has to be recognized that some efforts on underutilized crops (e.g. kiwi, before it became an important commodity) have done none of these things.

Below are some of the issues that we should consider in developing a monitoring system for assessing the impact of our promotion process.

- *Conservation aspects*

  1. How many accessions should be conserved (*in situ* and *ex situ*) to safeguard the representative genetic diversity of these species and to provide at the same time the variability needed by breeders and users in general?
  2. Given the new opportunities brought about by gene transfer, how broad should our efforts be in the conservation of the genepool of the species? Should we include the tertiary genepool in our collecting efforts?
  3. What should be the minimum level of knowledge on the ecogeographical distribution of the taxa and its genetic erosion status?
  4. How much local knowledge should be safeguarded?
- *Utilization aspects*

  1. What would be the level at which we would consider the economy of the species self-sustainable?
  2. What is the research threshold, that is the minimum amount of research addressing the economic development of the crop, beyond which we would consider the species 'properly addressed'?
  3. What would be the minimum information needed on nutritional aspects and processing aspects required for successful industrial applications?
- *Agronomic aspects*

  1. What should be the minimum know-how required for enabling a proper cultivation of the species?
  2. What basic information should be required for understanding the multiplication method and regeneration capacity of the species?
  3. What should be the minimum level of information on pests, diseases and other cultivation-related problems?
- *Policy-legal aspects*

  1. What basic policy framework should be used to enable the proper deployment of the species among the farmers?
  2. What should be the minimum level of policy attention required to ensure a sustainable use of these resources (particularly for wild species)?

The need for information

The availability of information has been always a major constraint in the promotion of underutilized species. This will continue to be so, particularly with regard to the monitoring of the points raised above. For this reason, it is important to stress the need for a change in attitude while reporting the agricultural statistics at both national and international levels. The agricultural statistics year book of FAO could, for instance, also be broadened in its scope by using less general figures (e.g. root and tuber crops) and by including crops that are not commodities but that are being marketed and for which information may already be available at national level. The possibility of accessing these data to guide workers at local level should be also addressed. Opportunities for strengthening informatics capacities, for instance one computer per village managed by local extension officers, should not be seen as far fetched.

Improving the availability of information on underutilized crops is one of the most important areas demanding our immediate attention. At the formal level, individual studies on underutilized crops continue to need support to ensure their publication. Further studies that bring together the sparse and often inaccessible literature are also needed. At local level, there is a need to gather and document information that is maintained within farming communities. The recognition of the value of this by researchers and scientists can often act as a powerful stimulus to improve a community's own valuation of the knowledge.

The importance of partnerships

The work on underutilized species is perhaps the most challenging endeavour in the history of plant genetic resources since the early 1970s, a period that witnessed a world race to rescue landraces of major crops (Pistorius, 1997). The Green Revolution was made possible by the use of that collected diversity but also thanks to the strengthening of national and international research to address the improvement of these major crops. Today, underutilized species will perhaps never command such a major undertaking as major crops did and they will thus need a different approach for their promotion.

Such an approach would have as its most strategic part, the so called 'filiere'. Filiere is a French word used to define the link between all stakeholders and activities starting from the collection, use enhancement, policy definition to marketing and commercialization. Such a chain of actors, which is needed at local, regional, national and international level, will allow us to cover research aspects but also marketing and policy issues usually dealt with in isolation. The filiere concept can be considered an evolution of the networking concept for plant genetic resources based on *more efficient partnership and participatory approaches*. The filiere would thus bring about greater participation of local actors to ensure the addressing of local needs properly. It would also ensure the wider representation and participation of stakeholders of the food processing and marketing sector as well policy makers who have traditionally been left aside from plant genetic resource activities (Padulosi, 1999c).

Such a partnership for underutilized species in many cases will have to be re-invented, and this should not discourage partners. The filiere approach will build up a more coherent system to meet the multiplicity of challenges in the promotion of underutilized species. Although the filiere will consist of, particularly, local, regional and national players, there will be a role for international organizations to ensure that lessons learnt in one region can benefit other regions. The strengthening of the links among international stakeholders involved in the promotion of underutilized species is indeed strategic to allow best use of existing capacities and promote synergism across regions.

Underutilized species constitute a category defined by their social value and status. For this reason people and farmers play an important role in reversing their decline in use and arresting their genetic erosion. Farmers and forest dwellers are the source of information for revealing the potentials of these species, their distribution and local use. Participatory research should therefore be actively pursued among stakeholders, particularly in the following areas:

- constraint analyses ('production to consumption') and development of strategic work plans for enhancing seed/germplasm selection and supply, production, processing, commercialization, marketing (*greater cooperation between private sector and extension workers*);
- characterization and evaluation work using formal descriptors and farmers' criteria (*closer cooperation between informal associations/NGOs and international and national research organizations*);
- development/strengthening the seed supply systems, both formal and informal (*closer participation of farmers in government-led efforts*);
- participatory plant breeding and selection activities (*bridging the gap between farmers' needs and breeders' objectives*).

Research Issues and Opportunities

Neglected and underutilized crops are essential to the livelihoods of millions of poor farmers throughout the world. As noted above, they are part of the (threatened) biological assets of the rural poor. In identifying research and development issues that should be addressed, it is essential to approach the problem from this perspective. Trying to convert an underutilized crop into some modern high value commodity may not only be inappropriate to a community's real needs and concerns, it may actually be counter-productive. Similarly it does nothing to address the problems for the remaining hundreds of underutilized crops.

One key strategic element involves the deliberate attempt to explore how conservation and use can be combined to secure the resource base of such crops. The approaches may differ, depending on whether the crop is seed propagated or clonally propagated, annual or perennial, outbreeding or self-pollinated. However the basic questions remain the same. What is the smallest size of *ex situ* collection that can cover substantial amounts of diversity and how can it be most economically maintained? How much diversity will remain in production systems and how can this be monitored? How can resources be secured through linkages and collaborations, involving producers, consumers, the formal and informal sectors, to ensure that both conservation through use and conservation for use can be sustained?

New technologies (e.g. molecular genetics and geographical information systems (GIS)) will certainly play their part in the process of developing conservation and use strategies. There are already a number of examples which show how useful they can be (Young *et al.*, 1999); but often only in small-scale research activities that need scaling up. Perhaps there needs to be some deliberate determination of the way in which these powerful tools can be best used for such crops.

As implied above, there is also much work to be done on the development of sustainable linkages between organizations, farmers and consumers. It will always be unlikely that any one organization will have the resources to support work on the scale needed for the individual underutilized crop. Hence the interest in networks and filieres. A major challenge is how to make these networks perform and to make them sustainable.

Strengthened community involvement in the management of underutilized crops and a deliberate attention to resourcing their needs for new materials (and securing access to existing ones) will provide a basis for some more work on key production issues. The first of these is obviously that of the development of improved materials. Participatory plant breeding approaches may not only be an important element of the work on these crops; it may be the only feasible approach to obtaining improved materials. Similarly, participatory approaches may be essential to resolving other production and marketing constraints.

Ultimately, we have to recognize that underutilized crops present their own range of problems and opportunities. These are important to many farmers in ways that are complementary to and different from their concerns for the major crops. Attempting to copy large crop solutions across to these species will help neither the improved conservation and use of the crops nor the interests of the farmers who grow them. Developing an agenda specific to the crops must be recognized as an important and continuing need.

References

Abbiw, D.K. (1990) *Useful Plants of Ghana: West African Uses of Wild and Cultivated Plants.* Intermediate Technology Publications and The Royal Botanic Gardens Kew, UK.

Al-Eiswi, D.M. and Takruri, H.R. (1989) A checklist of wild edible plants in Jordan. *Arab Gulf Journal of Scientific Research –B: Agricultural and Biological Sciences* 7, 79–101.

Alvarez-Buylla Roces, M.A., Lazos Chavero, E. and Garcia-Barrios, J.R. (1989) Home gardens of a humid tropical region in South East Mexico: An example of an agroforestry cropping system in a recently established community. *Agroforestry Systems* 8, 133–156.

Anonymous (1996) Portugal country report to FAO. *International Conference and Programme on Plant Genetic Resources (ICPPGR).* Food and Agriculture Organization of the United Nations, Rome, Italy.

Arora, R.K. and Nayar, E.R. (1984) *Wild Relatives of Crop Plants in India*, NBPGR Science Monograph No. 7. New Delhi, India.

Arora, R.K. and Pandey, A. (1996) *Wild Edible Plants of India: Diversity, Conservation and Use.* National Bureau of Plant Genetic Resources, New Delhi, India.

Ashraful, H., Paramesh, N. and Farid, H.U. (1997) Forest genetic resources conservation and utilization in Bangladesh. In: Hossain, M.G., Arora, R.K. and Mathur, P.N. (eds) *Plant Genetic Resource – Bangladesh Perspective. Proceedings of a National Workshop on Plant Genetic Resources, Dhaka, 26–29 August 1997.* Agricultural Research Council, Barnic Printers, Dhaka, India, pp. 104–131.

Bates, D.M. (1985) Plant utilization: patterns and prospects. *Economic Botany* 39(3), 241–265.

Becker, B. (1984) Wild plants in the nutrition of the population of arid regions in Africa: 3 case studies from Kenya and Senegal. *Gottinger Beitrage zur Land und Forstwirtschaft in den Tropen und Subtropen*, 6.

Bhag Mal, Paroda, R.S. and Sudhir Kochlar (1997) Underutilized crops and their implications in farming systems in India. In: Smartt, J. and Haq, N. (eds) *Domestication, Production and Utilization of New Crops.* International Centre for Underutilized Crops, Southampton, UK, pp. 30–45.

Bianco, V.V. (1989) Wild plants utilizable as vegetables and condiment herbs in Italy. *Int. Symp. Hort. Germplasm Cultivated and Wild. 1998, Beijing, China*, 2, pp. 55–64.

Bianco, V.V. (1992) Usual and specialty vegetable crops in Mediterranean Countries. *Acta Horticulturae* 318, 65–76.

Blench, R.M. (1997) *Neglected Species, Livelihood and Biodiversity in Difficult Areas: How should the Public Sector respond?* Natural Resources Perspective No. 23. Overseas Development Institute, UK.

Bruchner, S.B. (1989) *Useful Plants of Neotropical Origin and Their Wild Relatives*. Springer-Verlag, London, UK.

Burgess, M.A. (1994) Cultural responsibility in the preservation of local economic plant resources. *Biodiversity and Conservation* 3, 126–136.

Burkill, H.M. (1985) *The Useful Plants of West Tropical Africa*. Royal Botanic Gardens, Kew, London, UK.

Campbell, A. (1986) The use of wild food plants and drought in Botswana. *Journal of Arid Environments* 11, 81–91.

CGIAR (1994) *Challenging Hunger – The role of the CGIAR*. Consultative Group on International Agricultural Research, Washington, DC, USA.

Chikov, P. (1973) Herbal medicines in the Soviet Union. *World Health* September, 18-23.

Collins, W.W. and Hawtin, G.C. (1999) Conserving and using crop plant biodiversity in agroecosystems. In: Collins, W.W. and Qualset, C.O. (eds) *Biodiversity in Agroecosystems*. CRC Press, Boca Raton, Florida, USA, pp. 267–281.

Corbetta, F. (1991) *Piante spontanee mangerecce* [spontaneous edible plants]. Guide Pratiche Edagricole, Bologna.

De Groot, P. and Haq, N. (1995) Promotion of traditional and underutilized crops. ICUC/CSC Commonwealth Science Council Series No. CSC(95) AGR23. Technical Paper 311.

de Padua, L.S., Bunyapraphatsara, N. and Lemmens, R.H.M.J. (eds) (1999) *Plant Resources of South East Asia*, 21(1). PROSEA, Bogor, Indonesia.

Duke, A. (1990) New crops survey. In: Janick, J. and Simon, J. (eds) *Advances in New Crops. Proceedings of the First National Symposium on New Crops: Research, Development, Economics, Indianapolis, Indiana, 23–26 October 1988.* Timber Press, Portland, Oregon, USA, pp. 54–57.

Eyzaguirre, P., Padulosi, S. and Hodgkin, T. (1999) IPGRI's strategy for neglected and underutilized species and the human dimension of agrobiodiversity. In: Padulosi, S. (ed.) *Priority Setting for Underutilized and Neglected Plant Species of the Mediterranean Region. Report of the IPGRI Conference, 9–11 February 1998, ICARDA, Aleppo, Syria.* International Plant Genetic Resources Institute, Rome, Italy, pp. 1–20.

FAO (1996a) *Global Plan of Action for the Conservation and Sustainable Utilisation of Plant Genetic Resources for Food and Agriculture* and Leipzig declaration, adopted by the International Technical Conference on Plant Genetic Resources, Leipzig, Germany, 17–23 June 1996. Food and Agriculture Organization of the United Nations, Rome, Italy.

FAO (1996b) *Report on the State of the World's Plant Genetic Resources for Food and Agriculture*, prepared for the International Technical Conference on Plant Genetic Resources, Leipzig, Germany, 17–23 June 1996. Food and Agriculture Organization of the United Nations, Rome, Italy.

FAO (1997) *Human Nutrition in the Developing World*. FAO Food and Nutrition Series No. 29. Food and Agricultural Organization of the United Nations, Rome, Italy.

Fowler, C. and Mooney, P. (1990) *The Threatened Gene–Food, Politics and the Loss of Genetic Diversity*. The Lutworth Press, Cambridge, UK.

Ferguson, A.R. (1999) New temperate fruits: *Actinidia chinensis* and *Actinidia deliciosa*. In: Janick, J. (ed.) *Perspective on New Crops and New Uses. Proceedings of the Fourth National Symposium on New Crops and New Uses. Biodiversity and Agricultural Sustainability, 8–11 November 1998, Phoenix, Arizona*, pp. 342–347.

Guarino, L. (ed.) (1997) Traditional African vegetables. Promoting the conservation and use of underutilized and neglected crops 16. *Proceedings of the IPGRI International Workshop on Genetic Resources of Traditional Vegetables in Africa: Conservation and Use, 29–31 August 1995, ICRAF-HQ, Nairobi, Kenya.* Institute of Plant Genetics and Crop Plant Research, Gatersleben/International Plant Genetic Resources Institute, Rome, Italy.

Hammer, K., Knüpffer, H., Laghetti, G. and. Perrino, P. (1992) *Seeds from the Past. A Catalogue of Crop Germplasm in South Italy and Sicily*. CNR, Istituto del Germoplasma, Bari, Italy.

Harlan, J.R. (1989) Wild grass seed harvesting in the Sahara and Sub-Sahara of Africa. In: Harris, D.R. and Hillman, G.C. (eds) *Foraging and Farming: the Evolution of Plant Exploitation*, One World Archaeology-B, Unwin Hyman, London, UK, pp. 79–98.

Heywood, V. (1991) Conservation of germplasm of wild plant species. In: Sandlund, O.T., Hindar, K. and Brown, A.H.D. (eds) *Conservation of Biodiversity for Sustainable Development*. Norwegian University Press and Cambridge, pp. 189–203.

Heywood, V. (1999) *Use and Potential of Wild Plants in Farm Households*, FAO Farm System Management Series No. 15. Food and Agriculture Organization of the United Nations, Rome, Italy

Hmamouchi, M. (1999) *Les Plantes Medicinales et Aromatiques Marocaines* [medicinal and aromatic plants of Morocco]. Imprimerie de Fedala, Mohammedia, Morocco.

Jansen, P.C.M., Lemmens, R.H.M.J., Oyen, L.P.A., Siemonsma, J.S., Stravast, F.M. and van Valkenburg, J.L.C.H. (eds) (1991) *Plant Resources of South East Asia: Basic List of Species and Commodity Grouping*. Pudoc, Wageningen, The Netherlands.

Jardin, C. (1967) *List of Food Used in Africa*. Food and Agriculture Organization of the United Nations, Rome, Italy.

Juma, C. (1989) *Biological Diversity and Innovation: Conserving and Utilising Genetic Resources in Kenya.* African Centre for Technology Studies, Nairobi, Kenya.

Karagöz, A. (1996) Agronomic practices and socio economic aspects of emmer and einkorn cultivation in Turkey. In: Padulosi, S., Hammer, K. and Heller, J. (eds) *Hulled wheats: promotion of conservation and use of valuable underutilized species, Proceedings of the First International Workshop on Hulled Wheats, 21–22 July 1995, Castelvecchio Pascoli, Tuscany, Italy.* International Plant Genetic Resources Institute, Rome, Italy, pp. 172–177.

Kayimov, A.K., Sultanov, R.A. and Chernova, G.M. (1998) *Pistacia* in Central Asia. In: Padulosi, S. and Hadj-Hassan, A. (eds) Pistacia: *Towards a Comprehensive Documentation of Distribution and Use of its Genetic Diversity in Central & West Asia, North Africa and Mediterranean Europe. Report of the IPGRI Workshop, 14–17 December 1998, Irbid, Jordan.* International Plant Genetic Resources Institute, Rome, Italy.

Kunkel, G. (1984) *Plants for Human Consumption.* Koeltz Scientific Books. Koenigstein, Germany.

Lazaroff, L. (1989) Strategies for development of a new crop. In: Wickens, G.E., Haq, N. and. Day, P. (eds) *New Crops for Food and Industry.* Chapman and Hall, London, pp. 108–119.

Leaman, D.J., Fassil, H. and Thormann, I. (1999) Conserving medicinal and aromatic plant species: identifying the contribution of the International Plant Genetic Resources Institute. Study commissioned by the International Development Research Centre (IDRC). International Plant Genetic Resources Institute, Rome, Italy.

Madulid, D.A. (1979) *Rare and Vanishing Fruit Trees and Shrubs in The Philippines.* IUCN, Gland.

Malaisse, F. and Parent, G. (1985) Edible wild vegetable products in the Zambezian woodland area: a nutritional and ecological approach. *Ecology of Food and Nutrition* 18, 43–82.

Martin, F., Campbell, C. and Ruberte, R. (1987) *Perennial Edible Fruits of the Tropics*, Agriculture Handbook No. 642. Agricultural Research Service, United States Department of Agriculture Beltsville, Maryland.

Maundu, P.M., Ngugi, G.W. and Kabuye, C.H.S. (1999) *Traditional Food Plants of Kenya.* Kenya Resources Centre for Indigenous Knowledge, National Museum of Kenya.

Maxted, N., Hawkes, J.G., Guarino, L. and Sawkins, M. (1997) *Genetic Resources and Crop Evolution* 44, 337–348.

Mohamedien, S. (1995) Rocket cultivation in Egypt. In: Padulosi, S. (compiler), *Rocket Genetic Resources Network. Report of the First Meeting, 13–15 November 1994, Lisbon, Portugal.* International Plant Genetic Resources Institute, Rome, Italy, pp. 61–62.

Monti, L. (ed.) (1997) *Proceedings of the CNR International Workshop on Neglected Plant Genetic Resources with a Landscape and Cultural Importance for the Mediterranean Region, 7–9 November 1996, Naples, Italy.*

Myers, N. (1983) *A Wealth of Wild Species.* Westview Press, Boulder, Colorado.

Ogle, B.M. and Grivetti, L.E. (1995) Legacy of the chameleon: edible wild plants in the Kingdom of Swaziland, Southern Africa. A cultural, ecological, nutritional study. Part II- Demographics, species availability and dietary use, analyses by ecological zone. *Ecology of Food and Nutrition* 17, 1–30.

Padoch, C. and de Jong, W. (1991) The house gardens of Santa Rosa: diversity and variability in an Amazonian agricultural system. *Economic Botany* 45(2), 166–175.

Padulosi, S. (ed.) (1997) Oregano. *Report of the First International Workshop on Oregano, 9 May, Valenzano, Bari, Italy.* International Plant Genetic Resources Institute, Rome, Italy.

Padulosi, S. (ed.) (1999a) Priority setting for underutilized and neglected plant species of the Mediterranean region. *Report of the IPGRI Conference, 9–11 February 1998, ICARDA, Aleppo. Syria.* International Plant Genetic Resources Institute, Rome, Italy.

Padulosi, S. (1999b) Criteria for priority setting in initiatives dealing with underutilized crops in Europe. In: Gass, T., Frese, F., Begemann, E. and Lipmann, E. (compilers) *Implementation of the Global Plan of Action in Europe – Conservation and Sustainable Utilization of Plant Genetic Resources for Food and Agriculture. Proceedings of the European Symposium, 30 June-3 July 1998, Braunschweig, Germany.* International Plant Genetic Resources Institute, Rome, Italy, pp. 236–247.

Padulosi, S. (1999c) Partners and partnership. In: Swaminathan, M.S. (ed.) *Enlarging the Basis of Food Security: the Role of Underutilized Species. International Workshop, M.S. Swaminathan Research Foundation, 17–19 February 1999, Chennai, India* (in press).

Padulosi, S., Cifarelli, S., Monti, L.M. and Perrino, P. (1987) Cowpea germplasm in southern Italy. FAO/IBPGR *Plant Genetic Resources Newsletter* 71, 37.

Padulosi, S., Caruso, T. and Barone, E. (eds) (1996a) *Taxonomy, Distribution, Conservation and Uses of* Pistacia *Genetic Resources, Report of a Workshop, 29–30 June 1995, Palermo.*

Padulosi, S, Hammer, K. and Heller, J. (eds) (1996b) Hulled wheats: promotion of conservation and use of valuable underutilized species. *Proceedings of the First International Workshop on Hulled Wheats, 21–22 July 1995, Castelvecchio Pascoli, Tuscany, Italy.* International Plant Genetic Resources Institute, Rome, Italy.

Padulosi, S., Eyzaguirre, P. and Hodgkin, T. (1999) Challenges and strategies in promoting conservation and use of underutilized and neglected species. In: Janick, J. (ed.) *Perspective on New Crops and New Uses, Proceedings of the Fourth National Symposium on New Crops and New Uses. Biodiversity and Agricultural Sustainability, 8–11 November 1998, Phoenix, Arizona*, pp. 140–145.

Paroda, R.S. and Bhag Mal (1989) New plant sources for food and industry in India. In: Wickens, G.E., Haq, N. and Day, P. (eds) *New Crops for Food and Industry*. Southampton, UK, pp. 135–149.

Paroda, R.S. and Bhag Mal (1993) Developing a National Programme for Research on Underutilized crops in India. *Proceedings of the First International Crop Science Congress, Ames, Iowa, USA, 14–22 July 1992.*

Pimpini, F. and Enzo, M. (1997) Present and future prospects for rocket cultivation in the Veneto region. In: Padulosi, S. and Pignone, D. (eds) *Rocket: an old Mediterranean crop for the world. Report of the II International Workshop on Rocket, 13 December 1996, Padova, Italy*. International Plant Genetic Resources Institute, Rome, Italy, pp. 51–66.

Pistorius, R. (1997) *Scientists, Plants and Politics – A History of the Plant Genetic Resources Movement*. International Plant Genetic Resources Institute, Rome, Italy.

Prescott-Allen & Prescott Allen (1990) How many plants feed the world? *Conservation Biology* 4, 365–374.

Rehm, S. (ed.) (1994) *Multilingual Dictionary of Agronomic Plants*. Kluwer Academic Publishers, Dordrecht, The Netherlands.

Roche, L. and Dourojeanni, M.J. (1992) Criteria for selection of conservation areas for forest genetic resources. In: Kapoor-Vijay, P. and White, J. (eds) *Conservation Biology: A Training Manual for Biological Diversity and Genetic Resources*. Commonwealth Secretariat, pp. 61–70.

Samant, S.S. and Dhar, U. (1997) Diversity, endemism and economic potential of wild edible plants of Indian Himalaya. *Int. J. Sustain. Dev. World Ecol.* 4, 179–191.

Sankary, M.N. (1977) *Flora and Range Management of the Syrian Arid and Very Arid Areas*. Aleppo University, Aleppo, Syria.

Saw, L.G., La Frankie, J.V., Kochummen, K.M. and Yap, S.K. (1991) Fruit trees in a Malaysian Rain Forest. *Economic Botany* 45(1), 120–136.

Simons, A.J. (1996) ICRAF's strategy for domestication of non-wood tree products. In: *Domestication and commercialization of non-timber forest products in agroforestry systems. Proceedings of an International Conference held in Nairobi, Kenya, 19–23 February 1996*. Food and Agriculture Organization of the United Nations, Rome, Italy, pp. 8–22.

Smith, R.W. (1997) *Preface to the Proceedings of the International Conference on Domestication, Production and Utilization of New Crops*. Smartt, J. and Haq, N. (eds) International Centre for Underutilized Crops (ICUC), University of Southampton, UK.

Swaminathan, M.S. (ed.) (1999) Enlarging the basis of food security: the role of underutilized species. *International Workshop held at the M.S. Swaminathan Research Foundation, 17–19 February 1999, Chennai, India.*

Terrell, E.E., Hill, S.R., Wiersema, J.H. and Rice, W.E. (1977) *A Checklist of Names for 3,000 Vascular Plants of Economic Importance*, US Department of Agriculture, Agricultural Handbook, Washington, DC.

UNEP (1992) *Convention on Biological Diversity*. United Nations Environment Programme.

UNEP (1995) *Global Biodiversity Assessment*. United Nations Environment Programme, Cambridge University Press, Cambridge, UK.

Uphof, J.C.T. (1968) *Dictionary of Economic Plants*. Stechert-Hafner, New York.

Vietmeyer, N. (1990) The new crops era. In: Janick, J. and Simon, J. (eds) *Advances in New Crops. Proceedings of the First National Symposium on New Crops: Research, Development, Economic, Indianapolis, Indiana, 23–26 October 1988*. Timber Press, Portland, Oregon, pp. xviii–xxii.

von Maydell, H.J. (1989) Criteria for the selection of food producing trees and shrubs in semi arid regions. In: Wickens, G.E., Haq, N. and. Day, P. (eds) *New Crops for Food and Industry*. Southampton, UK, pp. 66–75.

WCMC (1992) *Global Biodiversity: Status of the Earth's Living Resources*. World Conservation Monitoring Center, Chapman and Hall, London, UK.

Wickens, G.E. (1995) *Edible Nuts*. Non Wood Forest Product Series, no. 5. Food and :Agricultural Organization of the United Nations, Rome, Italy.

Wickens, G.E., Haq, N. and Day, P. (eds) (1989) *New Crops for Food and Industry*. Southampton, UK.

Wilson, E.O. (1992) *The Diversity of Life*. Penguin, London.

Yarnell, R.A. (1964) *Aboriginal Relationships between Culture and Plant Life in the Upper Great Lakes Region*. University of Michigan Anthrop Papers No. 23.

Young, J.A., Christensen, B.M., Schaad, M.S., Herdendorf, M.E., Vance, G.F. and Munn, L.C. (1999) A geographic information system to identify areas for alternative crops in northwestern Wyoming. In: Janick, J. (ed.) *Perspective on New Crops and New Uses. Proceedings of the Fourth National Symposium on New Crops and New Uses. Biodiversity and Agricultural Sustainability, 8–11 November 1998, Phoenix, Arizona*, pp. 176–180.

Zeven, A.C. and de Wet, J.M.J. (1982) *Dictionary of Cultivated Plants and their Regions of Diversity Excluding most Ornamentals, Forest Trees and Lower Plants*. 2nd edn. Pudoc, Wageningen, The Netherlands.

Zohary, M. (1952) A monograph study of the genus *Pistacia*. *Palestine Journal of Botany, Jerusalem Series* 5(4), 187–228.

31 An Initiative in Exploration and Management of Plant Genetic Diversity in Saudi Arabia

T.A. Al-Turki

KACST Herbarium, Natural Resources and Environmental Research Institute, King Abdulaziz City for Science and Technology, Kingdom of Saudi Arabia

Introduction

Over the past half century, with the rapid increase in global population, depletion of our genetic resources has been occurring at a rapid pace. Conservation of biodiversity and genes in wild species is imperative; hence major efforts to manage and conserve our biosphere's natural resources have been made, initially in the genebanks for cultivated plants and now for wild plants. Use of biotechnological tools for more effective conservation is also occurring. There is a need for further enhancing these efforts in the 21st century to conserve plant genetic resources of all habitats on which human as well as animal life depends. In these efforts, a genebank, whether it be of clones, seeds, tissues or pollen, is a 'bottled' miniature biosphere, protected and preserved in a contracted space, facilitating quick access and use (Kaplan, 1998).

Genebanking is a global effort, aiming at the well being of humanity and should embrace the biodiversity of all continents. The plant diversity of Saudi Arabia should also be preserved so as to contribute to global progress. The present initiative aims to aid the preservation of the genes and germplasm of all categories of plants in Saudi Arabia, both wild and cultivated.

King Abdulaziz City for Science and Technology (KACST), having appreciated the value of the country's genetic resources, is currently launching a project aiming at establishing a national plant genebank. This project intends to undertake collection, evaluation, preservation and documentation of genetic variation in wild and cultivated plants in the Kingdom of Saudi Arabia in the first phase, which would eventually extend to endemic and endangered plants.

Saudi Arabia: Landscapes

The Kingdom of Saudi Arabia lies between 15°45′ and 34°35′ N and 34°40′ and 55°45′ E (Chaudhary and Al-Jowaid, 1999). Occupying the lion's share of the Arabian Peninsula, the Kingdom, with an area of *c.* 2,200,000 km^2, contains significant diversity of arid vegetation. The diverse physiographic (Fig. 31.1) features coupled with their peculiar climates have influenced the vegetation remarkably.

Though there is a vast expanse of desert in the Kingdom, Saudi Arabia is not totally a desert. The mountain ranges bordering the western seashore, rising from 500 m or so, run into escarpments as high as 3000 m (Chaudhary and Al-Jowaid, 1999). This mountain system receives more rain than other parts of the country, and holds arborescent vegetation with high species richness. The shrubby and herbaceous life forms subtended by the arborescent vegetation also tend to be diverse.

Numerous small and large interlaced wadis are distributed throughout the country. Because of the

(eds J.M.M. Engels, V. Ramanatha Rao, A.H.D. Brown and M.T. Jackson)

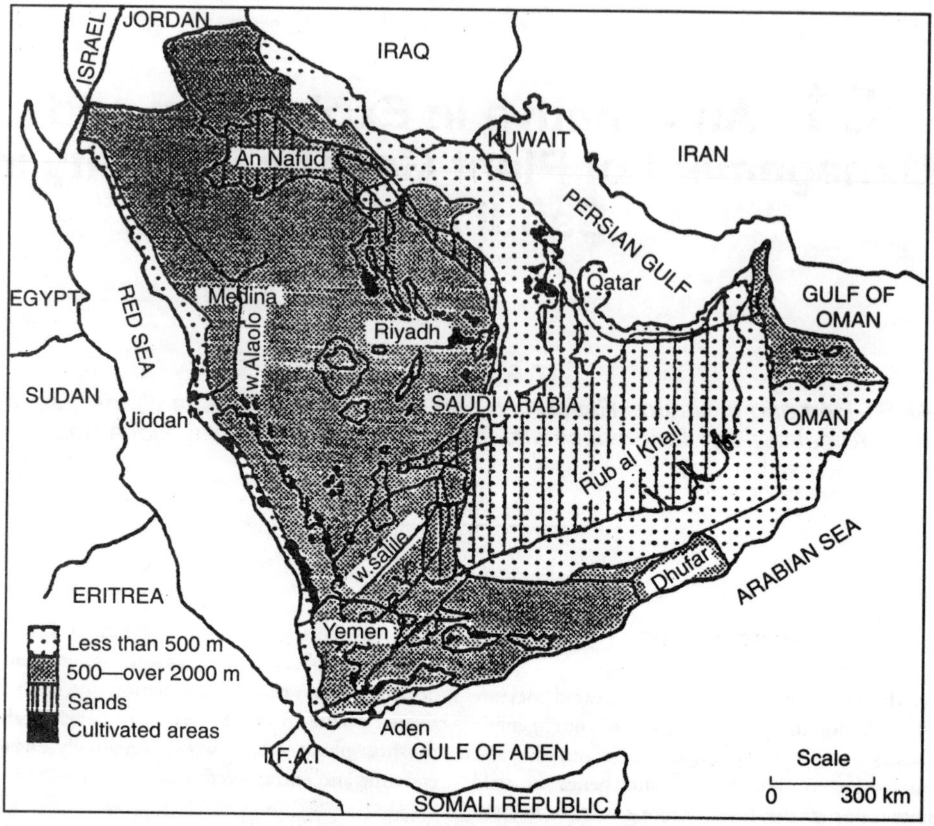

Fig. 31.1. Physiographic map of Saudi Arabia (after Abd El-Rehman, 1986).

slow seepage of moisture preserved beneath the adjoining landmasses, the lower landscapes forming the wadis hold a better soil moisture regime and shelter more species than the adjoining areas. These wadis provide shelter for a large segment of the flora of the country.

The inland sabkhas, coastal salt marshes, and the littoral tidal zones host diverse halophytic communities. The relief features edging the Arabian shield are also peculiar niches for particular kinds of plants. The Rub' al-Khali, the Nafud and the Dahna are vast sand expanses, which experience extreme aridity (Abd El-Rehman, 1986). Although sparse and species poor, flora comprising species characteristically adapted to the extremes of the xeric climate inhabit these deserts.

Surrounded by the Mediterranean, Near East, and Abyssinian (Ethiopian) and Indian centres, Arabia is an important plant diversity centre (Miller and Cope, 1996; Al-Farhan, 1999). The Peninsula provides some of the least man-modified landscapes and life forms within the Irano-Turanian phytochorion (Boulos *et al.*, 1994). Saudi Arabia is also an ancient cultural centre housing many agricultural crops and practices.

The Plant Genetic Resources of Saudi Arabia

Until recently, floristic botany in the Arabian Peninsula received low priority. The hostile desert environment, the extremely hot climate and the recurring droughts in many places were responsible for this, in addition to the paucity of trained botanists in the region. Migahid and Hammoda's (1978) *Flora of Saudi Arabia* is out of date and since then, a large number of new taxa and extension of distribution of some taxa have been reported (see Table 31.1). A comprehensive floristic treatment on a par with modern botanical practices is yet to emerge (Miller and Cope, 1996), though

Table 31.1. List of new taxa reported from Saudi Arabia in the past few years.

Family	Others	Endemics	Total	References
Asteraceae	7	5	12	Al-Farhan *et al.*, 1997; Ghafoor, 1997; Ghafoor and Al-Turki, 1999; Tan, 1995
Chenopodiaceae	9	1	10	Al-Turki and Ghafoor, 1996; Al-Turki and Swarupanandan, 2000; Al-Turki *et al.*, 2000a, b
Convolvulaceae	—	2	2	Al-Farhan, 1991; 1993
Cruciferae	2	8	10	Hedge and Tan, 1987
Fabaceae	4	—	4	Al-Farhan *et al.*, 1997
Flacourtiaceae	1	—	1	Al-Farhan *et al.*, 1997
Labiatae	1	3	4	Al-Farhan *et al.*, 1997; King, 1988; Hedge, 1982
Malvaceae	1	—	1	Al-Farhan *et al.*, 1997
Plumbaginaceae	1	—	1	Abedin *et al.*, 1985
Solanaceae	2	—	2	Abedin *et al.*, 1991
Total	28	19	47	

excellent floristic accounts exist for some regions (Mandaville, 1990). The pictorial guide to the flora of the Kingdom by Collenette (1999) was a significant task, but a truly scientific flora for the Kingdom is only now emerging (Chaudhary, 1999). Except for random efforts, provincial checklists (Al-Turki, 1997), illustrated guides (Heemstra *et al.*, 1990) and floras (Mandaville, 1990) are far from the main focus. According to a recent compendium, the total number of flowering plant species in the country is estimated to be about 2500 (Collenette, 1998, 1999).

Zohary (1973) reviewed the potential of wild plant resources of the Middle East and South-west Asia. He discussed plants used by man through the ages and the 19th century nomads, who collected and utilized the locally occurring wild plants for food, fruits, condiments, healing, fuel, industrial uses, fencing, shade, pot herbs and so on. The genepools enumerated by Zohary (1973) still exist and need the attention of scientists working in the field of agriculture and biotechnology.

Since the oil boom that started in the first half of the 20th century, plant-based resource utilization of the natives has expanded significantly. In one way or another, the number of cultivated plant varieties and species in the Kingdom has increased by at least twofold since the 1930s, making it extremely rich and catering to the interests of the people.

Cultivated plants and their wild relatives

Cereals

Saudi Arabia is an active importer of food material; however, in the case of wheat, it is self-sufficient and is an exporter. The cultivation of imported varieties has resulted in gradual disappearance of local races of wheat, but some of them are still grown in some parts of the Kingdom. The genus *Aegilops* (goatgrass), said to be the progenitor of cultivated wheats (*Triticum* sp.), is represented by three variable hardy species (*Aegilops kotschyi* Boiss., *Aegilops peregrina* (Hack.) Maire & Weiller, and *Aegilops vavilovii* (Zhuk.) Chennav.) in the Kingdom (Chaudhary, 1989; Collenette, 1998). *Aegilops* species grow in wheat fields and probably hybridize with them. *Triticum dicoccum* or emmer was cultivated in the middle ages and is the oldest known wheat, which cannot be easily threshed due to sub-persistent glumes. Even after the introduction of improved varieties, many landraces of wheat and barley exist here (Boulos *et al.*, 1994). Along with the northern mountains of Yemen, the Asir mountains in south-western Saudi Arabia have been identified to be an important location for collecting wild wheats. Other important cereal crops widely cultivated along the up-country include maize and sorghum. The International Plant Genetic Resources Institute (IPGRI) notes that the south-western Arabian Peninsula is a primary centre of diversification of *Sorghum*.

Fruits

A variety of fruit crops are grown in the country, supplementing the diet of the population. Date, mango, orange, plantains, grapes, lemon, fig and so on are cultivated on a large scale in different parts of the country, especially along Najran and Asir. Out of the *c.* 5000 named cultivars of date palm (*Phoenix dactylifera* L.; Arabic: Nakhl, Balh) (Anonymous, 1914; Al-Ghamdi, 1996), *c.* 400 cultivars (Sawaya, 1986) exist in the Kingdom. However, only 50–60 of these are considered economically important. A variety of products derived from date palm are exported in large quantities. A seedless date and another extremely sweet variety, 'sukkri', are elite cultivars preferred by people. *Ziziphus spina-christi* L. (Shajrat Sidr or Nabek), one of the few truly Arabian native tree species (Mandaville, 1990), is grown in towns and villages along with the many exotics now introduced and well adapted to the Saudi Arabian environment and bears pleasant tasting edible berries (Anonymous, 1992), which are stored by Bedouins. The pomegranate, *Punica granatum* L. (Arabic: Rumman), is another fruit represented by cultivated and wild varieties.

Fodder and forages

The rangeland spanning 14,000 km^2 is a megarefugium of wild resources of fodder and forage plants. The Arabian Peninsula is considered important for the genepool of lucerne (*Medicago sativa* L. Fabaceae). Forage crops in the wild that cattle, camels and sheep prefer need to be identified, evaluated and selected.

Introduced and naturalized plants

Originally introduced from Ethiopia (Boulos *et al.*, 1994), qahwa arabia (Arabic coffee) is a hospitality drink of Arabian countries. Arabic coffee (*Coffea arabica* Benth., *Rubiaceae*) is widely cultivated on the terraced mountain slopes of the highlands of Yemen and south-western Saudi Arabia (in Jizan, Al-Baha and Abha provinces). It is a traditional cash crop of individual holdings at above 1500 m. Coffee cultivation is integrated with sheep farming, where the sheep yield the fertilizer.

In recent years, a large number of smaller trees and shrubs have been introduced into the country as hedge plants or as avenue trees, including species of *Atriplex* (*Chenopodiaceae*), *Prosopis*, Australian *Acacia* and *Pithecellobium* (*Mimosaceae*), *Tamarindus* (*Caesalpiniaceae*), and many others. At least some of them might have acclimatized and naturalized in semi-urban environments. These are also genepools of value for the future.

A wide variety of vegetables like onion, okra, aubergine, potato, tomato, carrot, cabbage, turnip, different kinds of pumpkins, cucumbers and gourds, chilli, sugarbeet and a host of pulses and leafy vegetables from different parts of the world are being cultivated in different parts of the country, in open fields as well as under controlled conditions. These provide an excellent kaleidoscopic mixture of different cultivars from Egypt, Yemen, Pakistan, India, Bangladesh, Sri Lanka, Indonesia and the Philippines, in addition to native Saudi Arabian cultivars. While the introduced species have contributed to the richness of the plant resources of the country, the fact remains that they have also been responsible for the neglect of many native cultivars and wild plants. Efforts to locate these germplasm should be an important aspect of biodiversity conservation in the Kingdom.

Wild plants

Endemic and endangered plants

Approximately 2243 native flowering plant species are known from the Kingdom (Al-Farhan, 2000). Roughly 20% of the species (500 spp.) are found only in limited habitats, 656 exist only as small populations, 100 are endangered or of vulnerable status, *c.* 40 are endemic and 30 are on the verge of extinction (NCWCD, 1998; Al-Farhan, 2000). Floristic explorations and critical studies of selected plant groups by experts continue to add new species and new records of which a significant number (nearly 40%) fall to the endemic and rare categories or with restricted distribution (Table 31.1). El-Sheikh *et al.* (1989) and NCWCD (1998) list the existing and potential threats to plant biodiversity. Wood cutting, over-grazing, off-road driving, agriculture, urbanization and drought are the main threats, and Collenette (1999) notes that four endemic and 18 non-endemic species have become extinct in the Kingdom.

Recent explorations by the KACST herbarium staff have drawn our attention to the dwindling populations of a number of rare and threatened plants of the Kingdom. *Mimusops laurifolia* (Forssk.) Friis (*Sapotaceae*), *Faidherbia albida* (Delile) A. Chev. (*Fabaceae*) and *Kleinia pendula* (Forssk.) Sch.-Bip. (*Compositae*), *Breonadia salicina* (Vahl) Hepper & Wood (*Rubiaceae*), species of *Salicornia* (*Chenopodiaceae*), are typical examples (Fig. 31.2).

1. *M. laurifolia* (Forssk.) Friis is a rare tree of the ravine forests and scrubs between 600 and 1100 m msl. This species is also found in a number of countries, such as Eritrea, Ethiopia, Somalia and Yemen. It is restricted to areas with precipitation in the cooler months (Friis, 1980). In Saudi Arabia, the tree is known to have scattered distribution of sparse populations in the wadis of the Jizan Province, like Wadi Jora. Very recently, two trees of the species have been located by the author and his team in the hill folds of Jabal Shada (Al-Baha Province, a mountain escarpment), in a seasonal streamlet. Regeneration of the tree has not been observed *in situ*.

2. *F. albida* (Delile) A. Chev. (*Fabaceae*), another large tree mimicking the rainforest trees in size (to 4 m girth), inhabits the wadis cutting the mountain ranges of the Taif region (Wadi Al-Quyrat, Wadi Nail and Wadi Sharan: Al-Turki and Mehmood personal observation). The species is again restricted to one or two sites within the country with densities as low as 10–15 per population.

3. *K. pendula* (Forssk.) Sch.-Bip. (*Compositae*) is another curious plant with beautiful red flowers and a leafless stem. The author, during various field trips, has confirmed the endangered species status given to it by El-Sheikh *et al.* (1989). The plant has so far been known only from two localities in Jabal Fayfa (Jizan Province), one of the mountain ranges of the south-western corner of Saudi Arabia. Very little is known about the range of distribution of the species and there is every possibility that it may become extinct, unless protected.

4. *B. salicina* (Vahl) Hepper & Wood (*Rubiaceae*) is a tall tree up to 20 m in height and 2 m in diameter growing along high escarpments from 500 to

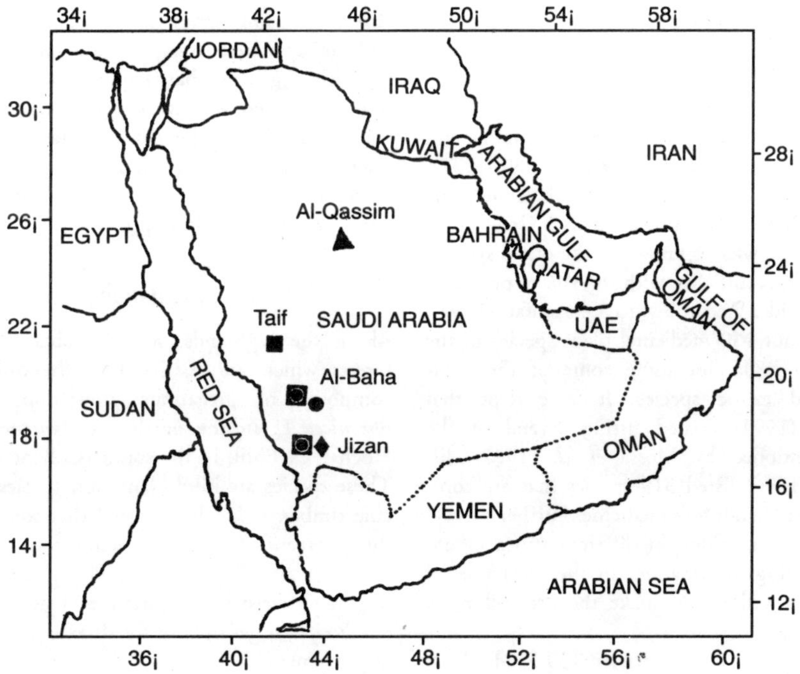

Fig. 31.2. Distribution map of some rare plants in Saudi Arabia: ◙ *Mimusops laurifolia* (Forssk.) Friis (*Sapotaceae*); ■ *Faidherbia albida* (Delile) A. Chev. (*Fabaceae*); ◆ *Kleinia pendula* (Forssk.) Sch.-Bip. (*Compositae*); ● *Breonadia salicina* (Vahl) Hepper & Wood (*Rubiaceae*); ▲ *Salicornia* sp. (*Chenopodiaceae*).

2000 m. Though common in Yemen, its population is very thin in Saudi Arabia, where it is classified as endangered (Collenette, 1999). Our recent explorations indicate its presence in the Al-Baha Province too (at Jabal Shada), again being represented by a sparse population comprising just three or four trees (Al-Turki and Swarupanandan, personal observation).

5. *Salicornia* (*Chenopodiaceae*) is an oil-seed crop used in saline agriculture (Glenn *et al.*, 1991). In the course of the explorations for halophytes in the country, we have come across two species of *Salicornia*, probably new, one of which inhabits inland sabkhas of Al-Qassim (Central Province). The normal vigorous form of the latter species is a robust shrublet, measuring as much as 130 cm in diameter and 60 cm or more in height. The natural home of the plant being inland sabkha (salt marsh), it is highly salt-tolerant, and may grow in the coastal environment too. It is a promising species which may stand in lieu of *Salicornia bigelovii* (a crop cultivated with saline water: see Glenn *et al.*, 1991), or may be a good genetic material for hybridization with the latter (Al-Turki and Swarupanandan, 2001).

Medicinal plants

With roots running prior to the philosopher and physician Avicenna (Ibn-e Senna: 980–1037), Arabia has a rich tradition of herbal medicine (Ghazanfar, 1994) which blended with Greek practices to become what is known as the unani system. The tradition is still alive with the village people of the countryside. According to Ghazanfar (1994), there are about 260 medicinal plant species in the Arabian Peninsula, including some of the commonly used exotic species. It is evident that Ghazanfar (1994) missed listing several of the species mentioned by Mossa *et al.* (1987). El-Shanwani (1996) listed 319 species that are commonly used in Saudi folk medicine. Further surveys and contacts with elders of different tribes should yield interesting information on the local uses of native plants and would make the list still more lengthy.

Al-Yahya *et al.* (1990) screened 150 Saudi plants for their pharmocological, phytochemical, pharmacognostical, antimicrobial and toxicological properties, providing valuable information on the richness of medicinally active constituents. Active constituents responsible for the action of plants mentioned by Mossa *et al.* (1987) and El-Shanwani (1996) include: mucilage (*Plantago* sp.), phenols (*Mentha* and *Thymus* spp.), tannins (*Acacia* spp.), coumarins (*Apium graveolens* L., *Ammi visnaga* (L.) Lam), anthraquinones (*Cassia senna* L.), flavonoids (*Citrus limmon* L.), anthocyanin (*Vitis vinifera* L.), glucosilinates (*Raphanus sativa, Sinapis alba* L.), anti-inflammatory volatile oils (*Matricaria recutita* L.), saponins (*Glycirrhiza glabra* L., *Chenopodium ambrosoides* L.), glycosides (*Datura* spp.) and alkaloids (*Salsola kali* L.). Some species of the gymnosperm genus *Ephedra* yield ephedrine which is useful in cough syrup. Species of *Datura* (*Solanaceae*) contain alkaloids useful in cases of megalomania and species of *Withania* and *Solanum* (*Solanaceae*) yield alkaloids useful in many other situations. Several species of *Commiphora* (Burseraceae: 7–8 species) are also important in this context.

Plants for oils, gums, resins and perfumes

In arid environments, there are a number of herbs which contain aromatic oils or substances useful in the manufacture of perfumes or insect repellents. *Artemisia* (*Compositae*), *Lavandula* (*Labiatae*), *Ruta* and *Haplophyllum* (*Rutaceae*) and many others are promising genera that require proper chemical screening. There are several plants which can form a basis for their industrial exploitation: *Artemisia judaica, Balanites aegyptiaca, Commiphora* spp., *Ruta chaleppensis, Haplophyllum tuberculatum, Eruca sativa* and *Brassica* spp.

Trees of plantation potential

Along the highlands, Saudi Arabia has conifer forests which are restricted to the hilltops and composed of gregarious growth of *Juniperus phoenicea* L. in the north and *Juniperus procera* Hochst. ex Endl. in the south (Collenette, 1999). These species are ideal plantation species for valuable timber. It has been noted that some parts of the *Juniperus* forest are facing serious dieback (Gardner and Fisher, 1994; Fisher, 1997) which highlights the need for careful evaluation of different *Juniperus* germplasm for disease resistance and improvement.

Plants for ecorestoration

Deserts are unequalled segments of the arid parts of the biosphere containing many bushy shrubs and low trees useful in arid land reclamation. Species of *Tamarix* L. (*Tamaricaceae*), *Ziziphus* Mill. (*Rhamnaceae*) and *Prosopis* L. (Arabian mesquite, *Mimosaceae*) are important. Several species of *Ficus* suited to the arid environment, *Dodonaea angustifolia* L. *f.* (*Sapindaceae*), already used as a hedge plant in urban gardening, can also be considered for revegetation purposes. Both the western and eastern coasts of the country are feathered with scattered pockets of mangrove forests mostly composed of *Avicennia marina* (Forssk.) Viern. (*Verbenaceae*) and *Rhizophora mucronata* Lam. (*Rhizophoraceae*). Coastal land protection and reclamation are very important and some of the species mentioned here might become important for this purpose in the future.

Much of the land in the country is affected by aridity, which is worsened by high soil salinity. Nearly 40% of the total land cover of the Kingdom is classified as saline, out of which approximately 4% is at the upper extreme (Ali, 1999; Al-Turki *et al.*, 2000a). Shrubs and trees useful in arid saline land reclamation and the woody species protecting the coastal mangroves, are all greatly attuned to the saline environment. The genepools of many of these plants originated there and, by and large, are halophytes. A preliminary short listing shows that there are as many as 190 species of halophytes in the Kingdom, dominated by members of *Chenopodiaceae* (Al-Turki *et al.*, 2000a). Some of these extreme drought and salt tolerant species should be investigated to improve the productivity of saline habitats.

Ornamental plants

Ornamental plants are another ignored segment of the biodiversity of the country. Many succulents belonging to the families *Asclepiadaceae*, *Compositae*, *Euphorbiaceae* and *Crassulaceae* are very beautiful and are worth introducing to garden domestication. Tens of species of *Caralluma* and genera lately separated from it, species of *Huernia, Stapelia, Duvalia*, and curious species of the Asteracean genus *Kleinia*, the spiny shrubs of *Euphorbia*, hauntingly elegant species of *Aloe* with differing spikes and colour of flowers, some of the beautiful *Portulaca* and so on are excellent materials for rock gardens. Similarly there are several other plant species that could become ornamental plants directly or play a role as rootstocks for other flowering plants. Work needs to be done in this area.

Preliminary Efforts in Establishing the Genebank

The initial work involved gathering information on plant species and their distribution. Riyadh, the capital city of the Kingdom, has five herbaria and by consulting the herbarium specimens and literature resources concerning localities, the distribution of plants in the Kingdom available with the herbaria were assembled. Additionally, the experts at the herbaria also helped to deal with the taxonomy and identification of the plants. The infrastructure facilities available in these institutions were put to use to establish a viable national genebank.

Initially, we have collected 83 samples of seeds of 52 species of plants belonging to 35 genera from the Kingdom and documented the details (Table 31.2). Equal numbers of seeds from 50 plants from each of the population of a given species were collected. In the case of rare plants whatever was available from the surviving plants was gathered. The seeds collected include a large number of halophytes, and include species like *Arthrocnemum macrostachyum* (Moric.) K. Koch., *Bienertia cycloptera* Bunge, *Chenopodium ambrosoides* L., *C. glaucum* L., *C. murale* L., *Halocnemum strobilaceum* (Pall.) M. Bieb., *Halopeplis perfoliata* (Forssk.) Asch., *Salicornia* spp., *Suaeda aegyptiaca* (Hasselq.) Zohary, *Suaeda maritima* (L.) Dumourt., and *Suaeda vermiculata* Forssk. mostly belonging to *Chenopodiaceae*. The non-halophytes are represented by species like *Dodonaea angustifolia* L. *f.* (*Sapindaceae*), *Faidherbia albida* (Delile) A. Chev. (*Fabaceae*), *Kleinia pendula* (Forssk.) Sch.-Bip. (*Compositae*), *Mimusops laurifolia* (Forssk.) Friis (*Sapotaceae*), *Rumex nervosus* Vahl (*Polygonaceae*), *Solanum incanum* L. and *Solanum surattense* Burm. f. (*Solanaceae*).

Samples were collected from habitats like coastal lagoons, inland sabkhas, wadis, sand dunes, low- and high-elevational plateaux and mountain escarpments. Species collected include many of the endemic and endangered taxa mentioned in previous sections. The seeds are stored at 5°C, processed for preservation at −18°C, and viability evaluated following conventional methods (Ellis *et al.*, 1985). The genebank is planned to be multi-organizational and is expected to cater to the conservation requirements of the Kingdom. The genebank will be managed by the scientists of KACST, and other organizations involved would be King Saud

Table 31.2. Seed sample collections of various plant families from Saudi Arabia.

Family	Genera	Species	Samples
Amaranthaceae	1	1	1
Capparidaceae	2	2	3
Caryophyllaceae	1	1	1
Chenopodiaceae	8	17	33
Convolvulaceae	2	2	3
Compositae	1	2	2
Cucurbitaceae	1	1	1
Cupressaceae	1	1	2
Labiatae	2	4	7
Leguminosae	6	7	8
Polygonaceae	1	2	3
Portulacaceae	1	1	1
Resedaceae	1	3	8
Salvadoraceae	1	1	1
Sapindaceae	1	1	1
Sapotaceae	1	1	1
Scrophulariaceae	1	1	1
Solanaceae	2	3	5
Zygophyllaceae	1	1	1
Total	35	52	83

University, the National Commission for Wildlife Conservation and the research stations of the Ministry of Agriculture and Water located in different parts of the Kingdom. The germplasm collected by staff of the agricultural research stations at Riyadh, Hofuf, Abha, Baha, Jizan, Al-Jouf and Qassim will be brought to the main seed storage facility under construction at KACST Experimental Station at Riyadh. The collections at the moment are seed-based and lodged at the Natural Resources and Environmental Research Institute of the KACST, where facilities for preservation of samples at −18 °C exist. The samples are to be replicated in more than one centre in the future.

Future Plans

Our future plan envisages the conservation of germplasm of underused leguminous nitrogen fixers, wild relatives of crop plants, halophytes, plants useful in desert reclamation, native medicinal herbs, edible geophytes like *Allium* and other succulent ornamentals, and species belonging to the threatened and endangered categories. Special focus is envisaged on halophytes having economic potential as a source of raw materials in the pharmaceutical (*Salsola tetrandra* Forssk.) and paper (*Juncus* spp.) industries and as range plants (*Kochia* spp.) for livestock. Presently we have established a storage facility for seeds for a few months without causing physical damage or loss to their genetic variability according to the methods and techniques outlined by Roberts (1975) and Hanson (1985). In future we plan to conserve germplasm using the following techniques:

1. Storage of recalcitrant seeds using liquid nitrogen (Villiers, 1975; Justice and Bass, 1978; Stanwood and Bass, 1981).
2. Pollen storage (Shivanna and Rangaswamy, 1992; Connor and Towill, 1993; Hoekstra, 1995).
3. Tissue culture, including cell/callus culture (D'Amato, 1975; Anonymous, 1985) and meristem/shoot tip cultures (Morel, 1975).

Apart from our taxonomic database for the herbarium collections, a genebank database will be established where information on accessions will be stored and will be searchable. Once the database is organized, it should be available for use by scientists from the region and abroad. Links to other genebanks will also be established. A safety duplication of the material conserved in the genebank is also planned in different institutions in the future, against the risk of losing them. Any regional collaboration on the initiative will be highly appreciated.

Conclusions

The absence of a genebank in the Arabian Peninsula has considerably hampered biotechnological progress and preservation of biodiversity. It is surprising that no collaborative effort has been made by Gulf Council Countries and international organizations such as IBPGR, IPGRI and FAO for establishing a genebank in the region for germplasm conservation.

The facility developed for long-term storage of seeds collected so far is enough for initial collections but is in need of trained personnel for biotechnological experimentation and databasing of germplasm and herbarium collections. The genebanking of the entire flora of Saudi Arabia will be phased into a number of long-term stages involving several experts from within the Kingdom and developed countries in the fields of plant taxonomy, seed technology and physiology. Information on the local uses of native plants will be gathered through surveys and contacts with elders of different villages and tribes and incorporated in the database.

The genebank information system would be supplemented with details of usefulness of taxa and their products, genetic variability, economic highlights, availability of the taxa in the wild and in the genebank. It will be made accessible on the Internet and linked to other germplasm resources so that information on our materials is available to the world community. Data on seed samples will be stored in a sophisticated management system which will be divided into three categories: sample identification, evaluation and inventory management.

Acknowledgements

The author thanks and acknowledges various help received from: Dr A.H. Al-Farhan (KSU), Dr Abdul Rahman Khoja (NCWCD-Taif), Mrs Sheila Collenette (UK), Mr Riyadh Basahy (Jizan), Mr S.F. Mehmood, Dr K. Swarupanandan and Mr Abdul Ghafoor (NRERI-KACST).

References

Abd El-Rehman, A.A. (1986) The deserts of the Arabian Peninsula. In: Eveanari, M., Noy-Meir, I. and Goodall, D.W. (eds) *Ecosystems of the World*, Vol. 12(B). Elsevier, Amsterdam, pp. 29–55.

Abedin, S., Mossa, J.S. and Al-Yahya, M.A. (1985) Contributions to the flora of Saudi Arabia, part 1. Some new records. *J. Pharm. Univ. Karachi* 4(1), 1–7.

Abedin, S., Al-Yahya, M.A., Chaudhary, S.A. and Mossa, J.S. (1991) Contributions to the flora of Saudi Arabia, part 2. A revision of the family Solanaceae. *Pakistan Journal of Botany* 23(2), 257–282.

Al-Farhan, A.H. (1991) Two new *Convolvulus* taxa from the Arabian Peninsula. *Botanical Journal of the Linnean Society* 106, 259–263.

Al-Farhan, A.H. (1993) A new species of *Convolvulus* (Convolvulaceae) from Saudi Arabia. *Brittonia* 45(2), 169–171.

Al-Farhan, A.H. (1999) A phytogeographical analysis of the floristic elements in Saudi Arabia. *Pakistan Journal of Botany* 2(3), 702–711.

Al-Farhan, A. (2000) An evaluation of the current status of the flora of Saudi Arabia. Country report presented at the Second Arabian Plants Subject Group Meeting, Abudhabi, May 2000 (Draft).

Al-Farhan, A.H., Thomas, J. and Alallah, M.I.H. (1997) Some new records for the flora of Saudi Arabia. *Kuwait Journal of Science and Engineering* 24, 123–130.

Al-Ghamdi, A.S. (1996) Date palm (*Phoenix dactylifera* L.) germplasm bank in King Faisal University, Saudi Arabia: 1. Justification, implementation and organization. In: *Proceedings of the Third Symposium on the Date Palm in Saudi Arabia*, vol 1. Date Palm Research Centre, King Faisal University and Mars Publications, Riyadh, Saudi Arabia, pp. 505–516.

Ali, M.A. (1999) Soils and summary of meteorological data. In: *Flora of the Kingdom of Saudi Arabia*, Vol. 1, Ministry of Agriculture and Water, Riyadh, pp. 10–11.

Al-Turki, T.A. (1997) A preliminary checklist of the flora of Qassim, Saudi Arabia. *Fedd. Repert.* 108 (3–4), 259–280.

Al-Turki, T.A. and Ghafoor, A. (1996) The genus *Chenopodium* L. in Saudi Arabia. *Fedd. Repert.* 107(3–4), 189–208.

Al-Turki, T.A. and Swarupanandan, K. (2001) Morphology and taxonomy of the Eurasian species of *Salicornia* Linn. (Chenopodiaceae) and two new species from Arabia (in press).

Al-Turki, T.A., Ghafoor, A., Omer, S. and Davy, A.J. (2000a) A preliminary contribution to the halophyte vegetation and flora of Saudi Arabia (in press).

Al-Turki, T. A., Omer, S. and Ghafoor, A. (2000b) A synopsis of the genus *Atriplex* in Saudi Arabia. *Feddes Reportorium* 111 (5–6), 255–267.

Al-Yahya, M.A., Al-Meshal, I.A.R., Mossa, J.S., Al-Badar, A.A. and Tariq, M. (1990) *Saudi Plants: a Phytochemical and Biological Approach.* GDRGP, KACST, Riyadh.

Anonymous (1914) Origin of the date palm. *Journal of Heredity* 5, 498–508.

Anonymous (1985) Tissue culture for crops, stress-resistant from cell culture. *Summary of First Annual International Plant Biotechnology Network (IPBNet).* Fort Collins, Colorado.

Anonymous (1992) *Natural History of Saudi Arabia: an Introduction.* Ministry of Agriculture and Water, Riyadh.

Berjak, P. and Pammenter, N.W. (1997) Progress in understanding and manipulation of desiccation-sensitive (recalcitrant) seeds. In: Ellis, R.H. and Black, M. (eds) *Basic and Applied Aspects of Seed Biology.* Kluwer Academic Publishers, Amsterdam, The Netherlands, pp. 689–703.

Boulos, L., Miller, A.G. and Mill, R.R. (1994) Regional overview: Southwest Asia and the Middle East. In: Davis, S.D., Heywood, V.H. and Hamilton, A.C. (eds) *Centres of Plant Diversity: a Guide and Strategy for their Conservation.* WWF and IUCN Publications Unit, Cambridge, UK, pp. 293–348.

Chaudhary, S.A. (1989) *Grasses of Saudi Arabia.* Ministry of Agriculture and Water, Riyadh.

Chaudhary, S.A. (1999) *Flora of the Kingdom of Saudi Arabia*, vol. 1. Ministry of Agriculture and Water, Riyadh.

Chaudhary, S.A. and Al-Jowaid, A.A.A. (1999) *Vegetation of the Kingdom of Saudi Arabia.* Ministry of Agriculture and Water, Riyadh.

Collenette, S. (1998) *A Checklist of Botanical Species in Saudi Arabia.* International Asclepiad Society, UK.

Collenette, S. (1999) *Wild Flowers of Saudi Arabia.* National Commission for Wildlife Conservation and Development, Riyadh.

Connor, K.F. and Towill, L.E. (1993) Pollen handling protocol and hydration/dehydration characteristics of pollen for application to long-term storage. *Euphytica* 68, 77–84.

D'Amato, F. (1975) The problem of genetic stability in plant tissue and cultures. In: Frankel, O.H. and Hawkes, J.G. (eds) *Linnean Society Symposium*, Ser. 2. Academic Press, London, pp. 333–348.

Eberhart, S.A., Roos, E.E. and Towill, L.E. (1991) Strategies for long-term management of germplasm collections. In: Falk, D.A. and Holsinger, K.E. (eds) *Genetics and Conservation of Rare Plants.* Oxford University Press, Oxford, UK, pp. 135–152.

El-Shanwani, M.A.A. (1996) *Al-nibatat al-mustakhdima fi al-tibb al-sha'abi al-Saudi* [Plants used in Saudi folk medicine]. General Directorate of Research Grants Program, KACST, Riyadh.

El-Sheikh, A.M., Chaudhary, S.A., Hassan, H.M. and Al-Farraj, M.M. (1989) Endangered or rare plant species in Saudi Arabia. In: Abu-Zinada, A.H., Goriup, P.D. and Nader, A. (eds) *Wildlife Conservation and Development in Saudi Arabia.* NCWCD, Riyadh, pp. 193–199.

Ellis, R.H., Hong, T.D. and Roberts, E.H. (1985) *Handbooks for Genebanks*, No. 2. *Handbook of Seed Technology for Genebanks*, Vol. 1. *Principles and Methodology.* International Board for Plant Genetic Resources, Rome, Italy.

Fisher, M. (1997) Decline in the Juniper woodlands of Raydah reserve in southwestern Saudi Arabia: a response to climatic change? *Global Ecology and Biogeography Letters* 6, 379–386.

Friis, I. (1980) The taxonomy and distribution of *Mimusops laurifolia* (Sapotaceae). *Kew Bulletin* 35(4), 785–792.

Gardner, A.S. and Fisher, M. (1994) How the forest lost its trees: Just so. Story telling about *Juniperus excelsa* in Arabia. *Journal of Arid Environments* 26, 199–201.

Ghafoor, A. (1997) A new species of *Anthemis* L. (Asteraceae-Anthemideae) from Saudi Arabia. *Feddes Repert.* 108 (5–6), 319–323.

Ghafoor, A. and Al-Turki, T.A. (1997) A synopsis of the genus *Anthemis* L. (Compositae-Anthemideae) in Saudi Arabia. *Candollea* 52, 457–474.

Ghafoor, A. and Al-Turki, T.A. (1999) A new *Anthemis* (Asteraceae) from Saudi Arabia. *Edinburgh Journal of Botany* 56(1), 55–59.

Ghazanfar, S.A. (1994) *Handbook of Arabian Medicinal Plants.* CRC Press, Boca Raton, Florida.

Glenn, E.P., O'Leary, J.W., Watson, M.C., Thompson, T.L. and Kuehl, R.O. (1991) *Salicornia bigelovii* Torr.: an oilseed halophyte for sea water irrigation. *Science* 251, 1065–1067.

Hanson, J. (1985) Procedures for handling seeds in genebanks. *Practical Manual for Genebanks.* No. 1. International Board for Plant Genetic Resources, Rome.

Hedge, I.C. (1982) Studies in the flora of Arabia. 2. Some new and interesting species of Labiatae. *Notes RBG Edinb.* 40(1), 63–73.

Hedge, I.C. and Tan, K. (1987) Two remarkable new Cruciferae from Saudi Arabia. *Plant Systems Evolution* 156, 197–206.

Heemstra, H.H., Hassan, H.O. and Al-Minwer, F.S. (1990) *Plants of Northern Saudi Arabia.* Range and Animal Development Research Centre, Ministry of Agriculture and Water, Sakaka, Al-Jouf, Saudi Arabia.

Hoekstra, F.A. (1995) Collecting pollen for genetic resources conservation. In: Guarino, L, Ramanatha Rao, V. and Reid, R. (eds) *Collecting Plant Genetic Diversity*. CAB International, Wallingford, UK.
Justice, O.L. and Bass, L.N. (1978) *Principles and Practices of Seed Storage*. Agriculture handbook no. 506. Government Printing Office, Washington, DC.
Kaplan, J. (1998) Conserving the world's plants. *Agricultural Research* 46(9), 4–9.
King, R.A. (1988) Studies on the flora of Arabia 19. *Teucrium* in the Arabian Peninsula and Socotra. *Notes RBG Edinb.* 45(1), 21–42.
Mandaville, J.P. (1990) *Flora of Eastern Saudi Arabia*. Kegan Paul International, London.
Migahid, A.M. and Hammouda, M.A. (1978) *Flora of Saudi Arabia*, 2nd edn. Riyadh University, Riyadh.
Miller, A.G. and Cope, T.A. (1996) *Flora of the Arabian Peninsula and Socotra*. Edinburgh University Press, Edinburgh, UK.
Morel, G. (1975) Meristem culture techniques for the long-term storage of cultivated plants. In: Frankel, O.H. and Hawkes, J.G. (eds) *Crop Genetic Resources for Today and Tomorrow*. Cambridge University Press, Cambridge, UK, pp. 327–332.
Mossa, J.S., Al-Yahya, M.A. and Al-Meshal, I.A. (1987) *Medicinal Plants of Saudi Arabia*, vol. 1. King Saud University Libraries, Riyadh.
NCWCD (1998) *Species Status and Conservation Strategy*. B. *Endangered, Vulnerable and Rare Plant Taxa in the Kingdom of Saudi Arabia*. National Commission for Wildlife Conservation and Development, Riyadh.
Roberts, E.H. (1975) Problems of long-term storage of seed and pollen for genetic resources conservation. In: Frankel, O.H. and Hawkes, J.G. (eds) *Crop Genetic Resources for Today and Tomorrow*. Cambridge University Press, Cambridge, UK, pp. 269–296.
Sawaya, W.J. (1986) *Dates of Saudi Arabia*. Ministry of Agriculture and Water, Riyadh.
Shivanna, K.R. and Rangaswamy, N.S. (1992) *Pollen Biology: A Laboratory Manual*. Springer-Verlag, Berlin, Germany.
Stanwood, P.C. and Bass, L.N. (1981) Seed germplasm preservation using liquid nitrogen. *Seed Science and Technology* 9, 423–437.
Tan, K. (1995) Seven new species of *Echinops* (Asteraceae). *Ann. Bot. Finnici.* 32, 117–126.
Villiers, T.A. (1975) Genetic maintenance of seeds in imbibed storage. In: Frankel, O.H. and Hawkes, J.G. (eds) *Crop Genetic Resources for Today and Tomorrow*. Cambridge University Press, Cambridge, UK, pp. 297–315.
Zohary, M. (1973) *Geobotanical Foundations of the Middle East*. Gustav Fischer Verlag, Stuttgart, Germany.

32 Mushroom Breeding and Cultivation Enhances *ex situ* Conservation of Mediterranean *Pleurotus* Taxa

G. Zervakis[1] and G. Venturella[2]

[1] *National Agricultural Research Foundation, Institute of Kalamata, Kalamata, Greece;* [2] *Università degli Studi di Palermo, Dipartimento di Scienze Botaniche, Palermo, Italy*

Introduction

More than 360 plant species are reported to have their primary or secondary centres of origin in the Mediterranean region; such estimates do not take into account additional hundreds of useful species harvested directly from the wild and not yet domesticated (Eyzaguirre *et al.*, 1999). Such a unique array of biodiversity is, however, in danger of disappearing forever, threatened by pollution, anthropogenic activities, urbanization, modern agricultural practices, land erosion, over-exploitation of natural habitats of high ecological value, and so on.

The fungi are essential, yet little understood and often overlooked components of healthy ecosystems. Ignorance of the fungi, which is the second most diverse group of organisms after the insects, is revealed by the fact that only 5–10% of the existing fungal biodiversity has been discovered and described (Hawksworth, 1991). For most of the species identified very few data are available on their biology and most importantly on their ecological and/or economic value. Research in both the field and the laboratory has stressed the essential functions of these organisms in the recycling of organic matter and the release of nutrients into the soil, in preventing nutrient depletion, improving soil structure, and facilitating reforestation through mycorrhiza formation, while some of these fungi produce basidiomata that are edible and nutritious. Given the many benefits that have been derived from a few fungal groups that are sufficiently understood, it is likely that the fungi are the most under-exploited group of organisms and possibly they possess the largest store of potential uses and applications for sustainable development.

Although the international plant genetic resources movement has devoted much of its attention towards the collection and conservation of major food crops for ensuring fairly good representation of the genetic variability of the staples in world genebanks, the same cannot be said about its efforts to safeguard edible mushroom biodiversity. These species, like the minor plant species, do not contribute much as basic food; yet they do contribute significantly to diversifying the human diet, to provide important chemical compounds for the 'neutraceutical' industry (Chang, 1996), and most importantly have the potential to act as sources of additional income for farmers by adding value to agricultural by-products of low economic value. Therefore, their conservation is essential, as has already been the target in the case of 'underutilized' plant species by the Food and Agriculture Organization (IPGRI – Underutilized Mediterranean Species project).

Fungal Biodiversity and Conservation

The topic of fungal conservation has only recently attracted the attention of mycologists. In the 1970s, only a few people worried about the fate of fungi, especially on the effects of mass collecting on some popular edible mushrooms. In the 1980s, reports started to appear on the prominent decrease of some groups of macrofungi in some countries (Arnolds, 1988; Fellner, 1989). The influence of mushroom picking appeared to be small in comparison with the effect of changes in the environment. Systematic mapping in some countries revealed essential information on the geographical distribution of macrofungi and showed the very limited occurrence and distribution of many species (Krieglsteiner, 1991; Nauta and Vellinga, 1992), hence the need to protect these sites. Today there is a widespread notion that many fungi are threatened and that efforts for their conservation are necessary. Air pollution presents a severe problem, whereas in the recent past national and international agricultural policies have led to a fast and irreversible loss of old and poorly managed pastures and their associated fungi. The same policies put at risk the few remains of old-growth forests with their characteristic wood-inhabiting fungi due to uncontrolled exploitation for timber. In addition, peat bogs with their few but strongly specialized fungi have been drained, and the existence of coastal sand dunes with their striking gasteromycetes is jeopardized because of the strong pressure of recreation and expansion of towns.

About 300 species of macrofungi are considered to be endangered throughout Europe, on the basis of published national and regional lists (Ing, 1993). They are arranged in four categories depending on the degree and geographical extent of endangerment:

A: widespread losses, rapidly declining populations, many national extinctions, high-level concern;
B: widespread losses, evidence of steady decline, some national extinctions, medium level concern;
C: widespread, but scattered, populations, fewer extinctions, low-level concern; and
D: local losses, some extinctions but mainly at edge of geographical range.

The evidence of losses is biased towards central, western and northern Europe, whereas for the southern and eastern regions of the continent data are scarce and fragmentary. It is striking that Red Lists for fungi do not exist in southern Europe. The lack of reliable information on distribution of fungi in Europe has repeatedly been indicated in the past, and it means that any conclusions drawn about decline are at most tentative and may even be misleading. Hence, any list to cover this continent as a whole must be a compromise, especially during a period of rapid climatic change to which the fungi are among the first organisms to respond.

In Italy and Spain, fungal biodiversity has been more extensively studied in comparison with other Mediterranean countries. As one might expect, a high species diversity is demonstrated throughout this area, and regional inventories have been published while Red Lists are under preparation (Venturella, 1991; Perini *et al.*, 1993; Venturella *et al.*, 1997; Ivancevic, 1998). The French programme for inventory and mapping of the fungi began in 1991 and is currently under way (Redeuilh and Courtecuisse, 1999). In Greece, significant progress on inventorying the country's fungal biodiversity has been made during the last two decades, and has resulted, among other things, in producing two annotated checklists of macrofungi belonging to Basidiomycotina and Ascomycotina (Zervakis *et al.*, 1998, 1999). However, since mapping of the mycoflora is far from being complete and sound data are missing as regards the occurrence of macromycetes in the past, a Red List still seems to be possible only in a distant future.

Analysis of Red Lists shows that many edible fungi can be considered to be decreasing and threatened in parts of Europe. A similar decline may be occurring in parts of North America and Asia as well, but data on long-term changes in these regions are not available. Harvesting of mushrooms may directly affect production of sporocarps in subsequent years, either by damaging or exhausting mycelia, shifting competitive relations with other species, or by causing reproductive failure due to decreased spore production. In addition, harvesting may indirectly affect productivity owing to trampling (compacting soils) or raking the forest, causing long-term damage to mushroom populations. However, data analysis is complicated by strong, weather-induced natural fluctuations in sporocarp production (Ohenoja, 1984; Egli *et al.*, 1990).

Strongly endangered groups are lignicolous fungi of large logs in old-growth forests and soil-inhabiting fungi of peat bogs, wetlands and low-productivity grasslands. Their decline is primarily attributed to changes in land use. Among forest fungi, ectomycorrhizal species are the most threatened group in densely populated or heavily industrialized regions. Various authors have reported a correlation between a decline of sporocarps of ectomycorrhizal fungi and ambient concentrations of air pollutants such as SO_2 and NH_2 (Fellner, 1989; Termorshuizen and Schaffers, 1991). In addition, mycelia of ectomycorrhizal species disappear soon after windthrow, clear-cutting, or burning of a forest (Ohenoja, 1988; Egli *et al.*, 1990). The associated tree species also play an important role; for example, *Tuber aestivum* grows with *Fagus* and *Quercus* and *Suillus luteus* and *Lactarius deliciosus* with two-needled *Pinus* species, whereas the wood-inhabiting *Kuehneromyces mutabilis* and *Flammulina velutipes* are only found on wood of deciduous trees. Similarly the age of trees exerts a major influence as certain species such as *S. luteus* fruit with young trees, while others like *Cantharellus* spp. are mainly found with mature trees. Furthermore, most macromycetes are restricted to certain soil types: chanterelles are mainly found on acid, sandy soils with a low nutrient status and thin litter layer, whereas trufflers prefer shallow, basic, mineral soils above calcareous bedrock. On the other hand, some saprotrophic species (e.g. *Lepista nuda*, *Macrolepiota rachodes*) thrive in nutrient rich, deep litter layers (Arnolds, 1995). Consequently, while creating appropriate environmental conditions can enhance the productivity of some edible mushrooms, there is no ideal management strategy which will stimulate the yield of all edible species.

In conclusion, the ecological data in floras and other taxonomic works on fungi are generally very superficial. There is a great need for accurate information on the ranges of habitat types, substrates and host specificity of species in different parts of Europe, preferably in a semi-quantitative form, and by monitoring in permanent plots. This information may be gathered by mycocoenological studies as well as during annotated floristic and geographical studies. Special attention should be paid to experimental studies on the impact of various anthropogenic factors on the mycoflora.

Systematics of Edible Mushroom Species

Floristic, geographical and ecological studies are impossible without a thorough taxonomic and nomenclatural basis. On the other hand, geographical and ecological studies may provide valuable information on taxonomy. There is continued need for both classical revisions of many critical groups of fungi and for advanced research on the genetic basis of our species concepts. Increasing discrepancies in the application of species concepts by various groups of workers in different parts of Europe are a subject of concern.

Systematics is an area of fundamental importance to biological science and forms the basis of information exchange. Morphology of the fungal fruit-body is the traditional basis for classification in fungi, but macroscopic features alone are rarely adequate to support taxonomic conclusions, while microscopy and physiology provide more reliable characters. Mating compatibility studies, based on the biological species concept, are widely used among higher fungi and offer valuable data on speciation processes. Recently, isozyme and molecular phylogenetic techniques added a powerful tool for understanding species concepts and systematics in this group of organisms.

An indicative case where a multi-task taxonomic approach was required to solve the perplexing relationships among species was the mushroom genus *Pleurotus* ('oyster-mushroom'). Since an essential prerequisite for mushroom breeding and cultivation is the use of properly identified biological material, numerous studies were conducted in the past to elucidate taxonomy within the genus by employing morphological, physiological and compatibility criteria (Eger *et al.*, 1979; Hilber, 1982), without arriving at a common conclusion as regards the taxonomy of related taxa, e.g. *Pleurotus columbinus*, *Pleurotus ostreatus*, *Pleurotus pulmonarius* and *Pleurotus sapidus*. Controversies were attributed to inaccurate original assignments, to the plasticity and the consequent overlapping of many morphological features caused by environmental factors, and to the lack of sufficient and sound physiological characters.

In addition to traditional taxonomic criteria, the results of matings between homokaryotic isolates were used, and therefore the biological species concept (Dobzhansky, 1970) has been applied to various groups of *Basidiomycotina* for species delimitation, for establishing systematic relationships among different taxa and for drawing more accurate conclusions about speciation processes (Boidin, 1986; Petersen, 1995; Zervakis and Balis, 1996), although its generalized use in fungi has proved to be problematic in certain cases (Brasier, 1987). The importance of adopting such a concept is obvious, as the ability of individuals to interbreed provides information on the type of reproductive system and the structure of mating factors reveals the degree of gene flow between populations and the existence of genetic and/or sterility barriers. Thus, studies on mating behaviour, apart from contributing to phylogeny, are valuable for confirming the validity of existing morphologically based taxa and for identifying reliable anatomical taxonomic characters. On the other hand, although the use of incompatibility criteria has provided valuable data, their full exploitation is yet to be accomplished, as it requires a detailed knowledge of the geographical distribution and a thorough understanding of the speciation mechanisms in this particular fungal group.

However, since nowadays a variety of morphological, physiological, ecological, compatibility, biochemical and molecular criteria is available, handling of multiple taxonomic characters and their final synthesis depends on the research area and the problem. Since the characters examined exhibit variation appropriate to the question posed, they have a clear and independent genetic basis, and the data are collected and analysed in such a way that it is possible to compare and combine phylogenetic hypotheses derived from them (Moritz and Hillis, 1990). If the problem of systematics is the object of study, then integration of the analyses of several different types of characters is strongly recommended (Kohn, 1992). For example, the relationship between molecular characters and other types of characters (particularly morphological characters) can produce an interesting dialectic (Doyle, 1992).

Therefore, *Pleurotus* systematics were recently addressed by modern biochemical (Zervakis and Labarère, 1992; Zervakis *et al.*, 1994) and molecular (Vilgalys and Sun, 1994; Iraçabal *et al.*, 1995; Vilgalys *et al.*, 1996; Zervakis *et al.*, 2001) approaches. Furthermore, numerical analysis and cladograms derived after the treatment of results from starch gel electrophoresis and polyacrylamide gel electrophoresis isoelectric focusing of dikaryotic and monokaryotic cultures, and restriction fragment length polymorphisms (RFLP) or sequencing of ribosomal DNA, contributed further to understanding *Pleurotus* phylogenetics and systematics in co-evaluation with traditional taxonomic criteria including morphology and mating tests.

Pleurotus *mushroom breeding and cultivation*

Well-understood taxonomic relationships and the application of modern molecular techniques are of great importance in today's successful mushroom breeding, for proper identification of biological material and strain authentication. However, the principal requirement for breeding is the availability of large amounts of variation in traits that can be selected for. Mushroom strains differ in traits and in gene combinations that can be generated by conventional methods of crosses, by protoplast fusion technology, and by transformation with genes cloned using recombinant DNA technology. As the organisms themselves are the only natural source of genetic diversity, the extinction of a species would mean a loss of the genetic information it contains, which amounts to a loss of a vast number of unique genes. For these reasons it is essential to establish a germplasm bank, through the collection of wild isolates, and characterization of available edible mushroom strains. In addition, it is also very important to study the ecological, morphological and physiological properties to determine mating types, substrate degradation efficacies, mycelial growth rates and cultivation parameters. Each strain's biochemical potential, for example in cellulose, hemicellulose, and lignin degradation, should be assessed for serving as criteria to identify high-quality strains suitable for breeding programmes. Furthermore, the development of DNA fingerprinting with the PCR and random amplified polymorphism DNA (RAPD) methodologies have contributed immensely to strain authentication and genetic manipulation.

Breeding has contributed much to the spread of the cultivation of *Pleurotus* species, whose world commercial production approaches annual rates of 1 Mt and is the third most cultivated mushroom (Chang, 1996). *Pleurotus* species are particularly known as efficient decomposers of a large range of lignocellulosic wastes and for producing edible basidiomata of high organoleptic qualities. These properties significantly favoured the increase in their commercial cultivation since growth substrates are cheap and abundant resulting from agricultural, forest and related industrial activities.

In the Mediterranean region, *Pleurotus* cultivation is relatively widespread and accounts for about 10–20% of the total mushroom production of the area, the rest being attributed to the button mushroom *Agaricus bisporus*. Its cultivation is relatively more popular in France and Italy, probably due to the mycophilic tradition of those countries and to the fact that these mushrooms also occur in nature and are among the favourites for collecting. In contrast, *Pleurotus* cultivation in Greece is still in its infancy and production is low (Philippoussis and Zervakis, 2000).

However, the trends indicate a significant increase in the future since the cultivation of *Pleurotus* species could be readily integrated within a sustainable agricultural production scheme. The entire cultivation process is by definition 'environmentally friendly' by exploiting and valorizing agroindustrial wastes of low economic value (which are usually incorporated into soils, burned or disposed of irrationally), some of which are potentially hazardous for the ecosystem. In addition, it requires relatively low inputs of energy (especially if the biological material used is carefully selected as regards adaptation to the local climatic conditions) and labour. The fact that *Pleurotus* mushrooms are widely accepted as a food of high organoleptic properties and maintain relatively high prices for the grower in the market is another important consideration. Furthermore, the spent substrate could be used for fodder through the enrichment of the initial waste with the fungal protein and at the same time by improving digestibility by preferentially consuming lignin and hemicellulose and leaving cellulose fairly intact as an energy source for ruminants (Tripothi and Yadar, 1992). The substrate could also be used as a plant fertilizer and soil conditioner beneficial to overall soil fertility and stability as well as vegetable yield (Levanon and Danai, 1995).

A special group of *Pleurotus* mushrooms appears in nature in close association with *Umbelliferae* plants, growing as weak parasites on their roots and stems. Species like *Pleurotus eryngii* and *Pleurotus nebrodensis* are typical constituents of the Mediterranean mycoflora and are particularly popular as choice edible mushrooms. Collection by local people is widespread and the prices they obtain in regional markets is very high compared with cultivated mushrooms (Venturella, 1995). Therefore, various cultivation methodologies have been recently developed in order to both safeguard the respective natural genetic resources and to provide an additional source of income for the indigenous farmers. Research is being conducted on the elucidation of the taxonomy within this group aiming at obtaining biological material to be readily exploited for breeding purposes (Zervakis and Venturella, 1998; Venturella *et al.*, 2001).

In Sicily, *P. eryngii* and *P. nebrodensis* strains were developed from selected local wild isolates (Venturella and Ferri, 1996). Their main features are the fast growth-rates of the mycelium and the trend to precociously differentiate basidiomata from a colonized substrate presenting a compact mycelium with uniform development. The growth substrate is prepared with wheat straw and residues of sugarbeet processing, pasteurized in thermoresistant polypropylene sacks with a capacity of 4 kg, inoculated with spawn and then incubated at 25°C. At the end of the incubation period the sacks are placed in rows in ridges dug in the ground (dimensions: 3 m × 1 m × 0.2 m). After opening, the sacks are covered with 2–3 cm of soil, and the beds are covered with a black shading net fitted on metal arches. Periodically beds are lightly irrigated to maintain the right degree of humidity. These cultivation tests aim to examine the suitability of certain mountainous and plain areas (i.e. situated at different altitudes) located *ex situ* with respect to the natural sites of occurrence. The tests are also expected to show the production efficiency and the quality properties of the different strains selected, as well as other cultivation parameters related to growing mushrooms outdoors. The tests are also to assess the suitable environmental conditions (depending on the season and the strain characteristics) required in order to reach yield values that will permit large-scale cultivation of *P. nebrodensis* and *P. eryngii*.

Similarly, in the southern Peloponnese (Greece), a project aiming to valorize locally abundant agricultural wastes (olive press-cake, groundnut shells, wheat straw, corn cobs, etc.) through the cultivation of *Pleurotus* mushrooms is currently under way. Special emphasis is placed on the exploitation of the cultivation potential of indigenous species like *P. ostreatus*, *P. pulmonarius* and *P. eryngii*. A well-insulated production chamber has been constructed with a total covered area of 350 m^2, equipped with up-to-date environmental control units and software.

Cultivation methodologies include substrate chopping, mixing and wetting, pasteurization in bulk at 60°C, inoculation with spawn developed from local wild strains, incubation at 28°C, and mushroom production at a temperature range of 16–22°C (depending on the strain used) at various levels of aeration, illumination and relative air humidity in order to assess the optimal environmental parameters that will permit high yield and fine product quality. Of paramount importance is the development of strains and cultivation techniques suitable for exploiting the climatic conditions of the area and the production potential of the biological material used.

Conclusions

The use of mushrooms as human food is increasing day by day. In particular, the production of *Pleurotus* could be further augmented to support the rising demand for the product. The first and probably most important aspect is to assess the genetic heterogeneity within the populations of various *Pleurotus* species. In fact species biodiversity is extremely important in nature and in crop species since it is the pool of genetic diversity which permits the species to react to environmental change.

As is well known, *ex situ* conservation actions include the establishment of genebanks, which should be further supported by adequate financial resources for infrastructure and running expenses, and the existence of specialized personnel needed for research and management activities. Despite the obvious economic importance of such measures for the much neglected group of microorganisms, mushroom breeding and cultivation (through its notable socioeconomic importance) could contribute in a very different and interesting way as an alternative means of promoting *ex situ* conservation of valuable biological material. In addition, promotion of integration between *ex situ* and *in situ* protection could be accomplished by proper diffusion of relevant information to the public, financial support for conducting inventory studies and for discovering the largely unknown wealth of the Mediterranean ecosystems, as well as the promotion of collaboration among institutions and organizations within this important region.

Acknowledgement

Financial support by the Assessorato Agricoltura e Foreste della Regione Siciliana, Programma Operativo Plurifondo 1994/1999 – Misura 10.4 – Ricerca Applicata, indagini e sperimentazione di interesse regionale (Progetto FUNGIS) is gratefully acknowledged.

References

Arnolds, E. (1988) The changing macromycete flora in the Netherlands. *Transactions of the British Mycological Society* 90, 391–406.

Arnolds, E. (1995) Conservation and management of natural populations of edible fungi. *Canadian Journal of Botany* 73 (Suppl. 1), S987–S998.

Boidin, J. (1986) Intercompatibility and the species concept in the saprobic Basidiomycotina. *Mycotaxon* 26, 319–336.

Brasier, C.M. (1987) The dynamics of fungal speciation. In: Rayner, A.D.M., Brasier, C.M. and Moore, D. (eds) *Evolutionary Biology of the Fungi*. Cambridge University Press, Cambridge, UK, pp. 231–238.

Chang, S.T. (1996) Mushroom research and development – equality and mutual benefit. In: Royse, D.J. (ed) *Mushroom Biology and Mushroom Products*. Pennsylvania State University, University Park, pp. 1–10.

Dobzhansky, T. (1970) *Genetics of the Evolutionary Process*. Columbia University Press, New York.

Doyle, J.J. (1992) Gene trees and species trees: molecular systematics as one-character taxonomy. *Systemic Botany* 17, 144–163.

Eger, G., Li, S.F. and Leal-Lara, H. (1979) Contribution to the discussion on the species concept in the *Pleurotus ostreatus* complex. *Mycologia* 71, 577–588.

Egli, S., Ayer, F. and Chatelain, F. (1990) Die Einfluss des Pilzsammelns auf die Pilzflora. *Mycol. Helv.* 3, 417–428.

Eyzaguirre, P.B., Padulosi, S. and Hodgkin, T. (1999) IPGRI's strategy for neglected and underutilized species and the human dimension of agrobiodiversity. In: Padulosi, S. (ed) *Priority-setting for Underutilized and Neglected Plant Species of the Mediterranean Region*. IPGRI, Rome, Italy, pp. 1–19.

Fellner, R. (1989) Mycorrhiza-forming fungi as bioindicators of air pollution. *Agriculture, Ecosystems and Environment* 28, 115–120.

Hawksworth, D.L. (1991) The fungal dimension of biodiversity: magnitude, significance, and conservation. *Mycological Research* 95, 641–655.

Hilber, O. (1982) Die Gattung *Pleurotus* (Fr.) Kummer unter besonderer Berücksichtigung des *Pleurotus eryngii*-Formenkomplexes. *Bibliotheca Mycologica* 87. J. Cramer, Vaduz.

Ing, B. (1993) Towards a red list of endangered European macrofungi. In: Pegler, D.N., Boddy, L., Ing, B. and Kirk, P.M. (eds) *Fungi of Europe: Investigation, Recording and Conservation.* The Royal Botanic Gardens, Kew, UK, pp. 231–237.

Iraçabal, B., Zervakis, G. and Labarère, J. (1995) Molecular systematics of the genus *Pleurotus*: analysis of restriction polymorphisms in ribosomal DNA. *Microbiology-UK* 141, 1479–1490.

Ivancevic, B. (1998) A preliminary red list of the macromycetes of Yugoslavia. In: Perini, C. (ed) *Conservation of Fungi in Europe, Proceedings of the 4th Meeting of the European Council for the Conservation of Fungi.* Vipiteno, Italy, pp. 57–61.

Kohn, L.M. (1992) Developing new characters for fungal systematics: An experimental approach for determining the rank of resolution. *Mycologia* 84, 139–153.

Krieglsteiner, G.J. (1991) *Verbreitungsatlas der Grosspilze Deutschlands (West).* Band 1, 2. Ulmer Verlag, Stuttgart, Germany.

Levanon, D. and Danai, O. (1995) Chemical, physical and microbiological considerations in recycling spent mushroom substrate. *Compost Science and Utilization* 3, 72–79.

Moritz, C. and Hillis, D.M. (1990) Molecular systematics: context and controversies. In: Hillis, D.M. and Moritz, C. (eds) *Molecular Systematics.* Sinauer Associates, Massachusetts, USA, pp. 1–10.

Nauta, M. and Vellinga, E.C. (1992) Towards a distribution atlas of macrofungi in the Netherlands. *Mycologist* 6, 6–10.

Ohenoja, E. (1984) Fruit body production of larger fungi in Finland. *Ann. Bot. Fenn.* 21, 349–355.

Ohenoja, E. (1988) Effect of forest management procedures on fungal fruit body production in Finland. *Acta Bot. Fenn.* 136, 81–84.

Perini, C., Barluzzi, C. and De Dominicis, V. (1993) Fungal communities in Mediterranean and submediterranean woodlands. In: Pegler, D.N., Boddy, L., Ing, B. and Kirk, P.M. (eds) *Fungi of Europe: Investigation, Recording and Conservation.* The Royal Botanic Gardens, Kew, UK, pp. 77–92.

Petersen, R.H. (1995) There's more to a mushroom than meets the eye: mating studies in the Agaricales. *Mycologia* 87, 1–17.

Philippoussis, A. and Zervakis, G. (2000) Cultivation of edible mushrooms in Greece: presentation of the current status and analysis of future trends. In: Van Griensven, L. (ed.) *Science and Cultivation of Edible Fungi.* Baalkema, The Netherlands, pp. 843–848.

Redeuilh, G. and Courtecuisse, R. (1999) Current and forthcoming computer developments around the French mycological inventory program. In: *Book of Abstracts of the XIII Congress of European Mycologists.* Alcala de Henares, Spain.

Termorshuizen, A. and Schaffers, A. (1991) The decline of carpophores of ectomycorrhizal fungi in stands of *Pinus sylvestris* L. in the Netherlands: possible causes. *Nova Hedwigia* 53, 267–289.

Tripothi, J.P. and Yadar, J.S. (1992) Optimization of solid substrate fermentation of wheat straw into animal feed by *Pleurotus ostreatus* – a pilot effort. *Anim. Feed Sci. Techn.* 37, 59–72.

Venturella, G. (1991) A check-list of Sicilian fungi. *Bocconea* 2, 5–221.

Venturella, G. (1995) I Pleuroti di potenziale interesse colturale in Sicilia. In: Fenech, L., Schimmenti, E., Tudisca, S. and Venturella, G. (eds) *La Fungicoltura in Sicilia.* Università degli Studi di Palermo, pp. 35–43.

Venturella G. and Ferri, F. (1996) Preliminary results of *ex situ* cultivation tests on *Pleurotus nebrodensis. Quad. Bot. Ambientale Appl.* 5 (1994), 61–65.

Venturella G., Perini, C., Barluzzi, C., Pacioni, G., Bernicchia, A., Padovan, F., Quadraccia, L. and Onofri, S. (1997) Towards a Red Data List of fungi in Italy. *Bocconea* 5, 867–872.

Venturella, G., Zervakis, G. and La Rocca, S. (2000) *Pleurotus eryngii* var. *elaeoselini* var. nov. from Sicily. *Mycotaxon* 76, 419–427.

Vilgalys, R. and Sun, B.L. (1994) Ancient and recent patterns of geographic speciation in the oyster mushroom *Pleurotus* revealed by phylogenetic analysis of ribosomal DNA sequences. *Proceedings of the National Academy of Sciences USA* 91, 4599–4603.

Vilgalys, R., Moncalvo, J.-M., Liou, S.-R. and Volovcek, M. (1996) Recent advances in molecular systematics of the genus *Pleurotus.* In: Royse, D.J. (ed.) *Mushroom Biology and Mushroom Products.* Pennsylvania State University, University Park, pp. 91–102.

Zervakis, G. and Balis, C. (1996) A pluralistic approach on the study of *Pleurotus* species, with emphasis on compatibility and physiology of the European morphotaxa. *Mycological Research* 100, 717–731.

Zervakis, G. and Labarère, J. (1992) Taxonomic relationships within the fungal genus *Pleurotus* as determined by isoelectric focusing analysis of enzyme patterns. *Journal of General Microbiology* 138, 635–645.

Zervakis, G. and Venturella, G. (1998) Towards the elucidation of the systematics of the *Pleurotus* taxa growing in Umbellifers. In: *Proceedings of the Sixth International Mycological Congress (abstract).* Jerusalem.

Zervakis, G., Sourdis, J. and Balis, C. (1994) Genetic variability and systematics of eleven *Pleurotus* species based on isozyme analysis. *Mycological Research* 98, 329–341.

Zervakis, G., Dimou, D. and Balis, C. (1998) A check-list of the Greek macrofungi, including hosts and biogeographic distribution: I. Basidiomycotina. *Mycotaxon* 66, 273–336.

Zervakis, G., Lizon, P., Dimou D. and Polemis, E. (1999) Annotated check-list of the Greek macrofungi: II. Ascomycotina. *Mycotaxon* 72, 487–506.

Zervakis, G., Venturella, G. and Papadopoulou, K. (2001) Genetic polymorphism and taxonomic relationships of the *Pleurotus eryngii* species-complex as resolved through the analysis of random amplified DNA patterns, isozyme profiles and ecomorphological characters. *Microbiology - UK* (in press).

33 Conservation and Use of Underutilized Crops: an Indian Perspective

V. Joshi,[1] P.L. Gautam,[1] Bhag Mal,[2] G.D. Sharma[1] and S. Kochhar[1]

[1]*National Bureau of Plant Genetic Resources, New Delhi, India;* [2]*International Plant Genetic Resources Institute (IPGRI), New Delhi, India*

Introduction

The dependence of humankind on plant resources is inevitable. Since the dawn of agriculture, domestication and gathering of desired plant species have helped in the evolution of useful plant species and these resources have been exploited to our advantage. So far, out of the estimated 75,000 species of edible plants (Gautam and Singh, 1998), only about 150 have been widely used. Of these about 30 species provide 90% of the world's food. Therefore, there has been focused attention by the researchers on exploiting alternative plant species or underutilized species for multifarious uses. Many of these occur in extreme environments and marginal or waste lands, and are being lost through rapid loss of natural habitats, especially in the tropics. The plants on which we depend in such adverse conditions/regions can be called the life support species (Paroda, 1988).

Underutilized plants constitute the lesser known species in terms of trade and research, often well adapted to marginal and stress conditions. Generally they possess promising nutritional and industrial importance for a variety of purposes for humankind. Their cultivation is restricted to specialized geographical pockets in different agroecological regions, mainly by poor farming communities which derive their sustenance and livelihood from such plants. But their commercial importance and market value is still unknown to the public. Therefore, research on underutilized crops holds promise to attain sustainability, profitability and diversification in agriculture and to restore the balance of trade, reduce India's dependence on imports and to make us more competitive in agricultural exports. This will make the global economy more sound and in many cases benefit the environment as well, since our depleting resources can be replaced with renewable ones.

The underutilized plant species of economic importance are the key to sustainable agriculture in most of the developing countries facing resource constraints as well as rapid depletion of natural resources due to ever-increasing population pressure. Poverty at both rural and urban levels leads to various health and nutritional problems respectively. This can be improved upon through wider cultivation and inclusion of underutilized crops in our food habit. Food and Agriculture Organization (FAO) statistics reveal (Swaminathan, 1999) that while about 800 million children, women and men are currently suffering from protein-calorie undernutrition, over 2 billion suffer from hidden hunger and there is a high frequency of low birth-weight children caused by the deficiency of micronutrients in the diet, particularly iron. Such micronutrients are in plenty in *Panicum miliaceum* (proso millet), *Paspalum scrobiculatum* (kodo millet), *Chenopodium* (chenopod), *Amaranthus* (amaranth), *Fagopyrum* (buckwheat) and so on. These underutilized plants

 Managing Plant Genetic Diversity
(eds J.M.M. Engels, V. Ramanatha Rao, A.H.D. Brown and M.T. Jackson)

can help to make diets more balanced and hence can play an important role in combating silent hunger (Swaminathan, 1999). These crops could also help in poverty alleviation by providing income generating opportunities to farmers by linking the development of these crops to market opportunities. But many of these crops are in danger of loss through competition resulting from the introduction of higher value commodities into the farming system.

Underutilized crop species have a potential that needs to be exploited to our advantage. Most of these crop species do not require high inputs, can be grown in marginal and degraded lands and at same time contribute to increased agricultural production, crop diversification and a better environment. In addition, such efforts will help in conserving and using the genetic resources of underutilized crop species. Favourable policy conditions lie at the core of alleviating poverty and sustainably managing natural resources. To achieve sustainable agriculture, one has to look at the immediate needs of farmers and the future needs of society, which requires a broad analysis of people and the environment.

Underutilized Plant Genetic Resources from Higher Plants

India, located between 8° and 38° N and 68° and 93.5° E, shows altitudinal variations ranging from below sea level to more than 3500 m above mean sea level and exhibits extreme diversity for edapho-climatic conditions, agroclimatic regions and ecosystems. The Indian sub-continent is a reservoir of several plant species, which have known uses of future potential for the benefit of human beings. India occupies a unique position among the major gene-rich countries of the world with a bounty of 49,000 species of higher plants known to occur. About 30% of these species are endemic: 17,500 species (Gautam and Singh, 1998) occurring in 16 major vegetation types of the country.

Several systematic efforts have been made to compile information on lesser known food plants including wild resources used by farmers and tribal communities in different regions of the country. Ethnobotanical investigations have been made to record a large number of wild plant species used by native tribal and aboriginal people, to meet their varied requirements (Jain and Sinha, 1987; MEF, 1994). Immensely rich landrace diversity occurs in major agri-horticultural crops, which is attributed to the farmers' conscious or unconscious selections, inherited and perpetuated over generations. Also, an enormous richness occurs in terms of crop relatives (326 species). Arora (1985) and Arora and Pandey (1996) observed that, out of a rich floristic wealth of higher plants in the country, over 1000 wild food plants are used and human association with these has been much older than with crops that emerged with settled agriculture. Over 50 of these species have been domesticated by native communities. Others hold promise as nutritional food supplements for future domestication and exploitation.

Concerted research efforts have been made in the recent past on domesticated and cultivated underutilized species, particularly for crop diversification for food and commerce (Bhag Mal, 1988, 1994; Paroda, 1988; Paroda and Bhag Mal, 1989, 1992; Wickens *et al.*, 1989; Bhag Mal and Joshi, 1991; Zhou, 1992). Several international agencies, such as the Overseas Development Agency (ODA), International Plant Genetic Resources Institute (IPGRI), United States Agency for International Development (USAID), and International Centre for Underutilized Crops (ICUC), also encouraged research on these underused species in order to broaden the range of plant species under cultivation. This has helped to raise concern and awareness for safe conservation and sustainable use of genetic resources of underutilized plant species.

However, given the resources, it is not possible to work on all the useful species and priorities need to be identified. The choice of crop species including their relative priority has to be clearly highlighted. In this context, it may be added that the underutilized food crop plants given research priority in India include pseudocereals and minor millets, minor grain legumes, fodder and energy plantation crops and some high value industrial plants (Table 33.1).

Coordinated Activities on Underutilized Crops: Indian Scenario

There are several underutilized and introduced plant species that have naturalized well in the country and hold great potential to be exploited for various purposes. For strengthening of research on underutilized plants, an All India Coordinated Research Project on Underutilized Plants (AICRPUUP) was initiated in 1984 with its headquarters at the National Bureau of

Table 33.1. Current status of germplasm collections in various underutilized crops in India (1999).

Crop	Active holdings	Holdings in the National Gene Bank	Areas showing rich diversity in India	Centres holding active germplasm collections
Pseudocereals				
Grain amaranths	3000	1952	Himalayan belt	SHM, AKL
Buckwheat	300	102	Central north	SHM
Chenopods	100	2	Gujarat Ghats	SHM
Job's tears	50	—	Northern India	SHL
Oil seeds				
Perilla	50	6	North-eastern states	SHL
Paradise tree	2	—	Orissa	AKL, MTU
Minor pulses				
Rice bean	500	983	North-eastern states	SHL, BHO, LDH
Faba bean	500	172	North-western plains	DLI, HSR
Adzuki bean	180	151	Western Himalayas	SHM
Winged bean	300	5	Deccan Plateau	AKL
Vegetables				
Kankoda	100	—	Deccan Plateau	RAH
Atriplex	50	—	Saline areas of Gujarat and Rajasthan	JOD
Other crops				
Jojoba	100	74	—	JOD
Guayule	50	—	—	JOD, HSR
Tumba	300	—	Rajasthan, Gujarat	JOD, MND
Casuarina	50	—	Southern states	MTU
Bamboo	50	—	Northern India	BSR

SHM, Shimla; AKL, Akola; MTU, Mettupalayam; JOD, Jodhpur; HSR, Hisar; MND, Mandor; BSR, Basar, Arunachal Pradesh; RAH, Rahuri; BHO, Bhowai; DLI, Delhi and LDH, Ludhiana.

Plant Genetic Resources (NBPGR), New Delhi. At present the project operates at 21 centres in different agro-climatic zones of India. Concerted research efforts have been made in this project in respect of germplasm build-up, maintenance and evaluation of some prioritized underutilized crop plants. Salient achievements in germplasm assembly and conservation and utilization, are briefly discussed in this chapter. The objectives of this programme are outlined below.

1. To find new plant resources for food, fodder and industrial uses.
2. To build up an extensive germplasm collection and to characterize, conserve and enhance them.
3. To identify/develop high yielding varieties of these crops for different farming situations.
4. To evolve an appropriate package of agronomic practices for their economic cultivation.
5. To disseminate knowledge about potential species for their popularization.

The underutilized crops identified in the Indian programme have been prioritized according to their suitability to the existing cropping systems or potential to supplement new systems. This project has identified priority plant species belonging to four major crop groups: food crops, fodder and energy plants and industrial plants.

Germplasm collection and assembly

Efforts through well-organized multi-crop explorations have resulted in sizeable germplasm collections over the years in a few selected species. Useful germplasm has also been introduced from other countries through NBPGR, which maintains links with over 70 countries. Thus, through exploration and introduction efforts, over 6000 accessions have so far been built up for selected underutilized crops from indigenous and exotic sources (Table 33.1). Information about the introductions is published in the Plant Introduction Reporters of NBPGR. Crop inventories for winged bean and jojoba have also been published.

Extensive collections were made of grain amaranth (*Amaranthus* sp.), and over 3000 accessions were assembled from Uttar Pradesh, Himachal Pradesh and North Eastern Hill (NEH) region, Gujarat and Maharashtra; 200 collections of buckwheat (*Fagopyrum* sp.) were also made from the same region. Over 500 collections of rice bean (*Vigna umbellata*) have been made from NEH region, Nepal and Sikkim. Forty-eight collections of *Jatropha curcas* were made from coastal Maharashtra and Malwa region of Madhya Pradesh. In winged bean (*Psophocarpus tetragonolobus*), some collections were made from the peninsular region of India.

The NBPGR took the lead role in collecting and characterization of most of the crops followed by a few individual centres for particular crops, namely Mandor for *Citrullus colocynthis* (tumba) (330), Rahuri for *Momordica dioica* (94), Mettupalayam for *Casuarina* (22) and Basar for bamboos (37 species).

There are some areas in India which still need to be surveyed for collecting germplasm of underutilized crop species. A new programme, National Agricultural Technology Project (NATP), has been initiated which is being operated through the Indian Council of Agricultural Research, Government of India. In this project, priority areas for various explorations and germplasm collection have been identified throughout the country. Under NATP, germplasm of rice bean, buckwheat, amaranths, broad bean (*Phaseolus multiflorus*), *Perilla* sp., *Buchanania lanzan*, *Aesendra butyracea*, *Lepidium sativum* and *Cleome viscosa* will be collected and conserved in our National Gene Bank (NGB).

Germplasm evaluation and documentation

Systematic evaluation and characterization of indigenous and introduced germplasm have been carried out in important underutilized crops which include pseudocereals and minor grain legumes, minor fruits, vegetables, oilseed crops and industrial crops. The pseudocereals have a high degree of acceptability in the poor farming households and sustainable agriculture although, presently, the area covered under these crops is meagre (Singh and Thomas, 1978; Bhag Mal, 1994). This group of underutilized crops includes amaranth, buckwheat, chenopod and minor millets. Rice bean, faba bean, adzuki bean, moth bean, *Lathyrus*, guar and horsegram constitute the category of minor grain legumes and are also grown for green fodder. Crop-wise research accomplishments are briefly described here.

Food crops

AMARANTH (*AMARANTHUS* SPECIES). Amaranth is a multipurpose crop with good potential for exploitation as grain, vegetable and fodder. Considerable variation has been observed for different nutritional constituents. The seed protein content has been reported to vary from 8.86% to 19.6% (Misra *et al.*, 1985; You *et al.*, 1987; Girenko *et al.*, 1988). Analysis of foliage of 61 accessions of *Amaranthus* species revealed the best combination of high carotene, appreciable level of protein and low nitrate and oxalate contents among some vegetable types as well as among a few grain types (Prakash and Pal, 1991). Its grains have high protein with high lysine and a good balance of other essential amino acids. Being an excellent source of iron and β-carotene, it can help in reducing iron and vitamin A deficiency, especially among rural populations. The presence of a higher amount of folic acid can help in increasing the blood haemoglobin. Amaranth is used as an important ingredient in food by people from the entire Himalayan region, and to some extent in the states of Gujarat, Maharashtra, Karnataka and eastern parts of Uttar Pradesh (UP). Amaranth can be utilized to manufacture products such as baby cereal foods, candies and snacks, protein drinks, and particularly hypoallergenic foods (Williams and Brenner, 1995), thus value-addition can make this crop more attractive. Grain amaranths are broad leaved, C4 plants with small seeds. The C4 photosynthetic pathway is particularly efficient at high temperature, in bright sunlight and dry conditions. Once the growth of the plant is well under way it can even withstand acute drought conditions. There are three species of cultivated amaranth, namely, *Amaranthus hypochondriacus*, *Amaranthus caudatus* and

Amaranthus cruentus, which are used as grain and leafy vegetables in the Himalayas. These species are sensitive to day length and are known to be short-day species.

The evaluation of germplasm collections of grain *Amaranthus* over several years led to the identification of accessions that offer opportunities for its utilization and improvement. IC 42258–1 (Annapurna) originating from Pauri Garhwal, UP, has been released for cultivation by the Central Varietal Release Committee (Joshi and Rana, 1991). Two varieties, GA-1 from Gujarat and Suvarna from Bangalore have also been released. A catalogue on 800 accessions evaluated for 29 descriptors has been published (Joshi, 1981). The highest content of protein (15.3%), sugar (6.4%) and iron (10.66 ppm) was observed in Suvarna variety.

BUCKWHEAT (*FAGOPYRUM* SP.). Buckwheat is the most important crop of the mountain regions above 1600 m a.s.l. both for grain and greens (Singh and Thomas, 1978). It occupies about 90% of the cultivated land in the higher Himalayas with solid stands. It is a short duration crop (2–3 months) and fits well in the high Himalayas where the growing season is short. In the higher Himalayas, up to 4500 m, this is the only crop grown (Joshi and Paroda, 1991). The tender shoots are used as a leafy vegetable, the flowers and green leaves are used for extraction of rutin used in medicine (McGregor and McKilllican, 1952), the flower produces honey of good quality. A total of 577 accessions have been collected including *Fagopyrum esculentum* (284), *Fagopyrum emarginatum* (55), *Fagopyrum tataricum* (198), *Fagopyrum tataricum* var. *himalianum* (30), *Fagopyrum giganteum* (5) and *Fagopyrum cymosum* (5) in 27 multi-crop explorations covering diverse agroecological regions of the Himalayas. A set of 408 accessions has been evaluated for 31 descriptors and a catalogue has been published (Joshi and Paroda, 1991). A wide range of variation was found for several traits, which can be useful in improvement of this crop. On the basis of multi-location testing at Shimla, Almora and Ranichauri, the varieties Himpriya and VL-7 have been released for cultivation (Table 33.2).

Table 33.2. List of improved varieties released in different underutilized plants in India.

Crop/variety	Year	Economic product	Yield (t ha^{-1})	Recommended areas/regions
Amaranth				
Annapurna	1986	Grain	2.25	Northwest hills
GA-1	1991	Grain	2.50	Gujarat, Maharashtra
Suvarna	1994	Grain	1.95	Karnataka State
Buckwheat				
Himpriya	1991	Grain	1.50	High-altitude region
VL-7	1992	Grain	1.00	Mid-hills of UP
PRB-1	1997	Grain	2.00	Hills
Winged bean				
AKWB-1	1991	Green pods	10.50	All winged bean areas
Rice bean				
RBL-1	1987	Grain	1.50	Punjab State
RBL-6	1991	Grain	1.80	NW and NE regions
PRR-2	1997	Grain	1.50	North-west hills
Faba bean				
VH 82-1	1994	Grain	4.20	Northern plains
Guayule				
Arizona-2	1986	Rubber	1.35	Arid and semiarid regions
HG-8	1991	Rubber	1.50	Arid and semiarid regions
Jojoba				
EC 33198	1986	Oil	0.50	Arid regions and coastal areas

CHENOPOD (*CHENOPODIUM* SP.). Chenopod is an important vegetable and grain crop for the hill region and is generally consumed mixed with other cereals. The grain is sometimes used as a staple food, consumed in the form of porridge, pudding, and also cooked with rice. Four species, namely *Chenopodium album, Chenopodium quinoa, Chenopodium nutaliae* and *Chenopodium pallidicaule*, are known to be cultivated. *C. album* is the most widely distributed and is grown in the Himalayan region. The Himalayan grain chenopod is comparable to Andean quinova in nutrient composition and is much better than wheat, barley, maize and rice (Tapia *et al.*, 1979; Pratap, 1982; Cusack, 1984; Wood, 1985). Grain protein quality equals that of milk and contains high lysine (6 g 100 g^{-1} protein), methionine (2.3 g 100 g^{-1} protein) and cysteine (1.2 g 100 g^{-1} protein). The grains are also used in local alcoholic preparations (Pratap and Kapoor, 1985). Out of the four domesticated species, *C. album* is the most widely distributed and is grown in the Himalayan region. The crop is also suited to the mixed farming system, particularly the multiple cropping system (Pratap and Kapoor, 1987).

Germplasm comprising 84 accessions of *C. album* was evaluated for 12 descriptors at NBPGR Regional Station, Shimla, and a wide range of variation has been observed for different agromorphological traits. Analysis of foliage of ten species of *Chenopodium* at the National Botanical Research Institute (NBRI), Lucknow, revealed a wide range of variation for protein (26–64 g kg^{-1}), carotene (78–190 mg kg^{-1}), vitamin C (0.5–2.4 g kg^{-1}), nitrate (2.6–5.0 g kg^{-1}) and oxalate (9–39 g kg^{-1}).

RICE BEAN (*VIGNA UMBELLATA*). Rice bean is a promising multipurpose legume crop with good potential as food, fodder, green manure and a cover crop. The dried seeds are usually cooked and eaten with or without rice, young immature pods and leaves are used as a vegetable. The seeds contain a high amount of protein (20.9%) and limiting amino acids, tryptophan (0.79–1.10%) and methionine (0.45–1.18%), which rank it as one of the best among pulses (NAS, 1979). Though it is suited for the lowland humid tropics, some of the cultivars are also adapted to subtropical conditions in the plains. It does well in sandy loam to heavy soils. It is also reported to be moderately drought tolerant (Duke, 1981). In India, its distribution is mainly confined to the tribal regions of the hilly areas of north-eastern hills and the Western and Eastern Ghats (Arora *et al.*, 1980; Chandel *et al.*, 1984).

Field evaluation of 530 accessions of rice bean for 36 descriptors at NBPGR, New Delhi, revealed a wide range of variation for a number of traits and considerable variation existed between collections from different geographical areas (Chandel *et al.*, 1988). Analysis of the biochemical constituents of rice bean seeds (15–16) revealed variation for crude protein (17.8–25.2%), ash (3.8–4.1%), calcium (315–450 mg 100 g^{-1}), phosphorus (197–393 mg 100 g^{-1}), iron (1–5 mg 100 g^{-1}). RBL-1 has been released for cultivation.

ADZUKI BEAN (*VIGNA ANGULARIS*). Adzuki bean is a legume crop and has a wide variety of uses. The dried seeds are used for human food, either cooked whole or made into a meal used in soups, cakes or confections. In Japan, it is used largely as human food in the form of meal or paste. In India, it is used as a pulse, either whole or split (Thomas *et al.*, 1974). Sprouted beans are used as a vegetable. The crop is also reported to be grown for forage and also as green manure in China and Japan. The seeds and leaves have medicinal properties (Sacks, 1977). The seeds contain 19.9 g protein per 100 g of seed (Duke, 1981). Adzuki bean is a short-day plant and requires almost the same climatic conditions as soybeans. It can be grown on all types of soil from light to heavy clay but does not grow well on extremely acidic soil. The crop is more tolerant to heavy rainfall than other grain legumes. It is also reported to be grown as a rainfed crop (Thomas *et al.*, 1974).

Thirty-one accessions were evaluated at NBPGR Regional Station, Shimla, and a wide range of variation was observed for several important traits. The observation row trial comprising 13 promising lines was conducted at three locations, Shimla, Palampur and Ranichauri, which revealed HPU-1 to be an early maturing (110 days) accession.

FABA BEAN (*VICIA FABA*). Faba bean has been one of the main sources of protein for people in the Middle East and North Africa since ancient times. In India, it is reported to be under cultivation as a minor crop in the Himalayan hills, Bihar, eastern Uttar Pradesh, Punjab, Haryana, Jammu and

Kashmir. Today, faba bean is becoming important as a source of protein (26.2%). The nutritive value of the crop is quite high and is regarded as a substitute for meat or skimmed milk. The green pods are used as a vegetable, and the seed is used dried, fresh or canned. It prefers a cooler climate and is grown as a winter annual in warm temperate and subtropical areas. It grows best on rich loamy soils and is tolerant to acidic soils. It is unable to withstand drought. It tolerates annual rainfall of 23–209 cm, annual mean temperature of 5.6–27.5°C and pH of 4.5–8.3 (Duke, 1981).

Evaluation of germplasm at Haryana Agricultural University, Hisar (Chhabra, 1983), indicated a wide range of variation for a number of important traits. Considerable genetic variation has also been reported for seed protein (Akbar *et al.*, 1992). Haryana Agricultural University, Hisar Centre, provided the lead function for faba bean breeding and VH 82-1 with seed yield of 420 kg ha^{-1} was released for cultivation in the Northern Plain Region.

WINGED BEAN (*PSOPHOCARPUS TETRAGONOLOBUS*). Winged bean is rich in protein (29.8–37.4% in seeds and 10.9% in tubers) and oil, and has potential as a multipurpose crop. Its grains and roots are edible and the plant is used as fodder, green manure and cover crop. The immature pods, leaves, young sprouts and flowers are consumed as a vegetable or in soups, seed oil is used for cooking and the oil cake as animal feed. Protein in seeds is comparable with that of soybeans in digestibility. Seeds are rich in the antioxidant, tocopherol, which improves human utilization of vitamin A, often deficient in the tropics. Seeds are reported to contain trypsin and chymotrypsin inhibitors. It is largely cultivated as a backyard or a garden crop in most of South-east Asia and is consumed locally. In India, its cultivation is confined to humid, subtropical parts of the north-eastern region, Bengal, Bihar, Western Ghats. The crop can be grown in a range of soil types and is found typically on well drained acid soils (pH 4.3–7.5). It is reported to tolerate annual precipitation of 70–410 cm and annual mean temperature of 15.4–27.5°C (Duke, 1981).

Evaluation of 88 accessions for 25 descriptors revealed considerable variability for different yield attributes, and the information has been catalogued (Chandel *et al.*, 1984).

Industrial crops

The use of agricultural commodities as industrial raw materials will rise as non-renewable resources become increasingly scarce and expensive, as businesses modify their manufacturing systems to use renewable materials and minimize waste generation; and as consumers become more aware, emphasis on concerns for a better environment will grow. The ultimate success of a new product, thus, will depend on generating strong demand from consumers, who are becoming more aware of environmental and other concerns. Research and development is just one step in the process of introducing new products to the market place. An equally important step is developing an appropriate marketing strategy that moves a new product from the laboratory towards commercial success. A number of plant species have been identified for research (Paroda, 1979), including *J. curcas, C. colocynthis* and *Cuphea* spp.

PURGING NUT (*J. CURCAS*). This is an industrial oil yielding species that is adapted to marginal lands, for example sandy, clayey, gravelly and eroded soils, and produces a semi-drying oil that can be used in fuel mixtures, as an illuminant and for making soaps and candles. Fifteen accessions were evaluated at Gujarat Agricultural University, Sardar Krushinagar. A wide range of variation was observed for some important yield and related traits (217–409 cm plant height, 30–45 branches per plant, 10–25 clusters per plant, 69–817 g per 1000 seeds. An initial varietal trial conducted at Gujarat and Orissa revealed genotype Chhattrapati to be the highest seed yielder (263 kg ha^{-1}).

TUMBA (*CITRULLUS COLOCYNTHIS*). This perennial creeper grows on sandy undulated plains and sand dunes and is quite drought hardy. The seeds contain 30–34% pale yellow oil, which contains an alkaloid, a glucoside and a saponin. Dried pulp of unripe fruit, freed from rind, constitutes the drug 'colocynth'. The roots have purgative properties and are used in jaundice, rheumatism and urinary diseases. An advanced varietal trial of eight entries conducted at four locations revealed GP 59 as the highest fruit and seed yielder. Germplasm comprising 40 accessions of tumba were evaluated at Rajasthan Agricultural University, Mandor. Data recorded on different traits revealed a high degree of variation for several traits. Initial varietal trials conducted at five locations revealed seed yield to vary from 101.00 to 323.50 kg ha^{-1}, the maximum being exhibited by GP 161 and followed by GP 172 and GP 45.

CUPHEA SPP. These species adapted to a temperate climate are good sources of industrial oils. Germplasm evaluation of 30 accessions of different *Cuphea* species at Shimla revealed a wide range of variation for different traits. The range of variation for number of fruit (capsule) was 38.6–44.9, 1.7–2.4 cm for capsule length and 13.6–67.0 g for seed yield per plant.

Further work on all the above species is in progress.

Germplasm conservation

High priority should be accorded to the conservation and sustainable utilization of many neglected and threatened plants that the local rural populations inhabiting the extreme environments depend on. In view of large-scale over-exploitation of land use practices, ecosystems or parts of them are under threat such as in the humid tropical and temperate forests because of the fragility of these ecosystems. In diversity-rich areas conservation of such plant species assumes great priority for current and future use. Their conservation can be accomplished through: (i) the farming community (*in situ* conservation through the continued use of local cultivars); and (ii) conservation through the formal sector (*ex situ* conservation through genebanks). Domestication for cultivation may have to be initiated in areas where the climatic conditions are similar to the niche of wild species.

Valuable germplasm of underutilized plant species is maintained by the NBPGR, New Delhi, and its Regional Stations, located in different agro-climatic zones of the country. Also, a few selected centres of AICRPUUP have the responsibility for maintaining the germplasm of particular species. The current status of germplasm holdings is given in Table 33.1. The majority of the germplasm collections is presently maintained through periodic regeneration. Facilities for long-term storage exist at NBPGR and currently 177,913 accessions of different crops including 3447 underutilized crops are maintained in this facility.

Germplasm utilization

Varietal development

Variability found in the germplasm of a few selected native and introduced species has been successfully utilized in developing improved varieties through selection procedures. Based on multilocation testing over the years, promising varieties have been identified/released in grain amaranth, buckwheat, winged bean, rice bean, faba bean, guayule and jojoba. Details about the year of identification/release, yield potential and the recommended areas are given in Table 33.2.

Breeding efforts

Concerted breeding efforts have also been initiated in a few selected potential crops at specified research centres for developing better varieties. Hybridization programmes are currently underway for amaranth, rice bean and tumba (*C. colocynthis*), while mutation breeding programmes are being carried out in rice bean, winged bean and tumba. Extensive work has been undertaken by making 26 sets of 46 interspecific crosses involving watermelon, matira, kalingda and tumba, belonging to different species of *Citrullus*. Evaluation of F_2 and F_4 line progenies was undertaken in grain amaranth. Highest seed yield of 10.28 kg ha^{-1} was recorded in the cross PRA 9401 × IC38269 followed by 9.17 kg ha^{-1} in Annapurna × PRA8801 in the F_2 generation.

Linkages

There is a need to generate sufficient information and resource base on underutilized crops of high potential to achieve the goal of popularizing them as regular crops. This is expected to help to meet the ever-increasing demand for food and other agro-products as well as to sustain the ecological balance by maintaining species/varietal diversity of the cultivated crop species. It also helps in conserving the cultural heritage and sustainable farming systems in remote areas and on marginal lands since most of the new crops are linked to such subsistence farming systems in other regions in India. The research and development programmes should, therefore, be based on an integrated multidisciplinary crop improvement concept for obtaining desired results quickly.

The underutilized crop species offer exciting prospects for crop diversification, increasing agricultural productivity and meeting the needs of industry and the community. Increased emphasis on these crops will continue in the context of current farming systems research and will help to supplement agricultural production. However, concerted efforts on underused crops need the urgent attention of the planners and policy-makers, administrators, researchers, extension workers, farmers and industry. Effective and close linkages between the national programmes on underutilized crops within the Asia-Pacific region and also with national programmes in other regions of the world need to be established.

During the last meeting of the South Asian national coordinators on plant genetic resources (SAN-PGR) (1–3 September 1998) the prospects and priorities of research on underutilized crops were assessed. One of the recommendations that emerged from this meeting was that the existing informal crop networks on underutilized crops, for example, sesame, *Lathyrus*, buckwheat and safflower, should be developed into formal networks. Further, the following four new crops/crop groups were identified as priorities for establishing crop networks in the South Asia region: (i) minor millets – finger millet, kodo millet, foxtail millet, little millet; (ii) minor legumes – black gram, rice bean, lablab bean, horsegram; (iii) amaranth; and (iv) cucumber. This recommendation was based on certain agreed criteria, namely: (i) regional importance; (ii) potential to contribute to food and nutrition security; and (iii) not addressed by any of the CGIAR Centres. The national programmes and IPGRI should initiate appropriate action to establish the above-mentioned networks. The national programmes agreed to provide all relevant information (diversity, distribution, collection, maintenance and ongoing activities) on these crops to IPGRI so that an appropriate strategy for action could be developed.

Conclusion

It is imperative that agricultural production should be sustained/enhanced through the use of alternative plant species which are underutilized and under-exploited. The pressing need to increase agricultural production of commercial crops and the resulting strain on the environment is an enormous task even for the most resourceful countries. To be able to grow more underutilized crops, there is a need for appropriate policies to be in place, which should be part of the national policy for alleviating poverty and sustainably managing natural resources. Without appropriate policies, simply improving the agricultural technology and natural resources management will not translate into adequate food supply, improved nutrition and the conservation of natural resources. To achieve sustainable agriculture, one has to look at the immediate needs of farmers and the future needs of society, which requires a broad analysis of people and the environment. A major problem in promoting underutilized or under-exploited crops is the difficulty in obtaining sufficient quantity of quality seed or planting material. Therefore, a balanced seed industry has to be developed that takes care of this need. Further, improved agrotechnologies on these crops will allow us to diversify the agricultural sector, including agroindustry. Such diversification will enhance the development of trade and utilization of several under-exploited plants and their products. Therefore, underutilized plant species, commonly raised by the poor farming households under subsistence agriculture, demand a change in the strategy for their conservation for sustainable utilization.

As a part of a specific strategy for sustainable management of genetic resources of underutilized species, the following actions need to be taken.

1. Priorities on future exploration and collection based on gaps in areas and diversity required.
2. Inventory and database for information documentation and dissemination and to help better monitoring and management.
3. *Ex situ* conservation through long-term seed storage facility primarily at NBPGR/ICAR, including field genebanks, should be complemented with *in situ* and on-farm conservation of these valuable materials.
4. Breeding, agronomic and quality studies for the development/promotion of improved varieties of various underutilized crops.
5. Capacity building for taxonomic classification and genepool studies of new materials should be promoted. Adequate importance should be given to human resource development and training in the relevant areas.
6. Links should be developed with other areas such as product development and marketing to be able to exploit the potential of the species on which work is being done.

7. Development/promotion of underutilized crops should be able to respond to emerging PGR issues, such as access and benefit sharing, equity, agreements for exchange of material and, above all, transparency. Addressing the problems of conservation and sustainable use of underutilized crops for the present and future on a scientifically sound basis is a common concern for all of us and must be seen in a broader context, building upon synergies and shared experiences.

References

Akbar, M.A., Tomer, Y.S., Gupta, P.C., and Singh, N. (1992) *Faba Bean* (Vicia faba L.): *A Potential Feed and Food Crop*. Technical Bulletin, HAU Press, Hisar, India, p. 62.

Arora, R.K. (1985) Genetic resources of less known cultivated food plants. *NBPGR Science Monograph 10*.

Arora, R.K. and Anjula Pandey (1996) *Wild-edible Plants of India: Diversity, Conservation and Use*. NBPGR, New Delhi, India.

Arora, R.K., Chandel, K.P.S., Pant, K.C. and Joshi, B.S. (1980) Rice bean – a tribal pulse of eastern India. *Economic Botany* 34(3), 260–263.

Bhag Mal, (1988) Underutilized plants programme in India – concepts and future perspective. In: Paroda, R.S., Promila Kapoor, Arora, R.K. and Bhag Mal (eds) *Life Support Species: Diversity and Conservation. Proceedings of the CCS/ICAR International Workshop on Maintenance and Evaluation of Life Support Species in Asia and the Pacific Region, 4–7 April*. NBPGR, New Delhi, India, pp. 153–159.

Bhag Mal (1994) *Underutilized Grain Legumes and Pseudocereals – Their Potentials in Asia*, RAPA Publication 1994/ 14. FAO, Bangkok, Thailand.

Bhag Mal and Joshi, V. (1991) Underutilized plant resources. In: Paroda, R.S. and Arora, R.K. (eds) *Plant Genetic Resources – Conservation and Management*. Malhotra Publishing House, New Delhi, India, pp. 211–229.

Campbell, C.G. (1997) *Buckwheat*, Fagopyrum esculentum *Moench*. IPGRI, Rome, Italy.

Chandel, K.P.S., Pant, K.C. and Arora, R.K. (1984) Winged bean in India. *NBPGR Sciences Monograph* 8.

Chandel, K.P.S., Arora, R.K. and Pant, K.C. (1988) *NBPGR Science Monograph* 12. NBPGR, New Delhi, India.

Chhabra, A. (1983) Assessment of genetic variability and selection of parents for hybridization programme in broad bean. MSc thesis, Haryana Agricultural University, Hisar, India.

Cusack, D.F. (1984) Quinova: grain of the Incas. *Ecologist* 14, 21–31.

Duke, J.A. (1981) *Handbook of Legumes of World Economic Importance*. Plenum Press, New York, p. 345.

Gautam, P.L. and Singh, A.K. (1998) Agro-biodiversity and Intellectual Property Rights (IPR) related issues. *Indian Journal of Plant Genetic Resources* 11(2), 129–151.

Girenko, M.M., Borodkin, A.S. and Voskresenskaya, V.V. (1988) Variation in the main economically useful characters in amaranth. *Sbornik Nauchnykh Trudoy po Prikladnoi Botanike, Genetike i Selekisii* 118, 59–67.

Jain, S.K. and Sinha, B.K. (1987) Ethnobotanical aspects of life support species – some emergency and supplementary foods among aboriginals in India. In: Paroda, R.S., Promila Kapoor, Arora R.K. and Bhag Mal (eds) *Life Support Species: Diversity and Conservation. Proceedings of the CCS/ICAR International Workshop on Maintenance and Evaluation of Life Support Species in Asia and the Pacific Region, 4–7 April*. NBPGR, New Delhi, India, pp. 173–180.

Joshi, B.D. (1981) *Catalogue on Amaranth Germplasm*. NBPGR Regional Station, Shimla.

Joshi, B.D. and Paroda, R.S. (1991) Buckwheat in India. *Science Monograph* 2, NBPGR, Shimla, p.117.

Joshi, B.D. and Rana, R.S. (1991) Grain Amaranths: The Future Food Crop. *Science Monograph*. 3. NBPGR Shimla, p.152.

McGregor, W.G. and McKillican, M.E. (1952) Rutin content of varieties of buckwheat. *Sci. Agri.* 32, 48–51.

MEF (1994) *Ethnobiology in India – a Status Report*. All India Coordinated Research Project on Ethnobiology, Ministry of Environment and Forests, New Delhi, India.

Misra, P.S., Prakash, D., Pandey, R.M. and Pal, M. (1985) Protein and amino acid composition of grain amaranth seed. *Fitoterapia* 5, 318–320.

NAS (1979) *Tropical Legumes: Resources for the Future*. National Academy of Sciences, Washington, DC.

Paroda, R.S. (1979) Plant resources of Indian arid zone for industrial uses. In: Goodin, J.R. and Northington, D.K. (eds) *Arid Land Plant Resources*. Texas Technical University, Texas.

Paroda, R.S. (1988) The need for life support species: an Indian perspective. In: Paroda, R.S., Promila Kapoor, Arora, R.K. and Bhag Mal (eds) *Life Support Species: Diversity and Conservation. Proceedings of the CCS/ICAR International Workshop on Maintenance and Evaluation of Life Support Species in Asia and the Pacific Region, 4–7 April*. NBPGR, New Delhi, India, pp. 153–159.

Paroda, R.S. and Bhag Mal (1989) New plant sources for food and industry in India. In: Wickens, G.E., Haq, N. and Day, P. (eds) *New Crops for Food and Industry*. Chapman and Hall, London, pp. 135–149.
Paroda, R.S. and Bhag Mal (1992) Developing national programme for research on underutilized crops in India. *Proceedings of the International Crop Science Congress, Ames, Iowa, USA, 14–22 July.*
Paroda, R.S., Kapoor, P., Arora, R.K. and Bhag Mal (eds) (1988) *Life Support Species: Diversity and Conservation.* NBPGR, New Delhi, India.
Prakash, D. and Pal, M. (1991) Nutritional and antinutritional composition of vegetable and grain amaranth leaves. *J. Sci Food Agric.* 57, 573–583.
Pratap, T. (1982) Cultivated grain chenopods of Himachal Pradesh: distribution, variations and ethnobotany. PhD thesis, Dept of Biosciences, Himachal Pradesh University, Shimla, India.
Pratap, T. and Kapoor, P. (1985) The Himalayan grain chenopods I. Distribution and ethnobotany. *Agriculture Ecosystems and Environment* 14, 185–199.
Pratap, T. and Kapoor, P. (1987) The Himalayan grain chenopods III. An under-exploited food plant with promising potential. *Agriculture Ecosystems and Environment* 19, 71–79.
Sacks, F.M. (1977) A literature review of *Phaseolus angularis* – The Adzuki bean. *Economic Botany* 31, 9–15.
Singh, H.B. and Thomas, T.A. (1978) *Grain Amaranth, Buckwheat and Chenopods.* Indian Council of Agriculture Research, New Delhi, India.
Swaminathan, M.S. (1999) Enlarging the basis of food security: role of underutilized species. In: *Proceedings of the International Consultation organized by the Genetic Resources Policy Committee (GRPC) of the CGIAR, M.S. Swaminathan Research Foundation, Chennai, India, 17–19 February 1999*, p. 22.
Tapia, M., Gandarillas, H., Alandia, S., Cardozo, A., Mujica, A., Ortiz, R., Otazu, V., Rea, J., Salas, B. and Zanabria, E. (1979) *La Quinova Y La Kaniwa: Cultivos Andinos.* CIID and IICA, Bogota, Colombia.
Thomas, T.A., Patel, D.P. and Bhagat, N.R. (1974) Adzuki bean, a new promising pulse for the hills. *Indian Fmg.* 23(12), 29–30.
Wickens, G.E., Haq, N. and Day, P. (eds) (1989) *New Crops for Food and Industry*. Chapman and Hall, London.
Wilkes, G. (1984) Germplasm conservation towards the year 2000: Potential for new crops and enhancement of present crops. In: Yeatman, C., Kefton, D. and Wilkes, G. (eds) *Plant Genetic Resources: A Conservation Imperative*. Amer. Assoc. for the Advancement of Sciences, Washington, DC.
Williams, J.T. and Brenner, D. (1995) Grain Amaranths (*Amaranthus* species). In: *Underutilized Crops: Cereals and Pseudocereals.* Chapman and Hall (& ICUC), London, pp. 129–186.
Wood, R.T. (1985) Tale of the food survivor. Quinova. *East-West Journal* 63–67.
You, S.X., Sun, H.L., Chang, B.Y., Chen, Z.P. and Zuo, J.W. (1987) The nutritional composition of grain amaranth and its potential for utilization. *Acta Agronomica Sinica* 13(2), 151–156.
Zhou Ming-De (1992) *Less Utilized Crop Genetic Resources of East Asia.* IPGRI Office for East Asia, Beijing.

34 Underutilized Edible Plants from South Africa: a Perspective

T.V. Jacobs
Department of Botany, University of Transkei, UNITRA, Umtata, South Africa

Introduction

Recent discoveries of hominid fossils indicate that southern Africa may have been one of the cradles of humanity. Today the indigenous peoples of southern Africa are: the San (Bushmen), the Khoi (Hottentots) and the Bantu, the latter comprising the South Sotho, North Sotho, Tswana, Zulu, Cape Nguni, Swazi, Transvaal Ndebele, Tsonga/Tonga, Venda, Ambo, Herero and Mpukushu (Liengme, 1983). For thousands of years the hunter-gatherers of South Africa have survived on the bounty of grasslands and forests. Initially it was a matter of trial and error. Anything that was palatable was eaten. One can reasonably assume that in the process of finding food for survival, many lives may have been sacrificed. Later on, the wisdom of the ages was readily available and information on useful and harmful plants was handed down from generation to generation.

Africans, both rural and urban, enjoy meat. With escalating prices, rural populations are in dire straits and find it hard to make ends meet. As pointed out by Wehmeyer and Rose (1983), with the decreasing availability of animal protein, the rural population is forced to rely on protein from wild plants.

Traditional Use of Food Plants

Information on the use of food plants traditionally stayed with elderly women in rural society. This information was readily shared with friends and relatives. With the arrival of colonizers, some travellers such as Burchell (1822), Campbell (1822), Kay (1833), Ayliff (1858), Mayr (1906), Ashton (1939), Bryant (1939) and Chapman (1971) gleaned some of the information and recorded it. As indicated by Liengme (1983), since 1957 no less than 25 books and articles have appeared on the San food habits. Notable among them are Lee (1965), Maguire (1978), Marshall (1965), Silberbauer (1965) and Story (1959). A great deal of literature is available on Bantu foods, notably Quin (1959), Franz (1971) and Rose and Jacot Guillarmond (1974).

As indicated above, literature abounds on the traditional use of food plants. However, reports are rather old, mostly fragmentary, dealing with the food habits of one tribe or the other. Hence it was felt that a comprehensive survey, dealing with the traditional use of food plants among the rural populations of South Africa might yield some valuable information relevant to our times.

Gathering of information on food habits

During the current study, students in different parts of South Africa conducted interviews with elderly women in rural communities. Apart from oral interviews, some respondents also completed a questionnaire. Specimens collected have been identified and lodged in the Kei Herbarium, Umtata, South Africa. In some cases relevant information was obtained from available literature (Drummond, 1984; Pooley, 1993, 1998; Fabian and Germishuizen, 1997).

 Managing Plant Genetic Diversity
(eds J.M.M. Engels, V. Ramanatha Rao, A.H.D. Brown and M.T. Jackson)

Use of a questionnaire

In addition to 376 oral interviews conducted in various parts of South Africa, a group of 285 members of the rural population completed a questionnaire. Subjects were mostly women from rural communities. Both in the oral interview and in the questionnaire, the following information was collected:

- relishes or pot-herbs, or 'imifino' plants, used by the local population;
- part of the plant used;
- method of preparation;
- plants that yield edible roots, tubers or corms;
- plants that yield edible fruits.

The questionnaire was translated into local languages and responses were recorded in the presence of the interviewer. While the mean number of plant names recorded during oral interviews was seven, the mean number of plant names on the completed questionnaire was nine.

Humans and the Food Chain in Southern Africa

An examination of the food habits of the indigenous peoples of South Africa indicates that 101 plant species are being used as food. Leaves and stems of 65 plant species are cooked and eaten as relishes or pot-herbs (Table 34.1). In the preparation of pot-herbs, several plant species are cooked together to achieve the desired flavour. It is eaten with the traditional main course, maize meal. Roots, tubers and corms from 26 plant species (Table 34.2) are collected at the appropriate time of the year and eaten. Fleshy fruits and nuts mainly from ten trees are collected and eaten as food supplements.

Plants that produce edible fruits

Over 100 South African shrubs and trees yield edible fruits. However, most of them have very little commercial or nutritive value. They are mostly collected and eaten by children who roam the veld. Some edible fruits, which have nutritive or commercial importance, have been selected for consideration.

SCLEROCARYA BIRREA (A.RICH) HOST. SUBSP. *CAFFRA* (*ANACARDIACEAE*), MARULA. *Scleros* (Gk) hard, *karya* (Gk) walnut refers to the hard stone in the fruit. A medium sized deciduous, dioecious tree found in low altitudes along the Natal and Transvaal coast, Mozambique, Swaziland and tropical Africa (Pooley, 1993). A tree may produce up to 90,000 fruits (up to 40 mm in diameter) in one season. Fruits are rich in vitamin C. Nuts contain 25% protein, 50–60% oil, vitamins E and C, magnesium, potassium and calcium (Swart, 1982). Ripe fruits and nuts are eaten. A potent alcoholic beverage is obtained when sugar is added to the fruit pulp. A liqueur named Amarula is commercially produced.

ADANSONIA DIGITATA L. (*BOMBACACEAE*), BAOBAB. Baobab is a comparatively short, about 15 m in height, but grotesquely fat tree, 15–30 m in circumference, occurring in low altitudes in hot dry woodlands. It looks like a tree planted upside down. Around 10 cm long, the ovoid fruit contains many seeds which are embedded in a white powdery pulp which contains appreciable quantities of tartaric acid and potassium bitartate (Palgrave, 1977). Both the seed and the pulp have a tart flavour. A drink can be made from dried pulp mixed with water (Swart, 1982). Fresh leaves are cooked and eaten.

DOVYALIS CAFFRA (HOOK.F. & HARV.) HOOK.F. (*FLACOURTIACEAE*), KEI APPLE. A shrub or small tree up to 8 m, found in the dry regions of the Eastern Cape and Transvaal. The 50 mm long spines make it a good hedge plant. Plants are dioecious and the female tree fruits profusely, fruits being round and yellow (40 mm in diameter). The pulp is rich in vitamin C. Fresh fruits are eaten or the juice is extracted. It is excellent for the preparation of jam and jelly.

CARISSA MACROCARPA (ECKL.) A. DC (*APOCYANACEAE*), BIG NUM-NUM. A shrub or small tree up to 5 m found on the Natal and Transkei coast. Large forked spines make them excellent hedge plants. The red oval fruits up to 50 mm long are rich in vitamin C, calcium, magnesium and phosphorus (Pooley, 1993). The fruit has a delicious flavour and can be used to make jam or jelly.

BEQUAERTIODENDRON MAGALISMONTANUM (SONDER) HEINE & HEMSLEY (*SAPOTACEAE*). A small to medium sized tree up to 10 m in height, on rocky outcrops and hills. Fruits measure 1.5 to 2.5 cm, red and fleshy with a high vitamin C content. They have a sweet but astringent flavour and are good for making jam, jelly and wine (Palgrave, 1977).

MIMUSOPS CAFFRA E. MAY EX A. DC. (*SAPOTACEAE*), COASTAL RED MILKWOOD. A shrub or medium sized tree up to 15 m in height in the dune forests of Natal and the Eastern Cape. Fruit are red, oval, up to 20 × 15 mm, produced in profusion. The mealy pulp is agreeably sweet and starchy and is used in the production of jelly and an alcoholic beverage.

PHOENIX RECLINATA JACQ. (*ARECACEAE*), WILD DATE PALM. Erect or gracefully reclining palms up to 10 m in height that are found along watercourses in the Eastern Cape and Natal. The fruits, oval, up to 15 mm, orange-brown when ripe, appear in large hanging branches. The fruit tastes like cultivated date *Phoenix dactylifera*. 'Palm heart' and young leaves are eaten. An intoxicating drink (mjemane, ubusulu) is produced from the sap (Pooley, 1993).

CORDYLA AFRICANA LOUR. (*FABACEAE*), WILD MANGO. A large, deciduous tree up to 25 m in height, restricted to riverine forests in the eastern Transvaal and Natal. Unlike other legumes, pods of this tree resemble small mangoes, rather spherical, up to 8 × 6 cm, yellow when ripe, with seeds embedded in a fleshy pulp (Palgrave, 1977). Some consider it one of the best wild fruits. It contains about 76 mg of ascorbic acid per 100 g of pulp. Seeds are also edible (Swart, 1982).

PARINARI CURATELLIFOLIA PLANCHON EX BENTH. (*CHRYSOBALANACEAE*), MOBOLA PLUM. A large evergreen tree up to 13 m high, usually found on sandy soils in open deciduous woodlands. The oval fruit is round, 5 × 3.5 cm, russet-yellow and pitted. It has a pleasant tasting yellow flesh, which may be made into porridge, a non-alcoholic beverage or an intoxicating liquor. It can be dried and stored for a long period.

TYLOSEMA FASSOGLENSE (SCHWEINF.) TORRE & HILLC. (*FABACEAE*), GEMSBOK BEAN. A shrub with prostrate, trailing or climbing stems which arise from a large underground tuber with an average mass of about 14 kg, but known to reach 200 kg. It occurs naturally in the North-west, Northern and Gauteng provinces (Fabian and Germishuizen, 1997). Pods contain 2–6 flattened seeds up to 15 × 10 mm. Roasted beans are tasty, having a slight coffee flavour. They contain up to 36% protein and 46% oil. The tuber has a high moisture content and nutritive values and may be eaten raw or cooked (Swart, 1982).

Plant Species that Need Expansion and Improvement

Table 34.3 lists plants identified by this study as showing promise for domestication and large-scale cultivation.

Four of the 65 plants commonly used as potherbs are fit for domestication and large-scale cultivation: *Chenopodium album* L., *Physalis viscosa* L., *Solanum nodiflorum* Jacq. and *Urtica urens* L. According to Wehmeyer and Rose (1983), *C. album* has an energy value of 200 kJ 100 g^{-1} and the calcium content is 239 mg 100 g^{-1}. The energy value of *P. viscosa* is 226 kJ 100 g^{-1} and has a very high carbohydrate content, 72 g 100 g^{-1}. The energy value of *S. nodiflorum* is 219 kJ 100 g^{-1} and the calcium content is 213 mg 100 g^{-1}. Finally *U. urens* has an energy value of 232 kJ 100 g^{-1} and has a very high calcium content, 528 mg 100 g^{-1}.

The selected plants also possess appreciable quantities of magnesium, iron, sodium, potassium, copper, zinc and phosphorus, together with thiamine, riboflavin, nicotinic acid, vitamin C and carotene.

Two edible tubers have been singled out for large-scale cultivation: *Dioscorea cotinifolia* Kunth. and *Plectranthus esculentus* N.E.Br. *D. cotinifolia* (wild yam) produces several tubers around 90 × 50 mm. It is quite delicious, when boiled or roasted. *P. esculentus* (African potato, wild potato) produces 20–30 finger-like tubers, which taste like sweet potato. They were planted and stored in some rural areas in the past (Pooley, 1998).

Wild fruits that have potential for large-scale cultivation and commercial exploitation are *Cordyla africana* Lour. and *Sclerocarya birrea* (A. Rich.)

Table 34.1. Plants used as relishes or pot-herb, 'imifino'.

Plant name	Family	Local name (X, Xhosa; Z, Zulu)
Aloe boylei Bak.	*Aloeaceae*	Grass aloe, incothobe (Z)
A. cooperi Bak.	*Aloeaceae*	Cooper's aloe, isiphuthymane (Z)
A. maculata All.	*Aloeaceae*	Soap aloe, ingcelwane (X), icene (Z)
Amaranthus hybridus L.	*Amaranthaceae*	Pigweed, umtyutyu (X), umbhido (Z)
A. spinosus L.	*Amaranthaceae*	Spiny pigweed
Aneilema aequinoctiale Kunth.	*Commelinaceae*	Idangabane elikhulu (Z)
Aponogeton junceus Lehm.	*Aponogetonaceae*	Lijo-tsa-liho-hoana (SS)
Asclepias eminens Schltr.	*Asclepiadaceae*	Large turret flower
Asystesia gangetica (L.) Ander.	*Acanthaceae*	Isihobo (Z)
Bidens pilosa L.	*Asteraceae*	Black jack, umhlabangubo (X), uqadolo (Z)
Boerhavia diffusa L.	*Nyctaginaceae*	Spiderling
Carpobrotus dimidiatus L. Bol.	*Mesembryanthemaceae*	Dune vygie, umgongozi (Z)
Celosia trigyna L.	*Amaranthaceae*	Silver spinach
Centella asiatica (L.) Urb.	*Apiaceae*	Marsh pennywort, icudwane (Z)
Chenopodium album L.	*Chenopodiaceae*	Fat hen, white goose foot
Coccinia adoensis (Rich) Cogn.	*Cucurbitaceae*	
C. hirtella Cogn.	*Cucurbitaceae*	
C. palmata (Sond.) Cogn.	*Cucurbitaceae*	Wild cucumber, uthangazane Iwehlathi (Z)
Colocasia esculenta	*Araceae*	
Cleome gynandra L.	*Capparaceae*	Umzonde (Z)
C. monophylla L.	*Capparaceae*	Spindlepod, isiwisa esiluhlaza (Z)
Commelina benghalensis L.	*Commelinaceae*	Idlebendlele (Z)
Corchorus asplenifolius Burch.	*Tiliaceae*	Ubangalala, igusha (Z)
Cucumis hirsutus Sond.	*Cucurbitaceae*	Wild cucumber, uthangazane (Z)
C. zeyheri Sond.	*Cucurbitaceae*	Iselwa-lenja (Z)
Cyphia elata Harv.	*Lobeliaceae*	Ikhonti (X)
Gunnera perpensa L.	*Gunneraceae*	Wild rhubarb, uxobo (Z)
Heliophila rigidiscula Sond.	*Brassicaceae*	Grassland blue cross
Hibiscus calyphyllus Cav.	*Malvaceae*	Yellow wild hibiscus
Hypoestes aristata (Vahl) Solan	*Acanthaceae*	Ribbon bush, uhlonyana (Z)
Ipomoea alba L.	*Convolvulaceae*	Moonflower
I. crassipes Hook.	*Convolvulaceae*	Ubhoko, uvimbukhalo (Z)
I. obscura (L.) Ker-Gawl.	*Convolvulaceae*	Wild petunia, usiboniseleni (Z)
Lactuca indica L.	*Asteraceae*	Wild lettuce
Lobelia flaccida (Pers.) A.DC	*Lobeliaceae*	Ubulawu (X), isidala (Z)
Nemesia denticulata (Bent) For.	*Scrophulariaceae*	
Nymphaea nouchali Sav.	*Nymphaeaceae*	Blue waterlily, intekwane (X), izubu (Z)
Orbeopsis lutea (N.E.BR) Leach	*Asclepiadaceae*	Yellow carrion flower
Oxalis corniculata L.	*Oxalidaceae*	Yellow sorrel, isithate (Z)
O. latifolia H.B.K.	*Oxalidaceae*	Pink garden sorrel
O. smithiana Eckl.	*Oxalidaceae*	Narrow-leaved sorrel, izotho (X)
Papaver aculeatum Thunb.	*Papaveraceae*	Orange poppy, sehlohlo (SS)
Pentarrhinum insipidum E.May	*Asclepiadaceae*	African heartvine
Physalis viscosa L.	*Solanaceae*	Wild gooseberry
Portulaca oleracea L.	*Portulacaceae*	Purslane
Rumex crispes L.	*Polygonaceae*	Ubuklunga (X,Z)
R. lanceolatus Thunb.	*Polygonaceae*	Smooth dock, idolonyana (X) idolekeyana(Z)
Rubus pinnatus Wild.	*Rosaceae*	Igundube (X)
Raphanus nasturtio aquatica L.	*Brassicaceae*	Iwatane (X)
Sida dregei Burtt Davy	*Brassicaceae*	Spider leg, umdiza wethafa (Z)
Sida cordifolia L.	*Malvaceae*	Flannel weed

Table 34.1. *Continued*

Plant name	Family	Local name (X, Xhosa; Z, Zulu)
S. rhombifolia L.	*Malvaceae*	Uvevane (X,Z)
Solanum nodiflorum Jacq.	*Solanaceae*	Purple sobosbo berry
S. retroflexum Dun.	*Solanaceae*	Sobosbo berry, umsobo wesinja (X)
Sonchus oleraceus L.	*Asteraceae*	Inlaba (X)
S. nanus Sond. Ex Harv.	*Asteraceae*	Thistle, sethokajane (SS)
Sparrmania rcinocarpa Kuntze	*Tiliaceae*	Isibundane (Z)
Strelitzia nicolai Regel & Koer.	*Strelitziaceae*	Ikamanga (X)
Tridax procumbens L.	*Asteraceae*	Tridax
Typha capensis (Roxbr.) N.E.BR.	*Typhaceae*	Bulrush, umkhanzi (X), ibhuma (Z)
Vigna frutescens Harv.	*Fabaceae*	Wild sweetpea
V. luteola (Jacq.) Benth.	*Fabaceae*	Yellow wild sweetpea, isikhwali (Z)
V. unguiculata (L.) Walp.	*Fabaceae*	Wild cowpea, umcwasibe (Z)
Urtica urens L.	*Urticaceae*	Imbabazane (X)
Zantedeschia aethiopica (L.) Sg	*Araceae*	White arum lily, intebe (X,Z)

Table 34.2. Plants that yield edible roots, tubers and corms.

Plant name	Family	Local name (X, Xhosa; Z, Zulu)
Albuca canadensis (L.) Leigt.	*Aizoaceae*	
Aponogeton distachyos L.F.	*Aponogetonaceae*	
Argyrolobium tuberosum Eckl.	*Fabaceae*	Little russet pea, olubomvu (Z)
Asclepias aurea Schltr.	*Asclepiadaceae*	Golden star flower, umanquandra (Z)
A. gibba (E.May) Schltr.	*Asclepiadaceae*	Humped turret flower
Brachystelma circinatum RB	*Asclepiadaceae*	Bird cage brachystelma
B. foetidum Schltr.	*Asclepiadaceae*	Foetid brachystelma
B. gerrardi Harv.	*Asclepiadaceae*	
Coccinia rehmanii Cogn.	*Cucurbitaceae*	Wild cucumber, uselwa-lwenyoka (Z)
Colpoon compressum Berg.	*Santalaceae*	
Ceropegia woodii Schltr.	*Asclepiadaceae*	String of hearts
Cyphia elata Harv.	*Lobeliaceae*	Igonsi (Z)
Cyrtanthus breviflorus Harv.	*Amaryllidaceae*	Yellow fire lily, injobo (Z)
C. stenanthus Bak.	*Amaryllidaceae*	Ebomvu (Z)
C. tuckii Bak.	*Amaryllidaceae*	Isiwesa (Z)
Dioscorea cotinifolia Kunth.	*Dioscoreaceae*	Wild yam, umtane (X), intana (Z)
D. rupicola Kunth.	*Dioscoreaceae*	Inkwa (Z)
Gladiolus eckloni Lehm.	*Iridaceae*	Sheathland gladiolus
Gunnera perpensa L.	*Gunneraceae*	Wild rhubarb, uklenya (Z), uxobo (X,Z)
Hypoxis argentea Harv. Ex Bak.	*Hypoxidaceae*	Small yellow star flower, inongwe
Ipomoea albivenia (Lind.) Swt.	*Convolvulaceae*	Umgwiligwili (Z)
Phragmites australia (Cav.) Std.	*Poaceae*	
Plectranthus esculentus N.E.BR.	*Lamiaceae*	Wild potato, umhlaza (Z)
Rhyncosia totta (Thun.) DC	*Fabaceae*	Yellow carpet bean
Satyrium longicauda Lindl.	*Orchidaceae*	Blushing bride satyrium, unokleshe (Z)
S. macrophyllum Lindl.	*Orchidaceae*	Unokleshe (Z)

Table 34.3. Plant species that show promise for domestication and large-scale cultivation.

Plant name	Family	Reason for selection
Chenopodium album L.	*Chenopodiaceae*	High energy value
Cordyla africana Lour.	*Fabaceae*	High ascorbic acid content
Dioscorea cotinifolia Kunth.	*Dioscoreaceae*	Nutritious tubers
Physalis viscosa L.	*Solanaceae*	High carbohydrate content
Plectranthus esculentus N.E.BR.	*Lamiaceae*	Nutritious tubers
Sclerocarya birrea (A. Rich.) Host. subsp. *caffra*	*Anacardiaceae*	Fruit rich in vitamin C, alcoholic beverage
Solanum nodiflorum Jacq.	*Solanaceae*	High energy value
Urtica urens L.	*Urticaceae*	Very high calcium content

Host. subsp. *caffra. C. africana* (wild mango) grows in the sand forests in Maputaland, Swaziland, eastern Transvaal and Natal. It is also found in Mozambique and Zimbabwe. The pulp of the fruit is rich in vitamin C and may be eaten fresh or cooked. Seeds, up to eight, are also edible. *S. birrea* should be considered South Africa's number one wild fruit. Apart from the highly nutritious fruit pulp and nut, which are enjoyed by both humans and animals, ripe fruits have great potential for commercial exploitation as an alcoholic beverage and liqueur.

Conclusion

The South African population is growing rapidly. Prices of commodities, including those of staple food materials are steadily on the rise. More and more people are losing jobs and find it difficult to maintain their families on a healthy diet. At this juncture, rural people have to return to the practice of collecting wild herbs, tubers and edible fruits, which grow abundantly in the veld. Although they are seasonal, the leaves of some plants such as *S. nodiflorum* and *S. retroflexum* can be dried and preserved for use during the winter months (Wehmeyer and Rose, 1983). Every housewife should possess information on edible herbs, roots and fruits. Moreover, she should know how to dry and preserve them for future use. Malnutrition is an ugly spectre in many parts of Africa. By encouraging the use of veld foods and also by promoting further work and awareness of the use of such nutritious material, scientists can help in alleviating some of the food and nutrition problems in southern Africa.

References

Ashton, E. (1939) A sociological sketch of Sotho diet. *Transactions of the Royal Society of South Africa* 27, 147–214.

Ayliff, J. (1858) *Kafir Laws and Customs*. Wesleyan Mission Press, Mt Coke, South Africa.

Bryant, A.T. (1939) *A Description of Native Foodstuffs and their Preparation*. National Food Research Institute, Pretoria, South Africa.

Burchell, W. (1822) *Travels in the Interior of Southern Africa*. Longmans, London.

Campbell, J. (1822) *Travels in South Africa, Second Journey*. F. Westley, London.

Chapman, J. (1971) *Travels in the Interior of Africa, 1849–1863*. Balkema, Cape Town, South Africa.

Drummond, R.B. (1984) *Arable Weeds of Zimbabwe*. Agricultural Research Trust of Zimbabwe, Harare, Zimbabwe, Africa.

Fabian, A. and Germishuizen, G. (1997) *Wild Flowers of the Northern South Africa*. Fernwood Press, Vlaeberg, South Africa.

Franz, H.C. (1971) Traditional diet of the Bantu of the Pietersburg District. *South African Medical Journal* 45, 1323–1325.

Kay, L. (1833) *Travels and Researches in Caffraria*. John Mason, London.

Lee, R.B. (1965) The subsistence ecology of the Kung Bushmen. PhD thesis, University of California, Berkeley, California.

Liengme, C.A. (1983) A survey of ethnobotanical research in South Africa. *Bohalia* 14, 621–629.

Maguire, B. (1978) The food plants of the Khu Bushmen of the north-eastern South West Africa. MSc thesis, University of the Witwatersrand, South Africa.

Marshall, L. (1965) The Kung Bushmen of the Kalahari desert. In: Gibbs, J. (ed.) *Peoples of Africa*. Holt, Rinehart and Winston, New York, pp. 24–78.

Mayr, F. (1906) The Zulu Kafirs of Natal. *Anthropos* 1, 453–457.

Palgrave, K.C. (1977) *Trees of Southern Africa*. Struik Publishers, Cape Town, South Africa.

Pooley, E. (1988) *A Field Guide to Wild Flowers of Kwazulu Natal and the Eastern Region*. National Flora Publications Trust, Durban, South Africa.

Pooley, E. (1993) *The Complete Field Guide to the Trees of Natal, Zululand and Transkei*. Natal Flora Publications Trust, Durban, South Africa.

Quin, P.J. (1959) *Food and Feeding Habits of the Pedi*. Witwatersrand University Press, Johannesburg, South Africa.

Rose, E.F. and Jacot Guillarmond, A. (1974) Plants gathered as foodstuffs by the Transkeian peoples. *South African Medical Journal* 48, 1688–1690.

Silberbauer, G.B. (1965) *Bushmen Survey Report*. Bechunaland Government, Gabarone, Botswana.

Story, R. (1959) Some plants used by the Bushmen in obtaining food and water. *Memoirs of the Botanical Survey of South Africa*, 30. National Botanical Institute, Pretoria.

Swart, W.J. (1982) Marula food and drink. *Farmer's Weekly* 82, 35–37.

Wehmeyer, A.H. and Rose, E. (1983) Important indigenous plants used in the Transkei as food supplements. *Bothalia* 14, 613–615.

35 'Mining the Gold': Finding Allelic Variants for Improved Crop Conservation and Use

S. Kresovich, A.J. Luongo and S.J. Schloss
Institute for Genomic Diversity and Department of Plant Breeding, Cornell University, Ithaca, New York, USA

Current State of Plant Genetic Resources Conservation

As both a science and vocation, plant genetic resources conservation has achieved a higher global profile and made great progress over the past 50 years. A representative sampling of advances includes:

- improved understanding of the genepools of cultivated crops and their weedy and wild relatives;
- integration of advanced systematics to better understand species relationships and population structure;
- improved strategies to conserve genetic resources (*ex situ* and *in situ*);
- increased accessibility to both genetic resources and associated information;
- integration of plant pathology to provide 'cleaner' collections and safer global movement of genetic resources;
- integration of breeding and molecular mapping theories and strategies to discover and better understand the genetic basis of useful traits;
- integration of genomics to better discover and characterize both neutral and adaptive variation; and
- application of geographic information system technologies for managing information on global plant diversity.

These achievements set the stage for a future that should prove to be challenging and rewarding.

After generations of characterizing genes one or a few at a time, researchers now have access to the complete genome sequences of many bacterial species and several eukaryotes, including the completion of the first plant genomic sequence for *Arabidopsis thaliana*. The breadth of this intensive effort is expected to grow exponentially in the future. For example, there presently exists a global effort to understand the function of all plant genes by the year 2010. If this effort reaches any level of success, the effect on plant genetic resources conservation should be significant. As such, *ex situ* genebanks are likely to face new kinds of demands from the user community. In the future, users may want a series of genes with a specific function regardless of the sources of such genes. Other users may want to receive DNA sequences or markers instead of seeds or other traditional means of carrying DNA. They may want a series of specific alleles rather than accessions where alleles are segregating. A future-oriented analysis of these possible trends and their implications is very important in order to be able to predict, and thus prepare for, a changing role for genebanks and curators.

 Managing Plant Genetic Diversity
(eds J.M.M. Engels, V. Ramanatha Rao, A.H.D. Brown and M.T. Jackson)

Highlighting New Theories and Technologies

Realizing that genes are genes

Comparative genomics involves the study and identification of the molecular bases of differences and similarities between organisms at different taxonomic levels. Over the past decade, geneticists and plant breeders have made great progress in understanding genome organization and diversity as a result of putting their investigations in an evolutionary context (Gale and Devos, 1998). That is, taxa that are evolutionarily related (e.g. grasses, legumes and solanaceous crops) have strikingly similar genome organization. Moreover, at the fundamental level of the gene, many sequences are highly conserved across families. Therefore, users of genetic resources will acquire useful genes from repositories independent of their source.

Comparative genetics provides the potential for trait extrapolation from a species where the genetic control is well understood and for which there are molecular markers to a species that has a limited amount of information. For example, rice is regarded as a model for cereal genomics because of its small genome. The similarity of cereal genomes, in general, means that the genetic and physical maps of rice can be used as reference points for exploration of the much larger and more difficult genomes of the other major and minor cereal crops (Wilson *et al.*, 1999). Conversely, decades of breeding work and molecular analysis of maize, wheat and barley can now find direct application in rice improvement. Comparative genetics can also be used to locate desirable alleles in genepools close to the target crop so that transfer can be achieved by conventional methods. Once the research base has been developed for the legumes and solanaceous crops, the same applications are expected (SGRP, 1999).

Understanding the complexities of genetic diversity

In order to emphasize the importance of representative genetic resources collections and the characterization and utilization of them for studies involving genetic rather than phenotypic data, we present an example through which the genetic changes underlying the domestication of maize came to be understood. Maize was domesticated from the wild Mexican grass teosinte (*Zea mays* ssp. *parviglumis* or ssp. *mexicana*) between 5000 and 10,000 years ago (Beadle, 1939; Galinat, 1971, 1983; de Wet and Harlan, 1972; Iltis, 1983; Wang *et al.*, 1999), in only a few hundred years (Eyre-Walker *et al.*, 1998). The domestication of maize involved early seed gatherers and agriculturalists selecting seed from wild teosinte plants that had desirable characteristics and storing it for planting the following year. After generations of this selection process, maize landraces were created. Those teosinte plants lacking desirable traits were left to remain as wild relatives of maize. Two important variables of this domestication process were that the initial crop population was probably extremely small, and the selection pressure that was applied to the initial population by the early agriculturalists was very intense (Eyre-Walker *et al.*, 1998). This selection pressure was in the form of phenotypic selection for morphological and agronomic traits that made growing maize more productive and stable.

One possible result of intense selection pressure on a relatively small population over a short time period is vast phenotypic differences between the progenitor–descendent pair, but these differences can be due to small changes in just a few genes. Maize has between 20,000 and 30,000 genes and, of these, as few as five loci are responsible for the morphological and physiological differentiation between maize and teosinte (Beadle, 1939; Doebley *et al.*, 1990, 1995). Not all of these loci have been studied, but those that have been characterized have shed a great amount of light on the process of crop plant domestication, the molecular evolution of the genes, and the biochemical pathways involved.

Because the domestication of maize occurred quickly, it is somewhat surprising that there are significant morphological differences between maize and teosinte (Iltis, 1983). Some of the most obvious physical differences between this progenitor–descendent pair include number and size of lateral branches, locations and numbers of reproductive organs, and ear size and morphology. Teosinte plants have a main stalk as well as elongated lateral branches extending from most nodes. Both the main stalk and the lateral branches are tipped by male inflorescences called tassels, which distribute pollen to neighbouring plants. The small female inflorescences, or ears, are borne in clusters of one to five on short secondary branches located in nodes of leaves along the primary branches. The structure of the maize plant is dramatically different from that of teosinte. Maize has one main stalk, and the lateral branches are severely truncated to as little as 1 cm in length. These lateral branches, instead of appearing at most nodes, occur at only three or four nodes, and are tipped not by tassels but by greatly enlarged ears. The remaining nodes have axillary buds whose development is arrested.

In addition to these gross morphological differences between teosinte and maize, the ears differ in size, shape and structure. Teosinte has small ears with kernels that are encased in hardened stone-like fruitcases, which decrease the utility of the kernels as food. On the other hand, maize has greatly enlarged ears with kernels that are not encased. The physiological basis for this difference is arrested development of the fruitcase early in maize ear development. As a result, maize kernels are exposed on the ear and are easily accessible for consumption. These differences in plant and ear architecture between teosinte and maize reveal the unique evolutionary changes that must have taken place during the domestication process. Most of these evolutionary changes are the result of a modification of a single organ, such as the fruitcase or lateral branch (Doebley and Wang, 1997). These differences in developmental regulation provide evidence for the hypothesis that many of these gross phenotypic changes are controlled by a small number of genes. Therefore, the traditional method of observing phenotypes and classifying accessions based solely on phenotypic or geographic data may be misleading in the case of maize and teosinte because these two species are much more genetically similar than their phenotype would predict.

A simple function such as repression of organ development suggests a common developmental signal for each characteristic which, when turned on (or off, as in maize) at a particular time and place in development, can repress the growth of any or all organs of the plant. Doebley and colleagues have studied the genetics of this regulatory phenomenon in detail and have uncovered three major genes involved in the domestication process, *teosinte branched 1* (*tb1*) (Doebley *et al.*, 1995; Doebley and Wang, 1997; Wang *et al.*, 1999), *terminal ear 1* (*te1*) (White and Doebley, 1999), and *teosinte glume architecture 1* (*tga1*) (Dorweiler *et al.*, 1993). All three genes are hypothesized to be developmental regulatory factors, and the differential expression of each produces the variable phenotypes observed in the progenitor–descendent pair, teosinte and maize.

The *tb1* gene is a regulatory locus on maize chromosome 1 (Doebley *et al.*, 1995; Doebley and Wang, 1997; Wang *et al.*, 1999) that was first characterized as a quantitative trait locus (QTL) controlling lateral branch size and morphology between maize and teosinte. This gene, when expressed at high levels in lateral buds, represses the growth of lateral branches as a result of the accumulation of high levels of messenger RNA (mRNA). In contrast, teosinte has low accumulation of *tb1* mRNA, resulting in extended lateral branches. The differential rates of mRNA accumulation are a result of varying levels of *tb1* expression in these two species. The gene is highly expressed in maize, whereas it is expressed at extremely low levels in teosinte.

Wang and colleagues (1999) looked for evidence of domestication selection pressure on this gene in the form of reduced sequence diversity as compared with neutrally evolving genes within the same genome. They sequenced a 2.9 kb region of the gene, including 1.1 kb of the 5′ non-transcribed region (NTR) and most of the predicted transcriptional unit (TU) from a geographically and genetically diverse sample of maize and teosinte. The sample included plants from *Z. mays* ssp. *mays*, ssp. *parviglumis*, ssp. *mexicana*, and *Zea diploperennis*, some of which were obtained from the International Maize and Wheat Improvement Centre (CIMMYT) in Mexico. These lines were sampled in order to represent a number of stages between the progenitor teosinte and the descendent maize. Many of the accessions utilized in the study do not have phenotypic characteristics that would be obviously or immediately useful to maize breeders or to curators, but they were collected anyway, and provided the genetic link needed between maize and its wild ancestor for the *tb1* sequence diversity analysis. This is an example of the importance of genetic rather than solely phenotypic characterization of accessions. Moreover, cases like this demonstrate the use of genetic resources in non-breeding applications and their value in enhancing our understanding of the organism and, ultimately, its effective conservation and use.

The second characterized gene, *terminal ear 1*, is a regulatory locus thought to be responsible for the switch from lateral branches tipped with tassels in teosinte to lateral branches tipped with ears in maize (White and Doebley, 1999). This gene was mapped as a QTL on the long arm of chromosome 3 along with a second gene, *tassel-replaces-upper-ear 1* (*tru1*) (Doebley *et al.*, 1995). It has been difficult to determine which, if either, of these genes had a role in the domestication of maize, but one clue is that *te1* encodes a protein with conserved RNA-binding domains. It is hypothesized that it may function through RNA-binding activity, a common method for regulatory loci. A 1.4 kb segment of this gene was sequenced from 25 *Zea* and *Tripsacum* individuals in order to look for evidence of past selection as was seen for *tb1*. The sequence encompassed the first and second exons and part of the third exon, as well as two small introns.

The panel included a number of US inbred maize lines, members of the subspecies *parviglumis*, *mexicana* and *huehuetenangensis*, *Z. diploperennis*, *Zea luxurians*, and *Tripsacum dactyloides* and *Tripsacum floridanum*. These lines of teosinte were also obtained from the collection at CIMMYT. To the authors' surprise, these tests indicated that *te1* did not experience a selective sweep as would be expected under the hypothesis that it was involved in the domestication process, but it is actually undergoing neutral evolution. In addition, the phylogenetic tree did not contain a monophyletic maize clade as would be expected if the gene went through selection pressure for a specific sequence. This is further evidence that, at least for the sequences studied, *te1* was not heavily selected during maize domestication. However, the upstream regulatory region of the gene was not sequenced as it was for *tb1*. There is still a possibility, therefore, that this gene could show similar directional selective pressure during domestication as *tb1*. The sequence diversity at this locus provided information as to the age of the *Zea* genepool. Assuming a substitution rate of 5.9–6.5 $\times$ 10^{-9} substitutions per synonymous site per year for a neutral gene such as *te1* (Gaut *et al.*, 1996), an estimate of 1.2–1.4 million years was consistent with estimates derived from known neutrally evolving genes, such as *Adh1* (Gaut and Clegg, 1993). Negative results for selective pressure on *te1* suggest that *tru1* may be the key domestication gene in the QTL on chromosome 3, but a similar experiment for this gene remains to be undertaken.

The third characterized gene hypothesized to be involved in the domestication of maize from teosinte is called *teosinte glume architecture 1* (*tga1*). This gene, located near the centromere on the short arm of chromosome 4, is thought to control the structure of the fruit casing in which maize or teosinte kernels are housed (Dorweiler *et al.*, 1993). On teosinte ears, individual kernels are encased in stone-like fruitcases, so the kernels are not easily accessible for herbivory by humans or other animals. In maize, the same components of the fruitcase are present, but their development is disrupted, so that the kernels are exposed on the ear.

A single allelic difference at this genetic locus made it possible for early agriculturalists to extract the kernels from maize ears for consumption. The teosinte allele causes high levels of silica and lignin, both hard substances, to be deposited in the fruit casing, while the maize allele conditions slower growth of the fruit casing and a decrease in the deposition of silica and lignin. In this early work on maize evolution, the methods the researchers had to rely on to yield clues about the involvement of this gene in the domestication of maize were not as informative as molecular evolution studies. However, the researchers were able to map the gene as a QTL and to intogress specific maize or teosinte alleles into known genetic backgrounds in order to understand the gene's function. For example, when the maize allele was inserted into a teosinte background, the fruitcase appeared much more like the maize fruitcase, with the kernels partially exposed on the ear. Additionally, the hard glume covering the kernel became significantly softer and more pliable. After inserting the teosinte allele into a modern maize background, they saw the reverse trend. The developing kernels were constricted by the glume, which became hard and encased the kernel as would be seen in a teosinte plant. This process caused a majority of the maize kernels to crack as they matured, a significant detriment to the health of the crop. The nature and number of phenotypic changes caused by this locus led researchers to believe that it is a regulatory locus controlling the expression of a set of other genes (Dorweiler *et al.*, 1993).

The allelic variants at *tga1* between teosinte and maize are seen as some of the most important changes to take place in the evolution of maize, because they made it possible for maize to be the harvestable grain crop on which we rely so heavily today. The molecular evolution of this gene is currently being studied in more detail in order to understand, as we do now for *tb1*, the molecular mechanism behind the gross morphological changes in ear structure as wild teosinte was transformed to modern maize.

As noted earlier, the domestication of maize was a relatively fast and extremely intense process in which a short, many-tillered, low yielding plant species was transformed into a tall, single-stalked, tremendously high-yielding plant species. This transformation has been observed in many other crop plants as well (Paterson *et al.*, 1995; Smartt and Simmonds, 1995) but none of the other domestication stories are as profound and elegant as the maize story. This crop underwent extreme selection pressure by early agriculturalists, and as a result there exist today extensive morphological differences between the progenitor, teosinte, and the descendent, maize. The most amazing part of the story is that these differences, though prominent, are possibly all due to a common developmental signal controlling the expression of the four or more genes discussed here. The fact that as few as five loci could control the differentiation between teosinte and maize is astounding, and the three genes already characterized, *tb1*, *te1* and *tga1*, are thought to be important factors in a cascade of regulatory action which may control numerous other genes involved in the domestication process. The story of maize domestication has become a model for evolutionary biologists undertaking similar studies for other crops, and it is entirely likely that the maize story will become even more fascinating as we come to understand the individual steps involved in gene regulation and expression in the future.

The same type of study that was used to uncover the maize story has been applied recently to understand major changes in fruit size during the domestication of tomato (Frary *et al.*, 2000). Tomato, along with other fruit and grain crops, underwent rapid and dramatic phenotypic changes during the domestication process (Smartt and Simmonds, 1995). Along with other features, the tomato fruit was transformed from a tiny (< 1 cm diameter) berry-sized home for hundreds of seeds with very little fruit tissue to a gigantic (> 15 cm diameter) edible fruit. Previous studies have determined that this size change is controlled by as many as 28 QTL (Grandillo *et al.*, 1999). However, there are just six QTL which account for the majority of fruit size variation in tomato evolution. Included in this group is one QTL, *fw2.2*, which changes fruit weight by up to 30% between wild and domesticated *Lycopersicon* species. All of the sampled wild tomato species exhibit the small-fruit allele at this locus, while modern cultivars contain the large-fruit allele, demonstrating high correlation between the morphological phenotype and the QTL genotype (Alpert *et al.*, 1995).

The gene underlying *fw2.2*, *ORFX*, was recently isolated using a map-based cloning technique (Frary *et al.*, 2000). Complementation tests revealed that when large-fruited modern cultivars were transformed with the small-fruit allele from *Lycopersicon pennellii*, the segregating progeny from the primary transformants separated into two groups, those containing the transgene and exhibiting a reduction in fruit size and weight, and those not containing the transgene and exhibiting no change in fruit weight. This result coincided with a second line of data on the expression pattern of *ORFX*. Transcripts of the gene were detected in all pre-anthesis floral organs of both large and small-fruited near-isogenic tomato lines, but mRNA levels in the carpels of small-fruited NILs (near isogenic lines) greatly exceeded levels in the carpels of their large-fruited counterparts. Therefore, it is possible that the *ORFX* gene represses the growth of carpels in which high levels of its mRNA accumulate. This observation is consistent with the example of *tb1* in maize, which operates as a repressor of organ growth in the lateral branch primordia of maize as a result of high levels of mRNA accumulation.

An additional analysis performed in this study was a sequence alignment between the alleles from small- and large-fruited NILs. An 830 bp fragment containing the *ORFX* open reading frame, 55 bp from the 3′ untranslated region (UTR), and 95 bp from the 5′ UTR, was sequenced for each NIL. Only three non-synonymous nucleotide substitutions were identified along the entire sequence, and depending on the position of the protein start codon, all of these substitutions could be located in the 5′ UTR. These nucleotide changes, if located in the 5′ UTR and not in the coding sequence, provide evidence that the differential mRNA levels in the wild and cultivated tomato fruits are not due to changes in coding sequence, but rather to modified expression of the *ORFX* gene. Regulatory genes such as *ORFX* of tomato and *tb1* of maize have apparently played a major role in the domestication of these crops from their wild ancestors and account for tremendous quantities of phenotypic variation in crop genepools.

The maize and tomato domestication stories should provide a warning to all curators and users of genetic resources that major phenotypic differences between accessions do not always mean that there are equally extensive genetic differences. In addition, significant contributions to agronomically desirable traits may result from the regulation, both spatial and temporal, of gene expression rather than differences in amino acid sequences or protein structures. The challenge for curators now is to interpret how this knowledge affects not only genetic resources conservation, but where and how to look for useful alleles for genetic diversity characterization and plant breeding.

Discovering new genes in new ways

An important ongoing activity associated with genetic resources collections is the molecular characterization of accessions and individuals with neutral genetic markers (Dean *et al.*, 1999). In addition to revealing redundancies and the structure of diversity in collections, genotyping provides the foundation for using the diversity within collections for other genetic analyses. In the past, plant breeders have been constrained by the need to have appropriate mapping populations. However, alternative population-based methods are now being utilized for mapping genes in humans and *Drosophila*, and the application of this approach is another way to utilize both genetic resources collections and natural populations.

Association mapping, or linkage disequilibrium mapping within a sample of known pedigree, exploits related individuals that differ for a particular trait in order to establish which region of the genome is associated with the phenotype among the population members. The underlying assumption of this method is that the gene responsible for the phenotype is identical by descent, and that markers that flank it will be common among accessions that share the phenotype. Although there are not yet any examples of association mapping in plant populations, an example from mapping in humans involves the identification of regions that contribute to susceptibility to type 1 diabetes, a disorder in which the β cells of the islets in the pancreas are destroyed by the immune system (Eckenrode *et al.*, 2000). Population samples consisting of siblings with diabetes have been used for a genome-wide scan to identify multiple participating loci (Davies *et al.*, 1994; Hashimoto *et al.*, 1994). Additionally, similar samples have been used to fine-map and evaluate regions containing candidate genes (Eckenrode *et al.*, 2000). In order to apply this method to mapping genes using plant genetic resources collections, the following prerequisite resources will be required: (i) a dense set of molecular markers; (ii) passport and phenotypic data; (iii) information on population structure; and (iv) a sample with contrasting genotypes for the trait of interest. Association mapping could potentially be applied to locate genes underlying adaptive traits such as disease resistance using either natural populations or genetic resources collections.

A central effort in genomics continues to be the identification of genes controlling traits of interest, and clearly there are options for using genetic resources collections in gene discovery. The next phase of genetic resources characterization will be the discovery and description of allelic variation, and natural populations and genetic resources collections will prove to be some of the most valuable sources of alleles for adaptive traits. Although the methods to survey allelic diversity are not yet widely available, they serve as prototypes for the potential use of genetic resources collections in allele discovery. One such method is a locus-specific microarray, which would allow a scan of variation at a locus from many accessions on single chip (Chee *et al.*, 1996). Combined with advances in prediction of protein structure from sequence data (Skolnick *et al.*, 2000), this could become a powerful tool for searching genebanks for specific alleles. The link between diversity at the DNA level and phenotypic level is gene expression. With the advent of microarrays it is now possible to observe a genotype in action at a particular developmental stage or in response to an environmental stimulus (Richmond and Somerville, 2000). This will be a useful tool for dissection of more complex traits, such as drought tolerance or nutritional pathways, to identify the diversity in expression that provides the most direct link to phenotype.

We are now able to make use of extensive biological resources (such as markers or mapping populations) that have an immediate impact on plant breeding. Advances in utilizing plant genetic resources to their fullest potential are founded on the concept that while a high-yielding line often contains a greater number of positive alleles at the loci associated with yield, there are almost always some loci for which the inferior parent contributes a superior allele to the offspring (Tanksley and McCouch, 1997). Therefore, to provide plant breeders with a broader array of useful allelic diversity, it is logical to complement our classical approach of identifying good phenotypes in collections or natural populations with the new-found possibility of looking for good genes whenever and wherever they exist, independent of the genetic background in which they currently reside.

Genetic resources are generally thought of as a reservoir of adaptive traits that are introgressed individually into breeding programmes. While this is the predominant use for the collections, genomics will enable us to utilize our genetic resources more broadly. Some of the most tantalizing stories in genetic resources utilization involve the positive contribution of wild relatives with poor agronomic type to quantitative traits affecting yield in wide crosses. For example, alleles from the wild rice *Oryza rufipogon* were associated with a 17% increase in grain yield per plant in a cross with the cultivated *Oryza sativa* (Xiao *et al.*, 1998). It is clear that many agronomically desirable alleles may be masked by agronomically undesirable phenotypes.

Framework Needs to Take Advantage of Future Opportunities

Genebank managers and curators now have a vast array of genetic and genomic tools available for their use. In the short term, the tools of structural genomics will be linked with bioinformatic tools to provide a framework for progress in comparative genomics and diversity. As a complement, functional genomics tools will become more accessible so that curators ultimately will ultimately be able to catalogue useful genetic diversity in their collections.

Moreover, there are some important resources necessary for the kind of studies discussed in this chapter and for future utilization of genetic resources collections. These include detailed molecular maps, without which QTL mapping cannot be performed; mapping populations as well as natural populations with which to create these maps; sequence-based markers such as expressed sequence tags or single nucleotide polymorphisms with which researchers can understand not only the exact location but also more detailed characterization of genes of interest; and bacterial artificial chromosome libraries for map-based cloning of genes. These biological resources are not only useful to evolutionary biologists trying to understand the origins of our food crops and their place in global agriculture, but also to systematists trying to understand the intertwining relationships between all forms of life on earth, to biochemists trying to understand the biochemical mechanisms behind phenotypic traits, and to plant geneticists and breeders trying to use all of this knowledge to improve agronomic traits of crops of interest in order to feed our ever-expanding population. We have reached a time when selected cases exist in which genetic resources collections are effectively becoming genebanks in the truest sense of the word. Advances in science and technology in the 21st century will support this transition.

Acknowledgements

The authors thank the U.S. Agency for International Development, Division on Agriculture and Food Security, for providing support for conference attendance.

References

Alpert, K.B., Grandillo, S. and Tanksley, S.D. (1995) Fw-2.2 – a major QTL controlling fruit weight is common to both red-fruited and green-fruited tomato species. *Theoretical and Applied Genetics* 91, 994–1000.

Beadle, G. (1939) Teosinte and the origin of maize. *Journal of Heredity* 30, 245–247.

Chee, M., Yang, R., Hubbell, E., Berno, A., Huang, X.C., Stern, D., Winkler, J., Lockhart, D.J., Morris M.S. and Fodor, S.P.A. (1996) Accessing genetic information with high-density DNA arrays. *Science* 274, 610–614.

Davies, J.L., Yoshihiko, K., Bennett, S., Copeman, J.B., Cordell, H.J., Pritchard, L.E., Reed, P.W., Gough, S.C.L., Jenkins, C., Palmer, S.M., Balfour, K.M., Rowe, B.R., Farrall, M., Barnett, A.H., Bain, S.C. and Todd, J.A. (1994) A genome-wide search for human type 1 diabetes susceptibility genes. *Nature* 371, 130–136.

Dean, R.E., Dahlberg, J.A., Hopkins, M.S., Mitchell, S.E. and Kresovich, S. (1999) Genetic redundancy and diversity among 'Orange' accessions in the U.S. national sorghum collection as assessed with simple sequence repeat (SSR) makers. *Crop Science* 39, 1215–1221.

De Wet, J.M.J. and Harlan, J.R. (1972) Origin of maize: the tripartite hypothesis. *Euphytica* 21, 271–279.

Doebley, J. and Wang, R.L. (1997) Genetics and the evolution of plant form: an example from maize. *Cold Spring Harbor Symposia on Quantitative Biology* 62, 361–367.

Doebley, J., Stec, A., Wendel, J. and Edwards, M. (1990) Genetic and morphological analysis of a maize-teosinte F_2 population: implications for the origin of maize. *Proceedings of the National Academy of Sciences USA* 87, 9888–9892.

Doebley, J., Stec A. and Gustus, C. (1995) *Teosinte branched 1* and the origin of maize: evidence for epistasis and the evolution of dominance. *Genetics* 141, 333–346.

Dorweiler, J., Stec, A., Kermicle, J. and Doebley, J. (1993) *Teosinte glume architecture 1*: a genetic locus controlling a key step in maize evolution. *Science* 262, 233–235.

Eckenrode, S., Marron, M.P., Nicholls, R., Yang, M.C.K., Yang, J.J., Guida Fonseca, L.C. and She, J. (2000) Fine-mapping of the type 1 diabetes locus (*IDDM4*) on chromosome 11q and evaluation of two candidate genes (*FADD* and *GALN*) by affected sibpair and linkage-disequilibrium analyses. *American Journal of Human Genetics* 106, 4–18.

Eyre-Walker, A., Gaut, R.L., Hilton, H., Feldman, D.L. and Gaut, B.S. (1998) Investigation of the bottleneck leading to the domestication of maize. *Proceedings of the National Academy of Sciences USA* 95, 4441–4446.

Frary, A., Nesbitt, T.C., Frary, A., Grandillo, S., van der Knaap, E., Cong, B., Liu, J., Meller, J., Elber, R., Alpert, K.B. and Tanksley, S.D. (2000) fw2.2: A quantitative trait locus key to the evolution of tomato fruit size. *Science* 289, 85–88.

Gale, M.D. and Devos, K.M. (1998) Plant comparative genomics after 10 years. *Science* 282, 656–659.

Galinat, W. (1971) The origin of maize. *Annual Review of Genetics* 5, 447–478.

Galinat, W. (1983) The origin of maize as shown by key morphological traits of its ancestor, teosinte. *Maydica* 28, 121–138.

Gaut, B.S. and Clegg, M.T. (1993) Molecular evolution of the *Adh1* locus in the genus *Zea*. *Proceedings of the National Academy of Sciences USA* 90, 5095–5099.

Gaut, B.S., Morton, B.R., McCaig, B.C. and Clegg, M.T. (1996) Substitution rate comparisons between grasses and palms: synonymous rate differences at the nuclear gene *Adh1* parallel rate differences at the plastid gene *rbcL*. *Proceedings of the National Academy of Sciences USA* 93, 10274–10279.

Grandillo, S., Ku, H.M. and Tanksley, S.D. (1999) Identifying the loci responsible for natural variation in fruit size and shape in tomato. *Theoretical and Applied Genetics* 99, 978–987.

Hashimoto, L., Habita, C., Beressi, P., Delepine, M., Besse, C., Cambon-Thomsen, A., Deschamps, I., Rotter, J.I., Djoulah, S., James, M.R., Froguel, P., Weissenbach, J., Lathrop, G.M. and Julier, C. (1994) Genetic mapping of a susceptibility locus for insulin-dependent diabetes mellitus on chromosome 11q. *Nature* 371, 161–164.

Iltis, H. (1983) From teosinte to maize: the catastrophic sexual transmutation. *Science* 222, 886–894.

Paterson, A.H., Lin, Y.R., Li, Z.K., Schertz, K.F., Doebley, J.F., Pinson, S.R.M., Liu, S.C., Stansel, J.W. and Irvine, J.E. (1995) Convergent domestication of cereal crops by independent mutations at corresponding genetic loci. *Science* 269, 1714–1718.

Richmond, T. and Somerville, S. (2000) Chasing the dream: plant EST microarrays. *Current Opinion in Plant Biology* 3, 108–116.

SGRP (System-wide Genetic Resources Programme of the CGIAR) (1999) *Genebanks and Comparative Genetics*. Report to the Technical Advisory Committee of the CGIAR.

Skolnick, J., Fetrow, J.S. and Kolinski, A. (2000) Structural genomics and its importance for gene function analysis. *Nature Biotechnology* 18, 283–287.

Smartt, J. and Simmonds, N.W. (1995) *Evolution of Crop Plants*. Longman, London.

Tanksley, S.D. and McCouch, S.R. (1997) Seed banks and molecular maps: unlocking genetic potential from the wild. *Science* 277, 1063–1066.

Wang, R.L., Stec, A., Hey, J., Lukens, L. and Doebley, J. (1999) The limits of selection during maize domestication. *Nature* 398, 236–239.

White, S.E. and Doebley, J.F. (1999) The molecular evolution of *terminal ear 1*, a regulatory gene in the genus *Zea*. *Genetics* 153, 1455–1462.

Wilson, W.A., Harrington, S.E., Woodman, W.L., Lee, M., Sorrells, M.E. and McCouch, S.R. (1999) Inferences on the genome structure of progenitor maize through comparative analysis of rice, maize, and the domesticated Panicoids. *Genetics* 153, 453–473.

Xiao, J.H., Grandillo, J.M., Ahn, S.N., Yuan, L.P., Tanksley, S.D. and McCouch, S.R. (1998) Identification of trait-improving quantitative trait loci alleles from a wild rice relative, *Oryza rufipogon*. *Genetics* 150, 899–909.

36 Geographic Information Systems (GIS) and the Conservation and Use of Plant Genetic Resources

L. Guarino,[1] A. Jarvis,[1] R.J. Hijmans[2] and N. Maxted[3]
[1]*International Plant Genetic Resources Institute, Regional Office for the Americas, IPGRI c/o CIAT, Cali, Colombia;* [2]*International Potato Center (CIP), Lima, Peru;* [3]*School of Biological Sciences, University of Birmingham, Birmingham, UK*

Introduction

The management of genetic resources is a complex, multi-faceted process. It involves a number of distinct stages, which are nevertheless linked and interrelated, from the selection of priority taxa, to the design and implementation of complementary conservation strategies and the development and exchange of the results of germplasm use (Maxted *et al.*, 1997a). These different components generate various types of data, including information on the identity (passport data) and characteristics of germplasm (characterization and evaluation data), which are crucial for the effectiveness of the process as a whole. In the end, these data refer to stands of wild or cultivated plants found growing in specific, known places. The location of these places on the surface of the earth is also included in the passport data of genebank and herbarium documentation. This means that the data associated with germplasm are 'geo-referenced', and therefore amenable to 'spatial analysis', or the description and modelling of patterns and relationships in geographical data (Bailey, 1994).

Each of the different components of the process of conservation and use of genetic resources not only generates, but also requires, data. For example, germplasm collection results in data on the distribution, the phenology, the ethnobotany and, once characterization and evaluation have been carried out, the genetic diversity of the target taxon. However, the collector will clearly benefit from considering – before venturing out into the field – any such data that may already exist in the literature and the documentation systems of genebanks and herbaria. Because they are geo-referenced, the data coming out of the genetic resources management process can be analysed not only on their own, but also in conjunction with other location data, from whatever additional source. Thus our collector, presented with scanty data on previous collections, could use climatic, vegetation and soil data from the study region to estimate the distribution and phenology of the taxon concerned.

In this chapter, we describe how spatial analysis of the geo-referenced data generated by the process of conservation and use of genetic resources, using geographic information systems (GIS), can feed back to enhance and facilitate the process, and indeed add value to the germplasm collections. We start with a brief introduction to GIS technology. We then examine a number of key stages of the genetic resources management process and discuss how GIS may be used to increase their efficiency and effectiveness. This chapter thus follows the structure of Guarino *et al.* (1999), while updating the information presented there with new examples and additional references (including relevant Internet resources).

 Managing Plant Genetic Diversity
(eds J.M.M. Engels, V. Ramanatha Rao, A.H.D. Brown and M.T. Jackson)

Geographic Information Systems

A GIS may be formally defined as a database management system which can simultaneously handle spatial data in graphics form, i.e. maps, or the 'where', and related, logically attached, non-spatial, attribute data, i.e. the labels and descriptions of the different areas within a map, or the 'what'. Simpler definitions have been given, perhaps the most jargon-free being 'a tool for managing information of any kind according to where it is located' (Treweek, 1999). The main elements of a GIS are as follows (Guarino, 1995; Guarino *et al.*, 1999):

- data input, verification and editing;
- data storage, retrieval and management;
- data manipulation and analysis; and
- output.

Data input

Data can be entered into a GIS by digitizing paper maps and their associated attribute information *de novo* using a digitizing table or scanner, or by importing existing digital datasets, including remote sensing images. Genebank curators can import data from the database of their documentation system into a GIS, using the latitude and longitude fields in the passport data to provide the link to digitized maps and remote sensing images. Some regional or global scale datasets are available from organizations such as the Food and Agriculture Organization of the United Nations (FAO), the United Nations Environment Programme's Global Resource Information Database (UNEP/GRID), the International Soil Reference and Information Centre (ISRIC), the World Conservation Monitoring Centre (WCMC), the US Geological Survey and the international agricultural research institutes (IARCs) of the Consultative Group on International Agricultural Research (CGIAR). As an example, the spatial data holdings of the IARCs can be searched in the CGIAR Spatial Data Catalogue,[1] and a wide range of global datasets ranging from maps of terrestrial ecosystems to human population density are available at the UNEP/GRID web site[2] and from the Eros Data Centre of the US Geological Survey.[3]

[1] www.griada.no/cgiar.htmls.mdindex.htm
[2] www.grid2.cr.usgs.gov
[3] edcwww.cr.usgs.gov/earthshots/slow/

Data storage

There are two main types of GIS data: vector and raster. Vector files store geographic data as points, lines or polygons. Polygons represent areas of different sizes and shapes where a particular attribute is equal throughout. In contrast, in raster (or grid) data, an area is divided into an array of regularly shaped cells. Each individual cell is assigned a value for the variable being studied. Examples of vector data include maps of roads, or administrative zones. Typical raster datasets include satellite images, elevation (so called Digital Elevation Models) and interpolated climate data.

Each type of data has advantages and disadvantages. However, modern GIS packages can handle both types of data, and analyses can be made across the two data types.

Data manipulation and analysis

The spatial processing system (to manipulate the 'where') and database management system (for the 'what') of a GIS allow the user to bring together diverse datasets, make them compatible, and combine and analyse them. A distinction has been made in describing the analytical capabilities of GIS between deterministic and statistical methods (Bailey, 1994). Deterministic functionalities provided by GIS include network analysis, three-dimensional modelling and projection algebra. Spatial analysis includes the tabulation or mapping of basic summary statistics for data in areas of interest. Relevant tools include query facilities, Boolean operations on attributes, map overlay facilities and buffer creation. More complex statistical approaches to the investigation, comparison and modelling of geographical patterns, include spatial correlation, pattern and trend analysis, and interpolation techniques such as kriging (Bailey, 1994).

Output

GIS allows visualization of spatial data (maps) on a computer screen and the production of printed maps, allowing such manipulations as selecting areas or layers for output, and changing scale and colour. However, the outputs of GIS analyses are not just maps, but can also include tables, graphs and animations. Three-dimensional visualization is one of the key topics in current GIS development, and is revolutionizing cartographic representation.

Use of GIS in Plant Genetic Resources Conservation and Use

In this section, we discuss how different types of spatial analysis can be applied to locality, characterization and evaluation data in a GIS environment to enhance the efficiency of genetic resources management. We concentrate on the following five components of the process:

1. Ecogeographic surveying;
2. Field exploration;
3. Design, management and monitoring of *in situ* reserves;
4. Germplasm evaluation; and
5. Use of genetic resources.

Ecogeographic surveying

Maxted *et al.* (1995) described the process of collating information on the taxonomy, genetic diversity, geographic distribution, ecological adaptation and ethnobotany of a plant group, as well as on the geography, ecology, climate and the human setting of study regions. The sources of the information for such 'ecogeographic surveys' will include herbarium specimens, germplasm accession passport data, experts, the formal and grey literature, field notes and maps.

It is only on the foundation of basic information such as (but not limited to) this that sensible conservation decisions can be made, for example regarding when, where and how to collect germplasm, and where genetic reserves might best be established and how they would need to be monitored and managed. This is because biodiversity is not evenly distributed over the surface of the earth. Some places will, therefore, be of a higher priority for collection or conservation than others. One of the main objectives of ecogeographic surveys is thus the identification of those geographic areas which are:

- likely to contain specific desired traits (adaptations), taxa or habitats of interest;
- highly diverse (whether environmentally, taxonomically or genetically);
- complementary to each other;
- currently missing or under-represented in conservation efforts;
- threatened with genetic erosion.

Areas likely to contain germplasm of interest

Although data on the geographic distribution of species are often scanty, even this limited information can be used to identify areas where a species has not been previously recorded but where it might still be expected to occur. GIS tools such as BIOCLIM[4] (Busby, 1991; see also GARP,[5] Genetic Algorithm for Rule-set Production, an extension of the BIOCLIM approach), DOMAIN[6] (Carpenter *et al.*, 1993) and FLORAMAP[7] (Jones and Gladkov, 1999; Jones *et al.*, 1997) attempt this type of extrapolation using climate data. They first estimate conditions at sites where a species has been recorded. Although the details of the methodologies differ, the software then derive a climatic 'envelope' for the set of collecting sites and display all other areas that have some level of similarity to the collecting sites. An example of the output provided by FLORAMAP is shown in Fig. 36.1. Afonin and Greene (1999) describe a similar methodology. Model-based approaches (e.g. Walker, 1990; Stockwell and Noble, 1992) have also been used to investigate the so-called 'potential' distribution of a species. An instructive, large-scale example of the application of GIS to the problem of 'objective prediction of the full distribution of a species from incomplete point distribution maps, based on its ecological preferences' is the work being carried out by the Royal Botanic Gardens, Kew,[8] in the context of Madagascar's Environmental Action Plan. These techniques have mainly been used at the species level. However, they can also be applied to infraspecific entities such as botanical varieties, groups of similar accessions (e.g. based on multivariate analysis of morphological or molecular characterization), or even farmer-recognized landraces.

A species being targeted for conservation may

[4] www.dino.wiz.uni-kassel.de/model_db/mdb/bioclim.html
[5] www.kaos.erin.gov.au/general/biodiv_model/ERIN/GARP/home.html
[6] www.cgiar.org/cifor/research/intro_d.html
[7] www.ciat.cgiar.org/floramap
[8] www.rbgkew.org.uk/herbarium/madagascar/plant_dis.html

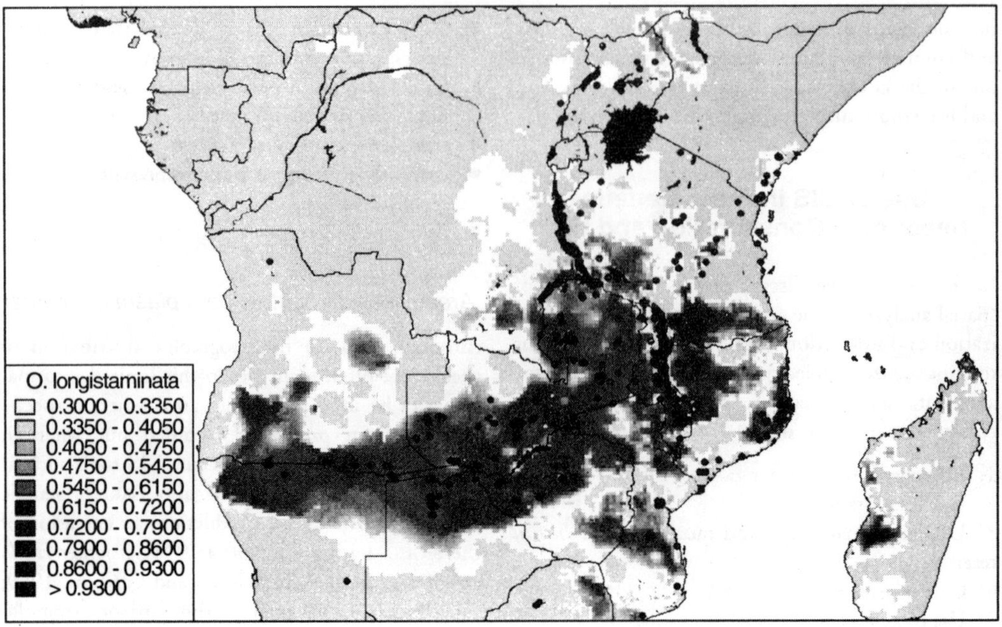

Fig. 36.1. Results of FLORAMAP analysis of the distribution of *Oryza longistaminata* in southern Africa. The shading shows areas with increasingly high levels of climatic similarity with the sites where the species has been encountered (the dark dots, which represent germplasm accessions and herbarium records), and therefore in theory increasing probability of finding the species. The climate variables used were monthly rainfall totals, monthly average temperatures and monthly average diurnal temperature range. Map used courtesy of Kihika Kiambi (IPGRI). See text for more information on the FLORAMAP software.

be closely associated with a specific habitat, vegetation, land use or landform. Remote sensing imagery such as aerial photographs, satellite imagery and radar images, combined with other sources of data in a GIS environment, and properly ground-truthed, can be useful in locating these features. For example, Landsat™ satellite images were used by Veitch *et al.* (1995) to map heathland fragments in southern England, and by Afonin and Greene (1999) to locate herbaceous meadows in the western Caucasus. Isolated areas of cultivation can also be spotted, assuming they are larger than the resolution of the remote sensing system (about 30 m in the case of Landsat images, and 15 m for SPOT images), and in some cases even the crops being grown can be identified.

Obtaining, manipulating and analysing satellite imagery can be a complex, costly and time-consuming task, but some useful software tools are available. An example is WINDISP, which was originally developed for the FAO Global Information and Early Warning System (GIEWS), and is 'a public domain, easy to use software package for the display and analysis of satellite images, maps and associated databases'. It can be downloaded from the Internet.[9] Some data on ecosystems and land use derived from satellite sources are available pre-processed and free on the Internet.[10]

Sometimes material with very specific characteristics needs to be targeted for conservation, perhaps because a need for it has been expressed by the user community or it is specifically threatened. Here the problem is one of using environmental data to identify areas where material with the required adaptation

[9] www.fao.org/giews/english/windisp/windisp.htm
[10] For example see www.edcdaac.usgs.gov/glcc/glcc.html for global datasets derived from AVHRR 1 km remote sensing imagery.

might be expected to occur. GIS can be used to superimpose appropriate thematic coverage from different sources to identify such areas. Thus, for example, CIAT's *An Atlas of Cassava in Africa* (Carter *et al.*, 1992), an excellent hardcopy product of a GIS project, divides the cultivation area of cassava on the basis of production environments, based on agricultural censuses, specialized surveys and crop experts.

If characterization and evaluation of germplasm have already been carried out, geostatistical methods such as kriging and other forms of interpolation can be used to describe the spatial pattern of variation in genetic, morphological and agronomic traits (or combinations of traits) among populations. The effects of natural selection on a broad scale can then be explored by seeking correlations with environmental and other factors, and disentangled from the effects of isolation by distance and micro-environmental variation (e.g. Epperson, 1993; Monestiez *et al.*, 1994). Differential systematics can also be used (Kirkpatrick, 1974), involving the combination of different character contour maps into a single map of a systematic function, the ridges of which reveal areas where maximum change over distance occurs. These have been called genetic boundaries (see also Monmonier, 1973; Pigliucci and Barbujani, 1991). Such analyses can guide the user to areas where specific traits are prevalent, and also help the conservationist identify areas that are relatively homogeneous but different from each other for the characters being studied. This is discussed further in the section on complementarity.

High diversity areas

Some geographic areas show greater taxonomic or genetic diversity than others. Diversity is often the most important consideration at all stages of genetic resource work, yet remains a difficult parameter to map and analyse. Diversity studies usually begin by dividing the target area (or strata within the target area, e.g. climate zones) into a number of smaller zones, for each of which a measure of diversity can be calculated. Different geometric, political or socio-economic spatial units have been used (e.g. see references in Csuiti *et al.*, 1997), ideally areas of equal shape and size (to reduce the area effect on diversity measures), such as square grid cells, are best. For example, Nabhan (1991) used presence/absence of species to investigate patterns in the taxonomic diversity of wild *Phaseolus* in different grid cells covering the Sierra Madre, Mexico. Measures of diversity based on morphological characters or molecular markers can also be used, as done by Pickersgill (1984) to calculate the morphological diversity of cultivated *Capsicum* spp. in different grid cells within Central and South America. Ferguson *et al.* (1998) calculated genetic diversity in wild *Lens* spp. using randomly amplified polymorphic DNA (RAPD) markers for different grid cells in the Mediterranean basin. To support this type of analysis by genebanks, IPGRI and CIP are collaborating in the development of software called DIVA,[11] which calculates diversity indices for all the cells in a user-defined grid given latitude, longitude and characterization data for a set of accessions, and maps the results.

A complementary method for mapping diversity is the 'point-based' approach used by the Spatial Intraspecific Diversity (SID) software described by Nelson *et al.* (1999).[12] SID reads the coordinates of each accession point in turn, and draws a circle of user-defined radius around it. All accession points lying within this circle are then used to calculate diversity within the defined radius, using either the Shannon-Weaver or Simpson index. The result is assigned to the spatial location of the observation for which the calculation was done. The advantage of this method lies in the ease with which point-based values of diversity can be interpolated and modelled. Figure 36.2 shows sample SID and DIVA outputs for the same dataset.

Various studies have investigated the use of different environmental parameters as surrogates for species diversity (Gaston, 1996; Faith and Walker, 1996). One example of the application of GIS to this kind of investigation is provided by Miller (1986), who showed that variation in elevation, calculated by GIS from topographical data for each of a large number of polygons in the southern Appalachian region, is a useful predictor of the richness of rare species in those polygons. This approach is being applied at the infraspecific level

[11] gis.cip.cgiar.org/gis/tools/diva.htm
[12] www.gis.ciat.cgiar.org/sid

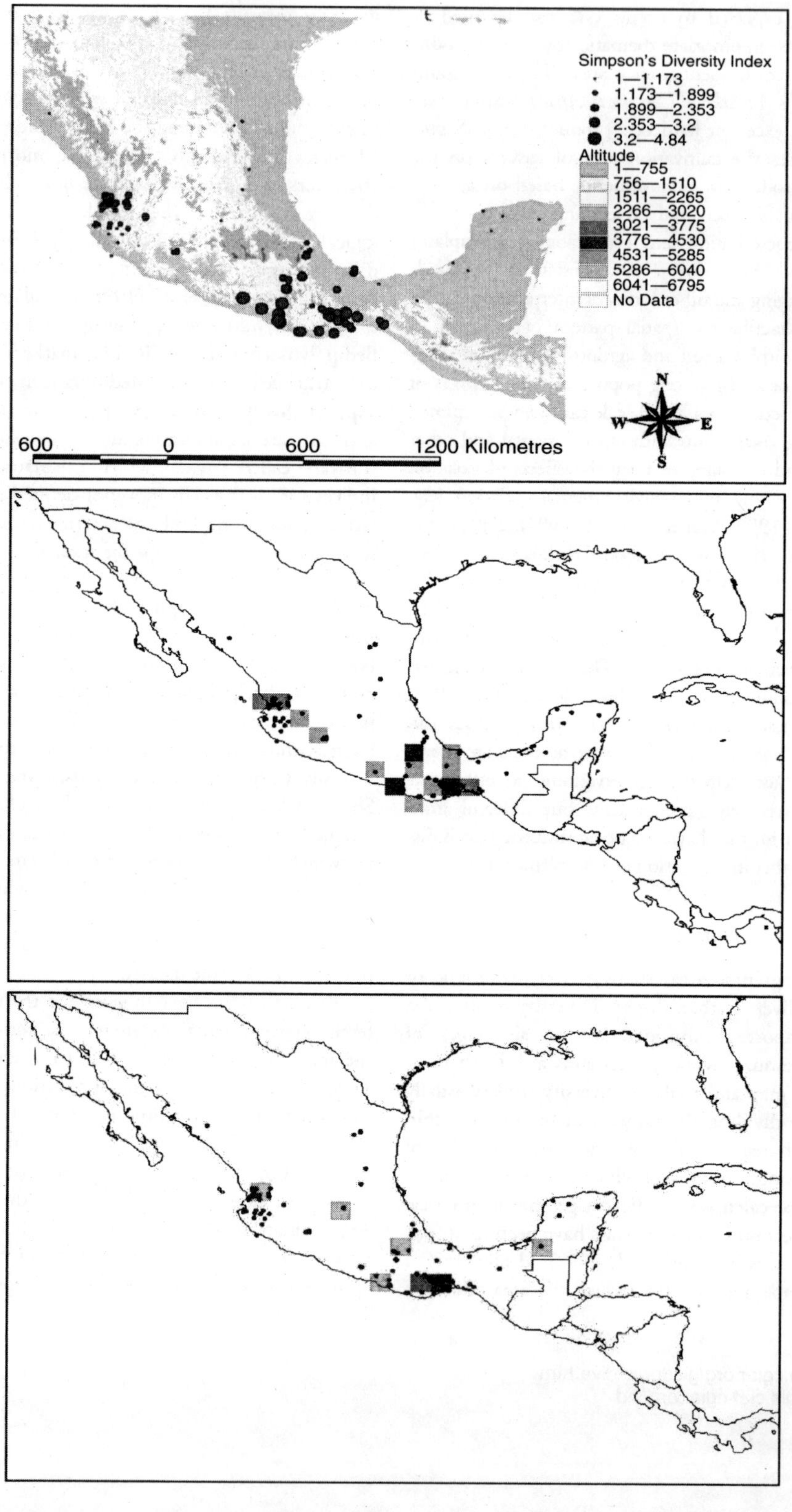
Simpson's Diversity Index
1—1.173
1.173—1.899
1.899—2.353
2.353—3.2
3.2—4.84
Altitude
1—755
756—1510
1511—2265
2266—3020
3021—3775
3776—4530
4531—5285
5286—6040
6041—6795
No Data
N
W
E
S
600
0
600
1200 Kilometres

in a study of the relationship between environmental and human diversity and genetic diversity in cultivated groundnut in Ecuador and Guatemala. The study is a collaboration among the national programmes of the countries involved, USDA, IPGRI and CIAT (International Centre for Tropical Agriculture).

Complementary areas

Analyses of diversity can be refined further. One possible enhancement may be necessitated by the fact that two areas may have equal richness or diversity of taxa or morphotypes, but the ones in one square may be similar to each other (e.g. closely related), while those in the other may be more different. Other things being equal, the second area would be the higher conservation priority (Humphries *et al.*, 1995). The procedure described by Vane-Wright *et al.* (1991), and available in their WORLDMAP software,[13] allows the diversity measure to be weighted for the distinctness of taxonomic units, calculated from a phylogeny based on cladistic analysis. Another possible refinement might be to use the potential distribution of species, say as predicted by FLORAMAP, to calculate diversity values, to counteract the kinds of sampling biases described by Hijmans *et al.* (2000a).

It is not enough to simply identify areas that are highly diverse to maximize the amount of diversity protected for a given amount of effort, because all the higher diversity areas might actually contain the same diversity, as well as the same amount of diversity. One approach to the optimal targeting of conservation effort is to use multivariate statistics to classify or ordinate spatial units according to the species found there, on the basis of characterization or evaluation data, or in terms of environmental conditions (Booth *et al.*, 1989; Pollak and Corbett, 1993; Corbett, 1998). Spatial units can then be targeted for conservation separately in a stratified manner from each distinct cluster of similar units.

More sophisticated methods are also available. Iterative procedures (e.g. Rebelo and Sigfried, 1992) can be used to choose the smallest number of spatial units such that each species, morphotype, etc. will be present in at least one (or two, three, etc.) unit(s) in the set. A recent study of 19 different techniques (Csuti *et al.*, 1997) found that this and various other heuristic techniques can all be very efficient at solving the problem of selecting potential biodiversity conservation sites. However, except when dealing with large, complicated datasets, the study recommends the use of a linear programming approach called a 'branch-and-bound' algorithm. The DIVA software tool supports complementarity analysis as described by Rebelo and Sigfried (1992), and is being used on a variety of datasets, notably the geographic distribution of wild potato species.

Under-conserved areas

The WORLDMAP software allows the user to select a grid square so that a subsequent run identifies those grid squares which are complementary to the selected one. If the selected grid square is the most diverse, the process approximates that of Rebelo and Sigfried (1992). On the other hand, if the selected grid square includes an existing protected area (a global dataset on protected areas is available from WCMC, the World Conservation Monitoring Centre[14]) the result is 'gap analysis.' The concept and some applications are discussed

[13] www.nhm.ac.uk/science/projects/worldmap
[14] www.wcmc.org.uk/cis

Fig. 36.2. (*opposite*) Comparison of diversity analysis using point-centred and grid-based approaches. The dataset consists of locality data for ten species of *Stylosanthes* in Mexico, collated from 119 herbarium specimens by Susana Gama of UNAM as part of a BADC-funded IPGRI research project. The first map shows the results of point-centred species-level diversity analysis using the SID software (Simpson's diversity index). The dots show the location of specimens. The size of the dots is proportional to the species diversity in a one-degree circle around each accession point. The second map shows the results of analysing the same dataset using the DIVA software. The dark grid cells are the most diverse (Shannon-Weaver diversity index), light grid cells the least diverse, and others intermediate. The third map shows the results of complementarity analysis as described by Rebelo and Sigfried (1992), and implemented by DIVA. The dark grid cell is the most diverse. The other grid cells are the additional ones that would have to be conserved to retain at least one population of each species. See text for more information on SID and DIVA.

in detail by Scott *et al.* (1993).[15] The process involves the use of 'digital map overlays in a GIS to identify individual species, species-rich areas, and vegetation types that are not represented or under-represented in existing biodiversity management areas.' These 'existing biodiversity management areas' could be protected areas, but also areas where germplasm collection has already been adequately carried out. Identifying ecogeographical gaps in existing *ex situ* germplasm collections in this way is increasingly important as the large-scale, coarse-grid collecting of the past 25 years is replaced with a more targeted, fine-grid approach.

Threatened areas

Areas under imminent threat of genetic erosion are clearly going to be high priorities for conservation (Guarino, 1999). But how can we predict where genetic erosion is going to take place before it actually happens? Remote sensing has been used to document and model changes in the extent and characteristics of forests (e.g. Skole and Tucker, 1993; Gastellu-Etchegorry *et al.*, 1993),[16] wetlands (e.g. Sebastini *et al.*, 1989), deserts,[17] different land use types (e.g. Zheng *et al.*, 1997) and cropping,[18] and must surely have a role to play in predicting genetic erosion.

Hutchinson and Weiss (1999) have recently described what that role might be. They present a model where different potential causes of genetic erosion are associated with specific indicators. They then list observable phenomena by which these indicators may be assessed, and finally the means by which information on the values of these observables might be obtained, including such diverse sources as remote sensing imagery, agricultural censuses, news reports and fieldwork. A similar conceptual framework has been successfully applied to early warning of crop failure and famine in Africa (for example by USAID's Famine Early Warning System, FEWS[19]). Thus, if environmental degradation is postulated as a possible cause of genetic erosion for a particular genepool, indicators might include desertification, soil erosion and the extension of agriculture into increasingly marginal areas. Observables might include decreasing peak NDVI (Normalized Difference Vegetation Index, a measure of the development of vegetation) values over the course of years, increasing outmigration and the appearance of new fields in marginal areas, all of which can be managed in a GIS environment.

However, the link between these kinds of changes and genetic changes in specific crops or wild species of interest has still not been sufficiently investigated. We may have observables, but we are not sure in most cases whether they are in fact associated with genetic erosion, because this has rarely been measured. An exception is perhaps habitat fragmentation, because the link between this phenomenon and genetic diversity, particularly in tree species, has been the subject of a number of theoretical and practical investigations (e.g. Templeton *et al.*, 1990; Ledig, 1992). The use of remote sensing to monitor and predict genetic erosion in the same way as is being done for food security and famine remains a theoretical, though no less interesting, and tantalizing, possibility.

The somewhat different problem of identifying areas of past genetic erosion is exemplified by ongoing work by CIP and IPGRI on cultivated potatoes in Peru. In this study, isozyme data from the CIP *ex situ* potato collection are being used to estimate levels of genetic variation in each 20 km $\times$ 20 km^2 in a grid covering the whole potato-growing area of the country. Data on the extent of potato cultivation and the biophysical environment in each square are used as the independent variables in a multiple regression model, with genetic diversity as the dependent variable. Negative deviations from the model are then mapped to identify possible areas of genetic erosion, and correlations sought with a number of socio-economic variables (such as accessibility and population growth) to identify possible causative factors for any lower than expected level of genetic diversity.

Synthetic analyses: a case study

GIS tools can be used not only to map the different kinds of areas described above, but also, crucially, to show where they coincide. We have used a dataset on the distribution of 15 wild *Gossypium* species in Africa to illustrate a possible methodology for the

[15] See also www.gap.uidaho.edu/gap for an application of this methodology.
[16] edcwww.cr.usgs.gov/earthshots/slow/Rondonia/Rondonia
[17] www.medalus.leeds.ac.uk/medalus.html
[18] edcwww.cr.usgs.gov/earthshots/slow/Mozambique/Mozambique
[19] www.info.usaid.gov/fews

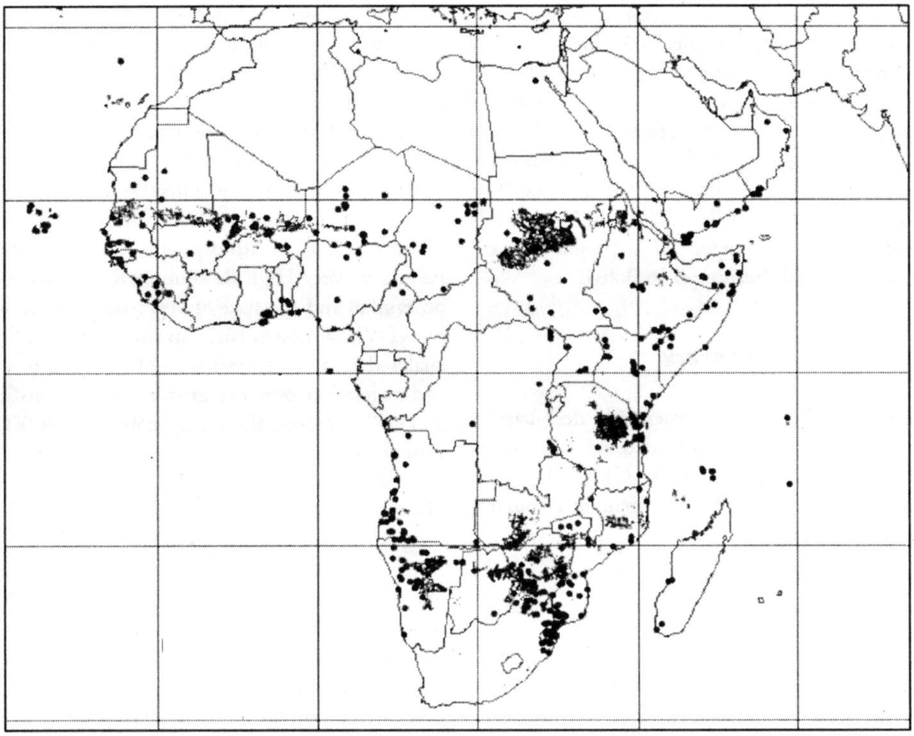

Fig. 36.3. High priority areas for conservation of wild *Gossypium* in Africa are shown in the shaded areas. Dots are locations of herbarium specimens. See text for details of the analysis.

integration of different datasets and analyses in a GIS environment in support of a large-scale ecogeographic study (Holubec, 1998). The methodology involves four separate analyses, the results of which are then combined to produce a single map of priority areas for conservation interventions.

1. Probability of high diversity. First, the SID software was used on the localities of 607 herbarium specimens to calculate species diversity within a 1-degree radius of each collecting point. The locality data of the top 20% diversity points were then imported into FLORAMAP. The FLORAMAP output was a probability surface displaying areas with climatic conditions most suitable to harbour a high diversity of *Gossypium* species.
2. Proximity to existing accession points. Next, a map was prepared which prioritizes areas most distant from existing collections, thus targeting geographical gaps in existing collections.
3. Proximity to roads. Third, an accessibility map was developed, with ease of access running from 1 (directly on a road) to 0 (20 km from the road). A CIAT dataset on road networks was used.
4. Genetic erosion risk assessment. Fourth, an index of genetic erosion risk was produced, using a variety of data sources. The first component of the index is severity of soil degradation, using a dataset from the International Soil Reference and Information Centre.[20] The threat of habitat loss was crudely modelled based on population growth. Those areas undergoing the highest rate of population growth were assumed to present the greatest risk of genetic erosion. This analysis was made using a dataset on population density[21] for 1970 and 1990. The final component uses an ILRI dataset on cattle density for the African continent (Kruska *et al.*, 1995). It is assumed that a high density of cattle produces a significant risk to wild species of cotton. These three components were put together with equal weighting to form a single index of genetic erosion risk.

Figure 36.3 shows the results of combining these four components into a single map, which highlights

[20] www.isric.nl/GLASOD.htm
[21] www.grid.cr.usgs.gov/clearinghouse/datalist.html

those areas most in need of conservation because of a combination of high diversity and high risk of genetic erosion. Such a methodology can be adapted to account for the specific objectives of a conservation programme. It could be refined further by, for example, overlaying the results of complementarity analysis or giving higher priority to specific taxa. It is currently being applied by IPGRI, USDA and national partners in support of a programme of exploration for wild *Arachis* spp. in Bolivia.

Fieldwork

The ecogeographic study is essentially desk-based. It needs to be refined and expanded, and its results implemented, through fieldwork. GIS technology can contribute to this phase of genetic resources conservation through:

- the development of field aids; and
- providing information on the optimal timing of field visits.

Field aids

Maps are crucial to fieldwork. Unfortunately, published maps showing a specific factor of interest to the conservationist (or combination of factors) may not be available. An example of the use of GIS to develop mapping products for use in the field is provided by Greene *et al.* (1999a). A number of moisture and temperature maps synthesizing some 60 monthly climate variables at 500 m resolution were prepared in support of a joint USA/Russia forage collecting programme in the Caucasus. Even when published maps are available they may be unwieldy to use: a single soil map of the study region in the Caucasus was synthesized from four different map series. These maps allowed the identification of ecogeographic gradients in the field, the slope of which influenced decisions about sampling frequency. They were also used to monitor, during the course of the fieldwork, which combinations of environmental conditions were being adequately sampled and which were not.

Timing

Rainfall can be unpredictable in both space and time, especially in the arid and semiarid tropics, making it difficult sometimes to time field exploration work to coincide with the most appropriate stage of vegetation development. Satellite imagery can provide data on the state of vegetation with relatively short lag-times. Use of such data could allow collectors to be more precise in timing their visit, adding greatly to the cost-effectiveness of collecting trips. Meteosat and NOAA/AVHRR (National Oceanic and Atmospheric Administration Advanced Very High Resolution Radiometer) data on rainfall and the state of vegetation (as measured by NDVI), although fairly low in resolution, can be analysed to allow surveillance of the state of crops and vegetation over the growing season. Justice *et al.* (1987) describe the annual course of NDVI in a variety of East African vegetation types, and how this measure relates to the phenology of rainfall and plant growth. The USA/Russia forage collecting programme mentioned above used Landsat MSS (multi-spectral scanner) and TM (thematic mapper) imagery not only to locate the primary target of meadow patches, but also to evaluate the stability of their phenology across years (Afonin and Green, 1999). The patches with the most predictable phenology were considered the highest priority for field visits, because of the greater likelihood of finding seed of the target species for collection.

FAO's WINDISP software package has various features that are of relevance in this context, for example, tools to compare two images and analyse trends in a time-series of images. Much analysis may not be necessary, however, as there are sources of relevant pre-packaged data. It is possible for a collector in, say, Zimbabwe to determine the location of potential collection areas from the satellite imagery available on the Zimbabwe Meteorological Services Department's Vegetation Index Report web pages[22] or indeed FAO's GIEWS[23] and then visit the areas a matter of days later.

Design, management and monitoring of in situ reserves

The design of protected areas, including questions of optimal size and shape, zonation and networking (e.g. Given, 1994; Hawkes *et al.*, 1997) is a spatial problem, and spatial analysis in a GIS environment has been applied to them, although mainly in the

[22] www.weather.utande.co.zw/navigator/latest-vegetation.idc
[23] www.fao.org/giews/english/giewse.htm

context of ecosystem or animal conservation. Thus, Howard (1996) discusses how spatial information on species richness, distribution and abundance of an endangered species, disturbance and distribution of timber resources within a forest can be used to develop a zoning plan for a forest reserve, including different use areas, buffer zones and a core. This type of application presents the challenge of integrating demographic, socio-economic, cultural and other data on, and from, the human population with data on the biophysical environment and on the target taxon. For example, Fox *et al.* (1996) used GIS to map areas where the objectives of protecting red pandas and those of meeting the claims and grazing-land needs of communities living in the area either coincided or came into conflict. Harmsworth (1998) described an attempt to manage within a GIS information on the cultural values of different features of the landscape, flora ahd fauna, with a view to developing resource and environmental management plans more in tune with the requirements of local people.

A particularly comprehensive example of the application of GIS to the design of an *in situ* reserve is the work of Kremen *et al.* (1999) on the Masoala National Park in Madagascar. They defined a set of 13 design criteria relating to ecological and socio-economic sustainability, which, for example, included that special consideration should be given to rare and threatened habitats and species, that the limits of the park should follow natural features to make them easy to respect and that buffer areas should be large enough to meet the subsistence need of the human population surrounding the park. Information needs were then identified, combining maps, satellite imagery, biodiversity surveys and population studies. Data from these different sources were then analysed in a GIS environment to develop a park proposal balancing human and wildlife needs, including a core rainforest area, three satellite marine reserves and a large surrounding multiple-use zone. The proposal was discussed with local people and, once their approval was obtained, formed the basis of a national decree establishing the national park.

Given the rapid rate of environmental and socio-economic changes (including the threat of global climate change), it is important for reserve design to take into account possible future scenarios. GIS combined with computer modelling can be used to optimize reserve selection and management based not only on current climatic and vegetation patterns, but also under different assumptions of change. Menon and Bawa (1997) used GIS to document land use change, deforestation and habitat fragmentation in the Western Ghats of India, and investigate the socio-economic drivers of these processes to develop predictive models. These were then used to estimate the current and future effectiveness of the protected area network for biodiversity conservation.

Once the *in situ* reserve has been established, some form of management intervention may need to be implemented, and its results on the frequency, abundance, demography and genetic diversity of target species monitored (Maxted *et al.*, 1997b). GIS can assist in defining and implementing the management plan. Although most examples of such applications relate to ecosystem management (e.g. Kessell, 1990), Liu *et al.* (1995) used a spatial population model in a GIS to simulate the potential effects of a range of forest management plans on the size of the population of a bird species. Working in a GIS environment can ease the application of such analytical techniques of plant demography as transition matrix sensitivity analysis (Silvertown and Lovett Doust, 1993), thus hopefully stimulating what is still their incomplete integration into the armoury of plant genetic resources conservation.

Germplasm regeneration and evaluation

A key step in both germplasm multiplication and evaluation is to decide what sites are best for which accessions. The environment of the regeneration site must not only provide suitable conditions for reliable flowering and seed production, but also any necessary triggers for the different stages of plant development (Sackville Hamilton and Chorlton, 1997). Unless controlled pollination is practised, pollinators must also be present. On the other hand, pests and diseases must not be prevalent and populations of species related to the one(s) being regenerated should not be found nearby. Climate data and distribution maps for pests, diseases, pollinators and wild relatives can be overlapped using GIS to identify potential sites for regeneration. If the choice of locations for regeneration is limited, which is usually the case, such data can be used to estimate the relative suitability of different available sites for the regeneration of different accessions, by comparison with the environment at their site of collection (e.g. using FloraMap). They can also help to determine the unique needs of individual accessions so that growing procedures can be customized in sub-optimal sites.

As for germplasm evaluation, both ecologically optimal and stress locations are usually necessary. Classification of testing locations using GIS-derived climatic data (e.g. Pollak and Pham, 1989) can be used to determine the suitability of different sites for the evaluation of specific traits, or combinations of traits, of interest. If a site has already been chosen, GIS can be used to determine the extent of the region where similar conditions apply, and hence the relevance and potential impact of the evaluation work (Chapman and Barreto, 1996). CIMMYT's *Africa Maize Research Atlas* on CD-ROM (Hodson *et al.*, 1999) provides an interesting illustration of how GIS can support germplasm evaluation work. It provides a compendium of data layers and an easy-to-use GIS exploration tool, allowing researchers to 'characterize regions of interest and target and predict the potential impacts of promising germplasm'.

Use of genetic resources

The major aim of conserving germplasm is, of course, for it to be used. Adequate access policies, good linkages between genebanks and users and well-developed breeding and other use programmes are necessary conditions for genetic resources to be used to the full; but they are not sufficient. The lack of data on accessions and the large size of collections can still present significant bottlenecks to the use of conserved material. GIS can help to alleviate the problem in the first instance by improving the quality and quantity of the locality data associated with collections. Given adequate data on the origin of material, GIS can also help to identify particularly interesting germplasm and reveal the structure of diversity within collections. Finally, GIS can be used to assess the potential impact of the products of germplasm use.

Better locality data on accessions

Typically, a large proportion of germplasm accessions, especially older ones, lack latitude and longitude data, though elevation data are more common. GIS can help the collector and genebank curator in completing such passport data. Gazetteers, both published and on-line (e.g. the *Getty Thesaurus of Geographic Names*[24] and the *GEOnet Names Server*,[25] which provide the coordinates for about 3.5 million global geographic features, such as populated places), maps and collectors' notebooks can be used to determine latitude and longitude for accessions where only a locality name is available. With localities marked on paper maps, latitude and longitude may be estimated by using digitized maps of the area and employing the common facility of GIS software to return coordinates when a specific point is selected, say by moving the cursor to it.

Even when passport data are available, they are sometimes inaccurate. Rapid data visualization using a GIS can reveal obvious errors in latitude and longitude data resulting in accessions falling into the sea, a lake or the wrong country. Ecogeographic outliers will also be suspect. Chapman and Busby (1994) describe a method of spotting outliers which involves plotting the climate profile of each specimen and looking for entries that are out of step with the rest. Similarly, FLORAMAP can be used to identify accessions with climatically discordant localities and to carry out other data checking analyses (Jones and Gladkov, 1999). Data can be checked further by comparing locality descriptions in the passport data with GIS datasets of administrative boundaries. A methodology for such data-checking is presented by Hijmans *et al.* (1999).

Germplasm targeting

Next, once locality data are better and more complete, such data as altitude, major soil type, land use and vegetation, if missing from the field documentation, can be estimated by overlaying the locations of sites on different thematic base maps of the appropriate scale and reading off the map attributes. Interpolated climate surfaces can also be used to characterize conditions at each collection site given its latitude, longitude and altitude, for example using FLORAMAP. Steiner and Greene (1996) describe this process as 'retro-classification' of accessions and give an example of its usefulness using the *Lotus* collection of the US National Plant Germplasm System (NPGS) (see also Greene *et al.*, 1999b).

Retro-classification of the collection site can guide the use of the material by focusing the attention of users on the most promising material for their specific

[24] www.shiva.pub.getty.edu/tgn_browser/
[25] www.164.214.2.59/gns/html/index.html

purposes. For example, although they have been developed in the context of introduction of tree species, the climate matching programmes described by Booth (1990; 1999) for various countries are valuable tools for targeting the introduction of germplasm to specific areas for evaluation and use based on the conditions where it was collected. Similarly, Chapman and Taba (quoted by Chapman and Barreto, 1996) overlaid germplasm locations on soil maps to identify candidates from an extensive maize collection that might be adapted to alkaline soils. A geographical approach was also used by Beebe *et al.* (1997) to identify edaphic adaptation in beans as part of a breeding programme.

Structure of collections

'A core collection is a selected and limited set of accessions derived from an existing germplasm collection, chosen to represent the genetic spectrum in the whole collection and including as much as possible of its genetic diversity' (Brown, 1993). The core collection can be used to evaluate traits that are expensive or time-consuming to measure on the whole collection, for example complex yield and quality traits and general combining ability with local germplasm. These studies can then be used to identify material in the rest of the collection that might require further investigation.

Different workers have employed passport, characterization (taxonomic, morphological, molecular, etc.) or evaluation data, often in different, usually hierarchical, combinations, to classify accessions into groups and identify duplicates. However, the importance of location data is widely acknowledged and agroclimatic conditions at the collection site is often one of the more important types of criteria used in the classification process. The case of the *Phaseolus vulgaris* core collection described by Tohme *et al.* (1995) illustrates this – and the role of GIS – well. First, regions were prioritized based on the history of the crop. Then, interpolated surfaces for four parameters (length of growing season, photoperiod, soil type and moisture regime) were used to define 54 distinct environments, and each 10-minute grid cell was assigned to one of these classes. Passport data were then used to match each landrace accession to an environmental class. Finally, accessions in each environmental class were stratified according to characterization data (growth habit, grain colour and grain size) and selections made at random from within each stratum within each environmental class.

Environmental data on accessions derived from analysis using GIS can be used to investigate the structure of genetic diversity collections without going as far as developing core collections. Thus, analysis of various datasets using FLORAMAP, including *Passiflora* spp. in the Andes (Sergio Segura, personal communication, 2000), wild rice in southern Africa (Kihika Kiambi, personal communication, 2000) and *Stylosanthes* in South America (Jones and Gladkov, 1999), has revealed correspondences between the groupings of accessions by clustering using the climatic characteristics of collection sites and those derived from isozyme and DNA marker data. Indeed, Chapman and Barreto (1996) have suggested that as the linkage between genetic resources and environmental data is explored, 'geographic trait loci' may be identified, where particular parts of the genome are linked to adaptation to different conditions.

Assessing impact

GIS has a role to play not only in facilitating the use of germplasm, but also in predicting the likely impact of that use. Thus, a breeding programme could weigh the relative effects of achieving a range of different improvement objectives, and then concentrate on the ones likely to have the greatest impact with a particular priority target group. An example is provided by recent work at CIP on frost tolerance in potato (Hijmans *et al.*, 2000b). A crop growth simulation model was used in conjunction with high-resolution monthly climate surfaces for the Altiplano of southern Peru and northern Bolivia to predict the likely effect on yield in different areas within the region of increasing the frost tolerance of a standard variety by specific amounts. This type of analysis is strongly dependent on the availability of the kind of detailed, within-country, geo-referenced crop distribution data that are being assembled by CGIAR Centres (e.g. Huaccho and Hijmans, 1999; Hyman, 1999).

Prospects

The application of GIS to estimating the impact of improved germplasm opens the way for a fully integrated approach to the use of GIS in plant genetic resources management, in which GIS-based analysis guides and facilitates the process from start to finish and back again. First, the likely impact of a particular potential use product in a given area is assessed using GIS; for example a more frost-tolerant potato variety. Next, retro-classification of past collection sites is carried out with GIS to identify germplasm in existing national and international collections which could be useful in developing the proposed product. In the potato case, interpolated surfaces for the occurrence or severity of frost in the Andes could be used. If sufficient suitable material is not available in genebanks, climatic data are used to identify areas where further germplasm collection should be done. This could be approached by superimposing maps of frost severity and potato cultivation, or by running the collecting localities of frost-tolerant accessions through FloraMap. Once promising material has been obtained, GIS is used to identify appropriate evaluation sites. The best material coming out of the evaluation programme is then used to develop the improved product, the actual performance of which is finally fed back into the impact model to start the cycle all over again.

The potential is clearly there for GIS-based analysis of data on germplasm accessions to add greatly to the value of the data, and thus of the germplasm to which they pertain. It should significantly enhance the cost-effectiveness of conservation efforts, and facilitate use of germplasm by breeders and others. However, it must be admitted that GIS technology has not been taken up by plant genetic resources conservation programmes to the extent that one might have predicted on the basis of its potential. Part of the reason is that many such programmes, particularly in developing countries, have significant resource constraints, and GIS hardware, software and data are perceived as being expensive, difficult to obtain and complex to use.

The perceived 'barriers to entry' may still be too high, but the current revolution in GIS technology is putting both the data and the analytical tools within the reach of many. However, if the potential of GIS in the field of plant genetic resources conservation and use is to be fulfilled, the plant genetic resources conservation community needs to take positive action itself at the international, regional and national levels, and not simply wait for the technology to come to its aid. At the international level, the CGIAR Centres must show the same leadership in the application of GIS-based approaches to plant genetic resources conservation and use as they have been showing in their application to natural resources management. An important first step will be the integration of more GIS capability into the International Crop Information System (ICIS) being developed by CIMMYT (Fox and Skovmand, 1996) and the on-line database of the System-wide Information Network for Genetic Resources (SINGER[26]). In fact, Web-based tools, where maps and accession points can be manipulated on-line with easy-to-use, browser-based, graphical interfaces will help wider adoption of GIS by plant genetic resources conservation programmes.

IPGRI has been collaborating with other CGIAR Centres, in particular CIAT and CIP, in implementing a three-pronged strategy of support to GIS adoption by national and regional programmes. This entails the following activities:

- development, testing and dissemination of a menu of basic methodologies;
- development, promotion and distribution of easy-to-use, inexpensive software tools (and data) specifically aimed at carrying out the analyses alluded to in the previous point; and
- awareness-building.

The approach that has been taken is to implement a series of case studies to develop methodologies, whenever possible based on the use of specialized software such as FloraMap, SID and DIVA. The hope is that the results of such case studies will be used to build awareness of the potential of GIS technology at both the technical and decision-making levels within the plant genetic resources conservation community.

At the regional level, plant genetic resources networks have a potentially pivotal role to play in the process of GIS adoption. If GIS technology is still not within the reach of many individual national programmes, it should be possible to find a mechanism whereby the members of a regional network can share the costs, as well as benefits, of GIS adoption. GIS provides an ideal environment within which to manage and analyse data from a number of countries in a common framework, and thus develop both regional and national conservation strategies based on a broad vision of the problems and opportunities.

At the national level, genebanks need to forge linkages with institutions which already have GIS capacity, for example meteorological services, the geography departments of universities, mapping and survey departments and national statistical and natural resources institutes. This will guarantee

[26] www.singer.cgiar.org/

access to data and expertise in the longer term and, therefore, sustainability. It will also eventually place genebank managers in a position to lobby decision-makers with tangible evidence of the gains in effectiveness that can accrue from the use of this technology, so that both national and international resources may be accessed for the eventual acquisition of in-house GIS capability.

It is certainly possible to visualize in a not too distant future GIS being used routinely and in an integrated fashion in support of plant genetic resources conservation and use as outlined above. However, before we are swept away by our enthusiasm, it is worth reminding ourselves of the fact that none of the analyses that have been described here would be possible without adequate data on germplasm, in particular accurate locality data in coordinate form. At the risk of seeming prosaic, the future of GIS adoption, with the improvements in efficiency of plant genetic resources programmes which it will bring, is likely to be more dependent on these programmes having solid documentation components, that is, effective data management systems and lots of data to manage, than on methodological or software innovations.

Acknowledgements

We have greatly profited from discussion of the issues presented here with David Williams (IPGRI), Karen Williams (USDA) and Glenn Hyman (CIAT). Thanks are due to USDA and CIAT for facilitating AJ's IPGRI-CIAT internship. We are grateful to Kihika Kiambi (IPGRI) for letting us use the results of his *Oryza* work and for making valuable comments on the paper. Vojtech Holubec (Research Institute of Crop Production, Czech Republic) kindly made available to us the *Gossypium* dataset he originally put together in the late 1980s as part of an IBPGR-funded study at Texas A&M University. Nelly Giraldo (IPGRI) helped with the figures. FLORAMAP is one of the first truly user-friendly GIS software-cum-data tools to be aimed at the plant conservationist, and we would like to thank its originator, Peter Jones (CIAT), for showing the way.

References

Afonin, A. and Greene, S.L. (1999) Germplasm collecting using modern geographic information technologies: directions explored by the N.I. Vavilov Institute of Plant Industry. In: Greene, S.L. and Guarino, L. (eds) *Linking Genetic Resources and Geography: Emerging Strategies for Conserving and Using Crop Biodiversity.* CSSA Special Publication No. 27. ASA and CSSA, Madison, Wisconsin, pp. 75–85.

Bailey, T.C. (1994) A review of statistical spatial analysis in geographical information systems. In: Fotheringham, S. and Rogerson, P. (eds) *Spatial Analysis and GIS.* Taylor & Francis, London, pp. 13–44.

Beebe, S., Lynch, J., Galwey, N., Tohme, J. and Ochoa, I. (1997) A geographical approach to identify phosphorus-efficient genotypes among landraces of common bean. *Euphytica* 95, 325–336.

Booth, T. (1990) Mapping regions climatically suitable for particular tree species at the global scale. *For. Ecol. Manag.* 36, 47–60.

Booth, T. (1999) Matching germplasm to geography: environmental analysis for plant introduction. In: Greene, S.L. and Guarino, L. (eds) *Linking Genetic Resources and Geography: Emerging Strategies for Conserving and Using Crop Biodiversity.* CSSA Special Publication No. 27. ASA and CSSA, Madison, Wisconsin, pp. 63–74.

Booth, T.H., Searle, S.D. and Boland, D.J. (1989) Bioclimatic analysis to assist provenance selection for trials. *New For.* 3, 225–234.

Brown, A.H.D. (1993) The core collection at the crossroads. In: Hodgkin, T., Brown, A.H.D., van Hintum, Th.J.L. and Morales, E.A.V. (eds) *Core Collections of Plant Genetic Resources.* John Wiley & Sons, Chichester, UK, pp. 3–19.

Busby, J.R. (1991) BIOCLIM – a bioclimate prediction system. In: Margules, C.R. and Austin, M.P. (eds) *Nature Conservation: Cost Effective Biological Surveys and Data Analysis.* CSIRO, Melbourne, Australia, pp. 4–68.

Carpenter, G., Gillison, A.N. and Winter, J. (1993) DOMAIN: a flexible modelling procedure for mapping potential distributions of plants and animals. *Biodiversity and Conservation* 2, 667–680.

Carter, S.E., Fresco, L.O., Jones, P.G. with Fairbairn, J.N. (1992) *An Atlas of Cassava in Africa: Historical, Agroecological and Demographic Aspects of Crop Distribution.* CIAT, Cali, Colombia.

Chapman, A.D. and Busby, J.R. (1994) Linking plant species information to continental biodiversity inventory, climate modeling and environmental monitoring. In: Miller, R.I. (ed.) *Mapping the Diversity of Nature.* Chapman and Hall, London, pp. 179–195.

Chapman, S.C. and Barreto, H.J. (1996) Using simulation models and spatial databases to improve the efficiency of plant breeding programs. In: Cooper, M. and Hammer, G.L. (eds) *Plant Adaptation and Crop Improvement.* CAB International, Wallingford, UK, pp. 563–587.

Chatelain, C., Gautier, L. and Spichiger, R. (1996) A recent history of forest fragmentation in south-western Ivory Coast. *Biodiversity and Conservation* 5, 37–53.

Corbett, J.D. (1998) Classifying maize production zones in Kenya through multivariate cluster analysis. In: Hassan, R.M. (ed) *Maize Technology Development and Transfer: A GIS Application for Research Planning in Kenya.* CAB International, Wallingford, UK, pp. 15–25.

Csuti, B., Polasky, S., Williams, P.H., Pressey, R.L., Camm, J.D., Kershaw, M., Kiester, A.R., Downs, B., Hamilton, R., Huso, M. and Sahr, K. (1997) A comparison of reserve selection algorithms using data on terrestrial vertebrates in Oregon. *Biological Conservation* 80, 83–97.

Epperson, B.K. (1993) Recent advances in correlation studies of spatial patterns of genetic variation. *Evolutionary Biology* 27, 95–155.

Faith, D.P. and Walker, P.A. (1996) Environmental diversity: on the best-possible use of surrogate data for assessing the relative biodiversity of sets of areas. *Biodiversity and Conservation* 5, 399–415.

Ferguson, M.E., Ford-Lloyd, B.V., Robertson, L.D., Maxted, N. and Newbury, H.J. (1998) Mapping the geographical distribution of genetic variation in the genus *Lens* for the enhanced conservation of plant genetic diversity. *Molecular Ecology* 7, 1743–1755.

Fox, J., Yonzon, P. and Podger, N. (1996) Mapping conflicts between biodiversity and human needs in Langtang National Park, Nepal. *Conservation Biology* 10, 562–569.

Fox, P.N. and Skovmand, B. (1996) The International Crop Information System (ICIS) – connects genebank to breeder to farmer's field. In: Cooper, M. and Hammer, G.L. (eds) *Plant Adaptation and Crop Improvement.* CAB International, Wallingford, UK, pp. 317–326.

Gastellu-Etchegorry, J.P., Estreguil, C., Mougin E. and Laumonier, Y. (1993) A GIS based methodology for small scale monitoring of tropical forests – a case study in Sumatra. *International Journal of Remote Sensing* 14, 2349–2368.

Gaston, K.J. (1996) Species richness: measure and measurement. In: Gaston, K.J. (ed.) *Biodiversity: A Biology of Numbers and Difference.* Blackwell Science, Oxford, UK, pp. 77–113.

Given, D.R. (1994) *Principles and Practice of Plant Conservation.* Chapman and Hall, London.

Greene, S.L., Hart, T. and Afonin, A. (1999a) Using geographic information to acquire wild crop germplasm: I. Map development and field use. *Crop Science* 39, 836–842.

Greene, S.L., Hart, T. and Afonin, A. (1999b) Using geographic information to acquire wild crop germplasm: II. Post collection analysis. *Crop Science* 39, 843–849.

Guarino, L. (1995) Geographic information systems and remote sensing for the plant germplasm collector. In: Guarino, L., Ramanatha Rao, V. and Reid, R. (eds) *Collecting Plant Genetic Diversity. Technical Guidelines.* CAB International, Wallingford, UK, pp. 315–328.

Guarino, L. (1999) Approaches to measuring genetic erosion. In: Serwinski, J. and Faberova, I. (eds) *Proceedings of the Technical Meeting on the Methodology of the FAO World Information and Early Warning System on Plant Genetic Resources, 21–23 June 1999, Prague, Czech Republic.* Research Institute of Crop Production, Prague, Czech Republic and FAO, Rome, Italy, pp. 26–28.

Guarino, L., Maxted, N. and Sawkins, M. (1999) Analysis of geo-referenced data and the conservation and use of plant genetic resources. In: Greene, S.L. and Guarino, L. (eds) *Linking Genetic Resources and Geography: Emerging Strategies for Conserving and Using Crop Biodiversity.* CSSA Special Publication No. 27. ASA and CSSA, Madison, Wisconsin, pp. 1–24.

Harmsworth, G. (1998) Indigenous values and GIS: a method and a framework. *Indigenous Knowledge and Development Monitor* 6, 3–7.

Hawkes, J.G., Maxted, N. and Zohary, D. (1997) Reserve design. In: Maxted, N., Forde-Lloyd, B.V. and Hawkes, J.G. (eds) *Plant Genetic Conservation: The* In Situ *Approach.* Chapman and Hall, London, pp. 132–143.

Hijmans, R.J., Schreuder, M., De la Cruz, J. and Guarino, L. (1999) Using GIS to check co-ordinates of genebank accessions. *Genetic Resources and Crop Evolution* 46, 291–296.

Hijmans, R.J., Garrett, K.A., Huaman, Z., Zhang, D.P., Schreuder, M. and Bonierbale, M. (2000a) Assessing the geographic representativeness of genebank collections: the case of Bolivian wild potatoes. *Conservation Biology.*

Hijmans, R.J., Condori, B., Carillo, R. and Kropff, M.J. (2000b) Estimating the potential impact of frost tolerant potato cultivars in the Altiplano (Peru and Bolivia). *Proceedings of the Third International Symposium on System Approaches for Agricultural Development (SAAD3).* CD-ROM publication. CIP, Lima, Peru.

Hodson, D.P., Rodriguez, A., White, J.W., Corbett, J.D., O'Brian, R.F. and Banziger, M. (1999) *Africa Maize Research Atlas* (v. 2.0). CIMMYT, Mexico DF, Mexico.

Holubec, V. (1998) Extinction threat of wild African *Gossypium* species in their center of diversity. In: Damania, A.B.,

Valkoun, J., Willcox, G. and Qualset, C.O. (eds) *The Origins of Agriculture and Crop Domestication.* ICARDA, Aleppo, Syria, pp. 286–290.

Howard, P.C. (1996) Guidelines for the selection of forest nature reserves, with special reference to Uganda. In: Bennun, L.A., Aman, R. and Crafter, S. (eds) *Conservation of Biodiversity in Africa. Local Initiatives and Institutional Roles.* Centre for Biodiversity, Nairobi, Kenya, pp. 245–262.

Huaccho, L. and Hijmans, R.J. (1999) A global geo-referenced database of potato production for 1995–1997 (GPOT97). *Production Systems and Natural Resource Management Department Working Paper* 1. CIP, Lima, Peru. www.cipotato.org/data/potato_atlas/gpot97.htm

Humphries, C.J., Williams, P.H. and Vane-Wright, R.I. (1995) Measuring biodiversity value for conservation. *Annual Review of Ecology and Systematics* 26, 93–111.

Hutchinson, C.F. and Weiss, E. (1999) In: Serwinski, J. and Faberova, I. (eds) *Proceedings of the Technical Meeting on the Methodology of the FAO World Information and Early Warning System on Plant Genetic Resources. 21–23 June 1999, Prague, Czech Republic.* Research Institute of Crop Production, Prague, Czech Republic and FAO, Rome, Italy, pp. 19–25.

Hyman, G. (1999) Crop distribution mapping: applications and techniques for broad-scale analysis of crop geography. In: Pande, S., Johansen, C., Lauren, J. and Bantilan, F.T. Jr (eds) *GIS Analysis of Cropping Systems. Proceedings of an International Workshop on Harmonization of Databases for GIS Analysis of Cropping Systems in the Asia Region.* Cornell University, Ithaca, USA and ICRISAT, Patencheru, India, pp. 91–96.

Jones, P.G. and Gladkov, A. (1999) *FloraMap: A Computer Tool for the Distribution of Plants and Other Organisms in the Wild.* CIAT, Cali, Colombia.

Jones, P.G., Beebe, S.E., Tohme, J. and Galwey, N.W. (1997) The use of geographical information systems in biodiversity exploration and conservation. *Biodiversity Conservation* 6, 947–958.

Justice, C.O., Holben, B.N. and Gwynne, M.D. (1987) Monitoring East African vegetation using AVHRR data. *International Journal of Remote Sensing* 7, 1453–1474.

Kessell, S.R. (1990) An Australian geographic information and modeling system for natural area management. *International Journal of GIS* 4, 333–362.

Kirkpatrick, J.B. (1974) The use of differential systematics in geographic research. *Area* 6, 52–53.

Kremen, C., Razafimahatratra, V., Guillery, R.P., Rakotomalala, J., Weiss, A. and Ratsisompatrarivo, J.-S., (1999) Designing the Masoala National Park in Madagascar based on biological and socio-economic data. *Conservation Biology* 13, 1055–1068.

Kruska, R.L., Perry, B.D. and Reid, R.S. (1995) Recent progress in the development of decision support systems for improved animal health. In: *AfricaGIS'95 Documents. Proceedings of the AfricaGIS'95 Conference, 6–10 March 1995, Abidjan, Ivory Coast,* pp. 524–538.

Ledig, F.T. (1992) Human impacts on genetic diversity in forest ecosystems. *Oikos* 63, 87–108.

Liu, J.G., Dunning, J.B. and Pulliam, H.R. (1995) Potential effects of a forest management plan on Bachman's sparrows (*Aimophila aestivalis*): linking a spatially explicit model with GIS. *Conservation Biology* 9, 62–75.

Maxted, N., van Slageren, M.W. and Rihan, J.R. (1995) Ecogeographic surveys. In: Guarino, L., Ramanatha Rao, V. and Reid, R. (eds) *Collecting Plant Genetic Diversity. Technical Guidelines.* CAB International, Wallingford, UK, pp. 255–285.

Maxted, N., Forde-Lloyd, B.V. and Hawkes, J.G. (1997a) Complementary conservation strategies. In: Maxted, N., Forde-Lloyd, B.V. and Hawkes, J.G. (eds) *Plant Genetic Conservation: The* In Situ *Approach.* Chapman and Hall, London, pp. 15–39.

Maxted, N., Guarino, L. and Dulloo, M.E. (1997b) Management and monitoring. In: Maxted, N., Forde-Lloyd, B.V. and Hawkes, J.G. (eds) *Plant Genetic Conservation: The* In Situ *Approach.* Chapman and Hall, London, pp. 144–159.

Menon, C. and Bawa, K.S. (1997) Applications of geographic information systems, remote sensing, and a landscape ecology approach to biodiversity conservation in the Western Ghats. *Current Science* 73, 134–145.

Miller, R.I. (1986) Predicting rare plant distribution patterns in the southern Appalachians of the south-eastern USA. *Journal of Biogeography* 13, 293–311.

Monestiez, P., Goulard, M. and Charmet, G. (1994) Geostatistics for spatial genetic structures: study of wild populations of perennial ryegrass. *Theoretical and Applied Genetics* 88, 33–41.

Monmonier, M. (1973) Maximum-difference barriers: an alternative numerical regionalization method. *Geogr. Anal.* 3, 245–261.

Nabhan, G.P. (1991) Wild *Phaseolus* ecogeography in the Sierra Madre Occidental, Mexico. *Systematic and Ecogeographic Studies of Crop Gene Pools* No. 5. IBPGR, Rome, Italy.

Nelson A., LeClerc, G. and Grum, M. (1997) The development of an integrated Tcl/Tk and C interface to determine, visualize and interrogate infraspecifc biodiversity. Internal document. GIS Laboratory, CIAT, Cali, Colombia.

Pickersgill, B. (1984) Migrations of chili peppers, *Capsicum* spp., in the Americas. In: Stone, D. (ed.) *Pre-Columbian Plant Migration. Papers of the Peabody Museum of Archaeology and Ethnology*, Vol. 76. Harvard University Press, Boston, Massachusetts, pp. 105–123.

Pigliucci, M. and Barbujani, G. (1991) Geographical pattern of gene frequencies in Italian populations of *Ornithogalum montanum* (Liliaceae). *Genetic Research* 58, 95–104.

Pollak, L.M. and Corbett, J.D. (1993) Using GIS datasets to classify maize-growing regions in Mexico and central America. *Agronomy Journal* 85, 1133–1138.

Pollak, L.M. and Pham, H.N. (1989) Classification of maize testing locations in sub-Saharan Africa by using agroclimatic data. *Maydica* 34, 1133–1138.

Rebelo, A.G. and Sigfried, W.R. (1992) Where should nature reserves be located in the Cape Floristic Region, South Africa? Models for the spatial configuration of a reserve network aimed at maximizing the protection of diversity. *Conservation Biology* 6, 243–252.

Sackville Hamilton, N.R. and Chorlton, K.H. (1997) *Regeneration of Accessions in Seed Collections: A Decision Guide.* Handbook for Genebanks No. 5. IPGRI, Rome, Italy.

Scott, J.M., Davis, F., Csuti, B., Noss, R., Butterfield, B., Groves, C., Anderson, H., Caicco, S., Dérchia, F., Edwards, T.C., Ulliman, J. and Wright, R.G. (1993) Gap analysis: a geographic approach to protection of biological diversity. *Wildlife Monographs* 123, 1–41.

Sebastini, M., Sambrano, A., Villamizar, A. and Villalba, C. (1989) Cumulative impact and sequential geographical analysis as tools for land use planning. A case study: Laguna La Reina, Miranda state, Venezuela. *Journal of Environmental Management* 29, 237–248.

Silvertown, J.W. and Lovett Doust, J. (1995) *Introduction to Plant Population Biology.* Blackwell Scientific Publications, Oxford, UK.

Skole, D. and Tucker, C. (1993) Tropical deforestation and habitat fragmentation in the Amazon: satellite data from 1978 to 1988. *Science* 260, 1905–1910.

Steiner, J.J. and Greene, S.L. (1996) Proposed ecological descriptors and their utility for plant germplasm collections. *Crop Science* 36, 439–451.

Stockwell, D.R.B. and Noble, I.R. (1992) Induction of sets of rules from animal distribution data: a robust and informative method of data analysis. *Math. Comput. Simul.* 33, 385–390.

Templeton, A.R., Shaw, K., Routman, E. and Davis, S.K. (1990) The genetic consequences of habitat fragmentation. *Ann. Missouri Bot. Gard.* 77, 13–27.

Tohme, J., Jones, P., Beebe, S. and Iwanaga, M. (1995) The combined use of agroecological and characterization data to establish the CIAT Phaseolus vulgaris core collection. In: Hodgkin, T., Brown, A.H.D., van Hintum, Th.J.L. and Morales, E.A.V. (eds) *Core Collections of Plant Genetic Resources.* John Wiley & Sons, Chichester, UK, pp. 95–107.

Treweek, J. (1999) *Ecological Impact Assessment.* Blackwell Science, Oxford, UK.

Vane-Wright, R.I., Humphries, C.J. and Williams, P.H. (1991) What to protect? Systematics and the agony of choice. *Biological Conservation* 55, 235–254.

Veitch, N., Web, N.R. and Wyatt, B.K. (1995) The application of geographic information systems and remotely sensed data to the conservation of heathland fragments. *Biological Conservation* 72, 91–97.

Walker, P.A. (1990) Modelling wildlife distributions using a geographic information system: kangaroos in relation to climate. *Journal of Biogeography* 17, 279–289.

Zheng, D., Wallin, D.O., Hao, Z. (1997) Rates and patterns of landscape change between 1972 and 1988 in the Changbai mountain area of China and North Korea. *Landscape Ecology* 12, 241–254.

37 Predicting Germplasm Differentiation Using GIS-derived Information

S.L. Greene,[1] M. Gritsenko,[1] G. Vandemark[1] and R.C. Johnson[2]
[1] *USDA ARS, Washington State University, Prosser, WA 99350, USA;*
[2] *USDA ARS WRPIS, Washington State University, Pullman, WA 99164, USA*

Introduction

Recognizing the influence that environment has on selection and gene flow, plant collectors have long capitalized on ecogeographic information to sample genetically diverse plant populations. The underlying premise has been that environmental descriptors can be used to detect plant population structure, and are much easier to obtain than phenotypic or genotypic descriptions of plant populations. Traditional sources of landscape and climate descriptions include maps, local expertise and on-site observation. In the past decade, plant collectors have increasingly capitalized on strategies developed by geographers and natural resource managers to characterize the environment. Three basic technologies are used: global positioning systems (GPS), remote sensing devices and geographic information system (GIS) software. GPS provides quick and accurate pinpointing of position. Remote sensing devices, such as satellites, describe the landscape using a wide array of attributes, at various spatial and temporal scales. GIS software allows us to organize, understand and ask questions about the spatial data that are collected. Maps can be produced from secondarily derived information. For example, geostatistical methods can be applied to a set of individual points (i.e. monthly rainfall and elevation data from meteorological stations) to produce a map that estimates the contiguous landscape coverage of an environmental attribute such as rainfall. Another common map involves the reduction of complex data using classification analysis. For example, monthly rainfall, snow and relative humidity can be grouped into similar classes to produce a single map that classifies an area into moisture zones. The widely recognized linkage between plant population adaptation to the surrounding environment makes this type of information particularly valuable for those interested in sampling, conserving and using plant genetic diversity. Guarino (1995) and Guarino *et al.* (1999) provide an introduction to GIS technology and discuss how geo-referenced data can be used in collecting and managing plant genetic resources.

As with other technologies, GIS-derived data and analysis need to be correctly applied to genetic resource questions to ensure valid answers emerge. Genetic structure is influenced by factors other than environment, such as reproductive biology, effects of migration, biotic selection pressure (i.e. pathogens) and human selection pressure (Loveless and Hamrick, 1984; Hedrick, 1986; Slatkin, 1987). Greene and Hart (1999) briefly review factors that contribute to genetic differentiation in plant populations that can be attributed to ecogeographic influence. Another consideration for effectively using environmental information to understand population structure is to ensure spatial and/or temporal scales match between plant and environmental attributes. Hart (1999) described

 Managing Plant Genetic Diversity
(eds J.M.M. Engels, V. Ramanatha Rao, A.H.D. Brown and M.T. Jackson)

the significance of scale in acquiring meaningful measures of the environment. Accurately measuring or estimating environmental attributes is challenging in its own right, and the cost can be high, depending on the level of GIS expertise needed and availability of datasets. However, costs are decreasing substantially as more data are available at a lower cost or freely available over the Internet, and GIS software packages become more accessible to non GIS-specialists.

Considering the appeal and complexities of using secondarily derived environmental data as surrogate information for population genetic structure, we felt it was important to validate the usefulness of GIS-derived information in predicting intraspecific differentiation in plant populations. Our hypothesis was that GIS-derived ecogeographic description would assist in distinguishing accessions of the same species, collected from the same area. To test the predictive value of GIS-derived information we classified 35 accessions of red clover collected in the Caucasus Mountains using both morphological data collected in a common environment and random amplified polymorphic DNA (RAPD) marker data. GIS-derived environmental attributes such as elevation and climate were used to classify the same 35 accessions. We then examined how well the GIS-derived classification corresponded to classifications based on morphologic and genetic attributes obtained from the collected germplasm.

Collecting Germplasm in the Caucasus Mountains, Russia

A joint expedition between the United States Department of Agriculture National Plant Germplasm System, and the Vavilov Institute of Plant Industry (VIR) was carried out in 1995. The target region covered an area that extended 600 km east–west and 250 km north–south, between 43–45° N and 38–42° E in the Western Caucasus Mountains in southern Russia. This is a floristically rich area laying adjacent to Vavilov's Southwest Asiatic centre of diversity for cultivated plants. Prior to the collecting trip, maps were developed using primary and secondarily derived ecogeographic information (Greene *et al.*, 1999a). VIR has made subsequent explorations to the same region in 1997 and 1999 (A. Afonin, St Petersburg, Russia, personal communications, 2000). VIR scientists continue to use and refine the GIS database developed in 1995.

Producing a GIS database and maps

Greene *et al.* (1999a) described the procedures used to develop the Caucasus Mountain maps and GIS database. We will briefly review here how remotely sensed elevation data and long-term weather station data were used to develop the secondarily derived topographical and climate zone maps that were used to guide the collecting of germplasm. Elevation data were originally acquired through aerial surveillance and represented spot height data at a 500-m horizontal interval. The database was acquired from academic sources and used to develop a digital elevation model (DEM). A DEM is a digital representation of elevation over a two-dimensional surface by the regular array of height referenced to a common datum.

The Caucasus DEM data were reported in metres above sea level. Slope and aspect maps were developed at a 250-m horizontal interval. Slope data were reported in degrees, and similar to elevation, represented a quantitative attribute. The aspect data were reported using an eight-class nominal scale centred on the cardinal directions, and thus represented a qualitative attribute. Climate data were obtained from 111 meteorological stations distributed throughout the collection area. Accumulated rainfall, absolute monthly minimum and maximum temperatures (i.e. coldest and hottest day of each month), average relative humidity and average wind speed were reported as long-term monthly averages. Initially, monthly climate surfaces were created reflecting the effects of station data and neighbouring terrain for each climate attribute. General trends were examined in the monthly maps. Monthly accumulated rainfall and monthly average minimal temperature were aggregated into logical seasons: winter (November to February), spring transition (March to April); early growing season (May to June), late growing season (July to August) and autumn transition (September to October). Humidity was averaged through the early and late growing season (March to June, July to October), maximum temperature averaged over the growing season (May to August), and wind speed averaged over summer (May to September) and winter (November to February).

Seasonal attributes were then classified to produce maps depicting temperature and moisture zones. For the temperature zone map, all minimum and maximum temperature groups were layered in a single unweighted six-variable map file, and a cluster analysis carried out that resulted in a 14-zone temperature map across the collection area. Seasonal groups of rainfall, relative humidity, and wind speed were layered in a single unweighted nine-variable map file and a cluster analysis carried out on the attributes that resulted in a 15-zone moisture map across the study area.

Germplasm sampling strategy and post collection site analysis

The moisture and temperature zone maps, and elevation slope and aspect maps were used in the 1995 collecting trip in the Caucasus Mountains. The collection trip focused on wild populations with an emphasis on sites not recently disturbed by natural or anthropogenic forces. Generally, 50 or more individual plants were sampled from each population. Targeted species were sampled as they were encountered. Local collecting site information, including soil pH was recorded using the data form of Steiner and Greene (1996). Latitude and longitude positions were obtained using GPS.

A total of 550 accessions, representing 37 genera and approximately 109 species, were collected from 149 sites during the 3-month period. This paper focuses on 35 accessions of red clover (*Trifolium pratense* L.) that were collected from 35 sites. The red clover accessions were separated into two classes, based on collection site data (Greene *et al.* 1999b). The first group was designated as unadapted. Based on work by Cocks and Ehrman (1987), which suggested that ecotypic differentiation occurred within 20 years for forage species, and computer simulations carried out by Haywood (1991), which indicated that 10 to 20 generations was sufficient for ecotypic differentiation to occur, we defined populations as unadapted if passport data suggested the population had not been locally present for 20 years (Greene *et al.*, 1999a). Populations sampled from these types of sites may not reflect genetic adaptation to the collection site. These included accessions collected along roadsides, streams, construction areas, overgrazed meadows, or areas otherwise disturbed. The first group also contained accessions collected from sites representing microhabitat that was anomalous to the broader climatic regime described by the GIS database. Populations collected from this class of sites would probably not correlate with GIS-derived information. The second group was designated as adapted and consisted of accessions that had been collected at sites having no evidence of recent demographic disturbance, and that were influenced predominately by the broader climatic regime described by the GIS database. Populations sampled from these types of sites were more likely to reflect adaptation to the site.

Evaluating Caucasus Red Clover Accessions

In 1998, original seed of 35 Caucasus red clover accessions, and five US cultivars, were planted in Prosser, Washington, and Pullman, Washington. Approximately 60 plants were evaluated at each location. The following attributes were measured: plant height, growth habit, winter survival, leaf mark, percentage plants flowering in the seedling year, stem pubescence, flower length, corolla tube length, length of the shortest corolla lobe, length of the longest corolla lobe, standard colour, corolla tube colour, keel colour, days to bloom from 1 Jan in the second year, and bloom duration in the second year.

In addition to the morphological data, data from RAPD markers were obtained. Leaf tissue was harvested and combined to form a bulk leaf sample that represented 18 individual plants. DNA was extracted by the method of Doyle and Doyle (1989). A series of polymerase chain reactions (PCR) were conducted and amplification products were resolved, stained with ethidium bromide and visualized with a source of UV light. Each primer-plant sample combination was repeated at least twice, and negative control reactions lacking DNA were included for each primer. All gels were scored for both polymorphic and monomorphic bands. It was assumed that bands of the same molecular weight in different individuals were identical.

Based on a preliminary analysis, the following set of morphological attributes was selected to develop a morphological distance matrix: growth habit, flower length, pubescence, leaf mark, % flowering first year, winter survival, days to bloom, duration of bloom and standard colour. Euclidean distance coefficients were calculated to generate a set of morphological distance matrices for the two groups of accessions (unadapted, adapted). The RAPD data were used to construct a genetic distance matrix. Since red clover is a perennial outcrossing species, we used the procedure used by Nebauer *et al.* (1999) to calculate genetic distances in the perennial cross-pollinated species *Digitalis obscura* L. Lynch and Milligan (1994) recommended this procedure to minimize bias in the calculation of genetic distance using RAPD markers.

Genetic distance matrices were generated using 45 out of a total of 83 RAPD bands. The correspondence coefficient used was the simple matching (SM) coefficient. The RAPD data were used to generate two genetic distance matrices that were estimated for accessions collected from unadapted sites and adapted sites.

Distance matrices were also developed for the following GIS-derived attributes: DEM-derived elevation, slope, and aspect; moisture zone and temperature zone; and latitude and longitude coordinates. Distance was estimated using all ecogeographic attributes (i.e. elevation, slope and aspect and climate zones); elevation and climate; elevation and latitude; climate and latitude; and geographic distance. Euclidean distance coefficients were calculated to generate a set of distance matrices based on accessions from unadapted and adapted accessions.

We examined the correlation between the distance matrices generated from morphological, RAPD marker and GIS-derived data using Mantel's Test (Mantel, 1967). Correspondence between the morphological and RAPD distance matrices and each of these with geographic distance matrices were examined. A total of eight different GIS data matrices representing various combinations of GIS-derived attributes were compared against the matrices obtained from morphological and RAPD marker data. Calculation of distance matrices and matrices comparisons were carried out using the NTSYS-pc program version 2.02k (Rohlf, 1993).

Correspondence between Morphological and RAPD Marker Data

The morphological data grouped the 35 accessions into five distinct groups (Fig. 37.1). These groups remained distinct for both the adapted and unadapted accessions. Sixty-six per cent of the accessions fell into a single class that was characterized by a

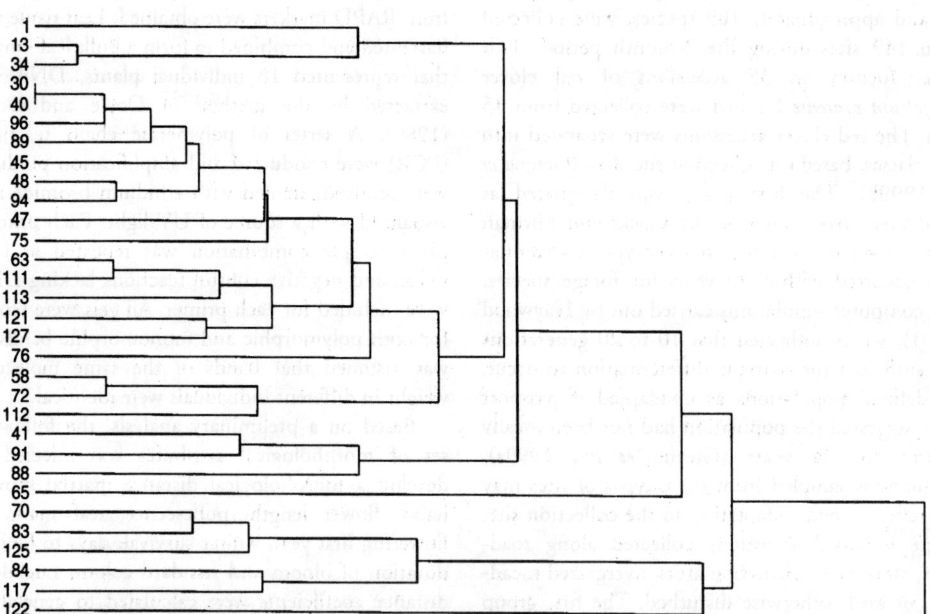

Fig. 37.1. Dendrogram resulting from cluster analysis of morphological attributes of red clover accessions collected in Russia and evaluated in Prosser, Washington.

mostly upright growth habit, good winter survival (85.0%), normal or delta leaf marks, less stem pubescence, first year bloom and had flowers that were smaller with pale, purplish standards. This group was among the earliest and longest blooming germplasm. A single accession made up the second class, and was distinguished by having the lowest winter survival (60%), and the earliest and longest bloom period. The third class of accessions had the greatest level of pubescence, slightly larger and darker flowers and was medium in maturity and bloom duration. The fourth class contained only a single member and was distinguished by an upright growth habit, the greatest winter survival, lowest percentage of first year flowering, the latest bloom date and one of the shortest durations of bloom. The fifth class of accessions was characterized by having an extended to basal leaf mark, large flowers, later bloom date and the shortest bloom duration. The RAPD marker data did not cluster the accessions into classes that were as distinct as those formed by the morphological data. When the distance matrices for adapted accessions were compared, the Mantel test was significant at $P < 0.01$, indicating that the morphological and RAPD data classified the 35 accessions in the same manner. However, for the unadapted group, there was no significant correlation between the morphological and RAPD distance matrices. This suggested that accessions we considered to be in genetic equilibrium with the environment grouped in a similar fashion, using either morphological or molecular marker data. However, for accessions not in equilibrium, morphological and molecular markers did not correspond.

Comparing Geographic Distance with Distance Estimates Based on Morphological and RAPD Markers

There was no correlation between geographic distance and RAPD marker distance or morphological distance for either the adapted or unadapted sets of accessions. Other studies have reported a lack of correspondence between geographic distance and distance measures based on RAPD markers. Although Nebauer *et al.* (1999) identified clusters of *D. obscura* (based on RAPD distance) that had been collected from the same geographic region within Spain, the correlation between genetic distance and geographic distance was not significant. In wild barley (*Hordeum spontaneum* Koch), Nevo *et al.* (1998) found no association between interpopulation genetic distance estimated using RAPD markers and geographic distance in accessions collected from Israel, Turkey and Iran. Steiner (1999) did not find a significant correlation between geographic distance and RAPD market distance in sets of birdsfoot trefoil (*Lotus corniculatus* L.) germplasm representing Old World distribution, and regional-scale distribution in the Caucasus Mountains. Generally, patterns of molecular marker and morphological variation have been associated with ecogeographic variation, not geographic distance *per se* (i.e. Berdahl *et al.*, 1999; Bennett, 2000; Beebe *et al.*, 2000).

Associating GIS-derived Ecogeographic Distance Estimates with Estimates Based on Morphological and RAPD Markers

There was a significant correlation between the morphological distance matrix and the various GIS-derived matrices. Distances that included elevation were generally the strongest, while distances that included only climate had the weakest correlation. The strongest correlation was with elevation and elevation and latitude. Group correspondence between morphological attributes and elevation and latitude was evident from scatter plots of the first and second eigenvectors (Fig. 37.2). For accessions we had identified as adapted, all distance matrices based on GIS-derived data were significantly correlated with the distance matrix based on morphological attributes. For accessions we had identified as unadapted, with the exception of the matrix containing only moisture and temperature zone data, the GIS distance matrices were significantly correlated with the morphological distance matrix. Assuming that the traits we measured contributed to fitness, this suggested that populations we had classified as unadapted were in fact adapted to the sites they were collected from, in that the morphological characteristics of accessions we had previously defined as unadapted were in fact correlated to specific environmental conditions.

The GIS-derived distance matrices did not generally correlate with the genetic distance matrix obtained from RAPD markers for either the adapted or unadapted populations. For accessions classified as unadapted, there was no association between any of the GIS-derived distance matrices and the genetic distance matrix based on RAPD markers. However, when adapted accessions were examined, there was a significant but weak correlation between the distance

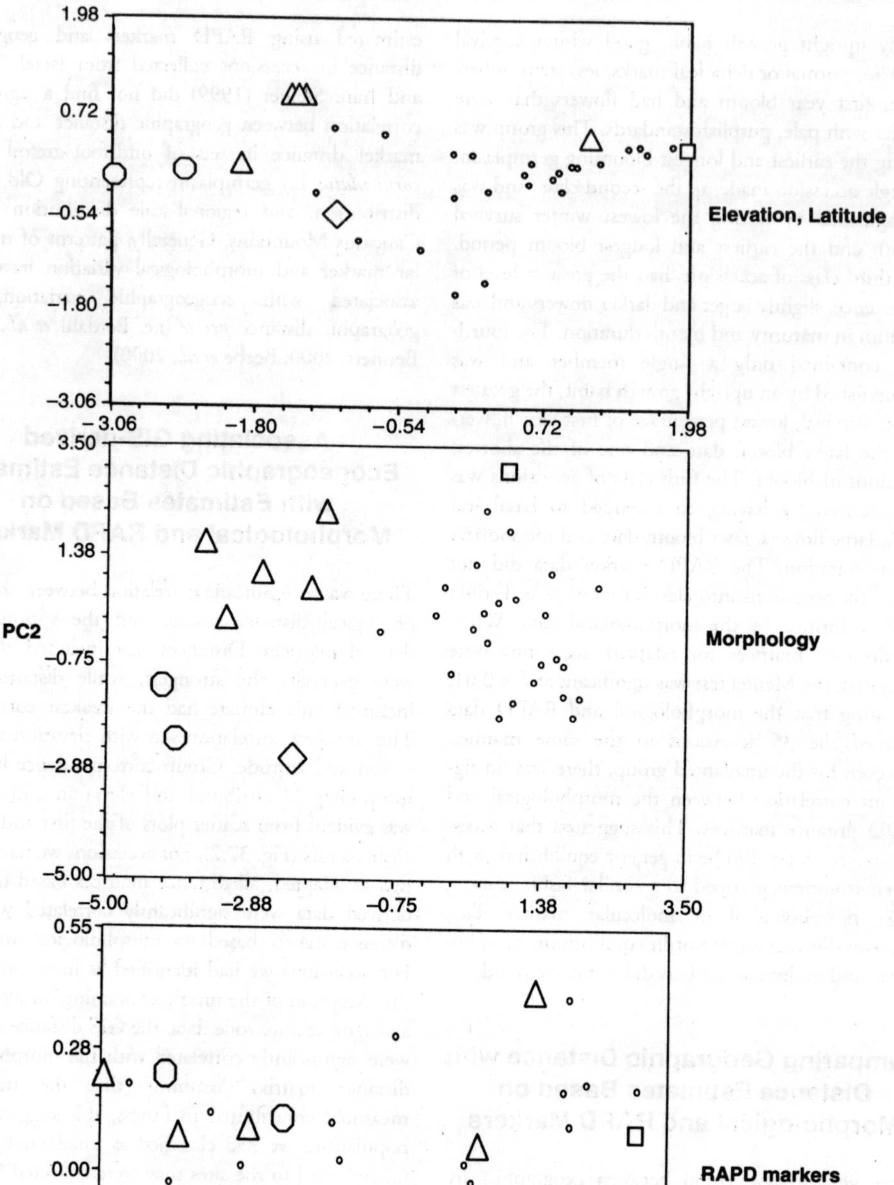

Fig. 37.2. Scatter plots of first and second eigenvectors for distance matrices based on DEM elevation and latitude, morphology and RAPD markers using all red clover accessions.

matrices based on elevation and latitude, and moisture and temperature zone and latitude. The lack of correlation between RAPD markers and ecogeographic data has been reported in a number of papers; for example, Heaton *et al.* (1999) reported no association between RAPD markers and ecological and phenotypic differences in *Manillkara zapota* (L.), a variable outcrossing tropical tree. Schierenbeck *et al.* (1997) found no association between elevation and genetic variation among populations growing at different elevations for four tropical tree species. However, Steiner (1999) found a significant correspondence between genetic distance based on ITS sequences and distance measures based on climate, elevation and latitude. The correspondence was greater in perennial *Trifolium* species than annual species (Steiner, 1999). Heaton *et al.* (1999) suggested a number of reasons for the lack of association between DNA molecular markers and geography. Our results suggested that the prevalence of neutral markers may be a factor considering the high correspondence between the morphological distance matrix and GIS-derived distance matrices, compared with the low correspondence observed between the RAPD distance matrix and GIS-derived matrices. However, the fact that we observed weak but significant correlation between the RAPD matrix and climate and elevation attributes when we examined adapted accessions, yet saw no association with the unadapted accessions, suggested that the RAPD markers may have some association with adaptive alleles. RAPD markers identified adaptive DNA differentiation in wild emmer wheat growing in two contrasting microhabitats (Li *et al.*, 1999). This is certainly a subject where further research will help to clarify the situation.

Conclusion

Our hypothesis was that GIS-derived ecogeographic description could assist in identifying variation among accessions of the same species, collected from the same area. This type of information is particularly valuable for plant collectors interested in obtaining germplasm adapted to unique environments, or in efficiently sampling the genetic variation of a species within a given region. As a test of our hypothesis, we examined the correlation between GIS-derived ecogeographic information and morphological and molecular marker variation for 35 accessions. We found that GIS-derived data were correlated with patterns of germplasm variation measured by morphological and DNA-level characters. Of particular value was the DEM estimate of elevation, which was strongly correlated with germplasm variation. Elevation data are readily obtained, and can often be obtained freely from the Internet. Our study suggested that this type of information could be used as a straightforward guide for sampling diverse germplasm. Latitude was also a useful predictor of variation. Although more complex to develop, climate information was also correlated with morphological and molecular marker variation.

Generally, GIS-derived ecogeographic data were more strongly correlated to morphological variation. A positive correlation was observed for accessions that had been defined as unadapted and adapted. Although we examined a limited number of accessions, these results suggest that morphological adaptation may occur in a shorter time frame than our previous assumption (Greene *et al.*, 1999b). The GIS-derived data had a weaker correlation with genetic distances as estimated using RAPD markers. Only a weak association was observed, and only with accessions that we had defined as adapted. Finally, geographic distance was not associated with either morphological or RAPD variation.

Greene and Hart (1999) discussed the effective use of GIS-derived information in collecting plant germplasm. This present study reinforces the discussion. For example, the differential response of adapted and unadapted accessions in this study illustrates the importance of collecting populations that are in equilibrium with the environment from which they are collected if project objectives are to associate germplasm variation with environmental variation. This study also illustrated how morphological and DNA-level variations differ in how they correlate with environmental variation. The difference between morphological data and RAPD marker data is not surprising considering that populations that are recent emigrants may undergo relatively rapid changes in adaptive allele frequencies due to intense selection pressure, while neutral allele frequencies would be expected to change at a slower rate.

Provided researchers take into account the inherent limitations and challenges of using ecogeographic information, particularly secondarily derived information, we believe that the use of GIS-derived ecogeographic information promises to bring greater precision and efficiency in the acquisition of plant genetic resources.

References

Beebe, S., Skroch, P.W., Tohme, J., Duque, M.C., Pedraza, F. and Nienhuis, J. (2000) Structure of genetic diversity among common bean landraces of Middle American origin based on correspondence analysis of RAPD. *Crop Science* 40, 264–273.

Bennett, S.J. (2000) Genetic variation of five species of *Trifolium* L. from south-west Turkey. *Genetic Resources and Crop Evolution* 47, 81–91.

Berdahl, J.D., Mayland, H.F., Asay, K.H. and Jefferson, P.G. (1999) Variation in agronomic and morphological traits among Russian wild rye accessions. *Crop Science* 39, 1890–1895.

Cock, P.S. and Ehrman, T.A.M. (1987) The geographic origin of frost tolerance in Syrian pasture legumes. *Journal of Applied Ecology* 24, 678–683.

Doyle, J.J. and Doyle, J.L. (1989) Isolation of plant DNA from fresh tissue. *Focus* 12, 13–15.

Greene, S.L. and Hart, T.C. (1999) Implementing a geographic analysis in germplasm conservation. In: Greene, S.L. and Guarino, L. (eds) *Linking Genetic Resources and Geography: Emerging Strategies for Conserving Crop Biodiversity.* CSSA Special Publication No 27. ASA and CSSA, Madison, Wisconsin, pp. 25–38.

Greene, S.L., Hart, T.C. and Afonin, A. (1999a) Using geographic information to acquire wild crop germplasm for *ex situ* collections: I. Map development and use. *Crop Science* 39, 836–842.

Greene, S.L., Hart, T.C. and Afonin, A. (1999b) Using geographic information to acquire wild crop germplasm for *ex situ* collections: II. Post collection analysis. *Crop Science* 39, 843–849.

Guarino, L. (1995) Geographic information systems and remote sensing for the plant germplasm collector. In: Guarino, L., Ramanatha Rao, V. and Reed, R. (eds) *Collecting Plant Genetic Diversity.* CAB International, Wallingford, UK, pp. 315–328.

Guarino, L., Maxted, N. and Sawkins, M. (1999) Analysis of georeferenced data and the conservation and use of plant genetic resources. In: Greene, S.L. and Guarino, L. (eds) *Linking Genetic Resources and Geography: Emerging Strategies for Conserving Crop Biodiversity.* CSSA Special Publication No. 27. ASA and CSSA, Madison, Wisconsin, pp. 1–24.

Hart, T.C. (1999) Scale considerations in mapping for germplasm acquisition and the assessment of *ex situ* collections. In: Greene, S.L. and Guarino, L. (eds) *Linking Genetic Resources and Geography: Emerging Strategies for Conserving Crop Biodiversity.* CSSA Special Publication No. 27. ASA and CSSA, Madison, Wisconsin, pp. 51–61.

Haywood, J.S. (1991) Spatial analysis of genetic variation in plant populations. *Annual Review of Ecology and Systematics* 22, 335–355.

Heaton, H.J., Whitkus, R. and Gomez-Pompa, A. (1999) Extreme ecological and phenotypic differences in the tropical tree chicozapote (*Manilkara zapota* (L.) P. Royen) are not matched by genetic divergence: a random amplified polymorphic DNA (RAPD) analysis. *Molecular Ecology* 8, 627–632.

Hedrick, P.W. (1986) Genetic polymorphisms in heterogeneous environments: a decade later. *Annual Review of Ecology and Systematics* 17, 535–566.

Li, Y.C., Fahima, T., Beiles, A., Korel, A.B. and Nevo, E. (1999) Microclimatic stress and adaptive DNA differentiation in wild emmer wheat, *Triticum dicoccoides. Theoretical and Applied Genetics* 98, 873–883.

Loveless, M.D. and Hamrick, J.L. (1984) Ecological determinants of genetic structure in plant populations. *Annual Review of Ecological Systematics* 15, 65–95.

Lynch, M. and Milligan, B.G. (1994) Analysis of population genetic structure with RAPD markers. *Molecular Ecology* 3, 91–99.

Manly, B.F.J. (1994) *Multivariate Statistical Methods: a Primer*, 2nd edn. Chapman and Hall, New York, USA.

Mantel, N.A. (1967) The detection of disease clustering and a generalized regression approach. *Cancer Research* 27, 209–220.

Nebauer, S.G., Del Castillo-Agudo, L. and Segura, J. (1999) RAPD variation within and among natural populations of outcrossing willow-leaved foxglove (*Digitalis obscura* L.). *Theoretical and Applied Genetics* 98, 985–994.

Nevo, E., Baum, B., Beiles, A. and Johnson, D.A. (1998) Ecological correlates of RAPD DNA diversity of wild barley, *Hordeum spontaneum*, in the fertile crescent. *Genetic Resources and Crop Evolution* 45, 152–159.

Rohlf, F.J. (1993) *NTSYS: Numerical Taxonomy and Multivariate Analysis System*, v. 2.02k. Exeter Software, Setauket, New York.

Schierenbeck, K.A., Skupski, M., Lieberman, D. and Lieberman, M. (1997) Population structure and genetic diversity in four tropical tree species in Costa Rica. *Molecular Ecology* 6, 137–144.

Slatkin, M. (1987) Gene flow and the geographic structure of natural populations. *Science* 236, 787–792.

Steiner, J.J. (1999) Plant genotype, phenotype and ecogeography. In: Greene, S.L. and Guarino, L. (eds) *Linking Genetic Resources and Geography: Emerging Strategies for Conserving and Using Crop Biodiversity.* CSSA Special Publication No. 27. ASA and CSSA, Madison, Wisconsin, pp. 39–50.

Steiner, J.J. and Greene, S.L. (1996) Proposed use of ecological descriptors and their utility for plant germplasm collections. *Crop Science* 36, 439–451.

38 *In situ* Conservation of Forest Genetic Resources at Regional Level: Two Complementary Programmes Using GIS Approach

K.N. Ganeshaiah,[1] R. Uma Shaanker,[2] N. Barve,[3] M.C. Kiran,[3] K.S. Bawa[3,4] and V. Ramanatha Rao[5]

[1]Department of Genetics and Plant Breeding and [2]Department of Crop Physiology, University of Agricultural Sciences, Bangalore, India; [3]Ashoka Trust for Research in Ecology and the Environment (ATREE), Hebbal, Bangalore, India; [4]University of Massachusetts, Boston, Massachusetts, USA; [5]IPGRI Regional Office for Asia, the Pacific and Oceania, Serdang, Malaysia

Introduction

Conservation of genetic resources began strongly as an *ex situ* programme and there have been several suggestions of *ex situ* conservation of forest genetic resources as well (FAO, 1992; Yang and Yeh, 1992; Singh, 1996). *Ex situ* conservation of genetic diversity of the forest species requires that we have a good knowledge of their breeding behaviour, population structure, genetic differentiation of the provenances, pollination systems, seed storage requirements, life history details and so on (Marshall and Brown, 1998; Brown and Hardner, 1999). For most of the forest species these are hardly understood and even if these details are available, *ex situ* conservation of these species is difficult because of the vast areas required and prohibitive costs involved in maintaining and evaluating their populations (Shands, 1991). Thus *in situ* conservation of genetic resources of the forest species is emerging as a predominant strategy (Smith and Schultes, 1990; Bawa *et al.*, 1991; Ledig, 1992).

In situ conservation of genetic resources, especially of the forest species, requires identification of the 'Genetic Diversity Hot Spots' (GDHSs) for each of the species such that we protect these areas for conserving their genetic diversity (Uma Shaanker and Ganeshaiah, 1997; Ganeshaiah and Uma Shaanker, 1998, 1999). Such GDHSs and genetically diverse populations are identified for a number of species and conservation strategies are formulated accordingly (e.g. for sandalwood, *Phyllanthus* species and *Bambusa* spp.) at our centre. However, while this approach could be very effective for the conservation of genetic diversity of certain targeted species, it may not be an appropriate and useful protocol for conserving the entire plant genetic resources of a target region, or of a country.

Forest genetic resources of a region or of a country, constitute the total pool of genetic diversity at several layers from intraspecific to inter-community and to that at the ecosystem level (Fig. 38.1). Conserving this entire range of genetic diversity requires a pyramiding of the conservation efforts of all the individual species. Such an approach calls for identifying the genetic hotspots for all or most of the important species of the area (or the country)

 Managing Plant Genetic Diversity (eds J.M.M. Engels, V. Ramanatha Rao, A.H.D. Brown and M.T. Jackson)

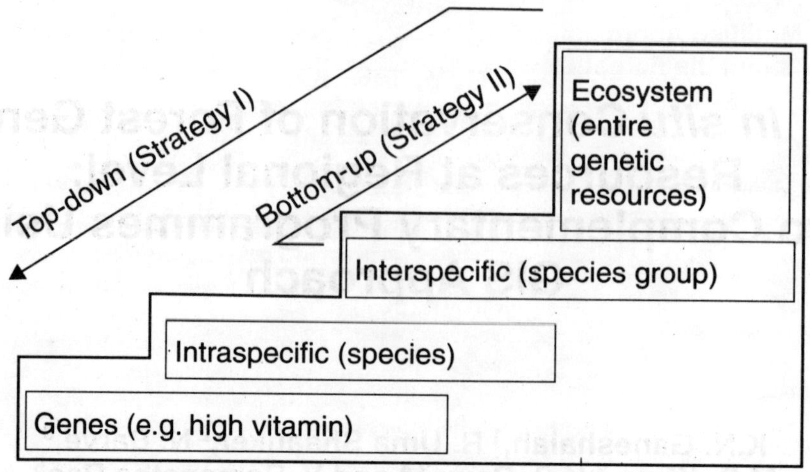

Fig. 38.1. The pyramid of the different hierarchies of the genetic resources of a region or of an ecosystem. The present approach focuses on each species and begins at the base of the pyramid. The two new strategies suggested focus the conservation efforts at the ecosystem level and proceed downwards (strategy I) or at a group of related species and proceed upwards (strategy II).

and pooling these hotspots to arrive at a final map depicting areas for protection. For instance, conservation of genetic resources of the entire Western Ghats requires that we develop such independent plans for about 4000–6000 species and join them meaningfully! Such a task, besides being Herculean and time consuming, might not be cost-effective: several overlapping areas are likely to be identified for different species and the final map may consist of several small, disjointed areas, conservation of all of which will be problematic if not impossible. Hence this approach may not be an effective and useful strategy especially considering the rapid rate at which we are losing our forest genetic resources. We obviously need an altogether different approach to conserve the forest genetic resources of a region or of a country.

Alternative Suggestions for *in situ* Conservation

Several methods are being proposed for *in situ* conservation of biological resources in general and of genetic diversity in particular. These range from identifying a network of the reserves that complementarily constitute most of the biological diversity of the region for conservation (Rebelo and Siegfried, 1992), use of landscape approach for identifying distinct habitat elements for conservation (Menon and Bawa, 1997) and prioritizing the conservation sites based on the alternative species (Peterson and Navarro-Sigüenza, 1999). While continued concern with species is essential, landscape-level issues also need much greater attention. Franklin (1993), for instance, argued that efforts to preserve biological diversity must focus increasingly at the ecosystem level because of the immense number of species, the majority of which are currently unknown. An ecosystem approach is also the only way to conserve processes and habitats (such as forest canopies, below ground habitats and hyporheic zones) that, with their constituent species, are poorly known. Similarly, our incomplete knowledge of the functioning of the habitats of various sizes has led to the continued debate on whether or not to conserve the species as part of the large ecosystem or as a part of several small habitats (Rebelo and Siegfried, 1992).

However, in these suggestions, two issues do not receive simultaneous and serious attention: (i) the possibility of augmenting the efforts to conserve the genetic resources with those to conserve the biological diversity; and (ii) the possible approaches and techniques to be employed to conserve the genetic resources at an ecosystem level. In what follows we have attempted to combine the elements of the suggestions by earlier workers but with a special focus on *in situ* conservation of the genetic resources along with the biological diversity of the target area.

Two Modified Approaches for Conserving the Forest Genetic Resources of a Target Region

An effective and comprehensive strategy for the conservation of the entire forest genetic resources of a target area needs to be essentially based more on identification of the genetic hotspots for the specific *groups* of plants and/or for the *habitats*, than for individual species. Hotspots thus identified could be combined together to arrive at a regional or national conservation plan which could help us to conserve the most diversity with least effort and cost. Accordingly, we propose two independent but complementary approaches for addressing the national or regional level conservation of the forest genetic resources. These two, 'top-down' and 'bottom-up' approaches, which aim at identifying the areas to be conserved at the ecosystem level and for specific groups of plants respectively, together offer the complementary sets of areas to be conserved.

Top-down approach

Consider a biodiversity rich region, for example the Western Ghats, one of the 18 (now 25) hotspots of diversity in the world, spread along the west coast of India along about a 1500 km stretch of mountain range (Myers, 1990). This area harbours about 4000 to 6000 species of angiosperms, almost a third of their richness in India, in diverse types of vegetation spread over a range of altitudes from 900 m to about 2000 m above sea level. It hosts at least about 1000 plant species that are economically very important for their medicinal purposes, timber and non-timber value, for their ornamental potential and/or industrial use (Gadgil *et al.*, 1997; Subhash Chandran, 1997). Thus, even if we aim at the conservation of the genetic diversity of merely the economically important species of this area, we need to have about 500 independent programmes. Even if such an effort is possible, it will not be able to address the conservation needs of the other plant species with less known or yet unknown potential. Obviously such species-based plans cannot be appropriate strategies. On the other hand, conservation of the entire genetic resources of these areas could be addressed by identifying the genetically diverse habitats and areas for conservation within them. We propose the following steps for this.

1. Stratification of the region

In the top-down approach we suggest a geographical information system (GIS) based stratification of the area to identify distinct bio-geographic habitats that are likely to harbour diverse genetic resources. The process of stratification can be based on a set of important climatic, physical, biological and stress factors that might together shape the structure of the genetic resources. For instance, we have attempted to stratify the Western Ghats based on different layers such as rainfall, temperature regimes, altitude range, vegetation types, soil types, diversity of plants in general and of the threatened species in particular and other social and economic threats (Figs 38.2–38.4; see the legends for details). Among these, however, some parameters such as temperature and altitude were found to be either highly correlated and or were found redundant. While we still are exploring the possibilities of incorporating other parameters, we found that stratification of the Western Ghats using even a subset of these parameters offers ecologically meaningful habitats.

2. Culling the minor habitats and filtering the large viable patches

Based on the three layers, namely rainfall, vegetation and density of threatened species, we identified 45 distinct habitats. Some of these, however, were very small, occupying less than 500 km^2. These small patches were culled as unimportant in the context of the conservation of the genetic diversity of the entire region and hence we were eventually left with 16 distinct types (Table 38.1). Though this culling process is likely to eliminate certain small but unique habitats (e.g. swamps), as a first approximation of distinct forest types we retained the minimal set containing representatives of the large forest types. Within each of these 16 habitats, however, a large number of patches were very small and hence are unlikely to hold viable sizes of populations of different species, especially of the rare and threatened species. We hence constructed the frequency distribution of the sizes of the patches (Fig. 38.5) for each of these habitats and identified those in the top 5% of the size category. These constituted the areas for further sampling for genetic conservation (Table 38.1).

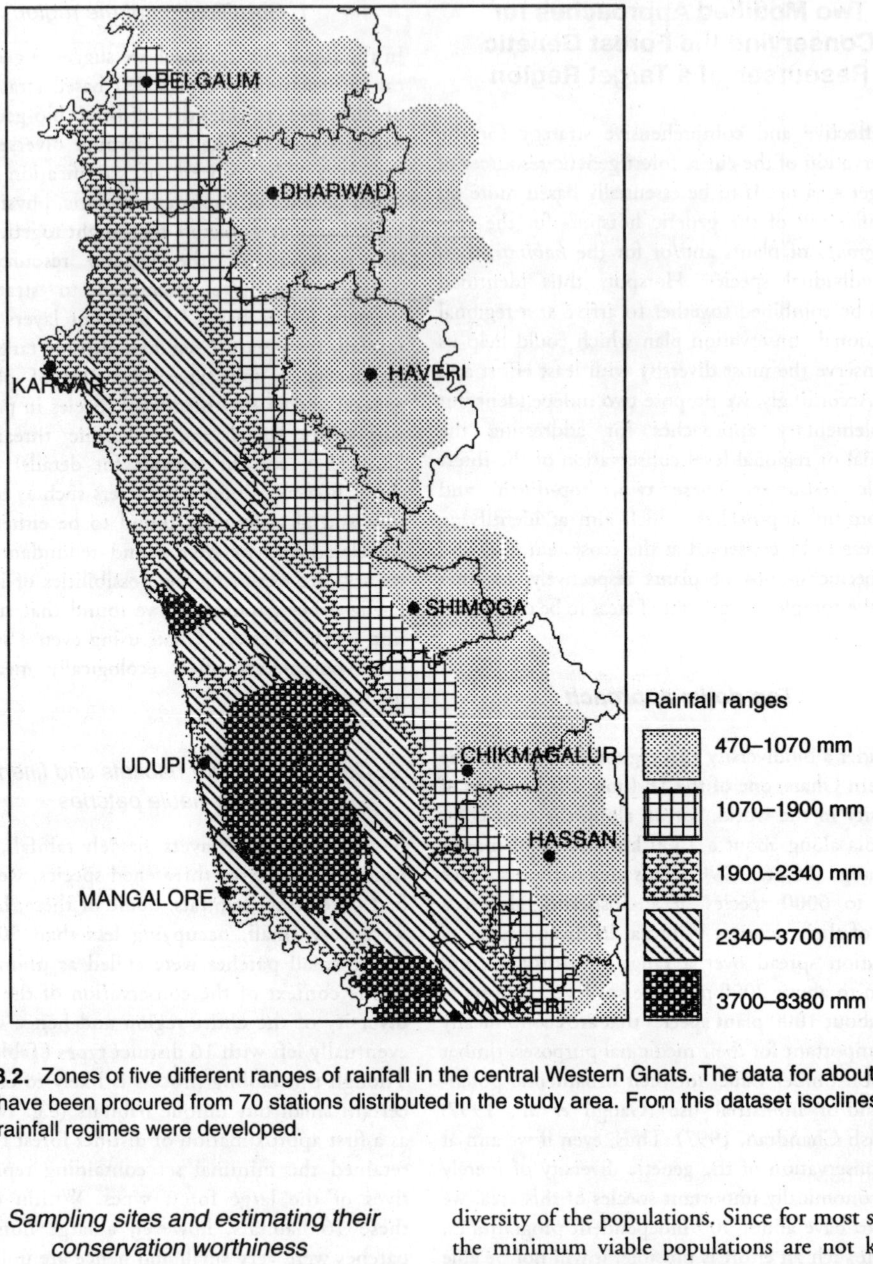

Fig. 38.2. Zones of five different ranges of rainfall in the central Western Ghats. The data for about 50 years have been procured from 70 stations distributed in the study area. From this dataset isoclines of equal rainfall regimes were developed.

3. Sampling sites and estimating their conservation worthiness

We propose that within these identified large patches, a certain number of sites or plots are identified across the altitude and latitude ranges for evaluating the floral diversity, population sizes or densities of economically important species and their genetic diversity. The genetic conservation value (GCVal) of a site can be computed as a function of the population size and the extent of genetic diversity of the populations. Since for most species, the minimum viable populations are not known, we suggest that the genetic conservation value of a site be computed as a sum of the genetic diversity, GDi, weighted with the population size P_i of the species (i) as given below.

$$GCVal_j = \sum_{i=1}^{S} Gdi \times P_i ,$$

where S = number of species in the jth plot.

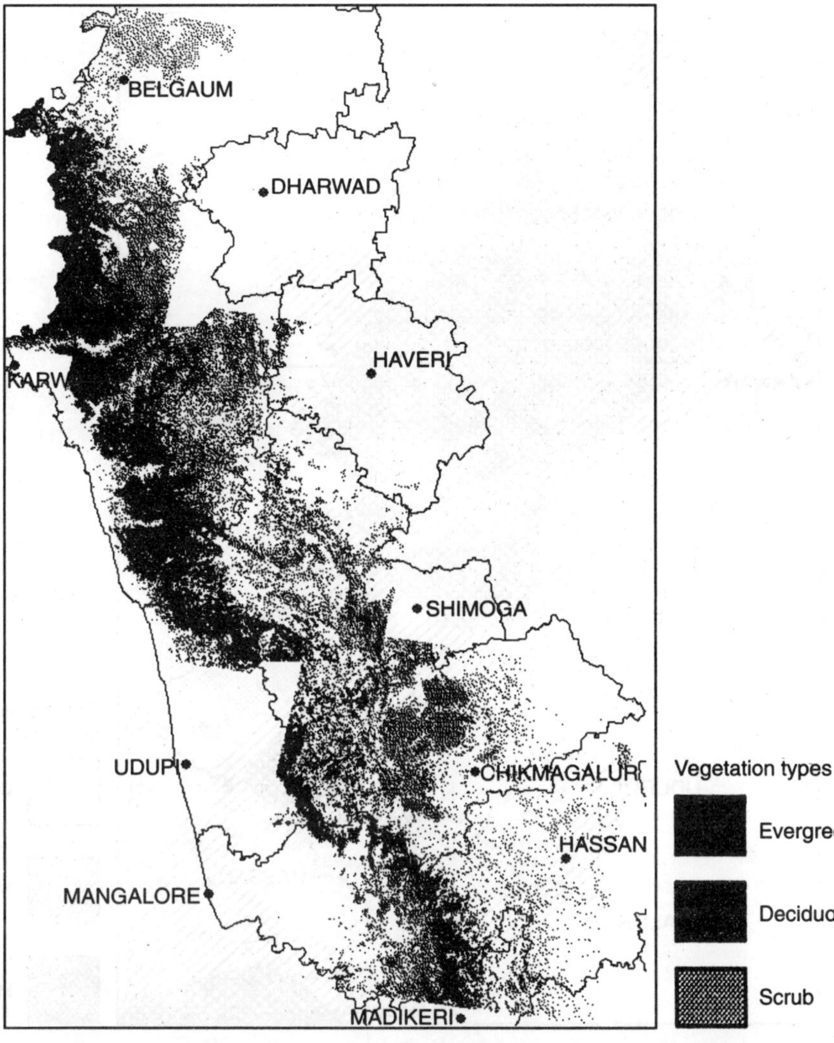

Fig. 38.3. Vegetation classes of the study area derived by a supervised classification of the satellite data from IRS 1C and IRS 1D for the year 1998. The ground data for the forest types (vegetation classes) were obtained from the forest survey done by the Forest Survey of India for the area during 1995–1997 (Shivaraj *et al.*, 2000) and from the vegetation maps prepared by the French Institute, Pondicherry, India, and published during the late 1980s.

The conservation worthiness of a site can be expressed as a direct function of the genetic conservation value and the species richness. Accordingly, plotting the sites sampled on these two estimates helps to identify the best set of sites for conservation. Those with the highest species diversity and genetic diversity index can be prioritized for conserving the genetic resources of the area (Fig. 38.6).

Bottom-up approach

In this approach, we propose that *in situ* conservation sites are selected together for a set of species that are either taxonomically related and/or share common use profiles. For example, the group-based approach can aim at identifying common sites for conservation of medicinally important plants or bamboos and/or rattans or a set of economically

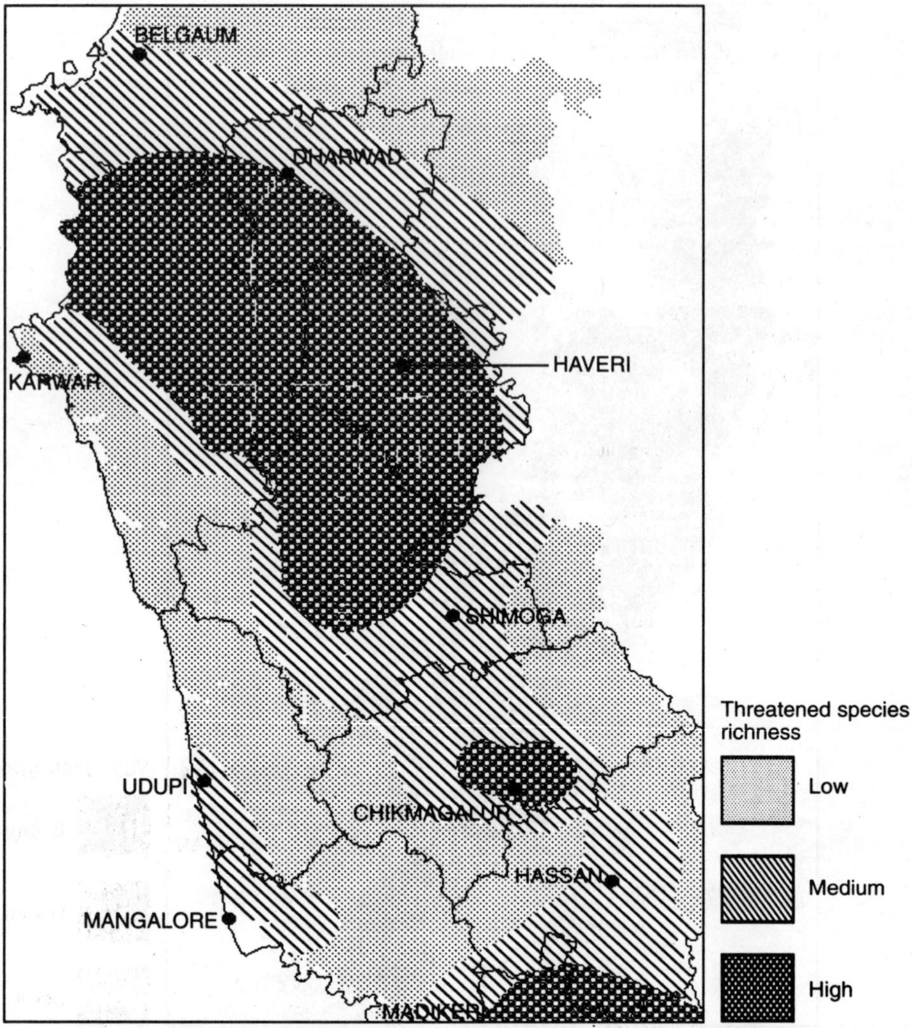

Fig. 38.4. Three regimes of the threatened species density in the study area. The threatened species data were obtained by an independent study done by the Karnataka Department of Forests during 1995 for the distribution of 400 threatened species in the Western Ghats (Ananda Rao, 1995). These data have been compiled from both flora records and field visits. To this, the distribution of the endemic species has also been added from Ahmedullah and Nayar (1986). The data from these two have been combined, the richness of the threatened and endemic species has been computed for different places and contours drawn. The three regimes identified from these contours indicate the sensitivity or vulnerability of an area for some of the unique species and perhaps also represent the extent of threat experienced.

important but endemic species. More often the species with common use profiles are likely to be taxonomically related to varied degree (as in bamboos and rattans). Even in the groups that contain taxonomically diverse ranges of species there are species subgroups that might be taxonomically related and/or share common habitats. For example, the *Triphala* (meaning three fruits) species complex, composed of *Terminalia bellerica* and *Terminalia chebula* along with *Phyllanthus emblica*, which forms

Table 38.1. Features of the 16 stratified habitats after removing[a] the small patches.

Class	Vegetation	Threatened species richness	Rainfall	Approximate area after removal of patches less than mean patch size (km^2)	Average patch size (km^2)
1	Scrub	low	very low	720.5065	0.2354
2	Scrub	medium	very low	462.7655	0.0863
3	Deciduous	medium	low	972.7485	0.3580
4	Deciduous	medium	medium	560.1004	0.2442
5	Deciduous	low	high	819.4231	0.1414
6	Evergreen	low	medium	1074.486	0.9019
7	Deciduous	high	low	1411.193	0.4641
8	Evergreen	medium	medium	704.3603	0.4589
9	Deciduous	medium	high	711.9055	0.2825
10	Deciduous	high	medium	1027.277	0.3521
11	Evergreen	low	high	1301.937	0.2909
12	Evergreen	low	very high	696.3894	0.3457
13	Evergreen	medium	high	708.979	0.1849
14	Evergreen	high	medium	912.1208	0.6417
15	Deciduous	high	high	922.3078	0.3800
16	Evergreen	high	high	567.0536	0.2302

[a]Classes less than 500 km^2 removed.
Patches less than average patch size also removed.

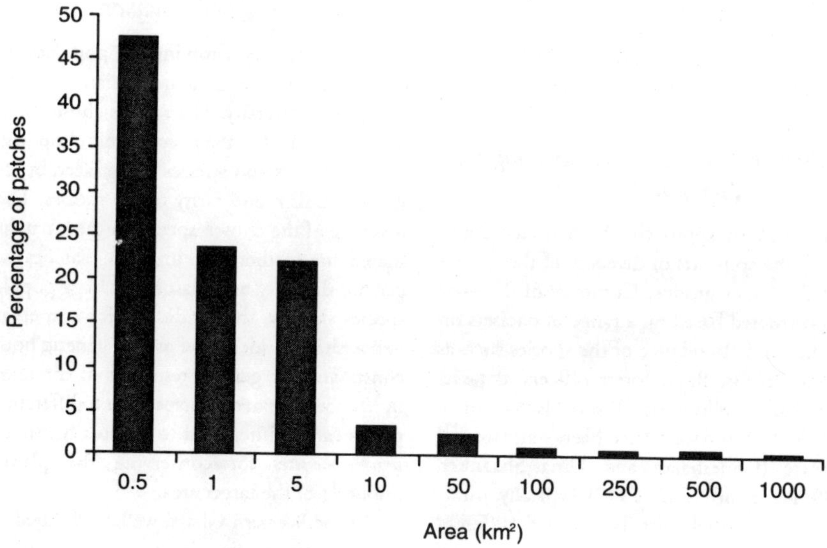

Fig. 38.5. Distribution of the patch sizes in all the layers. This analysis was performed for each of the stratified habitats and the largest areas were chosen for further studies.

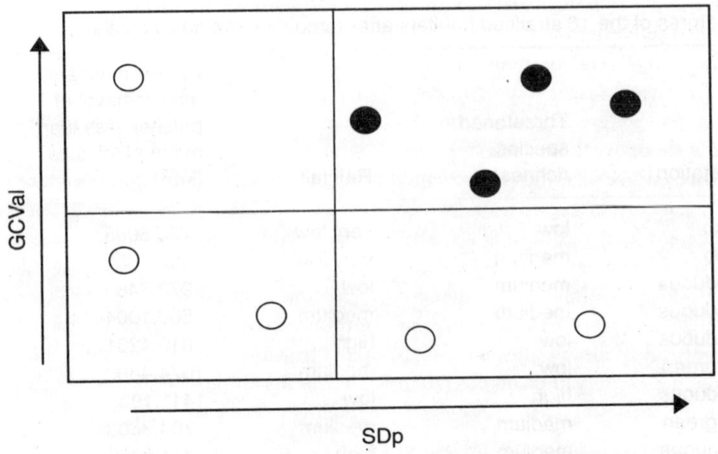

Fig. 38.6. A protocol for prioritizing sites for conservation at the ecosystem level. Those located on the top right-hand corner are those with high species richness as well as high genetic conservation value.

the important ingredients of Ayurvedic medicinal formulations, generally occur together in most of the *Combretaceae* rich deciduous habitats of South India. Thus, identifying a set of common conservation sites for such species might be not only a possibility but could be an effective method of conserving their genetic resources. Therefore we propose that conservation sites be identified for groups of such plant species that could be later combined to arrive at the regional sites for conservation. We propose the following steps in this approach.

1. Identifying the hotspots and locating the viable patches

In executing such an approach, the first step could be to identify the contours of diversity of the species richness of the target groups. Contours of diversity could be constructed based on a range of datasets on the occurrence and abundance of the species such as forest working plans, flora, forest officers' datasets and the herbaria collections. Protocols for these steps using the GIS software have been standardized at our centre (Ganeshaiah and Uma Shaanker, 1998, 1999; Ravikanth *et al.*, 1999) Typically, information on the spatial distribution of different species of the group chosen is collated in a spatial database system, the study region is divided into grids of suitable sizes and the density of species in each grid counted. From these density data, contours of different species densities are drawn and hotspots of species richness are identified. From within these identified hotspots of species diversity, patches containing the viable populations of a reasonably large set of species need to be located for estimating the genetic diversity of the species.

2. Estimating the genetic diversity and identifying the genetic hotspots for conservation

This step involves sampling the populations of a set of species that demand immediate conservation of the genetic diversity. The species chosen could be on the basis of: (i) their economic importance; (ii) human threats and stresses being faced by them; and (iii) endemism and rarity of the species. The genetic diversity of the chosen species in these patches is estimated for further prioritization. Patches with high genetic diversity and reasonably large populations of species shall be the candidates for conservation and hence shall be identified as the 'genetic hotspots' for conserving the genetic resources of the target group *in situ*. Such genetic hotspots for a different group of plants can be integrated to eventually arrive at a *network of sites* for conserving the plant genetic resources of the target area.

In the Western Ghats, we have located such hotspots of species diversity for a few groups of plants such as bamboos, rattans, dipterocarps, and *Triphala* (see above) and a few other medicinal plants. The species richness contours thus drawn for rattans for instance, suggest that there are two distinct areas where the rattan species are highly concentrated:

one just north of and other immediately south of the Pálghat gap along the Western Ghats (Fig. 38.7). From the northern patch, several locations with reasonably large populations of rattans located in fairly well protected forest zones were identified. Samples of *Calamus strictus* were drawn from these sites and their genetic diversity estimated using the isozymes of leaf samples. The data on the presence or absence of the bands of different isozymes of nine different enzymes were coded in a binary system and the extent of genetic diversity of different populations and hence of different areas estimated. From these datasets, we have identified specific sites with high genetic diversity for the conservation of the rattan genetic resources in the Western Ghats.

Complementation of the bottom-up with top-down approach: merits and problems

For an *in situ* conservation agenda of a small set of forest species, the current approach based on individual species is perhaps the most effective. But the list of species for which the genetic resources need to be conserved *in situ* is becoming larger and for this reason, an ecosystem-based approach that addresses most of the species of the habitat is perhaps the best strategy. Though this approach compromises on the details of the individual species, it does provide an economically and logistically effective conservation programme of the entire genetic resources at the ecosystem level. Nevertheless, the

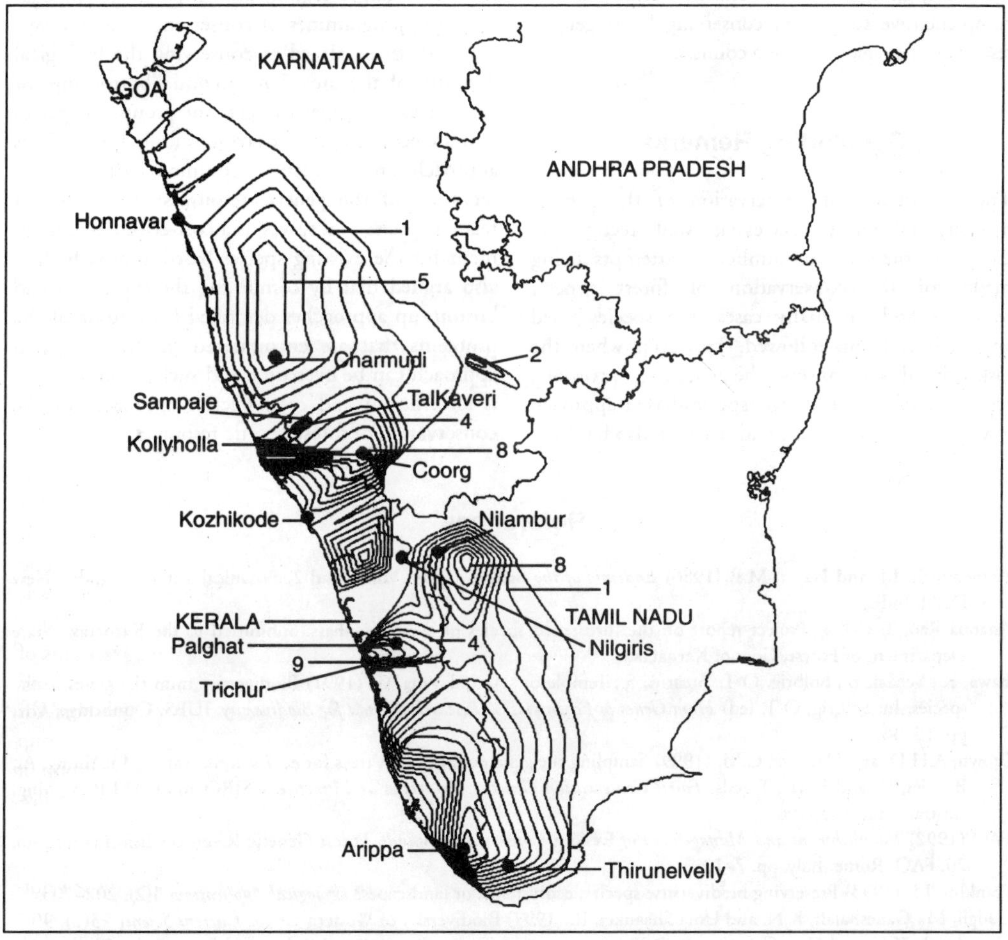

Fig. 38.7. Contours of rattan species diversity in the Western Ghats. Note that the two areas enclaved by the contours depicting high species diversity (near Coorg-Kozhikode and Trichur Nilambur) are the candidate areas for conservation of the genetic resources of the rattans. These were further sampled for the genetic diversity of focal species. Values attached to the contour lines indicate number of species of rattans in the area.

ecosystem approach does not ensure the conservation of the entire range of the genetic diversity of the species, a possibility that can be attained with the species-based approach. Similarly, it does not ensure the conservation of the rare alleles or genes that might be very important for the survival of the species during extreme and stress situations. These limitations of the ecosystem-based approach can be partially overcome by suitably combining the top-down with the specific group-based or bottom-up approach. By combining the sites identified through the two approaches and marking the overlaps, it might be possible to evolve a network of genetic conservation sites for a region or for a country. In this sense, we see the two approaches suggested here complementing each other and serving as an effective, economic and comprehensive strategy of conserving forest genetic resources of a region and/or a country.

Concluding Remarks

The importance of conservation of the genetic diversity of forest species is well recognized. Currently there are a number of attempts being made for the conservation of forest genetic resources and, in many cases, the species-based approach is being followed. However, when the task at hand is to conserve the genetic resources of a region or of a country, the species-based approach may not be very relevant and it could also be time-consuming and costly. Rather, such regional programmes of genetic conservation could become cost-effective by adopting an ecosystem approach and by amalgamating them with those of conserving the biological diversity of the area. We accordingly propose two complementary strategies to prioritize the areas and the biological material for *in situ* conservation of forest genetic resources of a region or of a country. First of these, the 'top-down' or ecosystem-based approach aims to identify the sites that could contain the genetic diversity of a wide range of forest species from diverse habitat types of the entire ecosystem; conserving such sites would conserve genetic diversity contained in them. This approach, based on the stratification of the entire target region, also attempts to amalgamate the programmes of conserving genetic diversity with those aimed at conserving the biological diversity of the area. The second, 'bottom-up' or group-based approach targets the groups of species that are economically or taxonomically linked. This approach aims to identify common areas for conservation of the genetic resources of a range of related species and serves as a cost-effective replacement for the existing species-based approach. It is also argued that by combining the top-down and bottom-up approaches described here, some of the problems that are encountered in the ecosystem approach can be overcome and such a combination is an efficient and cost-effective approach for the conservation of forest genetic resources.

References

Ahmedullah, M. and Nayar, M.P. (1986) *Endemic of the Indian Region*, Vol. 1 and 2. Botanical Survey of India, New Delhi, India.

Ananda Rao, T. (1995) Project report on the threatened species of Western Ghats. Submitted to the Karnataka State Department of Forest, Govt of Karnataka.

Bawa, K., Schaal, B., Solbrig, O.T., Stearns, S., Templeton, A. and Vida, G. (1991) Biodiversity from the genes to the species. In: Solbrig, O.T. (ed) *From Genes to Ecosystems: A Research Agenda for Biodiversity*. IUBS, Cambridge, UK, pp. 15–36.

Brown, A.H.D. and Hardner, C.M. (1999) Sampling the gene pool of forest trees for *ex situ* conservation. In: Young, A., Boshier, D. and Boyle, T. (eds) *Forest Conservation Genetics; Principles and Practice*. CSIRO and CABI Publishing, Australia, pp. 185–196.

FAO (1992) *Establishment and Management of* Ex Situ *Conservation Stands*. Forest Genetic Resources Information, no. 20. FAO, Rome, Italy, pp. 7–10.

Franklin, J.F. (1993) Preserving biodiversity: species, ecosystems, or landscapes? *Ecological Applications* 3(2), 202–205.

Gadgil, M., Ganeshaiah, K.N. and Uma Shaanker, R. (1997) Biodiversity of Western Ghats. *Current Science* 73(2), 99.

Ganeshaiah, K.N. and Uma Shaanker, R. (1998) Contours of conservation – A national agenda for mapping biodiversity. *Current Science* 75(3), 292–298.

Ganeshaiah, K.N. and Uma Shaanker, R. (1999) Planning conservation strategies. *GIS@development* September-October 3(5), 67–69.

Ledig, F.T. (1992) Human impacts on genetic diversity in forest ecosystems. *Oikos* 63, 87–108.

Marshall, D.R. and Brown, A.H.D. (1998) Sampling wild legume populations. In: *Genetic Resources of Mediterranean Pasture and Forage Legumes, Proceedings of an International Workshop*. Kluwer Academic Publishers, Dordrecht, The Netherlands, pp. 78–79.

Menon, S. and Bawa, K.S. (1997) Application of geographic information systems, remote-sensing, and a landscape ecology approach to biodiversity conservation in Western Ghats. *Current Science* 73(2), 134–145.

Myers, N. (1990) The biodiversity challenge: expanded hotspots analysis. *Environmentalist* 10, 243–256.

Nayar, M.P. (1996) *Hotspots of Endemic Plants of India, Nepal and Bhutan*. Tropical Botanic Garden and Research Institute, Palode, Thiruvananthapuram.

Peterson, A.T. and Navarro-Sigüenza, A.G. (1999) Alternate species concepts as bases for determining priority conservation areas. *Conservation Biology* 13(2), 427–431.

Ravikanth, G., Chaluvaraju, Uma Shaanker, R., Ganeshaiah, K.N. (1999) Mapping rattan species diversity in South India, *IPGRI Newsletter for Asia, The Pacific and Oceania* 28, 10.

Rebelo, A.G. and Siegfried, W.R. (1992) Where should nature reserves be located in the Cape Floristic Region, South Africa? Models for the spatial configuration of reserve network aimed at maximizing the protection of floral diversity. *Conservation Biology* 62(2), 243–252.

Shands, H.L. (1991) Complementarity of *in situ* and *ex situ* germplasm conservation from the standpoint of the future user. *Israel Journal of Botany* 40(5–6), 521–528.

Shivaraj, B., Narayani, B., Kiran, M.C., Ganeshaiah, K.N. and Uma Shaanker, R. (2000) Mapping the tree diversity of Uttara Kannada. *Journal of Indian Institute of Sciences* (in press).

Singh, U.V. (1996) Conservation of forest genetic resource – an *ex situ* management of secondary forests. *Indian Forester* 122(9), 787–794.

Smith, N.J.H. and Schultes, R.E. (1990) Deforestation and shrinking crop gene pools in Amazonia. *Environmental Conservation* 17(3), 227–234.

Subhash Chandran, M.D. (1997) On the ecological history of the Western Ghats. *Current Science* 73(2), 146–156.

Uma Shaanker, R. and Ganeshaiah, K.N. (1997) Mapping genetic diversity of *Phyllanthus emblica* : Forest gene banks as a new approach for *in situ* conservation of genetic resources. *Current Science* 73(2), 163–168.

Yang, R.-C. and Yeh, F.C. (1992) Genetic consequences of *in situ* and *ex situ* conservation of forest trees. *The Forestry Chronicle* 68(6), 720–729.

39 Managing Plant Genetic Resources and the Role of Private and Public Sectors: Oil Palm as a Model

N. Rajanaidu[1] and V. Ramanatha Rao[2]
[1]Malaysian Palm Oil Board, Bangi, Selangor, Malaysia; [2]IPGRI Regional Office for Asia, the Pacific and Oceania, Serdang, Darul Ehsaw, Malaysia

Introduction

It is now well recognized that biodiversity is essential for sustainable development and for the support of social and economic systems to meet food and nutritional needs. Biological diversity is defined as the variation present in all species of plants and animals, their genetic material and the ecosystems in which they occur (McNeely, 1992) and can be considered at three levels:

- genetic diversity (variation in genes and genotypes);
- species diversity (species richness);
- ecosystem diversity (populations/communities of species and their environment).

During the past few years, there has been increasing awareness of the holistic view of managing biodiversity, including agricultural biodiversity, conservation for sustainable utilization and development (Arora, 1997). While dealing with plant genetic resources, the emphasis is more on genetic diversity, with some focus on species diversity. Plant genetic resources (PGR) can include genotypes or populations of plants, representing modern and traditional cultivars, genetic stocks, wild species and so on. These are preserved in the form of living collections, seed, pollen, tissues and DNA.

Though the need for effective conservation and utilization of genetic resources is well recognized, its importance has been further highlighted in two global events. Firstly, in the Convention of Biological Diversity (CBD) agreed at the meeting held in Rio de Janeiro, Brazil and, secondly, the Global Plan of Action (GPA) stemming from the International Technical Conference on the Conservation and Use of Plant Genetic Resources for Food and Agriculture (Iwanaga, 1994; FAO, 1996a, b). Both conventions recognized the sovereign rights of countries over PGR within their borders. However, the onus to conserve and sustainably use PGR rests with countries and both CBD and GPA stress the importance of equitable sharing of these resources and technologies related to their utilization.

The genetic variation in plants was considered as an unlimited resource, its availability was taken for granted to meet all the current, and future needs. However, during the late 1960s, it was realized that the genetic variation available, particularly in landraces grown in the centres of diversity, would soon become extinct if it was not taken care of, due to genetic erosion. The problem became more acute with the wave of agricultural development demanded by the increasing population and the 'Green Revolution' was one of the main causes of genetic erosion, also affecting the centres of diversity, which are mainly located in the developing world. Many factors, like the extension of agricultural land and changes in land use, the introduction of modern techniques of agriculture, urbanization and industrialization, made the traditional cultivars obsolete and led to their rapid replacement with improved cultivars.

Generally speaking, conservation and sustainable utilization of PGR has always been considered as a national responsibility, that is, carried out and supported by the public sector. However, over the years, the multifaceted and multisectoral nature of germplasm activities has been recognized, and several different organizations and individuals may have a stake in their safe conservation and sustainable utilization. In resource rich areas of the world, agriculture has opted for large-scale monocultures. The major reasons for this are managerial and economic and not ecological in nature. However, obviously, diversity is essential as defence against risks and it is also a desirable ecological feature. Thus, conservation of genetic resources is an effort that needs to be supported by all sectors. Consequently, conservation efforts will need to involve all stakeholders concerned, including non-governmental organizations (NGOs), farmers' groups and the private sector, in order to improve the efficiency and reduce the costs. Like any activity, conservation of PGR requires funding, most of which is expected to come from governments as PGR have been considered as a public good. In many countries, it is becoming apparent that the public sector alone may not be able to support fully all the conservation operations. Consequently, other sectors that exploit commercially the PGR products are required to support this effort. Such support could include both collaboration in research and direct funding of PGR conservation activities, organized and coordinated by public sector institutions.

In this chapter, we briefly reflect on the costs of conservation and on the benefits to the private sector from publicly conserved germplasm. Using oil palm as an example, we argue for an improved management of PGR that is based on an increased collaboration between private and public sectors.

Economics of Conservation

The need for efficiently conserving and sustainably using PGR by all countries, as stressed by CBD and GPA, is well recognized. Most of the countries realize this responsibility and have developed, or are developing, strategies for conservation of PGR, *in situ* and/or *ex situ*. Conservation and utilization of plant genetic resources, like any other activity, requires investment, and economic questions thus become relevant. In recent years, this has gained further importance due to the economic slowdown in most of the countries around the world, giving rise to doubts regarding continued support for conservation efforts in many countries in general. This situation has led to the need to highlight the importance of PGR, in economic terms. There is thus the need to demonstrate that PGR are indeed valuable and that tangible benefits can be derived. Assessing the value of genetic resources collections requires being able to demonstrate their current use as well as to relate conserved germplasm to future plant improvement.

Generally, genetic resources are not traded, there are really no prices attached to them. Other things being equal, genetic resources that have the least cost of preservation ought to be ranked above those with greater cost (Brown, 1990). There have even been suggestions made that a genetic resource is worth preserving if it would yield products of commercial worth. If this norm had been applied 50 years ago we would have promoted the extinction of several hundreds of plant species that are worth millions of dollars of revenue today. Nevertheless, it is important to understand that the recognized value of a resource does not mean that it should be conserved wherever it grows. The genetic resources are considered important because of the belief that genetic resources are extremely valuable, so much so that we cannot afford the predicted rate of extinction during the next century (Brown, 1990). At the same time, one has to note that genetic resources have uncertain potential value and limited budgets necessitate identifying priorities for conservation actions, both in terms of plant species and methods of conservation. Moreover, cooperation of all concerned to pool the diminishing resources and to share the responsibilities and expertise would help us to conserve most genetic diversity cost-effectively.

In recent years, we hear and read much about cost-effectiveness. In fact, developing cost-effective conservation technologies and strategies is a major objective. However, usually when a reference is made to cost-effectiveness, the stress tends to be on cost and not on effectiveness. We need to note that our responsibility is the conservation and sustainable utilization of genetic resources and preservation of genetic integrity of conserved material effectively, minimizing the costs involved. The standard economic approach to the conservation issue alone will not help if it only considers static exchange of a fixed quantity of goods among consumers (or of inputs among producers) and plant germplasm is treated no differently than any other market good (Gowdy, 1993). So, it is important to recognize that any serious trade-offs between economic use and preservation of PGR may mean a dramatic loss of genetic diversity, or even the extinction of species/populations. Many of the standard economic theories tend to ignore this irreversibility and choices applied to specific species tend to overlook the total environmental context. In policy applications of the standard approach, efficiency is usually defined narrowly in terms of the optimal allocation of goods with market prices. In recent years, there have been several studies on determining the costs of conservation that emphasize the need for science-based cost-effective conservation techniques and also to provide justification for ongoing conservation efforts (Brown, 1990; Barbier *et al.*, 1994; Godden and Kambuou, 1996; Burstin *et al.*, 1997; Gollin *et al.*, 2000; Pimentel *et al.*, 1997; Cooper, 1998; Gollin and Smale, 1999). Most of these studies conclude that the conservation of biological diversity or PGR presents many problems related to policy, socio-economics, ethics and environmental issues. Significant benefits have been derived through the use of genetic resources for plant improvement and it seems to be fully justified that investments in conservation of PGR should continue.

Genetic resource conservation is a long-term effort with a large initial investment and continuing cost. As noted earlier, the funding support for the conservation of PGR has been mainly by the public sector. Some of the examples include USA, India and Japan. The annual operating budget of the National Seed Storage Laboratory in the USA is US$2,947,102 (Dr Steve Eberhart, 1999, personal communication). The Japanese government funds the entire budget of the genebank in Tsukuba, which was about US$1,578,430 in 1999 (Dr Shoji Miyazaki, 1999, personal communication). Even then, enhancement of agricultural production has received preferential support over PGR conservation mainly due to a few immediately tangible benefits, such as employment (Cohen *et al.*, 1991). What is probably urgently required is the development of a system of monitoring and costing conservation efforts, so that the efforts can be streamlined and made efficient and costs can be brought down.

There are not many examples of public and private sector collaboration in the conservation of PGR, barring the efforts of some of the European genebanks (for example, Lefort *et al.*, 1997). Generally speaking, the private sector has largely depended on the public sector for their raw germplasm. With the rapid innovations that are being made in the field of biotechnology, it is now probably more attractive to the private sector to be involved actively in the conservation and utilization of PGR. Such involvement could be not only in testing and utilizing them, but also financially supporting conservation efforts conducted by the public sector organizations. This is elaborated with oil palm as an example, in Malaysia, where the private sector contributes to all aspects of oil palm genetic resources activities.

A study carried out in 1996 indicated that private breeders were generally supportive of conservation efforts, especially the *ex situ* efforts, but were not prepared to pay for germplasm maintained in *in situ* conservation areas (Swanson and Luxmoore, 1996). This notion may be related to the limited ability of plant breeders to gain access to the material conserved *in situ*. Developing appropriate mechanisms to improve the access to material conserved *in situ* could help in encouraging the private sector to support conservation efforts. Even if the private sector is not willing to support financially the germplasm conserved *in situ*, their support to the public sector efforts on *ex situ* conservation will go a long way in a country's overall efforts to conserve its genetic resources.

It is becoming clear that substantial income could be generated by the efficient use of germplasm through marketing of genetically enhanced materials by the private sector, and hence it should be possible to increase investment in the field of plant sciences by private sector organizations involved in plant improvement. As noted earlier, there is a need for improving the efficiency of germplasm management and we argue here that this is possible by promoting collaborative efforts of public and private sectors.

Oil Palm Genetic Resources Programme at the Malaysian Palm Oil Board

The African oil palm (*Elaeis guineensis*) and the American oil palm (*Elaeis oleifera*) belong to the family *Palmae*. Both these species hybridize readily and produce fertile hybrids. The world production of oils and fats (including animal fats) was 109 Mt in 1999 and 34 Mt were exported. Palm oil production during this period was 20 Mt or 18% of the world production of oils and fats. Malaysian palm oil export figures amounted to 14 Mt or 40% of the world export and 12% of the world production. In this respect, Malaysia alone produced 10.5 Mt and exported 8.9 Mt valued at US$4.6 billion in 1999. It should be emphasized that African oil palm has the highest yielding capacity of 5 t ha^{-1} of oil and its yield is nearly 10 times that of soybean. World annual population growth is about 80 million and it is expected that only palm oil could fulfil the global demand for oils and fats.

The palm oil industry is a multi-billion dollar industry but, surprisingly, the whole industry, as noted earlier, is based on an extremely narrow genetic base (Arasu and Rajanaidu, 1975). The bulk of current planting material is based on four palms planted at Bogor Botanical Gardens in 1848. These four 'Bogor' palms formed the basic stock, which were later planted at Deli, Sumatra, Indonesia and came to be known as the 'Deli dura' population in oil palm breeding. 'Dura' (D) refers to genotypes that have thick-shelled fruits. Later, a limited number of tenera (T, hybrids between dura and pisifera forms) and pisifera (P, genotypes that have fruits without shell) samples were brought to South-east Asia from Africa.

It has been generally recognized that the narrow genetic base is the major obstacle to rapid selection progress in oil palm (Hardon and Thomas, 1968; Arasu and Rajanaidu, 1975; Ooi and Rajanaidu, 1979). This provided the initial impetus for the exploration and collecting of oil palm genetic resources from its centres of diversity. Early oil palm workers from different parts of the world mounted a number of expeditions to collect oil palm genetic resources. After the Second World War, workers in the Belgian Congo sampled oil palm germplasm at a number of sites (Vanderweyen, 1952; Pichel, 1956). Between 1961 and 1965, plant breeders from the Nigerian Institute for Oil Palm Research (NIFOR) collected genetic material at local markets and through village chiefs. This material was established at NIFOR. Evaluation of 72 prospected open-pollinated progenies was concluded and outstanding palms were selected for introduction into the current cycle of the breeding programme (Okwuagwu, 1985). Blaak (1967) sampled materials in the Bamenda Hills of Cameroon and some of those materials were planted at Lobe, Cameroon, and other places. In Côte d'Ivoire, the French oil palm workers systematically evaluated palms in the wild and the selected palms were progeny-tested for their breeding value (Meunier, 1969; Meunier and Baudouin, 1985). Earlier, Institut de Recherches pour les Huiles et Oleagineux (IRHO) selected 38 palms at Pobe and Dahomey and four palms at Bingerville in Côte d'Ivoire. These formed their basic genetic stock.

The Malaysian Palm Oil Board (MPOB) devised a comprehensive genetic resources programme on oil palm. The programme covers all aspects of germplasm collecting, conservation, evaluation/documentation, utilization and research on problems of conservation and utilization.

The African oil palm germplasm was collected systematically in Nigeria (1973), Cameroon (1984), Zaire (1984), Tanzania (1986), Madagascar (1986), Angola (1991), Senegal (1993), Gambia (1993), Sierra Leone (1994), Guinea (1994) and Ghana (1996). In addition, *E. oleifera, Jessenia, Oenocarpus, Bactris, Euterpe* and *Orbignya* species were sampled in Honduras, Nicaragua, Costa Rica, Panama, Colombia, Suriname, Peru and Colombia. The sampling method used for the collection was based on rainfall pattern, soil types, density of palm groves and so on. Normally, 5–10 palms per site were sampled. The number of sites visited in a country depended largely on the density of palm groves (for details see Rajanaidu, 1994).

Seeds were collected from healthy palms in the country of origin. The fruits were depericarped, washed, dried and inspected by the local quarantine officer before the issuance of a phytosanitary certificate. Subsequently, these seeds were dispatched to an intermediate quarantine station in a temperate country. The seeds were further inspected for pests and diseases before dispatching to Malaysia. A bunch was harvested from each of the sampled palms and the fruits from each bunch were planted separately. Seedlings derived from a bunch were considered as an 'open-pollinated' family. Very strict quarantine procedures were observed while importing oil palm germplasm from Africa and Latin America into Malaysia.

The germplasm from various parts of the world was planted in the form of open-pollinated families mainly at MPOB Research Station, Kluang, Johor, Malaysia. The materials were planted in the field using different experimental designs, such as completely randomized design (CRD) to evaluate inter-genotypic competition and randomized complete block design (RCBD) to evaluate productivity. Detailed data were collected on an individual palm basis from these plantings according to agreed descriptors developed jointly with the International Board for Plant Genetic Resources (IBPGR), the predecessor of the International Plant Genetic Resources Institute (IPGRI) (IBPGR, 1982). The characteristics that were studied included:

- fresh fruit bunches yield, FFB, weight and number of bunches produced per annum;
- oil and kernel content of bunches;
- height, measured from ground level to frond number 41;
- fatty acid composition of mesocarp oil;
- physiological parameters (such as harvest index, total dry matter);
- flower census, number of male and female flowers produced per annum;
- molecular data, including randomly amplified polymorphic DNA (RAPD), amplified fragment length polymorphism (AFLP) and restriction fragment length polymorphism (RFLP) data of leaf DNA analysis.

The information collected was stored in electronic databases, and information is available on request.

Oil palm germplasm collected in different parts of the world is maintained in the form of field genebank collections. At present, there is no long-term seed conservation at low temperature and low relative humidity conditions as is practised in other crops such as cereals and legumes, as oil palm seed is recalcitrant. The preservation of germplasm in the form of field genebanks is very expensive. They require a large amount of land and regular maintenance. However, currently this is the only method suitable for the conservation of the genetic resources of this crop. In the case of oil palm 148 palms are planted per hectare. Currently, the MPOB field genebank maintains more than 60,000 palms covering more than 400 ha. One major advantage of field genebank conservation is that the material is readily available for evaluation and breeding. This is critical in the case of perennial tree crops. The material conserved in the MPOB genebank is available for exchange on a bilateral basis.

Opportunities for Oil Palm Improvement through the Use of New Genetic Resources

The systematic collection, evaluation and conservation of oil palm genetic resources have generated elite genotypes for use in MPOB's breeding programme. Most of the valuable genetic resources were not previously available to oil palm breeders in Malaysia for exploiting the potential available in this genepool. Some of the most useful traits for future survival of the oil palm industry are present in the collection and these are:

- *High yield and dwarfness.* Current oil yield of commercial cultivars is 5–7 t ha^{-1}. The material such as the tenera palms yielding more than 10 t ha^{-1} oil have been found in some accessions collected from Nigeria, which can be used for increasing the productivity of the crop significantly. In addition, they are also short; their annual height increment is only 15–25 cm compared with 45–75 cm of current commercial planting material. The selected elite dura and tenera for exploitation are given in Table 39.1. A number of these palms collected at a few sites in East Central State of Nigeria possessed these traits.
- *High iodine value.* High iodine value (IV) confers high unsaturation to edible oils, which is an important consideration in oil palm breeding. A number of palms have been selected for high IV by evaluating collections from Nigeria. They have been multiplied to produce high IV planting material. The selfed material of the high IV palms is being evaluated. Some of the palms had an IV of > 70 while the current commercial varieties have a value of 52–54.
- *High kernel content.* The fast growing oleochemical industry in Malaysia requires a good source of lauric oil found in palm kernel. The prospect of breeding for high kernel to bunch (i.e. percentage of kernel in a bunch) is bright. Some Nigerian families have a mean kernel to bunch of more than 12% compared with 6% in current breeding progenies. It is possible to double the kernel content of future oil palm planting material. An economic analysis shows that breeding for the highest possible content of kernel in oil palm fruits yields the highest rate of return from the oil palm (Rajanaidu and Jalani, 1994).
- *High carotene content.* There is a considerable level of variation for carotene in *E. oleifera*, *E. guineensis* and their hybrids. Although the oil yield of pure *E. oleifera* is extremely low as compared with *E. guineensis*, the level of carotene is attractive enough to sell this oil in the form of capsules as health food. In addition, *E. oleifera* has a higher vitamin E (total tocopherol and

Table 39.1. High-yielding and dwarf Nigerian tenera palms.

Trial no.	Family	Palm no.	Oil (kg ha^{-1} per palm)	Oil (t ha^{-1} $year^{-1}$)	Height (cm $year^{-1}$)
0.149	28.17	12724	83.34	12.18	23.1
0.149	19.11	12279	75.94	11.24	21.5
0.149	13.05	12094	76.27	11.29	24.0
0.150	16.21	4352	70.39	10.42	24.9
0.150	19.13	3759	71.54	10.59	22.5
Current planting material				5.0	45–75.0

tocotrienols) content compared with *E. guineensis* (Choo and Yusof, 1996; Jalani *et al.*, 1997). An extensive screening of MPOB Nigerian palms for carotene was carried out and the variation for carotene in *E. guineensis* was in the range of 273–3512 p.p.m. Some of the crosses involving specific populations (No. 16 and No. 19) showed high carotene content (> 1000 p.p.m.), while the normal carotene level in current Deli duras is around 500 p.p.m. In general, the African germplasm has higher carotene content (> 900 p.p.m.). *E. oleifera* has slightly higher carotene than *E. guineensis* but, as noted earlier, the oil content is extremely low. A special breeding programme has been developed to exploit MPOB Nigerian duras and pisifera to develop high carotene planting materials.

Materials possessing the various new traits that have been described above are now available for utilization in oil palm improvement. Such improvement is expected to result not only in productivity increases but also in quality traits. Materials from other sources are being evaluated and more such useful traits may be identified soon. These traits for use would not have been available if there were no systematic collection, conservation and evaluation activities undertaken.

Malaysian Palm Oil Board and Funds for Oil Palm Genetic Resources

Prior to the Malaysian Palm Oil Board, since 1979 the Palm Oil Research Institute of Malaysia (PORIM) handled the genetic resources programme. MPOB is a public body, established in 2000. However, the funds for its operations are provided mainly by the palm oil industry (private) through a research cess levied on crude palm oil and palm kernel oil produced (Fig. 39.1). For instance in 1999, Malaysia produced 10.5 Mt of palm oil and the industry contributed US$30 million as a research cess. Of this, US$0.5 million was allocated for oil palm genetic resources activities and a further US$0.5 million is generated by MPOB through sale of seeds, pollen and lease of mother palms (Table 39.2). General conservation and research expenditures incurred by the oil palm genetic resources programme of MPOB are presented in Table 39.3.

The research priorities of MPOB are set by the Board, which comprises representatives mainly from the oil palm industry. A Programme Advisory Committee (PAC) assists the Board in decision-making; this consultative body, which scrutinizes the MPOB annual research programme, makes recommendations to the Board. Most of the members are from abroad, specialized in various disciplines. The PAC has approved MPOB's continued attention to oil palm genetic resources, including collection, conservation, evaluation and utilization and in recent years much attention has been paid to using

Table 39.2. Funds for genetic resource activities at MPOB, Malaysia in 1999 (in million US$).

Total research cess collection from industry	30
Biology division budget	10
Genetic resources budget	
Income generated (by MPOB through sale of seeds, pollen and lease of mother palms)	0.5
Net contribution by the Institute for genetic resources activities (from the cess collected)	0.5
Total for genetic resources	1.0

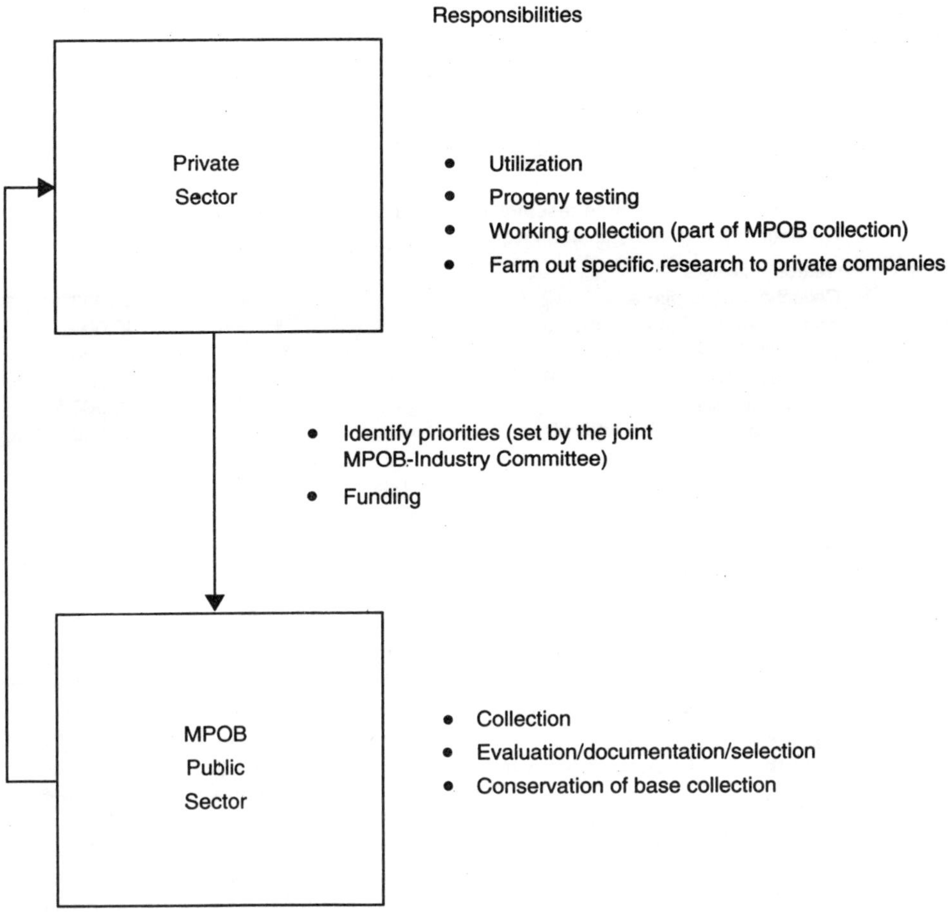

Fig. 39.1. Management of oil palm genetic resources at MPOB. Collaboration between private and public sectors.

advanced techniques for evaluation. Considerable research has been carried out on *in vitro* techniques for the conservation of oil palm genetic resources, but a workable technique is yet to be developed. Hence the germplasm continues to be maintained in the field genebank. Pollen storage technique is mainly used as a tool to facilitate asynchronous flowering genotypes. It has been found that pollen can be stored for a maximum of 2 years at around –4°C (Hartley, 1988). Research on various problems of conservation and utilization, including biotechnology and tissue culture (mainly for rapid multiplication and propagation of elite material) is ongoing (Rajanaidu and Arffin, 2000). For details of expenditures on several of the above-mentioned activities see Table 39.3.

Public–Private Sector Collaboration in Oil Palm Genetic Resources Activities

Collecting

MPOB has assembled the world's largest oil palm germplasm collection in collaboration with public and private organizations in different countries. Here are some examples in which both public and private sectors were involved in joint collecting missions. In 1973, 180,000 seeds were collected in Nigeria from 45 sites with the cooperation of NIFOR. In 1984, *E. guineensis* germplasm was collected in Zaire in collaboration with Unilever; 369 samples from 56 sites were collected. PORIM and Unilever jointly col-

Table 39.3. Expenditures on germplasm activities at MPOB, Malaysia.

	Cost in US$
General expenses	
Maintenance of germplasm 400 ha@US$700 ha^{-1} (fertilizers, weeding, harvesting)	280,000
Maintenance of vehicles	70,000
Personnel expenses (three research officers and five technicians)	100,000
Contingency and miscellaneous expenses	60,000
Research activities	
Collecting expenditure	50,000
Establishment of trials in the field	40,000
Vegetative data collection	10,000
Yield recording	50,000
Bunch analysis	70,000
Fatty acid composition analysis	80,000
Vitamin E analysis	40,000
Molecular screening (AFLP/RLFP) of populations	150,000
Data processing and analysis	50,000
Total	1,050,000

lected 95 samples from 36 sites in the western and eastern regions of Cameroon. Collecting of *E. guineensis* genetic resources was also carried out in 1986 in Tanzania and Madagascar with the cooperation of the Departments of Agriculture in Tanzania and Madagascar with partial financial assistance from IBPGR (Rajanaidu, 1986). Sixty samples from 13 sites were collected in Tanzania and 17 samples from four sites in Madagascar.

E. oleifera germplasm was sampled in Colombia, Costa Rica, Honduras, Nicaragua, Panama and Suriname with the cooperation of the Departments of Agriculture and private and public bodies in the above countries. In 1989, PORIM, with the cooperation of Instituto Colombiano Agropecuaria (ICA), collected germplasm of *Jessenia* and *Oenocarpus* in various parts of the country.

These activities needed about US$100,000 of which about 20% came from public sources (IBPGR), 15% from the private sector (mainly in kind) and 65% from Malaysian sources (again mainly from the private sector as part of the cess levied on palm oil production).

Maintenance, evaluation and conservation

The oil palm genetic materials collected in the above countries were maintained as a base collection with the objective of long-term conservation, evaluation and utilization. Part of the collection was distributed to the industry as a working collection to be evaluated in different parts of Malaysia in various ecological niches. These efforts not only help MPOB to evaluate materials in different sites but also help the users (the industry) to select materials suitable to the agroclimatic conditions in which they operate.

Evaluation of oil palm genetic resources efforts included the testing of the collected material at different sites. Such testing provided valuable information on genotype × environment interaction. This activity was carried out largely with the collaboration of the private sector (Rajanaidu and Rao, 1988). The private sector absorbs the costs for the establishment and maintenance of collections planted for evaluation. The information is shared between all the collaborating partners. The evaluation is carried out using proper experimental designs so that statistical analyses can be carried out and to ensure the evaluation of the economic potential of the material.

Oil palm, being a fairly big plant, is planted at 148 palms ha^{-1}. This makes the maintenance very expensive. The collection from Nigeria alone occupies more than 200 ha. Hence, the cooperation of the industry is vital to obtain land to test and conserve oil palm genetic resources. The private sector owns plantations throughout the country and every year about 5% of the land is replanted. Hence, it is possible to get suitable land for various experiments through their cooperation. In order to discuss and determine priorities, to develop action plans and to monitor the collaborative activities between MPOB and the private sector, a joint committee of oil palm breeders from both sectors has been established. The joint committee reviews the workplans and progress on a regular basis and takes decisions on allocation of funds, further experiments to be undertaken and the responsibilities of each partner. MPOB develops a list of material to be planted. The industry makes a selection from the list to be planted on their farms and the rest (generally about 50%) is planted by MPOB. The funds that the industry spends on these evaluations are in addition to the cess that is levied on production, part of which comes to MPOB.

Utilization of Genetic Resources by the Private Sector

The elite oil palm germplasm is being utilized by the industry in a number of ways for crop improvement, including:

- *Direct selection of individual palms.* About 3% of the tenera in the Nigerian collection had oil yields comparable with current planting materials. A third of these palms had an annual height increment significantly less than current commercial dura × pisifera (tenera) materials. Attempts are being made to clone these 1% palms by tissue culture technique by PORIM and the industry. Available data show that the outstanding families and individual teneras are normally found in the East Central State of Nigeria.
- *Progeny-testing of Nigerian elite palms.* Some of the outstanding Nigerian teneras were progeny-tested together with a range of Deli duras available in the industry and MPOB. The production of tenera × dura or dura × tenera hybrids and selfing of their dura and tenera parents is underway. The aim of the crossing programme is to progeny-test the Nigerian teneras to study the combining ability. The selfs will be used for seed production following the procedure of reciprocal recurrent selection (Jacquemard *et al.*, 1981).
- *Broadening the genetic base of Deli duras and teneras.* The overall genetic variability of current Deli duras and teneras could be broadened by crossing Deli duras with Nigerian duras, and teneras such as AVROS, La Me, Yangambi, URT and 27B could be mated with Nigerian teneras or pisiferas. These introgressed populations, with increased amounts of heritable variation for desired traits, will be the basis for further selection and breeding. Careful choice of the germplasm at this stage will increase selection efficiency and the probability of obtaining desirable segregants. Such selected palms can be expected to possess one or a combination of the desired traits; for example crossing Deli duras with African palms gave marked increases in the additive variance for bunch yield and, especially, its components, and weight and bunch number per palm. The Nigerian germplasm palms gave relatively high yields. The main collection as a whole, covering 165 ha on inland soils, gave fresh fruit bunch yields of 23–24 t ha^{-1} $year^{-1}$ in 8–10 years after planting. On coastal soils yields have been in excess of 25 t ha^{-1} $year^{-1}$ in the 8th year. Clearly the introgressing of such palms into existing breeding populations will maintain the yield potential, if not raise it, while broadening the genetic base.

Future Role of Public and Private Sectors in Oil Palm Genetic Resources Management

At present, the public sector is mainly involved in the collection, evaluation and conservation of oil palm genetic resources, whereas the development of planting material is with the private sector. In the future, it may be necessary to develop collaborative programmes between the public and private sectors from the outset. This is effective if the collecting team combines both scientific and business skills that could lead to sustainable conservation and utilization of oil palm genetic resources. Any successful venture could be marketed readily and arrangements could be made in such a way that the profits are shared between the public and private bodies in an equitable manner. For example, in the case of oil palm, MPOB, with the collaboration of the private sector, progeny tested the Deli duras with the Nigerian pisiferas. Outstanding commercial D × P planting materials with high yield and dwarf plant type were identified from these materials. Seed production and marketing of the these hybrid materials are undertaken by the private sector. MPOB receives a royalty for the supply of elite Nigerian pollen for the seed production. Thus, the benefits from the exploitation of oil palm genetic resources are shared by public and private sectors and the public sector is able to generate funds for the conservation of genetic resources. The farmers benefit as well since they can access the new varieties much faster than if the seed production was only done by the public sector. It should be possible to extend such joint ventures in other areas and other crops as well.

While it is possible to obtain funds for germplasm collection, the funds for long-term conservation of PGR are limited, especially in developing countries, which are endowed with the traditional landraces and wild relatives of crops. The MPOB has developed a mechanism to finance oil palm research and conservation activities. The funds are provided largely by the oil palm industry through a research cess levied on crude palm oil and palm kernel oil produced. This model could be adopted for other crops and in other countries.

Concluding Remarks

As indicated at the beginning of the chapter, conservation of genetic resources is an effort that needs to be supported by all sectors. One of the main reasons for the conservation of crop genetic resources is that landraces and other cultivated and non-cultivated materials offer a rich source of desired traits for the improvement of economic plants. However, it is not possible to foresee what kinds of germplasm will be required in the future and hence the need for the conservation of PGR. With advances in biotechnology, methods are becoming available for better accessing of desired types in large collections as well as improving methods for breeding, especially in the distantly related germplasm. This is expected to improve the efficiency of using germplasm that is being conserved in different genebanks around the globe. To efficiently conserve and sustainably use these resources, a multisectoral approach is needed.

It must also be noted that genetic resource conservation is a long-term activity with a large initial investment and recurrent expenses. Generally, this has been the onus of national governments and international public funding mechanisms. In Malaysia in the case of oil palm, a mechanism has been successfully established to jointly bear the burden of oil palm genetic resources and benefit from their use. In this mechanism, the public sector takes care of conservation of the genetic resources and the private sector funds the conservation programme. In turn, the private sector benefits from the MPOB's (public) breeding programme by using the varieties developed and marketing the seeds of improved varieties. Thus, the public and private sectors together share the responsibility of conserving the oil palm genetic resources, vital to the country. This win–win situation could be a model to other crops and to other countries.

With the increased role that the private sector might play in the future in the conservation and utilization of PGR, it is likely that the role of the public sector in conservation and utilization of PGR may diminish progressively, at least in more developed countries. However, keeping in mind the needs and responsibilities of countries and communities, it is important that the public sector, along with community-based organizations, continues to play a major role in the conservation of PGR. Such a role might increasingly be that of monitoring the use, sharing the benefits, establishing priorities for conservation actions and putting in place appropriate checks and balances. Given the complexities of intellectual property protection and the need to protect farmers' rights to which we all subscribe, this continued involvement is imperative. If proper care is not taken, biotechnology, as promoted by certain private companies, might prove too expensive for the developing countries (thus hindering development); its use might increase the gap between developing countries and industrialized nations. We should also think in terms of linking commercial benefits from the exploitation of PGR through biotechnology to conservation of PGR. Additionally, a spirit of understanding and sharing of ideas as well as technologies will help to do a better job of conservation and use. Only then will it be possible to realize the most obvious point that can be made about plants, animals, fungi and microorganisms; that they are the means by which we attain sustainability.

References

Arasu, N.T. and Rajanaidu, N. (1975) Conservation and utilization of genetic resources in the oil palm (*Elaeis guineensis* Jacq.). In: Williams, J.T., Lamoureux, C.H. and Sastrapradja, S. (eds) *South East Asian Plant Genetic Resources, Proceedings of the Third Southeast Asian Regional Symposium on Genetic Resources, Serpong, Indonesia, 22–24 August 1995.* Yayasan Kehati, Jakarta, Indonesia, pp. 182–186.

Arora, R.K. (1997) Biodiversity Convention, Global Plan of Action and the National Programmes. In: Hossain, M.G., Arora, R.K., and Mathur, P.N. (eds) *Plant Genetic Resources – Bangladesh Perspective, Proceedings of a National Workshop on Plant Genetic Resources, 26–29 August 1997.* Bangladesh Agricultural Research Council BARC-IPGRI, Dhaka, Bangladesh, pp. 28–35.

Barbier, E., Burgess, J. and Folke, C. (1994) Ecological and economic perspectives: convergence of divergence? In: *Paradise Lost: the Ecological Economies of Biodiversity.* Earthscan Publications, London, pp. 42–59.

Brown, G.M.J. (1990) Valuation of genetic resources. In: Orians, G.H., Brown, J.G.M., Kunin, W.E. and Swierzbinski, J.E. (eds) *The Preservation and Valuation of Biological Resources.* University of Washington Press, Seattle, Washington, pp. 205–233.

Burstin, J., Lefort, M., Mitteau, M., Sonnot, A. and Guiar, J. (1997) Towards the assessment of the cost of genebanks management: conservation, regeneration and characterization. *Plant Varieties and Seeds* 10, 163–172.

Choo, Y.M. and Yusof, B. (1996) *Elaeis oleifera* palm for pharmaceutical industry. PORIM TT No. 42, PORIM Information Series. PORIM, Bangi, Malaysia.

Cohen, J.I., Williams, J.T., Plucknett, D.L. and Shands, H. (1991) *Ex situ* conservation of plant genetic resources: global development and environmental concerns. *Science* 253, 866–872.

Cooper, J.C. (1998) The economics of public investment in agro-biodiversity conservation. In: Evenson, R.E., Gollin, D. and Santaniello, V. (eds) *Agricultural Values of Plant Genetic Resources.* CAB International, Wallingford, UK, pp. 43–54.

FAO (1996a) *Report on the State of the World's Plant Genetic Resources – International Technical Conference on Plant Genetic Resources, Leipzig, Germany.* FAO, Rome, Italy.

FAO (1996b) *Global Plan of Action for the Conservation and Sustainable Utilization of Plant Genetic Resources for Food and Agriculture.* FAO, Rome, Italy.

Gollin, D. and Smale, M. (1999) Valuing genetic diversity: crop plants and agroecosystems. In: Collins, W. and Qualset, C. (eds) *Biodiversity in Agroecosystems.* CRC Press, London, pp. 237–265.

Gollin, D., Smale, M. and Skovmand, B. (2000) Searching an *ex situ* collection of wheat genetic resources. *American Journal of Agricultural Economics* 82(4), 812–827.

Gowdy, J.M. (1993) Economic and biological aspects of genetic diversity. *Society and Natural Resources* 6, 1–16.

Hardon, J.J. and Thomas, R.L. (1968) Breeding and selection of oil palm in Malaya. *Oleagineux* 23(2), 85–90.

Hartley, C.W.S. (1988) *The Oil Palm* (Elaeis guineensis, *Jacq*). Longman Scientific and Technical, John Wiley & Sons, New York.

IBPGR (1982) *Descriptors for Oil Palm.* IBPGR, Rome, Italy.

Jacquemard, J.C., Meunier, J. and Bonnot, F. (1981) Etude Genetique de la production du croisement chex le palmier a huile *Elaeis guineensis. Oleagineux* 36(7), 543–552.

Jalani, B.S., Cheah, S.C., Rajanaidu, N. and Darud, A. (1997) Improvement of oil palm through breeding and biotechnology. *Journal of the American Oil Chemist's Society* 47(11), 1451–1455.

Lefort, M., Arbez, M., Chauvet, M., Dattee, Y., Guiard, J., Mitteau, M. and Sontot, A. (1997) The French strategy for the management of plant genetic resources. *Plant Varieties and Seeds* 10, 153–162.

McNeely, J.A. (1992) Nature and culture. *Nature and Resources* 28(3), 37–43.

Meunier, J. (1969) Etude des populations naturelles d'Elaeis guineensis en Côte d'Ivoire. *Oleagineux* 24(4), 195.

Meunier, J. and Baudouin, L. (1985) Evaluation and utilization of Yacoboue population of *Elaeis guineensis.* In: Soh, A.C., Rajanaiud, N. and Mohd Nasir Hassan Basri (eds) *Proceedings of International Workshop on Oil Palm Germplasm and Utilization.* ISOPB/PORIM, Kuala Lumpur, Malaysia, pp. 144–152.

Okwuagwu, C.O. (1985) The genetic base of the NIFOR oil palm breeding programme. In: Soh, A.C., Rajanaiud, N. and Mohd Nasir Hassan Basri (eds) *Proceedings of International Workshop on Oil Palm Germplasm and Utilization.* ISOPB/PORIM, Kuala Lumpur, Malaysia, pp. 228–297.

Ooi, S.C. and Rajanaidu, N. (1979) Establishment of oil palm genetic resources: theoretical and practical considerations. *Malaysian Applied Biology* 8(1), 15–28.

Pichel, R. (1956) L'Ameriolation du Palmier a huile du Congo Belge Conf. Franco-Brittanique Sur le palmier a huile. *Bulletin des Recherches Agronomiques* 14, 59–66.

Pimentel, D., Wilson, C., McCullum, C., Huang, R., Dwen, P., Flack, J., Tran, Q., Saltman, T. and Cliff, B. (1997) Economics and environmental benefits of biodiversity. *Bio-Science* 47(11), 747–757.

Rajanaidu, N. (1986) Collection of oil palm (*Elaeis guineensis*) genetic material in Tanzania and Madagascar. *ISOPB Newsletter* 3(4), 2–6.

Rajanaidu, N. (1994) *PORIM Oil Palm Genebank*. Palm Oil Research Institute of Malaysia, Bangi, Malaysia.

Rajanaidu, N. and Jalani, B.S. (1994) Potential sources of lauric oils for the oleochemical industry. In: Applewhite, T.H. (ed) *Proceedings of the World Conference and Exhibition on Lauric Oils: Sources, Processing and Applications* AOCS, Manila, Philippines, pp. 47–50.

Rajanaidu, N. and Rao, V. (1988) Oil palm genetic collections: their performance and use to the industry. In: Abdul Halim, Chew, P.S., Wood, B.J. and Pushparajah, E. (eds) *Proceedings of 1987 Int. Oil Palm Conference-Agriculture.* PORIM, Bangi, pp. 59–85.

Swanson, T.M. and Luxmoore, R.A. (1996) Industrial reliance upon biodiversity. Report of a Darwin Initiative Project. World Conservation Monitoring Centre, Washington, DC.

Vanderweyen, R. (1952) La prospection des palmeraies congolaises et ses premier results. *Bulletin d'Information de Institut National pour l'Etude Agronomique du Congo Belge* 1, 357–382.

40 The Community-based Conservation and Management of Genetic Diversity in Agroecosystems: the Role and Function of Law

S. Biber-Klemm
Faculty of Law, University of Basel, Basel, Switzerland

Introduction

In the last few decades the insight into the significance of biological diversity and the value of genetic and biochemical information increased significantly. On the ecosystem level, new scientific findings furthered the understanding of ecosystem functions and the crucial role of biodiversity to assure ecosystems regulatory functions, which provide the basis for economic activities such as agriculture (Barbier, 1996; Costanza *et al.*, 1997). At the level of the genes, the achievements in biotechnologies and the progress made in research technologies brought about profound changes in the agricultural and pharmaceutical industries and opened up new fields of application and uses of plant and animal (genetic) resources.

At the same time, awareness of the accelerated loss of diversity and insight into the threats of this process to the survival of humanity are growing. With a view to the plant genetic resources for food and agriculture (PGRFA), insight increased into the importance of their genetic diversity for maintaining food security in a changing world. This heightened awareness is mirrored in the efforts to conserve crop genetic information in *ex situ* genebanks, but also *in situ* in its original habitats. Efforts to this end in the political, institutional and legal field at international and national levels and within civil society are increasing.

On the international level, pertinent discussions and processes take place in different fora: (i) within the follow-up process of the Convention on Biological Diversity (CBD); (ii) in the revision process of the International Undertaking on Plant Genetic Resources (IU) of the Food and Agriculture Organization (FAO); and (iii) at the interface between conservation of PGRFA and trade, in the framework of the International Convention for the Protection of New Varieties of Plants (UPOV-Convention) and the Agreement on Trade-Related Aspects of Intellectual Property Rights (TRIPS).

These international conventions are at the core of intense political debates at the national and international levels. The debates focus on mechanisms to regulate access to genetic information and the sharing of benefits resulting from their use in industrial innovation and production processes. Closely connected are discussions on the means to alleviate the disparity in the legal protection of the intangible property resulting from formal breeding and innovation processes on the one hand, and from small-scale traditional breeding on the other hand.

This chapter looks at these problems from a legal point of view. It asks if it is possible to create specific legal instruments to support the conservation and maintenance of PGRFA within the framework of the above-mentioned international

(eds J.M.M. Engels, V. Ramanatha Rao, A.H.D. Brown and M.T. Jackson)

conventions. To this end, it analyses in a first step the causes leading to the loss of biodiversity in agroecosystems. It then describes the actual legal and institutional framework and the ongoing processes and main discussions at the international level. Against this background it identifies criteria and elements for a (legal) solution. On the basis of this information emerging legal strategies will be presented and the proposed solutions discussed.

Factual Background

Before entering into the legal and political discussion, it is important to explain the chapter's focus on the on-farm management of PGRFA and to present the elements of the conservation of agrobiodiversity which constitute the background for the following reflections.

Conservation of PGRFA: the importance of an integrated approach

PGRFA are sources for tailoring food production in order to meet future challenges. The Leipzig Declaration[1] summarizes the reasons for the importance of their conservation as follows: PGRFA have an essential importance for the food security of present and future generations. They are

> the basis of natural and directed evolution in the plant species most critical to the survival and well-being of human beings. All countries require plant genetic resources if they are to increase food supplies and agricultural production sustainably and meet the related challenges of changes in the environment, including climate change. (para 3)

With growing insight into the complexity of ecosystem networks and the importance of the interaction of all components of the system, awareness of the factors determining agricultural ecosystems has increased. Thus, for instance, the Subsidiary Body for Scientific, Technical and Technological Advice (SBSTTA) of the CBD defines the term agrobiodiversity in a broad sense, including 'the variety and variability of animals, plants and micro-organisms, at genetic, species and ecosystem levels, which are necessary to sustain key functions of the agro-ecosystem, its structure, and processes for, and in support of, food production and food security'.[2] PGRFA form one part of this system. According to the definition of the IU they include not only the man-made, cultivated varieties, but also their wild and weedy relatives (Art. 2.1 (a)).

With this goes the understanding of the interdependence between the maintenance of agroecosystems, the further evolution of PGRFA, and human intervention. Thus the Global Plan of Action (GPA)[3] points out that, unlike most natural biodiversity, PGRFA require continuous active human management. According to Brush (1994) *in situ* conservation of PGRFA includes hybridization within and between populations of wild, weedy and cultivated plants, competition among genotypes, natural and conscious selection at the local level and exchange of different genotypes among farmers and farms. Swanson points out that preservation of the diversity of PGRFA includes not only (static) conservation of the presently existing varieties – as is effectuated in *ex situ* genebanks – but prominently also maintenance of the underlying evolutionary processes, in order to conserve as many options for the future as possible (Swanson *et al.*, 1994; Swanson, 1995).[4]

From this it becomes clear that – as a supplement to *ex situ* conservation in genebanks – an integrative approach to conservation and maintenance of PGRFA diversity is necessary, which includes *in situ on-farm* conservation as an essential element. Against this background, the argumentative context and the close interrelationship between *in situ* on-farm conservation and the realization of farmers' rights becomes obvious (Brush, 1994;

[1] Leipzig Declaration on Conservation and Sustainable Utilization of Plant Genetic Resources for Food and Agriculture. Adopted by the International Technical Conference on Plant Genetic Resources. Leipzig, Germany, 17–23 June 1996.
[2] UNEP/CBD/SBSTTA/5/INF/10.
[3] Global Plan of Action for the Conservation and Sustainable Utilization of Plant Genetic Resources for Food and Agriculture. Adopted by the International Technical Conference on Plant Genetic Resources. Leipzig, Germany, 17–23 June 1996.
[4] Swanson points out that

> It must be recognized that human capital alone may not be capable of producing all important and valuable information. There is also a base biological dimension which generates information. This biological dimension is the evolutionary process which, through biological interaction and the process of selection, generates communities of life forms that contain substantial amounts of accumulated information (1994).

FAO, 1996 (sub Legal Texts and Conference Documents); Swaminathan, 1997). Accordingly, the IU points out the contributions of 'unnumbered generations of farmers in conserving, improving and making available plant genetic resources'.[5]

In the process of on-farm breeding of PGRFA, the knowledge of the farmers of specific characteristics plays an important role. Traditional knowledge – knowledge, innovation, and practices of indigenous and local communities embodying traditional lifestyles (Art. 8 (j) CBD) – can be characterized as a body of knowledge which has been built by a group of people, indigenous peoples or farming communities, through generations living in close contact with nature. In farming systems, it has a specific importance, given the high interdependence between domesticated species and humans. It encompasses a profound knowledge of ecosystem conditions and the breeding techniques adapted to local conditions, including the selection for tolerance or resistance to biotic and abiotic stresses. It further comprises information on the useful properties contained in the genotypes which have been selected and conserved.

Open access and exchange

Another key issue in plant breeding is open access to resources and exchange of crop varieties and related information. At the local level, in traditional societies and in subsistence farming systems, sharing and exchange of varieties is traditional and of high importance for the evolution of local varieties, and to assure food security. This exchange takes place in a reciprocal relationship. Castillo speaks of a 'culture of reciprocity', the basis of which is 'not money but trust, mutuality of benefit and social equivalence in the value of the item or service being exchanged' (Castillo, 1997).

At the international level, due to the movement and exchange of crops throughout history, crop species have spread from their primary centres of origin, and secondary centres of diversity have developed. Thus, with regard to crop diversity, a high interdependence between countries and continents exists. Even countries which hold large amounts of plant genetic diversity, will have to guard against the risks of over-reliance on too narrow a genetic base and thus continue to require access to diversity elsewhere for genes with useful traits. In the long term, breeding programmes would be at risk if access to the full range of genetic diversity was restricted (Cooper *et al.*, 1994; International Plant Genetic Resources Institute (IPGRI), 1996).

Open access to and sharing of genetic resources at the local, regional and global level are thus essential for maintaining the options to meet the needs of the growing world population. Accordingly, the original IU as adopted in 1983, declared plant genetic resources as a 'heritage of mankind' available without restriction. According to its Art. 5, samples were to be made available free of charge, on the basis of mutual exchange or on mutually agreed terms.

Background Problems

In general

From the above it can be concluded that traditional farming and on-farm management is important for the conservation and further evolution of PGRFA. Open access to and free exchange of crop germplasm are other essential elements. However, the present situation is characterized by the erosion of both genetic diversity, related traditional knowledge and the concept of free exchange (IPGRI, 1996; Girsberger, 1999; Biber-Klemm, 2000a, c). The resulting legal question is, how can a sound legal basis be created which prevents the loss of agricultural diversity and guarantees access to and exchange of genetic resources, forming a basis for mutual trust and supporting farmers and farming communities in the maintenance of on-farm crop evolution systems. In order to answer this question, it is important to understand the underlying reasons for erosion of diversity, knowledge and open access.

Three domains which influence this evolution can be identified. First, the specific characteristics of the information we are dealing with. Second, the effect of these characteristics is being intensified by actual economic and scientific evolutions. Third, the related processes taking place in both subsistence and industrialized societies.

[5] Annex II, Resolution 5/89, Para 4.

The characteristics of genetic information and traditional knowledge

In the questions on PGRFA and related traditional knowledge, we are essentially dealing with information: the genetic information as contained in the seeds of plant varieties and the information contained in traditional knowledge. Both genetic resources and traditional knowledge have a ubiquitous character which is specific to information of any kind: once it has been revealed and acquired by another person, it is impossible for the original owner to prove that it was exclusively his or hers. The information becomes independent of its original source. These characteristics correspond to the fundamental paradox of information: information is valueless until revealed, but its value, once revealed, cannot be appropriated. With a view to PGRFA and traditional knowledge these traits are due to the following qualities:

- PGRFA are self-propagating and also produced and traded as goods for consumption. The pertinent information, which results from the skill of generations of breeders, is contained in each seed. By use of modern biotechnology, access to this information can be gained on the basis of only a small quantity: one seed is sufficient to reproduce the information.
- Knowledge, once it has been revealed, is accessible to everybody. It is true that traditional knowledge has frequently been protected by customary laws within the cultural framework and structures of the communities (Dutfield, 1999; Greene and Drescher, unpublished). However, these customary laws are not sufficient to regulate conditions of exchange and trade in the broader context of the growing (intercontinental) interdependence and globalization of markets (see Dutfield, 1999; Girsberger, 1999).

In the context of formal research and development (R&D) processes, specific instruments to protect the informational values of new plant varieties or of inventions have been created: intellectual property rights. However as a rule, landraces and traditional knowledge presently cannot be protected by these instruments, as they do not correspond to the criteria requested by, for instance, patents or plant breeders' rights.[6] Thus, from the legal point of view, genetic diversity and traditional knowledge, once revealed or distributed, are in the public domain, accessible to everybody.

According to economic theories, in this informational character of biological diversity and related traditional knowledge lies an important reason for its loss. Because its main value cannot be appropriated, the value of biodiversity is ignored by land-owner decisions, and diversity of plant species is sacrificed in favour of a more lucrative use of land resources such as industrialized monocultures[7] (Swanson *et al.*, 1994; Swanson, 1995).

Due to this 'open-access' situation of plant genetic resources and of related traditional knowledge, scientists within industrial innovation systems could use and patent them for industrial utilization, often without further improvement and/or an additional 'inventive step'. This frequently happened without the consent of the holders of the resources and without compensation or sharing profits. Concerned people and communities speak of 'piracy of traditional knowledge'. Piracy of genetic resources and related traditional knowledge has been denounced for pharmaceutical resources, for example the neem tree, kava and turmeric. However, there has also been extensive documentation in the field of PGRFA. There patents have been applied for (and granted) even for 'in-trust' germplasm[8] (Rural Advancement Foundation International (RAFI), 1998; see also RAFI, 2000).

[6] In the case of patents, the innovation has to be new, to involve an inventive step or, synonymous, to be non-obvious and to be useful (Art. 27.1 TRIPS). The protective criteria for plant breeders' rights require that a plant variety be new, distinct, homogeneous (uniform) and stable. Traditional knowledge frequently lacks the criteria of novelty, the necessary inventive step, or the level of creativity seen as prerequisites for protection by industrial IPRs. Landraces as developed by informal breeding systems of farming communities fail to fulfil the criteria of stability and uniformity and thus fail to meet the protective criteria of the plant breeders' rights.

[7] Swanson describes this phenomenon as 'conversion process'. In this process, for the benefit of economic development, the more productive assets, e.g. a cultivated lucerne variety, are substituted for the less productive, e.g. diverse native grasses. That means that uses are changed from diverse to specialized ones. 'The biodiversity problem … is the result of the diffusion of this same process to the last unmodified habitat on earth. Diversity decline is the by-product of this scoping-in process by which the global biosphere is being homogenised for agricultural development' (1994).

[8] In-trust germplasm is placed as 'designated germplasm' by contract between the donor country or the International Agricultural Research Centre and the FAO under the auspices of the latter, and made freely available for agricultural research or breeding purposes. The donor organization is obligated not to claim legal ownership over the designated germplasm nor to seek any intellectual property rights over that germplasm or related information. The same obligations are to be passed on to all future recipients of designated germplasm (see Cooper *et al.*, 1994).

This situation leads to mistrust. If possible, the relevant information is kept secret and access is denied (examples in Biber-Klemm, 2000b). This is the contrary to the situation of mutual trust and open exchange evoked above.

The problem of the insurance value

According to Swanson *et al.* (1994) farmers in subsistence systems used to maintain a high diversity of crop species in order to assure their annual harvests and thus to guarantee a minimal level of production and to prevent food shortage, even at the cost of a higher average productivity. However, evolution in recent decades, and the results of the Green Revolution brought about other means to guarantee subsistence.

Swanson *et al.* (1994) enumerate the following points:

1. The substitution of new varieties for traditional: in getting accustomed to the new varieties and production methods, farmers' information and experience on the traditional varieties get lost. The new information now serves as insurance, the old varieties having lost this quality with the loss of the related information.
2. The substitution of other assets for biological ones: better connection of a rural area by roads and transport to population centres furthers production changes from 'food crops' to high yielding 'cash crops'. The centre additionally provides alternative insurances against food scarcity, such as, for example, the opportunity to make use of surplus human labour as another source of income.
3. The development of industry in the urban centres further de-links the income of the country and city dwellers. That means that even in times of crop failure, markets for the other products of farmers exist, which can be translated into income.

The consequence is that the individual's motivation to maintain diversity as an assurance is reduced. The availability of other means of insurance allows them to pursue the highest average yield, acquiring insurance by other means. From the perspective of the individual farmer, this concept makes sense. The problem is, though, that the farmer, in maintaining diversity for his or her own insurance, at the same time performed a task in the interest of the entire society: the conservation and maintenance of crop diversity as a 'global public good'. Consequently, in abandoning crop diversity as an individual insurance, the global interest in high crop diversity is also no longer supported.

The problem of negative incentives

According to Tanner (1996), in the industrialized countries the loss of traditional knowledge was closely linked to the evolution of the formal research and development of, for instance, pharmaceuticals in the industrial processes at the beginning of the 20th century. The preference for scientifically developed products, which were supported by intensive marketing measures, led to the underestimation and abandoning of traditional knowledge and, consequently, to its disappearance.

Similar results might have been brought about by the active promotion and spread of the blueprint approach to development, with its typical expressions of industrial agriculture and the closely related Green Revolution. These endeavours led to the preference of high-yielding varieties suited for industrial farming. They have increased food production but often replaced the older landraces (compare FAO, 1999 (sub Legal Texts and Conference Documents); Girsberger, 1999).

As the system of trade in raw materials has an important influence on agriculture, the development described is enhanced by the expansion of global markets, and the recent pattern of trade liberalization, which favours the highest possible yield at minimal cost. This intensifies the use of resources and has a double impact on biological diversity: on the one hand, agricultural (crop) biodiversity is homogenized by standardizing food production and consumption. On the other hand, conversion and degradation of habitats is promoted by intensifying production or by abandoning use.

Privatization of research and development and expanding intellectual property rights

The achievements of biotechnology have led to processes in the industrialized world which are relevant for the conservation of PGRFA. These processes are characterized by changes in the sponsorship and support frameworks for scientific R&D and in the regime of intellectual property rights (IPRs). This development is of importance for plant breeding, as it threatens the free exchange of PGRFA, local breeding efforts and the supply of the public good 'biological diversity'.

Changes in R&D

Changes in the R&D structure in the last few decades significantly influenced the target of research into PGRFA and the accessibility of biogenetic information. In the past, the development of science and technology was public sector driven, application was then usually developed by the private sector. However, in recent decades, R&D capabilities started to move out of public institutions and shifted into the private sector in both development and application.

As a result, priorities in research, development and distribution are increasingly determined by the interests of transnational private companies, and reflect industrial production strategies. The turn towards the private sector privileges planning and investment directed at short-term returns, and the exploitation of opportunities which neither lead to the production of public goods, nor address the special needs of local ecosystems. Moreover, high investments in R&D call for enhanced protection of the results of the research and the expansion of the commercial seed market (Serageldin, 1997; FAO, 1999 (sub Legal Texts and Conference Documents)). These developments lead to an expansion of the IPRs to protect the (innovative) results of private sector research and development.

Expansion of intellectual property rights

Property rights to protect informational values of human creativity and innovation are a common legal instrument in industrialized countries. In particular, patents[9] were, historically, the legal answer to the Industrial Revolution. Aimed at stimulating creativity and the scientific process, they were devised:

- on the one hand, to provide the owner (in the country where he or she applied for the patent) with an exclusive right to commercialize his or her creation for a limited time. The utilization and commercialization of the patented subject matter is only allowed with the permission (licence) of the patent owner. This allows the author of the invention to amortize his or her investments and creates a financial incentive.
- On the other hand, to oblige the inventor to disclose his or her invention, thus making possible further inventive steps on the basis of the former invention. The subsequent invention can in turn be patented if it corresponds with the protective criteria.

As has been mentioned above, the system of IPRs has grown over time in different ways (Drahos, 1999). First, in step with the increasing globalization of trade, the system of IPRs has undergone a process of regulatory globalization and harmonization. The latest example of this evolution is the conclusion of the TRIPS Agreement in the frame of the Uruguay Round of the General Agreement on Tariffs and Trade (GATT) negotiation. This agreement is the first global[10] legal instrument aiming at the harmonization of the regimes on intellectual property. Second, new rights were created. In our context, a case in point are the plant breeders' rights. Third, however, IPRs, in particular patents, have also expanded in scope. This is expecially the case with the patenting of biological materials:

- Some countries (USA and Australia) allow the patenting of plant varieties (Correa, 1999).
- The patenting of inventions relating to genes is allowed in several countries. These patents refer to a great variety of materials and processes. Thus patents can comprise entire plant cells, plants and plant varieties, including parent lines and hybrids (Correa, 1999). According to de Miranda Santos and Lewontin (1997), the patenting of gene sequences and biotechnological processes is performing the same function as varietal patents.[11]

This means that neither the patented genes nor the plant varieties are available for further breeding without previous consent from the holders of the IPRs. The right of farmers to save their seed, which

[9] Other types of intellectual property rights which could prove to be of relevance in our context are, e.g., trademarks, geographical indications, copy-rights, undisclosed information.
[10] Former international harmonization having been realized on regional levels.
[11] For details in particular regarding the situation in the USA, see de Miranda Santos and Lewontin, 1997; Barton, 1998; Erbisch and Velazquez, 1998; Correa, 1999.

is granted by the plant breeders' rights, is abolished by these developments (de Miranda Santos and Lewontin, 1997).

The tendency of this evolution of IPRs clearly goes in the direction of an expansion of the right of the industrial breeders. The described acceptance of broad claims, the uncertainty about the limits of the patent-holder's rights, and the aggressive enforcement of the rights can (even) lead to the monopolization of varieties and, by limiting research and breeding activities, to a perversion of the original idea of IPRs (de Miranda Santos and Lewontin, 1997; Correa, 1999).

Summary: resulting questions and elements for a legal solution

The analysis amply demonstrates the complex situation of the conservation of PGRFA at the crossroad between agriculture, trade, industrial breeding and environmental issues. This is particularly obvious with respect to the situation of the individual farmer and *in situ* on-farm conservation.

The impossibility of appropriating the value inherent in the diversity of crop species and related traditional knowledge, and the situation of open access lead to the loss of diversity and traditional knowledge. The unclear, unequal and highly politicized situation regarding the IPRs, fosters confusion and mistrust and leads to reluctance of countries and individuals or communities to grant access (Bragdon and Downes, 1998). The present situation is thus entirely opposed to the climate of mutual trust and of reciprocity which is the basis for open access and exchange and for innovative plant breeding.

Aspects of the Institutional Background and Legal Environment

From the above it becomes apparent that the key issues which require regulation in a consistent legal system are the closely linked questions on access to genetic resources and the sharing of the benefits resulting from their use. These questions are to be seen in context with the creation of incentives for the conservation and maintenance of PGRFA and related on-farm breeding systems. In doing so, a balance between the protection of industrial R&D by IPRs and the protection of the results of traditional invention systems has to be found.

In parallel to the technical and scientific evolution, the relevant international instruments are also in flux: the CBD and the IU on the side of the conservation and maintenance of PGR and traditional knowledge, the UPOV-Convention and the TRIPS Agreement on the side of the IPRs. In the following, a brief overview of the regulations and processes of importance in our context will be given.

The UPOV Convention and plant breeders' rights

Plant breeders' rights (PBRs) are IPRs to protect the results of formal plant breeding. In analogy to patents, they aim at creating incentives to develop new plant varieties. They are specifically devised to take account of the traditions between breeders. The national legislation on PBRs was, in 1961, harmonized by the International Convention for the Protection of New Varieties of Plants (UPOV-[12] Convention).

The protective criteria require that a plant variety be new, distinct, homogeneous (uniform) and stable. In particular the criteria of homogeneity and stability exclude the landraces and traditional varieties from protection. In analogy to the patent system, the PBR entitles its holder to exclusively market the variety, or to license it to other users. It does not, however, prevent others from using the variety for further breeding (the breeders' exemption). It originally also allowed farmers to store harvested material to be used as seeds in the next season (the farmers' privilege).

The Convention has been revised four times since it came into existence, with a trend for the strengthening of the rights granted, significantly expanding their scope. In our context the most important changes, as introduced by the last revision of 1991 (which came into force in 1998) are: (i) the extension of the rights of the holders to harvested material.[13] This has implications for the farmers' privilege, which now is provided for only

[12] The treaty is named after the French acronym for the International Union for the Protection of New Varieties of Plants.

[13] Art. 14.1, (1991) in comparison with art. 5.1, (1978).

as an optional exception; (ii) the newly introduced permission to cumulate patent and plant breeders' protection;[14] and (iii) the granting of a breeders' right only for 'essentially derived plant varieties' (for details see Barton, 1998; Bragdon and Downes, 1998; Girsberger, 1999). This expansion mirrors the evolution of the last decade in biotechnologies and in the seeds markets. It clearly leads to a broadening of the rights of the formal breeders.

The TRIPS Agreement

The TRIPS Agreement was adopted in 1993 as a result of the GATT/WTO (World Trade Organization) Uruguay Round of Multilateral Trade Negotiations. It sets international minimal standards to protect intellectual property and requires its members to establish this protection within given periods of time. Its aim is to set minimal standards for the national regulations on the protection of intellectual property by member states, and thus to 'reduce distortions and impediments to international trade and take into account the need to promote effective and adequate protection of intellectual property rights' (Preamble of the TRIPS Agreement).

The discussions with a view to rights to PGRFA and related traditional knowledge focus on Article 27(3) b. This article states that the Contracting Parties may except plants and animals from patentability. However, they are to provide for protection of plant varieties either by patents, or an effective *sui generis* system, or a combination thereof.

Opinions differ as to the interpretation of this article. One question is, whether the notion of *sui generis* is to be understood as referring to the system of protecting plant varieties of the UPOV-Convention. However, in the wording and the history of this article, nothing suggests this interpretation. Accordingly, countries have differing positions as to how it is to be implemented and further developed. Whereas industrial countries support the expansion of Article 27(3)b to include the obligation to protect biotechnological innovations, developing countries oppose patenting life forms. They propose to clarify Article 27(3)b of TRIPS in the sense that developing countries can opt for a national *sui generis* law that protects innovations of indigenous and local farming communities and that harmonizes this article with provisions of the CBD and the FAO's IU (World Trade Organization, 1999a, b).

Farmers' rights

The results of the informal breeding systems – traditionally bred varieties, landraces and their wild and weedy relatives – as a rule cannot be protected by the aforementioned IPRs: (i) they do not correspond to the protective criteria of plant breeders protection; and (ii) the respective systems may not be accessible for reasons of legitimization, difference in taxonomic systems, and lack of know-how and funding (Girsberger, 1999). Therefore, the information as contained in the landraces is freely accessible and can be used for further breeding in the industrial context without restriction.

The asymmetric treatment given to donors of germplasm and donors of technology in the process of plant breeding, in particular in connection with the expansion of the IPRs, has led to debates in the FAO. The discussions focused on the question of access to *ex situ* collections and of creation of rights to traditional PGRFA, the so-called farmers' rights. The farmers' rights were to balance the equity between the protection of the results of formal and informal breeding. These discussions finally led to the establishment of the – legally non-binding – International Undertaking on Plant Genetic Resources by the FAO Conference in 1983. The IU is the first comprehensive instrument on PGRFA. Its objective is 'to ensure that plant genetic resources of economic and/or social interest, particularly for agriculture, will be explored, preserved, evaluated and made available for plant breeding and scientific purposes' (Art. 1).[15]

In 1989, in an interpretative resolution to the IU, the FAO Conference established the recogni-

[14] In abolishing Art. 2.1. of the 1978 version.

[15] The International Undertaking is part of the FAO's Global System for the Conservation and Utilization of PGRFA, which comprises the Commission on Genetic Resources for Food and Agriculture (CGRFA), the Global Plan of Action for the Conservation and Sustainable Utilization of Plant Genetic Resources for Food and Agriculture (GPA), the International Fund for Plant Genetic Resources, the International Network of *Ex-Situ* Collections and the international network of *in situ* conservation areas and crop-related networks.

tion of farmers' rights.[16] The resolution recognizes that farmers, in particular of developing countries, have conserved, improved and made available plant genetic resources for the benefit of all of mankind. It defines farmers' rights as 'rights arising from the past, present and future contributions of farmers in conserving, improving, and making available plant genetic resources, particularly those in the centres of origin/diversity.' According to the resolution these rights have the purpose of ensuring full benefits to farmers, in particular to assist in the protection and conservation of their PGR and to allow farmers, their communities and countries to participate fully in the benefits derived from the improved use of PGR.

The IU has since been developed through agreed interpretations. However, relevant matters, such as conditions of access to PGR and the funding system, and thus also the implementation of farmers' rights, remained largely open. The conclusion of the CBD in 1992 initiated the revision process of the IU in order to adjust it to the accepted principles of the CBD. The aims of the revision were in particular: (i) to harmonize the IU with the CBD; and (ii) to 'consider' the issues of access to PGR and the realization of farmers' rights.[17]

In the debates on the implementation of farmers' rights, two basically different concepts were discussed: (i) the 'political' concept, which understands the notion of farmers' rights merely as a political slogan as opposed to the PBR (Bragdon and Downes, 1998). In this concept, farmers' rights are understood as a non-exclusive right, their major content being the right to compensation, financed by an international funding system (see e.g. Brush, 1994; Girsberger, 1999). (ii) The 'legal' concept, which has been reinforced by the conclusion of the CBD. This concept encompasses proposals to define farmers' rights as an alternative form of IPR, adapted to the specific requirements of conservation and maintenance of PGRFA and related traditional knowledge by farmers, farming families and communities (compare, e.g., Swaminathan, 1997; Shiva, 1999).

During its eighth session in April 1999, the Commission on Genetic Resources for Food and Agriculture, which is the body mandated to revise the IU, negotiated a new text on farmers' rights. According to Art. 15 of the Draft Version of the Revised IU,[18] 'the Parties agree that the responsibility for realizing farmers' rights ... rests with national governments.' The Parties are obliged to take measures to protect and promote farmers' rights 'in accordance with their needs and priorities.' The measures are to include, namely, the protection of traditional knowledge, the right to equitably participate in sharing benefits arising from the utilization of PGRFA and the right to participate at the national level in making decisions on matters related to the conservation and sustainable use of PGRFA (Art. 15, Draft Revised IU).

This solution has the advantage of disconnecting the two crucial and sensitive issues of benefit sharing and farmers' rights. However, it leaves the implementation of farmers' rights to a vast extent to the Contracting Parties and does not answer the question as to the nature and scope of their rights.

The Convention on Biological Diversity

The ongoing international debate on the terms for access to genetic resources and sharing the benefits from their use (and the growing self-confidence of indigenous peoples and farming communities as to the value of their knowledge) created pressure to include obligations in these issues into the CBD.[19] In particular the developing countries pressed for measures to ensure greater benefits for the countries of origin (Bragdon and Downes, 1998). Thus, the CBD decided against the principle of common heritage and confirms the sovereignty of states over their biological resources. The CBD is based on a broad notion of biological resources. The term is defined with reference to the utility of the resources. Accordingly, 'biological resources' include 'genetic resources, organisms, or any other

[16] Resolution 5/89 of the twenty-fifth session of the FAO Conference, Rome, 11–29 November 1989.
[17] CPGR-Ex1/94/3: Commission on Plant Genetic Resources, First Extraordinary Session, Rome, 7–11 November 1994, Revision of the International Undertaking: Mandate, Context, Background and Proposed Process.
[18] Commission on Genetic Resources for Food and Agriculture, Third Inter-Sessional Meeting of the Contact Group, Teheran, 26–31 August 2000. Composite Draft Text of the International Undertaking on Plant Genetic Resources (CGRFA/CG-3/00/2).
[19] Concluded in 1992 at the occasion of the United Nations Conference on Environment and Development, and entered into force in December 1993.

biotic component of ecosystems with actual or potential value for humanity' (Art. 2). Genetic resources, as viewed by the CBD, include genetic information found in both domesticated and wild species of animals and plants. The notion encompasses resources found in *in situ* conditions as well as stored *ex situ,* for example, in genebanks or botanical gardens. Thus the genetic information contained in PGRFA in principle falls under the regulations of the CBD.

This is of relevance in so far as the PGRFA therefore fall under the sovereign control of the state. This means that the contracting parties are – in the limits of international law – free to decide on the regulation of, for example, the ownership of the genetic resources situated in their territory. In particular the authority to determine access to genetic resources rests with the national governments and is subject to national legislation (Art. 15.1 CBD). The Contracting Parties are only obliged 'to endeavour to create conditions to facilitate access to genetic resources ... by other Contracting Parties' (Art. 15.2 CBD). So the original concept of the IU, which considered PGR as a heritage of mankind and, thus, as available without restriction (Art. 1), was not taken up in the CBD negotiations. The concept of the CBD, in contrast, is designed for contractual bilateral relations between donor countries and prospecting firms. It prescribes that access to genetic resources be granted under the condition of prior informed consent (PIC), mutually agreed terms, and fair and equitable sharing of research and development results as well as benefits arising from use of genetic resources (Art. 15 CBD).

The existence, value, and importance of *traditional knowledge* is implicitly recognized by the CBD. The pertinent Art. 8(j) obliges parties to encourage sharing of benefits arising from use of traditional knowledge. However, the obligation retains the reservation of national legislation and thus does not in itself confer rights to protect traditional knowledge.[20] As elaborated above, traditional knowledge from the point of view of international public law remains in the public domain.

The above makes clear that there are divergences between the CBD's regulations on access and benefit sharing and on traditional knowledge, and the specific requirements for PGRFA. However, the task to regulate rights to genetic information and traditional knowledge related to PGRFA, and the question of farmers' rights was left to the FAO by the Conference which adopted the Convention,[21] and there prompted the revision of the IU.[22] In the revision process of the IU, the antinomy between bilateral contracts versus open access was one of the main controversial issues (besides the related issues of benefit sharing, financial resources and farmers' rights). The draft solution, now elaborated by the negotiating commission, proposes the creation of a Multilateral System of Access and Benefit Sharing (Part IV). This Multilateral System is intended to mutually facilitate access to a limited number of PGRFA which are to be selected according to criteria of food security and interdependence[23] (Art. 11 and 12, Draft IU).[24] The details of the Facilitated Access have not yet been finalized by the negotiating body.[25]

From the institutional point of view, access to PGRFA and to the traditional knowledge related to it, thus follows a pattern which differs from the one established by the CBD for PGR and traditional knowledge in general. However, in both cases, the implementation is to a large extent transferred to the Contracting Parties. The question is whether, for the implementation of the legal protection of the different types of information – traditional knowledge related to PGRFA, information included *in* the PGRFA and traditional knowledge related to PGR in general – different instruments must be created at the national and international level, or whether, for example a general *sui generis* right to traditional knowledge can be created which covers all types of knowledge.

[20] In accordance with the principles of international public law, which as a rule recognizes only states and not individuals as subjects of rights and duties.
[21] CBD, Resolution III Paragraph 4.
[22] FAO Conference Resolution 7/93, on follow-up to Resolution 3 of the Nairobi Final Act.
[23] The conditions for the Facilitated Access within the Multilateral System and the content of the List still needing further discussions (state: September 2000).
[24] Composite Draft Text, FN 19.
[25] State in September 2000.

Conclusion

The analysis of the legal and institutional situation shows the imbalance between the protection of formal innovation on the one hand and informational values as contained in traditional knowledge and traditionally generated information on the other. However, both types of information are the result of intellectual accomplishments. It also illustrates the difficulties which arise in trying to find solutions. The protection of the informational values as contained in PGRFA and traditional knowledge is in itself complex. This complexity is amplified by the complicated pattern of stakeholders and interests involved (see Biber-Klemm, 2000a).

Whereas general principles and concepts for a solution seem to emerge as much in the domain of PGRFA as in relation to traditional knowledge related to other PGR, their implementation on the level of the authors of the information – individuals or communities – is far from clear. This is as true for the instruments at national level, as for their implementation and enforcement at international level in the context of international trade relations.

The design of legal instruments has to take the purposes and objectives of the attempted solution into account. Recent economic theories help to shed light on the specific problems that PGRFA and related traditional knowledge are confronted with in a system of increasingly globalized and liberalized markets. They thus can give further indications as to the elements that (legal) solutions ought to encompass.

Solutions: Background Reflections

The problem of open access

As described above, the open-access situation of biodiversity leads indirectly to land conversion, and thus to its loss. Whereas IPRs create incentives to invest in R&D in the steps of formal R&D and industrial production, no comparable incentives exist at the level of earlier production stages, that is, where basic information is created by informal R&D processes as effectuated by traditional breeders and indigenous people. Thus, the benefits from retaining diversity are appropriated at a level within the agricultural industry far removed from the individuals making decisions concerning land conversions (Swanson *et al.*, 1994; Swanson and Göschl, 2000).

Swanson and Göschl (2000) conclude that, in order to maintain basic investments and innovation processes, incentives should also be created at the level where the information is created, that is, at the level of the indigenous and farming communities. Swanson proposes the creation of an 'informational resource right' as a specific property right, tailored in analogy to IPRs, to protect the informational value of biogenetic information and to allow the appropriation of the value of the information at the level at which it has been generated.

The theory of the global public good

As developed above, the diversity of PGRFA is an important element to secure the nutritional needs of humanity and of future generations in a changing world. Agrobiodiversity is maintained by local subsistence farming systems, as a side-effect of their striving to prevent food shortages. The local farmers thus perform a service to humankind. This diversity presently is freely accessible to everybody and the information contained in the PGRFA has no marketable value. These elements correspond to the characteristics of the 'global public good'.[26]

The problem of a public good is that, while the entire society benefits from its use, no mechanisms for its generation such as market mechanisms, exist. Therefore, additional measures are needed for its provision (Swanson *et al.*, 1994; Kaul *et al.*, 1999). The public policy implication is that states, respectively international regimes, must play some role in the provision of such goods, otherwise they will be undersupplied (Stiglitz, 1999).

Stiglitz (1999) identifies two strategies to provide for the public good 'knowledge', that is, for informational values: (i) to increase the degree of appropriability of the returns by issuing IPRs and/or (ii) to grant direct government support. He points out that there are types of information, such as basic research and other fundamental forms of

[26] The notion of 'public good' is defined by two main qualities: (i) its benefits are non-rivalrous in consumption, i.e. their utilization by one person does not exclude the utilization by others; and (ii) its benefits are non-excludable, i.e. it is extremely difficult and costly to exclude others from the utilization of the good (Kaul *et al.*, 1999).

knowledge, which 'almost certainly' should remain in the public domain as parts of the 'global commons' and not be privatized by an intellectual property regime. However, in this case, public support is required. He proposes that the international community could claim the right to charge for the use of the global knowledge commons – as is the case for the utilization of other common goods.

Conclusion

The following conclusions can be drawn from the above:

1. Agrobiodiversity as a global public good needs to be supported by the international community. Support is possible either by creating rights which permit the privatization of the informational value or by subsidizing maintenance of biodiversity. In the case of support for the *in situ* conservation of PGRFA, this means either the creation of (property) rights to the informational value or the generation of a system to reimburse the efforts of the farmers in the maintenance of the public good 'diversity of PGRFA'.[27]
2. The equities between privatization and open access are to be weighed very carefully. A system to charge for access to the information belonging to the 'commons' has to be created.
3. Incentives are most effective if applied at the level where the decisions are taken, that is, at the level of the farmers and/or the farming communities, or the nation states.
4. If incentives are created, it is to be assured that the 'culture of reciprocity' (Castillo, 1997) and the customary laws governing the exchange of PGRFA, and the tradition of the knowledge between farming families and communities are maintained and supported.

Proposed Solutions

Elements and criteria

According to the concerns and the political, legal and factual situation described above, the key issue of a legal solution is to balance rights and compensations with respect to the different systems of innovation. The aim is to ensure the conservation of genetic diversity in agriculture and to support its community-based management. As has been elaborated above, a solution will have to be sought at both the national and the international level.

In particular – in addition to the criteria elaborated above – it ought to address the following objectives:

1. It should create a sound basis to secure access to genetic resources and their free exchange at local, regional and international levels.
2. The contributions of all stakeholders involved are to be adequately compensated. Equity in rights as well as in benefits is to be secured for individuals, communities and nation states.
3. A sound basis for partnership in R&D of PGRFA is to be created.
4. A solution should aim to minimize transaction costs while maximizing efficiency and effectiveness.

Overview

Solutions to create a legal basis for the issues of access to PGR and for the sharing of benefits resulting from their use are discussed for both; PGRFA and related traditional knowledge in the agricultural context in the sense of the IU, and for PGR and traditional knowledge in the more general context of the CBD.

Four main groups of proposed solutions can be distinguished. First, there is a series of proposals for non-compulsory codes of conduct for access and benefit sharing and models for solutions on the contractual level, the material transfer agreements (MTAs). A second group of measures focuses on the problem of allocating and documenting information and controlling lawful implementation by the purchasing states, by creating databases to register traditional knowledge. A third proposition is the creation of a funding system, particularly in the domain of farmers' rights. A fourth group of measures focuses on creation of *sui generis* rights to traditional resources (Posey, 1996; Cottier, 1998).

The solutions which are proposed in many variations in the steadily increasing literature follow in principle the two basic strands of thinking encountered in the theory of the global public good: (i) the concept of the creation of a funding system, which

[27] In conformity with the WTO rules.

is promoted in connection with the concept of the Multilateral System of Access and Benefit Sharing of the IU; and (ii) the creation of specifically tailored property rights to protect informally created innovation. The question is, whether these measures are appropriate for all types of information, or if in fact the present solution, which seems to exclusively favour the concept of the funding system for the implementation of the farmers' rights, is compelling. The thesis submitted is that a uniform solution will not meet the problems, but rather a combination of measures which encompass rights, instruments and institutions for their implementation and other supporting measures.

It is put forward that these measures need not follow the administrative and institutional separation presently existing between the 'agricultural' (IU) and the 'pharmacological' (CBD) sector on the international and national levels. It is put forward that instruments ought rather to be devised and applied according to factual criteria and differences between the types of information. Thus, in the following, the 'Traditional Intellectual Property Rights' (TIP rights) and the funding system will be discussed from this point of view.

Traditional intellectual property rights

It has been pointed out above that an unclear situation regarding the rights of individuals and communities to the informational values they generated leads to secrecy and to the erosion of open access (Correa, 1999; RAFI, 2000; Biber-Klemm, 2000b). The industrial IPRs were designed particularly to halt this tendency to secrecy, which was considered to harm the overall process of innovation and progress. In defining clear rights to protect the information and in creating the opportunity to license it to other users, the necessity to keep an invention secret was banned.

It is submitted that a specific right to protect traditionally generated information could also secure a sound basis for access, benefit sharing and open exchange as well as for contractual relations as within a Multilateral System on Access and Benefit Sharing. Cottier (1998) proposes the creation of so-called TIP rights within the framework of the TRIPS Agreement, designed to answer to the specific characteristics of traditional knowledge. In particular these rights ought not to rely on novelty, but encompass pre-existing traditional knowledge and know-how relating to plant and animal genetic resources.

The determination of the possible content and of the scope of such rights would have to be developed taking the economic and cultural significance, the potential holders of the rights and the different types of information into consideration. For instance, Drahos (1999) argues that devising such regulatory forms seems best left to indigenous people and the positive law of individual states. Such legislation is emerging in different states such as Peru (Biber-Klemm, 2000b), the Philippines and India (Cullet, 1999). However, protection on the national level cannot guarantee protection in international trade relations (see Biber-Klemm, 2000b). Therefore the integration into an international legal system, such as the integration in the revision process of TRIPS, is necessary, as well as the integration of control mechanisms into patent procedures.

In discussing the creation of specific IPRs to protect traditionally generated information, the different types of information are to be taken into consideration. Traditionally generated information exists either in the form of traditional knowledge which comes in addition to a resource, as is, for example, the case with knowledge on a specific quality of a crop variety or of a medicinal plant; or it is integrated into the genetic programme of cultivated varieties, which represents the experience and skill of generations of traditional breeders in adapting the variety to specific local circumstances. Traditional knowledge can be freely accessible within a community, or even be in the public domain, or it can belong to the spiritual heritage and tradition of a people and its use therefore be strictly regulated and restricted.

The scope of the rights ought to be designed accordingly; the creation of several types of rights can be imagined. A TIP right could, for example, encompass a mere right to compensation as could be imagined in the framework of the Multilateral System. However, TIP rights could also be designed in analogy to the existing IPRs, that is, procuring for their owner exclusive rights to decide over the utilization of the information. This second solution could prove to be appropriate for knowledge with a strong cultural or spiritual character.

Cottier (1998) proposes further limitation of the TIP rights to knowledge related to genetic resources and suggests that the entitlement be limited to commercial use and industrial production. Thus, the continued use of such knowledge by other communities and individuals would remain unimpaired. He points out that TIP rights should not be limited to local and indigenous communities embodying traditional lifestyles, as presently discussed in the CBD framework, but be open to rural communities and individuals all over the world.

It is obvious that the allocation of the right to individuals, communities or regions is primordial for its acquisition. To allocate the rights, it is proposed to create registration systems at the national and/or international level (Cottier, 1998). Special emphasis ought to be placed on elaborating measures at the interface between modern IPRs and TIP rights, the former needing adjustment to implement and enforce the latter. One solution could be a certification system, which would permit the control of the lawful acquisition of the information by the patent offices. Other aspects which ought to be internationally discussed and harmonized are, for example, the delimitation between discovery and invention, the requirements of the proof of prior art, and the option of joint industrial–traditional patenting.

The option of TIP rights is controversial. It is argued in particular with regard to landraces and related traditional knowledge that:

1. The economic relevance of traditional PGRFA, determined by the market, will be small, as breeders are primarily interested in the resources that have already been collected and characterized.
2. Additionally, the (small) compensation will be consumed by expenses caused by transaction costs and the protection of the rights.
3. The specific process of informal plant breeding and the specific characteristics of landraces make the distinction between specific innovators and innovations difficult.
4. Enforcing TIP rights seems uncertain, the potential holders lacking financial, legal and scientific means.
5. Traditional knowledge, being an integrated part of cultures and traditions of peoples, does not lend itself to a concept of private property.
6. The innovative system of traditional communities does not correspond to the concept of industrial IPRs.

Thus, it is feared that privatization of knowledge will hinder free exchange between groups and lead to conflicts over ownership, which could in turn jeopardize the further evolution of the landraces (Brush, 1994; Girsberger, 1999).

In favour of TIP rights it is argued that: (i) a clear legal basis for access and benefit-sharing arrangements, and for trade of information could be created; (ii) incentives to conserve and innovate could be allocated without intermediate steps; (iii) TIP rights are not only valuable between the parties of a contract, but grant the holder an absolute legal title which is also valuable vis-à-vis third parties; (iv) TIP rights would not only be of interest in the international market; they could also be instrumental in rewarding invention within local market systems (Cottier, 1998); and (v) they would permit the farming communities and the indigenous peoples to regain autonomy over their intellectual property (Drahos, 1999).

To ensure that TIP rights do not conflict with the objectives listed above, further research must seek to find a balance between aspects of protection and control of information on the one hand and accessibility and further development and innovation on the other. It is argued that this balance can be found in the definition of the rights to traditional knowledge and the determination of their scope. In particular the following items are to be carefully developed and evaluated:

1. The creation of the rights (*ipso iure* or by registration);
2. Their function, scope, duration, and effect;
3. Creation, scope and implementation of related rights to appeal and of dispute settlement mechanisms;
4. The integration of local customary law and the possible effects on traditional exchange of information at the local level and on access for scientific research;
5. The level of regulation.

Funding system

A key element in the creation of property rights with respect to informational values is to allocate the right to the authors of the information: individuals, groups or communities. Yet, in many cases – especially in the domain of so-called farmers' rights – it can be very difficult to determine the authors of the information and, thereby, the holders of the rights. Analogous situations exist in other domains, for example in the case of a publicly known folk remedy such as neem or kava. In these cases, compensation by way of a funding system could be envisaged, as proposed for the realization of farmers' rights (Brush, 1994; Girsberger, 1999) and provided for, for instance, in Art. 20 and 21 of the CBD and in the IU. This compensation should be funded by those having an interest in conservation of resources and those using the resources, that is, national governments, formal plant breeders and pharmaceutical industries. The financial means of the fund should be allocated on the basis of project proposals. Compensation through funds could be attributed to states, thus creating incentives to conserve biological diversity at the state level. An application system, open to projects handed in by stakeholders involved in conservation and sustainable use of the resources could be imagined, thus creating incentives for conservation and sustainable use of biological diversity at the level of the communities or individuals directly involved.

Conclusion

It has been discussed above that, in order to halt the genetic erosion of PGRFA, and to maintain evolutionary processes, incentives must be created to support on-farm conservation and management. It has been argued that, in order to stop the erosion of open accessibility, a clear legal basis for access and benefit sharing regulation is to be created. The analysis of the problem has then shown a complex pattern of influences.

From the background and the discussion presented, it becomes clear that no simple and uniform answer will meet the problem of protecting PGRFA and traditional knowledge. Thus, it is concluded that an 'either–or' solution will not solve the problems, but that action must be taken in several areas and in a multi-layered approach, which encompasses rights, instruments and institutions for their implementation and other supporting measures.

In general terms, the proposed solutions can be summarized as follows: first and principally, to allocate specific rights to traditional knowledge based on individual or community rights wherever this is possible. The aim would be to integrate a basic right to traditional knowledge in an international legally binding instrument such as the TRIPS Agreement, its detailed implementation being left to the competence of the nation states. Second, to create a legal basis and mechanism for a funding system which generates incentives for and permits compensation of the maintenance of biodiversity. The funding system could be applied in cases where the pertinent information cannot be allocated to specific authors, or is considered to have the quality of a public good and, thus, is not to be privatized.

Specific emphasis is to be given to the adjustment of existing IPRs. In particular, the rules for patenting plant varieties are to be specified. Further, measures are to be designed to implement and enforce the rights at the international and national level. Finally, additional supporting measures and incentives, such as labelling and auditing systems as well as tax reductions are to be elaborated to conserve and sustain use of biological resources and related traditional knowledge in the context of (international) markets and international trade.

Further research will have to be conducted in order to evaluate the political and practical feasibility of the proposed solutions and to elaborate the details of the measures, in accordance with the criteria above. It is to be emphasized that it is of great importance that these rights are elaborated and evaluated with utmost care and in cooperation with the people concerned. Of special importance is the creation and recognition of a community right to traditional knowledge at the international level.

However, the effective support of on-farm management and conservation of PGRFA, lies beyond the scope of legal solutions only. The basis of all measures must be the appreciation of the contribution of the local and indigenous communities and farmers to the conservation and development of PGRFA. In order to support informed decisions on the on-farm management and conservation of PGRFA, on access and benefit sharing, sound information and capacity-building with a view to the legal and institutional background are fundamental. The aim and function of legal instruments in this context would be to sustain self-confidence and mutual trust and, thus, to form a sound basis for autonomous decisions, cooperation and the participatory approach to research on and development of rural PGRFA.

References

Barbier, E.-B. (1996) Ecological economics, uncertainty and implications for policy setting priorities for biodiversity conservation. In: *OECD Proceedings, Investing in Biological Diversity, the Cairns Conference, Cairns, Australia, 25–28 March 1996.* OECD, Paris, France, pp. 115–140.

Barton, J.H. (1998) Aquiring protection for improved germplasm and inbred lines. In: Erbisch, F.H. and Maredia, K.M. (eds) *Intellectual Property Rights in Agricultural Biotechnology*, Biotechnology in Agriculture Series, No. 20. CAB International, Wallingford, UK, pp. 19–30.

Biber-Klemm, S. (2000a) Biotechnology and traditional knowledge – legal means to balance the equities. Biotechnology in the Global Economy, Proceedings of the International Conference on Biotechnology in the Global Economy, Harvard, 1999. In: *International Journal of Technology*, Vol. 2, Nos 1/2/3, 85–102.

Biber-Klemm, S. (2000b) *Intellectual Property Rights and Traditional Knowledge. Proceedings of the Workshop on Biodiversity and Biotechnology, Berne, 2000* (unpublished).

Biber-Klemm, S. (2000c) Incentives to bring about conservation and sustainable use of genetic resources in the framework of the world trade order. In: Cottier, T. and Mavroidis, P.C. (eds) *Intellectual Property: Trade, Competition and Sustainable Development, Proceedings of the World Trade Forum, Berne, 1999.* Michigan University Press, USA.

Bragdon, S.H. and Downes, D.R. (1998) Recent policy trends and developments related to the conservation, use and development of genetic resources. In: Engels. J. (ed.) *Issues in Genetic Resources* No. 7. IPGRI, Rome, Italy.

Brush, S.B. (1994) Providing farmers' rights through *in-situ* conservation of crop genetic resources. *Commission on Plant Genetic Resources, First Extraordinary Session, Rome, 1994. Background Study Paper* No. 3. FAO, Rome, Italy.

Castillo, G.T. (1997) Whose ethics and which equity?: Issues in the conservation and use of genetic resources for sustainable food security. In: International Plant Genetic Resources Institute (ed.) *Ethics and Equity in Conservation and Use of Genetic Resources for Sustainable Food Security. Proceedings of a Workshop to Develop Guidelines for the CGIAR, Foz do Iguaçu, Brazil, 1997*, pp. 19–31.

Cooper, D., Engels, J. and Frison, E. (1994) A multilateral system for plant genetic resources: imperatives, achievements and challenges. In: IPGRI (ed.) *Issues in Genetic Resources* No. 2. IPGRI, Rome, Italy.

Correa, C.M. (1999) Access to plant genetic resources and intellectual property rights. *Commission on Genetic Resources for Food and Agriculture, Background Paper* No. 8. FAO, Rome, Italy.

Costanza, R. *et al.* (1997) The value of the world's ecosystem services and natural capital. *Nature*, 387, 253–260.

Cottier, T. (1998) The protection of genetic resources and traditional knowledge: towards more specific rights and obligations. *Journal of International Economic Law* 1(4), 555–584.

Cullet, P. (1999) Revision of the TRIPS Agreement concerning the protection of plant varieties – lessons from India concerning the development of a *sui generis* system. *Journal of World Intellectual Property*, 2(4), 617–656.

De Miranda Santos, M. and Lewontin, R.C. (1997) Genetics, plant breeding and patents: conceptual contradictions and practical problems in protecting biological innovations. *Plant Genetic Resources Newsletter* 112, 1–8.

Drahos, P. (1999) Biotechnology, patents, markets and morality. *European Intellectual Property Review* 21(9), 441–449.

Dutfield, G. (1999) The Public and Private Domains: Intellectual Property Rights in Traditional Knowledge. In: Traditional Ecological Knowledge, WO 03/99, *Oxford Electronic Journal of Intellectual Property Rights* (www.oiprc.ox.ac.uk/EIWPO0399.html).

Erbisch, F.H. and Velazquez, C. (1998) Introduction to intellectual properties. In: Erbisch, F.H. and Maredia, K.M. (eds) *Intellectual Property Rights in Agricultural Biotechnology*, Biotechnology in Agriculture Series, No. 20. CAB International, Wallingford, UK, pp. 3–18.

Girsberger, M.A. (1999) *Biodiversity and the Concept of Farmers' Rights in International Law, Factual Background and Legal Analysis.* Peter Lang, Bern.

International Plant Genetic Resources Institute (IPGRI) (1996) Access to plant genetic resources and the equitable sharing of benefits: a contribution to the debate on systems for the exchange of germplasm. *Issues in Genetic Resources* No.4. IPGRI, Rome, Italy.

Kaul, I., Grunberg, I. and Stern, M.A. (1999) Defining global public goods. In: Kaul, I., Grunberg, I. and Stern, M.A. (eds) *Global Public Goods.* Oxford University Press, New York, pp. 2–19.

Posey, D.A. (1996) *Traditional Resource Rights: International Instruments for Protection and Compensation for Indigenous Peoples and Local Communities.* IUCN Biodiversity Programme, IUCN, Gland.

Rural Advancement Foundation International (RAFI) (1998) Plant breeders wrongs. (www.rafi.org.).

Rural Advancement Foundation International (RAFI) (2000) Biopiracy – RAFI's sixth annual update. *RAFI Communique*, Issue 65. RAFI, Winnipeg, Manitaba, Canada.

Serageldin, I. (1997) Equity and ethics: twin challenges, twin opportunities. In: International Plant Genetic Resources Institute (ed.) *Ethics and Equity in Conservation and Use of Genetic Resources for Sustainable Food Security. Proceedings of a Workshop to Develop Guidelines for the CGIAR, Foz do Iguaçu, Brazil, 1997*, pp. 1–6.

Shiva, V. (1999) Neem. A case study of intellectual property rights and traditional knowledge. *Council of Europe International Conference on Ethical Issues Arising from the Application of Biotechnology, Oviedo, Spain*. Council of Europe, Doc. SPK/07/PROV/SUM/E.

Stiglitz, J.E. (1999) Knowledge as a global public good. In: Kaul, I., Grunberg, I. and Stern, M.A. (eds) *Global Public Goods*. Oxford University Press, New York, pp. 308–325.

Swaminathan, M.S. (1997). Ethics and equity in the collection and use of plant genetic resources: Some issues and approaches. In: International Plant Genetic Resources Institute (ed.) *Ethics and Equity in Conservation and Use of Genetic Resources for Sustainable Food Security. Proceedings of a Workshop to Develop Guidelines for the CGIAR, Foz do Iguaçu, Brazil, 1997*, pp. 7–18.

Swanson, T.M. (1995) The appropriation of evolution's values: an institutional analysis of intellectual property regimes and biodiversity conservation. In: Swanson, T. (ed.) *Intellectual Property Rights and Biodiversity Conservation: an Interdisciplinary Analysis of the Values of Medicinal Plants*. Cambridge University Press, Cambridge, UK, pp. 141–175.

Swanson, T. and Göschl, T. (2000) Property rights issues involving plant genetic resources: implications of ownership for economic efficiency. *Ecological Economics* 32, 75–92.

Swanson, T.M., Pearce, D.W. and Cervigni, R. (1994) The appropriation of the benefits of plant genetic resources for agriculture: An economic analysis of the alternative mechanisms for biodiversity conservation. *Commission on Plant Genetic Resources, Background Paper* No. 1. FAO, Rome, Italy.

Tanner, J. (1996) Property rights, Innovationsdynamik und Marktmacht. Zur Bedeutung des schweizerischen Patent- und Markenschutzes für die Entwicklung der chemisch-pharmazeutischen Industrie (1907–1928). In: Ernst, A. and Wigger, E. (eds) *Die Neue Schweiz? Eine Gesellschaft zwischen Integration und Polarisierung*. Chronos, Zürich, pp. 273–303.

Legal Texts and Conference Documents

Agreement on Trade-Related Aspects of Intellectual Property Rights, adopted 15 December 1993 (TRIPS-Agreement), reprinted in 33 I.L.M. 1197 (1994).

Commission on Genetic Resources for Food and Agriculture: Third Inter-sessional Meeting of the Contact Group, Teheran, 26–31 August 2000. Composite draft text of the International Undertaking on Plant Genetic Resources, incorporating: the texts of articles 11, 12 and 15 negotiated during the Commission's Eighth Regular Session; and the texts of Articles 13, 14 and 16 negotiated during the First and Second Inter-Sessional Meetings of the Contact Group. CGRFA/CG-3/00/2.

Commission on Plant Genetic Resources: First Extraordinary Session, Rome, 7–11 November 1994: Revision of the International Undertaking: Mandate, Context, Background and Proposed Process. CPGR-Ex1/94/3.

Convention on Biological Diversity, adopted 5 June 1992 (Biodiversity Convention, CBD), reprinted in 31 I.L.M. 818 (1992).

Convention on Biological Diversity, Appendix: Resolutions adopted by the Conference for the Adoption of the Agreed Text of the Convention on Biological Diversity; Resolution 3 on the Interrelationship between the Convention on Biological Diversity and the Promotion of Sustainable Agriculture.

Convention on Biological Diversity, Subsidiary Body on Scientific, Technical and Technological Advice, Fifth meeting, Montreal, 31 January – 4 February 2000, Agricultural Biodiversity: Assessment of Ongoing Activities and Instruments. UNEP/CBD/SBSTTA/5/INF/10, 19 November 1999.

Convention on Biological Diversity, Subsidiary Body on Scientific, Technical and Technological Advice, Fifth meeting, Montreal, 31 January – 4 February 2000, Agricultural Biodiversity: Assessment of Ongoing Activities and Priorities for a Programme of Work, UNEP/CBD/SBSTTA/5/10, 23 October 1999.

FAO Global Plan of Action for the Conservation and Sustainable Utilization of Plant Genetic Resources for Food and Agriculture, adopted by the International Technical Conference on Plant Genetic Resources, Leipzig, Germany, 17–23 June 1996 (Global Action Plan (GAP)), www.web.icppgr.fao.org/gpaeng.htm

FAO Agricultural Biodiversity. Background Paper for the FAO/Netherlands Conference on the Multifunctional Character of Agriculture and Land, Maastricht 1999.

International Convention for the Protection of New Varieties of Plants of 2 December 1961, as revised at Geneva on 10 November 1972 (UPOV Convention 1961), reprinted in UPOV Publication No 295 and in The International Union for the Protection of New Varieties of Plants, www.upov.org/eng/convntns/1961/content.htm

International Convention for the Protection of New Varieties of Plants of 2 December 1961, as revised at Geneva on 10 November 1972, on 23 October 1978, and on 19 March 1991 (UPOV Convention 1991), reprinted in UPOV Publication No 221(E) and in The International Union for the Protection of New Varieties of Plants, www.upov.org/eng/convntns/1991/content.htm

International Undertaking on Plant Genetic Resources, adopted at the 22nd Session of the FAO Conference as FAO Conference Resolution 8/83 on 23 November 1983 (International Undertaking, IU), reprinted in International Undertaking on Plant Genetic Resources 1, FAO Doc. CPGR-Ex1/94/Inf.1 (September 1994).

International Undertaking on Plant Genetic Resources, Annex I, entitled 'Agreed Interpretation of the International Undertaking' adopted as FAO Conference Resolution 4/89 at the 25th Session of the FAO Conference on 29 November 1989 (Annex I), reprinted in International Undertaking on Plant Genetic Resources 7, FAO Doc. CPGR-Ex1/94/Inf.1 (September 1994).

International Undertaking on Plant Genetic Resources, Annex II, entitled 'Farmers' Rights', adopted as FAO Conference Resolution 5/89 at the 25th Session of the FAO Conference on 29 November 1989 (Annex II), reprinted in International Undertaking on Plant Genetic Resources 9, FAO Doc. CPGR-Ex1/94/Inf. 1 (September 1994).

Leipzig Declaration on Conservation and Sustainable Utilization of Plant Genetic Resources for Food and Agriculture. Adopted by the International Technical Conference on Plant Genetic Resources, Leipzig, Germany, 17–23 June 1996.

World Trade Organization Preparations for the 1999 Ministerial Conference: The TRIPS Agreement. Communication from Kenya on behalf of the African Group, WT/GC/W/302, 6 August 1999.

World Trade Organization Preparations for the 1999 Ministerial Conference: Proposal on Protection of the Intellectual Property Rights to the Traditional Knowledge of Local and Indigenous Communities. Communication from Bolivia, Colombia, Ecuador, Nicaragua, and Peru, WT/GC/W/362, 12 October 1999.

41 Evaluating the Benefits of Conserved Crop Germplasm in PNG

M. Milne,[1] D. Godden,[1] J. Kennedy[2] and R. Kambuou[3]
[1]*Department of Agricultural Economics, University of Sydney, New South Wales, Australia;* [2]*School of Business, La Trobe University, Bundoora, Victoria, Australia;* [3]*National Agricultural Research Institute, Port Moresby, Papua New Guinea*

Introduction

The greater commercialization of public breeding, and the higher profile of private breeders, supported by plant breeders' rights, has increasingly highlighted the value of genetic material used in plant breeding (Godden, 1984, 1991). There is increasing discussion as to the value of conserving stocks of genetic material in plant germplasm collections, and sources of future funding. Biologists have often encouraged the conservation of all or most material, arguing that it is potentially valuable (Evenson *et al.*, 1998). However, funding constraints are making this conservation effort extremely difficult, especially in countries like Papua New Guinea (PNG).

An economic estimation of the benefits and costs of conserving accessions in collections is required to determine optimal levels and sources of funding and the optimal economic organization of germplasm collections. The costs of maintaining the collections can usually be estimated from existing and historical data. However, empirical estimations of the benefits have been fraught with difficulties, despite recent methodological developments. While numerous studies have successfully documented the value of particular germplasm *ex post* and even estimated its value *ex post*, few have been able to determine this value *ex ante* (Godden and Kambuou, 1996). The focus of the present study is to assess the value of PNG's germplasm collections and to estimate the *potential* benefits of conserving this material.

In valuing PNG's crop germplasm collections, a stochastic dynamic programming model has been developed to determine the net benefit of maintaining existing collections held by the National Agricultural Research Institute (NARI) (formerly by the Department of Agriculture and Livestock (DAL)) in PNG (Kennedy *et al.*, 1997; Godden *et al.*, 1997). To analyse the relative costs of maintaining PNG's existing field germplasm collections, a spreadsheet model has been created (Godden *et al.*, 1998) and a solution derived for the problem of taro germplasm maintenance and breeding in PNG, given the best available current data (Kennedy *et al.*, 1997). Data on the transition probabilities for the two state variables, number of accessions and crop yields were not available for estimating the benefits. The following study was carried out in an attempt to provide the data for empirical analysis of the collections.

The following discussion on the valuation of benefits of germplasm collections in PNG begins with a summary of the different types of plant genetic resource values: *ex situ* and *in situ*. In the third section, methods of estimating the breeding value of genetic resources and collections are briefly reviewed. In section four the methods for the PNG case study are outlined. A description of the movement of PNG germplasm to national and international collections and assessment of the benefits of PNG material in these collections is given in the fifth section. In the final section, survey results are presented on the elicitation of subjective probabilities relating to: (i) the probability of maintaining existing collections (survey of curators); and (ii) potential yield gains from the maintenance of these collections (survey of breeders).

Values of Plant Genetic Resources

Both agriculturalists and environmentalists seek to preserve plant genetic resources. Agriculturalists usually emphasize the potential value plant genetic resources may have in a breeding programme and hence on farm production, while environmentalists tend to support the preservation of genetic biodiversity of all species (Evenson *et al.*, 1998).

Various taxonomies have been used to explain the value of genetic resources. Oldfield (1989) adopted a broad classification, distinguishing between direct values (consumptive use and productive use) and indirect values (non-consumptive use, option and existence). Swanson *et al.* (1994) divided values into non-use and use values of plant genetic resources and defined the following *use* values:

- *exploration value* for yield enhancement: the direct contribution to the value of products produced with plant genetic resources in the plant breeding industry for the purpose of enhancing yields of existing agricultural crops;
- *exploration value* for reduced variability: maintenance of a pool that may be explored for the ascertainment of traits that might contribute to the future stability of agricultural production;
- *portfolio or insurance value*: the value of avoiding output variability and providing for future sustainability through more stable systems of the diversity of plant genetic resources in current agricultural usage (across countries and across time); and
- *quasi-option value*: the value of maintaining potentially valuable traits in the event of environmental shifts. For example, if new bio-types of pests and pathogens are introduced, genetic resources not currently valued may increase in value (Evenson, 1996a).

Evenson (unpublished, 1996a) and Evenson *et al.* (1998) tended to define biodiversity values in terms of value to different interest groups. They considered 'consumer good' (existence) values and 'producer good' (use) values (Evenson, unpublished). Existence values were defined as those held by people who believe in the conservation of all genetic resources, for the sake of biodiversity and long-term sustainability (Evenson *et al.*, 1998). Use values were considered more important to utilitarians who focused on the benefits of genetic resources to humans now and in the future.

In reviewing the various classifications of genetic resource values, it is evident that there is no accepted terminology. 'Direct' and 'indirect' values are used to describe both major categories (Oldfield, 1989) and sub-categories (Evenson *et al.*, 1998) of genetic resource values (see Table 41.1). For the purposes of estimation, many studies have focused on the use values of plant germplasm resources and collections and in particular on the potential yield increases resulting from breeding. In Table 41.1, this value is given numerous names: productive use, breeding, crop improvement, yield exploration value. The following section reviews the different approaches that have been used in determining the genetic resource contributions to yield increases.

Valuation Methods of Genetic Resource Contributions to Yield Increases

Approaches to valuing plant genetic resources have been classified in numerous ways (Pearce and Cervigni, 1994; Evenson, 1996a; Evenson *et al.*, 1998). Pearce and Cervigni (1994) provide a summary on the usefulness of these approaches. To date there is no broadly accepted classification of genetic resource valuation methods or terminology. Similar terms are given for different techniques, and different expressions are adopted for the same method (Pearce and Cervigni, 1994; Evenson, 1996a).

Evenson (1996a) considered three methods for measuring the value of plant genetic resources: subjective, subjective probabilities and objective. In contrast, Pearce and Cervigni (1994) based their categorization on the attempt to capture non-use values and use values, mentioned in the previous section (see Table 41.2).

Indirect/objective methods

Objective methods have been defined as those used in analysing the relationship between production

Table 41.1. Sources of genetic resource values.

Oldfield (1989)	Indirect values			Direct values			
	Existence values	Option values	Non-consumptive use value (environmental services)	Consumptive use (recreational)		Productive use (commercially harvested)	
Evenson (unpublished)	Consumer good (existence value)			Producer good (use value)			
Evenson (1996a)	Conservationists, biologists			Agriculturalists		Bio-prospectors	
	Psychic green	Option value		Bio-support for agriculture Crop improvement Animal improvement		Bio-prospecting for pharmaceuticals and chemicals	
Evenson *et al.* (1998)	Non-use existence values			Use values			
				Direct use		Indirect use	
				Breeding	Recreation	Option	Diversity
Swanson *et al.* (1994)	Non-use values			Use values			
	Intrinsic			Yield exploration Stability exploration Portfolio Quasi-option			
Brown (1990)	Future non-consumptive use value			Direct productive value		Indirect productive value	

Table 41.2. Classifications of valuation methods for genetic resources.

Evenson (1996a)	Objective				Subjective		Subjective probability	
	Indirect		Direct		Contingent valuation			
	Hedonic specification	Returns to research	Breeding production function	Rent calculations				
	Qualitative and quantitative trait valuation	Quantitative trait valuation	e.g. Gollin and Evenson (1996)					
Evenson *et al.* (1998)	Hedonic approaches				Contingent valuation			
	Hedonic pricing	Mapping genetic flows e.g. Gollin and Evenson (1997)			Willingness to pay	Willingness to pay for on-farm diversity		
Pearce and Cervigni (1994)	Indirect				Direct			
	Hedonics	Household production functions			Contingent valuation	Experimental		
Norris and Kramer (1990)							Direct	Indirect

characteristics, genetic resource collections and breeding activities (Evenson, 1996a). Hedonic specification, returns to research and production functions fall under indirect objective methods. Indirect objective methods are considered indirect because they do not create a direct link between the size and evaluation state of plant germplasm collections (Evenson, 1996a). Preferences for the environmental good are revealed indirectly through the purchase of related marketed goods. Techniques for measuring the producer good value of genetic resources, usually require statistical regression to relate the measure of varietal improvement to factors expected to cause or produce varietal improvement (Evenson *et al.*, 1998).

Hedonic methods have been used to infer the value of individual traits or characteristics of a variety (Evenson, 1996b; Godden and Kambuou, 1996; Smale *et al.*, 1997; Evenson *et al.*, 1998; Gollin and Evenson, unpublished; Gollin *et al.*, 1998; Rao and Evenson, 1998). Traits contribute to the value of the variety through increasing yields or enabling higher yielding varieties to be planted in previously unsuitable environments (Evenson, 1996b).

The total contribution of genebanks and plant breeding programmes (in terms of public expenditures) to agricultural output has been well covered in the literature (e.g. Smale *et al.*, 1997; Azzam *et al.*, 1997). These studies do not attempt to isolate the factors of production involved in plant breeding and do not usually account for the number of breeding processes carried out before the new variety is released (Evenson, 1996a). Based on the statistical association between the area planted to new varieties and productivity, the returns to research literature has tended to value quantitative traits rather than qualitative traits. To capture the specific contribution made by genetic resources to plant breeding and yield improvement, a number of studies have disaggregated the different components of advancements responsible for yield gains (Godden, 1988; Godden and Brennan, 1994; Gollin and Evenson, 1998). The economic benefit of the new varieties is expressed as the product of the portion attributable to the introduction of new varieties and an appropriate shadow price (Pearce and Cervigni, 1994). The value of the stages of the breeding process, and the value of the inputs involved in that process, are not disaggregated.

The complete genealogy of the released variety is usually very complex (Pearce and Cervigni, 1994; Evenson, 1996a). In order to estimate the value of the genetic resource, all the genetic steps which led to developing the released variety would need to be measured in terms of yield value added and cost of the particular cross. That is, costs and benefits at each stage of the breeding process – collection, evaluation, pre-breeding and breeding – would need to be measured. Given the detailed information on actual breeding activities required, this type of analysis would be very data intensive and require an extensive database, such as the wheat database at the International Maize and Wheat Improvement Centre (CIMMYT).

Subjective/direct methods

Subjective/direct methods are used to elicit preferences directly through survey and experimental techniques, such as contingent valuation methods. The complexities of the breeding processes have limited the effective application of subjective methods for valuing direct or use values for genetic resources (Evenson, 1996a). Evenson *et al.* (1998) suggested that contingent valuation is more suited to estimating 'consumer goods' values than 'producer goods values. The average consumer would have little knowledge about the use of germplasm collections, and hence would find it difficult to assign a value.

Subjective probability methods

The management of plant germplasm collections is an economic decision problem. For decision making the only valid probability approach is the subjective one where decision-makers bear responsibility for their decisions and use their own strengths of conviction (Dillon, 1971). Subjective probabilities are beliefs held by individuals that reflect their degree of uncertainty about some idea or event. Compared to objective frequencies, subjective probabilities allow for incorporation of intuitive knowledge and recognition that the future may not be like the past (Dillon, 1971). They are not restricted to situations where a series of observations are available and are appropriate to depict events that only occur once or have never occurred (Norris and Kramer, 1990). Plant breeding is an uncertain process, and hence the value of germplasm in terms of future plant varieties that could be developed from it is similarly uncertain (Godden *et al.*, 1997).

Empirical studies of valuing PGR

Most empirical studies on the valuation of genetic resources and collections have used objective methods. The use of subjective probability methods for valuing genetic resources is limited (Evenson, 1996b; Evenson *et al.*, 1998; Kennedy *et al.*, 1997). Empirical work has focused on commercial crops, such as rice, wheat and maize, where detailed datasets are available (Evenson, unpublished, 1996b; Gollin and Evenson, unpublished; Evenson and Gollin, 1997; Gollin *et al.*, 1998). Gollin *et al.* (1998) developed a search model for traits of economic importance in *ex situ* collections of wheat genetic resources. They concluded that the underutilization of wheat germplasm collections was related to the costs and time lags in the breeding process and did not imply that accessions had no value.

Attempts have been made to value rare and pharmaceutical plants in the wild. Phillips and Meilleur (1998) undertook a statistical survey on the usefulness and economic potential of rare plants in the USA. They identified commercial relatives of endangered species and considered the likelihood of wild germplasm being incorporated into crops. Mendelsohn and Balick (1995) estimated private and social values for undiscovered pharmaceuticals in tropical forests. A complete collection and screening of all tropical plant species was valued at US$3–4 billion to a private pharmaceutical company and around US$147 billion to society.

Limited research has been performed on the value of subsistence crop germplasm to farmers' yields. Literature on yield gains of subsistence crops has tended to focus on research station trial outcomes in terms of changes in levels of inputs. Farm yield and production figures for subsistence crops are scarce, especially in developing countries.

Methods for PNG study

To value the net benefits of the maintenance of the PNG germplasm collections, a stochastic dynamic programming model was developed (Kennedy *et al.*, 1997). Decisions on germplasm maintenance and breeding expenditure are required at regular intervals (multistage decision making) depending on the current number of accessions and yield (state variables). It is expected that increasing maintenance expenditure will increase the likelihood of maintaining current or greater numbers of accessions. Greater breeding expenditure is expected to 'increase the probability of countering a yield reversion or obtaining a higher yield variety' (Kennedy *et al.*, 1997).

The number of accessions and future crop yields depends not only on these decision and state variables, but also on stochastic outcomes, such as the loss of germplasm due to volcanic eruptions or crop disease and pests. Since objective probabilities do not exist for these events, best 'guesstimates' of experts and sensitivity analysis are used. The model consists of transition probabilities for both state variables:

- probability of attaining, a number of years on, a certain number of accessions, given the current accession number and maintenance expenditure; and
- probability of attaining, a number of years on, a particular yield given current accession numbers, current yield and breeding expenditure.

The probabilistic state transition functions are the most difficult data to estimate but necessary in determining the optimal policy (Kennedy *et al.*, 1997). The current study attempts to elicit subjective probabilities for yields and collection size in order to carry out a complete empirical analysis for the value of germplasm collections in PNG.

As a preliminary step to determine both domestic and international benefits of PNG's germplasm collections, the movement of germplasm originating from PNG was tracked to other collections. Information was obtained through both direct correspondence and a questionnaire sent to likely international holders of PNG material. Where possible, estimates were obtained of existing yield advances achieved by the use of PNG material in breeding programmes and the potential importance of PNG material in current breeding. Given that only three respondents were successful in including PNG germplasm material in breeding programmes, a second survey was carried out to gain quantitative data on the potential benefits of germplasm collections of the study crops. The survey was designed to elicit subjective probabilities relating to expenditure, potential yield and size of collections. It was assumed that 'yield' was the only desirable trait in crops. Given the difficulty in making such estimates about future varieties (i.e. those not yet in the farmers' field), it was considered unreasonable to survey farmers.

An indirect approach was adopted to elicit probability density functions (PDFs) for predetermined intervals. Indirect methods refer to probabilities inferred from preferences or choices between possible bets, decisions or alternatives (Norris and Kramer, 1990). Indirect methods are useful when the resulting probabilities are not clear to the assessor (Winkler, 1967 in Norris and Kramer, 1990). In the case of PNG respondents, both the visual counter (assigning coloured chips to the likelihood of a given outcome) and the gamble method (betting odds until the assessor is indifferent between two offered bets) were attempted, with most respondents preferring the visual counter technique.

The elicitation of subjective probabilities from PNG curators and breeders was carried out face-to-face while respondents outside PNG were sent a written questionnaire. The questionnaire contained brief explanatory notes on the concepts of subjective probabilities. Respondents were not required to have PNG material.

Documentation of the movement of PNG germplasm

In tracking the movement of PNG germplasm to international collections, a sample of breeders and curators of taro, banana and sweet potato was established. Geographical coverage was wide and the number of PNG accessions in individual collections ranged from 1 to 274. The material had either been obtained directly from the PNG collections, collecting missions, international genebanks or from other researchers.

Table 41.3. Source of PNG taro germplasm in the national germplasm collection, Bubia, 1995.

Province	Number of accessions
Morobe	127
East New Britain	125
Central	91
East Sepik	55
West New Britain	33
Milne Bay	30
NSP	29
East Highlands province	23
West Highlands province	22
Western Province	22
Oro	7
ESP	6
Madang	3
Manus	1
Total	574

*Taro (*Colocasia esculenta*)*

In PNG, the national taro germplasm (301 accessions in February 1999) is held at the NARI Wet-Lowlands Mainland Programme, Bubia.[1] Duplicates are also held at a number of other stations throughout the country. Table 41.3 lists the origin of taro germplasm from around PNG held in the national taro collection at Bubia. A majority of the accessions were collected from Morobe, East New Britain and Central provinces (Kalabus, 1995). The working collection of Bubia also contains 28 introduced varieties from the Asia Pacific region.[2]

Table 41.4 contains the type of institution that has received PNG taro germplasm and the number of accessions obtained. So far, first and second round transfers have been documented, i.e. material that has moved from the place of origin to a germplasm collection (first round recipients) and transferred onwards (second round recipients). The outbreak of taro leaf blight in the Pacific in the last 5 years has restricted the movement of PNG taro germplasm.

*Banana (*Musa *spp.)*

The PNG national *Musa* germplasm collection is currently held at the NARI Dry-Lowlands Programme, Laloki[3] (298 accessions in 1999), and duplicated at the NARI Wet-Lowlands Islands Programme, Keravat. *Musa* germplasm material originating in PNG has been officially collected by international research institutes and breeding programmes since the 1960s, either directly from research stations or from independent fieldwork. Table 41.5 contains the recipients of banana germplasm from PNG, divided into virus indexing

[1] Formerly the Bubia Agricultural Research Centre.
[2] The taro working collection contains taro germplasm from Niue, Samoa, Fiji, Vanuatu, Hawaii, Cook Islands, New Caledonia, Federated States of Micronesia, Solomon Islands, Thailand and Indonesia. The accessions were sent to the working collection by LIPI in Indonesia, ERETA in Samoa and collected in the Solomon Islands by a former taro breeder at Bubia Station.
[3] Formerly the Laloki Agricultural Research Station.

Table 41.4. Distribution of PNG's taro germplasm to international recipients.

Type of institution holding germplasm material	Number of institutions in category	Number of accessions collected/received
Virus indexing centres	1	15
Breeding programmes (first round recipients)	1	N/A
Breeding programmes (second round recipients)	1	1
Research institutes (first round recipients)	5	23
Research institutes (second round recipients)	3	9
Botanical gardens	2	6

Table 41.5. Distribution of PNG's *Musa* germplasm.

Type of institution holding germplasm material	Number of institutions in category	Number of accessions collected/received
Virus indexing centres	3	278
Breeding programmes (first round recipients)	2	N/A
Breeding programmes (second round recipients)	21	259
Breeding programmes (third round recipients)	1	32
Research institutes (first round recipients)	3	107
Research institutes (second round recipients)	12	29
In vitro genebanks	2	274

centres (VICs), research institutes, breeding programmes and *in vitro* genebanks. Where information is available, the number of accessions collected or received is given.

*Sweet potato (*Ipomoea batatas*)*

In PNG, sweet potato germplasm is currently held at the NARI Highlands Programme, Aiyura[4] (1158 accessions in 1998), and the NARI Wet-Lowlands Islands, Keravat[5] (1062 accessions in 1998). Under the Secretariat for the Pacific Commission (SPC) and Pacific Regional Agricultural Programme (PRAP), PNG sweet potato varieties are being evaluated, indexed and tissue cultured at NARI Wet-Lowlands Islands, Keravat. The collection at Keravat also contains 73 accessions from overseas.[6]

Table 41.6 provides information on the type and number of recipients of sweet potato germplasm from PNG. Where information is available, the number of accessions collected or received is also given. Second round recipients have not yet been contacted directly. It is possible that they hold PNG material from other sources, hence the total number of PNG accessions in their collections may be greater. In Japan, PNG germplasm has been characterized and used as breeding materials in breeding programmes. However, to date, no PNG material has been used in breeding varieties released to farmers for cultivation.

Survey one

Valuation of PNG material in international collections commenced with questionnaires sent to 140 known[7] and likely holders of PNG taro, sweet potato and banana germplasm. From the 25 respondents, 13 had taro germplasm, 11 had banana germplasm, and one had sweet potato germplasm. Of the 25 respondents, 14 believed that

[4] Formerly the Highlands Agricultural Experiment Station, Aiyura.
[5] Formerly the Lowlands Agricultural Experiment Station, Keravat.
[6] The LAES collection contains sweet potato germplasm from Solomon Islands, Vanuatu, Samoa, Tonga, Australia, Indonesia, Philippines, Vietnam, Korea, Sri Lanka, Taiwan, India, Bangladesh, Burma, Nigeria, USA and Puerto Rico.
[7] International genebanks provided data on the request and transfer of PNG material from their international collections.

Table 41.6. Distribution of PNG's sweet potato germplasm.

Type of institution holding germplasm material	Number of institutions in category	Number of accessions collected/received
First round recipients	16	1001
Second round recipients	3	935
Third round recipients	27	40
Fourth round recipients	21	49
Pathogen testing facilities	1	94

they had no PNG germplasm material. Four recipients were unaware that material originating in PNG was in their collection.

The constraints to obtaining and using PNG germplasm were varied. Often, lack of knowledge about the material and access issues were the major constraints (Table 41.7). Most of the recipients of PNG germplasm material were not using it in breeding programmes. Only three respondents had breeding lines containing PNG material, all of whom were banana breeders (Table 41.8).

Survey two

Respondents to survey two included both international and PNG national curators and breeders of either banana, sweet potato or taro germplasm. From 70 questionnaires, 27 completed surveys were received. Table 41.9 provides a breakdown of respondents by location and type of collection. To reduce bias, estimates of subjective probability distributions, rather than point estimates, were collected from respondents (Evenson *et al.*, 1996). In similar studies, considerable time is spent in explaining and training respondents in the concept of subjective probabilities. Hence, face-to-face surveys proved more successful in obtaining complete probability distributions than the written survey. The use of the visual counter method with PNG respondents proved to be a successful technique in eliciting subjective probability data. Unfortunately, a number of written respondents identified only the most likely outcome for maintaining the collection rather than providing a probability distribution. To obtain complete probability distributions for input into the

Table 41.7. Constraints in obtaining and using PNG material.

Constraints	*Musa* respondents	Sweet potato respondents	Taro respondents
Knowledge about material	2	1	3
Access to material	3	1	1
Phytosanitary issues	1		2
Quality of material	2		
Low banana production	1		
Lack of funding	1		
No interest in material			1
Expense in obtaining material			1

Table 41.8. Breeding lines containing PNG *Musa* germplasm.

Type of institution holding PNG material	Number of PNG varieties held	% Yield increase from new varieties containing PNG material	Number of varieties distributed to farmers
International	1	10–15	0
National 1	80	> 15	6
National 2	94	5–10	0

Table 41.9. Survey respondents.

Collections	International		PNG	
	Curator	Breeder	Curator	Breeder
Taro (field)	4	1	4	1
Taro (tissue culture)	2			
Banana (field)	2	1	1	
Sweet potato (field)	2	3	2	1
Sweet potato (tissue culture)			1	
Aibika (field)			2	

dynamic programming model, a greater sample of complete responses is still required, especially from international breeders and curators. Ways of obtaining this information while still relying on written questionnaires are currently being explored.

Curator survey

The size of the respondents' collections ranged from 20 to 1200 accessions. The accessions were maintained for the purposes of biodiversity, breeding programmes, research and multiplication. Curators were asked about the likelihood of maintaining their current collections at different levels of expenditure over the next 10 years. As expected, most respondents suggested that with reduced expenditure, the probability of maintaining the collections decreased. Three respondents did not believe that they could maintain the current number of accessions in the collection without loss, even at current levels of expenditure.

Curators were then asked to consider relating different levels of expenditure to the likelihood of maintaining collections of varying sizes. In many instances, only the most likely occurrence was identified rather than a probability distribution. On average, it was considered more likely that a reduction in expenditure would lead to a proportional reduction in the size of the collection. However, three respondents thought that it was more likely that a 25% reduction in expenditure would lead to an even greater reduction in the size of the collection (25–50% rather than a 1–25% loss of material). The latter cases were all field collections. In contrast, with a 50% reduction in expenditure, five respondents still believed that they could maintain their collections with minimal losses (1–25% reduction in material).

Respondents were asked about the types of activities and resources that would be affected if expenditure was reduced. Losses in the quantity of labour and deferring of capital items ranked the highest. Other affected activities included research, maintenance, multiplication and isolation of field material. Reduction in the quantities of insecticides, herbicides, fertilizers and hand tools were mentioned under 'other resources'. In terms of the most valuable characteristics of accessions, 'pest and disease resistance' rated the highest followed by 'potential yield' and 'drought resistance'. Other characteristics of importance included palatability, cold tolerance and root quality traits. Given the outbreak of the taro leaf blight in the Pacific, and the prevalence of banana and sweet potato diseases, the response was as expected.

In considering the accessions that have been or are likely to be requested by a breeding programme, curators were asked to think about dividing up their germplasm collection into sub-collections where the first sub-collection S1 comprised the first 25% of the accessions they would discard, that is, those least likely to contribute to increasing yield; sub-collection S2 comprised the next 25% of accessions they would discard; sub-collection S3 comprised the 'second-best' 25% of accessions; and sub-collection S4 comprised the best 25% of accessions. Four respondents believed that material from all four sub-collections had been requested while five respondents felt that only the top 25% (S4) accessions had been distributed. In terms of likely requests, those respondents who believed that only their best material has been requested in the past, were more optimistic that less valued material (S2, S3 and S4) would be requested in the future.

Breeder survey

The number of accessions in breeding collections ranged from 40 to 500 and came from diverse sources. Three collections contained over 80% of internationally obtained material, two collections had over 70% from national collections, and three breeders had collected over 80% of their material on collection missions.

For a majority of breeders, yield potential characteristics were the most important traits followed by quality and pest and disease resistance. Quality of material was also given the highest ranking by three respondents. While 'pest and disease resistance' was ranked very highly by curators, it was not given a top ranking by any of the breeders.

Similarly to the curator questionnaire, breeders were asked to imagine dividing up their own germplasm working collection into sub-collections. Respondents were then asked to assess the likelihood of achieving a range of yield gains on research stations and farms in the next 10 years, with an ever-decreasing size in their working collection (where S1, the worst accessions, would be discarded first). The responses were not uniform. Four respondents thought that a reduction in the size of their collections would reduce the chances of yield gains, while two breeders believed that discarding less valuable material increased their chances of achieving yield gains. It was expected by 85% of the respondents that yield gains would remain the same or increase, under any change in the size of the working collection. One respondent believed that they could achieve maximum yield gains irrespective of the size of the collection. In comparing research and farm outcomes, the shape of the distributions remained the same but at lower yield gains for on-farm outcomes.

A similar set of questions was asked in regard to potential yields and a reduction in the size of the national germplasm collection. For most breeders, the most likely outcome was for yields to remain the same at most sizes of the collection, particularly for the whole collection. Four respondents believed that once the collection was reduced to the top 25% accessions, the yield gains would most likely be over 10%. In contrast, one breeder expected that yields would decrease with any reduction in the size of the national collection. Compared with the working collection outcomes, the changes in the size of the national collection increased the likelihood of reduced yield gains.

Breeders were then asked to consider the effect on potential yields if the national collection was lost and accessions were only obtained from international collections. Similar to the questions for the working and national collections, breeders were asked to state the probability of achieving a range of yield gains against a decreasing size in the international collection. Again the results were varied. An increasing number of respondents believed that yields would be more likely to decrease with a reduction in the size of international collections, especially on farms. However, for some respondents, the smaller the collection, the increased likelihood of achieving above 10% yield gains.

In comparing the requests for material in the past with the likelihood of requests in the future, most respondents said the number of expected requests would either be about the same or more. On average, the chances of achieving 10% yield gains from random accessions requested from other collections doubled when the number of accessions increased from 100 to 200. With 50 accessions, the possibility of increased yields ranged between 0 and 30%.

Finally, respondents were asked to indicate the probability of increasing yields by 10% at three levels of expenditure (see Table 41.10). Most breeders believed that at ever-decreasing levels of expenditure the chances of increased yields would fall. Three breeders believed that with an expenditure cut to 80% of the current level, there would be no likelihood of achieving a 10% yield gain. One breeder believed it was just as likely to achieve a 10% yield gain at the three levels of expenditure.

Some individual responses

From the seven completed breeder surveys, only three respondents provided complete probability distributions; one breeder per species. Despite the small sample size, some preliminary comments can be made. The banana breeder was the most responsive to changes in the size of the national collection. He believed that with any decrease in the size of the national collection it was most likely that expected potential yield gain would fall by up to 10%. The taro and sweet potato breeders thought it would be most likely, under any sized reduction in the number of accessions in the national collection, that potential yield gains would not be affected. All breeders believed that any change in the size of the international collection would not be likely to impact on their potential yield gains.

Breeders provided information on the origin of their material. It was expected that if a high proportion of material had come from a particular collection then by reducing the number of accessions in that collection, potential yield gains would be expected to fall. This was not evident from any of

Table 41.10. Subjective probability in achieving a 10% yield gain.

Respondents	20% increase in current expenditure	Current expenditure	20% decrease in current expenditure
1	45	35	20
2	36	35	29
3	46	27	27
4	74	26	0
5	50	50	0
6	100	0	0
7	33.3	33.3	33.3

the three responses. The responses from breeders and curators at the International Conference on Science and Technology for Managing Plant Genetic Diversity in the 21st Century (SAT21) are to be included in the analysis. The running of the model with this recent data is still work in progress.

Acknowledgements

The research reported in this paper was financially supported by the Australian Centre for International Agricultural Research. We would like to acknowledge the help of Grahame Jackson in providing invaluable contacts and information about international root crop curators and breeders. His help enabled us to establish a network of likely recipients of PNG material and respondents for the questionnaires. We would also like to thank John Brennan and Fred Stoddard for their comments on the design of the second questionnaire regarding the elicitation of subjective probabilities. The staff of NARI were invaluable in providing information on the national collections and participating in the surveys. Special thanks to Jimmy Risimeri, Norah Omot, Peter Gendua, Paul Van Wijmeersch and Janet Pomet.

References

Azzam, A., Azzam, S., Lhaloui, S., Amri, A., El Bouhssini, M. and Moussaoui, M. (1997) Economic returns to research in hessian fly (*Diptera*: *Cecidomyiidae*) resistant bread-wheat varieties in Morocco. *Journal of Economic Entomology* 90(1), 1–5.

Brown, G., Jr (1990) Valuation of genetic resources. In: Orians, G.H., Brown, G.M., Kunin, W.E. and Sweirzbinski, J.E. (eds) *The Preservation and Valuation of Biological Resources*. University of Washington Press, Seattle, pp. 203–228.

Dillon, J.L. (1971) An expository review of Bernoullian decision theory in agriculture: is utility futility? *Review of Marketing and Agricultural Economics* 39(1), 3–80.

Evenson, R.E. (1996a) Valuing agricultural biodiversity. In: *Conference Proceedings of the Global Agricultural Science Policy for the Twenty-First Century, Melbourne, 1996*, pp. 611–638.

Evenson, R.E. (1996b) An application of priority-setting methods to the rice biotechnology program. In: Evenson, R.E, Herdt, R.W. and Hossain, M. (eds) *Rice Research in Asia: Progress and Priorities*. CAB International, Wallingford, UK, pp. 327–345.

Evenson, R.E. and Gollin, D. (1997) Genetic resources, international organizations, and improvement in varieties. *Economic Development and Cultural Change* 45(3), 471–500.

Evenson, R.E., Gollin, D. and Santaniello, V. (1998) *Agricultural Values of Plant Genetic Resources*, CAB International, Wallingford, UK.

Godden, D. (1984) Plant breeders' rights and international agricultural research. *Food Policy* 9(3), 206–218.

Godden, D. (1988) Technical change embodied in new varieties of English winter wheat and spring barley. *Research and Development in Agriculture* 5(1), 108–113.

Godden, D. (1991) Induced institutional innovation: plant variety rights, patents and genetic engineering. *Oxford Agrarian Studies* 19(1), 3–19.

Godden, D. and Brennan, J. (1994) Technical change in southern NSW and British wheat. *Review of Marketing and Agricultural Economics* 62(2), 247–260.

Godden, D. and Kambuou, R. (1996) Making nature pay: economics of plant germplasm collections in Papua New Guinea. Paper presented at the 40th Annual Conference, Australian Agricultural and Resource Economics Society, University of Melbourne, Australia, February 1996.

Godden, D., Kennedy, J. and Kambuou, R. (1997) Economic modelling of plant germplasm collections in Papua New Guinea. Paper presented at the 41st Annual Conference, Australian Agricultural and Resource Economics Society, Gold Coast, Australia, January 1997.

Godden, D., Wicks, S., Kennnedy, J. and Kambuou, R. (1998) Decision support tools for crop plant germplasm maintenance in PNG. Paper presented at the 42nd Annual Conference of the Australian Agricultural and Resource Economics Society, Armidale, Australia, January 1998.

Gollin, D. and Evenson, R.E. (1998) An application of hedonic pricing methods to value rice genetic resources in India. In: Evenson, R.E., Gollin, D. and Santaniello, V. (eds) *Agricultural Values of Plant Genetic Resources.* CAB International, Wallingford, UK, pp. 139–150.

Gollin, D., Smale, M. and Skovmand, B. (1998) Optimal collection and search for crop genetic resources. In: Smale, M. (ed.) *Farmers, Gene Banks and Crop Breeding: Economic Analyses of Diversity in Wheat, Maize, and Rice.* Kluwer Academic Publishers, Norwell, Massachusetts, pp. 57–58.

Green, F.N. (1997) The value of the United Kingdom statutory seed collections from a genetic resource perspective. *Plant Varieties and Seeds* 10, 195–204.

Kalabus, E.O. (1995) *Cost estimated costings for the taro germplasm collection maintenance for 1994–1995.* Bubia Agricultural Research Centre, Lae, Papua New Guinea.

Kennedy, J., Godden, D. and Kambuou, R. (1997) Optimal management of plant germplasm collections in Papua New Guinea. Presented at the Fourth Conference, Association of Asian-Pacific Operational Research Societies, Melbourne, Australia, December 1997.

Mendelsohn, R. and Balick, M.J. (1995) The value of undiscovered pharmaceuticals in tropical forests. *Economic Botany* 49(2), 223–228.

Norris, P.E. and Kramer, R.A. (1990) The elicitation of subjective probabilities applications in agricultural economics. *Review of Marketing and Agricultural Economics* 58(2–3), 127–147.

Oldfield, M.L. (1989) *The Value of Conserving Genetic Resource.* Sinauer Associates, Sunderland, Massachusetts.

Pearce, D.W. and Cervigni, R. (1994) The valuation of the contribution of plant genetic resources. In: Swanson, T.M., Pearce, D.W. and Cervigini, R. (eds) *The Appropriation of the Benefits of Plant Genetic Resources for Agriculture: an Economic Analysis of the Alternative Mechanisms for Biodiversity Conservation.* Background Study Paper No. 1. Food and Agriculture Organization of the United Nations, Commission on Plant Genetic Resources, Rome, Italy.

Phillips, O.L. and Meilleur, B.A. (1998) Usefulness and economic potential of the rare plants of the United States: a statistical survey. *Economic Botany* 52(1), 57–67.

Rao, K.P.C. and Evenson, R.E. (1998) Varietal trait values for rice in India. In: Evenson, R.E., Gollin, D. and Santaniello, V. (eds) *Agricultural Values of Plant Genetic Resources.* CAB International, Wallingford, UK, pp. 151–156.

Smale, M., Hartell, J., Heisey, P.W. and Senauer, B. (1997) The contribution of genetic resources and diversity to wheat production in the Punjab of Pakistan. Working paper. CIMMYT Economics Program, Mexico.

Swanson, T.M., Pearce, D.W. and Cervigini, R. (eds) (1994) *The Appropriation of the Benefits of Plant Genetic Resources for Agriculture: an Economic Analysis of the Alternative Mechanisms for Biodiversity Conservation.* Background Study Paper No. 1. Food and Agriculture Organization of the United Nations, Commission on Plant Genetic Resources, Rome, Italy.

Winkler, R.L. (1967) The assessment of prior distributions in Bayesian analysis. *Journal of the American Statistical Association* 62(319), 776–800.

42 People, Plants and DNA: Perspectives on the Scientific and Technical Aspects of Conserving and Using Plant Genetic Resources

T. Hodgkin[1] and V. Ramanatha Rao[2]

[1] *International Plant Genetic Resources Institute (IPGRI), Rome, Italy;* [2] *IPGRI Regional Office for Asia, the Pacific and Oceania, Serdang, Selangor, Darvil Ehsan, Malaysia*

Introduction

The International Conference on Science and Technology for Managing Plant Genetic Diversity in the 21st century (SAT21) provided an opportunity to consider our current knowledge of how to conserve and use plant genetic resources. *The State of the World's Plant Genetic Resources for Food and Agriculture* (FAO, 1998), prepared for the International Technical Conference on Plant Genetic Resources, held in Leipzig in 1996, provides the most comprehensive overview of the global plant genetic resources conservation effort. Using this information, the Global Plan of Action (FAO, 1996) provided a clear agenda for conservation work. SAT21 provided a chance to review our technical capacity to carry out the work; to review the extent to which the methods, technologies and knowledge needed to conserve and use plant genetic resources were available; and to explore what new opportunities were on the horizon.

Conservation and use of plant genetic resources depend on a wide range of technical and scientific areas. Information to support conservation work comes not only from biological disciplines such as genetics, plant physiology, ecology, reproductive biology and biotechnology, but also from areas such as sociology, anthropology, economics and geography. A truly multidisciplinary approach is needed, so that knowledge and approaches from different areas can be put together in ways that enable genetic resources workers to identify what should be conserved, how it can best be conserved, and how the conserved materials can best be used. Soulé (1985) also emphasized the importance of a multidisciplinary approach and the need to draw on many different disciplines in developing solutions to the pressing problems of conservation. He suggested that conservation biology differed from most other biological sciences in that it was a 'crisis discipline' in which one must often act before all the facts were known. Often there is a need to carry out urgent and necessary interventions while still collecting information and seeking to understand the characteristics of the species one is trying to conserve.

Over the past three decades, the concerns of those involved in plant genetic resources conservation have dramatically expanded. There is a greatly increased recognition of the importance of embedding conservation in a wider social framework. This is reflected in an increasing concern with *in situ* conservation, an interest in an increasingly large number of useful plant species, often previously neglected by research, and a recognition of the importance of diversity within production systems, especially in marginal areas. The need to confront the economic issues involved in sustainable conservation is now accepted and there is a growing interest in ensuring that national resources management takes account of the role of biodiversity while developing national plans. Taken together, these reflect the objectives contained in the Convention on Biodiversity (UNEP, 1992) of ensuring that conservation contributes fully and effectively to world development.

The keynote papers at SAT 21 provided a broad overview of the major areas of conservation science identified by the organizers. Together with the presented papers, also published in this book, they provide perspectives on some of the most important areas for genetic resources workers today. Some subjects stand out as major preoccupations. These include the power of molecular methods and of geographic information systems (GIS), the potential of bioinformatics and the increasing concern with use, conservation in production systems and with crops that have been neglected or underutilized. The papers also raised issues with regard to the costs of conservation and increased responsibility of the private sector to support conservation efforts. In this chapter our objective is not to review the subjects themselves but rather to highlight some of the important questions that arise from considering recent scientific and technical developments in different areas of plant genetic resources conservation and use. We will try to identify gaps in the current research agenda as revealed by the conference and to identify some of the challenges for the future.

New Scientific Opportunities

There are two areas in which the rapid developments of the past few years are having a major effect on plant genetic resources work: molecular genetics and information technology. Molecular genetic methods now allow us to analyse diversity in plant genetic resources at the DNA sequence level and to obtain unparalleled amounts of information about the diversity present in plants. The information revolution enables us to manage and process the very large amounts of data generated by molecular tools and potentially provides all the access facilities of the Internet to such data.

Linked to both areas, and drawing extensively on improvements in information technology, the use of GIS is also having an important effect on our capacity to understand the distribution of diversity and the ways in which different factors affect it. GIS provides an ability to analyse distribution patterns and relate them to any other geographic information so as to plan conservation work in ways that take greatest advantage of known or predicted spatial factors.

Molecular genetics in conservation and use

The number of research reports providing some information on the ways in which diversity of particular markers is distributed in natural populations or a set of accessions is rapidly increasing. Many different molecular marker systems are used although the use of randomly amplified polymorphic DNA (RAPD) still seems to be most common, despite the weaknesses of this marker system (Karp *et al.*, 1997). The number and nature of the accessions used are also very variable. This *ad hoc* approach has definite weaknesses. Firstly, it is very difficult to compare different marker systems and determine which are the best for what purposes. Secondly, it is difficult to get a clear picture of how effective molecular markers are in helping us to understand the extent and distribution of diversity in crop genepools. As Karp noted (Chapter 4, this volume) we need to be clearer about the questions we ask. Deciding on the appropriate molecular techniques, or even whether molecular methods are needed at all, depends on formulating clear and rigorous questions or hypotheses for investigation. There also need to be a few carefully selected studies on specific crop genepools that would allow us to compare patterns of diversity as revealed by different molecular markers with those revealed by isozyme and morphological studies.

Do molecular markers tell us much about the patterns of diversity that might be observed for adaptive characters such as biotic or abiotic stress tolerance or for yield or quality related characters? Certainly they provide information about important diversity characteristics such as gene flow (Raybould *et al.*, 1996) and the breeding pattern of a population (Chase *et al.*, 1996). However, the evidence that the diversity patterns shown by molecular markers are associated with ecological characteristics is slight (but see, e.g., Owuor *et al.*, 1997). This is where a search for markers that might be useful for studying functional diversity, as described by Karp (Chapter 4, this volume), becomes important.

Genebank managers can already see the potential value of using molecular methods to support the operations of a genebank (Phippen *et al.*, 1997; van Hintum and Hodgkin, 2000). While it is doubtful that the methods will be useful for eliminating duplicates (except in the negative way of establishing that specific accessions are not duplicates), molecular markers can provide robust methods of checking accession integrity and of monitoring change during regeneration. This makes it unfortunate that, as far as we can tell, rather little work is being undertaken in this area. Studies are overdue on the ways in which different marker systems might best be used to monitor regeneration activities in genebanks.

The role that the genetic resources community might undertake in respect of conserving molecular genetic products has yet to be defined. For example, is there a need for genebanks to take some responsibility for DNA storage and what form might this take? Certainly, there are an increasing number of resources generated by the molecular community which are relevant to conservation work. Some genebanks might wish to store primers, probes and bacterial artificial chromosome libraries to facilitate their work. Similarly, there may be a need to store mapping populations generated to identify quantitative trait loci (QTLs) or create dense molecular maps, as was noted by a recent Consultative Group on International Agricultural Research (CGIAR) international workshop on genebanks and comparative genetics (SGRP, 2000).

A 'holy grail' for many workers using molecular genetics in genetic resources work has been the development of methods that might help to locate useful variants in large genebank collections (Kresovich *et al.*, Chapter 35, this volume). The ways in which this might be done are, in fact, unlikely to be either simple or direct. A substantial amount of sequence data will usually be required for the trait in question and, even when variants at a target sequence can be identified, a good phenotypic screening procedure will be needed to determine whether the variants have any functional significance. Furthermore, in the case of many abiotic stress responses, quite different metabolic pathways are involved at different stages of the plant response. In different situations, stress resistance seems to involve different gene systems and searching for a particular variant sequence of one specific gene may not always be the most useful approach for characteristics controlled by many genes.

In fact, locating useful genes in collections will require an integrated approach that brings together information from molecular studies and other areas. This might include using an extensive set of molecular markers for diversity studies, analysis of the extent of linkage disequilibrium and identification of areas of the genome where important genes may occur combined with more conventional approaches using passport data, GIS and core collections (Kresovich *et al.*, Chapter 35, this volume). These techniques could together provide an optimum set of candidate accessions for phenotypic and genetic analysis. In all cases, good phenotypic characterization methods will remain an essential component of any genebank search strategy.

Even where it is not directly applicable, the knowledge coming from plant molecular genetic studies is likely to continue to have a profound impact on genetic resources work. For example, molecular taxonomy will substantially improve our knowledge of the primary, secondary and tertiary genepools of many crops and evolutionary studies will help identify crop ancestors, past genetic bottlenecks and opportunities for introducing new useful variation. Studies of the genetic architecture of key yield related components (e.g. flower and seed production, maturity and photoperiod response) will enable us to focus on areas of the genome where diversity is particularly important. Analyses of genome synteny will assist in the identification and transfer of useful characteristics in crop relatives and help conservation of minor crops showing synteny with major ones. Understanding the potential value of these developments will be important as will engaging the interest of molecular genetics research workers in conservation issues.

The information revolution

Information management has always had a central place in plant genetic resources conservation. The need to identify, record and communicate information about accessions has led to the development of a substantial infrastructure with relatively highly developed database structure and information management systems. The development of the World Wide Web and electronic communication have been seized on enthusiastically and a number of directly accessible genebank databases are now available such as those of the CGIAR centres (through System-wide Information Network for Genetic Resources (SINGER): www.singer.cgiar.org) and the collections of a number of European genebanks (e.g for Brassicas see: www.plant.wageningen-ur.nl/about/Biodiversity/Cgn/collections/brasedb/

The information revolution offers much more than an ability to query databases at a distance. It provides opportunities for carrying out substantial investigations by linking databases from different sites so that, for example, geographic information can be linked with genebank passport data and molecular sequence data. Indeed, the information revolution is essential to the development of molecular genetics and for realizing the potentials of new areas of study such as genomics or proteomics.

Many genetic resources workers are likely to be interested in linking different databases together, and the ways in which this can be done will become an area of increasing research activity. Thus, databases from genebanks will be linked with those from botanical gardens and protected areas to support conservation planning. Sequence information from molecular databases will be compared with expressed sequence tagged sites (ESTs) obtained in diversity studies to target markers associated with potentially useful traits. Information from gene expression studies will be used to find potentially important genes that can then be studied in more detail to determine the different ways in which they vary.

Accessibility of the information continues to be an important issue at two levels. Firstly, it will be increasingly necessary for genebank managers and plant genetic resources workers to have Internet access. Already, molecular geneticists without such access find their research opportunities severely restricted. In many countries, access remains difficult and expensive, and opportunities for prolonged analysis of databases accessed through the Internet are strictly limited. Secondly, the data itself will need to remain (or be placed in) the public domain. The importance of this is widely recognized throughout the research community.

While the opportunities are considerable, they will remain ultimately dependent on the quality of data maintained. The first objective for genebanks will always need to be 'good accession level data, well curated, at source'. Only when this is the case and when the data become available electronically will the potential of bioinformatics be realized for genetic resources. Research on improved information management and documentation methods is still required and, while this may seem a disappointingly conservative conclusion, it reflects the needs identified by many genebank managers.

There are some practical concerns which need to be dealt with fairly soon. One is the way in which genebank workers should handle molecular information on accessions. The experience of CGN (Centre for Genetic Resources) in the Netherlands will be important in this regard; it is one of the first genebanks to confront the issue of having very large amounts of molecular information on its 2000 lettuce accessions (van Hintum and Hodgkin, 2000). Another issue will be the need to ensure that there are common vocabularies where different databases are being linked and that there are reasonably straightforward ways of linking them (Sobral, Chapter 16, this volume). There is also the need to develop improved procedures for handling indigenous knowledge which meets the needs of potential users while recognizing the particular status of the information and the concerns of those from whom it is obtained.

As Sobral (Chapter 16, this volume) noted, the information revolution will profoundly affect our understanding of the organisms we conserve. At its best it will provide a framework for a more integrative approach to biology where information from widely different sources can be brought together to help understand crop plant performance and diversity, and the forces that are responsible for the patterns we observe.

Understanding the distribution of diversity: the power of GIS

GIS are becoming the subject of increasing attention as their power to support the analysis of extent and distribution of genetic diversity is recognized. As Guarino *et al.* (Chapter 36, this volume) noted they could enhance the efficiency of surveying, field exploration, development of *in situ* management strategies, and the evaluation and use of genetic resources.

Surprisingly, given how rapidly the use of molecular methods is spreading, the number of plant genetic resources studies using GIS methods remains rather limited. There is a very real need for studies in different parts of the world that explore the use of GIS with a wider range of useful plant species. These should cover crops, forages, crop relatives and forest species, and include species with different biological characteristics.

Perhaps even more than with molecular methods or bioinformatics, the use of GIS in plant genetic resources work is in an experimental phase. Analytical techniques greatly outstrip our experience in using them. They are likely to be of considerable value in locating populations of species with known ecological preferences, for identifying areas of potential threat, or for finding populations with specific attributes such as drought or salinity tolerance. However, the limitations of the techniques and the ways in which they can best be put together have yet to be rigorously tested. More studies will be needed to determine the extent to which species characteristics fit neat ecological or environmental patterns so that one can extrapolate from known populations to find unknown ones. Similarly, the extent to which GIS can help in identifying patterns of intraspecific diversity needs to be further tested (e.g. see Greene *et al.*, Chapter 37, this volume). Work is also needed to test the power of time series analyses as a way of identifying patterns of change and areas where threats to existing populations may be greatest.

Bringing the new technologies together

A major concern in all three areas is that the availability of the new techniques is so unevenly distributed. Many developing countries cannot realistically expect to use the techniques in their work, despite the substantial amounts of diversity that they possess and for which they are responsible. Technology transfer, as it relates to genetic resources, must be a major priority involving collaboration between institutes, training and improved access to the different technologies.

From a research perspective, the greatest benefits are likely to come from investigations which bring together molecular methods, information sciences and GIS to focus on two different aspects of the distribution of diversity, which are themselves central to identifying what should be conserved. One involves the determination of general properties of diversity using anonymous or neutral markers. The other involves analysing distribution of variation in adaptive traits or in loci likely to be associated with adaptive variation. Such studies might include analyses of the distribution of adaptive and non-adaptive diversity using different molecular markers on defined populations of specific crop genepools in combination with GIS and in association with information from global genebank holdings. The genepools studied should include selected self- and cross-pollinated species, perennials and annuals, widely dispersed species and endemics. Hypotheses should be clearly formulated and the information from the different methods carefully evaluated.

At the same time, we need a more pragmatic approach to evaluate how best to use relevant elements of the new technologies to answer questions on individual populations, accessions and species which each have their own properties. Which methods are best for monitoring populations conserved *in situ*, for identifying those with most diversity or for determining if newly identified populations contain unique variation? The technologies are powerful, but also expensive and their use in many genebanks will continue to be limited to those areas where they can be of greatest benefit.

Improving Conservation Practices

The practice of conservation – ensuring that the *ex situ* conserved accessions and *in situ* plant populations in their care are maintained securely for future generations – remains the overriding preoccupation of most plant genetic resources workers. There are now some 6 million accessions conserved *ex situ* throughout the world. Of these, over 600,000 are maintained in the 11 genebanks of the CGIAR Centres and many countries now have genebanks with over 10,000 accessions which require very substantial resource inputs to retain them. Add to this the growing commitment to *in situ* conservation and, to many of us, the magnitude of the task must seem overwhelming.

Ex situ *technologies: securing the resources*

One of the most significant achievements of plant genetic resources work has been the establishment of procedures that provide for the safe, long-term maintenance of seeds. The procedures are generally well established, especially for our major seed propagated crops. For this reason, perhaps, seed conservation research has received rather little attention in recent years although some key questions remain. Thus, Engelmann and Engels (Chapter 9, this volume) noted the importance of continuing work on identifying optimum seed moisture content, on seed drying procedures and on ensuring that seed for long-term storage is harvested in the best possible condition.

A number of questions have been raised concerning the way in which accessions can best be managed. The possibility of combining similar accessions was discussed by van Hintum *et al.* (Chapter 11, this volume). Accession management has always been a thorny question and appropriate procedures will need to take account of genetic studies and management considerations. For example, Phippen *et al.* (1997) were able to show that combining very similar cabbage accessions would result in the probable loss of only 4% of the variation identified using RAPDs. Without more information on what RAPD diversity data measures, one may want to be cautious in deciding how to use these data. The 4% of diversity lost may be associated with agromorphological variation of value to breeders. The genebank manager will also need to balance the importance of this loss against the possibility that attempting to maintain all the accessions with reduced efficiency might result in greater losses of diversity.

More generally, we need much more empirical data on what is happening to the diversity maintained *ex situ*. The procedures that are being used by many managers represent a compromise between what is known to be optimal and what is possible, given available resources. This is bound to be the case, but the consequences of the alternative decisions that can be made by managers have yet to be studied. Decisions on frequency of regeneration, size of population used for regeneration, splitting or combining accessions, or balancing viability and regeneration are all likely to have an effect on the amount of diversity maintained, but the nature of that effect remains substantially unknown.

As Engelmann and Engels (Chapter 9, this volume) noted, an improved understanding of recalcitrance remains a priority for *ex situ* conservation. Molecular genetics would now seem to offer real opportunities to investigate gene expression during the seed maturation process and to reach a better understanding of the critical processes involved. Studies using genomic and proteomic approaches on suitable model species could provide a substantially improved understanding of the processes that lead to the development of mature seed in both normal and recalcitrant seeded species. Tissue culture methods and cryopreservation are likely to remain the appropriate procedures for storage of species with recalcitrant seeds but an improved understanding of recalcitrance would certainly lead to benefits in the management of conservation and exchange of these species.

In situ *technologies: conserving the processes of evolution*

In situ conservation is becoming increasingly important in the conservation of plant genetic resources. A major Global Environment Facility (GEF) funded programme of work has been undertaken in Turkey (Tan and Tan, Chapter 19, this volume) to conserve crop wild relatives. This led to the development of a substantial national programme to conserve crop wild relatives *in situ* and has provided much information on the financial and labour resources that might be needed and how such a programme might be implemented. *In situ* conservation of crops (on-farm conservation) has also received increasing support and is the subject of extensive research in a number of countries around the world (Pham *et al.*, Chapter 14, this volume; Chiwona, Chapter 15, this volume; Jarvis *et al.*, 2000).

In situ conservation requires some significant changes in perspective for those used to *ex situ* conservation. In the first place, the concern is less with the maintenance of specific characteristics or specific genotypes than with the maintenance of evolutionary potential. The target populations have to exist within ecosystems that are themselves dynamic and subject to change over time and to succession. On-farm conservation will depend for its effectiveness on the extent to which local varieties continue to meet the needs of farmers and communities and the approach taken needs to be one that is embedded in the community and reflects its values and concerns.

The research needed for *in situ* conservation of useful plants (both on-farm and in natural habitats) is substantial. It is a fairly young endeavour compared with *ex situ* conservation and the body of information that might help us in making choices about locations, populations, management procedures and appropriate interventions has not yet been fully developed.

For crop wild relatives a carefully selected set of detailed studies is required that complements the work done already in a few situations such as the wild wheats in Ammiad, Israel (Horowitz and Feldman, 1991) and beans in Costa Rica (Rocha *et al.*, Chapter 20, this volume). The exciting element of this research is that it will frequently have to be carried out alongside specific conservation activities and in full collaboration with communities living in the areas; a considerable challenge to research workers used to controlled experiments in defined environments. Since resources for the conservation actions needed will be extremely limited, methods also need to be found of identifying the key interventions required (which populations, of what species, in which areas) so that they, at least, can be implemented. Knowing what is most threatened and what is protected already (through being in protected areas or maintained *ex situ* or secured in other countries) will be an essential component of planning such interventions.

On-farm conservation requires a careful linking of research through multidisciplinary studies that will answer some important conservation questions (e.g. the amount of diversity in local varieties, its distribution and variation from year to year) with an understanding of the social and economic dimension and a concern with development. In obtaining the information that will allow the scientific basis of on-farm conservation to be established, participatory procedures must be used which fully involve local communities and farmers. In order to make on-farm conservation effective, we need to understand the ways in which farmer management affects the extent and distribution of diversity and to determine how production environment and farmer decision-making interact with crop biological properties to affect the diversity characteristics of local varieties.

Neither *in situ* nor *ex situ* conservation is adequate on its own. They need to be used in a complementary manner (Maxted *et al.*, 1997). However, what this actually means, in practice, has yet to be determined. The factors which might guide one to choose specific *in situ* or *ex situ* methods for a particular taxa, population or variety have been discussed in general but not, to our knowledge, tested in any substantial way and were not considered at SAT21 (although see Lamont and Palmer (2000) for an interesting approach with apple). It would be desirable to see investigations developed which might involve a number of partners working on a specific genepool, to test the concept in practice and determine how useful it really could be.

Aspects of Use

Any discussion about increasing the use of plant genetic resources usually involves some consideration about what 'use' should include. Many plant genetic resources, especially those of minor crops, are maintained as part of production systems and one aspect of the discussion clearly involves management of diversity in production; this has been termed conservation *through* use (FAO, 1996). The other aspect of use, more familiar to many plant genetic resources workers, concerns the use made of genetic resources conserved in genebanks (conservation *for* use).

The increasing interest in the use of plant genetic resources was fully reflected at SAT21 with substantial discussions of germplasm enhancement and of management of agrobiodiversity in production. Germplasm enhancement is now well recognized as an area where conservation workers can make a significant contribution (Ortiz, Chapter 26, this volume). Agrobiodiversity is a newer concern which reflects an increased recognition by the wider conservation community of the importance of diversity in production systems and a growing interest in the ways in which diversity can be used to enhance sustainable production (Sastrapradja and Balakhrishna, Chapter 12, this volume).

An increased concern with germplasm use also reflects the need to identify and quantify the contributions that are being made by plant genetic resources to agriculture and world food security. Data on the use of *ex situ* collections in research have also recently been provided by Dudnik *et al.* (2000). Much more information of this type is needed from a range of different situations to reflect the different national and crop perspectives and different types of use that can occur.

Genetic resources in production systems

The importance of maintaining genetic diversity within production systems has always been recognized by plant genetic resources workers. The dangers of a narrow genetic base have been described and the need to prevent its occurrence noted (e.g. National Research Council, 1972). However, much of the debate has tended not to become deeply involved in the issues or processes surrounding the deployment of diversity in production. A growing concern with on-farm conservation, focusing on specific crops and local varieties, and with sustainable production, has been matched with an increased interest in diversity in production systems (e.g. Wood and Lenne, 1999). This is now reflected, for example, in the development of an agrobiodiversity agenda for the Convention on Biological Diversity (UNEP, 2000).

One entry point for agrobiodiversity research that has already been discussed is the concern with on-farm conservation. The appreciation that communities and farmers often maintain large numbers of local varieties which meet their production needs has led to substantial programmes to investigate the genetic, ecological, socio-economic and other dimensions of this process. Over the next few years we can expect to develop an increasing understanding of the different circumstances in which farmers preferentially maintain their local materials, the diversity that is retained through this process and how it is managed (see, for example, Chiwona, Chapter 15, this volume; Pham *et al.*, Chapter 14, this volume; Louette *et al.*, 1997).

The next steps will involve going beyond the simple description of the management of local variety diversity of single crops and asking a much more complex and demanding, but at the same time significant set of questions. These will include investigating how best to use diversity within crops to improve production, not only using local varieties but also using other approaches (e.g. variety mixtures and multilines as illustrated recently for rice in China by Zhu *et al.*, 2000), and how best to combine different crops to optimize sustainability. The ways in which crop and non-crop components of agroecosystems interact also require investigation so as to explore the ways in which the different elements such as pollinators, pests, pathogens and weeds all interact within the agroecosystem. From this, we will begin to develop a better understanding of how diversity management in production systems can make the best contribution to increased sustainable production.

Neglected and underutilized crops: a need for action

Padulosi *et al.* (Chapter 30, this volume) make a clear case for increasing work on a range of crops that are locally and regionally important but have been neglected by research and conservation or are being displaced by changing agricultural production practices despite their value. The importance of an increased commitment to these crops is now recognized and new activities are being developed. However, the general problem, that there is a group of crops for which conservation is under-resourced, will always remain.

One way of approaching the conservation of these crops is by an increased emphasis on conservation through use. Where neglected crops can be maintained in production systems, this can constitute an important element of their conservation. Indeed, there is a recognition of the benefits for both conservation and use that come from linking producers and consumers with institutions involved in agriculture, marketing and research (the so-called 'filière' approach).

It might also be desirable to explore how to develop effective strategies for such crops that will make the most use of limited available resources. Cost-effective contributions to their conservation may include an increased emphasis on collating all available information on the target crop (since lack of information is often a key constraint), a deliberate adoption of core collection procedures for all *ex situ* conservation work (since the numbers conserved will nearly always be in the hundreds rather than thousands), and the development of monitoring procedures that can be used to chart production changes so as to determine if active interventions are needed.

Germplasm enhancement and pre-breeding

There is a widespread feeling that support for germplasm enhancement or pre-breeding has tended to decline with the increasing shift to commercial plant breeding of the last two decades. At the same time, the opportunities for making progress from the sort of work that has been classified as germplasm enhancement (Ortiz, Chapter 26, this volume) has substantially increased. For example, molecular genetic methods combined with increased opportunities for wide crossing using techniques such as embryo rescue enable us to locate and transfer new characters found in crop wild relatives (Tanksley and McCouch, 1997).

Ortiz (Chapter 26, this volume) rightly emphasizes the importance of the various 'base broadening' procedures that are being explored at present. Genetic resources workers need to collaborate with plant breeders in the development of procedures that allow effective testing of new materials and their introduction into improvement programmes in a systematic way. Both introgression and incorporation (*sensu* Simmonds, 1993) programmes will be needed. Programmes such as the Latin American Maize Program (Salhuana *et al.*, 1991), the HOPE scheme of Kannenburg (Kannenburg and Falk, 1995) or the sugarcane improvement programme in the West Indies (Kennedy, 2000) make relatively full use of genetic resources in sustained and effective ways. Genetic resources workers also need to re-examine the role they can play in evaluating genebank materials in a way that is useful to breeders. It has often been said that evaluation should not be undertaken by genebanks, but this is an oversimplification. Lack of evaluation is perceived as a real bottleneck to further use and some steps are needed to identify ways of overcoming the problem.

A common problem faced by genetic resources workers is that of identifying a limited set of accessions likely to be of most interest to a specific user from a rather large collection. Perhaps we need to explore the ways in which information from passport data, GIS analysis, the development of core collections, molecular studies and other approaches can be brought together to provide generalized search strategies for genebank managers.

New Demands and New Horizons

The economic imperative

The value of plant genetic diversity, as a component of biodiversity, has been estimated by different experts differently, while some maintain that such estimation is fruitless as biodiversity is invaluable. The responsibility of conserving plant genetic resources of a country, either *in situ* or *ex situ* ultimately lies with the country. Countries therefore have to determine what investment they will make and, like any other activity, economic aspects become important (see Milne *et al.*, Chapter 41, this volume). While it is important that such studies do not get bogged down in economic theories, it is necessary that a much more substantial economic analysis of costs and benefits from conservation work is undertaken. The world has recognized the value of these resources unequivocally but finding the funds needed for conservation work requires a much better understanding of the economic aspects of conservation work (see, e.g., Swanson, 1996). While there have been a number of recent contributions on the economics of conservation, (Burstin *et al.*, 1997; Gollin and Evenson, 1998; Gollin and Smale, 1999), rather little of this was reflected at SAT21, indicating that there is still a need to persuade plant conservation workers of the importance of incorporating economic perspectives in their work.

One specific aspect discussed at SAT21 concerns the roles of the private sector and of genebanks themselves in generating the funds needed for their operations. Rajanaidu and Ramanatha Rao (Chapter 39, this volume) suggest that genebanks could generate part of the funds they need themselves, which could be supplemented from public and private sources.

To date, the private sector has mostly been able to access germplasm from the public sector without any involvement in the process and without payment. With changing scenarios, it will become necessary that the private sector collaborates with the public sector in conserving the resources. It will be desirable that the public and private sector collaboration that is under way in some countries (Burstin *et al.*, 1997) develops even to the level of sharing the costs of conservation as has been done for oil palm in Malaysia (Rajanaidu and Ramanatha Rao, Chapter 39, this volume). This may eventually give a larger role to the private sector in conservation work, but the need for continued involvement of the public sector as the custodian of genetic resources has to be recognized, otherwise all different user needs will not be met. Some of the pitfalls of privatization have been described by Biber-Klemm (see Chapter 40, this volume) and it is imperative that appropriate checks and balances for accessing and using the genetic material and benefiting from it are maintained.

How well are we doing?

The paper by Brown and Brubaker (Chapter 24, this volume) represents one of the few attempts ever made in plant genetic resources work to identify ways of evaluating the effectiveness of germplasm conservation work in an organized way. There are many difficulties. It is impossible to say how much diversity exists within a crop or to decide on how much needs to be conserved by any means. It is often felt to be dangerous to provide uncertain or incomplete information that may be misinterpreted. The measurements that need to be made have not been clearly identified or generally agreed. However, Chapter 24 provides a solid basis for a debate of what indicators of effectiveness might be useful and how they could be applied.

More debate will certainly be needed on what indicators will be most useful and more experience is required on what information the different ones actually provide. It would be useful, in the first instance, to test the approaches proposed on specific collections or a specific crop. This will allow us to gain familiarity with the different measures, their accuracy, ease of collection and the ways in which they can be collected and interpreted.

Conclusions

There is a healthy and substantial quantity of research on relevant topics being carried out throughout the world; the amount of plant genetic resources research is growing and the findings are important. Research by national plant genetic resources programmes, by genebanks, universities and research institutes, will continue to be the mainstay of the global plant genetic resources conservation effort and to provide the necessary effective coverage of crops and issues which can steadily resolve many practical conservation problems.

SAT21 provided a good overview of the situation with few exceptions. One area which received rather little attention was the integration of social and cultural aspects into genetic resources work. There are some important developments in this area (e.g. Zimmerer, 1992; Tapia and De la Torre, 1997) that provide exciting prospects for a fuller understanding of diversity management by different communities but rather little of this was presented at the conference. Another area which received very little attention was the extent and rate of genetic erosion. There remains, in practice, almost no information on either quantity of genetic erosion (how many genes or alleles are being lost) or its rate. At some point this is an issue which the genebank community will have to face.

However, there are always some gaps in any meeting and, in this review chapter, we have also been rather selective and have concentrated on the dominant issues and trends. Thus, we have not explicitly discussed some topics on which valuable research is in progress, such as the conservation of useful agroforestry and forest species and medicinal and aromatic plants.

While the breadth of research is impressive, there are a number of research areas where there is an increasing need for a sustained, coordinated research programme which addresses a particular research area in some depth. It seems to be less easy to achieve this objective. This is partly because of the ways in which funding is presently allocated to research (often for short-term projects with limited and specific objectives). It is also because the development of such research projects depends very much on long-term planning and commitment, and on the development of extensive collaboration. However, the development of such programmes is not impossible; the global programme of work on *in situ* conservation of crop plants developed by a number of national partners and IPGRI provides one way. This constitutes a programme which addresses key research questions in a sustained way which, hopefully, can be maintained for long enough to yield the desired scientific outputs.

Together with the different topics identified in the body of this chapter where research is needed, it is possible to identify a few topics where there might be a particular benefit from a coordinated international research effort of some type. These are areas where depth as well as breadth is particularly important and where a focused research agenda is likely to be valuable in terms of achieving a qualitative advance in our conservation capability. Obviously, any selection will be subjective but we believe that a few areas can be identified where the need for a focused international research agenda is greatest and the benefits of such an approach are likely to be most substantial. These include:

- The investigation of the patterns of diversity shown by neutral molecular markers, molecular markers associated with functional genes and agromorphological characters, using a few selected crop genepools. This should integrate and critically assess the use of GIS and link it to the use of molecular methods of describing diversity. It would provide an understanding of patterns of genetic diversity of different types of characters in crop genepools and the way in which different forces affect them. The investigation should include crops, forages and crop relatives selected to cover different biological characteristics such as life history and breeding system.

- A collaborative investigation of the way in which different types and patterns of diversity contribute to sustainable production. The rhetoric that diversity is beneficial needs to be placed on a much firmer footing at the production system level. There is a need to critically explore the ways in which different forms of diversity can best be deployed in different situations, to define where and in what ways it is most beneficial, and how it can most effectively be maintained or introduced. This will require a quite new collaboration between a very wide range of participants in which expertise on intraspecific diversity from on-farm conservation work will be only one of the dimensions in a wider agrobiodiversity context.
- The continued development of ways of locating potentially useful accessions in genebank collections through a selected set of model studies aimed at finding material with specific attributes in existing collections. This will involve integrating molecular tools with bioinformatics and conventional studies using passport and characterization data in development of integrated search procedures. In a sense it will provide important entry points for molecular contributions from synteny and from character studies (molecular genetic control of specific characters such as disease resistance and abiotic stress tolerance).
- At a more practical level, there remains the need to continue to address genebank management issues in a coordinated way that ensures that we can understand how actual management practices affect accession integrity and genetic content. The most pressing need remains the development of procedures for monitoring and improving regeneration procedures. Evidence suggests (Burstin *et al.*, 1997) that regeneration costs the same as all other *ex situ* operations put together and it remains the stage at which accessions are most vulnerable to loss of integrity and to change in genetic constitution from drift and shift (Parzies *et al.*, 2000).

References

Burstin, J., Lefort, M., Mitteau M., Sonnot, A. and Guiar, J. (1997) Towards the assessment of the cost of genebanks management: conservation, regeneration and characterization. *Plant Varieties and Seeds* 10, 163–172.

Chase, M.R., Moller, C., Kessell, R. and Bawa, K.S. (1996) Distant gene flow in tropical trees. *Nature* 383, 398–399.

Dudnik, N.S., Thormann, I. and Hodgkin, T. (2000) The extent of use of plant genetic resources in research – a literature survey. *Crop Science* 41, 6–10.

FAO (1996) *The Global Plan of Action for the Conservation and Sustainable Utilization of Plant Genetic Resources for Food and Agriculture.* FAO, Rome, Italy.

FAO (1998) *The State of the World's Plant Genetic Resources for Food and Agriculture.* FAO, Rome, Italy.

Gollin, D. and Evenson, R.E. (1998) An application of hedonic pricing methods to value rice genetic resources in India. In: Evenson, R.E., Gollin, D. and Santaniello, V. (eds) *Agricultural Values of Plant Genetic Resources.* CAB International, Wallingford, UK.

Gollin, D. and Smale, M. (1999) Valuing genetic diversity: crop plants and agroecosystems. In: Collins, W. and Qualset, C. (eds) *Biodiversity in Agroecosystems.* CRC Press, London, pp. 237–266.

van Hintum, Th.J.L. and Hodgkin, T. (2000) Molecular diversity in a lettuce genebank collection (Abstract). *International Conference on Science and Technology for Managing Plant Genetic Diversity in the 21st Century, Kuala Lumpur, Malaysia.* IPGRI, Rome, Italy.

Horowitz, A. and Feldman, M. (1991) Evaluation of the wild-wheat study at Ammiad. *Israel Journal of Botany* 40, 501–508.

Jarvis, D., Sthapit, B. and Sears, L. (eds) (2000) *Conserving Agricultural Biodiversity* in situ: *A Scientific Basis for Sustainable Agriculture.* IPGRI, Rome, Italy.

Kannenburg, L.W. and Falk, D.E. (1995) Models for activation of plant genetic resources for crop breeding programmes. *Canadian Journal of Plant Science* 75, 45–53.

Karp, A., Kresovich, S., Bhat, K.V., Ayad, W.G. and Hodgkin, T. (1997) Molecuolar tools in plant genetic resources conservation: a guide to the technologies, *IPGRI Technical Bulletin* No 2. IPGRI, Rome, Italy.

Kennedy, A.J. (2000) Genetic base broadening in the West Indies sugar cane breeding programme by the incorporation of wild species. In: Cooper, H.D., Spillane, C. and Hodgkin, T. (eds) *Broadening the Genetic Bases of Crop Production.* CAB International, Wallingford, UK, pp. 283–294.

Lamont, E.J. and Palmer, P.D. (2000) Complementary methods of conserving cultivated *Malus* genetic diversity in the United Kingdom *ex situ* and on-farm (Abstract). *International Conference on Science and Technology for Managing Plant Genetic Diversity in the 21st Century, Kuala Lumpur, Malaysia.* IPGRI, Rome, Italy.

Louette, D., Charrier, A. and Berthaud, J. (1997) In situ conservation of maize in Mexico: genetic diversity and maize seed management in a traditional community. *Economic Botany* 51, 20–38.

Maxted, N., Ford Lloyd, B.V. and Hawkes, J.G. (1997) Complementary conservation strategies In: Maxted, N., Ford Lloyd, B.V. and Hawkes, J.G. (eds) *Plant Genetic Conservation: The* in situ *Approach.* Chapman and Hall, London, UK, pp. 15–39.

National Research Council (1972) *Genetic Vulnerability of Major Crops.* National Academy of Sciences, Washington, DC.

Owuor, E.D., Fahima, T., Beiles, A., Korol, A. and Nevo, E. (1997) Population genetic response to microsite ecological stress in wild barley, *Hordeum spontaneum. Molecular Ecology* 6, 1177–1187.

Parzies, H.K., Spoor, W. and Ennos, R.A. (2000) Genetic diversity of landrace accessions (*Hordeum vulgare* ssp. *vulgare*) conserved for different lengths of time in *ex situ* genebanks. *Heredity* 84, 476–486.

Phippen, W.B., Kresovich, S., Candelas, F.G. and McFerson, J.R. (1997) Molecular characterization can quantify and partition variation among genebank holdings: a case study with phenotypically similar accessions of *Brassica oleracea* var *capitata* L. (cabbage) 'Golden Acre'. *Theoretical and Applied Genetics* 94, 227–234.

Raybould, A.F., Mogg, R.J. and Clarke, R.T. (1996) The genetic structure of *Beta vulgaris* ssp *maritima* (sea beet) populations: RFLPs and isozymes show different patterns of gene flow. *Heredity* 77, 245–250.

Salhuana, W., Jones, Q. and Sevilla, R. (1991) The Latin American Maize Project: Model for rescue and use of irreplaceable germplasm. *Diversity* 7, 40–42.

SGRP (2000) *Annual Report of the CGIAR System-wide Genetic Resources Programme.* IPGRI, Rome, Italy.

Simmonds, N.W. (1993) Introgression and incorporation: strategies for the use of crop genetic resources. *Biological Reviews* 68, 539–562.

Soulé, M.E. (1985) What is conservation biology. *BioScience* 35, 727–734.

Swanson, T. (1996) Global values of biological diversity: the public interest in the conservation of plant genetic resources for agriculture. *Plant Genetic Resources Newsletter* 105, 1–7.

Tanksley, S.D. and McCouch, S.R. (1997) Seed banks and molecular maps: unlocking genetic potential from the wild. *Science* 277, 1063–1066.

Tapia, M.E. and De la Torre, A. (1997) *La Mujer Campesina y las Semillas Andinas.* FAO/IPGRI, Rome, Italy.

UNEP (1992) *Convention on Biological Diversity.* United Nations Environment Programme (UNEP), Nairobi, Kenya.

UNEP (2000) *Decision V/5 Agricultural Biodiversity: Review of Phase 1 of the Programme of Work and Adoption of a Multi-year Work Programme.* CBD, United Nations Environment Programme (UNEP), Nairobi, Kenya.

Wood, D. and Lenne, J.M. (eds) (1999) *Agrobiodiversity: Characterization, Utilization and Management.* CAB International, Wallingford, UK.

Zimmerer, K.S. (1992) The loss and maintenance of native crops in mountain agriculture. *GeoJournal* 27(1), 61–72.

Zhu, Y., Chen, H., Fan, J., Li, Y., Chen, J., Fan, J., Yang, S., Hu, L., Leung, H., Mew, T.W., Teng, P.S., Wang, Z. and Mundt, C. (2000) Genetic diversity and disease control in rice. *Nature* 406, 718–722.

Index

Page references in *italic* refer to tables or illustrations.